Tropical Rain Forest Ecology, Diversity, and Conservation

Tropical Rain Forest Ecology, Diversity, and Conservation

Jaboury Ghazoul and Douglas Sheil

OXFORD
UNIVERSITY PRESS

OXFORD

UNIVERSITY PRESS

Great Clarendon Street, Oxford OX2 6DP

Oxford University Press is a department of the University of Oxford.
It furthers the University's objective of excellence in research, scholarship,
and education by publishing worldwide in

Oxford New York

Auckland Cape Town Dar es Salaam Hong Kong Karachi
Kuala Lumpur Madrid Melbourne Mexico City Nairobi
New Delhi Shanghai Taipei Toronto

With offices in

Argentina Austria Brazil Chile Czech Republic France Greece
Guatemala Hungary Italy Japan Poland Portugal Singapore
South Korea Switzerland Thailand Turkey Ukraine Vietnam

Oxford is a registered trade mark of Oxford University Press
in the UK and in certain other countries

Published in the United States
by Oxford University Press Inc., New York

British Library Cataloguing in Publication Data

Data available

Library of Congress Cataloging in Publication Data

Data available

Typeset by SPI Publisher Services, Pondicherry, India
Printed in Great Britain
on acid-free paper by
CPI Antony Rowe, Chippenham, Wiltshire

ISBN 978-0-19-928588-4 (Pbk.)
       978-0-19-928587-7 (Hbk.)

1 3 5 7 9 10 8 6 4 2

Dedication

'How paramount the future is to the present when one is surrounded by children.'
   Charles Darwin

Despite what you may read elsewhere, this is really a book for my children: Zeki, Antar, Alligin, and Sanna. Hence the pictures. I hope this book remains relevant when they reach my age.

JG

'Curiouser and curiouser!'
   Alice in *Alice's Adventures in Wonderland* by Lewis Carroll (Charles Lutwidge Dodgson), 1865

To everyone (and their children) with curiosity and concern for tropical forests and their peoples. I wish you luck.

DS

Acknowledgements

In some ways this has been a team effort, and consequently we have many to thank. Most important is the special thanks we extend to our respective partners, Katharine (Gnome) Liston and Miriam van Heist, as well as Jaboury's children, for tolerating late nights, lost weekends, and foul moods, and for providing the support and enthusiasm for this project throughout.

Our interest and concern in tropical forests has deeper roots. We have been lucky to have been inspired, encouraged, and challenged by many people—too many to name them all. Jaboury's first tropical rain forest adventures, in Vietnam, were coincident with his first adventures with his, now, wife. The two have been lifelong projects ever since. He is forever indebted to his parents, Ramez Ghazoul and Jemila Topalian, whose support, encouragement, and interest allowed him basically to do whatever he wanted. Jaboury also wishes to thank Julian Evans for sowing the seed of this book and providing the encouragement to nurture and develop it. Douglas especially acknowledges the key role of his parents, Richard and Margaret Sheil, as well as Colyear Dawkins, Jo Eggeling, Tim Whitmore, Cass Clunies-Ross, Peter Savill, Howard Wright, and Jeff Sayer for their interest and support in helping him forge a career in ecology—a field he once thought was 'something you do in the holidays'.

Draft texts were reviewed by a variety of experts, and their comments have improved the final version substantially. These include Nico Bluethgen, Corey Bradshaw, Francis Brearley, Tom Brooks, Emilio Bruna, David Burslem, Robin Chazdon, Carol Colfer, Mauro Galletti, Manuel Guariguata, Dennis Hansen, Miriam van Heist, Chris Kettle, Lian Pin Koh, Robert Morley, Meine van Noordwijk, Gary Paoli, Lourens Poorter, Axel Poulsen, Mauricio Quesada, Lucy Rist, Navjot Sodhi, William Sunderlin, Cam Webb, Doug Yu, and Roderick Zagt.

We extend special thanks to Lucy Rist, who sought and obtained all the necessary permissions for images and tables, read through drafts and provided comments as necessary, all in a magnificently organized and efficient manner.

We also thank all those who provided or helped us track down many excellent photos and illustrations (and we wish we had space to use more), including George Beccaloni, Karin Beer, Robert Bitariho, Manuel Boissiere, Julia Born, Carsten Bruehl, Rhett Butler, Damien Caillaud, Robin Chazdon, Stuart Davies, Edmond Dounias, Eberhard Fischer, Carlos Garcia-Robledo, Dennis Hansen, Michael Heckenberger, Ernest Hennig, David Hughes, Christopher Kaiser-Bunbury, Alex Kendall, Smitha Krishnan, Lee Su See, Ariel Lugo, Andrew Macgregor, Erik Meijaard, Steven Paton, Rob Pickles, Stuart Pimm, Lourens Poorter, Axel Poulsen, Lucy Rist, Valenti Rull, Jose Sabino, Hanster Steege, Matthew Struebig, Indah Susilanasari, Alan Timmermann, Ellen Wang, Alex Wild, and Doug Yu. Clinton Jenkins kindly provided his modified Fuller projection map of forest cover and deforestation. Michael Adams kindly allowed us to use a detail from one of his paintings as the cover for this book.

We gratefully acknowledge CIFOR's contribution in supporting Douglas's research and writing (from 1998 to 2008) and for allowing him time, including a sabbatical, to work on this book.

Thanks to Wageningen University, the Netherlands, especially to Douglas's 2006 sabbatical hosts, the Forest Ecology and Forest Management group: Frans Bongers, Frits Mohren, Frank Sterck, Lourens Poorter, and Joke Jensen. Thanks also to Wageningen's Forest Policy and Conservation group, notably Freerk Wiersum. A special thanks to the excellent Wageningen University Library for help and support with tracking references during and after Douglas's sabbatical.

Help and support in chasing book content was also provided to Douglas by CIFOR, notably Robert Nasi, Markku Kanninen, Indah Susilanasari, Popi Astriani, Greg Clough, Sven Wunder, David Kaimowitz, Laura Snook, Manuel Guariguata, Manuel Ruiz-Perez, Kuswata Kartawinata, Ismayadi Samsoedin, and many other colleagues, but especially the CIFOR library staff.

We thank Helen Eaton, Philippa Hendry, and Ian Sherman at OUP for their guidance and patience.

Contents

CHAPTER 1

Tropical rain forests: myths and inspirations

'The true shape of this tree is largely unknown because prudent nature has set this evil plant apart from the dwellings of people, and placed it in unknown mountain ranges and wildernesses.'

Georg Everhard Rumpf, or 'Rumphius'[1] (reprint 1993)

Rain forests have long captured the imagination of those who live in temperate regions. For many centuries the wet tropics were a rich source of myth and magic. Fanciful images are more than a millennium old. From stories of the lands of spices, of magical birds-of-paradise, the feathers of which rendered their possessor impervious to weapons (and were therefore much prized during the medieval crusades), golden jungle cities and headless peoples with faces on their chests (Figure 1.1), and tribes of violent female warriors, to more modern images of Tarzan, King Kong and of lost lands where dinosaurs still roam. Into this mix over the centuries has come recognition of human cultures that collected shrunken heads, of human-like apes, and of ruins of lost jungle civilizations. No wonder that over time myths have become intertwined with realities to the point where they may not be readily distinguished. This enchantment still colours our perceptions and scientists have not been immune.

While many can now visit a rain forest themselves or experience it through the lens of television, the myths have not necessarily retreated. Even now it is not always clear which popular claims for rain forests are scientifically sound and which are misconceptions. Any sum-

mary of what rain forests represent must acknowledge the many myths which still colour feelings and judgements even among the most objectively minded researchers. In discussing rain forest we cannot always avoid clichés and superlatives, but we can at least seek to justify them.

The 17th-century quote at the start of this chapter notes the challenge of investigating tropical forest organisms. Tropical forests remain large, diverse, and hard to access, whereas most species within them are small, rare, and localized. But progress is being made. The tree mentioned in Rumphius' quote is now considered neither elusive nor evil: *Antiaris toxicaria* is actually widespread in Asia and Africa and, contrary to Rumphius' beliefs, its fumes will not kill those who linger close by. Yet *A. toxicaria* represents a minority of rain forest species in that we know something about it: we still know precious little about the form distribution and ecology of the several millions of other rain forest organisms.

Certainly significant progress has been made on some of the bigger mysteries too. Two centuries after Rumphius' time Alfred Russel Wallace also spent time in the equatorial islands of the **Moluccas** collecting and observing

---

[1] Georg Everhard Rumpf (more widely known as Rumphius) was a German botanist who lived in Ambon (one of the Moluccas Islands in present-day Indonesia) in the 17th century. Here he writes about the tree *Antiaris toxicaria* (Beekman 1993).

1

Levinus Hulsius's depiction of Sir Walter Raleigh's
headless men in Guiana (1599).

Figure 1.1 Headless men—one of the many tropical rain forest myths.

tropical species. Suffering from fever, probably malaria, he had an inspiration which he managed to scribble on paper and develop into an essay. In February 1858, he sent Charles Darwin his unpublished article 'On the Tendency of Varieties to Depart Indefinitely from the Original Type'. This essay spurred Charles Darwin hurriedly to summarize his lifetime's work into a short treatise, and the presentation and publication of these two essays was the public origin of the theory of natural selection.

It is often our understanding of evolution that allows us to interpret our ecological observations within a meaningful framework. What has emerged is in many ways far more wonderful than the myths that science has shattered. The adaptations and complexity of species and the nature of their interactions continue to provoke astonishment from the youngest children to the most eminent ecologist. The wonder is open-ended—there is

much more to discover. Species unknown to science abound in tropical rain forests. Remarkable too are the behaviours, relationships, and dependencies exhibited among the more familiar animals and plants which we are only beginning to appreciate. Lizards that run on water, ants that glide, plants that 'walk' are not myths but scientific facts—and no less fascinating for that.

Any scientifically-driven exploration of rain forests builds upon generally accepted assertions and seeks to advance beyond them. Scientific imagination and creativity are an essential part of this process. In some ways scientists create their own 'myths', albeit they work hard to support or falsify such myths through examining any implications which can be verified or contradicted by further observations and experimentation.

Scientists seeking to understand how nature functions may be overwhelmed by the rain forest—the complexity does not facilitate easy enquiry. But scientists are drawn to rain forests precisely because of such complexity—we are responding first and foremost not to our desire to understand (although this of course is also important) but to our primeval fascination with the unknown, to the intrinsic beauty (Figure 1.2), and to the revelation of new and frequently astonishing stories of nature. Instead of myths, we now have scientific stories that are no less enchanting, and perhaps more so owing to the new doors they open to our appreciation of nature's complexity. We ponder the big mysteries of evolution and the origins of tropical diversity, but we also amuse ourselves with quirky questions, answers to which remain elusive: why are two species of fruit bat the only male mammals to suckle their young? Why do sloths climb down from the forest canopy to defecate? We need to understand the past and present to be able to grapple effectively with challenges of the future: What are the implications of biodiversity losses and forest clearance for humans? How will climate change affect ecosystem dynamics and the long-term viability of rain forests?

Defining rain forests

Before embarking on a scientific rain forest tour we must begin with a definition: What are rain forests? Even this proves an elusive exercise. While many cases are clear-cut and uncontroversial, this is not always the case. Any definition of tropical rain forests is plagued by the need to consider transitional communities, mosaics, the range of distinctive formations often included, and the modified and successional communities that may (or may not) become more 'rain forest-like' over time. One map of forest types distinguishes several kinds of forest that might be encompassed within a more generic 'rain forest' category (Figure 1.3).

The German botanist A. F. W. Schimper coined the term tropical rain forest (in German, *tropische Regenwald*) at the end of the nineteenth century[2]. The broader term *tropical moist forest* includes more seasonal formations (Sommer 1976). It is this more general class of forests that is often used in large-scale overviews of

Figure 1.2 Borneo's rainforest. Why do such scenes convey such magnificence and beauty, and indeed are such feelings universal?

Photo by Julia Born.

[2] Originally published in 1898 as *Pflanzengeographie auf physiologischer Grundlage* (in German), his translated work along with 'rain forest' was published by Clarendon Press in 1903 as *Plant-geography upon a physiological basis* with revisions and co-authorship from William Rogers Fisher, Percy Groom, and Isaac Bayley Balfour.

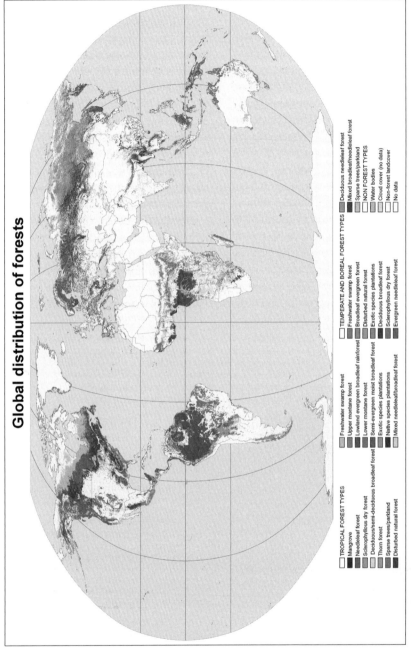

Figure 1.3 Map of forest distributions worldwide. (See colour plate 1.)

Map courtesy of WCMC-UNEP: http://www.unep-wcmc.org/forest/global_map.htm

forest cover or forest loss, as there is seldom a clear and easy way to distinguish intermediate forests.

Tropical rain forest definitions can be based on the climate (and potential vegetation) and/or the actual form of the vegetation present. Taking the first approach, true tropical rain forests are warm, over 20°C year round, and usually receive over 1,500 mm of rain every year with no pronounced cold or dry spells.[3] High mountain rain forests can certainly be colder, while rain forest type formations (often minus the epiphytes) can also develop with much less rain where plants can avoid moisture deficits by accessing ground water reserves, fog or other means. Closed forests can form in the lowland tropics on as little as 800 mm of rain a year, so the transition from wet to dry forests is gradual and any defining line is ultimately arbitrary. Any physiologically relevant definition would need to address overall water balance and seasonality. Another relevant definition is for '**megathermal**' climates (Morley 2000). These are the wet and frost-free climates, and scientists have used such criteria to determine the extent of rain forests in the ancient past.

This book

We have chosen to organize this book into three main sections reflecting 1) natural history and variety of life forms, 2) biogeography and ecology, and 3) human interactions and influences. Each section serves a different purpose, and each may be more closely read by different sectors of our audience. Needless to say, we hope that there is something of value in all sections for all readers.

Our first section provides a brief overview of the wealth and variety of life, and seeks to convey some inkling of why rain forests have captured the imagination and interest of explorers, collectors, and biologists over the ages.

Our second section provides evolutionary, biogeographical, and ecological perspectives. Here we review scientific advances and current knowledge about the origins of these forests, what shaped and continues to shape their distributions and formations, and how and why their communities are structured in the way they are. The diversity of tropical rain forest species is perhaps their most enthralling feature, but providing an answer to even the most basic question of how many species there are

remains absurdly elusive. We remain intrigued and largely perplexed not only by the extent of rain forest diversity, but also by its origins and maintenance, although ecologists and other biologists have many suggestions.

Our final section describes the interaction of humans with the rain forest. In tangible terms the human race has depended and continues to depend on rain forests for products and services. More prosaically, rain forests have contributed to the cultural and spiritual wealth of innumerable peoples and societies, including the myths that have arisen in modern Western perceptions. The dramatic decline in recent decades of rain forest formations and associated biota marks a concern for humans as a loss of commodities and services that might directly impact on livelihoods, but also as a diminution of our cultural heritage. We close this section with a discussion of human responses to this legacy.

We have attempted to evaluate past and current ecological debates, we have outlined major patterns and underlying processes, and we have considered future challenges and possible responses. While we have sought to present common consensus, we acknowledge that many views are contested. In the traditions of the scientific approach, we have sought to be objective in our writing, but we have also been unable to resist anecdotes, asides, favoured examples, and biases towards familiar locations. This reflects the interplay of our scientific training with our innate curiosity and enthusiasm for our subject, for which we ask the readers' forgiveness and indulgence. In the end, we have written this book as much for ourselves as for others.

As we look back on Rumphius' words with the benefit of four centuries of learning, we have come a long way. In another 400 years it is quite possible that our words may seem as quaint and inaccurate as his. It is also quite possible that our words become largely irrelevant: Rumphius might be able to rest in peace, secure in the knowledge that his evil tree and its domain will have been permanently removed from human concern. Thus, our book may serve a higher purpose—we hope that in addition to the scientific information it provides to the reader, it also generates a fascination for the rain forest biome that inspires and motivates action towards its conservation.

---

[3] Rain forest precipitation can exceed 10,000 mm in a year.

Section I

The natural heritage

CHAPTER 2

An exuberance of plant life

'The reader who is familiar with tropical nature only through the medium of books and botanical gardens...will think that I have unaccountably forgotten to mention the brilliant flowers, which in gorgeous masses of crimson, gold or azure must spangle these verdant precipices, hang over the cascade, and adorn the margin of the mountain stream. But what is the reality?'

Alfred Russel Wallace (1869 reprint 1996)

'The exuberance of plant life in the humid tropics continues to dazzle scientists from the more sober temperate biomes.'

Tim Whitmore (1998)

Viewed from space, Earth's wet equatorial lands are green and clouded. Coming closer, we see a lush abundance of plant life—favoured and nurtured by abundant light, moisture, and everlasting summer. Penetrating the sunlit canopy and descending to the ground we enter a still, damp, and dappled underworld of saplings and stems, roots and fungi, growth and decay.

From the ground, an observer is sometimes frustrated that '...all is out of reach of the curious and admiring naturalist. It is only over the outside of the great dome of verdure exposed to the vertical rays of the sun that flowers are produced, and on many of these trees there is not a single blossom to be found at a height less than a hundred feet. The whole glory of these forests could only be seen by sailing gently in a balloon over the undulating flowery surface above' (Wallace 1853).

But, as eyes grow accustomed to the dim light beneath, patience and curiosity bring rewards. Tropical forests are distinguished by abundant **epiphytes**, tangled climbers, massive **buttressed** stems, flowers and fruits growing on woody trunks. Every surface is an opportunity for life,

from pendulous **ferns**, creeping roots of epiphytic figs, **hyphal** networks stretching across the soil surface, to **moss** mats, **lichens**, and algal-films.

The impenetrable jungle beloved by Hollywood has little in common with the reality of pristine rain forests. Thickets occur only at forest edges and in disturbed areas. The **understorey** vegetation is usually sparse and open, creating a poet's vision of a 'green cathedral' (Slater 2003), a vaulted ceiling enclosing an expansive space full of marvellous detail. This detail bears scrutiny, and in this chapter we introduce the variety, structure and form of rain forest plants.

2.1 Botanical overview

A large proportion of the Earth's 300,000–500,000 higher plant species reside in the humid tropics (Pitman and Jorgensen 2002). We know little about the majority (Box 2.1).

High botanical diversity characterizes many tropical wet forests, and the figures are impressive. A complete survey of 100 m^2 of Costa Rican lowland forest[1] yielded

---

[1] Horquetas, in the Atlantic lowlands. To assess epiphytes, every tree was felled.

Box 2.1   Challenges in tropical botany

Taxonomy explores relationships through the classification of organisms. This is the front line in assessing tropical forest biota. Botanical taxonomy distinguishes and classifies plants so that we can enumerate species, identify populations, assess relationships, and develop ecological and evolutionary perspectives for management and conservation. Despite considerable taxonomic effort, respected estimates of global plant species numbers still vary widely. In a typical year, more than 2,000 species and 90 new genera are described, and these numbers show no sign of declining (Prance et al. 2000).

Consider Borneo: an estimated 15–35% of Borneo's flora remains undescribed (Beaman and Burley 2003). More than 1,000 species of orchids have been collected in Sarawak (Malaysian West Borneo), but almost a third (306) have been recorded only once (Beaman and Burley 2003). The botany of Brunei Darussalam in Borneo is relatively well-known, yet even here only 17 palms were reported before 1988; seven years later the number was 140 (Prance et al. 2000). All but one of the 17 species of

Begonia (Begoniaceae) collected in Brunei since 1989 have been new to science (Earl of Cranbrook and Edwards 1994). These examples and others (Figure 2.1) are typical of the wider wet tropics (Beaman and Burley 2003).

Figure 2.1 *Thismia ophiuris* (Burmanniaceae), a rare and easily overlooked saprophyte from Matang, Borneo.

Photo by Axel Poulsen.

---

233 **vascular** species (Whitmore et al. 1985). A non-destructive survey in 200 m² in Tesso Nilo, Sumatra, Indonesia, catalogued 217 species[2] (Gillison 2001). In the 'Choco' Pacific coast of Colombia, 970 vascular species were recorded across 0.9 ha, implying around 1,000 per hectare (Galeano et al. 1998). Impressive as they may be, numbers or lists of species fail to convey the full variety of forms, adaptations, and behaviours of tropical plants.

2.1.1 Angiosperms: the flowering plants

Angiosperms dominate terrestrial biomes, and their diversity, especially family-level diversity, peaks in rain forests. This was not always the case, as we shall see in Chapter 7. Flowering plants are generally divided into two

groups: those with one seed leaf (**monocots**) and those with two (**dicots**). Of course it is not quite so simple. *Idiospermum australiense*, the 'idiot fruit' tree of Australian rain forests, is a '**polycotyledonous**' exception with two to six seed-leaves (Edwards and Gadek 2002). Various other apparently 'dicot' lineages (the so-called 'ranalean' families) are now known to pre-date the inferred dicot-monocot division. These include widespread tropical forest families such as Annonaceae, Lauraceae, and the Piperaceae, along with localized oddities such as the monotypic Amborellaceae, namely *Amborella trichopoda*, a peculiar rain forest **shrub endemic** to New Caledonia, currently considered the most primitive of all angiosperms (Feild and Arens 2005). The sister group to the monocots is now called the 'eudicots'.

---

[2] This density of species is 10–times higher than is typical for temperate latitudes.

Larger pantropical monocot families include palms (Arecaceae), orchids (Orchidaceae), and grasses (Poaceae). The monocots typically have their flower parts in threes (for example, three or six petals)[3]. Dicot families in tropical rain forests include Melastomataceae, Myrtaceae, Fabaceae and Fagaceae. Dicot trees usually dominate the canopy, but local formations (Chapter 7) are sometimes dominated by monocots—bamboos (Box 2.2) or palms (Box 2.3)—or even by **gymnosperms** (see below).

2.1.2 Gymnosperms

Rain forest gymnosperms include conifers, *Gnetum* and **cycads**[4]. Most gymnosperms possess tough, simple leaves, sometimes reduced to narrow needles or even scales. Juvenile conifers are generally out-competed by faster growing angiosperms in productive sites, and they are seldom abundant outside **montane** or **edaphically** unusual sites (Coomes et al. 2005).

Prominent tropical conifers include Auracariaceae, Podocarpaceae and Pinaceae[5]. The Southern Hemisphere family Auracariaceae includes many emergents in the genera *Agathis* (13–21, depending on classification, mostly rain forest species in the Asia-Pacific) and *Auracaria* (19 mainly montane and temperate species) (Setoguchi et al. 1998). Most of the 105 species of the pantropical genus *Podocarpus* are montane trees with berry-like cones. Other Podocarpaceae include various genera with dis-

---

Box 2.2   Bamboos

Bamboos are primarily forest grasses (Poaceae, Section Bambusoideae [Bambuseae: woody, and Olyreae: herbaceous]). They include evergreen and deciduous species, and most possess woody, usually hollow, stems/culms (rattan canes, in contrast, are always solid) and below-ground networks of branching rhizomes (McClure 1966). Bamboos are widely distributed and include around 1,200 species in 90 genera (estimates vary from 1,000 to 1,500 species in 50 to 100 genera)[1].

Well known for the passage of decades between synchronized flowerings, and mass seeding, and die-back in many species, bamboos also spread vegetatively. Most tropical species form free-standing clumps (Figure 2.2), but some, like the Asian *Dinochloa* spp. and *Nastus* spp. have adopted a climbing or scrambling behaviour (Yap et al. 1995).

Tropical America and Southeast Asian forests are especially rich (Indonesian forests probably include at least 200 species), but bamboos are found far into temperate regions (Table 2.1). Madagascar has more species than the entire African continent.

Bamboos occur in all tropical forest habitats except mangrove. They are locally dominant in the wetter montane belt on African mountains and the south-

[1] They are difficult to collect and identify: the plants seldom flower and specimens are unwieldy. There are also various understorey grasses that are not (taxonomically) true bamboos but share similar forms.

---

[3] Pentastemonaceae Stemonaceae (two-species, *Pentastemona*), a family of forest herbs endemic to Sumatra, is unique among monocots in having five petals.

[4] Cycads include around 300 species patchily distributed at warmer latitudes. They are often associated with seasonal locations, typically on thin soil.

[5] Pines are an exclusively Northern Hemisphere group, and are also entirely absent (unless introduced) from sub-Saharan Africa. They are diverse in northern parts of Central America (Mexico, Belize, and Guatemala) where they often form a dominant component of montane forests, but reach to the equator only in Asia (North Sumatra). Many highly restricted conifer species occur scattered throughout the moist tropics—for example *Acmopyle sahniana* (Podocarpaceae), a montane tree in Fiji with fewer than 100 individuals.

western Amazon Basin (see Chapter 7). Small herbaceous bamboos occasionally dominate forest understoreys, such as *Pariana radiciflora* in Amazonian Ecuador (Poulsen and Balsley 1991). In Asia many species occur on river edges and forest margins. Malaysia's limestone *Dendrocalamus elegans* and ultramafic *Dinochloa obclavata* show clear soil preferences (Dransfield and Widjaja 1995).

Many distributions reflect human interventions: *Guadua aculeate* was once abundant in Central America but is now scarce and localized, while the exotic *Bambusa vulgaris* is naturalized and widespread across Jamaica (McClure 1966). Indeed various species, including *B. vulgaris*, are known only from cultivated and/or naturalized populations (Dransfield and Widjaja 1995).

Figure 2.2 Bamboo cathedral of Mpivie-St Anne, Loango National Park in Gabon, Central Africa.

Photo by Rhett A. Butler.

Table 2.1 Approximate regional numbers of woody bamboo (Bambuseae) species and genera.

Region	Species	Genera	Notes
Americas—tropical*	410	20	
-Brazilian Atlantic forests	62		27 endemic species
-Central America	73		
Asia—tropical and subtropical**	270	24	Majority are *Bambusa*
Asia—temperate	320	20	Majority are *Sasa*
Africa	3	3	All endemic
Madagascar	20	6	All endemic
Australia	3	2	
Pacific	4	2	

Data derived from Dransfield and Widjaja 1995; Judziewicz et al. 1999.

Box 2.3 Palms

'Palms provide one of the tallest trees (*Ceroxylon*), the longest woody climber (*Calamus*), the largest leaf (*Raphia*), the largest inflorescence (*Corypha*) and the largest seed (*Lodoicea*)[2] in the plant kingdom' (Purseglove 1972).

Palms (Arecaceae) are a dominant part of the understorey in many of the world's tropical forest regions. The two major palm leaf-forms are fan and pinnate. Persistent dead leaves (as opposed to dead leaf bases) are known only in fan-leaved palms, and climbing habit and stilt roots occur only in pinnate-leaved palms (Saakov 1983).

Most of the 2,200 described palm species occur in tropical forests where they can be locally abundant, reaching nearly 10,000 ha^{-1} in parts of the Peruvian Amazon (Vormisto 2002), 2611 ha^{-1} in Costa Rica (Wang and Augspurger 2004), and several hundred ha^{-1} in parts of Asia (LaFrankie and Saw 2005).

Asia is richest in palm species (Indonesia alone has 477), followed by the American tropics (Colombia has 277 species, Brazil 221). Continental Africa has only 50 species (Johnson and Group 1996), but Madagascar has at least 176 species (172 endemic).

Diversity and endemism on various islands are notable: Philippines has 109 endemics out of 157 total species; Western Samoa has 11 species all endemic; New Caledonia has 32 endemics of 33; Fiji 26 of 34; Vanuatu 12 of 17; and the Seychelles has six endemics, all in distinct monotypic genera (Johnson and Group 1996).

Rattans (sub-family Calamoideae) are diverse and abundant climbers in the paleotropics. They are most diverse in Asia. Various species provide commercially important cane used in furniture and basketry. In the Americas *Desmoncus* (sub-family Arecoideae) are used in a similar manner. Many rattans climb using a spined, whip-like *flagella* or **cirrus** on the end of the leaves. On reaching the canopy, they are often held only by their leaves, and as these senesce and die, new leaves are required to secure the plant. As palms produce new leaves only in conjunction with stem elongation, considerable lengths can arise through this running-on-the-spot mechanism, as well as by climbing among multiple tree crowns. A *Calamus manna* in Sabah has been measured at over 150 m.

[2] The palm *Lodoicea maldivica* (*coco-de-mer*) produces the largest nuts in the world, some weighing over 20 kg. Their only native home are the Seychelles islands of Praslin and Curieuse. For centuries the plants were known only from their giant seeds found floating on the Indian Ocean or washed up on beaches. The species was described from these seeds, which were falsely thought to be from the Maldives, hence the strikingly inappropriate specific name.

---

tinctly Asia-Pacific distributions (for example, *Dacrydium*, *Phyllocladus*, and *Sundacarpus*) (Farjon 1998).

Gnetaceae is a notable pan-tropical woody group of about 72 species in the single genus *Gnetum*. These climbers and shrubs (and two small Asian trees[6]) are commonly mistaken for angiosperms due to their remarkable similarities, including broad **reticulate**-veined leaves and fruit-like, fleshy seed-coats—a clear example of parallel evolution. People in parts of Africa eat *Gnetum* leaves, while in Asia people also eat the seeds.

2.1.3 Pteridophytes

Forest ferns, and fern allies, are diverse and often abundant. Around three-quarters of the world's roughly 20,000 fern species are found in the moist tropics, where they typically contribute 5–10% of species in local floras.

[6] *Gnetum gnemon* is common across large parts of Malesia and Papua.

Ferns benefit from moisture: they require free water, even if only as a film, for sexual fertilization (Page 2002), and their spores can disperse by wind over large distances to colonize suitable habitats.

Ferns on the forest floor typically have thin textured leaves, and can maintain sufficient photosynthesis at very low light. Fertile leaves are frequently narrower and more elevated than trophic leaves, allowing spores to take better advantage of the limited air movements.

Ferns possess fine **rhizomes** suited to exploiting tiny cracks and fissures. Some are well adapted to rocky surfaces, and others will readily grow in moss, such as tiny *Trichomanes* (Aspleniaceae) with leaves only two cells thick. Many have adapted to life as epiphytes. *Asplenium nidus* (Aspleniaceae) the common 'bird's nest fern' of Southeast Asia, collects moisture-absorbing litter within its central bowl. This mass is penetrated by the fern's own roots and provides habitat for others to establish and grow.

Some fern species growing in regularly flooded areas can live submerged for days, and have narrowed leaves to reduce the tug of the currents, such as *Microsorum pteropus* (Polypodiaceae) (de Winter and Amoroso 2003). A few fern species are shrubs: Java's *Oleandra neriiformis*

(Oleandraceae), for instance, has developed stiff stems and **spatulate** leaves and grows to 2 m high.

Tree ferns (*Cyathea* with 600–650 species; Cyatheaceae) are typical of montane forests, occurring also in some especially humid lowland forests (notably New Guinea and surrounding islands) (Figure 2.3). Stems in some species can reach 20 m tall and 20 cm in diameter. These stems are hard and durable but lack secondary thickening, although the stem may expand from stiff roots which envelop the trunk. Various 'fern allies' contribute greatly to higher order taxonomic diversity in any assessment of vascular plants. Ophioglossopsida include numerous terrestrial and epiphytic fern-like, frond-bearing forms. Selaginellopsida, the genus *Selaginella*, has about 700 species, including many fern-like forest forms. Lycophyta, the club mosses, were the Earth's dominant tall vegetation during the Carboniferous (Chapter 6). While the coal-forming giants are long gone, smaller *Lycopodiella cernua* is still common on forest edges on acidic soils throughout the tropics, but most species, such as *Huperzia*, *Lycopodium*, and *Lycopodiella*, are restricted to epilithic, epiphytic, and high mountain habitats. Diversity remains uncertain, but there appear to be at least 46 *Lycopodium* species in the Philippines alone (Amoroso

Figure 2.3 Tree ferns in Bwindi Impenetrable National Park, Uganda, East Africa.

Photo by Rhett A. Butler.

et al. 2001). The Psilotophyta are narrowly distributed, including *Tmesipteris* (generally epiphytes on tree ferns in the Australia-west Pacific region) and *Psilotum* (mainly Caribbean forest **herbs** and epiphytes).

Few herbivores exploit rain forest Pteridophytes, presumably due to their potent anti-herbivore defences. Exceptions like the Central American butterfly *Euptychia westwoodi* (Nymphalidae) which feeds on *Selaginella* have this resource virtually to themselves.

2.1.4 Bryophytes and allied taxa

Tropical forests contain around half the world's **bryophyte** species (Frahm et al. 1996). Though bryophytes are often viewed as 'archaic'—lacking 'true' vascular tissues or leaves—they have continued to evolve and diversify alongside seed plants (Gradstein and Pocs 1989). Their taxonomy remains challenging, and some forms classified until recently as thallose liverworts are now known to be epiphytic ferns (*Vittaria* spp.) (Frahm et al. 1996).

There are three principal types: the mosses (Musci) with perhaps 10,000 species worldwide; the **liverworts** (Hepaticae) with a similar number (6,000–10,000); and **hornworts** (Anthocerotaceae) with fewer than 100 species.

All species require moisture to breed and metabolize, but vary in their ability to tolerate desiccation. Bryophyte diversity and biomass peak in wet montane forests: one study at 3,700 m in Colombia reported 44 tonnes dry weight of bryophytes and associated suspended matter per hectare (Hofstede et al. 1993).

In epiphytic habitats, hornworts and most liverworts are restricted to shaded wet sites such as the lower trunk. Pendulous, feather and bracket mosses are specialized epiphytic forest forms, while *Dawsonia longifolia* (Polytrichaceae) in the mountains of Melanesia can grow to 50 cm as an understorey 'herb' and possesses sophisticated moisture conducting architecture. All these moisture-dependent groups are especially sensitive to environmental change, and deforestation is estimated to have caused at least 20 previously recorded bryophyte species to disappear from Singapore (Piippo et al. 2002).

2.2 Diverse forms

All green plants need light, moisture, and nutrients to survive, grow, reproduce, and disperse. These requirements have been met in an astonishing variety of ways that are most extravagantly realized in the tropical rain forests. Nonetheless, some forms and life strategies are characteristic and ubiquitous (Table 2.2).

Most forest plants, from giant trees to microscopic algae, are usually considered **autotrophic**, needing only light, moisture, atmospheric gases and mineral nutrients to grow. In fact most are also dependent on symbiotic partnerships with **mycorrhizal** fungi (Chapter 3). Parasitic plants are **heterotrophic**, extracting some or all of their nutrients (carbohydrates and minerals) and/or water from some plant or fungal (mycorrhiza) host. Carnivorous plants can extract nutrients from animals, but this is primarily opportunistic.

Table 2.2 Percentage relative abundance of forest life-forms by species in various well-studied sites. Definitions of life-forms and treatment of non-forest taxa vary amongst studies. The 'Others' category includes hemi-epiphytes, lithophytes, parasites, etc.

Life-forms	La Selva, Costa-Rica (McDade et al. 1994)	Guyana (Richards 1996)	Mt Cameroon, Cameroon (Cable and Cheek 1998)	Budongo, Uganda, Africa (Synnott 1985)
Trees	20	34	29	28
Shrubs (< 3 m)	17	18	10	14
Terrestrial herbs	34	—	32	24
Climbers	6	19	18	11
Epiphytic herbs	23	22	10	15*
Others	<1	7	3	8

2.2.1 An introduction to leaves

Leaves are the primary means of energy capture. Their form and arrangement varies considerably. Angiosperm leaves include simple, lobed, toothed, compound, and so forth, and modified leaves have become spines, tendrils, domatia, and traps.

Leaf size

Individual leaf size varies over five orders of magnitude (Table 2.3), and some tropical leaves are substantial (Figure 2.4). Ecologists use a simple scale to describe and summarize leaf and leaflet sizes (see Table 2.3). The ecological conditions that select for leaf sizes are often consistent enough over time and space to make it a useful criterion by which to classify forests. Hence Australian lowland forest is often referred to as '**notophyll** vine forest' (Webb 1959).

   Lowland forest trees commonly possess oval *notophyll* or *mesophyll* **laminas**, often in compound leaves. Smaller *microphylls* are predominant in heath and upper montane forests (Whitmore 1998). Leaves are often larger in humid understorey conditions with a reliable moisture supply. Some giant herbs, such as the bananas and aroids, possess extremely large leaves. The largest undivided dicotyledon leaves probably belong to Neotropical *Pentagonia* (Rubiaceae) that reach nearly 2 metres[7]. Large

palmate leaves, sometimes with long **petioles**, are (aside from palms) a distinctive, if minor, component of tropical forests. These are typical of trees colonizing open sites (pioneers, see Chapter 10), and various **lianas**. Streamside plants that need to withstand rapid floods typically possess narrow leaves.

Leaf form

Moisture can be a problem for leaves. Most higher plants cannot photosynthesize effectively with wet leaves (but see **CAM**, Chapter 10) as water films interfere with **transpiration** and gas exchange (Luttge 2004). Wetness also increases potential colonization by **epiphylls** (Chapter 3) or **pathogens**, and may leach away soluble nutrients.

Table 2.3 Standard names of leaf/leaflet (lamina) area-size classes (Raunkiaer 1934; Webb 1959). In the original Raunkiaer scheme (Raunkiaer 1934) *notophyll* and *mesophyll* were combined in a single *mesophyll* class which was later divided by Webb (1959).

Name	Area (mm2)
Picophyll	<2
Leptophyll	2–25
Nanophyll	25–225
Microphyll	225–2,025
Notophyll (small Mesophyll)	2,025–4,500
Mesophyll (large Mesophyll)	4,500–18,200
Platyphyll	18,200–36,400
Macrophyll	> 18 x l0

Figure 2.4 The leaves of *Coccoloba* spp. (Polygonaceae) in central Amazonia are large enough to shelter flocks of roosting bats.

Photo by Douglas Sheil.

[7] The biggest leaves of any plant belong to the African *Raphia* palms (swamp forest specialists): individual fronds can reach over 20 m.

Plants therefore employ various leaf forms and surface textures and coatings to reduce wetting (Ivey and DeSilva 2001; Farji Brener et al. 2002). **Drip-tips**, long tapering tips on many understorey leaves (Figure 2.6), appear to facilitate water shedding (Barker and Booth 1996; Rebelo and Williamson 1996; Farji Brener et al. 2002), although this does not seem to reduce colonization by epiphylls (Lucking and Bernecker-Lucking 2005). Many **bromeliads** possess a finely irregular surface of hydrophobic **cuticular** projections (foliar **trichomes**) that hold water droplets above the main surface (Pierce et al. 2001). As particulate matter often adheres to passing water drops, water-shedding mechanisms facilitate self-cleaning, and may thus reduce pathogens and **stomata**-blocking debris (Benzing 2000).

Numerous leaf venation patterns occur, but certain features are notable in rain forest. Firstly, reticulate-venation (characterized by a network of veins) is com-

mon amongst understorey monocots, and appears to reflect vascular redundancy, allowing long-lived leaves to function even when parts of their surface are lost or damaged (Givnish et al. 2005). Widespread brochidodromous venation (in which secondary veins loop and connect at the lamina margin), found in many tropical trees, similarly seems to provide some vascular redundancy and may increase the effectiveness of transporting water to the **leaf margin** (Roth-Nebelsick et al. 2001; Turner 2001). Probably related to these venation patterns, entire margined leaves are common in wet tropical forests, while serrated and toothed laminas are scarcer. A strikingly consistent empirical correlation between the proportion of woody dicot species with non-toothed leaf margins and mean annual temperature has been refined as a tool to assess past climates from fossil-leaf assemblages (Wolfe 1978; Royer et al. 2005).

Also common in the wet-tropics are swollen joints at the base of, or part-way along the leaf petioles, on both trees and herbs, including some ferns. The prevalence of this character, though likely to be associated with leaf angle adjustment and/or with leaf **abscission**, remains unexplained (Richards 1996).

Leaves vary within individual plants (Winn 1999). In some cases, this involves subtle physiological differences between light and shade leaves (Chapter 10). In others it is more marked, for example some Melastomataceae possess **dimorphic**, alternately large and small, leaves. More general are the leaf changes that often occur as plants mature: sapling leaves are often larger than adult leaves, and the simple leaves of juveniles are sometimes replaced by compound leaves in adults. In contrast, species with large adult leaves (many palms, for example) produce much smaller juvenile foliage. An especially prevalent syndrome of juvenile–adult leaf dimorphism, including reduced laminas and frequent red colouration of juvenile leaves, is found on various oceanic islands and may relate to herbivory (Hansen et al. 2004).

In about one-third of tropical tree species, leaf greening is delayed and juvenile foliage is transiently white, pale, red, pink, purple, or blue (see also Chapter 13). Similar juvenile colouration is occasionally found in herbs and other understorey plants. Young leaves are invariably softer than mature foliage, and sometimes even appear limp. In tropical forests most herbivore damage occurs on young foliage (Kursar and Coley

Figure 2.5 Drip-tips are most pronounced in very wet lowland forests, especially on long-lived understorey leaves and tree seedlings (here in Kutai, Kalimantan, Indonesia).

Photo by Douglas Sheil.

2003), and delayed greening may make leaves less visible[8] thereby reducing potential herbivory, or less nutritious thus reducing the cost of any losses, or the colours may reflect defence compounds (possibly against fungal pathogens). A study of saplings from eight *Shorea* (Dipterocarpaceae) species found those with delayed greening indeed suffered less herbivory (Numata et al. 2004). Pigments in young foliage may also reflect defence of vulnerable tissues against strong light, but evidence is lacking (Cai et al. 2005).

Senescent leaf colour can reflect **anthocyanin** (reddish) or **caratenoid** (yellow) pigments. Unlike autumnal temperate forests, brightly coloured senescent leaves are uncommon in rain forests, and can therefore be used to guide identification; for example, *Terminalia* (Combretaceae) and *Elaeocarpus* (Elaeocarpaceae) leaves wither to red, and the fallen leaves of some *Eleocarpus* and *Baccaurea* (Euphorbiaceae Phyllanthaceae) are yellow.

2.2.2 Trees, shrubs, and treelets

There are around 37,000 species of tropical forest trees (Odegaard 2000). Certain groups are especially important. Large Fabaceae are common in many forests. The Meliaceae too occur on all tropical continents and include various important timber species like the Neotropical *Swietenia* (true mahoganies), and their African (*Khaya* and *Entandrophragma*) and Australian-Asian (*Toona*) counterparts. Some families, including Dipterocarpaceae in Asia and to a lesser degree Chrysobalanaceae in South America, attain high regional diversity and dominance.

Trees range in size from unbranched treelets less than a person's height to New Guinea's 89-metre *Araucaria hunsteinii* (Araucariaceae) (Whitmore 1998). Tropical forests commonly contain many smaller trees, and considerable species diversity resides in these smaller stems. Important families include Rubiaceae (including wild coffee *Coffea* spp., understorey trees in Africa), Violaceae, Euphorbiaceae, Ochnaceae, Melastomataceae and palms

(Arecaceae). Shrubs (woody plants with several main stems) are usually scarce but can be locally common where forests are impacted by browsers or fire, as occurs on many edges.

Being tall: stems and wood

While leaves are often inaccessible, a tree's stem can provide clues to its identity (Keller 1996). Stem forms are strikingly varied, including round and straight (Asian *Agathis*; Araucariaceae); fluted (various *Alstonia*; Apocynaceae); or braided (such as Neotropical *Platypodium elegans*, Fabaceae, or the sinuous but durable stems of *Minquartia guinanensis* (Olacaceae Coulaceae) widely used as telegraph posts in the Central Amazon). Bark is diverse in colour and texture. It may be smooth, scaly, fissured, scrolled, or densely layered or fibrous, and may be shed in characteristic patterns. Colour varies from white (*Tristaniopsis*; Myrtaceae) to black (*Diospyros*; Ebenaceae). Neotropical *Calycophyllum* (Rubiaceae) has bark like burnished copper, while *Peltogyne paniculata* (Fabaceae) is distinctly red. Most smooth barks have a long-lived surface and commonly host lichens that may be specific: *Diospyros*, for example, commonly has a dark green microlichen (Whitmore 1998). Cutting stem surfaces can reveal distinctive layers, **exudates**, odours, and other clues[9].

Stem form reflects not only taxonomy, but mechanical and physiological factors (McMahon and Kronauer 1976; Niklas 1995; Midgley 2003). The form and structure of higher plants reflect a compromise between seeking favourable foliage distributions, mechanical stability, and gaining adequate water and nutrients. Few herbs are taller than a few metres, as they lack the strength provided by wood. The size of large forest trees is the evolutionary consequence of competition for light. As size increases, light capture typically improves, but the metabolic cost and/or physical ability of the plant to support and maintain itself become limiting. Forest trees grown in the open, without competition for light from neighbours, usually grow shorter and wider than when in the forest.

---

[8] It has been proposed that the ability to find such coloured, young, palatable foliage contributed to the development of colour vision (red-sensitivity) in higher animals, including old-world primates (Lucas et al. 1998; Osorio et al. 2004).
[9] *Dillenia* spp. (Dilleniaceae) hiss audibly when cut. The sap of most Anacardiaceae rapidly turns dark.

Most plants we call 'trees', 'shrubs', and 'treelets' undergo secondary growth via an active vascular **cambium**. This means the stem gains breadth and strength as the plant grows taller. Most stem tissue is wood (secondary **xylem**), providing structural support and facilitating effective water transport. As a young tree grows, the lower branches are usually shed ('self-pruning'). This reduces weight and maintenance requirements and may help to shed lianas and hemi-epiphytes.

Wood is a mechanically, anatomically, and biochemically complex and diverse material. Densities vary tenfold, reflecting the relative abundance of secondary and structural compounds (mainly **lignin**) and the abundance and size of air cavities within the cell walls. Density and strength are closely correlated, though the densest timbers can be brittle. One Amazonian plot with 268 tree species possessed wood densities (dry weight) ranging from 0.14 to 1.21 g cm^{-3} (per species mean = 0.65) (Fearnside 1997c). Tropical wood densities range only a little wider, from around 0.1–0.12 g cm^{-3} for balsa *Ochroma lagopus* (Malvaceae)—the fast-growing Neotropical canopy species beloved by model plane enthusiasts—to 1.3–1.4 g cm^{-3} for the tall, Neotropical *Piratinera guianensis* (syn. *Brosimum guianense*; Moraceae), 1.1–1.4 g cm^{-3} for the short Caribbean *Guaiacum officinale* (Zygophyllaceae) and 1.2 g cm^{-3} for Asian *Eusideroxylon zwageri* (Lauraceae). Some timbers contain hard mineral grains that quickly blunt any steel saw. Greenheart *Ocotea rodiaei* (Lauraceae) from Guyana and Surinam contains toxic **alkaloids** that can cause respiratory problems in those sawing it, but it is highly prized for marine construction due to its durability and resistance to seawater and marine organisms.

Roots

Roots anchor trees and access nutrients and water. A typical tree possesses millions of fine rootlets, each covered with even finer **root hairs**. Below-ground structural characterizations are rare. In many rain forests the root **biomass** is concentrated near the soil surface, and sometimes a dense root mat occurs. Tropical forests are intermediate in the global range of rooting depths[10]

(Schenka and Jackson 2005) with average root depths of 7.3 ± 2.8 m (standard deviation) (Canadell et al. 1996) in tropical evergreen forests and reaching an impressive 18 m in seasonally dry, evergreen forests near Pará in Eastern Amazonia (Nepstad et al. 1994).

Tree roots possess various structural forms (Jenik 1978). These influence and are influenced by the physical forces determining tree stability (Chiatante et al. 2001). Tropical forest tree roots are typically shallow (Figure 2.6) but occasional 'sinker roots' branch from lateral roots downwards and help anchor the tree. Deeper roots generally provide more reliable access to moisture than shallow roots in the same location. In Sarawak, Borneo, seedlings of sandy-soil specialist species (*Dryobalanops aromatica* and *Scaphium borneense*; Dipterocarpaceae and Sterculiaceae Malvaceae respectively) possess deeper tap roots than their clay specialist congeners (*D. lanceolata* and *S. longipetiolatum*), apparently reflecting the more severe moisture limitations in sandy soil (Yamada et al. 2005).

Swamp environments and flooded sites pose various problems. In shallow and tidal environments, knee roots (aerial loops) and **pneumataphores** (peg roots) help roots to respire in **anoxic** soils (for example, various **mangroves**) (Chapter 7). In more seasonal and deeply flooded forests like the Amazonian *varzea*, such roots are absent, but buttressing and **stilt roots** are common (Parolin et al. 2004a). In areas of dense sedimentation, many trees also grow adventitious roots that allow access to the aerated flood waters.

Buttresses—flanges joining tree roots and lower trunks (Figure 2.7)—are characteristic of tropical forests. They occur much more frequently on larger trees (Smith 1972; Richards 1996), on stems lacking deep tap roots (Crook et al. 1997), and in forests growing on soft, thin, and waterlogged soils (Richards 1996). Buttresses grow at least partly in response to forces borne by the individual tree (Chapman et al. 1998). Buttresses link forces from the sinker roots to the trunk and vice versa, and often extend further on the side of the trunk that faces the prevailing wind (Lewis 1988). They are usually in tension and will resonate when struck[11].

---

[10] Global rooting depth ranges from 0.3 m in tundra to 68 m for *Boscia albitrunca* (Capparaceae) in the central Kalahari.

[11] Territorial chimpanzees *Pan troglodytes* sometimes beat upon them and can be heard for considerable distances in East African forests (DS, pers. obs.).

Figure 2.6 Most tropical rain forest tree roots are shallow, as seen here exposed on a path at Bako, Sarawak, Borneo.

Photo by Julia Born.

Though the principal purpose of buttresses appears to be stability (Smith 1972; Fisher 1982; Chapman et al. 1998; Clair et al. 2003), various additional adaptive explanations include improved root respiration in anoxic soils, and protecting stems from bark-damaging herbivores (Sheil and Salim 2004). The stability explanation is consistent with simple tests that find buttressed trees hard to pull down (Crook et al. 1997), but does not explain morphological variation (Fisher 1982) or why they are rare outside the tropics (Smith 1972)[12].

Stilt roots[13] are frequent in swamp forests (Figure 2.7) and mangroves (mainly Rhizophoraceae), where they are viewed as adaptations to soft and waterlogged soils, improving stability and oxygen supplies (Jenik 1973). Occasionally they appear to 'prop up' leaning trees growing in very soft ground, such as *Protomegabaria stapfiana* (Euphorbiaceae) in West Africa (Jenik 1978). Stilts can also develop when a tree germinates on a raised substrate that subsequently rots away (Santiago 2000). Stilt roots are common on certain fast-growing,

[12] It has been suggested that the additional surface area of the lower tree stem makes buttresses metabolically expensive in environments where species require thick bark (Smith 1972).

[13] Distinct from stems which survive once erosion uncovers their roots.

A

B

Figure 2.7 Trees in many lowland forests, as here in Sabah, Borneo, are supported by various means, including stilts (A) and buttresses (B).

Photos by Julia Born (A) and Marlina Ardiyani (B).

dryland species where they appear to facilitate rapid juvenile growth, but may persist to support the mature trees. A benefit, especially for palms and other species lacking secondary thickening, is that vertical growth does not need to wait for full diameter size to be established. In Costa Rica, the stilt-rooted palms *Socratea durissima* and *Iriartea gigantean* (Arecaceae) were both taller, at equivalent diameter, earlier in development than *Welfia georgii*, a palm lacking stilts (Schatz et al. 1985).

Intriguingly, stilt roots may allow trees to 'walk'. Juvenile palms obstructed by physical obstacles will lean sideways; as the tree sends out new roots, the lower trunk and older roots rot away, and the crown 'walks' to new locations (Bodley and Benson 1980). Some *Pandanus* (Pandanaceae) likely do the same.

Apogeotropic roots, often called simply 'climbing roots' due to the manner in which they climb up stems, represent a peculiar rain forest adaptation to exploiting stem flows. These roots occur in several plant families and are found growing on nutrient poor soils. It appears that these roots provide a pathway in which nutrients entering the forest system as stem flow can be carried from one stem to another without entering the soil solution (Sanford 1987).

Crowns

Crowns constructed around a single leading apical shoot are called **monopodial**; those with multiple such shoots are called **sympodial**. Most young trees have monopodial crowns, as do various understorey treelets. Larger monopodial forms often provide longer, straighter stems with narrower canopies than sympodial forms, and are thus more commonly encountered as timber plantation species. In some large forest trees (such as *Araucaria*), the monopodial form can remain through maturity; others

undergo a metamorphosis to develop a sympoidal structure. Order of **branching** is often distinctive, from zero for tree ferns and palms, to 1–2 for *Dracaena* (Dracaenaceae/Ruscaceae) and *Pandanus* (Pandanaceae), to numerous as in many **legumes**. Branching angles are often typical of a given plant: many small boys in Africa recognize the dividing twigs of *Tabernaemontana* spp. (Apocynaceae) as perfect Y-shaped catapults.

In the main canopy, trees may exhibit the 'crown shyness' phenomenon where a gap is maintained between the extreme edges of canopies of adjacent trees (Figure 2.8). This has been observed in lowland rain forests worldwide (and in some other forest formations) but remains poorly understood. Possible explanations include mutual shading by adjacent trees causing abscission of leaves at the outer canopy edge, and mechanical abrasion of leaves and small branches among adjacent crowns (Putz et al. 1984; Rebertus 1988).

Deciduousness

Though true rain forests are mainly evergreen, most include some **deciduous** tree species. The proportion is

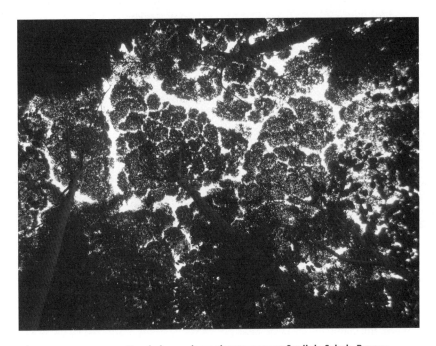

Figure 2.8 Crown shyness among canopy *Dryobalanops lanceolata* trees near Sepilok, Sabah, Borneo.
Photo by Julia Born.

low in the wettest and least seasonal regions and increases in regions with more pronounced dry seasons. For some plants, this ability is facultative: the plant only drops its leaves if conditions become dry (Borchert et al. 2002). In others it seems to be fixed. Exposed upper canopy species typically lose foliage while understorey species are more likely to remain evergreen. Leaf loss reduces water demand (Chapter 10), but deciduousness can also be a facultative response to heavy infestations of caterpillars, and some species, e.g. many *Ficus* spp. (Moraceae), have an obligatory leafdrop immediately before new leaf flushing.

Measurements of tropical forest deciduousness are scarce. One study, which considered canopy trees in three moist, semi-deciduous tropical forests in central Panama in sites with annual rainfalls of 2830, 2570 and 2060 mm yr^{-1}, recorded maximum levels of deciduous trees to be 4.8%, 6.3% and 24.3% (wetter to drier sites respectively) (Condit et al. 2000).

2.2.3 Climbers

About half of all vascular plant families include climbing species (Figure 2.9), and some, such as Convolvulaceae, Cucurbitaceae, and Vitaceae, are mainly climbers (Gentry 1991). Lianas are woody climbers that are, unlike tree stems, flexible in compression but strong in tension. Important families are Apocynaceae in Africa, Annonaceae, Araceae, and Arecaceae (the climbing rattan palm, Box 2.3) in Asia, and Araceae, Bignoniaceae, and Fabaceae in the Americas (Gentry 1991).

Lianas are abundant in most lowland tropical forests and account for about a quarter of woody stems and species in forests: one study recorded 1,597 lianas in a single hectare on Barro Colorado Island, Panama (Putz 1984). Densities generally decline with greater mean annual rainfall and increase with seasonality, but drop abruptly at latitudes over 24 degrees north (Schnitzer 2005). One explanation is that their vascular anatomy makes them vulnerable to freezing although structurally diverse in organization, their anatomy generally includes a high density of large water-bearing vessels (Schnitzer

2005). Sites in Africa and Madagascar have higher liana stem densities than Asia or the Neotropics (Gentry 1991; Schnitzer and Bongers 2002), and oceanic islands usually have few (Gentry 1991). Lianas and climbers often thrive in heavily disturbed forest (Kokou et al. 2002).

Lianas possess various climbing adaptations (Putz 1984; Nabe-Nielsen 2001). Adhesive climbers possess adhesive roots or tendrils[14], stem twiners twine tightly against their host (95% of species coil to the right[15]), branch twiners clasp their hosts using branches, tendril climbers use thin and mobile tendrils, and hook or thorn climbers use backward-pointing spines or hooks to attach to their hosts. There are also scramblers, which primarily lean on their host (Putz 1984; Nabe-Nielsen 2001). Hosts differ in their susceptibility to these processes. For example, few twiners can climb smooth **boles** over 10 cm diameter (Putz 1984). An additional characteristic of certain climbers such as *Monstera deliciosa* (Araceae) is that the seedlings, upon germination, grow towards darkness until they encounter the base of a tree up which they can then climb toward the light.

Larger trees often possess more lianas than smaller stems, even though only adhesive climbers can climb such stems directly. This appears to reflect prior infestation and later accumulation: one liana provides access for others to follow, though they can also sometimes move among crowns (Putz 1984; Nabe-Nielsen 2001). Nonetheless, areas with smaller trees often have high climber densities, reflecting a combination of 'climbability' and the influence of disturbance (Nabe-Nielsen 2001).

2.2.4 Epiphytes

Epiphytes range from terrestrial species that occasionally grow on trees (not considered here) to truly epiphytic species that are seldom found elsewhere (Benzing 1995). Forks and crevices on larger stems that accumulate humus, spores, and seeds are often especially rich sites (Figure 2.10). Forest canopies support around

---

[14] In contrast to climbers in general, these seem especially common in wetter, less seasonal climates.

[15] The yams, *Dioscorea* (Dioscoreaceae), are amongst the most common left-handed exceptions.

Figure 2.9 Lianas in Loango National Park, Gabon (Central Africa).

Photo by Rhett A. Butler.

24,000 or more vascular epiphyte species from 84 families (Kress 1986) or about 10% of global plant diversity. This diversity is almost exclusive to tropical forests (Zotz 2005). Over two-thirds are monocotyledons such as Orchidaceae and Bromeliaceae, around 20% are other dicots, and ferns and their allies comprise the rest (Kress 1986). Only five known gymnosperms are true epiphytes[16]. Lowland communities are dominated by

vascular plants, while bryophytes prevail at higher altitude.

In Ecuador, epiphytes represent about a quarter of all vascular plant species. African forests are commonly believed to have relatively few epiphyte species compared with the Americas or Asia (Nieder et al. 2001) where 10–15% is typical (Nieder et al. 2001). Indeed, plot-based records suggest epiphytes are around 8% of total vascular species in Ghana, and only 2–3% in central Africa (Nieder et al. 2001). The situation in Eastern Africa is less distinctive with epiphytes comprising as much as 20% of forest species in Budongo, Uganda (Synnott 1985). These differences across and within regions are probably related to moisture, seasonality, and climatic history[17].

True vascular epiphytes are usually evergreen. They include woody and herbaceous forms, and can possess diverse specialized features (Benzing 1995). Epiphytes can be long-lived—50 years or more for some orchids (Zotz 1995)—but tend to grow very slowly (Laube and Zotz 2003). Many species possess specialized adhesive roots that attach to tree branches where they forage for moisture and litter, in some cases creating structures to capture these (Figure 2.11). Some possess green roots, and some orchids (Orchidaceae) have lost their leaves altogether, such as *Microcoelia* spp. in Africa and *Campylocentrum* spp. in the Neotropics. Some appear able to move short distances by vegetative growth to gain a more advantageous position.

Epiphytes cannot use terrestrial soil as their water reservoir, and even very short periods without rain can cause water stress. Epiphytes are therefore often succulent and xeromorphic with tough leathery leaves. The Cactaceae, a family with many desert specialists, includes numerous wet-forest epiphytes, including the only indigenous **paleotropical** representative of the family *Rhipsalis baccifera*[18]. Many smaller plants are **poikilohydric** (able to survive desiccation), including most bryophytes and a few **filmy ferns** (Hymenophyllaceae) whose leaves are only a single cell thick, some larger ferns and even a few angiosperms. The

[16] These include epiphytic cycads in the Neotropics.
[17] Wet African mountains appear rich in epiphytes, which comprise 10% of vascular plants at Mount Cameroon in West Africa (Cable and Cheek 1998) and 8–9% in the Usambaras in Tanzania (Iversen 1991).
[18] Which reaches to Africa, Madagascar, the Comores, Seychelles, Mascarenes, and Sri Lanka.

Figure 2.10 In some wet forests—here, swamp forests in Papua—epiphytes such as these pendulous ferns colonize any available surfaces on which they can establish themselves.

Photo by Douglas Sheil.

Figure 2.11 *Scaphyglottis prolifera* (Orchidaceae) wrapping its adhesive roots around the branch of a tree.

Photo by Ernest Hennig.

epiphytic fern *Polypodium polypodioides*[19] (Polypodiaceae) can recover almost full vigour in the first 30 minutes of

rehydration after losing 97% of its normal water content (Stuart 1968).

Many detritus-catching ferns exploit their humus collection as a sponge—this wetness is also available to others. For example, various *Platycerium* (stag's horn ferns; Polypodiaceae) capture litter and water in a tank of leaves where other plants may germinate and grow. These detritus catchers form a continuum with water-tank forms, or **phytotelmata**. In **bromeliad** tanks, epidermal structures (**trichomes**) absorb nutrients from the detritus-enriched liquids that accumulate within them (Givnish et al. 1984).

Moisture, shade, organic matter, bark textures, fungi[20], and surface chemistry all appear to affect epiphytic distribution, and some very specialized associations occur though appear rare. The orchid *Cymbidiella pardalina* seems to grow only within the epiphyte *Platycerium madagascariense* (Polypodiaceae). Some associations are informative: for example tree fellers in Kalimantan will

---

[19] Also called the 'resurrection fern' on account of its desiccated appearance and rapid and visually dramatic recovery on wetting.

[20] Orchid seeds carry no food reserves and need specific fungi to establish and grow. As the fungi appear to gain little from this association, these orchids might better be viewed as parasites.

not cut a 'yellow meranti' (various species of pale-tim-bered *Shorea*, Dipterocarpaceae) holding *Drynaria* ferns (Drynariaceae), as they claim such trees are always rot-damaged, making the timber worthless.

Available surfaces are colonized by various nonvascu-lar plants: **cyanobacteria**, green algae, bryophytes, liv-erworts, lichens (Chapter 3), and small filmy ferns. On leaves these life-forms are called epiphylls, and they tend to be more common in the understorey. A small number of **cryptogams** occasionally colonize insects such as beetles (Gradstein and Pocs 1989), while sloths harbour algae in their fur.

2.2.5 Hemi-epiphytes

Hemi-epiphytes have both epiphytic and grounded life phases and occur widely in tropical forests, typically on around 10% or more of canopy trees. They are richer and more abundant in wetter regions, while some *Ficus* (Box 2.4) species are locally abundant even in drier forests.

Primary hemi-epiphytes (probably over 1,000 species) establish epiphytically, later sending down roots or shoots to the ground. They occur in at least 19 tropical families (plus three temperate), including the 'stranglers' Moraceae (*Ficus*) (Figure 2.12), Araliaceae, and Clusiaceae (Williams-Linera and Lawton 1995). Young plants invest considerable resources in connecting to the ground, and once the connection is established their leaves become less **xeromorphic** as moisture availability increases (Holbrook and Putz 1996). In some Clusiaceae, Araceae, and *Ficus*, roots are freely descending and thick. Some form distinctive **anastomozing** networks of strong **aer-ial roots** that envelop the supporting structure (Williams-Linera and Lawton 1995). Such fused stems can and do arise from multiple, genetically distinct individuals[21] (Thomson et al. 1991). Stranglers often kill their host and remain as free-standing trees, and some can colonize neighbouring stems in turn.

Secondary hemi-epiphytes (around 700 species glo-bally, all monocots of the Araceae, Cyclanthaceae, Marcgraviaceae) establish terrestrially, ascend their hosts as adhesive climbers, and, once a favourable location has been achieved, sever connection with the ground[22] (Kress 1986). The potential length of aerial roots can limit establishment height: *Anthurium clavigerum* (Araceae) in Panama, for example, rarely manages to reach more than 5 m above ground (Meyer and Zotz 2004). Different spe-cies show different host distributions, but these are sel-dom specific. Palms, tree ferns and older trees (>50 cm **dbh**) are common hosts (Williams-Linera and Lawton 1995).

2.2.6 Herbs

Rain forest herbs include diverse taxa and numerous forms. The understorey can include around half the total vascular plant species, and tropical rain forests would rank among the world's richest vegetation even if woody plants were ignored (Gentry and Dodson 1987). Herbs are often very patchily distributed, and typical densities range around only 0.5–2 m^{-2}, although sometimes more than 20 (Costa 2004). Annuals are largely absent. Toleration of low light and intermittent water availability have made many of these rain forest plants familiar house plants in temperate countries, including various Gesneriaceae, Commelinaceae, Zingiberaceae, ferns, and dwarf bamboo.

The wild bananas *Musa* spp. (Musaceae) of Asia are—like the related *Heliconia* (Heliconiaceae), *Ravenala* and *Strelitzia* (Strelitziaceae), and *Ensete* (Musaceae) of other regions—large perennial herbs. In *Musa*, tightly furled new leaves grow up through the centre of the non-woody stem before opening into a large oblong blade. Any given stem produces only a fixed number of succes-sive leaves before it produces an **inflorescence**. After the fruits mature and fill with hard seeds[23] the stem dies, but plants often **sucker**.

---

[21] A multi-branch sampling of 14 fig trees in Panama gave evidence for 45 distinct genetic individuals!

[22] As *Monstera acuminata* ascends a support, aerial roots are produced along the stem which eventually dies back from the base leaving only aerial feeding roots connecting the stem to the soil (Lopez-Portillo et al. 2000).

[23] The bananas typically carried and sold around the world are a sterile seedless form maintained only by human cultivators who plant suckers.

Box 2.4   Figs

Figs (*Ficus* spp., Moraceae) include almost 900 species, almost all of which occur in tropical forests (Janzen 1979; Weiblen 2002). *Ficus* is the only woody genus found in virtually all lowland tropical forests, though often at low densities. Some are climbers, some are shrubs or herbs, a minority are trees, but over half are hemi-epiphytes (Berg 1989). They are largely limited to tropical environments by their dependence on specialized wasps for pollination (Chapter 12). About half the species are functionally dioecious (Harrison and Yamamura 2003).

Figs possess a unique closed inflorescence called a syconium. For the fig-pollinating chalcid wasps, the syconium is a brood-site. Adult pollen-laden female wasps enter the syconium through a small gap and oviposit on hundreds of flowers within. The larvae feed on the ovules and flowers, pupate, and emerge as adults. After mating, the females, now covered in pollen from remaining flowers, leave the syconium in search of another flowering fig tree of the same species. The syconium ripens to form 'pseudo-fruits' containing the seeds.

Figs produce their ripe, sugar-rich syconia ('fruits') throughout the year, providing an important resource for frugivores when other foods are scarce. By providing fruit in times of fruit scarcity, they may

sustain vertebrates that are seed dispersers for many plant species at other times of the year (Shanahan et al. 2001b). They also seem to be an especially rich source of calcium for vertebrates (O'Brien et al. 1998). At least 1,274 bird and mammal species eat figs, along with a handful of reptiles and fishes (Shanahan et al. 2001b). Figs are good colonizers, even reaching distant sites such as new islands (Shanahan et al. 2001a).

Figure 2.12 A towering fig tree on the edge of a clearing in Bacan, Moluccas, Indonesia. Note the two people for scale.

Photo by Douglas Sheil.

Table 2.4 Global diversity of the genus *Ficus*.

Region	Number of subgenera	Number of species
Indo-Pacific	6	> 500
Borneo	6	>160
Papua New Guinea	6	> 150
Afrotropics	5	112
Neotropics	2	132
Global	7	> 800

From Harrison 2005.

Herbs typically possess softer leaves than woody plants, but the distinction is artificial. Some 'herbs' develop secondary lignified tissues (wood) in their basal parts (common in Acanthaceae, Gesneriaceae, and Rubiaceae) and are sometimes referred to as woody herbs or sub-shrubs. Patterned leaves are common (Box 2.5).

2.2.7 Carnivores

Carnivorous plants are a minor, yet ecologically fascinating, component of forest vegetation. Often absent on richer soils, they are favoured in wet, well-lit, nutrient poor habitats where **arthropod** prey (usually insects) are available. The nitrogen obtained through prey capture must compensate not only for nutrient scarcity, but also pay for maintaining the adaptations (Givnish et al. 1984). Suitable hunting sites include rocky outcrops, forest on wet sand, and clearings (pitcher plants are often common on roadsides). Example plants include sticky-haired sundews (*Drosera* spp., Droseraceae), and pitcher plants (Box 2.6).

2.2.8 Parasites

Parasitic plants are represented by more than 4,100 species (277 genera) spread across most major biomes (Press and Phoenix 2005), including most tropical forests[24]

(Nickrent 2002). *Mycotrophs* obtain carbohydrates and nutrients by parasitizing **mycorrhizal** fungi. **Haustorial** *parasites*, exclusively dicotoledons with one exception (Musselman 1980), use structures known as *haustoria* to penetrate host tissues. To gain water and nutrients, haustorial parasites require lower water potentials than their hosts—most readily accomplished by maintaining high transpiration rates. This requirement may explain why photosynthetic **hemiparasites** are rare in the understorey, and non-photosynthetic **holoparasitic** species predominate on the forest floor. Parasites almost certainly impact host productivity, viability and/or reproduction, but evaluations in tropical rain forests are lacking.

Mistletoes (Loranthaceae: around 1,400 species globally) are primarily tropical, green and shrubby haustorial hemiparasites. Most occupy above-ground, epiphyte-like positions on other woody plants on which they depend for water (Reid et al. 1995). Most rain forest species lack host specificity (Barlow 1981). Some send out extensive vegetative runners that can eventually form arboreal networks across multiple hosts (Reid et al. 1995).

Hemiparasitic trees include the scented sandalwoods (*Santalum*) of the Asia-Pacific region and *Okoubaka aubrevillei* (Santalaceae) of West Africa, which reaches nearly 40 m (Veenendaal et al. 1996). Only one gymnosperm is considered a parasite, the rare New Caledonian *Parasitaxus ustus* (Podocarpaceae) that attaches to the

---

Box 2.5   Patterned foliage

Patterned foliage is common in tropical rain forest understoreys, especially among herbs (Figure 2.13). The variegated form of the herb *Schismatoglottis calyptrata* (Araceae), from Sabah, Malaysia, possesses a greyish pattern due to air-spaces in the leaves. Though this patterning reduces photosynthetic efficiency, the blotchy form coexists widely with an unvariegated form, suggesting some benefit (Tsukaya et al. 2004).

Various advantages have been suggested for specific cases of patterns and speckling (Givnish 1990; Lev-Yadun 2003; Lev-Yadun et al. 2004). Patterns

may draw pollinators to flowers, prey to carnivorous plants, reveal homes or hiding places to beneficial animals, or help to conceal them. Broken outlines may camouflage plants from herbivores. Patterns may indicate, mimic or accentuate anti-herbivore properties. Bold patterns may undermine or challenge the camouflage of invertebrate herbivores, exposing them to increased predation. Most plausibly, if herbivores avoid leaves that are already damaged or colonized by other herbivores, then selection might favour colourations and/or forms that mimic

---

[24] Typically Balanophoraceae, Rafflesiaceae, Mitrastemonaceae, Opiliaceae, Loranthaceae, Lauraceae (*Cassytha*), and Olacaceae.

these states. Such damage mimicry may also be relevant to various leaf forms.

The mature leaves of *Amorphophallus titanum* and *A. gigas* (Araceae) reach up to 4 m, supporting an umbrella-like crown, and are the largest known plant structures that gain their rigidity solely from turgor (Hejnowicz and Barthlott 2005), but these structures remain fragile. It has been suggested that their typically blotchy colouration resembles tree saplings so as to discourage animals from running into them (Hejnowicz and Barthlott 2005).

Figure 2.13 Many understorey herbs possess strikingly patterned leaves: (A) *Alocasia* sp.; (B) *Pilea* sp.; (C) *Calathea lancifolia*; and (D) *Episcia* sp.

Photos by Douglas Sheil (A–B) and Jaboury Ghazoul (C–D).

Box 2.6   Pitcher plants

Pitcher-carnivory occurs among the paleotropical *Nepenthes* (Nepenthaceae) (Figure 2.14), some tank bromeliads (Bromeliaceae) in the Neotropics, and in *Heliamphora* spp. (Sarraceniaceae) endemic to the rock plateaux (tepuis) of Venezuela and neighbouring regions.

Pitchers possess various finely honed adaptations to attract and catch their prey. They attract insects using high contrast markings in visible and/or UV light, and sometimes provide nectar (Moran et al. 1999; Biesmeijer et al. 2005). The surface of the *Nepenthes* pitcher rim (peristome) possesses a regular microstructure with 'radial ridges of smooth, overlapping epidermal cells'. This surface develops a liquid film under humid conditions that resists adhesion by arthropod footpads (Bohn and Federle 2004). Despite a rough glandular surface, even the digestive chamber reduces the possibility of a sound footing as slippery, hood-like structures are placed over the glands (Gorb et al. 2004). Similarly, the pitcher bromeliads *Brocchinia reducta* and *Catopsis berteroniana* possess loose, thread-shaped 'crystalloids'. Experiments show that flies cannot easily walk on these surfaces, and quickly lose mobility, including the ability to escape by flight, owing to the waxes adhering to their feet (Gaume et al. 2004). The pitcher fluid itself is neither toxic nor caustic, but the presence of detergent-like compounds reduces surface tension and increases wetting. Prey typically drown and decompose.

Figure 2.14 *Nepenthes* pitcher plants (found from Madagascar, via Sri Lanka, through Melanesia to New Caledonia) include climbers, small herbs, and occasional epiphytes. This example in the wet forest of Brunei, Borneo.

Photo by Douglas Sheil.

---

roots of *Falcatifolium taxoides* (also Podocarpaceae). Other life-forms are scarce, but examples include *Dendrotrophe buxifolia* (Santalaceae), a small twining shrub from **Malesia**. Many parasites have low host specificity—*Santalum* spp. can infect grass and trees, and are known to damage underground cables—but some, such as *Rafflesia*, are specialized (Box 2.7).

2.2.9 Saprophytes

Chlorophyl-free **saprophytes** (actually *holoparasitic mycotrophs*, as only the fungi are truly saprophytic) are widespread but often hard to find as they appear only briefly to flower (see Figure 2.1), sometimes as multi-species aggregations. They are called *tumbuhan hantu* or

---

Box 2.7   One big flower: *Rafflesia*

All three Asian genera of the Rafflesiaceae family (*Rafflesia*, *Rhizanthes*, and *Sapria*), including the approximately 14 *Rafflesia* species, are parasites of *Tetrastigma* (Vitaceae) climbers (Nickrent 2002). Vegetatively, *Rafflesia* are little more than fibre clusters within the root tissues of their host. Yet *R. arnoldi* from Sumatra boasts the world's largest single flower reaching almost 1 m in diameter. Because *Rafflesia* derives its resources almost entirely at its host's expense, there is little selective pressure to moderate the size of its flowers (Figure 2.15). Flower buds take up to nine months to develop, and open for only a few days. The pollinators are probably carrion flies.

Figure 2.15 A flower of *Rafflesia keithii*, Sabah, Borneo.

Photo by Axel Poulsen.

'ghost plants' in Indonesia and neighbouring regions. One intensive study of herbs in Belalong Brunei (Borneo) recorded 10 ghostly species in five families, *Lecanorchis* (Orchidaceae), *Epixanthes* (Polygalaceae), *Burmannia* (Burmanniaceae), *Sciaphila* (Triuridaceae), *Petrosavia* (Melanthiaceae Liliaceae) (Earl of Cranbrook and Edwards 1994).

2.3   Conclusion

Tropical forests are generally rich in plant species and higher order taxa. Many species remain undescribed, and few have been studied ecologically. Many unusual and unique life-forms are found among a characteristic range. There is a danger that, by emphasizing the great diversity of forms harboured within them, rain forests will be seen as a bizarre collection of botanical oddities and freaks—an impression that would be unbalanced and unfortunate, as their conservatism and conformity should also be appreciated. Despite the taxonomic richness, most rain forest plants possess a form that reflects the success of a very limited number of life-history choices: trees, lianas, herbs, and epiphytes. This conformity goes further, as within each form the variations are seldom so marked as to readily tell any but the most familiar botanist what any particular plant actually is, or where it comes from. Indeed, as we shall see in Chapter 7, it is the conservatism of the vegetation in relation to the environment that allows a basic classification of rain forests separate to the species they contain. Having now glimpsed the diversity of plant life in the tropical rain forest we must consider where it comes from (Chapters 6 and 8), how it behaves (Chapters 10 and 12), and what will happen to it in the future (Chapter 17).

CHAPTER 3

The great unseen: fungi and microorganisms

'Today the tropical rain forests, which may contain more than half the species of plants and animals on Earth, grow on a mat of mycorrhizal fungi.'

E. O. Wilson (1992)

Fungi are the great unseen. They occur throughout the entire rain forest, from deep in the soil to high within canopy leaves. They perform multiple roles as **pathogens**, parasites, saprophytes and **mutualists**. Together with bacteria and other microorganisms, fungi are the least known components of tropical rain forest biodiversity, and yet control many of the most vital processes on which these rain forest ecosystems depend (Hawksworth and Colwell 1992). It is a reflection of how little is known that we treat fungi, **oomycetes**, bacteria, and slime moulds within a single short chapter even though they represent four different Kingdoms (Mycota Stramenopila Prokaryota and Protista respectively[1]). We dwell on fungi: we know most about them, and some at least are visually striking, comparatively familiar, and known to be functionally important. We address fungi as mutualists, **commensals**, saprohpytes, and pathogens, and briefly consider the vast and largely unknown role and diversity of other rain forest microbes.

3.1 Fungi

Worldwide, fewer than 100,000 species of fungi have been described, but most experts believe many more remain undescribed (Box 3.1) (Hawksworth 2001). Discovery must usually await the emergence of the fruiting body (Figure 3.1), a sporadic and unpredictable event, although genetic tools are providing new ways to reveal and distinguish species. Close to nothing is known in detail about the biology, ecological function, host range, or distribution of all but a few tropical species. This is striking, as fungi are immensely important as essential decomposers, as plant **symbionts** facilitating nutrient uptake, and as pathogens that influence plant distributions and arthropod populations (Table 3.1). Fungi often represent valuable food (Figure 3.2): in lowland rain forest in Sulawesi 40% of 1,250 beetle species feed directly on fungal fruit bodies or on fungal **mycelium** in wood (Hammond 1990). In Africa mountain gorillas sometimes chew on rotting wood, suggesting that the associated fungus may be a significant source of scarce nutrients like calcium.

3.1.1 Mycorrhizas

Mycorrhizas are mutualistic associations between fungi and plant roots, and are widespread in the tropics and elsewhere. Indeed, most (75-80%) vascular rain forest

---

[1] Note that names and classification vary depending on the authority used.

Box 3.1   Fungal diversity is mushrooming

Until quite recently few people had considered fungal diversity in the tropics. Most fungi are microscopic and undescribed, so defining and counting species remains a daunting challenge (Hawksworth 2004). Exacerbating this challenge is the realization that fungal diversity is far higher than previously thought: for example, 350 fungal species have been isolated from a small sample of leaves taken from only two Panamanian tree species (Arnold et al. 2000).

A figure of around 1.5 million extant fungal species globally (Hawksworth and Rossman 1997) is extrapolated from plant–fungi ratios of 1:6, based on a few well known temperate localities. Detailed work on tropical palms in Australia and Brunei has,

however, revealed plant–fungal ratios of around 1:33 (Frohlich and Hyde 1999; Frohlich et al. 2000) and current estimates of global diversity range from half-a-million to 10 million species (Hawksworth 2001). Owing to recent appreciation of the richness of previously overlooked groups such as lichens and **endophytes**, it is feasible that early estimates of 1.5 million species (Hawksworth 1991), derided by many as being gross overestimates, may in fact be gross underestimates (Arnold et al. 2000; Hawksworth 2001). With few available studies it is difficult to be confident about any of these figures, but it seems certain that tropical rain forests harbour a considerable degree of yet to be discovered fungal diversity (Hawksworth and Rossman 1997).

---

plants (including ferns and their allies) regularly form associations with mycorrhizal fungi (Fitter and Moyersoen 1996). Fungus–plant junctions allow transport of carbohydrates from the plant to the fungus, and mineral nutrients (especially phosphate) and water in the reverse direction, facilitated by the large surface area of the fungal mycelium that surrounds plant roots. These associations improve plant nutrient uptake and drought tolerance, and reduce vulnerability to plant pathogens, thereby greatly influencing the plant community (Kiers et al. 2000).

The two main mycorrhiza types are **arbuscular** mycorrhizas (AM), which have limited ability to degrade organic matter, and **ectomycorrhizas** (EM), which can access nutrients from labile organic material that is otherwise unavailable to plants. The common and widespread arbuscular mycorrhizas grow into the root **cortex** and form highly branched structures, arbuscules, within cortex cells. Arbuscules are responsible for nutrient exchange between host and fungus.

Ectomycorrhizas (Figure 3.3) are the next most abundant of mycorrhizal types, and evolved more recently. On a global scale, EM appear to be adapted to habitats rich in organic material (Fitter and Moyersoen 1996), although there is little evidence yet of such link in tropical forest systems specifically. Ectomycorrhizas are associated with a

select range of predominantly woody species (Pinaceae, Dipterocarpaceae, Fagaceae, *Eucalyptus*, Fabaceae [particularly Caesalpinoideae], Nyctaginaceae), including certain tree families and genera that form **monodominant** stands (Box 3.2). Until recently it was thought that the Dipterocarpaceae was only associated with ectomycorrhiza but now several AM–dipterocarp associations have also been identified (Tawaraya et al. 2003) (see also Table 3.2).

Additional forms of mycorrhiza include those associated with orchids (Pereira et al. 2005) and with the Ericaceae, ericoid mycorrhizas that can decompose humic material (Luteyn 2002). These associations are generally viewed as mutualistic, though orchid mycorrhizas sometimes behave as pathogens or saphrophytes and may gain little from the orchids.

Many higher plants can live in nutrient rich soils without mycorrhizas, but this is uncommon in tropical forests. Plants that apparently avoid mycorrhizal associations altogether are typically colonists of open ground (Proteaceae including the genus *Grevillea*), **haustorial** parasites (mistletoes), carnivorous species (e.g. *Drosera*, Droseraceae) and those like sedges (Cyperaceae) that live in waterlogged anoxic soils (Fitter and Moyersoen 1996). Plants without mycorrhizas usually possess finer roots, more root hairs, and relatively advanced chemical defences (Brundrett 2002).

Figure 3.1 The fruiting bodies of various tropical fungi.

Photos by Karin Beer.

Table 3.1 Key ecosystem processes mediated by microorganisms in tropical forests.

Process	Mechanism	Organism and group
Atmospheric/biotic nutrient cycling	N-fixation: autotrophic	Cyanobacteria on plants and soil: 14 genera
	N-fixation: heterotrophic	Yeasts and various bacteria in soil, wood and root surfaces
	N-fixation: symbiotic	As lichens, fungal associations with cyanobacteria. Bacteria (*Rhizobium* in legumes). In animal guts, bacteria (*Enterobacter*, *Klebsiella*) in termites
	Trace gas emissions	Mathanogenic bacteria and sulphur bacteria in the soil
	Denitrification	Bacteria: *Nitrosomonas* and many other genera
Carbon cycling	Photosynthesis	Algae, cyanobacteria (independently and in lichens)
	Respiration	All microorganisms
	Decomposition	Fungi and bacteria
Regulation of primary production	Competition for nutrients	Fungi and bacteria in soil, wood, and litter can compete with plants for limiting nutrients and reduce production
Regulation of secondary consumers	Pathogenic agents	Viruses and bacterial pathogens of insects, birds and reptiles
	Predators	Nematode trapping fungi, including wood-rotting basidiomycetes; slime moulds
Nutrient cycling	Decomposition, nutrient mineralization	Fungi and bacteria
	Mutualisms for nutrient uptake	Mycorrhizal fungi
	Leaching losses of N	Bacteria and fungi that convert ammonium and organic N to nitrate, which is more susceptible to leaching in most tropical soils
	Nutrient immobilization	Fungi and bacteria in soil and litter incorporate nutrients and prevent leaching losses during rains

Adapted from Lodge et al. 1996.

3.1.2 Endophytes

Plants harbour numerous non-pathogenic fungi or **endophytes** within their tissues. The ecological significance of endophytes is uncertain, as they appear to have no negative impacts on plant function, although some may provide plants with protection against insects and pathogens (Azevedo et al. 2000; Arnold et al. 2003). They are ubiquitous: they have been found in the leaves of every plant species examined so far. Evidence of host preference and spatial heterogeneity within leaves suggests that tropical endophytes may be extremely diverse, and perhaps with a diversity that is an order of magnitude higher than most comparable taxa (Arnold et al. 2000). For example, a single *Elaeocarpus* (Elaeocarpaceae) tree in Papua New Guinea yielded 200 endophytes, including numerous new species (Aptroot 2001).

Figure 3.2 Unidentified caterpillars feeding on a fungus in the Amazon.

Photo by Karin Beer.

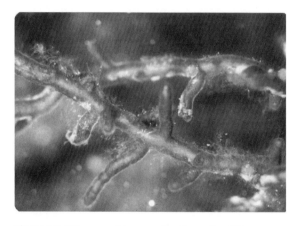

Figure 3.3 Ectomycorrhiza covering the roots of *Shorea leprosula*.

Photo by Lee Su See.

3.1.3 Saprophytes

Decomposition of wood (**lignin** and **cellulose**) depends on fungi and bacteria. Saprophytic fungi include a diverse range of macro- and microscopic species. The enzymes and acid compounds they secrete degrade large molecular complexes into simpler compounds. There is usually a succession of species, ranging from those that colonize fresh material to those that mop up afterwards.

Saprotrophic fungi form extensive **cords** and networks of aligned **hyphae** which can connect and translocate nutrients over several metres (Boddy 1993). Similar cords are used by some fungi as raised webs within the forest understorey for capturing litter before it reaches the ground. In tropical forests these fungal hyphae networks influence the local availability of nutrients (Guevara and Romero 2004).

Wood-decay **polypore** (and other) fungi can also be aggressive pathogens that attack living plants (Figure 3.4), often destroying roots or cambial tissue and ultimately killing trees or making them more susceptible to windthrow. Others simply colonize trees but do not spread until after tree death. Polypore fungi are certainly widespread, diverse and common: studies of Panamanian large wood-decay polypore fungi suggest that although the most abundant species colonize multiple hosts, fungal diversity in this group still exceeds that of the host trees (Gilbert et al. 2002; Ferrer and Gilbert 2003). Certainly most, if not all, tropical trees have associated

---

Box 3.2   Do mycorrhizal mutualisms maintain monodominance?

Although tropical forests are associated with tree diversity, they also harbour large areas dominated by one or a few tree species or genera. These taxa appear to be disproportionately associated with ectomycorrhizas as opposed to the more generally widespread arbuscular mycorrhizas. Thus it has been suggested that ectomycorrhizas facilitates monodominance by providing competitive advantages to trees that have them (Connell and Lowman 1989). Possible mechanisms by which ectomycorrhiza trees outcompete arbuscular mycorrhiza tree species include ectomycorrhiza fungi being more effective at obtaining nutrients in certain types of soils or in accessing nutrients locked in organic compounds. Ectomycorrhizas have also been shown to protect roots against herbivores and pathogens (Schelkle and Peterson 1996). Furthermore, they may form fungal networks that connect the root systems of adult trees with those of conspecific seedlings and saplings, allowing the former to subsidise the latter (Robinson and Fitter 1999; Onguene and Kuyper 2002; He et al. 2003).

The evidence for the ectomycorrhiza hypothesis is, however, equivocal and the theory remains inadequate (Torti and Coley 1999). Many monodominants possess ectomycorrhizas, but some such as *Gilbertiodendron dewevrei* and *Julbernardia seretii* (Fabaceae) from Ituri, Democratic Republic of Congo, and various Dipterocarpaceae also occasionally form associations with arbuscular mycorrhizas. Several well-known monodominants such as *Cynometra alexandri* (Fabaceae: Caesalpinioideae) in Uganda, *Mora excelsa* (Fabaceae: Caesalpinioideae) in Trinidad, and *Pentaclethra macroloba* (Fabaceae: Mimosoideae) in Costa Rica, appear to harbour only arbuscular mycorrhizas. Furthermore, many ectomycorrhiza species fail to attain dominance, such as *Manilkara* (Sapotaceae) species at Ituri, while some ectomycorrhiza taxa such as Dipterocarpaceae of Southeast Asia can form diverse stands themselves. In conclusion the ectomycorrhiza hypothesis as currently elucidated is insufficient to explain tropical monodominance.

---

wood-decay polypore fungi, and a substantial number of living trees of some tree species are also attacked.

3.1.4 Pathogens of plants

The majority of plant pathogens are fungi and oomycetes, although many bacteria are pathogenic too (see below). Pathogenic fungi and fungi-like microbes disperse as airborne spores and are particularly prevalent in the forest understorey[2] where conditions for germination prevail, notably high humidity, surface moisture, and cool temperatures (Gilbert 2005a,b). Foliar pathogens, mostly fungal, are perhaps the most familiar, causing necrosis, chlorosis, or leaf deformation. Pathogens also kill seeds and seedlings (damping-off diseases),

cause cankers in stems that block plants' vascular systems, decay wood in trunks and roots, and also attack flowers and developing fruits. The apparent ubiquity of these pathogens, and their impact on a number of life-history stages and plant parts, suggests that their collective impacts on plant performance and survival may be as great as, or even greater than, herbivores. Susceptibility to generalist pathogens varies among species, which could affect community composition (Pringle et al. 2007) (see also Chapter 8).

The main cause of damping-off disease, often a killer of tropical tree seedlings (see e.g. Hood et al. 2004) are *Phytophthora* and *Pythium* (Pythiaceae) oomycete species. On Barro Colorado Island, Panama, these pathogens infected 80% of examined species, and seedling

---

[2] In the Australian rain forest, spores have been reported to be 52 times more abundant in the understorey than canopy (Gilbert 2005a).

Table 3.2 Characteristics of the major types of mycorrhiza.

	Ectomycorrhiza	Arbuscular mycorrhiza	Ericoid mycorrhiza	'Orchid' mycorrhiza
Fungi involved	Basidiomycotina, Ascomycotina	Glomeromycota	Ascomycotina	Basidiomycotina (and others)
Plants involved	Taxonomically diverse (mainly woody)	Taxonomically diverse (widespread including pteridophytes)	Ericales (most examples known from temperate regions)	Orchidales
Ability to degrade organic matter	Labile organic matter	Limited	Humified material	Organic matter in soil and some wood-rotting species
Phosphorous for plant	Important	Important	Not known	Exclusive
Nitrogen	Important	Some	Important	Exclusive
Water	Some	Some	Not known	Not known
Pathogen	Sometimes	Sometimes	Not known	Not known
Other	Micronutrients Possible protection from toxins	Possible gain in micronutrients	Protection from high concentrations of Al, Fe, Cu and Zn, and to low pH	Net carbon transfer to plant
Net carbon transfer	To fungi	To fungi	To fungi	To plant

Derived and updated from Fitter and Moyersoen 1996.

Figure 3.4 Wood decaying fungi from Tapajos, Brazilian Amazon.

Photo by Douglas Sheil.

mortality of *Platypodium elegans* ranged from 35 to 81% (Augspurger 1983; Augspurger and Kelly 1984). Once considered fungi, oomycetes are more closely related to photosynthetic algae such as diatoms, and to the malaria parasite (Tyler et al. 2006). The best known example is *Phytophthora cinnamomi*, first isolated from cinnamon trees in Sumatra in 1922, but it is likely to have originated in or near New Guinea (Hardham 2005). It is now widespread and has become a major crop pest and threat to natural ecosystems. Pathology and persistence seem enhanced by moist conditions (Hardham 2005).

Seeds in the soil seed bank are also vulnerable to fungal diseases. Seeds of *Cecropia insignis* (Cecropiaceae) and *Miconia argentea* (Melastomataceae) in the soil suffer 39% and 47% annual mortality respectively (Dalling et al. 1998; Hood et al. 2004). Seed and seedling pathogenic agents are believed to be key players in the density-dependent processes that contribute to the maintenance of high species diversity in the tropics (Augspurger 1983; Gilbert 2002; Hood et al. 2004; Augspurger and Wilkinson 2007) (Chapter 8).

Foliar pathogenic fungi affect many larger plants but are rarely lethal and mainly reduce the photosynthetic capacity of leaves. They are remarkably widespread and abundant. Fungal pathogens accounted for 34% of leaf damage on 10 tree species in lowland tropical forest in Panama (Barone 1998), and in Los Tuxtlas, Mexico, 39 out of 57 understorey plant species and 45% of all examined leaves were damaged by fungal pathogens, although damage on each leaf was small, averaging less than 1% of leaf area (Garcia-Guzman and Dirzo 2001). All 21 tree species examined in several other localities in Panama and Brazil were affected by foliar diseases (Gilbert et al. 2002; Gilbert and Sousa 2002). While foliar pathogens appear ubiquitous across sites and seasons, there is, however, considerable variation in the degree of infection among species (Gilbert 2005). The probability of infection appears to be related to herbivory (Garcia-Guzman and Dirzo 2001), with leaf damage caused by herbivory rendering leaves much more susceptible to attack by foliar pathogens. It is possible, therefore, that species susceptibility is related as much to resistance to herbivory as it is to specific defences against fungal pathogens, although this remains to be explored.

Reduction in plant growth due to fungi can be substantial, as in the case of the tree *Erythrochiton gymnanthus* (Rutaceae) in Costa Rica, which suffered a 52% growth rate reduction due to infection by the petiolar pathogen *Phylloporia chrysita* (Hymenochaetaceae) (Esquivel and Carranza 1996). Pathogens of flowers are less common, as might be expected given the ephemeral nature of the resource, but in at least one case their effect on fruit set is substantial: with a reduction of as much as 75% in the understorey tree *Faramea occidentalis* (Rubiaceae) by the rust fungus *Aecidium faramea* (Pucciniaceae) (Travers et al. 1998).

3.1.5 Pathogens of insects

Many tropical forest insects are affected by **entomopathogenic** fungi (Figure 3.5). Neotropical forest soils may contain as many as 1,000–50,000 spores g^{-1} of *Metarhizium anisopliae* (Clavicipitaceae), a fungus that infects various insects, including leafcutter ants (Hughes et al. 2004). Despite constantly high exposure of leafcutter ants to pathogenic fungi, the large majority of ants remain uninfected owing to regular self-grooming and secretion of antibiotics (Hughes et al. 2002). Fungal spores that germinate and penetrate the host cuticle ultimately kill the insect when the fungus sporulates.

Entomopathogens often bring about bizarre behavioural alterations in their hosts; thus dying animals may climb as high as possible, presumably so that fungal spores may later be spread widely (Roy et al. 2006). Some of the most dramatic of fungal insect pathogens belong to *Cordyceps* (Clavicipitaceae) and related genera which infect many different insects, including moths, flies, and ants (Figure 3.5). The mycelium pervades the host entirely and eventually develops a club-shaped organ containing fungal spores and which is supported on the end of a stalk that may be as much as 10 cm in length.

The Laboulbeniales are perhaps the most unusual fungal group. This order of minute **ascomycetous** fungi is obligately associated with arthropods, mostly insects. One genus within the Laboulbeniales is *Rickia*, with approximately 145 species. The host range of species of *Rickia* encompasses distantly related arthropods such as mites (Acarina), millipedes (Diplopoda), mole crickets (Orthoptera), ants (Hymenoptera), and various beetles (Coleoptera). It is common to find the same fungus parasitizing both the host insect and the mites it carries (Weir 1998). The fungi attach to arthropod cuticles through which they absorb nutrients, although this appears to have little consequence for the host. Most of the 2,000 species have been described from temperate arthropods, but a recent study of Laboulbeniales from beetles collected in North Sulawesi indicates that tropical diversity may be high because host species are numerous and the parasites appear host-specific (Weir and Hammond 1997).

3.1.6 Insect symbionts

Along with various other microorganisms, fungi are essential to the digestive processes of many insects (Vega and Dowd 2005). A recent study found 650 yeasts in the digestive tracts of tropical beetles—adding almost 30% to the currently recognized yeast species (Suh et al. 2005). The high beetle diversity in tropical forests suggests that many more remain to be discovered.

Some termites (330 species), ambrosia beetles (around 3,400 species) and all 200 species of leafcutter ants are obligately dependent on species-specific fungi for food which they actively cultivate (Mueller et al. 2005). The

A

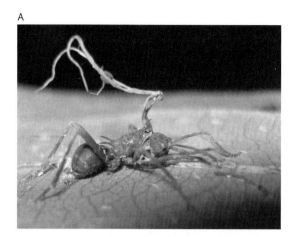

Figure 3.5 (A) The fruiting body of *Ophiocordyceps unilateralis* emerging from an infected *Polyrhachis armata* ant from Thailand.

Photo by David Hughes.

B

Figure 3.5 (B) *Cordiceps* sp. emerging from an infected ant from Guiana.

Photo by Douglas Sheil.

fungi grow on compost derived from leaves (ants), faeces (termites), or on the walls of galleries bored into wood (beetles), and appear to occur nowhere else. These fungal gardens were long thought to be free of any pathogens, as a result of the tending and weeding activities of the ant and termite mutualists, but now it appears that they are host to a variety of specialized microfungal par-

C

Figure 3.5 (C) Another unidentified fungal pathogen, this time on a wasp in Borneo.

Photo by Alex Wild.

asites that can potentially destroy fungal gardens, leading to colony mortality. To combat these pathogens, ants use antimicrobial 'pesticides' derived from bacteria grown on their own bodies (Currie et al. 2003)[3].

3.2 Lichens

Lichens are a complex symbiotic association of a fungus (mostly Ascomycetes), which absorbs nutrients and water, with a photosynthetic partner, an alga or cyanobacterium, which generates the carbohydrates that the lichen needs. Tropical forest lichens and algae are mostly epiphytic and hard to see, although straggly epiphytic *Usnea* (Parmeliaceae), Figure 3.6, are conspicuous in many wet, upper montane forests.

Foliicolous lichens live epiphytically on leaves and are among the most abundant epiphytes in tropical rain forests. They occur predominantly on the upper leaf surface of leaves in the forest understorey (Figure 3.7). Most derive no nutrients from the host plant, but by shading the leaf[4] it is assumed that they limit photosynthetic activity (Coley et al. 1993). Leaves may photo-acclimate by increasing chlorophyll concentra-

---

[3] These antimicrobial bacteria belong to the genus *Streptomyces*, a well-known genus of soil bacteria from which the pharmaceutical industry has derived several antibiotics for human use.

[4] Foliicolous lichens can cover a substantial portion of the leaf surface and sometimes the entire surface of individual leaves (Rogers et al. 1994).

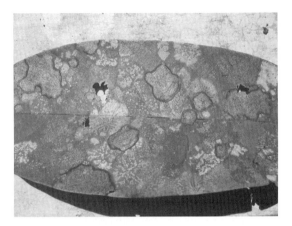

Figure 3.7 The long-lived leaves of understorey plants in non-seasonal rain forests (here in Sabah) are frequently colonized by a variety of epiphylls (mainly lichens, but also mosses and algae).

Photo by Julia Born.

Figure 3.6 Lichens of the genus *Usnea* grow hanging from tree branches, and are colloquially referred to as Old Man's Beard.

Photo by Eberhard Fischer.

tions (Anthony et al. 2002). These lichens have short life cycles corresponding to those of the leaves on which they reside, and they show marked structural diversity in response to differing light conditions.

Foliicolous lichens are highly diverse at very small scales: up to 38 lichen species have been counted on individual leaves of *Ocotea atirrensis* (Lauraceae) and 68 species on a 'large' leaf of *Carludovica palmate* (Cyclanthaceae) (Lucking and Matzer 2001). Even this hardly compares with two unidentified dicotyledonous leaves that, on an area of only 125 and 98 cm^2, supported 49 and 46 lichen species respectively (Lucking and Matzer 2001). Although diversity is high, foliicolous lichens have low **endemism** and most families and

genera are widely distributed. Between 73% and 87% of Brazil's Atlantic rain forest foliicolous lichen species are shared with Amazonian and Central American rain forests, and two-thirds show a wide intercontinental distribution (Caceres et al. 2000).

Lichens themselves harbour other algal species. Eighteen species of diatom belonging to nine genera have been found within a few strands of the common filamentous lichen *Coenogonium linkii* (Brachysiraceae), which grows on stems, hanging roots, and lianas in the understorey of neotropical lowland rain forests (Lakatos et al. 2004).

3.3   Other microorganisms

Microorganisms include a vast range of different and distinctive organisms[5], including various **prokaryotic** bacteria and **archaebacteria**, slime moulds, protozoa and amoeba. It is these microbial communities that form the slippery mats covering the surfaces of rocks and fallen trees in tropical forest understoreys. **Biofilms**, multispecies-structured communities of microbial cells

---

[5] The evolutionary distinctiveness of microorganisms dwarfs the diversity found within higher plants and animals. The genetic differences within and between slime moulds, fungi, bacteria, and archaebacteria are of a higher order of magnitude, resulting from divergences much earlier in Earth's history.

held together by their extracellular products, adhere to a variety of plant or rock surfaces, but their complexity, diversity, and function remain little understood (Andrews and Harris 2000). Biofilms are usually 20–30 μm thick but can be considerably more substantial in flowing streams. On plants biofilms may be beneficial, pathogenic, or neither. Saprophytic biofilms digest dead leaves, and many epiphytic biofilms simply use the plant for mechanical support. Bacterial films may contribute to nitrogen fixation, while others release substances that kill potential pathogens, and fungal biofilms on plant roots aid the absorption of water and nutrients. Bacterial pathogens also affect many insects, particularly moth or butterfly caterpillars in tropical forests (Martin and Travers 1989).

Among the most enigmatic microorganisms are **slime moulds**—a name that encompasses both the Dictyostelids and the unrelated Myxomycetes. As part of their life cycles both organisms aggregate to form macroscopic bodies that behave as a single organism. Dictyostelids aggregates ('pseudo-plasmodium') are often brightly coloured and occur on rotten wood. Dictyostelids are considered most diverse in the wet low altitude tropics (Swanson et al. 1999). Many remain undescribed, with 10 out of 35 taxa collected in Tikal, Guatemala previously unknown (Vadell et al. 1995). Tropical Myxomycete diversity is similarly unexplored but appears comparable with that of temperate forests. In Ecuadorian cloud forests distinct myxomycetes communities were found in dead wood, litter, and flowers, as well as amongst epi-

phyllic liverworts (Stephenson et al. 2004), while overall diversity and abundance declined with elevation and with reduced seasonality (Schnittler and Stephenson 2000; Novozhilov et al. 2001). As bacteriovores, slime moulds may have important roles in regulating bacterial populations and processing nutrients (Stephenson and Landolt 1998; Moore and Stephenson 2003).

The diverse microbes of the wet tropics have made these regions unhealthy for humans (see Chapter 15). Various organisms representing a wide range of taxa, from malarial protozoa through to a vast array of prokaryotes and viruses, along with evolutionary oddities such as *Giardia*[6], the rain forests are packed with pathogens, parasites, and opportunists. Virtually every multicellular rain forest organism has its associated microbes with which it has evolved in the dance of life: sometimes leading to conflict and sometimes to symbioses. We have much to learn about these dances.

3.4 Conclusion

There are several things that we can state with certainty about fungi and other microorganisms in tropical rain forests: they are widespread, abundant, and diverse; they are essential to the functioning of many ecological processes; and we know extremely little about them. They are likely to produce many surprises as our understanding improves.

---

[6] *Giardia* is a genus of protozoan parasites that colonize and reproduce in the small intestines of vertebrates, causing giardiasis. They belong to the Excavata, a eukaryote 'superclade'.

CHAPTER 4

More than monkeys: the vertebrates

'…the wild luxuriant tropical forest which stretched far away on every side; the rude uncultured savages who gathered around me—all had their influence in determining the emotions with which I gazed upon this "thing of beauty".'

Alfred Russel Wallace (1869 reprint 1996)

The image of rain forests portrayed by countless books and pictures is of lush forest teeming with immense abundance and variety of animal life. Such richness was immediately recognized and eloquently described by early explorers and naturalists, including Alexander von Humboldt, Henry Bates, Alfred Russel Wallace, and Charles Darwin. And yet a casual visit to a typical rain forest belies such expectations. Vertebrate animals are seldom conspicuous[1], although paw prints on soft ground, scrapes and scratches on tree trunks, and a plethora of strange noises betray signs of larger animal life. As familiarity grows, a visitor to Borneo, for example, learns to read the signs of the civet cat's nocturnal meanderings along the stream, the bearded pig's grovelling and wallowing on the forest floor, and the sun bear's search for beetle grubs hidden behind tree bark. Dawn and dusk are when birds and primates are most active, a daily routine heard also in Africa and in the Neotropics, where vociferous howler monkeys can be heard up to 16 km away. Amphibians render forests thick with cacophonous calls

through the night. Most mammals, including predators on the forest floor, insectivores in the canopy, and bats above it, are also active at night when they can be spotted in the beam of a powerful torch, itself a beacon to innumerable moths and beetles.

There are, however, marked differences in the distribution and abundances of animal species between and within the major forest blocks. Areas of exceptionally high avian species richness occur in the western Amazon near the Andes[2] where in less than 50 km^2 of forest 480 bird species have been recorded at Limoncocha, Ecuador, and 554 at Cocha Cashu, Peru[3]. In comparing distributions and structures of animal communities across continental forests it is also important to recognize that tropical forest communities are spatially heterogeneous and can only be understood through an appreciation of biogeographic, evolutionary, and ecological (as well as anthropogenic) factors.

Tropical regions have their own distinct forest faunas. Neotropical forests have more total species of birds, bats,

---

[1] This is not the case in East African forests where monkeys are not only frequently seen but are a nuisance in that they even come to houses, peer in the windows and steal (DS, pers. obs.).

[2] Molecular techniques reveal that recent (post-Pleistocene glaciations, 2–3 million years ago) evolutionary radiations in tectonically active areas gave rise to the exceptional species richness observed in the Andes (Moritz et al. 2000) (see Chapter 7).

[3] By comparison, the entire 244,110 km^2 of the United Kingdom has only 240 regularly visiting bird species.

Table 4.1 Number of vertebrate species from twelve tropical rain forest sites.

Site	Area (km3)	Birds	Mammals	Amphibians & reptiles
Tropical America				
La Selva, Costa Rica	15	410	117	134
Barro Colorado Island, Panama	15	443	115	133
Cocha Cashu, Peru	>50	554	139	129
Limoncocha, Ecuador	>50	480		
Manaus, Amazonia, Brazil	10	319	105	134
Africa				
Kibale, Western Uganda	766	305	87	83
Makoku, Gabon	2000	342	199	101
Asia				
Pasoh, Malaysia	8	212	89	~50
Ulu Endau, Malaysia	400	195	61	53
Temengor, Northern Malaysia	1489	154	101	72
Belalong Forest, Brunei (Borneo)	144	184	84	91
Danum Valley, Sabah, Malaysia (Borneo)	438	275	104	125
Australia-New Guinea				
Finnegan Uplands, NE Australia	281	82	37	57
Cooktown Lowlands, NE Australia	44	134	38	79
Gogol, New Guinea	10	162	27	57
Lakekamu, New Guinea	15	190	77	74

Various sources: McDade et al. 1994; Whitmore 1998; Kays and Allison 2001.

primates, and amphibians than forests elsewhere (see Table 4.1 for comparisons across major vertebrate groups), a pattern that is true for other taxa including higher plants (Chapter 2) and invertebrates (Chapter 5). There are 13 endemic families of birds in the Neotropics, while the rain forests of Africa and Asia each have only one (although Australia-New Guinea has ten). African rain forests have a large number and diversity of large terrestrial mammals, including great apes, such as gorillas and chimpanzees. Asian forests are notable for their variety of gliding animals. **Wallace's Line** (Chapter 6) separates the Asian faunas from markedly different New Guinean and Australian faunas to the east, where placental mammals are represented by bats and rodents alone. Thus primates are replaced by marsupials such as tree kangaroos (*Dendrolagus* spp.) and rain forest bandicoots (Peroryctidae). Madagascar is equally unique and is characterized by lemurs, tenrecs, chameleons, and an unusual carnivore fauna. There are also differences across **trophic**

guilds. In the Neotropics fruit-eating (**frugivorous**) primates are more abundant and leaf-eating (**folivorous**) primates scarcer than in the Old World (Table 4.2). It is likely that these patterns have both biogeographic and ecological causes.

4.1 Mammals

Rain forest mammals are diverse. Each rain forest region has its own set of unique and sometimes bizarre species. Placental mammals are widespread and dominate all rain forest regions except Australia-New Guinea, where marsupials, such as tree kangaroos (*Dendrolagus*, Macropodidae) and cuscus and possums (Phalangeridae), have filled niches occupied by monkeys elsewhere. Egg-laying mammals, Monotremes, are represented in rain forests by two species of echidnas (*Tachyglossus*, Tachyglossidae) in New Guinea and Australia.

Table 4.2 Species richness of vertebrate frugivores in the main tropical forest blocks.

		Frugivores		Folivores
Region	Birds	Bats	Primates	Primates
Neotropics	405	96	33	11*
Africa	149	26	32	9
Southeast Asia	143	66	11	25

* Unlike the African and Asian primates, Neotropical primate folivores are not specialists but also eat substantial amounts of ripe fruit when available.
Frugivore data from Fleming et al. 1987.

Figure 4.1 Brown-throated three-toed sloth *Bradypus variegatus* (mother and young) in the Amazonian varzea forests, Brazil.

Photo by Douglas Sheil.

Outside Australia-New Guinea placental mammals dominate the mammalian fauna[4] and are successful in the canopy where primates and bats are common. In terms of biomass, herbivorous sloths (Figure 4.1) and howler monkeys together account for 28% of mammalian biomass in rain forests in Surinam, and as much as 71% in Panama (Eisenberg and Thorington 1973). By contrast, insectivores and terrestrial herbivores are relatively scarce (Haugaasen and Peres 2005). While patterns have yet to be fully explained, insectivores, both arboreal and terrestrial, may suffer from competition with insectivorous birds that dominate avian communities, while scarcity of accessible food limits terrestrial herbivores. Our overview begins with primates and bats, and then considers ground-dwelling herbivores and carnivores.

4.1.1 Primates

The primates include 13 families, with 63 genera and over 270 species (Table 4.3) which occur across the tropics except for Australasia and the Pacific[5]. Most are forest dependent. The primates are divided into the 'lower' primates or prosimians (suborder Stersirhini), that include the lemurs in Madagascar and the lorises, pottos, and bush babies in Africa and Asia, and the 'higher' primates or anthropoids (suborder Haplorhini), such as marmosets, tarsiers, monkeys, and apes. Extant prosimians occur

[4] Although marsupial arboreal opossums (Didelphidae) occur in Neotropical forests, and, like many Neotropical monkeys, have prehensile tails.
[5] East of Sulawesi and Lombok, apart from some apparently introduced populations in Flores, Bacan (Mollucas), and more recently in Papua.

Table 4.3 Primate diversity by suborder.

Family	Examples Species	Distribution	Genera	
Suborder Prosimii				
Cheirogaleidae	Dwarf and mouse lemurs	Madagascar	5	9
Megaladapidae	Sportive lemurs	Madagascar	1	7
Lemuridae	Ring-tailed, crowned, ruffed, gentle lemurs	Madagascar	4	10
Indriidae	Indri, sifakas	Madagascar	3	6
Daubentonidae	Aye-aye	Madagascar	1	1
Lorisidae	Lorises, pottos, galagos	Africa, Asia	8	20
Tarsiidae	Tarsiers	Asia	1	5
Total Prosimii			*23*	*58*
Suborder Anthropoidea				
Callitrichidae	Marmosets, tamarins	Neotropics	5	37
Cebidae	Spider, woolly, howler, uakari monkeys	Neotropics	11	62
Cercopithecidae	Macaques, baboons, colobus, leaf monkeys, snub-nosed and proboscis monkeys	Africa, Asia	19	101
Hylobatidae	Gibbons	Asia	1	12
Pongidae	Chimpanzees, gorillas, orang-utans	Africa, Asia	3	5
Hominidae	Humans	Global	1	1
Total Anthropoidea			*40*	*218*
Total Primates			**63**	**276**

Adapted from Mittermeier and Konstant 2001.

only in the Old World tropics, despite the fact that North America was a centre of early diversification (Fleagle 1999). Five of the seven prosimian families exist only in Madagascar, but the tarsids and lorises extend through mainland Africa, India, and Southeast Asia.

Anthropoids have almost four times as many species as prosimians (Table 4.3) and are almost equally distributed between the New and Old World tropics. Most are diurnal. The New World monkeys (the Platyrrhini) share anatomical characteristics that, together with the fossil record, suggest descent from a single ancestral species which arrived in South America around 30–35 million years ago (Chapter 6) (Fleagle and Kay 1997). All Platyrrhini have tails, and in many genera the tail is prehensile, acting as an additional limb in their predominantly arboreal habitat. Asian and African monkeys (the Catarrhini) lack prehensile tails but have fully opposable digits, and use a broader variety of habitats. The Catarrhini are divided between the apes and the monkeys, based on various features including brain size, capacity for tool use, and learnt cultural behaviour.

The apes are divided into two groups: the great apes and the gibbons, both of which are forest dependent.

The great apes include the chimpanzee *Pan troglodytes*, the bonobo or pygmy chimpanzee *Pan paniscus*, gorilla *Gorilla gorilla*, and orang-utan *Pongo pygmaeus*. The chimpanzee, bonobo, and gorilla inhabit equatorial Africa, while the orang-utan, the world's largest (at over 100 kg) arboreal mammal, is found on the Southeast Asian islands of Sumatra and Borneo.

The 12 species of gibbon differ from great apes in being smaller, pair-bonded, and in not making nests. They occur in tropical and subtropical rain forests, from northeast India through southern China and Southeast Asia and into Indonesia. They move through the canopy by **brachiation**, using their long forearms to swing from one branch to another up to 15 m distant and at speeds of as much as 55 km h^{-1}. Gibbons advertise their territories with songs, a characteristic sound in many Asian rain forests.

Although some African rain forests report the highest primate biomass, there is variation. In Kibale Forest, Uganda, primate biomass reaches 3000 kg km^{-2} (Struhsaker 1975, 1997), but only 700 kg km^{-2} (Thomas 1991) in Ituri Forest, Democratic Republic of Congo.

Table 4.4 Regional variation in primate richness following a rainfall and seasonality gradient (mean ± standard deviation); NA = data not available.

Region	Localities	Species	Rainfall	Dry Months
Amazon	44	9.41 ± 2.91	2358 ± 624	3.22 ± 2.36
Orinoco and NE coast	14	4.64 ± 2.41	2277 ± 851	2.71 ± 2.20
North or west of Andes	6	3.33 ± 1.75	2396 ± 978	2.33 ± 3.20
Trinidad	1	2	2400	NA
Maracá Island, Amapá, Brazil	1	2	1600	6

From Kay et al. 1997.

Similar variation exists in Southeast Asia (Gupta and Chivers 1999) and South America (Peres 1997, 1999). In Asia and South America local primate species richness is to some degree correlated with primary productivity (Kay et al. 1997). This relation is particularly strong in the Neotropics where frugivory is common among primates, suggesting that local fruit productivity determined by rainfall may influence primate richness (Kay et al. 1997) (Table 4.4). In Africa there are fewer frugivorous primates and this pattern is less marked. Another pattern that is more apparent in Africa is that heterogeneous habitat mosaics support larger populations of primates (and other mammalian primary consumers) than homogenous forests (Oates et al. 1990; Thomas 1991; Fimbel 1994), to the extent that monkeys are almost absent in some monodominant forests (Thomas 1991).

Feeding behaviour

Many primates subsist on a variety of foods, but most species rely on relatively few resources for much of the year. Neotropical primates are mainly frugivorous and far less dependent on leaves than African and Asian species. They are often good seed dispersers: most New World monkeys (as well as lemurs and apes elsewhere) swallow seeds and defecate them some time later, often far from the parent tree. Paleotropical cercopithecine monkeys, on the other hand, store fruit in cheek pouches and spit

---

Box 4.1 Diets and adaptations

Gaining adequate nutrition from leaves alone demands specialization. Although leaves are available all year, they are difficult to digest, nutritionally poor, and often physically and chemically defended. They are, nevertheless, an important dietary source of minerals (especially calcium) and protein for various mammals. Fruit bats, for example, often chew leaves, swallow the nutrients, and spit out the bulky fibre.

Leaves also provide carbohydrates for organisms able to access them. The conversion of cellulose to sugars is dependent on specific gut micro-organisms. Folivory thus requires a digestive process that facilitates fermentation, notably a large compartmentalized stomach and extended gastro-intestinal tract harbouring the necessary symbiotic microorganisms. This solution has been adopted by large-bodied animals such as ruminants (deer, buffalo etc.) and to varying extents by other taxa including sloths, tree kangaroos, rhinos, tapirs, and okapi. A large body can be detrimental to canopy dwellers that support their weight on fine branches, so it is rare for birds to specialize on leaves (but see Hoatzin, Box 4.3). Folivory is not, however, uncommon among primates.

African colobus monkeys have, like ruminants, a particularly complex stomach with four chambers containing bacterial communities for fermentation of leaves. Bacterial fermentation is most effective in alkaline environments, and the proximal part of the stomach where fermentation takes place has a pH of 6–7. The need to maintain this high pH very likely determines the monkeys' foraging decisions. Colobines therefore supplement leafy diets with unripe fruit, but avoid ripe fruits, the fleshy portion of which is high in organic acids which would cause stomach pH to drop, and can even be fatal.

Generally, the longer leaf material remains in the gastrointestinal tract the more opportunity there is for nutrient extraction. Primates that use abundant but nutritionally low food such as leaves have relatively slow rates of food processing to maximize nutrient absorption. Rapid movement of food through the gut increases the total amount of food that can be processed in a given time, but is only possible for foods that are high in readily available energy, such as fruits. Larger animals can afford to process food more slowly as they have lower energy demands per unit body weight, and they are therefore able to make use of low-quality but abundant and predictable food such as leaves. Smaller species with higher metabolic rates can less afford to do this and seek high-quality food such as fruits and nuts. These resources are, however, unpredictable and patchy, but being rich in easily digestible sugars and starches they allow short unspecialized guts and small body size.

Animal food, although difficult to obtain, is nutritious and readily digested and requires a simple digestive tract. The exception is insect **chitin**, which is either excreted or, in specialist insectivores such as the tarsier, digested via microbial fermentation in an unusually large caeca. The African galago *Galago moholi* specializes on tree exudates and insects. The exudates and insect exoskeletons contain polysaccharides that are digested by microbial fermentation in the galago's complex and elongated gut.

Such categories of dietary adaptation reflect a dietary continuum from eating meat (difficult to catch but easy to digest) to folivory (readily available but relatively indigestible), with frugivory falling in between (fruit is available, but patchily and in limited quantities). Few animals specialize exclusively on a single form of food. Those that do, such as the folivorous proboscis monkeys (*Nasalis larvatus*) and faunivorous tarsiers (*Tarsius* spp., Tarsiidae) from Asia exemplify the extreme adaptations required to sustain them on such diets. All frugivores may sometimes supplement their diets with insects and/or leaves, and gut morphology may be highly variable between species depending on the typical diet. Faunivory and folivory require incompatible adaptations of gut and dentition, and consequently no mammals mix large quantities of animal matter and leaves, although many animals in both groups include fruit in their diets.

out seeds as they move through the forest, resulting in a less clumped seed distribution (Chapter 12).

The general paucity of fruit in equatorial Asian and African forests has probably limited the use of fruit by local primates, and several have become leaf-eating specialists. African colobus monkeys *Colobus* spp. and the Bornean proboscis monkey *Nasalis larvatus* have extended guts containing commensal bacteria to aid digestion of an almost purely leaf diet (Box 4.1). Folivorous primates, as well as other leaf-consuming animals, eat

soil at salt and clay licks which may help to detoxify plant defence compounds or supplement trace minerals that are lacking in folivorous diets (Box 4.2).

Insectivorous and faunivorous primates are smaller species, such as tarsiers in Asia, African galagos (Galagidae; commonly also known as bush babies), marmosets (*Callithrix*; Cebidae) in South America, and Madagascar's enigmatic aye-aye[6]. Mostly nocturnal, insectivorous primates supplement their diets with fruit and nectar (and also birds' eggs and geckos) in much the same way as frugivo-

---

[6] Originally described as a rodent, the aye-aye *Daubentonia madagascariensis* is actually a lemur and, at 80 cm from nose to tail, the largest nocturnal primate. The long middle finger is used to seek insects under tree bark. After tapping the bark it uses its sensitive hearing to detect the movement of insect larvae within the stem.

Box 4.2   Eating dirt: geophagy among animals

Consumption of soil, rock particles, or clay has been reported among forest ungulates, such as elephants, tapir, brocket deer, peccaries, and pacas, as well as primates including lemurs, marmosets, howler monkeys, colobus monkeys, langurs, gorillas, orang-utans, chimpanzees, and humans (especially pregnant women), and even reptiles, birds (parrots; Figure 4.2) and various invertebrates (Diamond 1999; Krishnamani and Mahaney 2000). These animals are often highly selective with respect to the soil consumed, and will return repeatedly to specific salt or clay 'licks'. The explanation for such widespread behaviour remains uncertain, and may vary with circumstances, but a predominantly herbivorous or omnivorous diet is a unifying characteristic of geophagous animals.

Rain forest plants often protect their leaves from herbivory with high concentrations of defensive secondary compounds, and folivores may ingest soil clays to adsorb and neutralize leaf toxins and secondary metabolites. Alternatively, ingested clays may buffer gut pH by adsorbing organic acids generated by bacterial fermentation. However, mineral supplementation has been the most commonly proposed explanation for geophagy. Salts and trace minerals, essential in cell enzyme systems and for synthesis of DNA and RNA, may be deficient in nutrient-poor diets, and animals may instead acquire them from soil sources. In addition, some clay minerals possess, as with milk of magnesia, established medicinal benefits.

Figure 4.2 Blue-and-yellow macaws *Ara ararauna*, Yellow-crowned parrots *Amazona ochrocephala*, and Scarlet macaws *Ara macao* feeding on clay at Tambopata rain forest in Peru.

Photo by Rhett A. Butler.

rous species will supplement theirs with grubs and insects. The smallest of all monkeys, the 80 g South American pygmy marmoset *Cebuella pygmaea*, and several other *Callithrix* marmosets, gouge holes in trees to feed on the exudates, though they also eat insects, fruit, and flowers.

Chimpanzees *Pan troglodytes* are primarily ripe fruit specialists, and the percentage of prey in their diet is usually very low (Newton-Fisher 1999). Nevertheless chimpanzees do, on occasion, prey on vertebrates (including other chimpanzees). In Kibale, Uganda, male chimpanzees hunt cooperatively and may kill 6–12% of the red colobus (*Procolobus*) population annually (Watts and Mitani 2002).

4.1.2 Bats

Bats, order Chiroptera, account for up to half of the mammal species in tropical rain forests. In Amazonian lowland forests, bat communities may include over 100 species in relatively small areas (Kalko 1998). Bats range in size from the giant Asian flying foxes *Pteropus* spp., with wingspans of 1.8 m, to the tiny bumblebee bat *Craseonycteris thonglongyai* of Thailand, perhaps the world's smallest mammal weighing about 1.5 g and only in 1974. The diets of tropical bats include fruit, nectar, insects, fish, frogs, and, notoriously, blood[7]. There are two major suborders: Megachiroptera are restricted to the Old World tropics and Pacific; Microchiroptera occur throughout the tropics. Megachiroptera are the true fruit bats and include the flying foxes, Pteropidae, of which there are also 15 species of nectar specialists. Microchiroptera are primarily insectivorous but also include fruit- and nectar-feeding species (Phyllostomidae). Fruit bats in both regions are important dispersers of many rain forest tree seeds.

Nectarivorous bats in both suborders are pollinators of many tropical rain forest plants. They rely on sight and smell to locate nocturnal flowers[8]. Sophisticated **echolocation** allows insectivorous bats to catch several thousand insects each night on the wing, which in some areas contributes a useful service in the control of malaria-carrying mosquitoes (Zinn and Humphrey 1981). South American fishing bats are able to detect the fins of small fish breaking the water surface using echolocation (Schnitzler et al. 1994). Other Neotropical forest bats prey on frogs, targeting them by their calls.

4.1.3 Ground-dwelling herbivores

Tropical rain forests support a low biomass of ground-dwelling herbivores (generally 1.5–5 times lower than arboreal herbivores) because the vegetation is mostly in the canopy and beyond reach (Coley and Barone 1996). They depend largely on fallen fruits, seeds, and flowers. Convergent evolution has given rise to the superficially similar mouse deer *Tragulus* spp. in Asia, duikers *Cephalophus* spp. and water chevrotain *Hyemoschus aquaticus* in Africa, and the agouti *Dasyprocta punctata* in South and Central America. Several species of small duiker antelopes coexist in African rain forests because each species is adapted to different types of fruits and seed. In the New World these niches are filled by rodents that have diversified to produce acouchis, pacas, and others (see 4.1.4 Rodents below).

Large tropical forest mammals include the okapi[9] *Okapia johnstoni*, elephant *Loxodonta africana*, pygmy hippo *Hexaprotodon liberiensis*, bongo *Tragelaphus euryceras*, and gorilla *Gorilla gorilla* of Africa, the tapir *Tapirus indicus*, rhinoceros *Dicerorhinus* and *Rhinocerus* spp., gaur *Bos gaurus*, and elephant *Elephas maximas* of Asia, and the tapir *Tapirus*

---

[7] Only three 'vampire' bat species feed on blood, and only one of these *Desmodus rotundus* feeds preferentially on mammals, the other two attacking birds. All three species occur only in tropical America, though *D. rotundus* is by far the most common and widespread. Although *D. rotundus* mainly feed on cattle and wild animals, they occasionally attack sleeping humans, sometimes causing infections and transmitting diseases such as rabies.

[8] Nectar-feeding bats pollinate many important tropical crops, including kapok, durian, mango, clove, guava, avocado, and breadfruit, and are the main pollinators of many Pacific Island forest trees.

[9] This reclusive diurnal species lives in some of the densest parts of the African rain forest. Long known by indigenous pygmy hunters, it was only discovered by the outside world in 1901, though descriptions of it were initially discounted. Its striped black and white flanks suggest a relationship to zebra, but it is actually a member of the giraffe family (Giraffidae).

terrestris of South America. Though typically uncommon, these mammals significantly influence the structure of the forest ecosystem. African forest elephants in the rain forest in the Congo Basin create clearings in the forest around water holes or mineral deposits. After several centuries these clearings may reach hundreds of metres across. Selective browsing and tree damage influence tree communities more widely. Javan rhinos *Rhinoceros sondaicus*, now confined to Ujong Kulon National Park on Java (Indonesia) and southern Vietnam but once ranging from the Brahmaputra River in Bangladesh to the coast of Vietnam, have also left their mark in the form of mountain paths that follow old rhino tracks.

Small herbivores specialize on nutritious foods such as fruits and nuts. However, the abundance of such food on rain forest floors is not sufficient to support large herbivores, which include a variety of lower quality plant materials in their diets. Salt licks provide essential minerals that are absent from low quality browse (Box 4.2), and their availability may even determine the size of local populations of species such as the Sumatran rhinoceros *Dicerorhinus sumatrensis* (Payne 1995). Other herbivores migrate in search of food. Borneo's bearded pigs *Sus barbatus*[10] (Figure 4.3) cover hundreds of kilometres in search of fruit falls, supplementing their diet by digging for worms, and during the synchronized mast fruiting events of Asian dipterocarp trees large groups of pigs congregate in areas of heavy fruiting (Curran and Leighton 2000). Most of the time food scarcity forces understorey herbivores, including bearded pigs, to avoid each other.

There are no large mammals in the Australian rain forests, although these, together with tropical rain forests of Southeast Asia and America, have lost several large mammals during Pleistocene and Holocene extinctions, apparently associated with the arrival of humans coupled with changing climate in the affected regions (Burney and Flannery 2005) (Chapter 15).

4.1.4 Rodents

The Rodentia is the largest mammalian order worldwide (over 2,000 species), and second to bats in the number of

Figure 4.3 The bearded pig *Sus barbatus* of Borneo, eating rambutan fruit.

Photo by Rhett A. Butler.

tropical rain forest species. In tropical forests rodents are important seed predators and dispersers, and provide prey for mammals, reptiles, and birds. The largest family is the Muridae to which rats and mice belong, but other large tropical families include squirrels (Sciuridae) and spiny rats (Echimyidae). Many rodents, and particularly rats, are predominantly ground-dwelling while squirrels are adapted for arboreal life, although there are many arboreal rats and some ground-dwelling squirrels. Arboreal rats have developed squirrel-like characteristics, including shortening of the muzzle to allow for better binocular vision, extended and strengthened toes, and long tails used for balance. Ground-dwelling

---

[10] *Sus barbatus* spend considerable time foraging for animal protein, digging and eating worms and insects. Indeed, these pigs are such good carrion feeders that some tribal people in Borneo bury their dead on raised platforms where pigs cannot disturb them.

squirrels have evolved in the opposite direction and towards the more typical rat shape, with an elongated snout to forage for ants and other invertebrates among the leaf litter (Earl of Cranbrook 1991).

Squirrels are widespread throughout tropical forests but are especially diverse in Southeast Asia (72 species in 23 genera), where they mainly occur in the canopy (giant squirrels *Ratufa* spp.) and among understorey branches and trunks (pygmy squirrels *Exilisciurus* spp.). There are many species of flying squirrels on islands of the Sunda shelf (11 island endemics), Borneo (14 species), and Peninsular Malaysia (11 species).

Spiny rats are the most diverse (78 species in 18 genera) and widespread of Neotropical rodents, and also among the most abundant (they give birth to 2–3 young every 2–3 months). They occur in almost all lowland tropical forests in Central and South America. Spiny rats feed on fruit and seeds (but will also take insects) and are important seed dispersers (and seed predators) of many forest plants (Adler and Kestell 1998). They are prey to ocelots, margays, jaguarundis, owls, and various snakes, so constitute an important component of the Neotropical forest food web. As their name suggests, spiny rats do have spines or bristles, but these are small and hidden by soft hair on their backs. The function of the spines is not known, but may help in repelling water in wet tropical forests.

4.1.5 Carnivores

Along with primates, carnivores like the tiger *Panthera tigris* and jaguar *Panthera onca* (Figure 4.4) are emblems of the rain forest. There is some ecological justification for this: by influencing herbivore densities, carnivores indirectly affect the wider community (Rao et al. 2001; Terborgh et al. 2001). Areas that have lost their carnivores can experience a dramatic increase in rodent numbers and a consequent increase in seed predation (Asquith et al. 1997; Terborgh et al. 2001) (see Chapter 13). Like primates, many forest carnivores are opportunistic in their dietary habits and also consume fruits. Indeed, smaller civets and mustelids are effective dispersers of many tropical rain forest seeds.

The order Carnivora dominates the mammalian carnivore fauna in all major tropical forest biomes outside Australia/Papua. The tiger is the largest rain forest carnivore at around 200 kg, and ranges widely through South and Southeast Asia. More common species include the leopard *Panthera pardus*, and several smaller cats that weigh no more than a few kilograms. Distributions are

Figure 4.4 Jaguar *Panthera onca* in Belize.
Photo by Rhett A. Butler.

sometimes hard to explain, and despite past coloniza-tion opportunities tigers and leopards are absent from Borneo (but occur on Java and Sumatra). The reason may be related to the naturally low biomass of potential prey species on Borneo compared to neighbouring regions (Meijaard 2004).

South America's cats are relatively recent arrivals from North America which crossed into South America on the formation of the land bridge between these two conti-nents 3 million years ago (see Chapter 6). The jaguar (120 kg) and puma *Puma concolor* (60 kg) are the largest. African rain forests have few cats, their **niches** being filled by genets of the civet family (Viverridae). Cats are entirely absent in Madagascar, where the fossa (Eupleridae) were originally mistaken for cats, illustrating a remarkable example of convergence.

Civets (Viverridae) and mongooses (Herpestidae) are important carnivore groups in the rain forests of Asia and Africa. Civets mark territories with a pungent secretion from anal glands[11]. Mongooses are best known for their agility in evading bites from the poisonous snakes that they occasionally predate, but their diet is diverse. In the Americas civets and mongooses are replaced by the Procyonidae, or racoons. Like the cats, the procyonids arrived relatively recently to South America a few million years ago (Brown and Lomolino 1998). Coatis *Nasua* spp. are among the most common members of this order and are opportunistic omnivores, foraging widely on both the ground and in trees. The olingo *Bassaricyon* spp. and kinkajou *Potos flavus* both eat figs and drink nectar.

Other mammalian carnivores include weasels and bears, but these groups are species poor in rain forests. Bears are omnivorous and feed on fruit, termites, wild bees, a variety of plants, and small vertebrates. African tropical wet forests lack bears, but hyenas, jackals, ser-vals, and other carnivores—even lions—will occasionally enter forests in search of prey.

In the Australian tropics, New Guinea, and South America before the Pliocene, mammalian carnivores were represented by marsupials, but most of these were extinct by the end of the Tertiary. The largest mammalian carnivore in Australia is now the tiger quoll *Dasyurus maculates* weighing less than 5 kg, and the largest in Papua is the New Guinea quoll *D. albopunctatus* at less than 1 kg. Both quolls feed on rodents, large insects, birds, and reptiles, although some fruit and even carrion are included in their diet.

4.1.6 Sloths, anteaters, and pangolins

Sloths and anteaters belong to the Neotropical order Xenarthra. Sloths have a wide distribution, occurring in tropical forests from southern Central America to north-eastern Argentina, though they are strictly arboreal[12] and do not live outside forest (Figure 4.1). They feed exclusively on tree leaves, and have a much lower metabolic rate than most other mammals of equivalent size. Consequently, they are slow moving and spend long periods asleep suspended in the forest canopy. To make room for their large guts they have reduced muscles, and are therefore not able to regu-late their body temperature efficiently through physiologi-cal means. Instead, they maintain their relatively low temperature (30–34°C) by basking in the sun, and at night body temperature can drop by as much as 12°C. Despite these apparent disadvantages, they are among the most common Neotropical forest mammals, and are important parts of the diets of ocelots, jaguars, and harpy eagles.

Like sloths, tamanduas *Tamandua tetradactyla* and *T. mexicana* (anteaters) occur widely though the Amazon basin and forests of Central and South America. They are mostly arboreal and use their extremely long tongue to feed on ants and termites, which they locate by scent. The silky anteater *Cyclopes didactylus* is a common arbo-real animal in the Amazon, but is rarely observed owing to its small size (about 1 kg) and nocturnal habits. Both the tamandua and silky anteater have prehensile tails and powerful front legs ending in curving claws to open ant and termite nests.

The Old World pangolins are classified in a single fam-ily and genus (Manidae: *Manis*) in the order Pholidota. Rain forest species occur in Africa (*Manis tricuspis and M. tetradactyla*) and Asia (*M. javanica*) (Figure 4.5) where

---

[11] These secretions are used by the perfume industry, and this has led to the introduction of the Malay Palm civet *Viverra tangalunga* to islands around Southeast Asia and New Guinea, causing havoc among the naïve indigenous fauna.

[12] Although also surprisingly good swimmers.

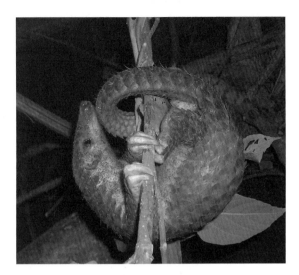

Figure 4.5 Malaysian pangolin *Manis javanica*.

Photo by Karin Beer.

they mostly feed on ants. These species have prehensile tails and are good climbers: the African species may spend most of their time in trees.

4.1.7 New mammalian species

New mammals continue to be discovered in remote forest regions. Indeed, around ten times as many new mammal species are being described as are bird species (Patterson 2000). Several new species described in the last few decades come from the Annamite mountains along the Laos and Vietnam border, and include a giant muntjac deer *Muntiacus vuquangensis* and the world's smallest deer, the leaf deer *Muntiacus putaoensis*. The enigmatic saola *Pseudoryx nghetinhensis*, another local novelty, appears most closely related to the bovids. Others such as the Vietnamese warty pig *Sus bucculentus* and Roosevelt's muntjac *M. rooseveltorum* previously thought to be extinct have recently been rediscovered.

The highland mangabey *Lophocebus kipunji* was described from Tanzania in 2005 (Jones et al. 2005). This is the first new monkey species (indeed a new genus) for Africa since the 1984 discovery of the sun-tailed monkey *Cercopithecus solatus* in Gabon. The Amazon basin, however, has yielded many new primates in recent decades, including two titi monkeys, *Callicebus bernardi* and *C. stephennashi* (van Roosmalen et al. 2002). These are just the latest additions to the *Callicebus* genus which comprised three species in 1963 but now, following taxonomic revisions of the genus and new discoveries, numbers 28! Indeed, 40 monkey species have been discovered since 1980, of which 13 are from the Brazilian Amazon. It is likely that this figure will be out of date by the time this book is in print.

4.2 Birds

Of the nearly 10,000 recognized species of birds in the world, well over half occur in the moist tropics. Many birds are attractive, easily observed, and widely valued, and species such as the resplendent quetzal and the hoatzin (Box 4.3) have come to embody rain forests for

---

Box 4.3 The Hoatzin: an enigmatic bird

The backwater swamps of South America's Amazon and Orinoco basins are inhabited by a strange bird. The chicken-sized hoatzin *Opisthocomus hoazin* has a taxonomic family (Opisthocomidae) to itself. It has been linked to Galliformes (chickens, pheasants, and grouse), but molecular evidence suggests an affinity to Cuculiformes (cuckoos), although this remains contentious and unresolved (Hughes and Baker 1999; Sorenson et al. 2003).

Hoatzin chicks have two claws on the first and second wing digits, a feature that has drawn comparisons with *Archaeopteryx*, the fossil proto-bird. When threatened, the chicks fall from the nest tree into the surrounding water—climbing back using their wing claws once danger has passed. Threatened adults produce a powerful smell to deter predators. The folivorous diet, supplemented by flowers and fruits, is also unusual among birds, and hoatzins are unique in fermenting the vegetable matter in a large double crop analogous to the foreguts of mammalian ruminants (see Box 4.1). The crop is already well-developed in newly hatched chicks, and acquires microbial populations of fermenting bacteria within the first two weeks of life, presumably by inoculation during feeding by adults. The large crop and slow digestion limit flight activity, making hoatzins relatively sedentary and inefficient fliers. Hoatzins live in loose family groups, and several nests can occur in a single tree. Non-breeding 'helpers' sometimes assist in incubating eggs and feeding chicks.

many visitors, thereby becoming important symbols for the conservation of the forests.

4.2.1 Distribution of major bird groups

The superficial similarity among bird taxa on different continents (e.g. sunbirds and hummingbirds, toucans and hornbills as in Figures 4.6 and 4.7) is frequently a result of convergent evolution rather than shared ancestry (Cibois et al. 2001). The bird communities of the Neotropics are, for the most part, evolutionarily distinct from those of the Paleotropics, and only a few, including swifts, swallows, pigeons, parrots, cuckoos, hawks, owls, falcons, and nightjars, are found in all major rain forest regions.

Neotropical rain forests have the greatest number of species, but fewer families than Old World tropical forests. The suboscines form a distinctive Neotropical group that includes ovenbirds, antbirds (Formicariidae and Thamnophilidae), manakins (Pipridae), and tyrant flycatchers (Tyrannidae). These birds, distinguished from the oscines by their more limited singing ability, dominate the understoreys of Neotropical forests, while the oscines dominate the canopies. Some large suboscine families, such as the ovenbirds (Furnariidae), consist of many divergent species that collectively perform the ecological functions of several Old World families.

Birds of African and Asian tropics are more similar to each other than to the distinctive tropical American avifauna. Important Old World families common to Africa and Asia include hornbills (Bucerotidae), bulbuls (Pycnonotidae), sunbirds (Nectariniidae), and barbets (Asian Megalaimidae and African Lybiidae). Pheasants

Figure 4.6 Rhinoceros hornbill *Buceros rhinoceros*, a large bird from Southeast Asia, including the Malay Peninsula, Sumatra, Borneo, Borneo, and Java. This hornbill eats fruit and insects.

Photo by Rhett A. Butler.

Figure 4.7 Swainson's toucan *Ramphastos swainsonii*, the second largest species of toucan, weighing around 600 g. This toucan is a native to Central American rain forests, ranging from southern Honduras southwards to northern Colombia.

Photo by Rhett A. Butler.

(Phasianidae) are primarily rain forest birds native to Southeast Asia. Notably these include the red jungle fowl *Gallus gallus*, the progenitor of the domestic chicken, along with several rather more spectacular species.

New Guinea Australia and neighbouring islands have several endemic families, including the spectacular bowerbirds (Ptilonorhynchidae), birds of paradise[13] (Paradisaeidae) (Figure 4.8), honeyeaters (Meliphagidae), and, uniquely, poisonous birds in the endemic genera

Figure 4.8 Lesser bird of paradise (*Paradisaea minor*) from the lowland rainforests of New Guinea. **(See colour plate 5.)**

Photo by Rhett A. Butler.

Pitohui and *Ifrita*[14] (Dumbacher et al. 1992). They lack, however, many common Old World groups, and bulbuls, babblers, barbets, and woodpeckers are absent, while hornbills, sunbirds, and pittas are greatly reduced in species. Islands also harbour many unique species and subspecies of mainland forms and make a large contribution to the diversity of birds globally (Mayr and Diamond 2001).

The forest understorey tends to be dominated by insectivorous birds, while frugivores dominate the canopy (Pearson 1977; Greenberg 1981; Terborgh et al. 1990). Nonetheless, many insectivores live in or use the forest canopy to varying degrees (Cohn-Haft et al. 1997; Naka 2004), and substantial differences in the structure of forest avifaunas exist at regional scales (Thiollay 2002; Naka 2004). Indeed many bird species, for example antwrens *Terenura*, *Herpsilochmus*, and *Myrmotherula*, are absent from large swathes (> 100 km) of apparently suitable habitat in American sites that lie well within the limits of their species ranges (Thiollay 2002). Such patchy, and unexplained, distributions have also been reported from New Guinea, Africa, and Amazonia (Diamond and Hamilton 1980; Beehler et al. 1995; Thiollay 2002).

4.2.2 Feeding guilds

The different avian feeding guilds (Table 4.5) are represented by characteristic families in each tropical region. Such categories neglect considerable variation within each class. For example, South American insectivorous ovenbirds include species that probe under bark, glean insects from leaves, hang from twigs to snatch insects from the underside of leaves, sift leaf litter, and even search for aquatic invertebrates (Pearson 1977). Variety and specialization of feeding strategies reduce competition among species and therefore allow more species to coexist (Box 4.4).

---

[13] Birds of paradise arriving in Europe from the Middle Ages up to the 18th century were fleshless skins from which the feet had been removed. Incredulous Europeans reasoned that, lacking feet, the birds could never land and, being without flesh, they were ethereal. Hence 'birds of paradise', and Linneaus' specific name *apoda* (footless) for the Great bird of paradise, *Paradisea apoda*.

[14] Though living on a different continent, these birds sequester batrachotoxins in their feathers, the same compounds found in Neotropical arrow-poison frogs. It is likely that eggs in the nest would also acquire toxin through contact with the nesting birds' feathers, and become unpalatable to predators. The poisons are apparently obtained from beetles in the little-studied genus *Choresine* (Dumbacher et al. 2004).

Table 4.5 Tropical forest bird species by dietary preferences in Africa and South America.

Foraging behaviour	Africa	%	South America	%
Frugivores	20	12.1	49	23.9
Nectarivores	1	0.6	12	5.9
Raptors and scavengers	15	9.1	13	6.3
Insectivores	129	78.2	131	63.9
Gleaning from foliage	77	46.7	55	26.8
Sallying from perches	21	12.7	21	10.2
Trunk and bark	13	7.9	31	15.1
Ground	13	7.9	21	10.2
Aerial and crepuscular	5	3.0	3	1.5
Total	294		336	

Adapted from Rickleffs and Schluter 1993.

Insectivores

Insectivores are the most common forest birds by species number (Table 4.5) and abundance. Their impact on forest canopy arthropods can be sufficiently high to benefit plants on which the arthropods feed (Van Bael et al. 2003). Niche differentiation among insectivores is marked by the variety of feeding strategies employed and spatial displacement of feeding locations (Pearson 1977), but even so competition may become intense on the arrival of winter migrants (Box 4.5).

A feature common to most rain forests is mixed flocks of insectivorous birds, with as many as 50 species foraging together. Flocks contain core species, represented by a single breeding pair that is a relatively permanent group member (Jullien and Thiollay 1998). Other birds join for varying periods. Flock membership may offer protection from predators through the combined vigilance of several species (Thiollay 1999).

Neotropical antbirds (Formicariidae and Thamnophilidae) specialize on following army ants, and prey upon the arthropods and small animals that are flushed out of the leaf litter and undergrowth by the ants. Various other species also follow army ants but more opportunistically. In Africa, too, various birds including thrushes and bulbuls often follow driver ants[15].

The swift family, the Apodidae, comprising swifts, swiftlets, and needletails, are found virtually worldwide but are most common and **speciose** in the tropics, where flying insects occur in abundance. Swifts are able to fly continuously for several weeks and spend almost all their lives on the wing above the forest canopy. Their feet are small and weak and cannot be used to perch, nor can they take off from the ground. They are social birds and nest in large colonies on cliffs and in caves in Southeast Asia and on forest trees in Africa. The cup-like nests of cave-nesting swiftlets in Southeast Asia are made from salivary secretions which set strongly enough to hold eggs and growing hatchlings[16].

Flower visitors

The availability of floral resources throughout the year in tropical rain forests has allowed the evolution of flower specialists, such as New World hummingbirds (Trochilidae) and Old World sunbirds (Nectariniidae), which in turn

---

[15] The ant fronts can be found by following the sound of the birds (DS, pers. obs.).

[16] The nests of the black-nest swiftlet *Collocalia maximus* and the white-nest swiftlet *C. fuciphagus* found within limestone caves in Borneo are collected at considerable risk as edible delicacies. The Chinese have prized these nests, and particularly the white nests, for at least 400 years, and eat them as 'bird's-nest soup'. The soup is consumed not so much for its taste but for its reputed effects in improving the voice, relieving gastric troubles, aiding renal functions, raising libido, alleviating asthma, curing tuberculosis, strengthening the immune system and improving concentration. They are served in wedding feasts. The nests are, as a result, extremely valuable: the 2003 retail price for white nests in Hong Kong (the main market) was about US$ 7 per gram (Hobbs 2004), although internet prices in 2009 averaged $3 per gram (JG, pers. obs.).

Box 4.4 Modes of coexistence

How do so many animal species coexist? Coexistence is achieved, in large part, by niche differentiation, that is, the separation of diets, activity periods, and foraging locations, among other things (Chapter 13). Hence tropical species tend to be more specialized, and tropical communities as a whole probably use a larger variety of resources and microhabitats. Being structurally and climatically complex, tropical forests provide many opportunities for specialization. At the most basic level, animal species occupy different parts of the forest at different times of day or night. Birds use different layers within the complex vertical structure of the forest (Karr and Roth 1971) (Table 4.6) and specialize on particular food types or sources. Among the foliage gleaning Neotropical insectivores, antbirds and furnariids dominate the understorey, while flycatchers and oscines use the canopy.

Nonetheless, given the number of co-occurring insectivorous species, such vertical stratification provides only a partial explanation of the observed diversity, and differentiation among species is also based on subtle differences in foraging strategy and food types (Karr 1971). Hummingbirds, for example, avoid competition by specializing on particular

floral types. This specialization is reinforced by the close match between their bills and the structural forms of the flowers they visit—similar patterns occurring among sunbirds. This co-evolved matching ensures exclusive access to nectar sources by the pollinating birds, benefiting birds and flowers alike (Endress 1994). Many birds adopt a generalized strategy of gleaning insects from leaves, but forage at different leaves: Amazonian cinnamon-throated woodcreeper *Dendrexetastes rufigula* and cinnamon-rumped foliage-gleaner *Philydor pyrrhodes* forage exclusively on different areas of palm fronds (Pearson 1977). Mammals too divide the forest amongst them in both space and time, thereby avoiding competition. The 35 bats of Barro Colorado Island in Panama form nine trophic guilds, and within each there is further specialization based on food size (Table 4.7). On the other hand, frogs, which are mostly opportunistic and generalist feeders, may coexist through the differentiation of their breeding sites rather than foraging preferences (see 4.3.1 Amphibians below). The fat-tailed dwarf lemur *Cheirogaleus medius* of Madagascar avoids competition entirely by hibernating in the dry season when fruit is scarce[1].

Table 4.6 Vertical zonation of birds at La Selva forest, Costa Rica.

Zone	Bird group
Above canopy	Vultures, hawks, swifts
Canopy	Toucans, cotingas, parrots, cacique birds
15–25 m	Woodpeckers, woodhewers, large trogons, jacamars, puffbirds
Understorey	Hummingbirds, antbirds, manakins, flycatchers, tanagers
Forest floor	Tinamous, great curassow, ground doves, wrens

Adapted from Whitmore 1998.

Table 4.7 The nine feeding guilds found amongst the 35 bat species at Barro Colorado Island, Panama.

Bat feeding guild	Habitat
Frugivores	Canopy
Frugivores	Near the ground
Frugivores	Scavengers
Omnivores	Nectar, pollen, fruit, and insects
Sanguivores	Blood of birds and mammals
Carnivores	Gleaning
Piscivores	Rivers and lakes
Insectivores	Waterbodies
Insectivores	Slow-flying hawking

Adapted from Whitmore 1998.

[1] It sleeps for seven months at a time and is the only known tropical mammal that hibernates.

Box 4.5   Seasonal coexistence of resident and migratory birds

In the tropics the number of insectivorous birds increases with the arrival of seasonal migrants from north temperate zones. These migrants arrive at a time when arthropod abundances are low. How do tropical communities support this increase in consumers? One idea is that residents are limited primarily by the abundance of large arthropods (termed 'breeding currency') needed to feed growing nestlings, while adult birds are limited by the general year-round availability of all arthropods. Thus two potential carrying capacities for birds in the tropics may exist, one determined by the availability of breeding currency and another higher one set by year-round arthropod biomass (Greenberg 1995).

A study in Jamaica examined the breeding currency hypothesis (Johnson et al. 2005). As predicted, resident bird abundance was correlated with large arthropod biomass, while total bird abundance in the non-breeding season correlated with year-round total arthropod biomass. Nevertheless, after controlling for the effects of arthropod biomass, migrant birds remained significantly more abundant than expected in disturbed forest locations. This last result supports another proposal that resident bird populations are limited by a scarcity of safe nesting sites and by high predation rates—which are higher in disturbed habitats—leaving feeding opportunities to be exploited by non-breeding migrants (Hutto 1980).

---

pollinate the flowers they visit. Hummingbirds are a common sight in Neotropical forests, where species occur in forest interiors (notably the comparatively dull hermit hummingbirds), and in canopies, gaps, and forest edges. An ability to hover, shared with many sunbirds, constrains body size—indeed the world's smallest bird species are the bee hummingbird *Mellisuga helenae* from Cuba and the little woodstar *Chaetocercus bombus* of Ecuador, both weighing less than 2 g.

Nectar is, however, deficient in lipids and proteins, and nectar-feeding birds frequently supplement their diets with insects. Leafbirds (*Chloropsis*, Chloropseidae), flowerpeckers (*Prionochilus* and *Dicaeum*, Dicaeidae), and spiderhunters (*Arachnothera*, Nectariniidae) of Asia also visit flowers for nectar, but they do so less regularly than either sunbirds or hummingbirds.

Frugivores
Fruit is patchily distributed in space and time and is deficient in nutritional fats and proteins[17]. Despite this, many

birds in all tropical forest regions rely on fruit. Indeed, fruit is so important that on Barro Colorado Island as much as 50% of forest bird biomass is supported by it (Willis 1980). Some toucans migrate between altitudes in the Brazilian Atlantic forest following fruit ripening (Galetti et al. 2000).

Tropical African forests have relatively few avian frugivores (though many fruit-feeding primates), and they are correspondingly poorer in bird-dispersed plant families such as the Burseraceae, Lauraceae, and palms. To make the most of their resources African hornbills track ripening fruit over very large areas and carry seeds over distances of more than 5 km[18] (Holbrook and Smith 2000; Holbrook et al. 2002).

Canopy frugivores such as guans (Cracidae), toucans (Ramphastidae), turacos (Musophagidae), and hornbills (Bucerotidae) tend to be larger than understorey frugivores like manakins (Pipridae), tanagers (Thraupidae), bulbuls (Pycnonotidae), flowerpeckers (Dicaeidae), and broadbills (Eurylaimidae) (Karr 1980; Wong 1984). Large size confers an ability to process large food items more

---

[17]   Exceptions include wild avocados of Central America that are a major source of food for quetzal.

[18]   Toucans, often considered the New World ecological equivalents to the Paleotropical hornbills, are by contrast extremely poor flyers and do not travel large distances in search of food. Consequently, dispersal distances of seed by toucans are far less than their erstwhile equivalents, the hornbills.

quickly and also reduces predation risk as a smaller variety of predators are able to tackle big birds. It has been speculated that the combination of low predation risk and high light source may have favoured the evolution of the typically bright colouration of many canopy frugivores, such as the remarkable birds of paradise (Paradisaeidae) in New Guinea and the quetzals (*Pharomachrus*, Trogonidae) in Central America, which use their colour display to signal for mates and challenge rivals.

Old World frugivores are generally larger than their New World counterparts, a pattern that holds for primates and bats as well as birds. Neotropical fruits also tend to be small: the maximum size of fruit that can be handled by a bird is a function of gape width (Wheelwright 1985). Reasons for larger-sized Asian and African frugivores remain uncertain: one possibility is the failure of large frugivores to colonize South America (Chapter 6). Another possibility is the lower reliability of fruit availability in Asia and Africa, where many of the forests are either markedly seasonal or support only low densities of fleshy fruited plant species, necessitating greater reserves and mobility to access larger areas. Contemporary body size differences can be misleading, as large mammals such as horses, gomphotheres, ground sloths, and other megafauna roamed South America perhaps even as recently as a few thousand years ago (Hansen and Galetti 2009), although it remains uncertain as to whether these were sufficiently abundant in Neotropical forests to have shaped seed evolution.

Megapodes

The megapodes (Megapodidae) have a largely Australia–New Guinea–Moluccas distribution[19], and all except one of the 19 species occur in rain forest. They are stocky ground-dwelling birds that scratch over the leaf litter in search of invertebrates and other edible titbits. They do not incubate their eggs with body heat but rely on external heat sources. On the Solomon Islands *Megapodius freycinet* lays eggs in sand warmed by geothermal heat, while other species may use hot springs and rotting tree stumps. The Moluccan megapode *Megapodius wallacei* simply relies on the heat of the sun to warm the black volcanic sand into which the eggs have been deposited, and large numbers congregate at suitable beaches. Several species bury their eggs within massive nest-mounds of decaying vegetation which, for the New Guinean scrubfowl *Megapodius affinis*, may be 12 m across and 5 m high. The Australian brush turkey *Alectura lathami* builds a somewhat smaller mound (up to 4 m wide and 2 m high) using vegetation gathered from the surrounding forest floor, and this can hold between 18 and 24 eggs. Warmth from the decomposing vegetation incubates the eggs, and male birds regulate the temperature to between 33°C and 35°C by adding or removing litter. On hatching, the young fully-feathered birds dig their way out of the mound and immediately lead an independent existence.

Raptors

Carnivorous birds such as hawks, eagles, and owls occur in almost all rain forests, often as top carnivores in the canopy, including Philippines' monkey eating eagle *Pithecophaga jefferyi*, African crowned eagle *Stephanoaetus* (*Spizaetus*) *coronatus*, and the Neotropical harpy *Harpia harpyja* and crested eagle *Morphnus guianensis*. The bones that accumulate beneath the nests of crowned eagles reveal a varied diet, including porcupines and tree hyrax. The harpy eagle is probably the main predator for sloths. Smaller raptors such as the Neotropical plumbeous *Ictinia plumbea* and swallow-tailed kites *Elanoides forficatus* (Accipitridae) prey upon large insects above the rain forest canopy. The double-toothed kite *Harpagus bidentatus* has been observed following monkeys and snatching any insects they flush out. Other accipiters (e.g. the sparrowhawks *Accipiter bicolor* and *A. superciliosus*) predate smaller birds in the understorey.

About 70 owl species are associated with tropical forests but are poorly studied. Scops owls *Otus* spp. and eagle owls *Bubo* spp. are the most abundant and speciose, although some are very rare, such as the Seychelles scops owl *Otus insularis* which has a population of about 360 birds (David Currie, pers. comm. with JG).

---

[19] Megapodes are also found in Philippines, Borneo, and the Nicobar Islands in the Indian Ocean.

4.3 Amphibians and Reptiles

4.3.1 Amphibians

Comprising frogs and toads, salamanders and caecilians, amphibians are a highly successful group that are widespread throughout the major tropical rain forest regions, but absent from most oceanic islands (Figure 4.9). South America has the greatest amphibian diversity, probably representing more than half the world's total species richness (Duellman 1999), although other tropical regions such as Sri Lanka (Box 4.6), Madagascar, Southeast Asia, New Guinea, and Central Africa are also speciose and taxonomic inventories remain incomplete. Amphibian densities vary. Typically, and very roughly, Central American rain forests attain 14–15 frogs per 100 m², 9–10 in African, 4–6 in South American, and 1–2 in Southeast Asian rain forests (see Box 4.6 and Tables 4.8 and 4.9 for regional **herpetofauna** comparisons), but there is considerable local variability as amphibian densities are heavily influenced by seasonality, humidity, altitude, and geology (which influences water drainage) (Stuart et al. 2004).

Tropical amphibians possess diverse breeding strategies (Haddad and Prado 2005). Oviposition and larval development in ponds or streams, as in temperate species, remains widespread, and large aggregations of chorusing males is a characteristic sound by still waters on tropical forest evenings. Many other species (e.g. microhylids and most *Bufo* species) oviposit in temporary pools, ponds or puddles, which presumably removes dangers from fish and aquatic insect predators, although overcrowding and desiccation present their own risks. Several species of poison-arrow frog (*Dendrobatidae)* and marsupial frogs (*Gastrotheca,* Amphignathodontidae) breed in pools of water collecting in tree holes or other pockets of water such as contained within many Neotropical bromeliads, their tadpoles feeding on mosquito larvae, frog eggs, or each other. In Borneo the tiny (up to 2 cm) Kinabalu dwarf toad *Pelophyne misera* from the montane forests of Mount Kinabalu sometimes lays eggs in *Nepenthes* pitcher plants.

Several frog species lay their eggs on vegetation overhanging water, into which the tadpoles drop on hatching. Others construct foam nests into which eggs are laid. These nests may be placed near to water pools or simply adjacent to depressions that are likely to flood. The

Figure 4.9 Two examples of frogs: a glass frog *Hyla granosa* from Ecuador (A) and the Bornean horned frog *Megophrys nasuta* (B).

Photos by Karin Beer.

Neotropical Dendrobatidae lay their eggs on land among humid leaf litter where they are tended by the adults. On hatching the tadpoles are carried by the mother to streams or water-bearing tree holes, leaf cups, and bromeliads. Finally, development has been made entirely independent of standing water by several common genera of

Box 4.6 Sri Lanka: an amphibian biodiversity hotspot

Recent surveys have uncovered over 100 species of Old World tree frogs (Rhacophorinae) in Sri Lanka's small remaining rain forests where only 18 were previously known (Meegaskumbura et al. 2002). These species lie within two taxonomic groupings: species of one lay eggs in foam 'nests' overhanging water courses, while species of the second group undergo direct development from eggs on land. Sri Lanka has lost more than 95% of its rain forest habitat, which has almost certainly caused the extinction of several amphibian species known only from 19th-century specimens. The persistence of these surviving spe-

cies has been attributed to their terrestrial development and fecundity (Lips et al. 2003; Stuart et al. 2004).

Sri Lanka's amphibian diversity now stands at 140 species, which is comparable to that of other tropical rain forest islands some of which are many times larger (Table 4.8). This confirms Sri Lanka's importance as an amphibian hotspot (similar survey effort applied elsewhere in Asia may also uncover many new species). Unfortunately, current global population declines and extinctions of amphibians may preclude a true appreciation of amphibian biodiversity.

Table 4.8 A comparison of known amphibian biodiversity among tropical forested islands

Island	Amphibian species	Area (000 km²)	Species/1,000km²
Sri Lanka	140	65.6	2.13
Philippines	96	299.8	0.32
Madagascar	190	587.0	0.32
Borneo	137*	746.3	0.18
New Guinea	225*	775.2	0.29

*Undescribed species are still regularly encountered in any significant survey of these regions.

amphibians (e.g. Neotropical plethodontid salamanders and *Eleutherodactylus* frogs) which, in very humid climates, lay clutches of large-yolked eggs on land; within the eggs larvae develop and hatch as tiny salamanders or froglets.

Caecilians are legless, burrowing amphibians comprising 33 genera and about 170 species distributed in humid regions in South America, Africa, the Seychelles, the Indian subcontinent and parts of Southeast Asia (Gower and Wilkinson 2005). They have not been recorded from Madagascar or Australasia. Larger species can reach 1.5 m in length (e.g. *Caecilia thompsoni* from South America). Most caecilians retain eggs and larvae within the oviduct from which developed offspring emerge (Kupfer et al. 2004). As generalist predators of soil invertebrates they

have reduced eyes, often covered by skin or bone, but possess sensory organs called 'tentacles' on the snout. Some caecilians are brightly coloured and patterned, perhaps as warning colouration, for they appear to produce repellent skin secretions. They share with snakes a single functional lung to accommodate their snake-like body plan, and one species, the 80 cm South American *Atretochoana eiselti*, lacks lungs entirely, making it by far the largest tetrapod to rely entirely on cutaneous gas exchange[20]. Despite their widespread occurrence, little is known about their ecology and behaviour, probably because they spend almost all their time in the soil.

Many amphibians harbour powerful toxins within their skins. The skin of an adult poison-dart frog

[20] The frog *Barbourula kalimantanensis*, recently rediscovered in Kalimantan, Borneo, has also been shown to be lungless (Bickford et al. 2008).

Table 4.9 Richness and density of tropical rain forest leaf litter amphibians and lizards across regions and altitudes (various sources).

Locality	Country	Elevation (m)	Richness	Density (100m^{-2})
SOUTHEAST ASIA				
Sarawak, Borneo	Malaysia	200	38	1.5
Nanga Tekalit	Borneo		28	1.2
Cuernos de Negros	Philippines	1000	12	5.1
Cuernos de Negros	Philippines	1250	12	11.3
Cuernos de Negros	Philippines	1500	4	15.0
AFRICA				
Lombe	Cameroon	<100	8	9.4
Kibale	Uganda	1500	19	1.9
CENTRAL AMERICA				
Osa	Costa Rica	100	35	11.6
La Selva	Costa Rica	100	34	14.7
Barro Colorado Island	Panama	100	27	14.6
Rio Canclon	Panama	100	20	6.3
Pipeline Road	Panama	300	14	12.6
Carti Road	Panama	300	7	29.8
Silugandi	Panama	<500	30	45.1
San Vito	Costa Rica	1200	27	58.7
Monteverde	Costa Rica	1500	15	6.7
SOUTH AMERICA				
Rio Llullapichas	Peru		13	15.5
Amazon	Brazil	200	15	4.8
Atlantic Forest	Brazil	1200	14	4.6

Phyllobates terribilis has enough batrachotoxin to kill 100 adult humans[21]. Apparently the poisons are derived from the arthropods upon which they feed, as frogs raised in captivity are not toxic. The Malagasy frog *Mantella laevigata* has adopted the same toxic defence strategy (Heying 2001). Many amphibians are well camouflaged, but toxic species are often brightly coloured, providing **aposematic** warnings to would-be predators.

Since the 1980s amphibians have undergone a dramatic global decline, and 32% of the 5,743 known amphibians are now threatened with extinction (Stuart et al. 2004) (see Chapter 18).

4.3.2 Reptiles

Rain forests are the sources of exaggerated reports of giant snakes, dragons, and other mythical creatures, and yet there is some truth in some of these stories. The record for the longest snake is held by a 10 m reticulated python *Python reticulatus* from Sulawesi, Indonesia, discovered in 1912. New Guinea has the Salvadori monitor lizard *Varanus salvadori*, the world's longest lizard[22] at 4.75 m. The world's largest snake by mass is the South American anaconda *Eunectes murinus* (Figure 4.10). The Neotropics are also home to the

[21] Embre and Choco Indians from Colombia use the skin toxins of *P. terribilis* and other poison-dart frogs to poison their blowpipe darts, and medical research using the alkaloid batrachotoxin from *P. terribilis* aims to develop muscle relaxants, heart stimulants, and anesthetics (Clarke 1997).

[22] The Indonesian island of Komodo, while not a genuine rain forest habitat, is home to the world's largest lizard, the 150 kg Komodo dragon *Varanus komodoensis*.

Figure 4.10 A relatively small anaconda, *Eunectes murinus*.

Photo by Rob Pickles.

remarkable basilisk lizard *Basiliscus basiliscus* of the iguanid family, which is able to run up to 20 m on water without sinking[23]. Projectile tongues, prehensile tails, independently swivelling eyes, changing skin colour, and a variety of head appendages make the African and Malagasy chameleons among the most distinctive forest reptiles. Madagascar has most of the estimated 130 species of chameleons, and both the smallest, the 2 cm pygmy stump-tailed chameleon *Brookesia minima* that dwells among the leaf litter, and the largest, the 80 cm Parson's chameleon *Calumma parsoni*, are found within its rain forests.

The distribution and abundance of reptiles are poorly understood compared with most other vertebrate groups. Reptiles, particularly lizards, are common in many tropical rain forests, especially among the leaf litter. Their abundance may stem from the fact that most species feed on arthropods, a very abundant food source, and their food requirements are low. Amazonian regions (of several thousand hectares) have 20–40 lizard species with at least 124 species occurring throughout the

Amazon (Duellman 1990; Da Silva and Sites 1995). Densities of herpetofauna appear far higher in Neotropical forests than Southeast Asian dipterocarp forests (Inger 1980a). One explanation may be that supra-annual mast fruiting of dipterocarp trees limits the availability of seed-feeding arthropods and therefore limits availability of food to insectivorous frogs and lizards. Central American rain forests have much higher densities of lizards than the Amazon (Vitt and Zani 1998). For lizards, African rain forests seem to have about half the species richness of Neotropical and Southeast Asian forests but are intermediate between the two in density (Hofer and Bersier 2001). Snake densities are more variable by site, although there has not been much standardized sampling (Hofer and Bersier 2001). Nevertheless, snake densities in African forests appear lower than in other rain forests and may reflect increased predator pressure, though this hypothesis remains to be verified (Janzen 1976).

Rain forest lizards may be grouped by their thermoregulatory behaviour. **Heliotherms** bask and forage

[23] Hence also known as the Jesus Christ lizard.

Box 4.7   Gliding and parachuting

A remarkably diverse set of animals move through the forest by gliding or parachuting from tree to tree. Such animals include lizards, squirrels, and even snakes and frogs, and although they are commonly referred to as 'flying' they are incapable of true flight (birds and bats being the exception).

Among mammals, gliding has evolved independently in several groups, including the 'flying' squirrels of the genera *Pteromys* and *Petaurista* (Rodentia: Sciuridae), the unrelated African bark-eating scaly-tailed squirrels (Anomaluridae), and the two species of colugos *Cybocephalus* (Dermoptera) from Southeast Asia. These mammals glide by means of a skin fold that stretches between the arms and legs, forming a simple aerofoil that catches the air as the animal leaps between trees.

In Asia the flying lizards of the genus *Draco* glide by expanding a skin membrane supported by elongated ribs that are folded against the side of the body when not in use (Figure 4.11). By contrast, the parachuting geckos of the Asian genus *Ptychozoon* have cuticular outgrowths that are not under muscular control but passively catch the air as the animal falls.

Southeast Asian *Chrysopelea* snakes are able to parachute by flattening their body to form a concave surface as they descend. They are remarkably versatile in flight and can even perform 90° turns.

Gliding or parachuting has evolved independently in at least two families of tree frog, Hylidae and Rhacophoridae. Frogs in the Neotropical hylid genus *Agalychnis* possess enlarged and extensively webbed feet and have a characteristic posture associated with enhanced gliding performance. The Bornean flying frogs of the genus *Rhacophorus* also have large webbed feet but are more accurately described as parachuters.

Gliding allows animals to move between trees more quickly and at less energetic cost than would

Figure 4.11 Five-banded flying lizard, Borneo.

Photo by Karin Beer.

otherwise be possible. Some African squirrels purposely trim vegetation that impinges upon their principal gliding routes (Emmons and Gentry 1983). Increased mobility increases food harvesting rates, which is particularly important when resources are patchy. Indeed, the abundance of gliding animals in Asian forests compared to their paucity in Neotropical forests has been linked to the relative scarcity of invertebrate prey in Asian dipterocarp forests (Inger 1980a, b). The prevalence of gliding in Asia has also been ascribed to the local paucity of lianas, improved efficiencies in more 'open' formations (Emmons and Gentry 1983) and 'taller' forests generally (Dudley and Devries 1990).

Comparison of mammalian glider faunas from Africa, Borneo, and Australia shows similar distributions: the largest gliders are less than 2 kg and mean glider mass is about 0.5 kg (Dial 2003). Smaller size is, however, also associated with smaller litters and increased predation. It may be for this reason that most mammalian gliders are nocturnal.

Box 4.8   Being arboreal

Movement within the forest canopy poses challenges for animals seeking to maximize foraging efficiency and avoid predation in this complex habitat. Climbing down one trunk and up the next is time consuming, energetically expensive, and exposes animals to terrestrial predators. Flying is a good alternative, but larger animals lack the agility required to fly within the canopy and are limited to its outer surface. Another efficient option is to glide or parachute (see Box 4.7) but this too is largely limited to smaller animals. These challenges of movement in the canopy have led to various anatomical and behavioural specializations. Forest vertebrates on different continents have developed similar forms and adaptations to canopy life. In some cases these involve relatively shorter limbs, sharp claws, suspensory habits, and prehensile tails (e.g. the Neotropical pygmy anteater *Cyclopes didactylus* and the Asian tree pangolin *Manis javanica*, Figure 4.5). Prehensile tails 'support alone the weight of the suspended body' (Emmons and Gentry 1983) and improve both reach and stability of the animal in feeding and locomotion in the canopy. Such tails evolved independently in six orders of mammal, two of reptiles, and one of amphibians.

Primates have developed a repertoire of movements that help them navigate the forest canopy.

The ability to raise limbs above the head and to swivel them in fully muscled ball-joints is a specific trait of arboreal species (the flexibility of the human arms and legs can be ascribed to the needs of more arboreal ancestors). For long-lived large species, such as chimpanzees, safety rather than efficiency appears to be the primary selective cause that maintains energetically costly climbing traits (Pontzer and Wrangham 2004).

The detailed form of the relationships underlying the anatomical, physiological, environmental traits and life-history requirements of arboreal animals remain only partially understood. A particularly striking example is brachiation: locomotion by which the forelimbs alone are used to move the suspended animal. Such swinging movements are adopted to some degree by woolly monkeys *Lagothrix* spp., spider monkeys *Ateles* spp., adolescent chimpanzees *Pan troglodytes*, and orang-utan *Pongo pygmaeus*. However, the masters of brachiation are gibbons (*Hylobates*). There are two distinct forms: continuous contact (analogous to walking) and ricochetal (analogous to running), both involving pendular motion that entails continuous exchange of kinetic and potential energy. This process minimizes energy losses, which mostly occur due to irregularities in the animals' trajectories (Bertram 2004).

---

in the sun, maintain relatively high body temperatures, and are more frequently found in forest gaps, though they enter the forest to forage (Vitt et al. 1998). Heliothermic species tend to be larger bodied active foragers that will seek out hidden prey, while the non-heliotherms are sit-and-wait foragers feeding on a wide variety of arthropods. Non-heliothermic lizards are associated with shaded habitats, maintain lower body temperatures, but use a wider variety of microhabitats than heliotherms. Many canopy species fall within this category, including the common flying lizards of the Asian genus *Draco* (see Boxes 4.7 and 4.8).

4.4   Fish

Fish are a neglected component of tropical forest biota. They are important prey for several mammals, including otters (*Lontra* and *Aonyx*), otter civets (*Cynogale*) and fishing cats (*Prionailurus*), and birds such as kingfishers. In Amazonian flooded forests, fish are also important seed dispersers (Chapter 12).

The Amazon basin contains about 3,000 species of fish (Val and de Almeida-Val 1995) or about 30% of all freshwater fish species and 10% of the global fish fauna (Groombridge and Jenkins 1998). Asian and African rain forest river systems also have very high fish species richness, although these are less well studied than the Amazonian communities. The Congo basin river system has over 700 species (WRI 2003). Indochina (Laos, Vietnam, and Cambodia) has over 930 fish species in 87 families, though many of these belong to the Mekong basin which contains much more than just rain forest habitat. Fish richness in Borneo is still impressive, with 290 species recorded in the Kapuas River, 147 species in the Mahakam River, and 115 species in the Baram River (Dudgeon 2000). These fish communities often support significant commercial and local fisheries (Crampton et al. 2004). The relatively low diversity and productivity of Australasian freshwater fish communities is because they are derived mainly from marine families, and consequently their colonization of floodplains has been limited (Coates 1993). This is in marked contrast to Amazonian fishes that have exploited a range of river, lake, and floodplain habitats and show considerable community structure across these habitats. Nevertheless, 375 freshwater species are known from New Guinea (Allen et al. 2000) where several centres of fish endemism have been identified. Asian fish communities are overwhelmingly dominated by the Cypridae (carps, minnows, danios), and differ in this respect from their Neotropical and African counterparts, which are represented mainly by the Cichlidae (cichlids and tilapias) and Characidae. The Characidae are absent from Asia, and the Cichlidae are represented by only two species in Sri Lanka (although there have been widespread introductions of African Tilapia) (Dudgeon 2000).

During the early 1990s between 10 and 20 million clown loaches *Botia macranthus* (Cobitidae), known from only a few locations in Borneo and Sumatra, were exported from Indonesia for the aquarium trade, and yet the biology of this species remains largely unknown (Kottelat and Whitten 1996). This paucity of information is typical. For example, a 1994 study of a **peat** swamp forest in Peninsular Malaysia yielded 47 fish species, including five new species, from a habitat type where only 26 had previously been recorded (Ng 1994a). In Southeast Asian peat swamp forests the so-called earth-

worm eels (Chaudhuriida) (Kottelat and Lim 1994) and some catfish (Ng and Lim 1993) actually live in the peat soils and are able to persist within the soil substrate even when the streams are dry. It is now recognized that 10% of Peninsular Malaysia's freshwater fish are endemic to peat forests (Ng 1994b).

The Amazon flood cycle has typical amplitude of about 10 m and floods large areas of forest for months. An estimated 70,000 km^2, or about 1%, of Amazonian forest floods each year (Saint-Paul et al. 2000). Flooding greatly increases the volume of water available, and also affects its spatial extent and oxygenation, connects lakes with the wider river system, and allows fish to migrate into forest habitat and between lake systems. Many Amazonian fish species migrate between the floodplains and river channels, tracking the rise and fall of flood waters. The floodplain forest becomes an important source of food during high-water, so many fish species feed on tree fruits and seeds and thereby disperse them (Goulding 1980; Forsberg et al. 1993; Mannheimer et al. 2003) (Chapter 12). As the waters recede most fish migrate back to the main river channels, but some, like the popular aquarium armoured catfish *Liposarcus pardalis*, are able to survive unfavourable environmental conditions during the low-water period by breathing air[24]. Many Asian fish also migrate along forest river systems, particularly during the wet season, when fish can exploit various terrestrially derived foods from riparian forests.

Fish communities are affected by water clarity. Fish such as the Characiformes, a very large group that includes the piranhas, hatchetfish, pencilfish, and tetras, occupy clear and shallow or surface waters. Catfishes (Siluriformes) in contrast rely on tactile and chemical receptors, and they, together with knifefish (Gymnotiformes) and electric eels which have electrical sensors, dominate turbid or deep waters (Tejerina-Garro et al. 1998).

Perhaps the fish most closely associated with the forest are those that live in ephemeral puddles on the forest floor. The 'annual' killifish (with more than a hundred species in tropical Africa and America) complete their life cycles within a year (sometimes just a few months)

---

[24] The starving piranha also stranded in these pools become increasingly aggressive, and it is this that has given them their dangerous reputation.

allowing them to live in shallow pools of water that dry up seasonally. Eggs remain viable until the following season (or longer). There are reliable accounts of these fish persisting in puddles formed by forest elephant footprints (J. Kingdon, pers. comm. with DS). Without forest shade, or in drought years, many pools may evaporate too quickly for the forest fish to survive. The West African fire-mouth panchax *Epiplatys dageti* (also a killifish) will move over moist ground to find more promising puddles, as do various walking catfish *Clarias* spp. of Asia that can breathe air.

4.5   Conclusion

Rain forests are immensely rich in vertebrate species, though animal diversity and density can be quite heterogeneous within each major tropical forest region. How animal diversity is determined remains a central question in ecology. Coexistence is at least partly facilitated by differences in the types of resources exploited, in the timing of resource-accessing activities, and in the location of those activities within the structure of the forest. Community differences among rain forest regions are due to biogeographic histories and regional ecologies. The absence of primates in Australasia is, for example, a biogeographic outcome, while the dominance of the Asian forests by dipterocarps, whose fruits are resource poor, is a likely ecological reason for the paucity of frugivorous mammals. Yet there are also many examples of convergent evolution responding to broadly similar environmental constraints and opportunities: toucans and hornbills, mouse deer and agoutis, being two examples. It is these biogeographic and ecological conditions that are the subjects of Chapters 6 and 7.

CHAPTER 5

The little things: invertebrates

'The high points of a rainforest trip are the moments when you enter the forest and when you get out of it.'

A. Fidler, *Call of the Amazon*

'The little things that run the world.'

E. O. Wilson (1987b)

Biting ants and mosquitoes, poisonous centipedes and scorpions, blood sucking leeches, stinging wasps, and a host of caterpillars sporting toxic spines and hairs, all these diminish the enjoyment of our visits to rain forests, but also add to their excitement. Rain forest invertebrates, in addition to being uncomfortable, can also be beautiful, are often functionally essential, and are invariably enigmatically complex in their interactions and behaviours. Invertebrates, and particularly insects, are also undoubtedly diverse and abundant. They have occupied almost every conceivable nook and cranny of the forest, such that if all plant material were to be dissolved from view it would still be possible to discern the structure of the forest, from the soil and leaf litter through to the stems and branches of trees and distribution of leaves in the canopy, from the invertebrates. It is therefore no wonder that estimates of global species richness, typically in the range of 5–10 million (Ødegaard 2000), have been derived by extrapolation from wholesale sampling of tropical rain forest insects (Box 5.1).

Box 5.1 How many species are there in the world?

Estimating global species richness has become something of a preoccupation among entomologists (Erwin 1982; Stork 1988; May 1990; Stork 1993). In 1982 Erwin surprised many by his calculation of 30 million species globally, extrapolated from 955 beetle species sampled from 19 *Leuhea seemannii* trees in Panama (Erwin 1982). His extrapolation relied on three qualifications: his estimate of the proportion of beetle host-specificity, the fraction of beetles found exclusively in the canopy, and the assumption that all tree species have the same number of insect associates (mainly herbivores). The last assumption is certainly not true, as rarer plants have fewer host-specific herbivores than more common species, and it is also increasingly clear that host-specificity is lower than previously thought (Novotny et al. 2002; Novotny and Basset 2005). There is a general consensus that global species richness stands at between 5–10 million, although we remain almost totally ignorant of the diversity of major invertebrate groups such as mites and nematodes (see also box 3.1).

Figure 5.1 (continues)

H

I

J

Figure 5.1 The variety of invertebrates of tropical rain forests. Beetles (A–C) dominate in terms of species and functional variety. Bugs (D [see colour plate 7]–F) are important herbivores, but some groups are also predatory. Orthopterans (G) include many forms of grasshoppers, katydids, and crickets which are primarily herbivorous. Butterflies (H [see colour plate 6]) are probably the most familiar and best known herbivorous group. Other common non-insect arthropods include spiders (I) and millipedes (J).

Photos by Carsten Bruehl (A–C, E–G) and Karin Beer (D, H–J).

Invertebrates include much more than just insects. The arthropod group, characterized by an exoskeleton, includes centipedes, millipedes, arachnids, and crustaceans (Figure 5.1). Other invertebrate groups are molluscs, primarily snails and slugs, annelid worms (mainly earthworms and leeches), as well as a variety of less common groups such as flatworms, velvet worms, and common but poorly known groups such as nematodes. Additionally, there are huge numbers of ecto- and endo-parasites of vertebrates and invertebrates. Even identifying and naming species is problematic, and a large proportion of scientifically described species are only recorded as single specimens from a single location.

An impression of our ignorance of tropical forest insects can be gained by comparing the proportion of species that cannot be named from each of a number of taxa. In one such study even common and well known taxa such as ants could not easily be assigned to known species (Lawton et al. 1998), while a large majority of species belonging to lesser known groups, including the relatively familiar beetles, remained unknown (Table 5.1). Still less is known about the ecology of tropical invertebrates, although collectively they are 'the little things that run the world' (Wilson 1987b), contributing to the decomposition of organic materials, pollination and seed predation, herbivory and other ecological processes that are essential to the normal functioning of tropical forests.

Table 5.1 Proportion of insects and other groups that remain unknown or difficult to identify to species from samples at the Mbalmayo Forest Reserve, Cameroon.

Group	Total species recorded	Percentage of species unknown
Birds	78	0
Butterflies	132	1
Flying beetles	825	50–70
Canopy beetles	342	>80
Canopy ants	96	40
Leaf litter ants	111	40
Termites	114	30
Soil nematodes	374	>90

Adapted from Lawton et al. 1998.

Extensive sampling of canopy invertebrates in the 1980s gave the impression of immense and previously undocumented species richness. The harvest of insects obtained by fogging the canopy with clouds of insecticide gives us more than just a rough approximation of species number—it provides an insight into the sheer abundance and variety of invertebrate types. It also impresses upon the collector the preponderance within the invertebrate community of a single group, ants, which accounts for 51–94% of the individual arthropods in these samples, and as much as 86% of the total biomass (Davidson et al. 2003). The forest floor is similarly rich in species, which again ants dominate in terms of abundance. Indeed, in tropical rain forests ant biomass alone has been estimated to be four times as much as that of all vertebrates (Holldobler and Wilson 1990), and ants have considerable functional importance as predators (Figure 5.2), herbivores, and mutualistic plant partners (Chapter 13). It is, however, beetles that contribute most in terms of forest species richness, particularly rich groups being rove beetles (Staphylinidae), weevils (Curculionidae), and leaf beetles (Chrysomelidae).

The structural and microclimatic complexity (and tree diversity) of forest canopies provides one explanation for such species diversity. However, species richness is actually fairly equitably distributed between the canopy and lower forest layers (Stork 1988; May 1990; Stork 1993; Stork and Blackburn 1993). Vertical stratification of certain groups of forest arthropods can, nevertheless, be marked (Basset 2001). Insect herbivores, such as chrysomelid and curculionid beetles and sap sucking bugs (Hemiptera and Homoptera), as well as the ants that tend them, are generally more abundant and species rich in the upper canopy where food resources for these taxa are more abundant and diverse (Basset et al. 1992). On the other hand, the abundance and diversity of pyralid moths and macro-moths from canopy and understorey samples in Borneo are almost identical, while geometrid moths are more diverse in the understorey (Willott 1999; Intachat and Holloway 2000; Schulze et al. 2001). Abundance and species richness of flies, parasitic wasps, and beetles are similar or even higher in the understorey of mixed rain forest in Surinam than in the canopy (de Dijn 2003). Vertical stratification of arthropods in the forest is complicated by seasonal changes, flowering phenology, and daily vertical migrations of many flying insects, as well as differences in arthropod abundance between tree species (Basset 1999; Barone 2000; Itioka et al. 2001; Wagner 2001).

In this chapter we present an overview of the main taxa of rain forest invertebrates, describing what they are and, where information is available, their relative abundances and distributions. Arthropods necessarily dominate much of this chapter as they are better known than other groups, and are ubiquitous, abundant, and functionally important (Table 5.2).

5.1 Insects

5.1.1 Butterflies and moths

Butterflies account for only 5–7% of the world's estimated 250,000 Lepidoptera species, the rest comprising various families of moths[1]. Most moths are herbivorous in their

---

[1] The richest tropical realm is, as for many other groups, the Neotropics, which has an estimated 35% of moth species, though many have yet to be described. Recent studies of certain Neotropical moth groups (Tineoidea, Tortricoidea, and Gracillarioidea) have shown that between 75% and 98% of species sampled were new to science. Estimates of moth species richness in Costa Rica are between 13,000 and 16,000 species, comparable to the species diversity of North America north of Mexico or Australia which have 380 and 150 times the area respectively.

Figure 5.2 Ants are ubiquitous and generalist predators of a wide range of invertebrate organisms. Here we see *Oecophylla* ants attacking a honeybee (A), *Harpegnathos* ants with an orthopteran nymph (B), *Leptogenys* ants swarming over a millipede (C) and an earthworm (D), and *Odontomachus* ant predating a small fungus fly (E), and unidentified ants predating a bug (F).

All photos from Borneo and by Carsten Bruehl.

Table 5.2 Some principal groups of arthropod forest biodiversity in Southeast Asia (adapted from Hill and Abang 2005). We exclude many specific groups associated with water, caves, birds' nests etc.

Life-style	Insect group
Burrowing (Fossorial)	Springtails (Collembola), Crickets (Gryllidae), Molecrickets (Gryllotalpidae), Termites (Isoptera), Ants (Formicidae), Bristletail (Thysanura), Tarantula Spiders (Theraphosidae)
Litter dwellers	Cockroaches (Dictyoptera), Ground beetles (Carabidae), Ants (Formicidae), Millipedes (Diplopoda, including Julidae, Polydesmidae, Harpagophoridae), Mites (Parasitidae)
Surface dwellers	Ground beetles (Carabidae), Ants (Formicidae), Ant-lions (Neuroptera), Centipedes (Chilopda, including Geophilidae, Scolopendridae, Scutigeridae), Spiders (Lycosidae, etc).
Foliage eaters	Katydids (Tettingoniidae), Stick insects (Phasmida), Butterfly and Moth caterpillars (Lepidoptera), Bagworms (Psychidae), Leaf Beetles (Chrysomelidae)
Sap suckers	Plant bugs (Aphididae, Cicadidae, Coccidae, Pentatomidae, Coreidae), Red Spider Mites (Tetranychidae)
Leaf miners	Leaf miners (moths), Microlepidoptera, Flies (Agromyzidae)
Gall makers	Gall Wasps (Cynipoidea, Chalcicoidea), Gall Midges (Cecidomyiidae), Gall mites (Eriophyiidae)
Fungus/Lichen eaters	Springtails (Collembola), Barklice (Psocoptera), Bagworms (Psychidae)
Wood borers	Termites (Isoptera), Jewel Beetles (Buprestidae), Longhorn Beetles (Cerambycidae), Ambrosia Beetles (Scolytidae), Goat Moths (Cossidae), Bark Moths (Metarbelidae), Carpenter Bees (Xylocopidae), Woodworm (Anobiidae)
Rotten wood eaters	Click Beetles (Elateridae), Stag Beetles (Lucanidae), Fungus Weevils (Curculionoidea)
Tree nesters	Some termites (Isoptera), Some ants (Formicidae), Wasps (Vespidae), Bees (Apidae)
Fruit eaters	Fruit Flies (Tephritidae), Some beetles (Scarabaeidae, etc.), Some caterpillars (Noctuidae)
Nectar feeders	June Beetles (Scarabaeidae), Butterflies and Moths (Adult Lepidoptera), Flies (Diptera)
Predators	Tiger Beetles (Carabidae), Rove Beetles (Staphylinidae), Ladybirds (Coccinellidae), Mantids (Mantidae), Lacewings (Neuroptera), Harvestmen (Opiliones), Predatory mites (Phytosedidae, Parasitidae), Spiders (Epeiridae Thomisidae, Attidae), Pseudoscorpions (Chelifers), Centipedes
Parasites of animals	Tick (Ixodidae), Mites (Acarina, Trombiculidae, Sarcoptidae, Demodicidae), Mosquitoes (Culicidae), Siphunculata, Tachinidae, Mallophaga, Hymenoptera-Parasitica
Pollinators	Bees (Apidae), Flies (Muscidae, Syrphidae), Thrips, (Thysanoptera), Weevils (Curculionidae) and other beetles, Moths and butterflies (Lepidoptera, various families).

larval stage, and their diversity in tropical regions may simply reflect a greater diversity of plant species, although little is actually known of the relationships between many moth species and their larval food plants (indeed most species are only known from the adult form). Moths are, nevertheless, important as herbivores, seed predators, and pollinators. The large hawkmoths (Sphingidae) are the primary pollinators for around 10% of Costa Rica's seasonally wet forest tree species (Bawa et al. 1985). In Southeast Asian forests tortricid and pyralid moths are predators of dipterocarp seed and may account for a substantial fraction of seed mortality. Other moth larvae feed on leaf litter, decaying wood, and fungi, and therefore have at least a minor role in decomposition and nutrient recycling.

Butterflies are amongst the most conspicuous and best known of rain forest insects. As with many other groups, they attain their highest richness in tropical rain forests, particularly in the Neotropics where a few hundred hectares can easily contain more species than occur in the whole of Europe or North America (Table 5.3).

Butterflies are divided into four main groups, the Papilionidae, Pieridae, Nymphalidae, and Lycaenidae, and the relative proportions of each in the Asian, African, and Neotropical regions are remarkably similar (Table 5.3). The proportion of butterflies in Madagascar is, however, strikingly different from other regions, in that lycaenids represent only 18% of all species compared with around 50% elsewhere. The other large tropical island, New Guinea, is also unusual in having a relatively high proportion of pierids.

Papilionids, commonly known as swallowtails owing to the 'tail' extending from the back edge of the hind wing of many species, have their highest diversity in the neotropics, but largest and most spectacular forms in the

Table 5.3 Comparative species richness of butterflies in Neotropical sites

Site	Area (ha)	Papilionidae	Pieridae	Nymphalidae	Lycaenidae	Total
Neotropics	780 million	169	347	1850–2500	2300–3000	5175–5975
La Selva, Costa Rica	1000	16	26	219	181	442
Jatun Sacha, Ecuador	600	25	23	306	345	699
Paquitza, Manu, Peru	3900	25	31	369	427	852
Cacaulandia, Brazil	2000	30	31	423	715	1199
Alto Jurura, Brazil	>500	38	37	467	496	1038
North America	1961 million	35	63	202	171	471
Europe	994 million					440
New Guinea	88 million	41	146	222	376	785
Malaysia	33 million	44	44	273	400	761
Africa	3036 million	80	145	1107	1397	2729
Madagascar	58 million	13	28	175	46	262
D. R. of Congo	234 million	48	100	607	551	1306
Australia	761 million	18	35	85	140	278

Adapted from DeVries 2001.

Old World. These 'birdwings' (*Troides* and *Ornithoptera*) range from Sri Lanka to New Guinea and Australia and attain highest diversity in Southeast Asia. Birdwing caterpillars feed on *Aristolochia* vines and incorporate the toxic compounds the vines produce into their bodies, making both the caterpillar and adult distasteful to predators.

The Pieridae are predominantly yellow or white butterflies, common throughout the tropics. Many species display 'mud-puddling' behaviour, where hundreds or thousands of males congregate at mud along streams, pools, or puddles from which they obtain sodium that is transferred to females during mating (Beck et al. 1999). Papilionid butterflies have also been observed doing the same.

The Nymphalidae are a diverse group of large and often brightly coloured or iridescent species such as the Morphinae and Apaturinae, or transparent species (Ithomiinae). Many form large and complex mimetic associations (Danainae, Ithomiinae, and Heliconiinae) in the forest understorey, while others have leaf-like camouflage and mainly inhabit the forest canopies (Charaxinae and Nymphalinae).

The small Lycaenidae, commonly called blues and coppers owing to the colouration of many of the species, account for about half of all tropical butterfly species. They display various life history strategies. Adults can feed on nectar, honeydew, carrion, fruit, or not at all. Caterpillars of different species feed on a wide variety of plants, but some also on other insects or their secretions. Many form complex symbiotic or parasitic associations with ants, living within the ant nest by producing chemical cues that alter ant behaviour. Some African and Southeast Asian species belong to mimicry complexes (see below). The almost exclusively Neotropical subfamily Riodininae, or metalmarks, are the most diverse lycaenid group. The caterpillars of about one third of the species live in close association with protective ants that are fed by special secretions.

Butterfly mimicry

Many tropical butterfly species closely resemble one another, even sharing conspicuous colour patterns, and fly together in the same forest habitat. This mimicry is an anti-predator strategy. Batesian mimicry, where a palatable species resembles a distasteful one and thereby avoids predation, is widespread, but is best exhibited by females of the African swallowtail *Papilio dardanus* which mimics different species across its range (Figure 5.3). Females of this species look completely different depending on which species they mimic.

Among Neotropical Ithomiinae and Heliconiinae and Asian Danainae (Nymphalidae), several co-occurring unpalatable species closely resemble and associate with each other. This 'Müllerian' mimicry reinforces the

Figure 5.3 A drawer of butterflies collected and arranged by Alfred Russel Wallace to illustrate Batesian mimicry, these specimens being from India, Africa, Southeast Asia, and Central America. The inedible model species are arranged in the far right column and the presumed edible mimics are to the left of them. Note that the males of two of the mimic species are not mimetic. All the writing is Wallace's. (See colour plate 2.)

Photo by George Beccaloni.

association of unpalatability with a particular colour pattern, and, by minimizing the number of visual cues, possibly makes it easier for a predator to learn which species to avoid (Kapan 2001). The tiger-pattern mimicry complex in Neotropical forests includes moths as well as butterflies (Figure 5.4) from several lepidopteran families. The Acraeinae butterflies in African forests form similar mimicry complexes[2].

5.1.2 Beetles

Beetles (Coleoptera) comprise around 350,000 recognized species from 160–170 families, depending on the classification system used, although almost two-thirds of species belong to only seven groups (Curculionidae, Staphylinidae, Scarabaeidae, Chrysomelidae, Buprestidae, Cerambycidae, and Carabidae). These are particularly well represented in the tropics. A study in Budongo Forest, Uganda, used canopy fogging to collect a total of 29,736 beetles (from 64 trees of four species) that were assigned to 1,433 species; 41.6% of these were represented by only a single specimen (Wagner 2001). Three years of light-trapping at Barro Colorado Island, Panama, yielded over 95,000 weevil (Curculionidae) specimens alone, from 1,239 species, 28% of which were represented by only a single specimen (Wolda 1992). It is this immense richness that has given beetles a prominent position in studies attempting to quantify global species richness (Box 5.1).

The weevils (Curculionidae) are by far the largest group with around 50,000 species. They primarily feed

[2] Relatively common among invertebrates, Müllerian mimicry rarely occurs in vertebrates, although it has been shown among dendrobatid frogs in Peru (Symula et al. 2001).

Figure 5.4 Müllerian mimicry complexes form where co-occurring noxious species evolve to resemble one another to reinforce visual unpalatability cues. Each pair of pictures shows a separate 'sub-ring' with the models (each a different species of *Melinaea*, Ithomiinae) above, and the mimics being different *Heliconius numata* (Heliconiinae) morphs, shown below. All of these species come from the same site in Ecuador. (See colour plate 3.)

Photo by George Beccaloni.

on leaves (as do the leaf beetles, Chrysomelidae), seeds, and wood. Their head is extended into a snout which, in some species, can be longer than the rest of the body.

The Scarabaeidae is a primarily phytophagous group that contains the dung beetles (Figure 5.5). Dung beetles recycle nutrients through the processing and transportation of dung, and disperse seeds that are embedded within vertebrate dung. Their role as seed dispersers varies as the abundance of species that bury no seeds, bury small seeds only, or bury small and large seeds, differs substantially among sites (Vulinec 2002). There are no

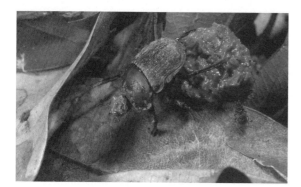

Figure 5.5 A dung beetle, here at Danum Valley, Sabah, Malaysia, rolling a dung ball to a safe location. Dung beetles are widespread in the tropical forests both in the understorey and canopy.

Photo by Carsten Bruehl.

obvious differences in species richness patterns across tropical regions among dung beetles (Davis 2000). The richest fauna of 87 identified species from Danum Valley in Sabah, Malaysian Borneo, probably reflects the comprehensive sampling, which includes 68,481 identified specimens (Davis 2000). The most common single species from this sample accounted for 27,383 individuals (40% of the total) while the majority of species were represented by less than five individuals.

Dung beetles of the forest canopy have been largely ignored, yet a significant portion of vertebrate dung remains arboreal and never reaches the forest floor, being instead caught on branches and leaves of the canopy or sub-canopy. This dung is utilized by dung beetles that are abundant only at heights exceeding a few metres above the forest floor and very rarely descend to the forest floor (Vulinec et al. 2007).

Various other species-rich, common, and widespread rain forest beetle groups include the Buprestidae, the larger forms of which have brightly coloured and metallic sheens to their carapaces and are much sought after for jewellery, hence their common name of jewel beetles. The family includes probably the largest arthropod genus, *Agrilus*, currently containing over 2,000 species, but as most tropical buprestids are scientifically undescribed the genus is almost certain to grow. The larvae of the long-horned beetles (Cerambycidae) are wood borers within living, dying, and dead trees. The adults feed on flowers, leaves, or bark. The ground beetles (Carabidae), including the tiger beetles, are predators with both ground-dwelling and arboreal forms. Although a large group, they remain mostly unknown: a study in Ecuadorian Amazonia yielded 2,329 individuals belonging to 318 species of which more than 50% were undescribed (Lucky et al. 2002). Furthermore, a high turnover in species composition between sampling dates and continued capture of new species with successive samples indicates that the species diversity of tropical arboreal carabid beetles may be far higher than current figures suggest (Lucky et al. 2002). Beetle taxonomists are unlikely to catalogue the majority of rain forest species any time soon.

5.1.3 The social insects: ants, bees, wasps, and termites

Social insects include all ants and termites and many species of bees and wasps. Although these account for less than 2% of all described insect species, ants alone amount to around one-third of insect biomass in Amazonian rain forest (Fittkau and Klinge 1973). They play a major role in pollination, nutrient cycling, soil turnover, herbivory, and predation (Chapters 12 and 13).

Ants

The exclusively social ants (Hymenoptera) are the most important group of social insects in terms of abundance and biomass (Box 5.2). They also make a major contribution to forest dynamics as herbivores and predators and form many diverse mutualistic associations with a variety of plants and insects (Chapter 13). The large majority of the estimated 11,000 ant species belong to the tropical rain forests where local diversity can be substantial (Holldobler and Wilson 1990) (Table 5.4). Sampling within 6 hectares of lowland rain forest at Kinabalu, Sabah, resulted in a collection of 524 ant species (Bruhl et al. 1998); 40 species[3] were collected from a single unidentified

---

Box 5.2   Explaining ant abundance in tropical forest canopies

Why are ants so abundant? What supports such an abundance of predatory and scavenging organisms? One possibility is that arthropod prey is more plentiful than sampling by insecticide fogging might suggest (Davidson et al. 2003). A high turnover of arthropods might also sustain high ant biomass, or, alternatively, ant workers may have little requirement for protein and might be sustained by abundant carbohydrate derived from plant exudates. It appears that canopy ants in Peru and Borneo act as indirect herbivores by tending and protecting sap-feeding and other insects, from which they collect exudates which provide both carbohydrate and protein requirements (Davidson et al. 2003) (Figure 5.6). Consequently, rain forest plants may experience considerable losses to ants acting as indirect herbivores, and this may sustain far larger ant populations than suggested by the abundance of ant prey alone (Davidson et al. 2003).

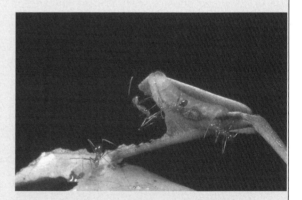

Figure 5.6 *Pheidole* sp. ants in Borneo tending a caterpillar from which the ants harvest secretions. In effect the ants are obtaining plant carbohydrates indirectly via the caterpillar's secretion, and the caterpillar in turn gains the ants' protection.

Photo by Carsten Bruehl.

---

[3] By comparison there are no more than 50 ant species native to the British Isles.

Table 5.4 Tropical rain forest ant communities.

Site	Sampled Area (ha)	Genera	Species	Species-rich genera
Malaysia: Kinabalu, Sabah	6	73	524	Camponotus
Malaysia: Pasoh	25	75	467	Camponotus, Polyrhachis
Brunei	1	57	232	Pheidole
Indonesia: Sulawesi	?	62	269	Polyrhachis
New Guinea	260	59	172	Pheidole, Polyrhachis
Australia: Kimberly rain forest	?	33	102	Pheidole
Peru	5	64	256	Pheidole
Peru	1000	78	520	Pheidole
Ghana	160	63	219	
Brazil: Atlantic Forest	?	71	272	

Adapted from Bruhl et al. 1998.

leguminous tree in the Tambopata Reserve in Peru (Wilson 1987a); 61 species from a sub-canopy tree in Sabah, Borneo (Floren 1995, quoted in Bruhl et al. 1998; and 95 species from a single *Goupia glabra* (Goupiaceae Celastraceae) tree in central Amazonia (Adis et al. 1998). The existence of so many ant species at the same sites, or even the same tree, begs the question as to how these species coexist. Samples from Sabah, Peru, and central Amazonia show that most species are rare: a single tree or area is dominated by one to three very common species, a further ten or so moderately common species, and the rest rare or represented by very few individuals[4].

Aggressive interactions among ant colonies at the edges of their territories are frequent. The nocturnal ant *Camponotus gigas* in Borneo maintains huge non-overlapping three-dimensional territories with ground area of up to 0.8 ha. These ants nest on the ground but forage in the canopy 50 or more metres above. To defend these territories from neighbouring colonies, the ants maintain outlying barracks housing large 'major' ants that patrol territory edges, attacking conspecifics and other ants deemed a threat (Pfeiffer and Linsenmair 2001). The ubiquity and agressiveness of ants has led to the evolution of many forms of ant-mimicry among a variety of other taxa (Box 5.3).

The ant faunas of the major tropical regions are similarly rich. The ant genera *Crematogaster*, *Camponotus*, and *Pheidole* are globally widespread except for some remote oceanic islands including Hawaii. The largest ant colonies belong to the swarm-raiding Neotropical army ants *Eciton*, and African driver ants *Dorylus*. The Neotropical leafcutter ants *Atta* and *Acromyrmex* also have huge colonies and are immensely important as herbivores that shape forest structure and composition (Chapter 13). Weaver ants *Oecophylla* are dominant elements of arboreal ant communities in Africa and Asia. Many other ant species have comparatively tiny colonies with no more than 150 individuals occupying short-lived nests in decaying twigs, tree hollows, or in amongst the leaf litter (Byrne 1994).

One of the more unpleasant camping experiences in the African rain forest is being awakened by viciously biting army ants (Figure 5.7). Though they do not sting, army ants have a powerful bite. Attempting to remove a biting ant often causes the head to break off with the mandibles still firmly embedded in the flesh. Army ants include the genera *Aenictus* (Africa and Australasia), *Dorylus* (Africa and Asia), and *Eciton* (America), although species that form large swarm-raiding colonies occur only in Africa and America. Some Asian species, e.g. *Pheidologeton*, have adopted similar behaviours, although their colonies are much smaller and make only occasional nocturnal surface raids from their underground nests (Berghoff et al. 2003). An army ant colony is nomadic, consisting of hundreds of thousands or mil-

[4] This pattern of species abundances is actually widespread among man invertebrate groups and other taxa.

Box 5.3   Myrmecomorphy

Given year-round abundance of ants in almost every tropical forest, and their aggressive anti-predator defences, it is unsurprising that other tropical arthropods mimic them. Ant mimicry, or **myrmecomorphy**, has evolved several times in spiders, bugs, and staphylinid beetles as well as in other groups, and is particularly common in tropical rain forests where, perhaps due to greater dependence on visually orientated predation, convergence to an ant-like form is remarkable. The adaptive significance may vary among species. Protection from predators is likely to be one reason. For example, all known staphylinids (rove beetles) that mimic ants live among army ant colonies, and the morphological and behavioural similarity to their forms may protect them from the

ants as well as from the variety of birds that follow army ant raids. Ants are also a readily available source of protein, and mimicry may allow some predators to prey on ants unmolested. Ant-mimicking spiders of the genus *Aphantochilus* feed on cephalotine ants, and appear to supplement their disguise by holding their victims before them when approached by other workers. This shielding behaviour has also been observed in the crab spider *Strophius nigricans* which is not myrmecomorphic, as an attractant and defence against *Camponotus* workers. Another crab spider *Amyciaea forticeps* attacks workers of the weaver ant *Oecophylla smaragdina*. By behaving like a dying or struggling ant worker it draws workers nearer to investigate. The spider then pounces.

lions of workers that swarm across the forest floor in a broad front, overpowering any animal unable to escape. Those that do escape the approaching horde are frequently predated by the attendant birds and other

Figure 5.7 Army ant colonies typically contain workers of many sizes, each specialized for a different task. Here a *Nomamyrmex esenbeckii* worker stands guard over a passing column, in Gamboa, Panama.

Photo by Alex Wild.

animals that accompany army ant colonies (see Chapter 13). Each evening the workers interlock their bodies to form a temporary bivouac that shelters the colony, including brood and queen, before dissembling the following morning to continue foraging. Although army ant colonies are huge, they possess only a single queen that lays as many as several hundred thousand eggs a month[5] (Holldobler and Wilson 1990).

A large part of army ant diets is the brood of other ant species (Gotwald 1995). It has been suggested that army ants may even maintain overall ant diversity by 'cropping' abundant species, thereby preventing dominance and maintaining ant diversity. The swarm raids of *Eciton burchelli* may, for example, selectively remove ant colonies of certain species, allowing other ant species to establish. In this way foraging army ants effectively initiate a process of succession in the ant community, and maintain a fairly constant fraction of the forest floor in each stage of succession, such that over large tracts of forest the relative abundance of ground-nesting ant species remains roughly constant.

[5] Consequently, army ant queens, at up to 5 cm in *Dorylus*, are the largest known ants.

Leafcutter ants (*Atta* and *Acromyrmex*) are the most prominent herbivores of Neotropical forests. Their subterranean nests typically harbour several million workers within structures that can exceed 6 m depth and reach a volume of 5 m³ (Holldobler and Wilson 1990). The workers harvest mainly leaf material, but also flowers, fruits, and seeds, from the vicinity of the nest. As much as 250 kg dry weight of plant material per year can be harvested by a single colony of *Atta vollenweideri* (Holldobler and Wilson 1990), while an *A. colombica* colony has been calculated to make an annual collection of 370 kg plant dry weight (Wirth et al. 1997). The ants do not feed on plant tissue but rather on specific symbiotic fungi that they cultivate on the plant material they collect (Chapter 3 and Chapter 13). Somehow these organisms made the transition from a hunter-gatherer lifestyle to horticulture long before humans.

When a leafcutter colony moves to a new site, or a queen leaves to establish a new colony[6], fungus is carried from the old nest site to establish new fungus gardens. The transportation by leafcutter ants of large amounts of leaves, flowers, and fruit, as well as their own bodies, to depths of 6 m greatly enriches the subsoil: the flow of 13 minerals through the underground refuse dumps of *Atta colombica* can be 16 to 98 times that in undisturbed leaf litter (Haines 1978).

Leafcutter ants clear the understorey vegetation in the immediate vicinity of the nest and thereby maintain understorey gaps (Wirth et al. 1997). As much as 80% of leaf damage in some Panamanian rain forests may be caused by *Atta* species (Vasconcelos and Cherrett 1997) possibly affecting plant species composition and ultimately contributing to the maintenance of biodiversity (Marquis 1984; Wirth et al. 1997).

While leafcutter and army ants dominate the New World tropics, it is the 12 species of weaver ants (*Oecophylla* spp.) that have become the most abundant social insects in the Old World tropics. The two most common and widespread species are *O. longinoda*, which occurs throughout the forested regions of the African tropics, and *O. smaragdina*, which ranges from India to the Australian wet tropics. Weaver ants are so called because of their leaf nests secured together by silk. The process of nest construction is highly complex, in that it requires a considerable degree of cooperation among the ants (Figure 5.8). Ants form living chains to bridge gaps between leaves and pull them together in an effort well beyond the ability of any single ant working alone. When several leaves have been manoeuvred into a tent-like configuration, some workers hold the leaves together while others use larvae as sources of silk to secure the leaves together. A single weaver colony can number over half a million individuals with nests extending through the crowns of several canopy trees. Some of the nests contain scale insects that are tended for their sugary secretions. Nests at the edge of the colony contain mainly ageing workers that defend the colony from incursions by other ants or intruders to which they deliver powerful bites[7].

Bees and wasps

Unlike ants, most bees and wasps (also Hymenoptera) are solitary rather than social. Some groups, such as the honeybees (*Apis* spp.) and stingless bees (Meliponinae), do form large and well organized colonies that contribute pollination services to many tropical forest plants.

Figure 5.8 The tent-like temporary nest structure, made from leaves stitched together, of the weaver ant *Oecophylla smaragdina* from Borneo.

Photo by Alex Wild.

---

[6] A single queen may produce as many as 200 million sterile females over a 10–15 years lifespan and only a few males.
[7] This defensive behaviour was first put to good use as long as 1,700 years ago by Chinese farmers who collected and introduced weaver ants to their citrus orchards to protect the crops from insect pests; the practice still continues.

Social bees attain their highest diversity in the tropics, and the several hundred species of stingless bees, together with honeybees, constitute numerically the most dominant bees of tropical forests. Some other bee groups (such as Euglossinae and Centridini) are largely confined to tropical and subtropical locations, but tropical rain forests overall are not especially rich in bee species. Tropical forests do, however, have a wider variety of bee-feeding ecologies than temperate bee communities, including the use of animal faeces, extra-floral nectar, resin and saps, floral fragrances, fruit and dead animals, in addition to the more usual floral resources (Roubik 1989).

Eight of the nine recognized honeybee species (*Apis*) occur within tropical rain forests (Oldroyd and Wongsiri 2006). They are predominantly Asian. Indeed, no honeybees occurred in New Guinea, Australia, or the Neotropics until introduced by humans. Honeybees are able to recruit large numbers of workers to rich floral resources, and their richness and abundance in Asian forests may

limit the richness of the Asian bee fauna overall, and may explain why Neotropical bee communities are richer[8].

The giant Asian honeybee *Apis dorsata* (Figure 5.9) suspends its large nest comb from tree limbs or cliffs many metres above the forest floor. Honey is regularly

A

B

Figure 5.9 (A) The giant Asiatic honeybee *Apis dorsata* pollinating coffee flowers in agroforestry systems within the Western Ghats rain forest; (B) multiple *Apis dorsata* colonies hanging on branches in the same region (see arrows).

Photos by Smitha Krishnan (A) and Lucy Rist (B).

[8] The recently introduced Africanized honeybee that has spread through South and Central America and is moving into North America has become a significant flower visitor in these regions, but it is not yet clear if it displaces native bee species (Paini 2004).

harvested from many of these nests. The trees in which they occur, and to which the bees return annually, are often conserved by local people to ensure continued honey harvests. *Apis dorsata* has considerable importance as a pollinator of several dipterocarp trees in Southeast Asia, as the bees recruit to mass flowering events over large distances (over 100 km) by colony migration (Itioka et al. 2001).

The stingless bees (Meliponinae, subfamily Apinae) are a predominantly tropical group. Species richness is highest in the Neotropics where they number around 300 species, compared to only 70 in Southeast Asia, 60 in Africa, and 14 in Australia. The largest genera are *Melipona*, which consists of over 50 Neotropical species, and *Trigona*, which is the largest and most widely distrib-

Figure 5.10 Stingless bees at their nest entrance tube emerging from the side of a tree, here in Mexico.

Photo by Douglas Sheil.

uted genus with more than 130 species. Species coexist by adopting different foraging strategies, some visiting scattered resources throughout the forest, while others recruit heavily to single flowering trees (Nagamitsu and Inoue 2005). Stingless bees lack a functional sting, but will readily attack intruders by biting particularly at the nose, ears, and eyes. Nests are usually located within tree cavities or in the ground, the entrance to which is lined with resin and developed into a tube (Figure 5.10) that may extend from a few millimetres to, in the case of *Trigona pectoralis*, as much as a metre.

Like bees, many tropical 'true' wasps[9] (Vespidae) are social and defend communal nests with powerful stings. Unlike bees, wasps are primarily carnivorous, obtaining mainly arthropod food for their larvae, although adult wasps might supplement their diet with nectar. Most rain forest wasps belong to the Polistinae and Vespinae (subfamilies within the Vespidae) which build exposed or enclosed nests of chewed plant fibres. Nest sizes vary considerably from a few centimetres to about one metre in diameter and are found in tree hollows, in the canopy, underground, or hanging under understorey leaves.

Termites

Termites (Isoptera) are almost as abundant in rain forests as ants, with estimated biomass in the Amazon exceeding 2000 kg ha^{-1} (Martius 1994), but are less conspicuous as they often remain concealed within their nests or under covered trails that they construct. Their nests are often conspicuous though, with many species forming large earthen structures rising from the forest floor or attached to trees (Figure 5.11). They are the main macro-decomposers in lowland rain forests, where they break up woody debris and make it available to other decomposers (Martius 1994).

Like ants, most termites form large social colonies where labour is differentiated among three castes: reproductives, soldiers, and workers. The abundance, species richness, and proportions of feeding groups of termites vary considerably across tropical rain forest regions. Regional and local species richness is highest in Africa, followed by South

[9] The term 'wasp' is used for a variety of other insects that are not closely related to vespids. These include an immense diversity of parasitic wasps and fig wasps.

Figure 5.11 A termite nest straddling downed branches in Borneo.

Photo by Julia Born.

America, Southeast Asia, Madagascar, and lowest in the Australian wet tropics[10] (Davies et al. 2003).

Termites feed on various components of soil organic matter, leaf litter, and dead wood, although the Macrotermitinae are obligate fungus-cultivators much like leafcutter ants. Four principal feeding groups have been determined, based on the position of their food substrates along a sequence of organic matter decomposition. First, termites in the Rhinotermitidae and Kalotermitidae clades feed on the least humified substrates such as undecayed wood. A second group feeds on wood in various stages of decomposition verging on humus. These belong to the higher termites (Termitidae) which are represented mainly by the Old World Macrotermitinae (which also include fungus cultivators) but also members of the New World Nasutitermitinae. The third group is the humus feeders, which consume a range of materials from humus associated with very rotten wood to humus particles in soil. The final group is the true soil feeders that favour the most humified and nutrient-poor organic material in soils. Soil feeding has been a major evolutionary development within the Termitidae and is represented by 67% of the Termitidae genera and around 50% of all known species (Brauman 2000).

Termites contribute to nutrient cycling by processing humus, soft litter, and dead wood, by building galleries through dead wood which accelerates decomposition, and by turning over the soil through their activities. Termites maintain a complex gut environment and biota. Their gut symbionts and chemistry[11] help digest cellulose and even lignin, and even fix small amounts of nitrogen (Bignell and Eggleton 1995). Excluding termites can slow wood decay (Takamura 2001). Nitrogen mineralization due to soil-feeding termites contributes to nitrogen fluxes in tropical ecosystems, probably surpassing those due to earthworms (Ji and Brune 2006). Calculations of the amount of organic matter consumed by termites ranges from 7 to 36 g m^{-2} year^{-1} (dry weight organic

---

[10] Differences in termite biogeography and biodiversity have been suggested to be a reliable indicator of tropical rainforest persistence through the Quaternary, the past 1.6 million years or so (Gathorne-Hardy et al. 2002).

[11] pH values in termite guts reach 12.5, considered to be the highest in the biological realm.

matter), equivalent to 0.9–3.4% of litter production in Mulu, Borneo, to 15.5–17.4 g m^{-2} year^{-1} (14.7–16.3% of litter production) at Pasoh forest, Peninsular Malaysia (Brauman 2000).

Termite guts produce methane, carbon dioxide, and even hydrogen gas in quantities that are globally significant: termite methane production was once estimated at around 20 Tg y^{-1} but is now thought to be 1.5 to 7.4 Tg^{-1}, or 1–4% of natural methane emissions[12]; CO_2 emissions are around 3500 Tg yr^{-1}, about 2% of the global flux (Chapter 9) (Sanderson 1996; Brune 1998; Sugimoto et al. 1998).

5.1.4 Other insects

The Diptera, or true flies, contains the mosquitoes (Culicidae), well-known on account of their medical importance as vectors of malaria, yellow fever, dengue, and other diseases. Of the 3,000 species of Culicidae worldwide, about 2,000 occur in tropical regions and 30% occur in the Neotropics. Costa Rica alone has over 200 species. Mosquito larvae develop in standing or slow moving water with specific species having specific preferences. Some favour arboreal water, or container habitats consisting of small pools of water trapped in bromeliad leaf axils, pitcher plant pools, tree holes, bamboo stems, or *Heliconia* leaf bracts[13]. The liquids held in these small containers are often rich in dissolved organic compounds and harbour a rich invertebrate community.

The Hemiptera, or true bugs, have piercing or sucking mouthparts, the rostrum, which has been adapted in different species to suit predatory or herbivorous lifestyles. Herbivorous sap-sucking bugs account for a considerable amount of plant herbivory, and many are protected by tending ants which gain plant derived exudates from the bugs in return (Box 5.2). In assassin bugs, and similar species, the rostrum is used to pierce the cuticle of other arthropods, or even to suck blood from birds and mammals.

The high fliers: fig wasps and chalcids
Although diminutive in size, fig wasps (Agaonidae) have considerable importance as the sole pollinators of figs

(*Ficus* spp.; Chapter 2, Box 2.4). As tropical rain forests are characterized by their rich plant diversity and relative isolation of conspecific trees, small parasitic and phytophagous insects, such as chalcid wasps and fig wasps, must often travel considerable distances to find suitable hosts. It appears that weak-flying insects, such as fig wasps, can cover long distances by actively flying up and above the canopy, where relatively strong winds facilitate their dispersal (Compton et al. 2000). On detecting the species-specific volatiles released by their host plants, these insects may fly down and into the canopy, where the lower wind speeds allow them to fly actively upwind to their hosts (Grison-Pige et al. 2002). In this way fig wasps can cover distances exceeding 30 km (Harrison 2003).

Arthropod serenades: Orthoptera and Cicadidae
About 40% of the known 2,200 species of Neotropical grasshoppers (Orthoptera: Caelifera) are canopy or sub-canopy species (Amedegnato 1997). These arboreal grasshoppers tend to be brightly coloured, with stout legs and body form, and protuberant eyes. Grasshoppers are diurnal and herbivorous. Crickets and katydids (Orthoptera: Ensifera and Tettigonida), on the other hand, are nocturnal and more catholic in their food preferences, including predatory and omnivorous species in addition to herbivores. However, it is the evening and nocturnal calls of crickets and katydids that are their most striking feature in tropical rain forests. Together with cicadas and frogs, these invertebrates ensure that dusk in many tropical forests is characterized by a cacophony of calls.

Tropical forests are indeed full of insect-generated sounds. Sound intensity is highest at dusk when male cicadas (Cicadidae) and crickets (as well as frogs) broadcast species-specific acoustic signals to attract females. These 'songs' can become incoherent against the background noise, and so different groups of animals segregate the timing of singing to precise time slots. Throughout Borneo, for example, the first half-hour of dusk is dominated by cicadas, with crickets

---

[12] Around 40% of global methane emissions are from natural sources.
[13] These structures are collectively known as phytotelmata: water bodies held by plants.

and frogs taking over in the second half-hour. These groups are also spatially separated: cicadas sing mainly from the canopy, while crickets use understorey perches little more than a few metres high. Crickets that sing at waterfalls have songs that are adapted to overcome the background noise of running water, while katydids and frogs alternate their calls to avoid confusion (Riede 1997). Other species avoid inter-specific interference by singing during the day or at night (not at dusk), but at these times forest atmospheric conditions are seldom as conducive for sound transmission.

Insect camouflage

Alfred Russel Wallace recognized and described many examples of protective camouflage among butterflies in Sumatra. He was particularly inspired by *Kallima paralekta* (Nymphalidae) which, when at rest, almost perfectly represents a dead leaf[14], and examples such as this contributed to his independent development of the theory of evolution by natural selection[15].

Several insect groups are masters of camouflage, which is most often a form of protective defence (Figure 5.12), but may also be offensive, as in crab spiders and orchid mantids that catch flower-visiting insects. Some ants, such as workers of the Neotropical genus *Basiceros*, actively wear camouflage by covering themselves with soil which is held in place by specialized hairs over most of the body. Camouflage can be chemical, such as the caterpillars of Lycaenid butterflies that reside undetected within ant nests, as well as other moths and other arthropods that raid *Apis dorsata* honeycombs.

The order Phasmidae (stick and leaf insects) is a small group of mostly large tropical insects that, as their name suggests, closely resemble sticks or leaves[16]. Some such as *Trychopeplus laciniatus* even have the appearance of lichen-covered twigs. Although often motionless, many will sway gently as they move, giving the impression of moving vegetation in a light wind. Others change colour

A

Figure 5.12 (continues).

[14] The Vietnamese jungle queen *Stichophthalma louisa* of the montane forests of Vietnam is perhaps even more perfectly camouflaged, as both sides of the wing resemble a dead leaf (Novotny et al. 1991).

[15] 'If an extraordinary adaptation as this stood alone, it would be very difficult to offer any explanation of it; but although it is perhaps the most perfect case of protective imitation known, there are hundreds of similar resemblances in nature, and from these it is possible to deduce a general theory of the manner in which they have been slowly brought about. The principal of variation and "natural selection"…offers the foundation for such a theory…' From *The Malay Archipelego*, first published in 1869 by Alfred Russel Wallace.

[16] Somewhat less insightful than Alfred Russel Wallace, Antonio Pigafetta published the first description of leaf insects in *The First Voyage Round the World by Magellan 1518–1521*: 'In this island [named as Cimbonbon by Pigafetta, and probably either Balambangan or Banggi islands, off the north coast of Borneo] are also found certain trees, the leaves of which, when they fall are animated and walk. They are like the leaves of the mulberry tree, but not so long; they have the leaf stalk short and pointed, and near the leaf stalk they have on each side two feet. If they are touched they escape, but if crushed they do not give out blood. I believe they live on air.'

Figure 5.12 Many tropical forest insects have become masters of camouflage, both for the purposes of protection and/or predation. The groups of praying mantids (A–C [see colour plate 4]), stick insects (D), various orthopterids (E–F), moths (G), and butterflies are particularly well represented in this respect. Insects commonly emulate living and dead leaves (B, E–G), bark surfaces (A), flowers (C), and sticks (D).

Photos by Carsten Bruehl (A–E and G) and Karin Beer (F).

Box 5.4   Invertebrate records from the tropical rain forests[1]

Given that rain forests hold by far the greatest number of invertebrate species of any terrestrial biome, it is perhaps unsurprising that rain forests include species that are among the largest of their kind.

The largest of all butterflies are New Guinea's Queen Alexandra's birdwing *Ornithoptera alexandrae* and Goliath birdwing *Ornithoptera goliath*, with wingspans of up to 28 cm. Africa's largest butterflies are the giant swallowtails *Papilio antimachus* (wingspan 23 cm) and the slightly smaller *Papilio zalmoxis*. Butterflies of the genus *Morpho* (Nymphalidae) are among the largest Neotropical species and are immediately recognizable by their bright iridescent and reflective blue colouration of the upper wing surface[2].

The longest insects in the world are the stick insects, with records held by *Phobaeticus kirbyi* from Borneo, with a body length up to 32 cm, and *Phobaeticus serratipes* of Peninsular Malaysia, with a body length of 27 cm. With legs outstretched, both these insects may measure over half a metre (57 cm) long (NHM 2008). The largest by body mass are the African giant scarabs, *Goliathus goliathus* and *G. Regius*, which reach 11 cm in length but weigh as much as 100 g. The Neotropical scarabs *Megasoma elephas* and *M. actaeon* (up to 12 cm), the Hercules beetle *Dynastes hercules* (17 cm), and the huge Amazonian cerambycid *Titanus giganteus* (up to 18 cm) are longer but less bulky.

The largest among spiders is the bird-eating goliath tarantula *Theraphosa blondi* with a leg-span of 25 cm, from the Guiana Shield of South America. When threatened this fearsome spider produces a hissing noise by rubbing the hairs on its legs together. Even tarantulas are prey to tarantula hawk wasps *Pepsis toppini* of Peru and Ecuador with a 15 cm leg-span, although the giant scoliid wasp *Megascolia procer* from Java, Indonesia, is undoubtedly bulkier.

The largest tropical forest invertebrates are found in the soil. Giant earthworms measuring 70 cm long are common in the rich montane rain forest on Kinabalu, Borneo (Blakemore et al. 2007b). Their presence is noted by their worm casts several centimetres high, found by the thousands within a few square metres. This giant is prey to the Kinabalu giant red leech *Mimobdella buettikoferi* which extends to about 25 cm in length. It is indeed fortunate that this leech feeds only on giant earthworms and is not, unlike its much smaller relatives on the lower slopes of Kinabalu mountain, a blood sucker. Other giant earthworms occur in Java, including one *Metaphire musica* that apparently emits a brief shrill ringing noise (Blakemore et al. 2007a), and French Guiana, where the recently discovered *Rhinodrilus saülensis* measures 80–120 cm (Lapied 2002).

[1] The information in Box 5.4 has been sourced from various websites, conference presentations, and directly from the Natural History Museum (London) collections.

[2] The underside, by contrast, is dull brown, and it is thought that the alternate flashing of bright blue and dull brown during flight confuses potential predators.

---

depending on the background vegetation or the time of day, adopting darker colours at night. The eggs of many stick insects resemble seeds and bear edible 'eliaosomes'—these are harvested by ants and the eggs dumped within the ant nest refuse, where they remain safe from predators (Compton and Ware 1991). Only around 30 of the 3,000 phasmid species, the Phyllinae, have adopted the appearance of leaves, and most of

these occur in Southeast Asian forests, although some species are found in India and the Fiji islands.

As with stick insects, many mantids (Mantoidea) resemble leaves, but the most spectacular forms of mimicry belong to the orchid mantids of the Southeast Asian rain forests that closely resemble orchid flowers. Such mimicry could be as much a trap for pollinators as concealment from predatory birds.

5.2   Other invertebrate groups

It is impossible to do justice to the diversity of tropical rain forest invertebrates in an overview, as indeed we remain largely ignorant of the myriad of forms above ground, let alone those below the forest floor. Some groups deserve a brief mention because they are widespread and abundant (spiders), because they are characteristic of some regions (land leeches), or because they are hard to ignore (Box 5.4).

5.2.1 Arachnids

Tropical rain forests are not especially rich in arachnids: only about 2% of the world's 94,000 arachnid species occur in Amazonia (Adis and Harvey 2000; Adis 2001). Tarantulas (Theraphosidae) are perhaps the spiders that are most widely associated with tropical rain forests, although they also occur in many other habitats including deserts. New World tarantulas are covered in **urticating** hairs that can be thrown in the direction of potential predators. Urticating hairs are also used to mark territories and are distributed at and around the burrow entrance. Old World tarantulas are more aggressive and their venom more toxic than New World tarantulas, but a bite is rarely more serious than a severe bee sting.

Spider density in the canopy averages about 3–4 per square metre, which is comparable to values from temperate forest (Russell-Smith and Stork 1995). Spiders generally account for less than 10% (and frequently less than 5%) of arthropods sampled from tropical forest canopies, although in the Amazonian *varzea* forest they amount to 12% (Adis et al. 1984), and in Australia's tropical rain forests they are much more common, amounting to 25% of all arthropods sampled[17] (Basset 1991).

In canopy samples from Sulawesi and Brunei in Southeast Asia the Theridiidae, which build webs that incorporate debris and leaf litter, are most abundant in terms of individuals, accounting for 20–30% of spiders. Other common groups are the web-spinning spiders such as Araneidae (10–20%) and Pholcidae (5–10%) and hunting spiders Salticidae (8–17%) and Oonopidae (7–11%) (Russell-Smith and Stork 1995).

Although most spiders lead a solitary existence, some species form aggregations and even social colonies. One common species that ranges from Panama to southern Brazil is *Anelosimus eximus* (Theridiidae) in forest gaps or edges. The colonies have a few to several thousand members, with overlapping generations and cooperative brood care (Ebert 1998). The web consists of a large sheet upon which fallen leaves are used as refuges. Hanging above the sheet is a tangle of non-sticky silk threads that ensnare prey. Prey is attacked by several spiders, allowing prey items many times the size of individual spiders to be tackled.

The golden orb spider *Nephila maculata* (Areneidae) is among the most common of large rain forest spiders in the Old World, with a range from Africa to India, China, Southeast Asia, tropical Australia, and the South Pacific islands. Another smaller species, *N. clavipes*, is widespread in the Neotropics. At 20 cm leg span with a 3–5 cm body, *N. maculata* is among the largest of orb weavers, and makes the largest and strongest web. The web can run from the top of a tree 6 m high and up to 2 m wide. Given the investment in such a structure, it is not surprising that webs are rarely dismantled and may last several months, if not years. The web is so strong that birds are occasionally caught. Trapped birds are not eaten but can cause considerable damage, and so the spider often leaves a line of insect husks on the web or builds smaller barrier webs around the main web to avoid such damage. The silk from these webs has long been used by Indonesians and New Guineans to make fishing lures, traps, nets, and bird snares, as well as bandages.

Several minor predatory arachnid groups are widely found within, and indeed almost confined to, tropical leaf litter. The 55 known species of Ricinulei form a highly developed and unusual group. They lack eyes and support a hood on the carapace that can be raised or lowered to cover the mouth and the chelicerae. They appear to be restricted to African and American rain forests. Whip-scorpions, Uropygi (Figure 5.13), are robust tropical predators that possess anal glands from which they spray chemical toxins to deter predators. Most of the 106 known species occur in rain forest. Rain forest leaf litters also harbour the 136 named species of

---

[17] This may be an overestimate as the sampling method used is inefficient at catching winged insects.

Figure 5.13 A whip scorpion (order Thelyphonida) from Borneo, though the group is distributed across tropical and subtropical regions worldwide.

Photo by Karin Beer.

Amblypygi and the 205 recognized species of Schizomida. These related groups possess extremely long front legs that are used as tactile organs. Scorpions, though more common in drier areas, can also be abundant in seasonal (and to a lesser degree aseasonal) rain forests.

Mites[18] and ticks (Acari) are external parasites of vertebrates and arthropods and are very common throughout the tropics—indeed they are the most common and diverse of all the arachnids. Owing to their small size (often less than 1 mm) they are rarely seen, but are encountered on probably every forest excursion.

5.2.2 Annelids: earthworms and leeches

Among the non-arthropods, tropical forest earthworms (Oligochaetae) have generally been thought to be less abundant in tropical forest soils and to contribute less to soil aeration, turnover, and breakdown of organic matter than either their temperate soil counterparts or other tropical soil macrofauna (termites and ants) (Anderson et al.

1983). Another review suggests that tropical forest earthworm abundance and biomass are not dissimilar to that of temperate woodlands, but considerably less than temperate and tropical grasslands (Fragoso and Lavelle 1992). The contribution of earthworms to decomposition and mineralization of tropical forest organic matter depends on the abundance of epigeic and anecic species, which fragment the leaf litter and relocate it into their burrows, versus endogeic worms, which mainly feed on soil organic matter (Fragoso and Lavelle 1992). Anecic earthworms appear to dominate the soil macrofauna of South American and some African forests, but may be less important in Central

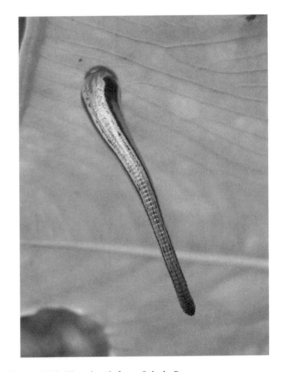

Figure 5.14 Tiger leech from Sabah, Borneo.

Photo by Karin Beer.

---

[18] Mites can also be found on plants and many are free-living. The Trombiculidae are a remarkable group of mites and include so-called 'chiggers', 'red bugs' or '*bêtes rouges*', whose larvae are skin parasites of vertebrates, including unlucky humans: they are reputed to be the most itch-inducing ectoparasite known. They are locally common in clearings in tropical America, and are associated with the large earth-mound nests of megapode birds in Asia and New Guinea. They are not recorded in Africa, and should not be confused with the African 'chigger'—a burrowing parasitic flea which has thrived since its inadvertant introduction (also from the neotropics).

American and Asian forests where endogeic species predominate (Fragoso and Lavelle 1992).

Land leeches *Haemadipsa* spp. (Achaetae) are common in tropical forests from Madagascar and India to Southeast Asia, New Guinea, and Australia (Figure 5.14). Although leeches are not known to transmit human diseases, their blood-sucking habit repels many erstwhile explorers and irritates forest scientists; while the bite is often made without pain, the resulting wounds are itchy, tend to bleed (due to anticoagulants in the leeches' bite), and easily become infected. Leeches are highly sensitive to movement and CO_2 which alerts them to an approaching host in whose direction they can move remarkably quickly. Although very common in many Asian rain forests, except those over limestone, surprisingly little is known about their basic ecology, including their diversity and their natural animal hosts[19].

5.2.3 Miscellaneous others

The mainly nocturnal millipedes (Myriopoda) and centipedes (Chilopoda) are common in virtually all rain forests. Millipedes (Figure 5.15) consume large quantities of leaf litter and therefore contribute to its breakdown, although unlike termites they cannot consume woody tissue. In seasonal forests they can reach very high densities. Centipedes are predatory, with a poisonous bite, feeding on insects, spiders and small crustaceans, and larger species even predate lizards, frogs, and birds. Some species attain considerable size (up to 26 cm for the Amazonian *Scolopendra gigantea*) (Figure 5.16).

Land crabs achieve high densities in mainland coastal rain forests, and in particularly wet rain forest environments far inland (Figure 5.17). *Gecarcinus quadratus* in coastal forests of Costa Rica's Corcovado National Park may achieve densities as high as 6 per square metre (Sherman 2003). These crabs forage on fallen leaf litter and collect it into their burrow chambers, which extend up to 150 cm deep. This litter relocation by land crabs

Figure 5.15 The pill millipede *Glomeris connexa* from Borneo is practically impregnable when rolled up in a ball.

Photo by Alex Kendall.

may affect profiles of soil organic carbon, rooting, and seedling distributions (Sherman 2003).

Land crabs (which need water bodies only for larval development) are especially rich in the forests of Christmas Island, where there are 14 species, more than

Figure 5.16 The giant centipede *Scolopendra gigantea* from the Amazonian forests.

Photo by Karin Beer.

---

[19] There are in addition a number of water margin species which move into the mouths or muzzles of vertebrates that come to drink at streams—it is these that give rise to some of the more outlandish travellers' tales of nasal leeches.

Figure 5.17 One of several crabs that occur mostly on the forest floor in tropical rain forests, this one being from Borneo.

Photo by Karin Beer.

Figure 5.18 A flatworm *Bipalium* sp. from Borneo.

Photo by Karin Beer.

any other rain forest. It is the annual migration to the ocean by 120 million red crabs *Gecarcoidea natalis* at the first rains of the wet season (December) that makes the island famous (Adamczewska and Morris 2001).

Other Neotropical crabs are tiny and live high above the forest floor in the pools of water that collect in the leaf axils of bromeliads. Other crustaceans such as isopods inhabit the moist leaf litter on the forest floor, while freshwater pools teem with water fleas and copepods and freshwater shrimps, which feed on suspended organic matter or scavenge in the mud and on rocks.

Onychophorans, commonly known as velvet worms, are an unusual and enigmatic group of soft-bodied terrestrial invertebrates that are widespread in the humid tropics. They are susceptible to desiccation, and are found in moist habitats such as leaf litter, in decomposing logs, and under stones or in cracks in the soil, where they predate other small invertebrates (Tait 1998). They are an ancient and largely unchanged group, very similar in appearance to their marine ancestors recorded from early Cambrian rocks (about 500 million years old). Far from being behaviourally primitive, these creatures form complex social groups that hunt collectively and are led by a single female (Reinhard and Rowell 2005).

Generally, rain forests are not especially rich in snails, although limited sampling may greatly underestimate true richness. A lack of calcareous nutrients from which to build shells, limited leaf litter in which to forage, and an abundance of predators combine to make tropical rain forests relatively hostile habitats for land snails. Hence species richness of tropical snail communities is not especially high compared to temperate moist forest localities: 60 snail species found in a 4 ha New Zealand temperate rain forest (Solem et al. 1981) is broadly similar to 61 species from 1.5 ha of alluvial soil forest at Danum Valley in Sabah, Malaysia (Schilthuizen and Rutjes 2001), 34 species from soil and leaf litter in French Guyana (Gargominy and Ripken 1998), 64 species from lowland tropical rain forest in Madagascar (Emberton 1997), and 97 species from Cameroonian tropical rain forest (de Winter and Gittenberger 1998). In limestone forests in Tabin, Sabah, Borneo, calcerous nutrients are more abundant and snail abundance is two- to tenfold higher than surrounding calcareous-poor areas, though species richness is only 1.3–2.5 times higher (Schilthuizen et al. 2003).

Land and tree snail communities can be very rich in species and endemics on tropical islands. The radiation of *Partula* and *Achatinella* tree snails on Polynesian Islands led to the evolution of hundreds of species, many restricted to single watersheds. Unfortunately many of the diverse and endemic *Partula* and *Achatinella* have become extinct in the wild through predation by the introduced snail *Euglandina rosea* (Cowie 1998).

Terrestrial flatworms (Platyhelminthes: Tricladida) glide over moist surfaces from the forest floor to the canopy (Figure 5.18). They are predatory and consume prey, such as snails, by enveloping them, secreting digestive juices, and sucking up the digested material.

5.3 Conclusion

Arthropods, and especially insects, account for the large majority of species and animal biomass in tropical rain forests. Though diverse and ubiquitous, our understanding of their variety and distribution is extremely limited and our knowledge remains patchy. It is, nevertheless, certain that invertebrates play key roles in many processes that shape rain forests, among which herbivory, plant reproduction, and nutrient cycling are explored in greater detail in later chapters.

Section II

Origins, patterns, and processes

CHAPTER 6

From the beginning: origins and transformation

'Geology teaches us that the surface of the land and the distribution of land and water is everywhere slowly changing. It further teaches us that the forms of life which inhabit that surface have, during every period of which we possess any record, been also slowly changing.'

Alfred Russel Wallace (1863)

Modern rain forests are the result of millions of years of interplay between evolving species and a changing Earth. Forests have come and gone. What can the past teach us about the present? For one thing, the diversity of so many terrestrial groups in wet tropical forests appears less mysterious when it is realized that this is where so many ancestral lineages were born, were sheltered, and have continued to diversify over time.

Modern rain forests pose riddle upon riddle: where do all these species come from, and how did they end up here? While some closely related species groups are widespread, others are localized. Some like tapirs (in Asia and the Tropical Americas) show oddly discerning intercontinental distributions. Clues to present distributions need to be sought and read through an understanding of how the world itself has changed.

The further back we go, the greater the uncertainties about what both 'tropical' and 'rain forests' mean. In this chapter we first summarize the ancient story of forests, especially those that appeared in wet and frost-free i.e. 'megathermal' climates, review the differences between regions, and reflect finally on events and processes influencing how regional rain forest biotas have come about.

6.1 Ancient forests

6.1.1 Mushrooms with a view

Vascular plants first appeared over 430 **Ma** (Silurian; Table 6.1). Fossils reveal that mysterious 'tree-like' organisms up to 9 m tall called 'Prototaxites' flourished from 420 Ma to the end of the Devonian some 60 Ma later. Long considered to have formed the 'earliest forests', they now appear to have been giant fungi or lichen[1] (Boyce et al. 2007).

6.1.2 Devonian to Carboniferous: the first tropical forests sow their own downfall

The earliest unambiguous fossil forests date to 385 Ma and occurred in swamps in North America. Two familiar

---

[1] The taxonomic affinities have been debated for 150 years as spores have never been observed.

Table 6.1 Names and estimated dates for geological time-periods, and significant events.

Era	Period	(Epoch)	Start (Ma)	Events
Cenezoic	Neogene	Quaternary — Holocene	0.012	Humans develop agriculture and metal work, and reach Americas, Pacific Islands and Madagascar.
		Pleistocene	1.81	Glaciations continue. Anatomically modern humans in Africa and Asia, develop use of fire and tools, reach New Guinea and Australia (40,000 ya). Last iceage ends.
		Pliocene	5.3	Climate cools, numerous glacial inter-glacial cycles.
	Paleogene	Tertiary — Miocene	23.0	Initially warm, but climate cools. Uplift of East Africa increases African aridity. Australia dries.
		Oligocene	33.9	South America and Antarctica separate. The Antarctic Circumpolar Current cools the continent. South America and Africa are islands.
		Eocene	55.8	Rapid warming, with cooling from mid-period, reducing forest areas, and leading to a more arid earth.
		Paleocene	65.5	India joins continental Asia. Gradually warmer and wetter. Believed to ends with a 'hydrate event' in which large volumes of methane were released into the atmosphere.
Mesozoic	Cretaceous		145	Gondwana fragments. India-Madagascar separates from Antarctica, and subsequently divides. Period ends mass extinction.
	Jurassic		200	Mainly arid in tropics. Pangea divides into Laurasia (North), Gondwana (South). India-Madagascar rift from Africa.
	Triassic		251	Equatorial areas mainly arid.
Paleozoic	Permian		300	Supercontinent Pangea forms. Ends in a major extinction (estimates suggest 70% of all land organisms eliminated).
	Carboniferous		359	Extensive swamp forests lead to coal formation.
	Devonian		416	Period ends with glaciation and extinction.
	Silurian		444	First vascular plants. Prototaxites 'forests'.
	Ordovician		488	First green plants and fungi on land. Ice age at end of period.
	Cambrian		542	Green algae, and non-vascular plants (similar to bryophytes) arise in aquatic habitats.

Dates based on The Concise Geologic Time Scale by Ogg and Gradstein, 2008.

tree architectures already co-existed (Figure 6.1) and some plants reached 8 m tall (Stein et al. 2007).

By 380 Ma Devonian swamp forests were widespread. By 360 Ma broad leaves with branched veins were common, this probably being advantageous in conditions of declining atmospheric CO_2 at the end of the Devonian. These warm, damp, multi-storeyed forests with vines were dominated by deep-rooted Archaeopterids (*Progymnospermopsida*, proto-gymnosperms) reaching 10–30 m, the first large trees to form forests (Algeo and Scheckler 1998). These forests harboured a variety of animals: arthropods included myriapods (Figure 6.2), collembolans, arachnids, and others, while their riverine channels sheltered tetrapod vertebrates. Copious litter allowed understorey fires to occur during droughts (Cressler 2001).

The Devonian forests changed the world and ultimately brought about their own demise. Deep roots and high atmospheric CO_2 (up to 10% concentration)

a   Gillboa tree

b   Archaeopteris

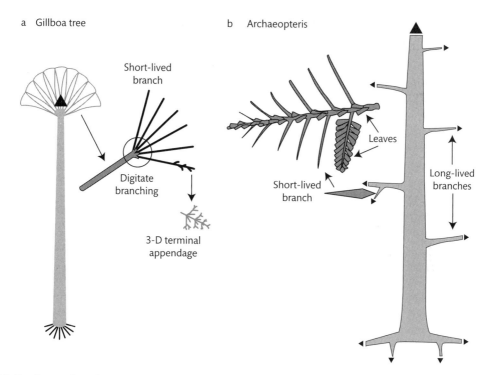

Figure 6.1 Two forms of tree from the Devonian: (A) A single-stemmed 'Gilboa' (a cladoxylopsid in the extinct Pseudosporochnales), perhaps the world's oldest tree, and (B) *Archaeopteris*, a pro-gymnosperm (extinct relative of seed plants) that possessed leafy twigs, long-lived roots, and branches that grew at the same time as the trunk. Both these basic architectures—single stemmed (e.g. tree-ferns, palms, and various monocotyledons) and branching with secondary thickening (most dicotyledonous and gymnosperm trees)—remain characteristic of modern forest trees.

From Meyer-Berthaud and Decombeix 2007.

promoted weathering and production of mineral carbonates, and the consequently increased nutrient inputs led to eutrophication of aquatic habitats—both processes removing carbon from the atmosphere. Thus, during the late Devonian, seas become anoxic and atmospheric CO_2 plummeted (by perhaps 95%) (Caplan and Bustin 1999). The result was a reverse greenhouse effect and severe cooling in which glaciers formed widely in the tropics (Crowley and Berner 2001), numerous extinctions, and a global demise of forest (Algeo and Scheckler 1998; Raymond and Metz 2004).

6.1.3 Carboniferous: the age of forest carbon

The Carboniferous period began with short fern-like zygopterids, sphenopsids (relatives of horsetails), and

Figure 6.2 A scene maybe little changed over 400 million years: a millipede on a *Selaginella*, Sarawak, Borneo.

Photo by Douglas Sheil.

Box 6.1   The inevitability of forests

Evolution of the tree growth habitat is a recurring element in Earth's history. Since the Devonian, trees have repeatedly evolved from non-tree ancestors (Petit and Hampe 2006). Given relatively benign and stable conditions with closely packed plants competing for light, the evolutionary outcome seems near inevitable: increasing size, strong stems, and ultimately tall forest trees.

This evolutionary potential remains contemporary. Studies of the Hawaiian flora suggest that lineages of erect woody plants have developed at least seven times from herbaceous ancestors in the last 5 Ma (Price and Wagner 2004). Apparently the metabolic machinery required for woody growth is present (or readily acquired) in many (angiosperm) herbs—suggesting these may share a woody ancestor.

---

short tree-like lycopods. Perhaps inevitably tall forests soon evolved once more (Box 6.1).

Coal from the extensive tropical swamp forests across the Pangea supercontinent put the 'carbon' in Carboniferous[2]. Lycopods, such as the thick-barked, 40 m *Lepidodendron*[3], lacked broad canopies but were nonetheless decked with epiphytes and a variety of climbers, twining vines, hooked scramblers, and tendril-bearing lianas (Krings et al. 2003). Atmospheric oxygen reached about 35%, permitting arthropods to attain monstrous sizes and evolve flight. High oxygen may have coaxed other fauna including vertebrates onto land (Beerling et al. 2002). Forest faunas included spiders, millipedes (up to 3 m long), scorpions, the first land snails, the earliest winged insects (including cockroaches, mayflies, and giant dragonflies), and eventually amphibians and primitive reptiles.

Gymnosperms became increasingly dominant and colonized progressively drier and fire-prone habitat up to the more arid end of the Carboniferous. Some speculate that the rise of more effective lignin and cellulose processing organisms, such as fungi and the wood roaches (Prototermite), reduced coal formation and aided seedlings to establish in drier habitats by processing litter (Taylor and Osborn 1996). Fossil evidence for surface root mats, like those considered characteristic of

some modern rain forests, dates back at least to the mid-Carboniferous, 305 Ma and around the time of the formation of the Pangea supercontinent. These forests already possessed deep roots and appear to have occurred not in swamps but on well drained mineral soil (Retallack and Germanheins 1994).

Ferns, fern allies, and gymnosperms continued to dominate terrestrial vegetation for the next 100 Ma. Many plant taxa somehow avoided the catastrophic extinction at the end of the Permian (251 Ma) which eliminated most forests (Eshet et al. 1995; McElwain and Punyasena 2007). Numerous animals, including many advanced reptiles, amphibians, and arthropods, became extinct. Wooded terrestrial ecosystems required more than 3 million years to re-establish (McElwain and Punyasena 2007). By this time, seed-plants had mostly replaced giant pteridophytes in the canopy, with spore-bearing plants relegated to the understorey.

6.1.4 The Mesozoic: browsing dinosaurs

During the Triassic and Jurassic, equatorial regions were monsoonal and often dry, equatorial forests occurred on coasts (Ziegler et al. 2003). For example, the fossils of Petrified Forest National Park, Arizona (North America) formed 220–215 Ma (Triassic) on the edge of Pangea near

---

[2] Organic matter (lignin or similar) accumulated in wet conditions across broad regions; it may be that (a) organisms able to break down such material effectively were still to evolve, or (b) climates encouraged peat formation (or both).

[3] Some authors assign these forms to the modern-day genus *Selaginella*, which would make it the oldest surviving genus of vascular plants. *Lepidodendron* possessed a scaled bark frequently observed in coal seams. These plants lacked deep roots but were shallowly anchored by hollow shoots (stigmaria) like modern day *Isöetes*.

the equator. These fossils reveal trees reaching to 60 m (mainly *Araucarioxylon arizonicum*, Araucariaceae) and an understorey of ferns, cycads, and giant horsetails. Fauna included predatory phytosaurs and large herbivorous dinosaurs, and small (presumably nocturnal) mammals[4].

Wetter sites occurred at higher latitudes. Indeed, fossil plant diversity generally appears higher at mid-latitudes in this period (Rees et al. 2004). These conditions apparently promoted speciation in some taxa. Beetles, for example, were already well established, and intense diversification in this period established numerous modern lineages (Hunt et al. 2007).

The late Cretaceous (80–65 Ma) supported extensive wet forests. These forests possessed the angiosperm dominance that many ecologists identify as characteristic of true modern rain forests. Some of these plant families evolved tolerance of colder and more arid climates and moved outside the wet tropics, but many never did (Donoghue 2008).

As Pangea divided so did mammals: the ancestors of today's egg-laying monotremes (e.g. the spiny *Echidna*) remained in Gondwana while other mammals (including the marsupials) headed north with Laurasia. Around this time primitive primates divided from other mammals leaving the cologos ('flying lemurs') and the slightly more distant tree-shrews (*Tupaia* spp.)—both now arboreal mammals found only in the wet forests of Southeast Asia—as our closest extant non-primate relatives. Plants adjusted to novel animal seed-dispersal opportunities seeds (Moles et al. 2005).

Flower power

Modern forests are dominated by flowering (angiosperm) trees. In contrast to most boreal and temperate forests, and to various tropical montane formations, gymnosperms are now a relatively minor component in lowland rain forests. This characteristic dominance arose in the Cretaceous.

Paleoclimatological evidence suggests extensive but dry tropical forests during the separation of the continents through the Cretaceous (Box 6.2). No unambiguous flowering plant (angiosperm) fossils appear until around 136 Ma (early Cretaceous) (Frohlich and Chase 2007). Genetic evidence suggests that many widespread tropical angiosperm families diverged only at around 110–100 Ma when Gondwana was breaking up and warm wet climates were extensive (Wikstrom et al. 2001; Davis et al. 2005).

The fossil record indicates that angiosperms appeared and diversified suddenly, and dominated wet terrestrial biomes within just a few million years (Anderson et al. 2005; Bell et al. 2005). This result is corroborated by genetic analyses (Wanga et al. 2009). To Charles Darwin this abrupt arrival and diversification posed 'an abominable mystery' and, despite some progress[5], their origins remain enigmatic today.

One proposal is that angiosperms evolved in a lost region. One candidate is the volcanic Ontong Java Plateau, an area of two million km² (approximately the size of western Europe) south of New Caledonia—fossilized wood fragments show that some or all was above sea level during the Early Cretaceous (Fitton et al. 2004). The wider geological activity in the southern Asia-Pacific region implies that other land-surfaces too have come and gone. A variety of phylogenetic and geological data lead to the suggestion that this region was a primary evolutionary centre for plants and much modern diversity (Heads 2009).

What were the early angiosperms like? Unfortunately the fossil record is uninformative. Extant basal lineages (such as *Amborella*, Austrobaileyales and Chloranthaceae) share traits[6] suggesting that their common ancestors grew in a warm, wet forest understorey (Feild et al. 2003, 2004) and were possibly insect pollinated (Hu et al. 2008).

The angiosperm radiation has been ascribed to many factors, including browsing dinosaurs and co-evolution with insect pollinators (Barrett and Willis 2001; Gorelick 2001), but no single theory is generally accepted. Whatever the cause, other groups diversified alongside angiosperms, including ants, bees, and various other

---

[4] By this time early birds were already stretching their wings. According to one theory, birds derived from forest-dwelling gliders. Feathered gliders/flyers certainly existed by the Jurassic, though their relationship to modern birds is uncertain.

[5] One striking result is that all extant gymnosperms and angiosperms appear monophyletic, and thus all species in one group are equally related to all species in the other.

[6] Such as woodiness, low and easily light-saturated photosynthetic rates, leaf anatomy suited to capturing understorey light, toothed spongy leaves (with thick parenchyma), small seed size, and clonal reproduction. Modern herbaceous,

---

Box 6.2   Changing lands

Continents have repeatedly separated and merged through geological time. The configuration of major land masses since 300 Ma, when the continents were unified into a single supercontinent 'Pangea', appears relatively clear. At around 200–160 Ma Pangea split into Laurasia in the north and Gondwana in the south[1]. Between lay the widening Tethys Ocean. Then Africa, India, and finally South America fractured from Gondwana, beginning jour-

neys that ultimately reconnected them with Laurasian lands. Australia–New Guinea was the last to break away from Antarctica.

The break-up of Gondwana from the late Jurassic (152 Ma) initiated the beginning of the isolation of the major rain forest blocks on separate continental fragments. This was particularly evident from the late Cretaceous (100 Ma), and this gradual isolation laid the foundations for their distinctive modern biotas.

[1] Laurasia comprised crust now divided between most of Eurasia, North America, and Greenland. Gondwana comprised crust now divided among Australia, Antarctica, South America, Africa (here including Italy), Madagascar, and India.

---

pollinating insects (see Box 6.3), birds, mammals, lizards, and even freshwater fish, and ferns (Grimaldi 1999; Novacek 1999; Schneider et al. 2004; Moreau et al. 2006).

6.1.5 The Tertiary: climate change

The K/T boundary (65 Ma; named for the Cretaceous (K) and Tertiary (T)) is physically marked by an anomalous worldwide iridium-rich band of sediments that is likely to reflect an extraterrestrial collision (coincident with extensive volcanism on the India-Seychelles landmass, the 'Deccan Traps'). The K/T boundary is associated with the disappearance of forests and, famously, the extinction of non-avian dinosaurs. Fossils from North America and elsewhere indicate that rain forests were abruptly replaced by herbaceous ferns, and recovered slowly. North American rain forests were re-established within 1.4 million years (Johnson and Ellis 2002) of the end-Cretaceous extinction, though evidence of intense insect damage suggests an unusual ecology (Wilf et al. 2006).

Early Tertiary climates favoured extensive rain forests (Box 6.4). Rain forest taxa, including angiosperms, mammals, and birds, diversified once more. Now free from competition with dinosaurs, mammals diversified, with many shifting from nocturnal to diurnal habits, and they soon filled a wide range of forest niches—though timing and details remain controversial (Bininda-Emonds et al. 2007).

Around 35–25 Ma the world entered a cooler period and forests retreated. The last 10–15 million years have seen C4-grasslands emerge as a major tropical biome (Linder and Rudall 2005). The evolution of grazing species along with fire created a powerful force restricting the distribution of forests (Retallack 2001).

Temperature was never stable. The last time that the Earth was 5°C warmer[7] than now was only 3 Ma. Variation then increased dramatically as the earth cooled. There have been over 20 glacial episodes in the last 2.5 million years. Under glacial conditions wet tropical climates cooled and dried, montane vegetation descended, and coastlines retreated seaward. Cooling lowered the tropical snowline (today c.4,500 m at the

---

non-woody plants are generally considered derived from a woody ancestry (water lilies, Nymphaceae, also an ancient lineage, are an exception).

[7] A temperature often mentioned in connection with current climate change.

Box 6.3   An amber window

Rapid decomposition means that rain forests typically yield poor macro-fossils. Amber is a notable exception, creating a small window on many details of rain forest life. Tropical forest trees are the source of most ambers (Langenheim 1995).

The oldest amber-like material appears to be Carboniferous, but extensive deposits begin much later in the Araucariaceae-dominated forests of the early Cretaceous 120–136 Ma found in the Lebanon and neighbouring regions. Cretaceous fossils include the oldest known examples of bees (Figure 6.3) and ants.

The best-studied are the Dominican ambers produced by *Hymenea* trees on Caribbean shores 20 Ma (Henwood 1993). Remains include numerous insects (including fig wasps), spiders, crustaceans, snails, onycophorans, nematodes, tardigrades, occasional vertebrates, plants, fungi, and slime moulds (Poinar and Poinar 1999). Careful treatments reveal microscopic bacteria, protozoans, and viruses, including the earliest malarial *Plasmodium* parasites inside a *Culex* mosquito (Poinar 2005) and the gut flora of extinct termites (Wier et al. 2002). Unexpected Dominican fossils include honey-pot ants, *Leptomyrmex*, now restricted to Australia, and *Chudania* leafhoppers, now known only from Asia and West Africa.

Dominican amber has also preserved an orchid pollinarium (*Meliorchis caribea* affinity) attached to an extinct stingless bee, *Proplebeia dominicana* (Ramirez et al. 2007). This discovery represents the

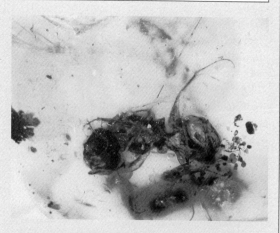

Figure 6.3 The oldest-known bee *Melittosphex burmensis* (3 mm long) from Cretaceous Burmese amber. Flowers preserved in the same deposits imply that insect–plant associations were already well established 100 Ma (Heads 2009), shortly after the assumed origin of the eudicots.

From Poinar and Danforth 2006.

first unambiguous fossil of Orchidaceae, now dated to the Miocene (20–15 Ma), and a unique direct fossil record of a plant–pollinator interaction. The amber fossil has been used to calibrate a molecular phylogenetic tree of the Orchidaceae, allowing the dating of the most recent common ancestor of extant orchids to the Late Cretaceous (84–76 Ma), supporting the hypothesis of an ancient origin for Orchidaceae (Ramirez et al. 2007).

---

equator) by as much as 1,500 m, lowering and compressing vegetation bands below (Flenley 1998).

The rise and fall of montane vegetation, and lowland oscillations between wet and dry vegetation, have been detected in pollen profiles from across the wet tropics. During these periods temperate species (e.g. oaks *Quercus*, alders *Alnus*, and gymnosperms) migrated into the tropics and amongst tropical highlands to an extent not seen previously (Morley 2000).

It is pertinent to reflect on how present climates relate to the past. Conditions in the last few thousand years

have been warmer and wetter, with more extensive natural rain forest cover, than during more than 90% of the past million years. Inevitably species have evolved and persisted in conditions and/or locations different from those in which they are currently found. While climatic differences had a considerable impact across the tropics, regional impacts were influenced by various factors, including relative elevation.

Glacials impacted on the distributions and adaptations of tropical species in many ways. Diminished wet tropical regions led to the concept of 'glacial **refugia**'.

Box 6.4   Paleo-biogeography and palynology

Rain forest paleo-biogeography draws on various disciplines. Geology and paleo-climatology set the stage: knowledge of plate tectonics, past sea levels and climate allow mapping of land and conditions over time.

Knowledge of ecosystems comes principally from plant microfossils (mainly pollen and spores), scarce macrofossils (leaves, wood, fruit, or seeds), and scarcer animal remains[2]. This fossil record is largely a record of swamp forests, including mangrove and peat formations, while dry land formations are seldom represented (Ziegler et al. 2003).

Palynology, primarily the analysis of pollen from sediments, is the principal approach for reconstructing plant communities. In samples from the Oligocene (34 Ma) to the present, pollen is normally attributed to extant taxa, but older sources demand more cautious taxonomic labels (Morley 2000). Indeed few plant genera have clear morphological variation among the pollen of their species.

Pollen and spore diversity can approximate plant diversity, although wind pollinated species tend to be over-represented, and the pollen from some taxa (e.g. Lauraceae) seldom fossilizes. Fossil pollen analysis reveals striking long-term diversity changes in South America (Figure 6.4), including a gradual rise then fall in plant diversity from 65 to 20 million years ago. The most floristically diverse periods appear to reflect climates favouring widespread rain forest cover. The increase in Eocene diversity is also observed in African and Indian pollen sequences, while the late Eocene decline is observed in tropical Asia (Morley 2000).

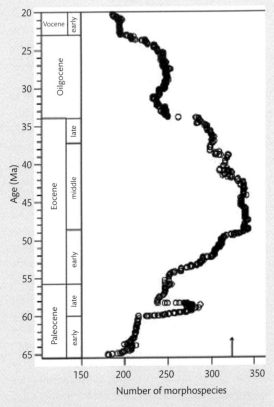

Figure 6.4 Changes in implied diversity (distinct pollen and spore forms or 'morphospecies') during the early to middle Cenozoic, in dated sediments from Colombia and Venezuela: the arrow (lower right) marks the Holocene (modern day) level.

From Jaramillo et al. 2006.

---

[2] An unusual example comes from the discovery of a 60 Ma giant snake fossil in Colombia *Titanoboa cerrejonensis* that would have dwarfed modern anacondas, reaching 13 m and weighing over a ton. According to the authors of the study this size is only possible if temperatures were several degrees warmer than today (Head et al. 2009).

Though largely discredited in the Neotropics (Colinvaux et al. 2000), refugia have had a major role in the history of understanding modern species richness patterns in other regions (Chapter 8).

Around 74,000 years ago, during the last inter-glacial, a vast volcanic event in Toba, North Sumatra, blasted more than 2,800 km³ of ash and dust into the atmosphere[8] (Chesner et al. 1991; Rampino and Self 1992). This may have caused global temperatures to fall rapidly leading into the last glaciation. Sea levels were 100 m lower as recently as 20,000 years ago. The current inter-glacial began 16,000–12,000 years ago, with tropical forests expanding once more (Figure 6.5).

Current plant and animal diversity is estimated to represent no more than 1–2% of the multicellular species that have existed over the past 600 million years (Erwin 2008)—and most of these survivors occur in the tropics (in forests and in coral reefs). In the following sections we will look more closely at how regional forest biotas arose.

6.2 The rain forest pick 'n' mix

Modern-day continental rain forests can be divided into five main biogeographical regions: the Americas, Southeast Asia, Africa, Madagascar, and Australia–New Guinea. Striking differences include the dominance of Dipterocarpaceae in Asia, the paucity of forest palms in Africa, the preponderance of Bignoniaceae (notable as

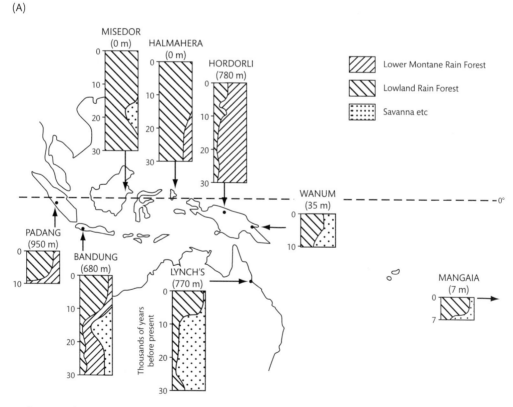

Figure 6.5 (continues)

---

[8] For comparison, the Krakatau eruption in 1883 ejected 21 km³ of ash and debris, and Mt St Helens in 1980 ejected only 1 km³.

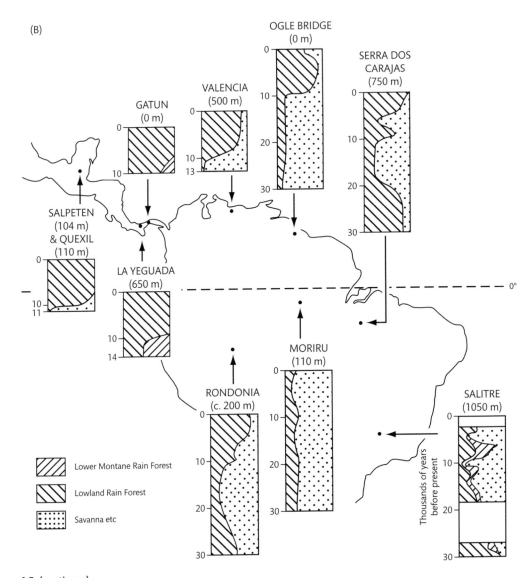

(B)

Figure 6.5 (continues)

(C)

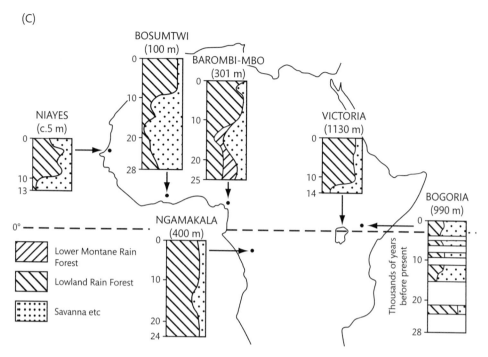

Figure 6.5 Selected lowland pollen diagrams from tropical regions covering the last 15–30 millennia (with human impacts omitted): (A) Southeast Asia and the West Pacific, (B) South and Central America, (C) Africa. Many sites show an increase in pollen indicative of lowland rain forest, an ascent of montane species, and a reduction of non-forest habitats since the end of the last glacial.

From Flenley 1998.

lianas) in the Americas, New Guinea's lack of monkeys, and Madagascar's lemurs. But there are also similarities (Table 6.2). While widely separated, these regions still share many related taxa. Explanations for these patterns reflect the interplay of local and global histories.

Here we examine how related taxa reached their modern-day tropical distributions. Such patterns are considered in the light of continental movements, changing climates, fluctuating sea levels, land bridges, and long-range dispersal. Some additional aspects, including diversification and local barriers, are considered further in Chapter 8. In the final section we shall briefly review the regional variation of modern rain forests.

6.2.1 Land routes

Gondwana divided

The 'Gondwana hypothesis' argues that many intercontinental tropical relationships arise as a result of

divergence among shared ancestral taxa present on the fragmenting Gondwana (Raven and Axelrod 1974). Gondwana encompassed many current major tropical regions, including South America, Africa, Madagascar, Australia–New Guinea, and the Indian subcontinent, as well as non-tropical land masses such as Antarctica which bears many similarities as revealed by the fossil record, and many widely distributed rain forest plants and animals now live on or near land that was part of Gondwana. Taxonomic relationships among various large-seeded tree taxa (e.g. *Chlorocardium* [Lauraceae] and Dipterocarpaceae) support a Gondwanan divergence (Baker et al. 2000a,b; Morley 2000; Chanderbali et al. 2001).

In other widely distributed plant taxa, such as *Ficus* (Zerega et al. 2005), the gymnosperm *Gnetum* (Won and Renner 2006), and the large-seeded *Ocotea* trees (Chanderbali et al. 2001), genetic evidence supports the view that intercontinental lineages diverged long

Table 6.2 Some key characteristics of the main rain forest regions. Rainfall is highly variable within each region, and the values given relate to the core rain forest areas.

	Neotropics	Africa	Madagascar	Southeast Asia	Australia–New Guinea
Main geographical feature	Amazon River basin and Andes Mountains	Congo River basin	Forests along eastern edge of island	Peninsula and islands on Sunda Shelf	Large, mountainous islands
Distinctive biological features*	Bromeliad epiphytes, high bird diversity, small primates	Low plant richness, many forest browsers	Lemurs, low fruit abundance	Dipterocarp tree family, mast fruiting of trees, large primates	Marsupial mammals, birds of paradise
Annual rainfall (mm)	2000–3000	1500–2500	2000–3000	2000–3000, often > 3000	2000–3000, often > 3000
Principal rain forest countries	Brazil, Peru, Colombia, Bolivia, Ecuador	Democratic Republic of Congo, Gabon, Cameroon	Malagasy Republic	Indonesia, Malaysia, Thailand, Laos, Cambodia	Papua New Guinea, Indonesia (West Papua), Australia
Geological history	Marine incursions. Andes rise (reversal of the Amazon). Land bridges.	Raised land surface. Few tall mountains. Repeated desiccations.	Isolated for 90 million years. Mostly dry. Wet mountains, and coastal strips and valleys.	Complex tectonic interactions. Large areas of shallow seas. Repeated division and connecting of many islands.	Isolation from other lands. Dessication (Australia). Mountain building (New Guinea).
Major or distinctive animal groups include	Sloths, Platyrrhine monkeys (include many small taxa), Caviomorph rodents, bears, hummingbirds, macaws, peccaries, leafcutter ants	Squirrels, Catarrhine monkeys, apes, pigs, elephants, elephant shrews (Macroscelidea) termites	Lemurs, tenrecs, fossa (Cryptoprocta ferox), chameleons, Mantellid frogs	Squirrels, Catarrhine monkeys, apes, pigs, bears, rhinos	Marsupials, monotremes, cassowary, birds of paradise, Megapodes cockatoos
Major or distinctive animal absences	Apes	Marsupials	Marsupials, cats, ungulates, apes	Marsupials (except Wallacea)	Ungulates, primates, squirrels, placental carnivores (introduced), pigs (introduced)
Major or distinctive plant families present	Bignoniaceae Caryocaraceae Chrysobalanaceae Fabaceae Lecythidaceae Melastomataceae Bromeliaceae Cactaceae Cyclanthaceae Passifloraceae Heliconiaceae Solanaceae Quiinaceae Marcgraviaceae Cannaceae	Fabaceae Meliaceae Irvingiaceae Rubiaceae Balsaminaceae Scytopetalaceae Melianthaceae	Arecaceae Ebenaceae Euphorbiaceae Erythroxylaceae Fabaceae Lauraceae Meliaceae Proteaceae Rubiaceae Sarcolaenaceae Sphaerosepalaceae	Arecaceae Dipterocarpaceae Fagaceae Rubiaceae Musaceae Pandanaceae	Myrtaceae Proteaceae Rutaceae Musaceae (Australia: Annonaceae Austrobaileyaceae Eupomatiaceae Himantandraceae Gyrocarpaceae Hernandiaceae Idiospermaceae Lauraceae and Monimiaceae)
Major or distinctive plant absences	Pittosporaceae Pandaceae Flagellariaceae	Bromeliaceae Winteraceae Few Arecaceae Elaeocarpaceae and various generalists like Campnosperma (Anacardiaceae)	Bromeliaceae Dipterocarpaceae	Bromeliaceae	Bromeliaceae Rafflesiaceae

Aadapted from Primack and Corlett 2005 and supplemented with additional sources.

after the continents themselves. Indeed, evaluations now challenge the ancient biogeographical story behind various taxa once viewed as 'classic' Gondwanan distributions (Box 6.5). Southern beeches *Nothofagus*, for instance, are now believed to owe much of their range, which extends from South America to Australia and into the New Guinean tropics, to relatively modern colonization (Swenson et al. 2001a; Swenson et al. 2001b). In addition, many tropical flora and fauna (including wet forest specialists such as caecilians) may in fact derive from Laurasia (the other fragment of Pangea).

In summary, distribution histories are often complex, and apparently more recent, than Gondwana alone can account for (Morley 2000, 2003). This has spurred a search for other dispersal routes.

Antarctic shortcuts and the Boreal tropical exchange program

During the Jurassic (at around 170 Ma) some sections of Antarctica reached a latitude of 50°S, and polar climates were in any case considerably milder: Ginkgo trees and cycads were plentiful. Evergreen and deciduous conifer-dominated forests occurred in Antarctica through the early and mid Cretaceous (Falcon-Lang and Cantrill 2001), being joined by *Nothofagus* through the later Cretaceous (from 80 Ma) as Gondwana fragmented[9].

Box 6.5   Araucariaceae, Gondwana, and Jurassic Park

The Araucariaceae (*Agathis* and *Araucaria*, see Chapter 2) now possess a primarily southern hemisphere distribution[3]. Phylogenetic methods suggest an ancient Australian origin, while tree cones and heavy seeds appear ill-suited to long distance dispersal—so no wonder the distribution of the Araucariaceae was long viewed as being classically Gondwanan.

New Caledonia, part of a larger land mass that rifted from Australia *c.*80 Ma (Ladiges and Cantrill 2007), is the prime Araucariaceae hotspot accounting for 18 of the world's 42 species. Perhaps, thought early botanists, this island is a botanical time-capsule presenting us with a real 'Jurassic Park' (or Cretaceous Creche) of ancient trees? Yet, all is not what it seems.

Jurassic fossils, and Cretaceous amber, reveal Araucariaceae in both hemispheres (Setoguchi et al.

1998; Kershaw and Wagstaff 2001). It is now considered unlikely that New Caledonia itself remained above sea level since 80 Ma[4] (although continuous dry land since around 40 Ma seems assured). Divergence within both the New Caledonian *Agathis* and *Araucaria* and relationships with other regional taxa imply more recent colonizations and radiations (Setoguchi et al. 1998; Stefenon et al. 2006).

An important lesson from this story is that taxa have often been more evolutionarily dynamic in time and space than can be inferred from modern distributions alone. Molecular tools and phylogenetic techniques, especially when supported by fossils and geological data, offer increasingly powerful means to address biogeographical hypotheses.

[3] Although *Agathis* reaches 18°N in the Philippines.

[4] A large area of thinned submerged continental crust lies off eastern Australia. This apparently once terrestrial land mass rifted from Australia 85–65 Ma, and interpretations imply terrestrial habitats between Australia and New Caledonia as recently as 40 Ma. It has thus been suggested that further examination of the evolution of the Araucariaceae should consider the history of a 'greater New Caledonia' region (Ladiges and Cantrill 2007).

---

[9] Antarctica maintained a stable polar latitude since 100 Ma.

At 55 Ma forests thrived at high latitudes, and the Antarctic fauna included both placental and marsupial mammals, but apparently no primates (Houle 1999). This may have been the period when marsupials reached Australia. Antarctica may have been sufficiently warm up to c.50 Ma for the exchange of hardier tropical taxa such as Malvaceae and Sapindaceae from South America to Australia (Morley 2003).

During these same periods the Northern Hemisphere was also warmer. English and Alaskan fossils reveal plants[10] now typical of Southeast Asian tropical coasts and forests, complete with woody lianas, buttressed trees, and numerous entire-margined drip-tip leaves (Wolfe 1972; Richardson et al. 2004). These forests spanned across neighbouring continents via a North Atlantic land bridge (Iceland–Faroes) or more northerly Bering Strait. This route helps explain the nearly 100 tropical angiosperm genera shared between America and Asia (Davis et al. 2002b).

By 33 Ma, cooling purged frost-sensitive vegetation from higher northern latitudes. Sensitive taxa migrated south or disappeared. Some persisted in Asia (for example, trees *Cinammomum* and *Lannea*). Seas and arid regions meant that few of these species found refuge in wet Africa—only eight Boreotropical genera are recognized in the modern African flora. Within the Americas, Boreotropical elements moved to the south of the North American Plate and eventually reached South America following the formation of the Panama Isthmus[11]. Some shared taxa have very limited modern distributions: Bonnetiaceae, a family of small trees, is found only in Malesia (*Ploiarium*) and in the Guyana Shield area of South American (*Bonnetia*); and Tetrameristaceae, which includes the peat swamp tree *Tetramerista* in western Malesia and the recently discovered *Pentamerista* trees in Guyana.

Animals too are believed to have taken a boreal route. Tapirs first reached North America from Asia and made their way down to South America later. North American Middle Eocene fossils also reveal (now extinct) Tarsiiform primates related to Asian species (Beard and Wang 1991; Gunnell 1995). Modern hummingbirds, though now solely American (and tropic centred), have an exclusively European pre-Pleistocene fossil record.

India transit: and the making of the paleotropical biota

The Indian plate, originally part of Gondwana, broke away in a series of steps beginning around 160 Ma. It separated sequentially from Antarctica (c.130 Ma), Madagascar (96–84 Ma), and finally Seychelles (65 Ma) before moving northward to collide with Eurasia's underbelly (c.60–50 Ma) (Conti et al. 2002). India remained close to Africa and Madagascar for much of the journey, and maintained species exchanges (including dinosaurs), with adjacent lands[12] throughout the late Cretaceous (Briggs 2003) (Figure 6.6).

By carrying many taxa from Africa and Madagascar to Asia, from where they subsequently entered Australasia, India is responsible for the existence of the 'Paleotropics' as a meaningful term—up until then the Asian and Southern flora had been relatively distinct (Morley 2000). Though the Indian flora was subsequently impoverished by aridity, its earlier legacy has come to dominate tropical Asia.

Various distributions resulted from India's role as a rafting landmass. Examples include the Asian dipterocarps (Box 6.6), *Durio* (Malvaceae), *Eugeissona* (Arecaceae), *Axinandra* evergreen trees (Crypteroniaceae), and the *Nepenthes* (Nepenthaceae) pitcher plants[13] (Meimberg et al. 2001). The Seychelles harbour the closest relatives of a recently discovered family of Indian frog (Biju and Bossuyt 2003), as well as a basal dipterocarp *Vateriopsis seychellarum* and *Nepenthes* pitcher plants. India may have brought a range of fauna to Eurasia, including haemadipsid leeches, Phylliinae leaf insects, lorises, and various amphibians. The mini-continent's role as a crucible of

---

[10] For example *Toona*, *Barringtonia* (now primarily a tropical swamp and beach forest plant), *Kandelia* (mangrove), *Myristica*, *Alangium*, *Cinnamomum*, *Firmiana*, *Macaranga*, and *Saurauia*, and species from the families Malpighiaceae and Annonaceae.

[11] Including *Ampelopsis*, *Saurauia*, *Symplocos*, and *Trigonobalanus* (Morley 2003).

[12] Caecilians, a good indicator of rain forests, still occur on the Seychelles and in continental Africa and Asia.

[13] Calibrated genetic phylogenies suggest that the ancestor of these Asian taxa (in Sri Lanka, Borneo, and the Malay Peninsula) arrived from Africa-Madagascar via the Indian plate (Conti et al. 2002).

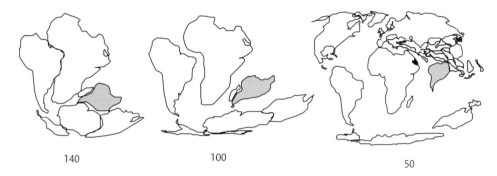

140   100   50

Figure 6.6 Paleogeographic reconstructions of continental plates during the early and middle Cretaceous and the early Tertiary, emphasizing the location of India (dates in millions of years). India moved rapidly (15 cm yr^{-1}) north until it collided with Asia 55–40 Ma. It is still moving slowly (8 cm yr^{-1}) (Briggs 2003) and its collision with Asia has played a major role in shaping Malesia.

Adapted from Briggs 2003.

Box 6.6   Roots of the Dipterocarpaceae

The modern Dipterocarpaceae possesses three sub-families. Molecular studies suggest they all (including the sister family Sarcolaenaceae, endemic to Madagascar) derive from a shared ectomycorrhizal Gondwanan ancestor at around 135 Ma (Ducousso et al. 2004).

The Monotoideae subfamily has three genera. Two occur in seasonal and open habitats in tropical Africa and Madagascar. The third, a monotypic genus represented by the large rain forest tree *Pseudomonotes tropenbosii*, was discovered in Colombia, South America in the mid 1980s (Londono et al. 1995; Morton et al. 1999).

The Pakaramoideae subfamily is represented by a single small ectomycorrhizal tree species *Pakaraimaea*

dipterocarpacea and was discovered in the 1970s, also in South America (Guyana and Venezuela).

The Dipterocarpoideae contains numerous important timber trees ranging from New Guinea through Malesia westwards to India and Sri Lanka and north to China, with an endemic monotypic genus in the Seychelles represented by *Vateriopsis seychellarum*, one species in Madagascar (*Monotes madagascariensis*), and African fossils in Somalia and Uganda. This characteristically Asian subfamily appears to have been carried from Africa on the Indian subcontinent (Morley 2000).

---

diversity in its own right is now clearest in India and Sri Lanka's endowment of endemic amphibians[14] (see Chapter 4).

A collision between Asia and Africa via the Arabian microplate (estimated 27–16 Ma) provided a land route which, owing to its aridity, may have favoured animal more than plant exchanges. Many present-day African lineages, including many large mammals[15], are derived from Eurasian immigrants. Elephants travelled out into Asia (Kappelman et al. 2003), as did monkeys which diversified and gave rise to apes in Asia. The ancestor

---

[14] Notably only one family of caecilian, Ichthyophiidae, is known in Southeast Asia; this also occurs in India along with another family, Uraetyphlidae, an Indian endemic.

[15] Notably, the Carnivora, Hystricognathi, Rhinocerotidae, Ruminantia.

we share with gorillas must have returned to Africa before 7 Ma.

American exchange: and the making of the Neotropical biota

Some limited contact between North and South American lands occurred (most likely via islands Morley 2000) in the Cretaceous, allowing the interchange of some dinosaurs, reptiles, and plants (e.g. *Gunnera* Gunneraceae which moved north) (Wanntorp and Wanntorp 2003).

The rate of interchange increased as the continents drew closer. Some immigrants from the North, such as *Guatteria* (now 265 or more canopy tree species distributed mainly in the Amazonian region) diversified rapidly after their arrival (only 30 species now occur in Central America) (Erkens et al. 2007). Transfers the other way included (*c*.5 Ma) *Titanis* the main South American predator—not a mammal but a large (200 kg) flightless bird.

The Panama isthmus bridged the two American continents only 3–4 million years ago. Upon formation of the isthmus, South American mammal families initially rose from 32 to 39, before declining to the current 35[16]. Immigrants included peccaries, bears, cats, coatimundis, dogs, deer, tapirs, shrews, squirrels, and (now extinct) mastodonts and American horses. Capybaras, porcupines, agoutis, spiny rats, monkeys, tamarins, and (extinct) glyptodonts moved in the opposite direction (Marshall et al. 1982).

In contrast to the mammals, there is no evidence for displacement of South American lowland plants by arrivals from the north. The extant flora of tropical Mexico and neighbouring regions include elements derived from tropical South America, tropical Central and North America (Burnham and Graham 1999).

In geological terms this interchange is relatively recent, and colonization processes are still underway. Many freshwater fish, for example, have been slow to move between the continents. Subtle patterns may also result: the Coryphoideae palms, which arrived from North America, maintain a northern-biased diversity pattern in South American forests (Bjorholm et al. 2006).

Coastal extensions and deep sea divisions

Sea levels are largely determined by global climate and by various geological processes. For most of the last 3 million years today's shallow seas were dry land. When sea levels were 40 m lower than now, some 9,500 years ago, dry-land bridges connected Malaysia, Sumatra, Java, and Borneo to continental Asia, while New Guinea was linked to Australia; and 17,000 years ago, when sea levels were 120 m lower, Southeast Asia gained more than 1.5 million km^2 of land (Voris 2000). These repeated and extended bridging periods during past glaciations explain why so many species are shared amongst these modern-day islands. A shared river catchment draining parts of east Sumatra, north Java, and western Borneo explains the close similarities among even their fish faunas.

Deep seas ensured that New Guinea–Australia never connected with the Southeast Asian mainland (Figure 6.7). Islands east and west of this line formed distinct land masses from the Late Eocene (40 Ma) (Figure 6.8). This explains the biogeographic division made famous by Wallace (Figure 6.9), and known as 'Wallace's Line'. One line cannot capture the biogeographic complexity of this region, though the implied barrier must have been very real for many animals: squirrels, monkeys, deer, and buffalo all failed to make it from Asia to New Guinea despite the relatively short linear distance, while in the opposite direction marsupials, spiny anteaters (echidnas), cockatoos, birds of paradise, and the Papua-Australian tree frogs (the Pelodryadidae) have never colonized mainland Southeast Asia. Rats and some others did, however, make the crossing.

The barrier is less marked for plants: various Asian plants including *Licuala* palms reached Australia, while dipterocarps reached only as far as New Guinea where the forests are predominantly 'Asian' in terms of plant origins (Morley 1998).

The plant movements were not one way. The Asian forests include taxa of Australasian provenance, especially within the montane and heath forest floras. The sharpness of this line for different taxonomic groups largely reflects probability for dispersal and age. The

---

[16] At least 16 Northern and 17 Southern mammal families were exchanged. Extinction and diversification rates both accelerated. Today 14 Northern families occur in South America, representing about 50% of all mammal genera.

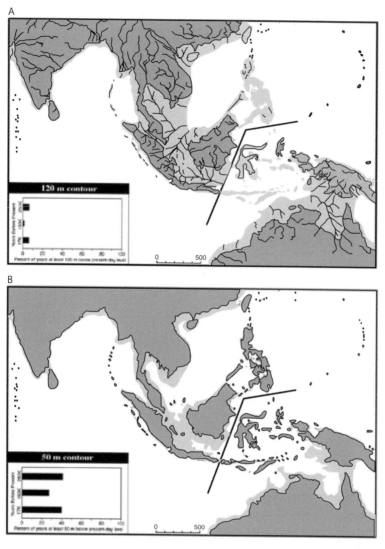

Figure 6.7 Maps of tropical Southeast Asia and Australia–New Guinea illustrating depth contours of 120 (A) and 50 m (B) below present level. The horizontal bar graph in the lower left corner estimates the percentage of time that sea level was at or below the illustrated level during the past 17,000 (bottom), 150,000 (middle), and 250,000 (top) years. Wallace's Line, which separates the zoogeographical regions of Asia and New Guinea, is drawn on the maps.

Adapted from Voris 2000.

antiquity of the flora compared to fauna contributes to the less marked floristic differences (Morley 2003).

Dynamic islands: stepping stones, conveyors, and mixers
Islands and archipelagos have risen and fallen throughout the Earth's history. Such islands can provide 'stepping stones', or perhaps 'conveyors', for dispersal. Ridges may have existed between South America and Africa up until the Oligocene, when Atlantic ridge islands were above water. Paleocene and Oligocene sediments from the now submerged Ninety-East Ridge reveal pollen of Australian affinity as well as a (now) Madagascan endemic (*Didymeles*, Didymelaceae) (Morley 2003). In some cases,

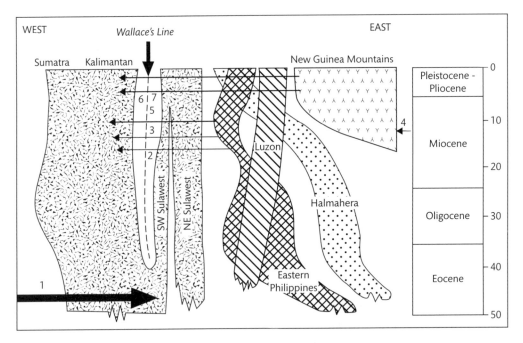

Figure 6.8 The proximity of Southeast Asian land masses through time. Southwest Sulawesi and Sunda were joined in the Middle Eocene and shared the same flora, whereas most of the islands of east Indonesia were not formed until the Middle Miocene. Halmahera and the Philippines have a much older history and would have supported a tropical flora throughout the Tertiary. The numbers on the figure refer to movements of various taxa: (1) dispersals from the Indian plate about 45 Ma; (2) Myrtaceae, 17 Ma; (3) *Camptostemon*, 14 Ma; (4) *Nothofagus*, Middle Miocene; (5) *Stenochlaena milnei*, 9.5 Ma; (6) *Podocarpus imbricatus*, 3.5 Ma; (7) *Phyllocladus*, 1 Ma.

From Morley 2000; data source from Hall 1995.

too, islands on dynamic plate fragments may, like India in miniature, have carried biota between distant regions. Such fragments may explain how some marine-intolerant freshwater fish moved between South and North America prior to land bridge formation.

In addition it is not unknown for species that colonize an island to evolve into new forms and later to recolonize the mainland. Such a history seems to account for some chameleons in Africa (via Madagascar), and short-faced bats and anole (*Anolis*) lizards in the Caribbean.

In tectonically complex island regions, such as Melanesia and the Caribbean, multiple plates interact mixing their biotas. Lands of both continental and oceanic origin rise, sink, fuse, divide, and exchange species. Islands combining multiple origins include Hispaniola and Cuba in the Caribbean, and Sulawesi in Asia—each of which possess rich 'collision biotas' from multiple sources.

6.2.2 Trans-oceanic movements

So far we have emphasized dry land routes for species migrations, but, despite the creative abilities of numerous evolutionary biogeographers, dry land routes prove insufficient to explain current biogeographical patterns. For example, the remarkable but well-established migration of the rain forest tree *Acridocarpus austrocaledonicus* (Malpighiaceae) from Madagascar to New Caledonia occurred approximately 8 Ma when no land route seems plausible (Davis et al. 2002a). Even within some ancient and widespread forest tree families like Annonaceae, Malpighiaceae, and Rhamnaceae, genetic studies suggest that various widely distributed lineages are too recently evolved to have reached their current distributions without trans-oceanic dispersal (Davis et al. 2004; Richardson et al. 2004). In any case, the biota of remote oceanic

Figure 6.9 Wallace's selection of fauna to illustrate the distinct species communities in Western Malesia and New Guinea. These are (clockwise from top left) (A) from Western Malesia: broadbills, gibbons, a 'long-tailed drongo shrike' (probably a racquet-tailed drongo, *Dicrurus remifer*), rhinoceros hornbill *Buceros rhinoceros*, and male and female argus pheasants *Argusianus argus*; (B) from New Guinea: a tree kangaroo *Dendrolagus inustis*, the fairy lory *Charmosyna papou*, the twelve-wired bird of paradise *Seleucides melanoleuca*, the common paradise kingfisher *Tanysiptera galatea*, and a crowned pigeon *Goura cristata (coronata)*.

From Wallace 1876.

islands such as Hawaii (discussed below) demands the recognition of numerous long-distance dispersal events.

Continental exchange: African–American species

Africa and South America broke physical contact *c.*96 Ma, and the Atlantic Ocean has been widening ever since. At least one-fifth of the 1,104 tree species in a 25 ha plot in the Ecuadorian Amazon are considered to belong to families and genera that arrived from the Paleotropics since South America became a discrete land mass (Pennington and Dick 2004; Renner 2004).

A widespread example is *Symphonia globulifera*, a small swamp tree that first colonized South America's Neotropical forests from Africa around 17 Ma (Dick et al. 2003) (Chapter 8). Some transfers have been west to east. Bromeliaceae and Cactaceae are diverse in the Americas (2,000–3,000 species each), but each has a single species in Africa (respectively, *Pitcairnia feliciana*, a terrestrial herb, and *Rhipsalis baccifera*, an epiphyte). Twelve angiosperm families and at least 110 plant genera now straddle the Atlantic with distributions essentially limited to the American and African tropics (Table 6.3).

Table 6.3 Selected plant taxa with apparently natural trans-Atlantic distributions.

Taxon	Direction: South America from (←) or to (→) Africa	When (estimate only)	Likely means of dispersal	Notes
Anacardiaceae *Spondias mombin*	→	Recent*	Water	Long considered a human introduction–this now seems unlikely (see Duvall 2006)
Arecaceae, *Elaeis oleifera*	←	Recent	Water	Neotropical *E. oleifera* readily hybridizes with the African oil palm *E. guinensis*
Arecaceae, *Raphia taedigera*	←	Recent	Water	One of 20–30 African/Malagasy species also occurs in Neotropics
Bromeliaceae, *Pitcairnia feliciana*	→	8 Ma	Wind	One of 260 species is African
Cactaceae, *Rhipsalis baccifera*	→	Recent	Water/raft or bird	One of 50 Neotropical species also occurs in Africa
Clusiaceae, *Symphonia globulifera*	←	3 events, all < 17 Ma	Water	A single swamp species
Malvaceae, *Ceiba pentandra*	→	Recent	Wind or anthropogenic	One in Neotropics and Africa
Rhizophoraceae, *Cassipourea*	←	Recent	Water	9 neotropical several in Africa
Rhizophoraceae, *Rhizophora*	→	23 Ma	Water	Mangroves, 2 in Neotropics and one in West Africa (*R. mangle* shared)
Platyrrhine monkeys	←	33–35 Ma	Water/rafts	No longer occur in Africa
Caviomorph rodents	←	25–55 Ma	Water/rafts	Widespread
Mabuya, lizards	←	9 Ma	Water/rafts	Widespread
Amphisbaenians (burrowing, usually limbless lizards)	←	40 Ma	Water/rafts	Widespread. Some lineages imply further dispersal events (e.g. *Cadea* spp. in Cuba)

Adapted from Renner 2004 using various sources (genetic and pollen).
*Recent here means within the last 1.5 Ma and perhaps.

Given sufficient geological time, even extremely unlikely events become increasingly probable. For spores, and seeds such as *Ceiba*'s fluffy seeds ('kapok'), strong winds are a plausible dispersal force. Birds, too, when occasionally blown far from their normal routes, can carry seeds far and wide. Such transfers may have benefited from areas of mid-oceanic land which might themselves have harboured forest.

Rafts of buoyant vegetation expelled by rivers, or rafts of pumice, are also plausible modes of dispersal. Aerial surveys off the Gabon coast have recorded vegetation rafts kilometres long and more than 100 m broad (Ted Barr, pers. comm. with DS). Floating islands with standing 15 m trees have been reported at sea (Houle 1999). Currents and winds might facilitate trans-Atlantic transfer in either direction allowing interchanges of flotsam (Carranza and Arnold 2003). It is difficult to find alternative explanations for how monkeys and rodents reached South America from Africa during the Oligocene/Miocene (Poux et al. 2006) when the Atlantic was about half its current breadth. Even now at its current width a raft crossing might take as little as two weeks[17] (Houle 1999).

Immobile and itinerant animals

In terms of biogeography, plants have been more mobile than animals. Aside from some birds and bats, most forest vertebrates appear poor at sea-crossings, although rafting rodents and buoyant tortoises have proved able voyagers, having reached all the main forest regions and many islands. Amphibians seldom make even short marine crossings, although frogs on oceanic islands such as the Comoros indicate that it can happen (Vences et al. 2003).

Small distances are less of a problem, and some fauna, including pigs and elephants, are remarkably good swimmers. Larger animals may be washed out to sea by floods, tsunamis or powerful currents—or may swim willingly—later to scramble ashore elsewhere, while smaller animals can raft on debris or volcanic pumice. Perhaps this is how Asian cattle and pigs reached Sulawesi, and stegodonts (small extinct elephants) reached Timor (Diamond 1987).

6.2.3 Communities: chance assembly versus dependent webs

Species coexisting in today's tropical forests have arrived at different times, often from different regions. Though rain forests are often presented as a result of 'ancient specialized co-evolved relationships', species origins suggest instead a fortuitous gathering of independent species. Reality lies in between and inevitably varies with location. Dispersal and mixing do not preclude the evidence that some species combinations (and interactions) are more likely to arrive intact, to arise *in situ*, or to persist or to perish than others—a theme we explore at greater length in Chapter 14.

Often a small number of dispersal events have been fundamental in defining modern distributions. Evidence often hints that the timing and order in which species arrived and occupied available niches has been influential to the resulting communities that have evolved (Box 6.7).

The ways to be a successful rain forest organism are many, but the principal adaptations are limited. Similar organisms need not be closely related but may have converged at the same form from a different starting point. Thus scaled pangolin and armoured armadillo possess a similar solution to a similar lifestyle. Sometimes the differences are harder to identify: the leaves of the rain forest gymnosperm *Gnetum* are hard to tell from those of angiosperms with which they live (Chapter 2). Genetic analyses of the ground-dwelling stick insect (order Phasmatodea) 'tree lobsters' of New Guinea, Australia, New Caledonia, and various neighbouring islands indicate that despite their geographical proximity these are not a discrete closely related group of species but rather a case of convergence (Buckley et al. 2008).

---

[17] A 1 kg primate might survive without water for 13 days (Houle 1999).

Box 6.7   Primate niches, prior occupants, and a 'just so' caution

The prosimians (primate-like animals) appeared in the late Cretaceous. Haplorrhines, the 'real' monkeys, arrived later in the Eocene 50–45 Ma, evolving from lemur-like precursors (Gebo et al. 2001; Bajpai et al. 2008). Haplorrhines spread widely in Africa and Asia where they outcompeted all but the specialized nocturnal insectivores (bushbabies, lorises, tarsiers, and pottos).

Regional primate communities appear to reflect colonization histories (Reed and Bidner 2004). Monkeys never colonized Madagascar where lemuroid primates—the wet-nosed Strepsirhines— had already radiated into largely empty forests after colonization from Africa 65–50 Ma. (Primates also failed to colonize New Guinea, where marsupials such as tree kangaroos and cuscus now occupy similar niches—though the order of events and whether the faunas ever met is uncertain.)

While Africa and Asia have few small monkeys, South America has many. The difference may reflect competition: squirrels only arrived in South America long after monkeys, while apparently no prosimians made it (despite some Caribbean Island fossils). In the Americas leaf-eating specialists are notable by their absence—perhaps because sloths already dominated this niche (Fleagle 1999; Corlett and Primack 2006). Furthermore, while South America lacks nocturnal prosimians, it possesses the only truly nocturnal monkeys, the owl-monkeys *Aotus* spp., which fill a similar role.

We end with a caution about 'just so' stories. The above ideas are pleasing conjectures—they may even be true. But they are only 'good science' if considered as testable hypothesis, and so far at least, short of finding a few unexpected fossils, no one knows how such ideas could be disproved. Still, as long as we remain sceptical such speculations are intriguing, and some related biogeographical ideas *are* more open to assessment. For example, evidence for the role of niche occupancy comes from the observation that while many of Asia's smaller continental islands possess several primate species, these assemblies are not random, with typically only one species representing any given genus (Harcourt 1999).

6.3   Regional themes and variations: a brief biogeography of quirks

Modern communities can and should be interpreted in the light of historical processes and current conditions. Table 6.2 has already outlined some of the principal features of the major rain forest regions. Here we briefly review the distinctive characters and idiosyncratic nature of each region, noting the role of history and highlighting aspects not considered elsewhere.

6.3.1 Continental Africa

Africa's wet forest flora is considered depauperate compared to the other tropical continents with fewer fami- lies, genera, and species. Despite moving only 15 degrees northwards since leaving Gondwana, the main story is one of climatic instability and arid episodes. Africa is higher and drier than the other continents, and the biota is marked by past, and present, aridity. Various otherwise widespread genera like *Campnosperma*[18] are now absent. It was not always so.

Africa possessed rich Late Cretaceous rain forests, and Cameroonian deposits from only 39–26 Ma show a diverse flora (Morley 2000). Pollen shows that various plants such as Casuarinaceae, Chloranthaceae, Winteraceae, and Sarcolaenaceae lingered in South Africa, only disappearing after the mid-Miocene (Burgoyne et al. 2005). A once rich palm flora has also been lost (Morley 2000; Pan et al. 2006). Fungus-growing termites, though now typical of

---

[18] Trees (Anacardiaceae) found through Southeast Asia, South America, and Madagascar.

savannahs, are likely to have originated in African rain forests before migrating to Asia and Madagascar (Aanen and Eggleton 2005). It appears that much of Africa's once rich biota of rain forest species has in fact been lost—and many current African species are relative newcomers.

Climate and forest cover changes have been complex, modulated by atmospheric changes, high terrain and uplift along the East African Rifts[19], fire, herbivory, and grasslands. Extensive forests have repeatedly expanded and contracted in response to expanding major grass savannahs (Jacobs 2004) and associated increase in large herbivores (Bobe 2006), as well as increased aridity during glacial cycles (Burgoyne et al. 2005).

Even today Africa's forest animals lack many specializations associated with wet closed forests elsewhere, and many are more willing to cross open habitats. Various forest taxa developed wholly terrestrial lineages—these included various ground-dwelling primates (Miller et al. 2005). Today Africa is much richer in larger (> 5 kg) mammal species and poorer in smaller species than South America, which has remained predominantly forested while Africa became open. Today Africa has a large number of specialized grazers and browsers, while South America has few[20] (de Vivo and Carmignotto 2004).

6.3.2 Madagascar

Madagascar's relative isolation has resulted in a distinctive biota. It split from Africa at about 160 My (for another 32 million years it remained connected to South America and Australia via India and Antarctica) and finally separated from India at about 88 My. The Paleocene and Eocene climates were relatively arid, such that Madagascar's oldest vegetation is the dry spiny forest of the southwest (Goodman and Benstead 2003). Wet rain forests are considered of more recent origins, though details remain speculative. Episodes of aridity, in conjunction with localized riverine and montane refugia, appear to have had a major impact on the fine-grained patterns of diversity across the island (Goodman and Benstead 2003). Humans colonized the island only in the last 2,000 years, but have had a dramatic impact in modifying cover and eliminating most of the larger fauna[21] (Goodman and Benstead 2003; Perez et al. 2005). Before people arrived in Madagascar there were no hoofed animals.

The dominant view is that the ancestors of most of Madagascar's modern flora and vertebrate fauna (including all living mammals) reached Madagascar from across the sea (Leigh et al. 2007). The age of these colonization events has resulted in high levels of endemism amongst species and genera but not families—thus while 96% of Madagascar's 4,220 species of trees and woody shrubs are endemic, only 161 of its 490 genera, and 7 of its 107 families, of such plants are endemic (Schatz 2000 cited in Leigh et al. 2007).

Phylogenies of the modern species suggest that Madagascar's four major extant orders of native mammals, including lemurs and tenrecs (Tenrecidae), each result from an adaptive radiation deriving from a single colonization event (Goodman and Benstead 2003). Data on invertebrates are inevitably incomplete, but Madagascar appears to be an important centre of diversity with, for example, 10% of the world's ant species (Fisher 2003) and some distinctive groups including the hissing cockroaches.

Wet forests are now largely limited to the east coast and mountains. Groups such as palms and orchids are more diverse than in continental Africa. Madagascar, like Africa, lacks some otherwise widespread families (Magnoliaceae, Chloranthaceae) but differs in possessing others (Winteraceae, Elaeocarpaceae), and shares several families with Asia (Nepenthaceae). Among the fauna, larger mammalian predators are a notable absence with the largest, the 17 kg giant fossa *Cryptoprocta spelea*, now extinct—leaving only the smaller fossa *C. ferox*.

---

[19] Which divided the lowland forest biota into the now very restricted biota of coastal East Africa and the more extensive forests of West and Central Africa.

[20] South America's extinct Pleistocene mammals were larger, so historical differences were somewhat less, but these fauna were seldom forest users.

[21] Post-human extinctions include a forest hippopotamus, various large lemurs, and two large eagles.

6.3.3 Americas

For most of the last 60 million years, having split from Antarctica, and before linking with the continent to its north, South America was a large warm wet island. Amazonian pollen shows many modern Amazonian genera already present by the late Miocene. Climate has played a role in shaping the rich biodiversity of Amazonia, but other key factors influencing the biota have been mountain building, changes in water-cover, and species exchange with North America.

In contrast to the high ground of Central Africa, the Amazon is predominantly a low alluvial basin prone to flooding and readily shifting rivers. The Amazon Basin has probably been submerged by occasional but vast fresh and marine water incursions since the Cretaceous. The Amazon itself originally flowed westward, entering the Pacific though modern Ecuador. At around 40 Ma the rising proto-Andes dammed the flow and an immense shallow lake developed, which first drained northwards through the Guyanas and later to the east. From 10 to 6 Ma, much of Amazonia was, once again, a vast freshwater wetland, and when sea levels were high a shallow sea probably opened to the Caribbean (Hoorn 2006; Hubert and Renno 2006). Consequently much of the Amazon's rich river fauna, including freshwater dolphins, stingrays, anchovies (Engraulididae), drums (Sciaenidae), and needlefish (Belonidae), are derived from marine Caribbean species (Lovejoy et al. 2006).

The rising Andes promoted spectacular diversification within montane forests (Burnham and Graham 1999) and divided the lowlands into Pacific and Amazonian forests. Many related taxa occur on both sides but have become distinct species.

Grasslands appeared 32 Ma, though they never spread to the extent seen in Africa. Various large mammals, including giant sloths, proliferated but remained largely absent from closed forests. The degree to which the Amazon forest was reduced in extent and fragmented during the last and more ancient glaciations has been strongly debated (Chapter 8).

Besides the extensive Amazon-Guyana lowland, other important forest formations occur. The Atlantic strip of south-eastern Brazil captures enough moisture to support patchy rain forests rich in endemic species—perhaps half of the tree species also occur in the Amazon or elsewhere in the Americas, probably reflecting periods of more extensive forest cover. Further north the forests of Central America have affinities to the Amazon, Guyana, and Pacific forests, though they tend to be more seasonal. To the west of the Andes lies a band of extremely lush Pacific coast rain forest that shares many genera with the Amazon but also possesses many endemic species.

Caribbean

Many Caribbean islands possess seasonal lowland and montane rain forests. Generally these have close affinity with mainland South and Central America but are less species rich. Individual histories have led to distinctive species communities.

After South America and Africa separated, an island chain (the Greater Antilles, including Cuba and Hispaniola) was dragged from the Pacific into the modern Caribbean by tectonic processes. It was a bumpy ride as islands rose and sank, divided and merged. Phylogenetic studies (e.g. in woody *Exostema*, Rubiaceae) show disjunctions between various regions of Cuba and Hispaniola consistent with past connections between these composite islands (McDowell et al. 2003). Cuba now has striking plant diversity, with half of over 6,500 vascular plants being endemic.

Some Caribbean endemics are relics of once more widespread taxa (e.g. the shrew-like Cuban and Hispaniolan solenodons, *Solenodon cubanus* and *S. paradoxus* Solenodontidae[22]). There have also been extensive diversification and specialization, and the region is rich in reptiles with 150 endemic *Anolis* lizards and 82 endemic dwarf geckos[23] (*Sphaerodactylus*).

---

[22] An ancient group that diverged from other insectivores in the Cretaceous. Fossils show that similar animals were common in North American forests 30 Ma.

[23] *Sphaerodactylus ariasae* and *S. parthenopion* from the Dominican Republic and the US Virgin Islands respectively are the smallest lizards in the world. The radiation of *Leptotyphlops* snakes include *L. bilineata*, the world's thinnest snake—no fatter than a pencil lead.

6.3.4 Asia

Today, Asia contains many of the world's most distinctive tropical forest formations. Biogeographically the region is complex. Multi-plate tectonic complexity, volcanism, India's transfer role (see above), islands, shallow seas, and ever-changing land bridges and sea channels have given the Asian rain forest region a convoluted biogeographical history.

The most important biogeographical feature of the region is Wallace's Line and associated transitions, already described above. But other features that have shaped the biogeography of the Asian region include the connection of the Sunda region with East Asia, currently via the Malay Peninsula and the Isthmus of Kra (modern-day Thailand). This was an important passage by which northern hemisphere rain forest elements were able to find low-latitude refuges during a period of cooling in the mid-Tertiary (45 Ma) (Morley 2000). Refuges in other regions were less accessible during this time, owing to wide oceanic barriers. This is thought to explain the present-day diversity of primitive angiosperms in the montane forest of Southeast Asia.

Despite a land connection, various evidence, including pollen and climate simulations, suggests that a climatically seasonal savannah corridor reached down through equatorial Southeast Asia during Quaternary glaciations, and reduced dispersal of rain forest species between Sumatra and Borneo (Figure 6.10). Pleistocene pollen indicates that the area around Kuala Lumpur in Peninsular Malaysia was dominated by a grassy pine savannah vegetation not found in Malaysia today (Morley 1998, 2000). Though many of the islands now support forest, when originally formed the newly isolated fauna reflected the prevailing habitat of the time: so many common forest taxa are absent from those islands which lacked forest at that moment. These records imply that during the last glacial there were several areas in the Sunda region that remained forested: west of Sumatra,

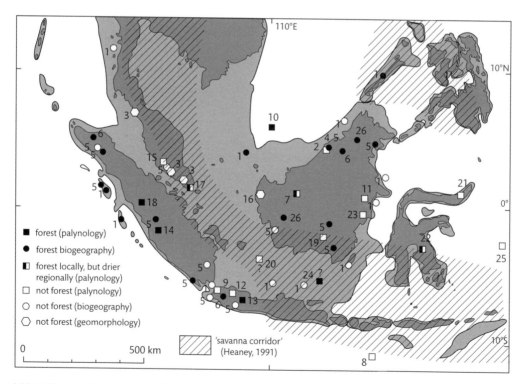

Figure 6.10 Evidence of forest and non-forest cover during the low sea level period of the last glacial.

From Bird et al. 2005.

north-west of Borneo, the Malacca Straits, and around Palawan. Other areas may have been covered by more open vegetation types like tree savannah, or open deciduous forest: on and to the east of the Malay/Thai Peninsula, the Java Sea area, including the Sunda Strait, and eastern Borneo (Meijaard and van der Zon 2003).

6.3.5 New Guinea and Australia

Australia, once well forested, is now dominated by deserts and aridity. In 100-50 Ma much of Australia was covered by low diversity temperate gymnosperm forests (mainly Podocarpaceae and Araucariaceae) (Morley 2000). In the Early Eocene (54-49 Ma) mangrove swamps which included *Nypa* palms (Figure 6.11) became widespread across south-east Australia. During the middle and late Eocene Australia remained densely forested, but as the climate cooled the mangroves became increasingly restricted, with *Nypa* palms persisting in Queensland (Morley 2000). The Early Oligocene (34-30 Ma) coincided with severe cooling, and temperate *Nothofagus* rain forest expanded at the expense of **megathermal** angiosperm taxa. By the Early Miocene (23-16 Ma) warm temperate rain forests of *Nothofagus*, *Elaeocarpus*, and Podocarpaceae were widespread.

Cover declined dramatically in the Middle Miocene (16-10 Ma) due to the drifting of the Australian Plate into the southern subtropical high pressure zone, which coincided with falling global temperatures. Increasing aridity caused the decline of rain forest elements, which were replaced by grasslands. A few taxa arrived from tropical Asia (e.g. *Pteris* ferns and Caesalpinoideae). Primitive angiosperms, such as members of Winteraceae and Chloranthaceae[24], which were widespread in the late Cretaceous, persisted through the Tertiary to the present day, indicating the presence of wet frost-free refuges along the eastern Queensland coasts, maintained by warm and moist onshore winds. During the cooler and drier Quaternary glacial periods these rain forests presumably found refuge in damp valley bottoms (Adam 1994).

Today Australia's rain forests are mainly restricted to north Queensland's coastal strip, and in New South Wales.

The forests are strongly seasonal yet predominantly evergreen, and include families such as Elaeocarpaceae Monimiaceae and Proteaceae that are scarce in other regions (Bowman and Prior 2005). The vegetation also includes species that appear similar to the ancestors from which many more xerophytic modern Australian species were derived, e.g. *Acacia*, *Brachychiton*, *Flindersia*, *Gardenia*, *Gunnera*, and *Gymnostoma* (*Casuarina*) (Whitmore 1998).

New Guinea is the youngest major rain forest region. From the Late Eocene to the end of the Oligocene (i.e. 35-23 Ma) New Guinea and most of the islands east of Sulawesi were submerged. New Guinea was uplifted as the Australian and Philippines plates collided at the end of the Oligocene. The main phase of angiosperm colonization of New Guinea occurred through the Early and Mid Miocene with Myrtaceae, Rubiaceae, and Arecaceae being important components of the vegetation from the Late Miocene to Pleistocene (Morley 2000). For much of the last 3 million years New Guinea and Australia have been one land mass, hence New Guinea having many biological affinities to Australia. Interestingly though, much of the tree flora is of Asian affinity (Good 1962), suggesting relatively ancient island hopping routes, possibly involving Sulawesi, prior to establishment of the Makassar Straits (Morley 2000).

Distinctive forest marsupials include the nocturnal cuscus or *Phalanger*, which has a long prehensile tail, and the tree-climbing kangaroos *Dendrolagus*. Large mammals, both carnivores and herbivores, are lacking (human impacts remain uncertain in the rain forest regions), but flightless birds, the cassowaries, are the dominant large animal in the forests.

Apparently New Guinea's bird fauna has twice as many fruit-eaters and nearly twice as many nectar-feeders as a comparable lowland forest in Peru (Cristoffer and Peres 2003). The lack of efficient predators may help explain the extraordinary diversification of brightly coloured birds, including birds of paradise, and conspicuous ground displays, as provided by bower birds. For those willing to take a long-term view, New Guinea's diverse flora and remarkable fauna, despite its youth, should be reassuring.

---

[24] Chloranthaceae is not closely related to any other family, and is among the earliest diverging angiosperm groups. It belongs to a sister group to the eudicots and monocots, but its origins remain unclear (Solds et al. 2008).

Figure 6.11 The monotypic palm *Nypa fruticans* characteristic of brackish swamps in Asia is one of the few extant angiosperms known to have existed in near identical form before the end of the Cretaceous. Though once widespread in Africa and the Americas, rapid sea level changes at the end of the Eocene (33 Ma) may have led to extinctions beyond Asia's extensive coasts and vast shallow seas (Morley 2000).

Photo by Douglas Sheil.

6.3.6 Oceanic and continental islands

Trans-oceanic dispersal is least controversial in explaining the biota of remote oceanic islands such as Hawaii. Such islands often possess numerous endemic species, but these typically derive from a limited number of colonist taxa (Cowie and Holland 2006).

Hawaii is amongst the best-studied examples. Analyses suggest that the 1,009 indigenous angiosperms derive from perhaps 263 distinct colonist species (Price and Wagner 2004); 89% of angiosperm species and 76% of pteridophytes are endemic. For wetter habitats, island age accounts for more variation in inter-island species differences than area, showing that colonization and diversification take time (Price 2004). The native vegetation evolved with little exposure to vertebrate herbivores, against which in consequence they are poorly defended. The islands were settled by Polynesians around 1,500 years ago, with the introduction of rats and pigs, reduction in lowland forest, and a catastrophic impact on the local biota—a process continued by the Western colonists who arrived in the last two centuries.

Young[25] remote Hawaii is unrepresentative of many oceanic islands. Fiji, for example, possesses an apparently older flora with 1,315 species in 137 families (including the endemic family Degeneriaceae). Much of this flora seems to have evolved more or less *in situ*, suggesting a long-term presence of dry land in the vicinity (Heads 2006).

Two generalizations about island floras: 1) widespread and abundant generalist species with abundant and readily dispersed propagules are more likely to cross oceans; and 2) taxa with lower dispersal rates when they

---

[25] Note that the Hawaiian archipelago is older than any of its extant islands (islands are born in a near continuous series only to age and disappear over a period of several millions of years); this raises the possibility that by hopping from older to younger islands, species might be much older than the current islands.

Figure 6.12 *Medusagyne oppositifolia*, an ancient and unique tree from the Seychelles. The long period of isolation has given rise to many unusual plants within the Seychelles archipelago.

Photo by Jaboury Ghazoul.

do occur tend to have higher rates of endemism. Regarding fauna, forest specialists, even birds and bats, seldom make sea crossings when compared with more generalist taxa.

The species richness of many oceanic islands is likely to have been considerably higher as recently as 1,000 years ago than what is observed today. Most extinctions almost certainly go unrecorded, but if the endemic *Partulina* and *Achatinella* snails of Hawaii reflect broader patterns, then the arrival of humans may have heralded the loss of as much as 50% of some groups (Hadfield 1986).

Seychelles

The Seychelles Islands are among the oldest islands in the world, and the only granitic islands located in mid-ocean. These islands are remnants of Gondwana from which they became detached when India drifted north toward Asia some 75 Ma. This long and continuous period of isolation has allowed the evolution of a unique assemblage of species, many of which appear to be evolutionary relics. Such ancient species include endemic palms such as coco-de-mer *Lodoicea maldivica* (Arecaceae), and the unique jellyfish tree *Medusagyne*

oppositifolia (Medusagynaceae, a monotypic family of uncertain affinity) (Figure 6.12), and among the animals seven species of caecilians.

The islands' long isolation and origin as fragments of Gondwana and India have resulted in a complex biogeography, with the plant community showing affinities to Madagascar and the Mascarenes, as well as to mainland Africa and Asia (Friedman 1994). The Seychellian species of *Impatiens thomassetii* (Balsaminaceae) and *Rothmannia annae* (Rubiaceae) are more closely related to species on the African continent than those on Madagascar or the Mascarenes, while several other species, e.g. *Amaracarpus pubescens* (Rubiaceae), are shared with Southeast Asia.

6.4 Conclusion

When climatic conditions allowed, vegetation cover with many structural similarities to modern rain forests has arisen repeatedly, following major extinction events in the Earth's past (e.g. after the Devonian, Permian, Cretaceous). The geographic origin of angiosperms remains unclear—but tropical forests acted as a cradle for the diversification of many plants and animals that we see in today's rain forests and elsewhere.

Various processes interact to result in present-day distributions. While plate movements and climatic determinants have played a central role, long-distance dispersal has also contributed greatly. We still have a great deal to learn about the histories of how current rain forest communities came to be the way they are, but some striking progress is being made.

CHAPTER 7

Many rain forests: formations and ecotones

'The greater number of trees, although so lofty, are not more than three or four feet in circumference. There are, of course, a few of much greater dimension...The woody creepers, themselves covered by other creepers, were of great thickness: some which I measured were two feet in circumference. Many of the older trees presented a very curious appearance from the tresses of a liana hanging from their boughs, and resembling bundles of hay. If the eye was turned from the world of foliage above, to the ground beneath, it was attracted by the extreme elegance of the leaves of the ferns and mimosae.'

Charles Darwin (1832)[1]

There is no single typical tropical rain forest. Each has its own floristic and structural characters, and it is often the structural characters that we first notice. Although the principal environmental features defining tropical rain forests are warm, wet, and frost-free climatic conditions, there is considerable variation in settings and resultant ecosystems. A tour of the world's forest regions might introduce us to Borneo's hill forests that many consider the archetype of tropical rain forests. Without leaving Borneo we could also visit low heath forest, densely populated with many small trees and saplings. On the other side of the world the hurricane-impacted forests of Puerto Rico, about half the height of the tall forests in Borneo, are festooned with fallen trees tangled in vines. Not too far away, on Barro Colorado Island in Panama, broad but not particularly tall trees are common. Moving westwards to western Amazonia we encounter understoreys packed with palms. Finally in Africa the scene changes yet again,

but here too there is variety with mixed canopy forest alongside single canopy monodominant forest at Ituri forest in the Democratic Republic of Congo.

7.1 Forest types

Tropical forests have been classified in numerous schemes addressing composition, structure, physiognomy, and environments. Since similar environments often possess similar vegetation, even on different continents, classifications based solely on selected environmental characteristics offer some advantages in a global overview (notably allowing generalization without reference to taxonomy). Our scheme (based on Whitmore 1998b) is presented in Table 7.1. The first division distinguishes ever-wet climates from those with a marked dry season. The second separates swamps from drylands.

---

[1] From *Journal of researches into the natural history and geology of the countries visited during the voyage of* HMS Beagle *round the world, under the command of Capt. Fitzroy, R.N.*1845. 2nd edn. John Murray, London, UK.

Table 7.1 Types of tropical rain forests and related formations. Note that 'seasonal rain forest' was previously called 'semi-evergreen rain forest', a name we now know to be misleading as such forests may be evergreen; and 'non-seasonal lowland rain forest' was previously called 'evergreen lowland rain forest', but many seasonal forests are also evergreen.

Climate	Water and drainage	Geology or soils	Location	Forest formation
Seasonally dry	Strong annual shortage			Monsoon forests (various formations)
	Slight annual shortage			Seasonal rain forest
Ever-wet (perhumid)	Dryland	Oxisols, ultisols	Lowlands	Non-seasonal lowland rain forest
			1200–1500 m	Lower montane rain forest
			1500–3000 m	Upper montane rain forest
			3000 m to tree line	Subalpine forest
		Podzolized sands	Mostly lowlands	Heath forest
		Limestone	Mostly lowlands	Forest over limestone
		Ultrabasic rocks	Mostly lowlands	Forest over ultrabasics
	Water table high (at least periodically)		Coastal salt water	Beach vegetation Mangrove forest Brackish water forest
		Oligotrophic peats	Inland freshwater	Peat swamp forest
		Eutrophic soils (organic and mineral)	Permanently wet	Freshwater swamp forest
			Periodically wet	Freshwater periodic swamp forest

(Location column for the 1200–1500 m, 1500–3000 m, and 3000 m to tree line rows is bracketed as 'Mountain')

Modified from Whitmore 1998b.

The third concerns geology and soils[2] (see also Chapter 9). Finally, altitudinal divisions are applied. This simple scheme is useful for broad generalization but inevitably neglects many factors that influence forest formations.

7.1.1 Non-seasonal lowland rain forest

Non-seasonal lowland rain forests include the world's most structurally complex and diverse vegetation. The classical form is tall, evergreen, and species-rich. The biggest trees may be more than a metre in diameter: tall, slender, and reaching over the main canopy (even to 70 m). These emergents are often deciduous.

Lowland forests in the wettest and most extensive forest regions tend to be the richest in tree and herb species. The understorey can be quite dark and open with a scattering of herbs, abundant saplings and seedlings, and locally abundant lianas. Vascular epiphytes are usually common, especially on larger trees, but bryophytes are relatively scarce. Buttresses are frequent on larger trees. Tree bark is often thin and smooth. **Cauliflory** and **ramiflory**, where flowers and inflorescences are produced directly from the stem and branches (Chapter 12), are often common. Leaves are commonly long-lived, tough, thick, and leathery. Large leaved understory plants can be common (Figure 7.1). Leaf margins are usually entire and drip-tips (Chapter 2) are frequent.

These rain forests dominate in areas without regular dry seasons (<100mm rain for one month or more), and are widespread in western Amazonia, Indonesia, Malaysia, and New Guinea, and occur locally in southwest Sri Lanka and Pacific Mexico. In Africa, suitable

[2] Another widely used scheme, developed by Holdridge, is based wholly on climate and so does not require information on local terrain, drainage, and substrate (Holdridge 1947).

Figure 7.1 *Lodoicea maldivica* palms in Praslin, Seychelles. Large, tough, and long-lived leaves are common among understory plants and saplings of canopy trees in forests free from seasonal moisture stress.

Photo by Jaboury Ghazoul.

conditions occur only in small blocks near the coast between Guinea and Liberia, and Cameroon and Gabon, including the Gulf of Guinea islands Bioko, São Tomé, and Principe.

7.1.2 Seasonal rain forest

Seasonal rain forests occur where there is moisture stress at some regular period during the year. Canopy height varies, although these forests are frequently as tall as, or taller than, non-seasonal rain forest in the same regions—most likely because soils are often better for plant growth. Emergents may be either absent or common, reaching 60 m in West Africa. Lianas are often abundant and buttresses common. Tree bark tends to be thicker and rougher, and cauliflory and ramiflory rarer than in less seasonal forest. Bamboos are sometimes common. Epiphytes are occasional to frequent, and include many ferns and orchids. The canopy usually contains a mix of deciduous and evergreen trees, leading to the names 'semi-evergreen' or 'semi-deciduous'. Nonetheless, many seasonal rain forests remain principally evergreen (Box 7.1).

Seasonal rain forests are the most extensive rain forest formations, and occur throughout the wet monsoonal tropics. They cover much of Central America, Brazil's Atlantic forests, and the south-east portion (*c*.1.8 million km²) of the Amazon Basin. They also occur on the margins of the main rain forest blocks in Asia, including East

---

Box 7.1   Seasonal or evergreen rain forests?

Older accounts imply that seasonal rain forests necessarily comprise a significant proportion of deciduous tree species in the canopy. Though often true, we now recognize many exceptions. The designations 'semi-evergreen' or 'semi-deciduous' as synonyms for seasonal rain forests are thus no longer tenable in an environmentally determined scheme (such as Table 7.1).

Consider some examples. Much of the eastern Amazon Basin is highly seasonal, but deep rooting canopy trees remain in leaf throughout the dry season (Nepstad et al. 1994; Myneni et al. 2007). In Africa's seasonal rain forests many formations are

evergreen; as droughts are short and variable in timing and degree, many canopy trees possess tough, drought-tolerant sclerophyll foliage (Veenendaal et al. 1996). The northernmost Asian rain forests below the Himalaya and in south China are in highly seasonal locations but are predominantly evergreen (probably due to the wet fogs and cool temperatures that reduce water demand during the winter dry season) (Zhang et al. 2006). The main Australian tropical rain forests are seasonal, although again the trees are predominantly evergreen (Bowman and Prior 2005).

Java and Bali and the monsoon systems of India's Western Ghats, the Andaman Islands, and part of Sri Lanka, Indochina, Southern China, the Philippines, Madagascar, and various tropical islands. Most of the African wet tropics are distinctly seasonal, although the dry seasons are less predictable than elsewhere.

7.1.3 The wet–dry and the latitudinal ecotones

Gradients in rainfall, seasonality, and latitude are interwoven. Moving from wet, non-seasonal equatorial forests into more seasonal and/or drier climates, there is generally a decline in floristic diversity and a reduction in structural variety (fewer vascular epiphytes, lianas and buttresses). Where seasonal droughts regularly last six months or more, forests are commonly known as 'dry tropical forests' and are distinct from rain forests in that woody cover is short, sometimes bushy, and dominated by xerophytic and usually deciduous vegetation that is often well adapted to fire.

The transition from rain forests to dry forests can be gradual. As rainfall declines, seasonal drought usually increases and a deciduous element usually predominates[3]. The canopy tends to become more open, while the abundance of grasses and other herbs increases.

Terrestrial vertebrates also often increase. Finally, an abrupt boundary maintained by fire and/or grazing often marks where woodland ends and grass savannah begins.

7.1.4 Montane rain forests and 'cloud forests'

Montane rain forests present a progression of formations (Figure 7.2). Mean air temperatures decline with altitude while diurnal variation increases. Moisture, cloud base, edaphic conditions, latitude, prevailing winds (Pendry and Proctor 1996; Ashton 2003), and various landscape aspects all affect zonation (Box 7.2). Montane forests are widespread. They are found in Southeast Asia (Pendry and Proctor 1997; Ashton 2003), the Caribbean Islands, the Andes (Kelly et al. 1994; Kessler 2001), and in Brazil's Atlantic forests. In Africa significant montane rain forest areas occur as localized patches and isolated islands[4]. Some montane rain forests occur even in arid regions, as mountains often capture significant moisture. For example, East Africa's Mt Kilimanjaro rises from arid plains but at 2,000 m receives over 3,000 mm yr^{-1} of rain (Hemp 2002).

Predictable vegetation patterns can be observed rising up the slopes of wet tropical mountains (Figure 7.2,

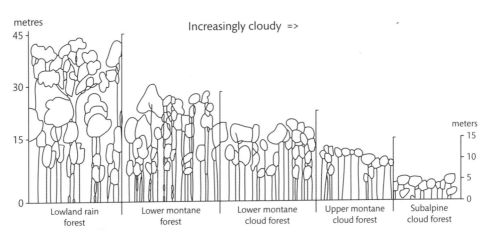

Figure 7.2 Generalized altitudinal change in forests with altitude on wet tropical mountains.

From Bruijnzeel and Hamilton 2000.

[3] Though not on very poor soils, or in Australia.
[4] For example, Mt Nimba, Mt Cameroon, the Cameroon highlands, the Gulf of Guinea Islands, and some of the larger East African mountains (Viurungas, Ruwenzoris, Elgon, the Eastern Arc).

Box 7.2   *Massenerhebungseffekt*: the 'mass elevation effect'

Mountain size and location influence forest zonation. All else being equal, forest zones sit higher on larger mountains than on smaller ones, especially in moist environments (Bruijnzeel and Hamilton 2000) (Figure 7.3). Explanations involve temperatures and clouds. Air temperatures on large mountains are typically higher than on smaller peaks, because they have larger areas warmed by the sun. On extensive plateaux, such as the East African highlands, 'lowland-like' daytime temperatures can occur even above 1,000 m. Persistent cloud cover also tends to be lower and narrower on smaller mountains. This reduces cloud shade and moisture which influences soil and vegetation.

Tropical air temperatures are reduced with proximity to oceans (and large water bodies). This explains why montane formations also tend to be lower near coasts versus those further inland.

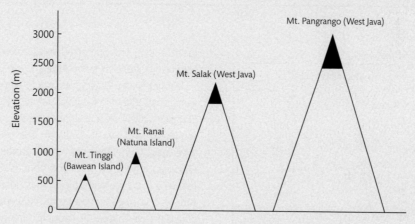

Figure 7.3 Mossy vegetation occurs lower on smaller mountains, and on small islands in Southeast Asia (Bruijnzeel and Hamilton 2000). This is the 'mass elevation effect' or *Massenerhebungseffekt* in its original German.

Table 7.2). The canopy gets progressively lower and lianas become increasingly scarce then disappear and are replaced by dense moss and other bryophytes (Figure 7.4), straggling lichens, bamboos, and tree ferns. Gymnosperms (*Podocarpus, Dacyrodes, Araucaria*) also can become common or even dominant. Tree species richness generally declines with altitude, although diversity in certain plant groups (notably epiphytes[5]) sometimes surpasses the lowlands (Heaney 2001), and local endemism is often high. The understorey is often dominated by a few herb species, such as the distinctive monocarpic herbs *Mimulopsis* spp. (Africa and Madagascar) and *Strobilanthes* spp. (Asia).

The transition from mesophyll-dominated lower montane to microphyll-dominated upper montane rain forest is often relatively abrupt at around 2,500 m, coinciding with a persistent cloud base[6]. Soils are often waterlogged and peat may form, sometimes with *Sphagnum* mosses.

[5] In the wettest forests, half of all leaf biomass may be held by epiphytes (Luttge 2004).
[6] The term 'cloud forest' is commonly used in South and Central America, but rarely elsewhere. In Asia these are 'upper montane rain forests', sometimes 'elfin forest' or 'mossy forest'. In Africa, 'Afromontane forest' is used, as well as 'upper montane forest'.

Table 7.2 Characters of structure and physiognomy that define the principal montane forest formations

Formation	Tropical (seasonal or non-seasonal) lowland rain forests	Tropical lower montane rain forest	Tropical upper montane rain forest (cloud forests)
Canopy height	25–45 m	15–33 m	1.5 (18 m)
Emergent trees	Characteristic, to 60 (80) m tall	Often absent, to 37 m tall	Usually absent, to 26 m tall
Pinnate leaves	Frequent	Rare	Very rare
Principal leaf size class of woody plants	Mesophyll	Mesophyll	Microphyll
Buttresses	Usually frequent and large	Uncommon, small	Usually absent
Cauliflory	Frequent	Rare	Absent
Big woody climbers	Abundant	Usually none	None
Adhesive bole climbers	Often abundant	Frequent to abundant	Very few
Vascular epiphytes	Frequent	Abundant	Frequent
Non-vascular epiphytes	Occasional	Occasional to abundant	Often abundant

From Whitmore 1998.

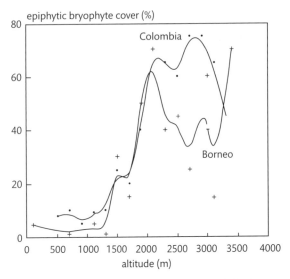

Figure 7.4 Cover of epiphytic bryophytes in tropical rain forests on the Sierra Nevada de Santa Marta, Colombia (after van Reenen and Gradstein, 1983) and on Mt Kinabalu, Borneo (after Frahm 1990). Lower cover values at 2,600–3,000m on Mt Kinabalu are due to a region of ultrabasic geology dominated by *Leptospermum* trees that have bark unsuitable for epiphytic bryophytes (Frahm and Gradstein 1991).

In areas bathed in cloud, bryophytes, filmy ferns, and liverworts cover both stems and rocks. Reaching higher still, the canopy surface becomes less irregular before the forest changes into a dwarf forest, usually at 2,800–3,200 m; this is sometimes called 'elfin woodland' due to

its shrunken and contorted trees. Such formations are especially extensive in the Andes.

Ultimately the forest gives way to alpine vegetation, at as high as 3,900 m in Papua and the Andes (Bruijnzeel and Hamilton 2000), with grasslands, heath, tundra, bogs, and meadows, and ice at 4,500 m. Fire often lowers such 'tree lines'[7]. Sometimes, as occurs at around 2,000 m on the Hawaiian Islands, an arid zone bounds the tree line. The reason is that mountains pierce the arid air (and temperature inversion) generated by the descending air at these latitudes (see Chapter 9). Latitude, altitude, and temperature variations also interact in more subtle ways (Box 7.3).

7.1.5 Freshwater swamp forests

Freshwater swamp forests occur in permanently or regularly flooded environments (freshwater swamp forest and freshwater periodic swamp forest). These also divide into two categories: white-water (sediment-rich 'eutrophic') and black-water (relatively nutrient-poor, often acidic 'oligotrophic') drainage systems (e.g. Amazonian and Orinoco tributaries). Intermediate, mixed or 'clear-water' drainages and rivers also occur (Amaral et al. 1997).

The key limitation is waterlogging. Flooding may be tidal, irregular, or seasonal. Where flooding is relatively minor, forests may be tall, complex, and generally similar to dryland forests. Canopy height and complexity usually decline as flooding increases. In some places, unstable and

---

[7] Typically associated with a coldest mean monthly temperature of -1° C.

Box 7.3   Altitude and latitude

Low temperatures influence vegetation, and frost appears to determine the boundary between montane and lowland forests: sub-zero temperatures kill most lowland species (Ohsawa 1995). In contrast sum periods of adequate warmth affect all plant growth and appear to determine the natural boundary for the upper tree-line (Ohsawa 1993).

Latitude–altitude interactions have been examined from the Asian equator northwards, avoiding the complications of deserts and dry winds. On the equator temperature changes are primarily diurnal, with seasonal variation increasing at higher latitudes. Montane rain forests typically occur above 1,000 m near the equator and drift lower as latitude increases. Eventually the forest reaches sea level at around 35°N where frost and temperatures limiting the growth season intersect in a single, subtropical lowland forest community that is similar to montane rain forests[1] (as in central China and Japan) (Ohsawa 1991, 1993).

According to one hypothesis, the altitude–latitude effect helps generate tropical diversity. Seasonal temperatures on high mountains at high latitudes can vary widely, so local species must be adaptable. In the tropics, high altitude temperatures vary little across the year, and so species specialize in a narrower range of conditions, making altitude a greater barrier to biological dispersal, and resulting in less mobile species. This may encourage more localized species distribution and higher species turnover along gradients (Janzen 1967). Indeed reptile and amphibian faunas separated by altitude overlap less in the tropics than in the temperate zone. The pattern for plants remains untested, though relatively narrow banding in the tropics has been noted (Ghalambor et al. 2006).

[1] Even in sub-tropical latitudes a 'rain forest element' is often distinguishable in the flora. Neither 'sub-tropical' nor 'extra-tropical' rain forests possess accepted definitions. 'Paratropical forests' has proved useful for cool, frost-free, evergreen rain forests with mean temperatures of 20–25 °C (Morley 2000).

short-lived alluvium banks may be covered only in grasses and herbs. Forests subjected to prolonged, deep flooding tend to possess lower basal area and more open canopies, and the trees often have above-ground aerial roots in the form of stilts or knees (Figure 7.5). The aerial portions of these roots contribute to the aeration of the root system. Exposed ground in seasonal swamp forests often possess little ground cover (Figure 7.5). Buttresses become common only when floods are minor and short-lived (Wittmann and Parolin 2005). Palms (Araceae) can be locally dominant with pandans (Pandanaceae) common too in parts of Asia, Africa, and New Guinea. Epiphytes, including ant plants, can be common. Vegetation is usually less diverse than in dryland forests, but these swamps often posses a rich and sometimes abundant aquatic fauna.

South American swamp forests cover substantial areas. The floodplains of the river Amazon cover more than 6 million square km[8] and the river remains tidal 900 km from the sea. In Iquitos, Peru, river levels vary by 20 m each year. Vegetation includes permanently, periodically, and occasionally flooded forests. The *igapó* forests (*mata de igapó*) are black-water systems draining the Guyana Shield Region (e.g. Rio Negro and lower Rio Branco) and possess a distinct ecology, including virtually no mosquitoes!

Seasonally flooded white-water *varzea* forest (*mata de várzea*) covers large areas (mainly focused on the Silmões-Amazonas, Purus, and Madeira Rivers) that are filled with lakes, pools, and extensive channel networks even during dry periods. The *varzea* range from *chavascal* (flooded for as long as 8 months with limited tree cover), to *restinga baixa* (flooded for maybe 4–6 months) and the *restinga alta* (flooded for 2–4 months and seldom to more than 2–3 metres deep). Variation in flooding depth and length of time, flow rates, and vegetation age lead to complex gradients and zonations (Parolin et al. 2004a,b). In seasonally flooded forests, trees commonly drop their leaves during some of the

[8] The Orinoco floodplain covers just over 1 million square km.

Figure 7.5 Exposed ground in seasonal swamp forests reveals little litter or other ground cover, and the stilt roots of a sapling in the foreground at Mamberamo, Papua, Indonesia.

Photo by Douglas Sheil.

flooded period[9], and most foliage possesses thick cuticles. The tree floras are rich in Myrtaceae, Euphorbiaceae, and Fabaceae (Mimosoidae) (Godoy et al. 1999). These forests possess a rich and specialized aquatic fauna, including a large number of frugivorous fish—and dolphins adapted for swimming amongst forest trees.

In Africa, about a third of the Congo basin is a braided mosaic of open water, herbaceous vegetation, and swamp forests. Most are white-water, 'varzea-like' forests, dominating the central region at the confluence of the Likouala, Likouala-aux-herbes, and Sangah rivers, and elsewhere such as the Ogooué Delta. Black-water, 'igapó-like' forests are restricted to north-east Gabon, small patches in Cameroon, and Congo, mainly around Lake

Mai Ndomebe. Swamp forests with more constant water levels are common throughout the region and often dominated by palms (*Phoenix reclinata, Raphia* spp.) and wild oil-palm (*Elaeis guineensis*) (Vande weghe 2004).

Swamp forests in the Niger Delta have largely been lost. Similarly the alluvial plains of Asia once carried extensive swamp forests, but much has been cleared to grow rice. Extensive white-water swamp forests still occur in New Guinea.

7.1.6 Heath forests

Heath forests occur on well-drained, often podzolized sandy soils[10]. Organic matter often accumulates on the soil surface (Moyersoen et al. 2001) and streams are usually acidic and rich with organic compounds. The canopy is usually low with abundant pole-sized trees and saplings. Larger trees and lianas are typically scarce. On the poorest soils the canopy is only a few metres high with frequent open patches; this is often associated with localized peat formation on wet sites in Borneo where it influences vegetation patterns (Miyamoto et al. 2003). Nutrient capture and retention strategies are often highly developed, often including dense, conspicuous root mats (Franco and Dezzeo 1994). Leaves are often waxy and greyish (occasionally red), and are smaller (microphylls predominate) and harder (more sclerophyllous) than in other lowland forests. Epiphytes and ant-plants, or myrmecophytes, can be common. Carnivorous plants often occur in exposed wet sites.

Vertebrates are scarce, but many plants are nonetheless dispersed by animals. In larger heath forest regions some specialist animals are found, such as in Amazonia where 37 bird species are associated with these sandy soils (Borges 2004).

Heath forests represent about 6% of forests in the Brazilian Amazon, where they are called *caatinga Amazonica*[11]. They are found in the upper Rio Negro and Rio Orinoco, and extend into Venezuela and Guyana on sands eroded from the highlands. The trees have a convoluted form and the vegetation includes many habitat endemics. The main forms are *campinara* (closed forest usually around 20 m tall) and *campina*

---

[9] In the *igapó* more trees appear to possess long-lived leaves.

[10] These areas are very slow to recover if cleared, presumably due to their limited nutrients (Pereira et al. 2003).

[11] The term *caatinga* is also used for a distinct semi-arid bushland in north-east Brazil.

Figure 7.6 Heath forest, or *kerangas*, at Bako National Park in Sarawak, Borneo.

Photo by Julia Born.

(low, open and more species-poor with a canopy of 8 m or less [Williams et al. 1996]). The water table is close to the surface and moss, lichens, and ferns can be abundant on the ground. Heath forests occasionally grade into more seasonal habitats or montane vegetations (Dezzeo et al. 1997).

Borneo's extensive heath forests are called *kerangas*[12] (Figure 7.6). In Sarawak these contain at least 948 tree species, less than the nearby evergreen lowland forests but more than the peat swamp forests that share numerous floristic similarities (Brunig 1974, quoted in Newbery 1991). There are small heath areas in Sumatra (especially Bangka Island), peninsular Malaysia, Thailand, and a few coastal sites in Indochina. In Africa, these formations are rare but can be found on coastal sands in the gulf of Guinea, in Gabon, Cameroon, the Ivory Coast, and possibly still on the Madagascar coast.

7.1.7 Peat forests

Peat swamp forests occur in wet environments lacking a marked dry season. Peat can occur on coasts behind mangroves or beach ridges (basinal peat swamps), inland on very nutrient-poor (podzolic) soils with poor drainage, in wet watersheds similar to temperate blanket bogs and in montane environments (Morley 2000). Inland peat formations, such as occur in the localized *igapó* Neotropical black-water catchments and the relatively species-rich *kerapah* in Borneo, tend to be associated with heath forests. There are relatively few vertebrate species, but the avifauna includes some specialists, and seasonal fruiting trees often attract frugivores from neighbouring vegetation (Gaither 1994).

In Asia, most coastal peats are less than 5,000 years old. Interior peats can be older but are still geologically recent (<10,000 yr). Under favourable conditions peat swamps rise and grow as organic material accumulates. The balance between inputs and decomposition creates a gentle but dynamic topography, and large shallow pools are often present (Shimamura and Momose 2005). The ground surface is irregular and pneumatophores may be common.

Asian peat domes can be 100 km across and more than 20 m deep, with distinctive forest types forming concentric bands. The outer forests reach 50 m tall and are a valuable

---

[12] *Kerangas* means land unsuitable for growing rice, in the indigenous Iban language.

timber resource. Soil nutrients generally become scarce towards the centre, where vegetation on the thickest peat is often less than 12 m high with tiny notophyll and microphyll leaves and abundant myrmecophytes and carnivorous herbs. However, 45 m tall forests have been reported on deep peat in Borneo, most likely reflecting an unusual hydrology and nutrient transfer (Page et al. 1999).

Undeveloped tropical peat lands are thought to cover 300,000–450,000 km², mostly in equatorial Asia. Peat forests are found in Borneo, Sumatra, New Guinea (east and west), peninsular Malaya, southern Thailand, and Mindanao. They occur locally in the Americas, in Guyana, Colombia, and in parts of the Amazon basin such as Tupinambara Island. Africa has only minor formations in the Niger Delta and in scattered locations surrounding marshy clearings (rather than in closed forests), associated with 'esobes' or elephant pools in the wetter parts of the western Congo basin (Vande weghe 2004). Agriculture is seldom viable in areas where peat exceeds depths of 1 m deep, but even so these forests have been badly impacted by human activities. After droughts peat forests are especially vulnerable to fire as the peat itself is flammable (Page et al. 2002).

7.1.8 Mangroves

Mangrove forests are sometimes included with rain forests. Mangrove vegetation occurs in a narrow band on many tropical, and some sub-tropical coasts, in estuaries, and in some cases far up tidal rivers[13]. Extensive formations occur where sediments accumulate.

Globally, mangrove vegetation includes 50–110 tree species (depending on definition). Mangrove forests usually possess a low diversity of trees (richest in eastern Asia), and many stands are composed of only 1–3 species, with distinctive communities zoned by substrate age, salinity, and tidal conditions. Understorey and climbers are typically absent, though a few ferns and

climbers may colonize the shore edge. Epiphytes (notably ant plants) and stem parasites can be common.

Establishment and stability on unstable substrate is a challenge for vegetation (Chapter 2). Plants must tolerate regular flooding with brackish or marine water. The tough-leaved trees possess various adaptations, including breathing roots (pneumatophores) and stilt roots (Figure 7.7). Salt tolerance involves different exclusion and excretion mechanisms in different taxa, though few species can endure the increased salinity from evaporation on very shallow tidal swamps.

Mangroves harbour a rich aquatic fauna, and provide important fish spawning grounds, and they often protect vulnerable, low-lying sedimentary land from erosion and storm damage. Mangrove forests have been widely cleared for land, salt works, shrimp ponds, and as a raw material (the dense wood makes good charcoal). Increasing die-back is associated with coastal pollution and declining freshwater inflows in many tropical rivers.

7.1.9 Other rain forest formations

Various minor rain forest types include coastal beach forests and forests over special soils. In some tropical regions the soils are rich in nickel, cobalt, copper, or other metals[14]. Such ultrabasic or ultramafic soils are extensive in New Caledonia, Cuba, parts of Southeast Asia and New Guinea, while distinct but related copper and cobalt rich soils occur in parts of the Congo. In some cases a specialized vegetation has developed (see Chapter 10). These forests, though variable, are often somewhat shorter and more open than are neighbouring forests on more typical soils (Proctor 2003). These formations often include localized species, and thus are often rich in endemics.

Rain forests over limestone are primarily a lowland phenomenon[15]. Most occur in Southeast Asia (including some raised massifs), bordering the Caribbean, and on

---

[13] Not all tropical islands have mangroves. No mangrove species are native to the Hawaiian Archipelago—although they have been introduced.

[14] These areas often represent oceanic crust thrust to the surface by tectonic activity.

[15] Notably in Borneo, China, Vietnam, Thailand, and Madagascar. Caribbean karst exposures, such as occur in the Greater Antilles, are generally too dry to support rain forest, though some local sites in Dominica and Jamaica, with easterly aspects backed by mountains, may carry more luxuriant vegetation. Small disturbed seasonal rain forest formations on karst also occur in Trinidad and Tobago (Eyre 1998). In Madagascar the near impenetrable karst forests are called 'Tsingy', meaning 'only on tip-toe' in Malagasy.

A

B

Figure 7.7 (A) The stilt roots (background) and pneumatophores (foreground), and (B) knee-roots of mangroves.

Photos by Jaboury Ghazoul.

oceanic islands. The soils tend to be rich in nutrients, but dry conditions caused by the permeable limestone geology often result in relatively low canopies. Rock outcrops, steep slopes, and diverse pockets of soils provide a variety of habitats. Lianas and herbs are often abundant. Reliable ground water and deep roots may allow forest trees to grow in seasonally dry regions. In jagged **karst** and sharp **makatea** (uplifted coral reef), terrain access is difficult and water scarce, thus discouraging conversion to agriculture and human settlement.

Gallery forests sometimes only a few trees wide may occur along rivers and drainage lines in drier regions. These often contain species also found in more extensive neighbouring forests. 'Gallery' systems can form extensive and locally significant formations on flood plains[16].

[16] For example, the Tana River in Kenya provides a coastal habitat for numerous highly restricted forest plants and wildlife, including two endemic primates, the Tana River Red Colobus (*Procolobus rufomitratus*) and the Tana River Crested Mangabey (*Cercocebus galeritus galeritus*) (Wahungu et al. 2005).

A

B

Figure 7.8 Inselbergs (A) and inselberg vegetation (B) on the Seychelles. Many species on inselbergs are adapted to very dry and extremely nutrient poor conditions, often subsisting on little more than bare rock.

Photo by Jaboury Ghazoul.

Rain forest **inselbergs** (rock outcrops; Figure 7.8) and the much larger table mountains (**tepuis** or mesas; Figure 7.9) occur mainly in the Guiana Shield of South America and in parts of West Africa, and on the ancient granitic islands of the Seychelles. They present a diverse range of dry micro-habitats. Some possess forest cover (Porembski and Barthlott 2000), and rock pools may occur. The principal ecological gradient is the decreasing soil depth culminating in bare rock (Parmentier et al. 2005). Along this gradient the proportion of species shared with the rain forest decreases. The open rock surface is a harsh environment, but nonetheless **cyano-bacterial** crusts can develop and support a diverse arthropod community (Vaculik et al. 2004). Lichens and mosses can also be prevalent. Annual herbs are often richly represented along with various drought-adapted plants[17], and common vegetation includes grasses (Poaceae), sedges (Cyperaceae), and Velloziaceae. Habitat endemics can be common: 65% of species found on Guiana Shield tepuis occur nowhere else (Berry et al. 1995), and 10% are endemic to the tepuis themselves, including genera of the Rapateaceae, Bromeliaceae, Xyridaceae, Eriocaulaceae, and Cyperaceae families. Inselbergs also share many species with surrounding rain forests and can play a significant role in the ecology of local wildlife.

Figure 7.9 The tepuis of the South American Guiana Highlands. These horizontally bedded, sandstone table mountains rise as high as 3,045 m (Mt Neblina on the Venezuela–Brazil border). The remarkable karst-like form in these quartzitic rocks suggests millions of years of weathering under high precipitation (Doerr 1999). Isolated and island-like, each mountain is home to unusual habitats with many endemics (Steyermark 1979; Williams et al. 1996).

Photo by Valenti Rull.

[17] In total, about 330 species of vascular desiccation-tolerant plants are known. Most are ferns (6 families) and monocotyledons (4 families). A few are dicotyledons (3 families). Nearly 90% of these occur on tropical inselbergs, especially in rain forest climates (Porembski and Barthlott 2000).

Many oceanic islands possess rain forests (Mueller-Dombois 2002). Diversity tends to decline with geographic isolation, small size, and (usually) younger biotas. The isolated Hawaiian rain forests are dominated by only two tree species, *Metrosideros polymorpha* (Myrtaceae) and *Acacia koa* (Mimoseae). Rich formations do occur: a single hectare of forest on tiny La Réunion in the Indian Ocean includes 43 tree species with a diameter over 8 cm (Strasberg et al. 1995). Islands often possess unusual ecologies: Hawaii's forests posses no native ants, amphibians, or conifers, and a bat (the Hawaiian hoary bat, *Lasiurus cinereus semotus*) is Hawaii's single native mammal[18].

At higher latitudes rain forest vegetation may persist in sites where mild oceanic climates allow. A good example is the 35 km^2 Norfolk Island located in the subtropics (29°3′ S) but possessing distinctly tropical vegetation[19].

7.2 Secondary forests

Secondary forests are 'forests regenerating largely through natural processes after significant disturbance of the original forest vegetation at a single point in time or over an extended period, and displaying a major difference in forest structure and/or composition with respect to nearby primary forests on similar sites' (FAO 2003). In practice, the structure and composition of secondary forests gradually, through age and history, grade into formations that are indistinguishable from 'primary' formations—the debate over just where to draw the line makes many definitions vague and hard to apply objectively.

Secondary and degraded forests (affected by logging, fire etc.) amount to around 4.5 million km^2, almost half of all tropical wet forests (FAO 2007). Around 1.2% (235,000 km^2) of tropical humid forests are in some stage of long-term (decadal) regrowth (Asner et al. 2009). They are a major source of forest products, including timber, and provide various goods and services, such as carbon sequestration and biodiversity conservation (see below). Prior land use history influences the nature and rate of regrowth—and age is a crucial factor.

Secondary forests can be natural or human in origin. If we go back far enough, most forests have regrown at some point (Chapter 6). Much of the Central African forests (southern Cameroon, south Central African Republic, Gabon, and Congo) appear to be only 2,500 years old (Maley 2002). Mexico's southern Yucatan forests date from the collapse of the Maya only six or seven centuries ago. European settlement of Puerto Rico was accompanied by near complete forest clearance, though widespread abandonment of agricultural land from the 1940s initiated forest regrowth to 45% forest cover by 2000, more than three-quarters of which is less than 50 years old (Helmer et al. 2008). During this period of regrowth, the mixing of native and alien tree species has generated biogeographically novel forest compositions (Lugo and Helmer 2004). Across the tropics, deliberate modifications through cutting and enrichment, as well as inadvertent impacts, have led to a rich variety of modified forests.

In some conditions, succession may stall or deviate from forest altogether. This is especially likely in environments with regular vegetation fires, which may be converted to grassland. The fire-adapted grass *Imperata cylindrica* is estimated to cover 5 million km^2 of humid land worldwide (notably in Southeast Asia, Africa, and South America); the very similar *I. brasiliensis* is more common in Central America. *Imperata* grasslands are relatively resistant to forest regeneration (MacDonald 2004). A similar pattern explains the *Saccharum spontaneum* grassland around the Panama Canal (Hooper et al. 2005).

Land-use context and history influences regrowth. The nature of any remnant vegetation and the distance to old-growth forests greatly influence plant and animal communities. Remnant vegetation can facilitate recovery by providing a refuge for organisms and by attracting seed dispersers that bring in new colonists.

While biomass and species richness generally increase over the first century of recovery, forests do not necessarily become increasingly similar to more natural old growth forests. Without immigration of shade-tolerant species, pioneer species appear able to

---

[18] Some tropical Pacific islands, like Tahiti, lack even bats.

[19] Vegetation includes lianas, endemic palms, tree-ferns, and even a parrot (*Cyanoramphus novaezelandiae cookii*). The plants include 174 native species, of which 51 are endemic.

perpetuate themselves. *Cavanillesia platanifolia* still comprises 60% of forests that regrew since the Spanish conquest four centuries ago in Darién, Panama (Whitmore 1998). Large areas used to cultivate *Uncaria gambir*, a source of tannin and tonic, in Peninsular Malaysia and in Singapore over a century ago, are still dominated by *Adinandra* or *Ploiarium* (Whitmore 1998b). Such regrowth on extensive areas of degraded lands has no natural equivalent.

7.3   Rain forest landscapes

In the past, rain forests were often considered as more or less homogenous formations. Site level observations and studies were emphasized. A deeper appreciation of scale and landscape is now changing how we view, examine, and manage tropical forests.

Landscapes are more than physical spaces: they are contexts in which species and communities interact and function. Increasingly we see how extent, context, variation, localized features, and spatial configuration are important aspects of forest ecology.

7.3.1 Landscape variation

Many variables influence rain forest landscapes: the most obvious are weather, drainage, and soil. Edaphic variation can be considerable at landscape and local scales, reflecting drainage, geology, and geomorphological processes. Soils of slopes and lower valleys, for example, are generally moister, better structured, and richer in plant nutrients than drought-prone ridge tops.

Mountains provide complex landscapes where geology and topography shape hydrology, exposure, insolation, soil depth, and disturbance regime, while climate, terrain, and vegetation combine to yield local climatic gradients. Unusual conjunctions of forests types may occur, such as fire-adapted (seasonal) montane pine forests alongside wet epiphyte-laden forests in the mountains of Mexico and Guatemala. On taller mountains cold night air sometimes flows downhill, leading to fingers of cold-tolerant vegetation along lowland valley bottoms[20]. In Monteverde, Costa Rica, montane trees taking the full brunt of the oceanic winds are only 15–20 m tall, while the floristically similar 'leeward' forests reach 25–30 m. Soils too are important, for example on Gunung Dulit in Borneo, where heath forest-type formations occur from the lowlands right into the upper montane forest zone (Richards 1996).

In some landscapes inherent dynamics are crucial to understanding vegetation patterns. In swamp forests, drainage patterns, flood periodicity, depth and water-flow rates all contribute to vegetation and habitat variety, but these landscapes are also dynamic, with new channels, sediment banks, oxbow lakes, and terraces regularly forming and old land disappearing (Godoy et al. 1999) (Figure 7.10).

7.3.2 Contexts and boundaries

Forest edges are associated with physical changes in temperature, humidity, and air movements, changes that often extend some distance into the forest. Forest edges often have their own biota: Asian forest rhinos, for example, are generally associated with forest–grassland mosaics. At the same time some mobile 'interior' species appear vulnerable to forest edges and seek to avoid them (Newmark 1991). Edges do not only occur with dry land: rivers, lakes, swamps, shorelines, beaches, and rocky exposures all offer various edge habitats that plants and animals may exploit.

Many species move in and out of forest ecosystems. Some are generalists and can survive in a wide range of habitats, while others are dependent on multiple habitats. For example, in some Central American forests the bees that pollinate many forest trees are dependent on open non-forest areas outside the forests (Gordon et al. 1990). In seasonal environments many species migrate to follow food or other resources.

Context also governs the movement, richness, and abundance of species. Remote oceanic island rain forests are low in species, as there are few opportunities to receive new species from across the sea. Such factors also operate within more localized contexts. For

---

[20] This is why mountain tea estates occasionally suffer localized frost damage as low as 1,000 m.

A

B

Figure 7.10 (A) Meandering river in Mamberamo West Papua. (B) Despite gentle terrain, swamp forests along meandering rivers often form complex dynamic landscapes featuring diverse elements in a context of environmental gradients, histories, and discontinuities.

example, species with traits for desiccation tolerance on exposed inselbergs can be more botanically isolated in wet forests than in drier habitats where savannah species can be exchanged with the surroundings (Burke 2003). This appears to be why in West Africa specialized inselberg vegetation gets richer with decreasing rainfall (Porembski et al. 1996). The vulnerability of species in small forest patches is strongly influenced by their ability to use the surrounding habitat (or 'matrix'): not surprisingly, those that can move and feed in the matrix are more likely to persist—for example, many large forest cats such as tiger, puma,

leopard, and jaguar are wide ranging and also prosper in open habitats.

7.3.3 Local features

There are many small-scale localized forest features that contribute to supporting the local biota. Animal species require a variety of suitable sites to find mates, breed, nest, bury eggs, complete larval stages, and seek dry season food. Animals often travel great distances to feed at mineral licks, drink at open pools, or breed at special sites. Many insects and even some vertebrates require

Figure 7.11 A perennial waterfall in Maliau (Sabah); such sites often possess moisture and light-loving flora that may be scarce or absent elsewhere.

Photo by Karin Beer.

specific types of water bodies or phytotelmata to complete their life cycles. These may need to be still or flowing water, oxygen rich or stagnant, shaded or sunlit, rich in prey or free of competitors. Suitable water-breeding sites may be as varied as streams, pools, wallows, elephant's footprints, tree hollows, tank epiphytes, pitcher plant pitchers, or *Heliconia* flowers.

Some forest features have their own specific ecologies. Many caves possess an endemic biota, while also harbouring species (bats, swiftlets, oilbirds) that interact with the surrounding rain forests. Large epiphytic *Asplenium* ferns can contain a biomass of invertebrates similar to that found in the whole of the rest of the tree crown on which they are growing (Ellwood and Foster 2004).

Forest gaps, caused by tree or branch falls, support many gap specialist plants and animals, if only temporarily as gaps fill quickly. Less ephemeral are waterfalls (Figure 7.11), escarpments, or river banks, where the more open forest structure provides habitat for forest edge or gap-loving species.

7.4 Forest structure

The mountain gorillas *Gorilla beringei beringei* of Bwindi Impenetrable National Park in south-west Uganda are unusual amongst gorillas in the effort they invest in climbing for fruit. These large animals need to solve a basic puzzle each time they wish to reach the fruits held many metres above, as the soaring *Chrysophyllum* trees are too broad-stemmed to climb directly; instead the gorillas must search for robust smaller climbable stems that are close enough to access the lower branches of the larger fruiting trees. Questions related to navigating forest structure—tree stem sizes, densities, and physical properties—are thus a key part of forest life for these gorillas. Diverse challenges of this type confront many species that exploit the forest's unevenly distributed resources.

Our earlier classification of large-scale rain forest formations was concerned with the external environment. Looking more closely, much complexity and variation result from the vegetation structure and changes within it. Rain forest structure can be viewed in many ways. Any basic outline must begin by considering the principal structural elements: the trees.

7.4.1 Stem packing

In almost every old-growth rain forest free of major external disturbances, small trees greatly outnumber large stems. Overall stem densities (stems over 10 cm diameter) are remarkably consistent at 400–700 per hectare, and basal area at 25–45 m^2 per hectare (Table 7.3). This consistency reflects geometrical constraints on canopy packing and the limited space and light available for foliage[21]. The overall spatial patterns of tree stems in most rain forests differ little from random (Lieberman 1994; Sheil and Ducey 2002).

Nevertheless, forests do differ. Montane and young regrowth forests typically lack very large stems, so total stem densities can be high. There are also regional differences: the forests of Southeast Asia often possess higher sapling densities than Neotropical or African forests, which have more understorey shrubs and treelets (Primack and Corlett 2005). The reasons for these differences are uncertain, but perhaps reflect the canopy dominance of the Dipterocarpaceae in Asia, a family that maintains a sizeable seedling bank (LaFrankie et al. 2006). In African forests lianas are more abundant than in other regions—but this may simply reflect climatic conditions and/or past disturbance history.

7.4.2 Canopies, strata, and spaces

The three-dimensional **structure** and arrangement of the rain forest canopy and canopy space provide an important habitat in itself and influence the environment beneath. The vertical structure of the forest has provided fuel for a long-lived controversy concerning characteristic vegetation layers or **strata** within rain forests: are they real or not? One quantified comparative study found continuous trends in vegetation structure by height in a tall Bornean forest, but detected a clear discontinuity

---

[21] Reflecting this packing effect, forests on slopes often possess a higher stem density (for a given stem size class) than more level formations.

Table 7.3 Stem densities and basal area for some intensively studied large plots.

Location	Americas			Asia				Africa		
	Yasuni, Ecuador	BCI, Panama	Luquillo, Puerto Rico, Caribbean	Lambir, Sarawak	Pasoh, Peninsular Malaysia	Palanan, Luzon, Philippines	Sinharaja, Sri Lanka	Korup, Cameroon	Ituri Mono-dominant, DR Congo	Ituri Mixed forest, DR Congo
Plot size (ha)	25	50	16	52	50	16	25	50	2 x 10 ha	2 x 10 ha
Altitude (m)	215–245	120–160	333–428	104–244	70–90	100–180	424–575	300–390	700–850	700–850
Annual rainfall (mm)	3081	2551	3548	2664	1788	3379	5016	5272	1785	1674
Dry months (<100mm)	0	3	0	0	1	0	0	3	3–4	3–4
Principal disturbance	Gaps	Gap	Hurricane	Gap & landslides	Gap	Typhoon	Gaps	Gap	Gap	Gap
Stems/ha >1cm	4581	8214	4171	6119	6094	4124	6707	6581	6843	8112
>10cm	429	677	875	636	702	537	531	492	358	438
>60cm	17	23	11	26	8	28	15	11	33	22
Basal area (m² ha⁻¹) >1cm	32.1	46.1	38.3	43.4	33	39.8	31	32	37.5	33.2
>10cm	27.8	40.1	34.4	37.8	27.3	36.1	25.7	26.1	32.6	26.3
>60cm	10.8	10.7	5.3	13.9	4.1	15.6	7.8	6.8	17.5	11.4
Tree species >1cm	1104	301	138	1182	814	335	205	494	283	259
>10cm	820	227	86	1003	678	262	167	307	168	164
Tree species (ha⁻¹) >1cm	655	169	73	618	495	197	142	236	159	149
>10cm	251	91	42	247	206	100	72	87	53	64

Compiled from data sourced from Losos and Leigh 2004.

between the overstorey and understorey in a Costa Rican forest. Canopy spaces within the shorter Neotropical forest were also smaller[22] (Dial et al. 2004). The generality of these differences and the processes that give rise to them await further study.

Canopy heights are far from uniform, and reflect differences of substrate on which the trees grow as well as topography, history, and local dynamics. At the Pasoh 50 ha plot in Malaysia, for example, canopy height averaged 34.7 m but varied from 11.6 m to 60.9 m. Canopy height differences here were strongly influenced by soil type with heights being generally lower on soils with poor drainage in alluvial or riverine areas.

7.4.3 Microclimates

Compared to neighbouring open habitats, rain forests provide cooler, more humid, darker, and in some respects more varied and dynamic environments. Below the canopy, weather and vegetation interact to cause significant microclimatic variation in both time and space. Patterns vary across forests, but some trends are relatively consistent. During the day canopy foliage often reaches temperatures several degrees above ambient. Temperature declines and humidity increases with depth beneath the canopy. At night the understorey becomes fractionally warmer than the canopy and overall climatic differences are less marked (Figure 7.12).

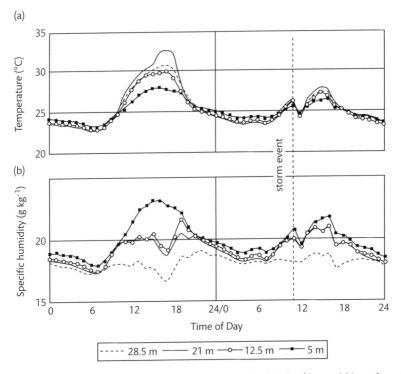

Figure 7.12 Two days of (a) air temperature and (b) humidity recorded at four heights (three within and one just above the forest canopy) in Amazonian Venezuela. The dashed line on the second day indicates a rainstorm. Microclimatic conditions above the canopy track external conditions. During the day, temperatures are lower under the canopy but humidity is higher. At night, the temperature gradient reverses as the canopy radiates heat. Humidity is high and varies little. Rain on the second day reduces temperature and humidity at all levels, but trends are soon re-established.

From Szarzynski and Anhuf 2001.

[22] Such structural differences may also underlie the frequency of gliding ability among Asian forest animals through the forest (See Chapter 4, Box 4.7).

Wind speeds decline beneath the forest canopy (Kruijt et al. 2000; McCay 2003). Vertical movements are associated with turbulence, local convection, and night-time temperature inversions[23]. Significant downward winds are rare (Kruijt et al. 2000). In still air, levels of ambient CO_2 increase in the understorey and, during daylight, decline in the canopy. Foliage and other vegetation elements have significant local effects. For example, transpiring epiphytes provide slightly cooler local conditions amongst tree branches than in epiphyte-free canopies (Schmidt et al. 2001).

7.4.4 Illumination

Canopy
Canopy illumination changes with the sun's position and cloud cover. In the wet equatorial tropics, total monthly sunlight can vary by 50% between cloud-free and overcast months. Further from the equator, day length and solar elevation become seasonal (van Schaik et al. 1993).

Within rain forests there is a strong vertical gradient in light, varying from full daylight above the canopy to sometimes below 1% daylight on the forest floor (Figure 7.13). Predictability also varies: while the upper canopy is guaranteed high light, the shaded understorey may experience occasional periods of strong sunlight as sun-flecks pepper the forest floor, or may be suddenly opened by a tree fall with longer lasting impact on light conditions.

The canopy intercepts most incident light, but patterns vary between forests. Observations in Sumatran dipterocarp forests suggest almost half of incident radiation reaches over one-third of the distance from the canopy top to the forest floor (Torquebiau 1988). In Panama 50% of leaf area can be concentrated in the top 5 m of the canopy (Wirth et al. 2001), where light extinction can reach 94% (Leigh 1999). Similarly, in Congolean *Gilbertiodendron dewevrei* monodominant forest light extinction reaches 95% just below the canopy at 34 m above the ground (Vierling and Wessman 2000).

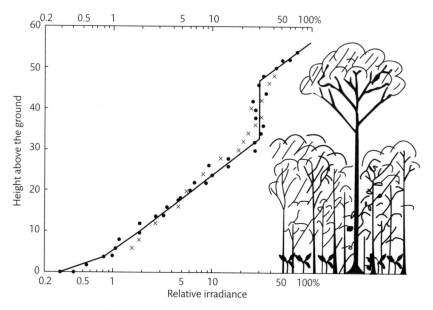

Figure 7.13 Vertical distribution of ambient light at Pasoh forest, Malaysia. Ambient light within the forest declines precipitously below the canopy, and falls to very low levels at the forest floor.
From Yoda 1978.

---

[23] Most fungi release their spores at night to make use of vertical air flows for spore dispersal.

Table 7.4 Leaf Area Index (LAI) and/or percentage light transmission (*T*) to the understorey in various forests.

Location	LAI	*Transmission* (%) and height of detector if above ground	
Evergreen dryland forests	7–9		
Malaysia, Pasoh (lowland)	8.0	0.44	
Sabah, Borneo, Danum (lowland)		1.58	
Gabon (lowland)		0.7 (25 cm)	
Costa Rica, La Selva		1.5 (139 cm), 1.9 (60 cm), 0.9–1.5 (70 cm)	
Sebulu Borneo (lowland)	8.0		
Brazil, near Manaus dry	5.7	1.1	
Singapore		0.6	
Brazil, Rio Negro upland		1.1	
Brazil, Rio Negro white water		1.1–2.2	
Brazil, Rio Negro black water		2.9–9.2	
New Guinea (2,500m)	5.5		
Puerto Rico (elfin forest 1000m)	2.0		
Cuba (Rosaria)	8.5		
Puerto Rico, El Verde		0.4–1.5	
Australia, Queensland (tropical)		0.4–2.0	
Congo (Monodominant *Gilbertiodendron*)		1.2	
Amazonian caatinga (waterlogged sand) (Rio Negro/Orinoco)	4.5–5.0	1–3	
Wallaba (Guyana) white sand		~1.5	
Seasonal		*Wet season*	*Dry season*
India		10	54
Mexico		5–6.4	20–55
Costa Rica (20–60 y secondary)		5–10	30–40
Australia Queensland		8	?
Temperate forests (evergreen)		10–60	
Temperate forests (deciduous)		0.3–4.0	

Compiled from Leigh 1999, Coomes and Grubb 2000; with additions.

Note, a common confusion relates to 'cover' versus 'closure'. *Canopy cover* is a property of an *area*: the proportion of the forest floor covered by a vertical projection of tree crowns. *Canopy closure* is a proportion of the sky covered by canopies when viewed from a point (Jennings et al. 1999).

The higher the density of foliage the lower the light energy that gets through; this concept allows ecologists to use **Leaf Area Index** (LAI), defined as the ratio of total one-sided leaf area per unit horizontal ground area, as a measure of shading. Some LAI and transmission figures are given for various forests in Table 7.4. In dry land lowland rain forests LAIs are often 6–8 (compared to 5–7 in temperate broad-leaved forests) (Leigh 1999). Forests that cast the deepest shade are typically the wettest and tallest, as drier or more stressful environments usually possess more open canopies with more light reaching the forest floor. How the forest canopy influences moisture, nutrient flows and atmospheric processes is considered in Chapter 9.

Light levels below the canopy decrease as an approximately exponential function of total Leaf Area Index. Nonetheless, leaf angles and arrangements mean that tree species composition has a significant influence (Canham et al. 1994; Kabakoff and Chazdon 1996) (Figure 7.14).

As canopy shade increases, there is a simultaneous shift in the ratio of red to far-red wavelengths (660 and 730 nm), a relationship that apparently holds in all forests including secondary formations (Figure 7.15). At La Selva, Costa Rica, the median red/far-red ratio is about 1.2 in the open, 0.99 in understorey sunflecks, and 0.42 in understorey shade. These differences have significance, as the ratio of red to far red wavelengths is used by many plants as a seed germination cue (Chapter 12).

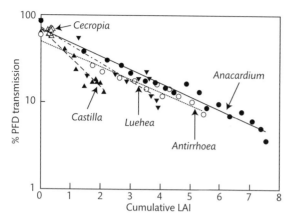

Figure 7.14 Light extinction as a function of cumulative leaf area index (LAI) through crowns of five Neotropical tree species (*Anacardium excelsum, Luehea seemannii, Antirrhoea trichantha, Castilla elastica,* and *Cecropia longipe*) in the Parque Natural Metropolitano, a seasonal tropical forest near Panama City, Republic of Panama. Mean percent PFD (photon flux density—a measure of irradiance) transmission at a given depth from the canopy surface is plotted against cumulative LAI above that depth.

From Kitajima et al. 2005.

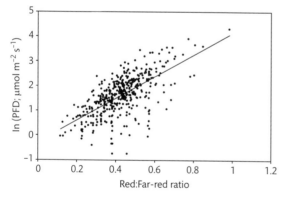

Figure 7.15 Variation in (log) photon flux density at 1.5 m height in secondary forest in Costa Rica is closely correlated to red:far-red ratios.

From Capers and Chazdon 2004.

Understorey

The forest understorey is a twilight world. Limited light reaches the forest floor directly, as most is scattered, transmitted through leaves, or reflected off various surfaces. Irregular canopy cover diversifies light levels below, but understorey vegetation too casts shade. As under-

storeys tend to be densest where canopy cover is least, ground-level illumination seldom correlates closely to tall tree cover (Montgomery and Chazdon 2001; Montgomery 2004).

Temporal variation in the amount of light reaching the forest floor at any particular location spans seconds to decades: from ephemeral sunflecks to daily and even multi-decadal cycles that reflect shifts in community composition and structure. Sunflecks possess considerably more energy than the ambient light levels, and require distinct physiological responses by plants that seek to make use of them (see Chapter 10). They shift as the sun arcs across the sky, lasting only as long as vegetation geometry and weather permit (Chazdon and Fetcher 1984; Benzig 1995; Niinemets and Valladares 2004). Leaf phenology contributes to seasonal variation below the canopy (Wirth et al. 2001) (see Figure 7.16).

7.4.5 Gaps

Gaps in the forest canopy create variation in the environment below. Air temperatures usually become closer to external conditions as gaps get larger (Brown 1993) and diurnal and seasonal changes become more marked (Figure 7.17). Gap soil moisture dynamics also differ from that in closed forest. Less rain is intercepted by tree crowns and therefore more reaches the soil. The density of plant roots may also be lower, but although gaps are wetter after rain, increased exposure to sunshine and wind can promote surface desiccation.

Much research has focused on illumination in relation to canopy gaps. Within larger gaps the microclimate is typically most extreme (i.e. drier and hotter) near the centre and least at the more shaded edges (Lieberman et al. 1989) (see Figure 7.18). Influence on temperature and humidity spreads laterally beyond the gap edge into the forest. Gap shape and orientation also influence conditions: for example light only reaches the ground when gap geometry, sun location, and cloud cover allow (Brown 1993). Due to the interplay of these factors, microclimatic variation tends to be highest in small rather than large gaps, and near gap edges rather than the centres.

Gaps are dynamic and ephemeral (see also Chapter 11). Neighbouring canopies rapidly grow into the open space, and understorey saplings, climbers, and other plants surge up from below. Ground-level illumination is

(a)

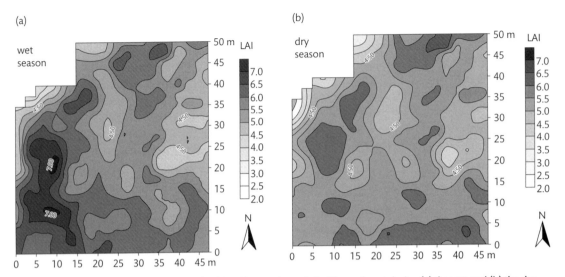

(b)

Figure 7.16 Contour map of Leaf Area Index (LAI) in old-growth semi-deciduous forest during (a) the wet and (b) the dry season (Barro Colorado Island, Panama). Local changes resulted in a 50% or more increase in light reaching the forest floor at 29% of the measuring locations, and a doubling or more at 13% of the locations.

From Wirth et al. 2001.

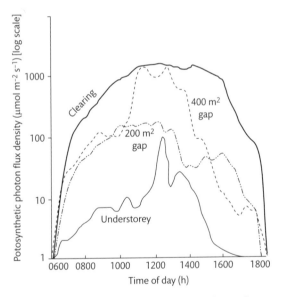

Figure 7.17 Daily patterns of photosynthetic photon flux density (μmol m^{-2} s^{-1}) in a Costa Rican forest during one sunny day in a 0.5 ha clearing, in 400 m^2 and 200 m^2 gaps, and in the closed understorey.

From Chazdon and Fetcher 1984.

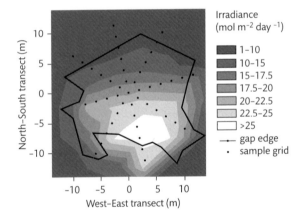

Figure 7.18 Contour plot illustrating sample-based estimates of potential daily irradiance within a 316 m^2 gap, Mabura Hills, Guyana.

From Houter and Pons 2005.

therefore quickly returned to near-understorey levels (Montgomery 2004). Gap formation in the upper canopy is only one cause of variation in understorey light events, which can be altered by branch loss, as well as climber and understorey plant death. Gaps are also heterogeneous and messy, often being cluttered with debris and surviving plants which provide additional layers of light level variation (Farris-Lopez et al. 2004; LaFrankie and Saw 2005). Such structural dynamics and their biological implications are examined further in Chapter 11.

7.5   Conclusion

Tropical rain forests vary greatly in many different respects. They are difficult to classify in neat satis-factory categories. Here we have only touched on the nature of this variation. Differences occur at all scales, both in the physical environment and in the biological communities that result—and variation is often continuous rather than discrete. Nevertheless, some characteristic differences in forest formations can be distinguished and generalized, and these often map to differences in climate, drainage, soils and disturbance history.

Additional aspects of rain forest variation and structure are considered in other chapters. Influences on water and nutrient availability are discussed in Chapter 9 and 10, and the ways in which plants are adapted to forest environments and the manner in which forest dynamics unfold are considered in Chapter 11.

CHAPTER 8

So many species, so many theories

'Thus, the nearer we approach the tropics, the greater the increase in the variety of structure, grace of form, and mixture of colors, as also in perpetual youth and vigour of organic life.'

Alexander von Humboldt (1849)

'What explains the riot?'

Charles Darwin (1850)

The uneven distribution of species richness across the globe was recognized long ago by early explorers and naturalists such as Alexander von Humboldt, who was probably the first to describe the global latitudinal diversity gradient that culminates in the profusion of life's varieties at the equatorial tropics. Why such a 'riot' of life, wondered Charles Darwin, and scientists have been wondering ever since. Why is it that 50 hectares of rain forest in Borneo contain as many tree species as there are in the entire north temperate region? How is it that nearly 300 tree species might be found in a hectare of Amazonian forest (Valencia et al. 1994) when an equivalent area of temperate forest typically possesses less than ten?

The search for explanations—the evolutionary, biogeographic, and ecological processes—that account for such biodiversity patterns is a central quest of tropical forest studies. Essentially, we still seek to answer the basic questions: Why are there so many tropical species, how do they coexist, and why does diversity vary among sites and regions? In this chapter we examine these patterns and some of the many explanations given for them. We mostly focus on plants, especially trees, as these have been the focus of especially intensive studies, but we also include examples from some animal groups.

While tropical rain forests are renowned for high species richness, this is neither uniformly distributed nor is it always the case (Figure 8.1, Table 8.1, Boxes 8.1, 8.2). Regions vary in the number of species they support. The Neotropics hold more species in most groups than the Asian tropics, with the African rainforests generally being species poor—at least in comparative terms. Some regions are sufficiently distinct to warrant special mention, not necessarily because of their richness, but rather due to the unique nature of their communities. The large islands of Madagascar and Australia–New Guinea stand out as having a disproportionate number of endemic taxa, resulting in distinctive rain forests (Chapter 6). Comparative measures of richness are additionally dependent on the scale of the observation: species richness within a plot in central Amazonia will, for example, often be much higher than an equivalent plot in Panama, but the cumulative richness of many such plots increases more rapidly in the Panamanian forest. This has obvious relevance to conservation and the distribution and location of protected areas. It also raises questions about the causes of these differences, an issue that is explored in further detail towards the end of this chapter.

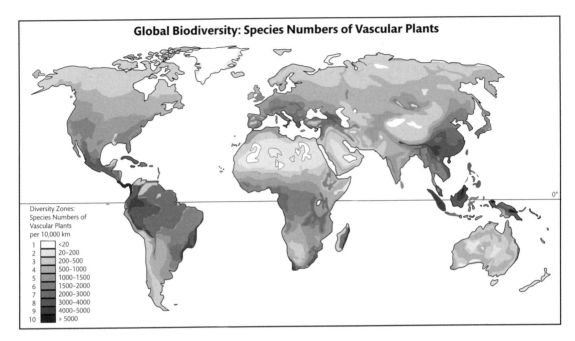

Figure 8.1 World map of species richness of vascular plants.

From Mutke and Barthlott 2005.

Table 8.1 Estimated numbers of plant species in the tropical rain forests of the world.

Area	Number of species	Rainforest (10⁶ ha)	Comments
Neotropics	93,500	400	Mainly Amazon basin; also includes some other forest types
African tropics	20,000	180	
African mainland	16,000		Includes West Africa, Congo region, and montane areas
Madagascar	4,000		A preliminary estimate; the flora is not well known
Asia-Pacific region	61,700	250	
Malesia	45,000		Malaysia and Indonesia, including New Guinea (separation precluded by the mixing of the floras)
Indochina and adjacent areas	10,000		An estimate as no flora has been completed for the region
Southern India	4,000		Mainly the Western Ghats
Sri Lanka	1,000		Sinharja and neighbouring localities
Australia	700		Queensland rain forest
Pacific Islands	1,000		Isolated rain forests with low species richness
Total	175,200		

Adapted from Primack and Corlett 2005.

Box 8.1   Elevational diversity gradients

Altitudinal gradients are often considered to be analogous to latitudinal gradients, but species diversity often peaks at mid-elevations (Rahbek 1995, 1997). This appears true of palms in New Guinea (Bachman et al. 2004), bryophytes and vascular plant groups in the tropical Andes (Kessler 2000, 2001; Kromer et al. 2005), epiphytes in Chiapas, Mexico (Wolf and Alejandro 2003), and Costa Rica (Cardelus et al. 2006), trees (Lieberman et al. 1996) and ferns (Watkins et al. 2006) in Costa Rica. Similar patterns have been noted for small non-**volant** mammals in the Philippines (Heaney 2001), Malaysia (Md. Nor 2001), Mexico (Sanchez-Cordero 2001), and Costa Rica (McCain 2004). Insects, however, show no clear pattern (McCoy 1990; Lees et al. 1999; Brehm et al. 2003; Novotny et al. 2005).

Rainfall, soil moisture, and evapotranspiration peak at intermediate elevations (although actual patterns are more complex than this might suggest) offering an explanation for high mid-elevation species richness through primary productivity (Brown 2001; Heaney 2001; Sanchez-Cordero 2001). There have been few direct tests of such an explanation. Although the mid-elevation richness peak is a widely observed phenomenon, the actual elevations of richness maxima vary among taxa to a degree that no single explanation seems universally plausible (Herzog et al. 2005). There are exceptions though. Bats, for example, show a continuous diversity decline with increasing altitude (Heaney 2001; Sanchez-Cordero 2001), and other groups even show mid-elevation troughs in species richness (Patterson et al. 1998).

A further difference between latitudinal and altitudinal gradients is the continued dominance, despite declining temperatures, of broad-leaved evergreen trees along tropical elevation gradients (whereas deciduous trees dominate at temperate latitudes). Low tropical climatic seasonality may explain this difference (Takyu et al. 2005).

8.1   Restricted range species and widespread generalists

8.1.1 Restricted range species

In addition to high species richness, many tropical areas have exceptionally high endemism, a measure of biological uniqueness. Endemics are defined as species (or taxa) restricted to particular areas. Such a definition can be applied in the context of a geopolitical region (e.g. Mexico) or a geographical unit (e.g. New Guinea, Mt Cameroon, the Congo Basin, or the Guiana Shield). Definitions of endemism based on political boundaries have value for states seeking to identify national conservation priorities, but are less useful to ecologists seeking to explore biological uniqueness across regions. More useful is the designation of 'restricted range species' which are those that occur entirely within a predetermined areal extent (e.g. species ranges less than 50,000 km²). Uniqueness is a function of scale, and varies depending on the designation of what constitutes a restricted range.

Endemism is not, of course, synonymous with diversity. A broad global correlation between richness and endemism has been reported for plant families (Williams et al. 1994) and mammalian species (Ceballos and Brown 1995), but the locations of areas with the highest values of endemism and richness are often distinct (Williams and Gaston 1994; Williams et al. 1994). For plant families, important endemic centres are New Caledonia, southwest China, the Guianas (Guyana, Suriname, and French Guiana), Madagascar and Costa Rica, as well as some non-tropical locations such as South Africa and central Chile (Williams et al. 1994). Mammalian endemism is highest in Australia, Indonesia, Madagascar, Mexico, New Guinea, and the Philippines (Ceballos and Brown 1995).

'Island' habitats such as mountain tops are generally species poor but often possess a high proportion of localized endemics, as isolation allows for speciation

---

Box 8.2 Low diversity and mondominant rain forests

Not all rain forests conform to the stereotype of being rich in tree species. Rain forests of oceanic islands are typically quite low in species. Swamp forests, in all regions, are often dominated by single species groves—a condition termed **monodominance**, e.g. *Mitragyna stipulosa* in Africa and *Mauritia flexuosa* palms in Amazonia. The peat swamp forests of western Borneo are frequently dominated by *Shorea albida* (Dipterocarpaceae), while lower montane forests in Papua and New Caledonia include dense cover of *Nothofagus* (Fagaceae), and similarly some wet mountains in Africa, patches in the Western Amazon, and indeed elsewhere are dominated by one species of bamboo.

Young secondary regrowth forests too are often dominated by only a handful of species. Mangroves are low in tree diversity. Forests in drier climates are low in species, some such as the *Cordia elaeagnoides* (Boraginaceae) and *Celaenodendron mexicanum* (Euphorbiaceae) forests in Mexico being dominated by very few species. Even in wet dryland forests, local dominance by one species is not uncommon, e.g. *Pentaclethra macroloba* a mimosoid tree dominates in parts of Costa Rica, *Parinari excelsa* (Chrysobalanaceae) in African montane forests, *Eusideroxylon zwageri* (Lauraceae) patches in lowland equatorial Asia.

One group of low diversity rain forests often considered together are those dominated by large caesalpinoid (legume) trees in African and Neotropical forests. Such forests occur mainly in wet but markedly seasonal regions. These include the *Gilbertiodendron dewevrei* dominated forest in Ituri, Congo, *Cynometra alexandri* forests in Uganda and Congo, *Dicymbe corymbosa* and *Mora* spp. as well as the Wallabas (*Eperua falcata* and *E. grandiflora*) in Guyana, and *Peltogyne gracilipes* in the Brazilian Amazon (Connell and Lowman 1989; Hart et al. 1989; Torti et al. 2001; Henkel et al. 2005). Some would also include *Julbernardia seretii* in Central Africa and *Talbotiella gentii* in drier forests in Ghana, West Africa and *Prioria copaifera* in Central America, along with the local groves of *Microberlinia bisulcata* and *Tetraberlinia* spp. in Cameroon.

Many reasons have been proposed for the occurrence of such monodominant forests (Connell and Lowman 1989; Hart et al. 1989) (see also Box 3.2) but as yet no single explanation seems generally accepted (Torti et al. 2001; Henkel et al. 2005; Mayor and Henkel 2006). The cause is not simply climate soil or drainage, as low diversity forests often occur in close proximity to richer forests on the same substrate (Hart et al. 1989). These forests do tend to occur in areas of ancient continuous forest cover without known catastrophic disturbances, suggesting that these sites may simply be the result of low disturbance and competitive exclusion (Connell and Lowman 1989; Sheil and Burslem 2003). Then monodominance is simply a result of the slow competitive processes that exclude all but the most durable species. This implies that the dominant species are competitively superior but very slow to disperse. In any case, explanations for these areas of such relative species poverty, as well as those of remarkable richness, must be part of any comprehensive explanation of variations in community richness (Connell and Lowman 1989).

---

and low immigration encourages uniqueness (Chapter 2). The slopes of Mt Kinabalu in northern Borneo represent one of the richest concentrations of endemic plant species in the world, and possibly the world's greatest concentration of orchids: more than 750 species in over 60 genera of which, at elevations above 2,700 m, around 90% are endemic to the mountain (Wood et al. 1993). Several species of *Rhododendron*, *Lithocarpus*, *Magnolia*, *Rhamnus*, *Ficus* and a variety of pitcher plants, including the spectacular *Nepenthes lowii*, also occur only on Mt Kinabalu (Cockburn 1978).

The isolated high plateaux of the tepuis[1] on the Guiana Shield in South America are similarly rich in plant species (Chapter 7). The endemic flora evolved over millions of years since the tepuis highlands were initially dissected and isolated from each other approximately 90 Ma (Berry et al. 1995). In contrast Mt Kinabalu is young (*c*.1.5 Ma) and its endemics evolved more recently from lower elevation ancestors (Barkman and Simpson 2001).

On larger scales, around half of the estimated 30,000 vascular plants of the tropical Andes are endemic to that area, with orchids again (as in Kinabalu) being particularly well represented (Jørgensen and León-Yánez 1999; Kessler 2000, 2002). Animal endemics are similarly abundant in mountain regions: the moist forests of the Andes harbour the highest bird diversity and endemism of any place on Earth, and even higher levels of amphibians and reptile endemism (Table 8.2).

Real islands are often centres of endemism (Chapter 2). Particularly high numbers of endemic plants are found on continental islands such as Madagascar, New Caledonia, and the Seychelles, presumably owing to divergence of an already rich continental biota from mainland taxa. Endemism on island archipelagos has been shown to correlate with island isolation, age, and elevation range for birds, mammals, and plants (Adler 1994; Carvajal and Adler 2005; Ricklefs and Bermingham 2007), but as yet there is no robust empirical model that can predict the proportion or number of endemic taxa by these variables (and others such as latitude and productivity) across tropical islands.

8.1.2 Widespread species

Some species, especially plants, are widely distributed (even though they may have narrow habitat requirements) and show little morphological divergence across very broad continental or even pantropical ranges (Pitman et al. 1999; Dick et al. 2003b). Examples include *Ceiba pentandra* (Malvaceae), a tall pioneer tree, and *Symphonia globulifera* (Clusiaceae), a shade-tolerant slow-growing medium-sized tree, both of which have native ranges that encompass Mesoamerica, the Caribbean, South America, and tropical Africa. In Amazonia many trees, and often the majority at any one site, belong to species with wide distributions (Pitman et al. 1999). It is this that gives Amazonia low beta-diversity. Molecular studies indicate that some widespread species are, actually, cryptic[2] sister species and that considerable genetic heterogeneity is masked by morphological similarity (Box 8.3).

8.2   Alpha- and beta-diversity

Different terms are used to distinguish the scales at which diversity is evaluated: **alpha-diversity** for point or within-plot or habitat samples; **beta-diversity** for the rate at which species accumulate as observations are

Table 8.2 Endemism among plants and vertebrate groups in the Andes

Taxonomic group	Species	Endemic species	% endemism
Plants	30,000	15,000	50.0
Mammals	570	75	13.2
Birds	1,724	579	33.6
Reptiles	610	275	45.1
Amphibians	981	673	68.6
Freshwater fishes	380	131	34.5

Conservation International 2007

---

[1] These are the characteristically flat-topped mountain summits of the Guyana Highlands, lying at altitudes of between 1,500 and 3,000 m, that are collectively termed the Pantepui.
[2] That is, not detected by morphologically-based characters.

Box 8.3 Are widespread species really widespread 'species'?

Why have widely distributed species not diverged across their range? Molecular approaches show that many widespread species are, in fact, **cryptic species** with separate evolutionary origins. The distribution of *Trema orientalis*[1], for example, spans Africa and Asia, but molecular analysis reveals that individuals from Africa are probably derived from the African sister species *T. africana*, while those from Asia are more closely related to *T. aspera* from Australia (Yesson et al. 2004). Similarly, *Symphonia globulifera* has considerable cryptic genetic divergence among its lineages, probably reflecting multiple Neotropical colonization events from Africa (Dick and Heuertz 2008). More surprisingly, substantial genetic differentiation among Mesoamerican populations contrasts with almost none among Amazonian populations (Dick et al. 2003a; Dick and Heuertz 2008). The patterns of genetic differentiation (high in Mesoamerica, low in Amazonia) reflect patterns of low species beta-diversity among widely separated Amazonian locations, compared to much closer Panamanian plots (Condit

et al. 2002). This is unexpected given the long fossil record of *S. globulifera* in Amazonia, and therefore the long period of time over which differentiation might have occurred. Habitat differences that might enhance divergence through isolation do exist across the Amazonian Basin, but they are relatively small compared to the mountain ranges that dissect the Mesoamerican range of *S. globulifera*.

Other examples of cryptic divergence include the tall pioneer tree *Ceiba pentandra*, which is naturally common over much of equatorial Africa and the Neotropics and shows little variation across its range. African populations of *C. pentandra* do have different numbers of chromosomes, a characteristic referred to as 'ploidy' (Dick et al. 2007). Such patterns are also found in other life-forms: genetic characterization suggests the widespread epiphytic fern *Asplenium nidus* is in fact a complex of several related species based on differences in chromosome number (ploidy varies from diploid to 16-ploid) (Murakami and Yatabe 2001).

[1] *Trema* is a pantropical genus of around 15 species of small pioneer trees from the family Ulmaceae.

---

made over larger landscape or regional scales; and **gamma diversity** to describe the total regional species richness[3] (Whittaker 1975).

Most lowland rain forests have high alpha-diversity, with values as high as 300 tree species (stems over 10 cm diameter) or over 1,000 vascular plant species per hectare in some tropical rainforests (Gentry 1988b; Valencia et al. 1994). Given these values for a single hectare, it is surprising to learn that the entire Amazonian lowlands of Ecuador, covering about 7 million hectares, contain around 4,000 vascular plant species, just four times as

many as are found in a single hectare (Condit et al. 2002). Thus while alpha-diversity is often extremely high, regional diversity values (gamma-diversity) are not much greater, suggesting little turnover of species (beta-diversity) across large areas of the western Amazon. Panama, and Central America in general, has comparatively low alpha-diversity but much greater richness at larger scales owing to high species turnover (beta-diversity), something that is also true for the inundated Amazonian *igapó* (black-water) forests (Duivenvoorden 1996; Fine et al. 2005) (Box 8.4).

[3] These terms can be confusing as they are applied variously at different scales. Thus there can be substantial variation in soil and topography within a single hectare, causing fine-scale floristic patterns which can be interpreted as either alpha- or beta-diversity depending on the scale of observation (Poulsen et al. 2006). Some authors do not even distinguish between beta- (variation within a region or gradient) and gamma- (variation amongst distinct regions or gradients) diversity measures, and instead refer only to beta-diversity measured at different scales.

Box 8.4   What explains beta-diversity? Dispersal vs. environmental models

One-hectare plots in the western Amazon have 2 to 10 times as many woody plant species as do one-hectare plots in Panama, but both regions support similar number of species (around 3,500–5,000) (Condit et al. 2002). Larger differences in species composition among Panamanian plots give steeper species accumulation curve with sequential plot sampling (higher beta-diversity) than in Peru or Ecuador (Figure 8.2). These observations have been explained on the basis of environmental heterogeneity (i.e. niche specialization) and dispersal limitation. Dispersal limitation partially accounts for community differences among plots separated by 0.2 to 50 km, but is not effective in explaining the often considerable dissimilarity among plots separated by less than 100 m (Condit et al. 2002). Significant predictors of species similarity at this scale include four environmental variables: elevation, precipitation, age of the forest stand, and geology, demonstrating the importance of habitat heterogeneity at local scales in shaping species composition (Duivenvoorden et al. 2002). Even so, 59% of the variation in species among Panamanian plots could not be explained by either distance or environment. Biogeographic explanations relating to Panama's complex geological history[2] may account for some of this variation.

Other studies have reiterated the importance of both dispersal processes and environmental variability in determining beta-diversity among tropical plants. Surveys of terrestrial ferns and Melastomataceae, a family of shrubs and small trees, at 163 sites in four Amazonian regions (Colombia, Ecuador, northern and southern Peru) found that differences among sites in these two plant groups were correlated with both geographic distance (indicating community differentiation by dispersal limitation) and environmental distance (i.e. habitat heterogeneity) (Tuomisto et al. 2003). Ferns, with their wind-dispersed spores, are less likely to be dispersal-limited than Melastomataceae, whose berries are mostly dispersed by birds. This is, indeed, reflected by the greater importance of habitat heterogeneity in structuring fern communities compared to the Melastomataceae, where dispersal limitation and habitat heterogeneity had near equal importance (Tuomisto et al. 2003).

Though probably sensitive to scale and context, dispersal limitation has been found to be more important than environmental heterogeneity in explaining the distribution of herbs and trees in Panama (Svenning et al. 2004, 2006; Chust et al. 2006) and palms in Ecuador and Peru (Vormisto et al. 2004; Normand et al. 2006). Normand et al. (2006) emphasize the importance of scale and the predominance of environmental factors, particularly soil moisture, in explaining beta-diversity of palms at local scales (25 m^2 grain size and 0–500 m in extent), in contrast to dispersal (distance) factors at regional scales (500 m^2 grain size and 0.3–143 km extent), reiterating the conclusion of Condit et al. (2002). Environmental heterogeneity (particularly differences in soil nutrient content, drainage and canopy openness) is more relevant than dispersal for ferns in Costa Rica (Jones et al. 2006), and trees in Borneo and south-west Amazonia (Phillips et al. 2003; Slik et al. 2003; Paoli et al. 2006).

Dispersal limitation may be more important in structuring juvenile tree communities, but as trees mature, niche-related processes become increasingly important, such that the distribution of mature trees is determined more by environmental heterogeneity (Paoli et al. 2006).

[2] Prior to the formation of a land bridge, Panama consisted of an archipelago that may have allowed the divergence of populations which, together with the subsequent mixing of independently derived plant communities on formation of the land bridge (Dick et al. 2003a), may explain much of the remaining variation. The complexity of this example underscores our limited understanding of the processes that determine tropical forest diversity.

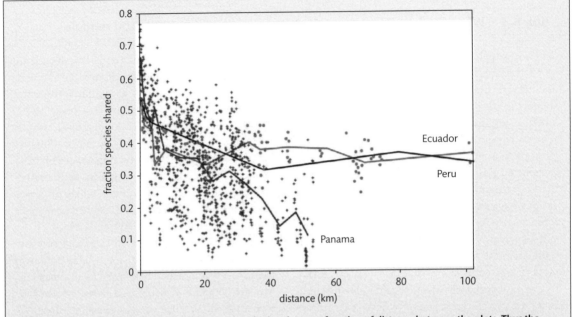

Figure 8.2 Sorensen similarity index between pairs of 1-ha plots as a function of distance between the plots. Thus the fraction of species shared among plots declines with the distance between plots more rapidly in Panama than at either of the western Amazonian locations, indicating higher beta-diversity.

From Condit et al. 2002.

8.2.1 Environmental factors underlying alpha-diversity

Species-energy theory (Wright 1983; Cousins 1989; Wright et al. 1993) suggests that the energy available to communities (generally represented as potential eva-potranspiration) limits the number of species they can contain. More energy supports more biomass and allows more individuals of each species to be sustained, hence larger and more viable populations. The theory is broadly supported outside the tropics, with current climatic conditions (a surrogate for available energy) being more consistently related to species diversity than many other potential explanatory variables (Currie 1991; Francis and Currie 2003; Hawkins et al. 2003). Most studies that explore energy-diversity relationships do not, however, extend into the tropics and so, despite the strength of the relationship at mid-latitudes, it remains unclear whether it retains its explanatory power in the tropics.

Positive relationships between local species richness and surrogates of primary productivity (such as rainfall) have been described for several tropical forest regions (Gentry 1988a; Phillips et al. 1994; Clinebell et al. 1995; Swaine 1996). Amazonian species diversity, for example, follows a rainfall gradient that increases from east to west (Figure 8.3). The strength of this correlation, and indeed the explanatory value of annual rainfall, has been questioned as some sites in central Amazonia have species richness that at least equal those of western Amazonia (ter Steege et al. 2000)—although sites on the Guiana Shield and eastern Amazonia are indeed consistently poorer. Species richness has also been positively (Gentry 1988a; Costa 2006; DeWalt et al. 2006) and negatively correlated with soil fertility (Huston 1980; Swaine 1996). Rainfall and soil fertility are, however, interrelated: high rainfall decreases soil fertility through leaching (Huston 1980, 1994; Swaine 1996), but minimizes drought stress allowing for higher primary productivity (DeWalt et al. 2006). Alternatively, high rainfall is known to favour natural plant enemies (especially fungi) through which density-dependent mortality favours rare species thus favouring diversity (Givnish 1999).

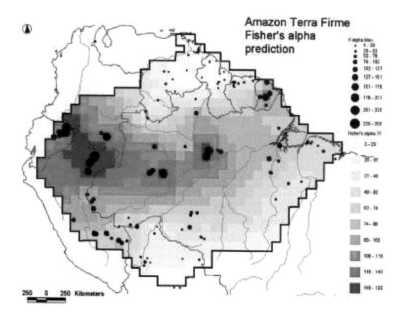

Figure 8.3 Amazonian average tree diversity (Fischer α-diversity statistic) based on 275 plots in *terra firma* forest in the Amazon and Guiana Shield rain forests. Diversity is highest in a narrow band from western Amazonia to central Amazonia, which broadly correlates with the rainfall gradient.

From ter Steege et al. 2003.

Correlative patterns are not shared across tropical continents. While dry season length is clearly linked to alpha-diversity in Amazonia, there is little apparent association in African forests (Parmentier et al. 2007) (Figure 8.4), the explanation probably being related to Africa's more arid climatic history (Chapter 6).

8.2.2 Environmental variability and beta-diversity

It seems intuitive that regions encompassing many heterogeneous environments will support greater species variability across their range (beta-diversity) as new communities adapted to different environmental conditions are encountered. Over relatively small distances (hundreds of metres to a few kilometres) edaphic, topographic, and climatic differences appear most important in determining beta-diversity (Tuomisto and Ruokolainen 2006) (Box 8.4). Thus floristically rich areas are often associated with steep topographical gradients and diverse geological, edaphic, and climatic conditions. They include the Pacific Coast, Choco region of Colombia and Ecuador, Mesoamerica, the tropical eastern Andes, Brazil's Atlantic forests, the Western Ghats in India, the

Eastern Arc Mountains of Tanzania, Mt Kinabalu in northern Borneo, as well as the large islands of Madagascar and New Guinea.

Explaining regional diversity patterns of animal taxa is more difficult than for plants, because animal taxonomic groups often respond quite differently within the same contexts. For example, the diversity of frogs in tropical Australia is explained by either ecological or biogeographic factors depending on the specific group concerned (Williams and Pearson 1997; Williams and Hero 2001). Species richness of non-microhylid rain forest frogs is thought to be the outcome of historical biogeography related to quaternery fluctuations in the size of the rain forest areas. Richness of microhylid frogs appears instead to be due to speciation driven by altitudinal gradients and localized isolation resulting from poor dispersal abilities. Patterns of species richness, and causes that underlie these patterns, may therefore vary among functional groups within broadly defined taxa (Williams and Hero 2001).

Species richness of Neotropical butterflies has been explained by environmental heterogeneity arising from disturbances (Brown 2005). Such disturbances include landslides, shifts in river courses, and flooding events. Rapid turnover in successional habitats over relatively small geographic scales best explains the distribution of butterfly species richness patterns across tropical South America.

8.3 Why are there more species in the tropics? Global and regional perspectives

At the broadest scale, species diversity of most plant and animal groups increases dramatically from polar to tropical regions. Such latitudinal diversity gradients[4] have been described for many taxa including angiosperms,

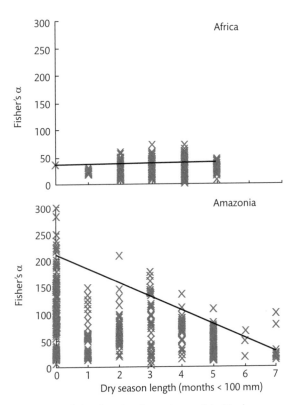

Figure 8.4 Alpha-diversity of trees is correlated to dry season length in Amazonia, but not African rainforests.

From Parmentier et al. 2007.

such as palms (Bjorholm et al. 2005), invertebrates, such as ants and butterflies, as well as mammals, birds, and others (Hillebrand 2004). Terrestrial latitudinal diversity gradients are paralleled by similar gradients among marine taxa, and the most parsimonious theories for latitudinal diversity gradients would need to account for both marine and terrestrial patterns, but such consideration lies beyond the scope of this book. The various models advanced to describe terrestrial patterns provide enough scope for debate.

Explanations for terrestrial species diversity fall into two main groups: biogeographic and evolutionary hypotheses that emphasize speciation (Table 8.3) and extinction; and ecological explanations that emphasize species coexistence (Mittelbach et al. 2007). Some models have been developed from a global perspective, others from a more regional perspective. None are necessarily exclusive.

8.3.1 Evolutionary and biogeographic explanations of species diversity

Environmental stability and geographical area
Alfred Russel Wallace (1878) was the first to suggest that the great diversity of tropical vegetation was a result of the slow accumulation of species over the aeons of relatively stable tropical rain forest existence. This idea, elaborated as the 'museum hypothesis', proposes that low extinction rates in climatically stable tropical environments allowed the persistence of species and accumulation of newly evolved forms which together account for species richness (Stebbins 1974). Lending support to this hypothesis is the observation that species richness of terrestrial vertebrates and vascular plants is negatively correlated with the amplitude of past climatic change, over time scales of 10–100 thousand years[5] (Jansson 2003). Further evidence is derived from phylogenetic relationships of temperate and tropical taxa. Angiosperms arrived suddenly and diversified in the tropics and spread to cooler regions during the Cretaceous (Chapter 6), and many temperate species owe their origins to still diverse

---

[4] Emphasizing latitudinal species richness gradients has been criticized for reducing the study of large-scale patterns to one dimension, thereby limiting opportunities for more comprehensive spatial understanding (Hawkins 2004).
[5] The time-scale of the Pleistocene glaciations.

Table 8.3 Major models of evolutionary processes that promote diversification of rainforest faunas.

Name	Geographic mode	Isolating barrier	Evolutionary mechanism	Main authors
Species-pump: forest refugia model	Allopatric	Dry forests or savannah formation during dry periods associated with Pleistocene glaciations	Isolation, drift, selection	(Haffer 1969, 1974, 1997a; Diamond and Hamilton 1980)
Species-pump: geological instability model	Parapatric or allopatric	Cycles of changing environmental conditions and gradients	Isolation, divergent selection	(Graham et al. 2006)
Paleogeography hypothesis	Allopatric	Continental seas, plateaux, flooded plains	Isolation, drift, directional selection	(Emsley 1965)
Riverine model	Allopatric	Major rivers	Isolation, drift	(Wallace 1852; Herschkovitz 1977)
Gradient model	Parapatric or allopatric	None necessary, but steep environmental gradients invoked	Divergent selection, with or without gene flow	(Endler 1977; Smith et al. 1997)

Adapted from Moritz et al. 2000 and Haffer 1997a.

tropical lineages (Jablonski et al. 2006). Thus tropical lineages have had more time to diversify (Ricklefs 2005). Making the transition from the tropics to temperate zones was likely to have been impeded by the need to evolve tolerance to cold or freezing conditions (Wiens and Donoghue 2004; Hawkins et al. 2006).

Furthermore, larger areas generally support more species[6]. The larger land surface area of the tropics[7] can therefore potentially support a greater total number of accumulated species (Terborgh 1973; Rosenzweig 1995). Thus both large area and long duration of tropical rain forests may have facilitated the evolution and accumulation of more species than could have evolved or persisted elsewhere. Indeed, tree species richness correlates strongly with biome area only when area is integrated over time periods stretching to the Eocene (55 Ma) (Fine and Ree 2006).

Speciation speed

The warmer tropics may drive higher speciation rates, in that temperature accelerates biological processes (Rohde 1992). This hypothesis argues that temperature increases mutation rates and shortens generation times, which leads to faster rates of population divergence and higher speciation rates. While the rate of molecular evolution of warm-bodied, fast developing endotherms is higher than 'cold-blooded' ectotherms (Martin and Palumbi 1993), minimal variation in body temperature within endothermic taxa (mammals and birds) stands at odds with the pronounced latitudinal diversity gradients that they too exhibit (Hillebrand 2004).

Environmental instability: species pumps and refugia in the Amazon

It has long been argued that areas of high species diversity and endemism of Amazonian birds, butterflies, primates, lizards, flowering plants, and various other taxa coincide (Haffer 1969; Mallet and Turner 1998; van der Hammen and Hooghiemstra 2000), and that these endemic-rich regions reflect the locations of wet forest refugia that became isolated within a matrix of dry savannah during repeated Pleistocene (glacial) cycles of Amazonian forest contraction (Haffer 1969; Simpson and Haffer 1978; Bonaccorso et al. 2006). Rather than just a series of 'museums' that allowed for species persistence, it was hypothesized that these geographically isolated refugia facilitated allopatric speciation (Haffer 1969, 1997a). Repeated cycles of contraction and expansion were therefore proposed to have driven a 'species-pump' with the consequent accumulation of species in the tropics.

While initially attractive, the Pleistocene refuge species-pump hypothesis is now considered untenable for Amazonia, largely owing to doubts that forest refugia existed (Box 8.5). In many cases it can also be shown that taxonomic groups evolved long before Pleistocene climate changes: fossil evidence indicates that many modern plant genera were already present in tropical South America in the early Eocene (at least 52 Ma) and that Amazonian plant species richness comparable to the modern era far precedes Pleistocene glacial cycles (Wilf et al. 2003). Molecular methods have also been used to date species radiations to well before the Pleistocene, and there is no molecular signal for a Pleistocene speciation burst (Moritz et al. 2000). A molecular phylogeny of *Inga*, a genus of predominantly small rain forest trees, reveals that the 300 or so species evolved between 1.8 and 7.2 Ma (Richardson et al. 2001). Geological events (e.g. uplift of the Andes and formation of the isthmus of Panama [Chapter 6]) are more congruent explanations of speciation among *Inga* than Pleistocene refuges. In addition to these fundamental criticisms, the spatial overlap of endemicity centres of unrelated taxa has been challenged (Knapp and Mallet 2003), and species richness

---

[6] Larger areas offer more opportunities for speciation by being more likely to encompass barriers to gene flow, and by being able to support larger populations with correspondingly greater genetic variation. Larger areas can also support larger population sizes that reduce extinction likelihood, and are more likely to contain refugia for species during periods of unfavourable environmental change.

[7] The tropics also have larger climatically similar areas than other parts of the globe, due to the declining surface area of latitudinal bands approaching the poles, a relatively constant mean temperature between 20°N and 20°S, and because regions of similar climate on either side of the equator abut.

Box 8.5 Amazonian refugia in the Pleistocene: a beautiful theory killed by ugly facts[3]

While glacial cycles ravaged temperate regions during the Pleistocene, the effects on tropical regions have been a subject of debate. A prevailing paradigm of recent decades is that periods of glaciation in high latitudes were paralleled in the tropics by periods of aridity that caused tropical wet forests to contract to wet forest refugia surrounded by dry savannahs. A number of hypotheses seeking to explain high tropical species richness have been founded on this paradigm (see main text for details). But does the evidence for such cycles of forest contraction and expansion stand up to scrutiny? For Africa and Australia there is widespread agreement that it does, but for the Amazon the long-running debate has only recently been resolved.

Two main lines of evidence[4] from palynology and geomorphology have been scrutinized with respect to Amazonian Pleistocene refugia. Pollen cores from Carajas, Lake Pata, and marine deposits off the mouth of the Amazon River, suggest that central Amazonian rainforests remained intact during the Pleistocene[5]. Grass pollen from cores need not imply aridity, but could be explained by seasonal river flood formations (Bush 1994; Colinvaux et al. 2000).

Geomorphological evidence arrayed in support of refugia includes fossil dune systems at multiple locations along the edges of the Amazonian forest region, feldspar-rich arkosic sands within marine fan deposits of the Amazon River, white sands in the Rio Negro drainage basin, and layers of stones within sediments in the Amazon Basin itself. Dune systems imply periods of aridity, but reported arid dune systems at some locations (e.g. the Pantanal) have been discredited, while at other locations dates of dune formation either do not correspond to the Pleistocene or are not specific to this period (Colinvaux et al 2001).

Feldspar minerals degrade rapidly in humid environments, and their presence in sediments normally indicates prior arid conditions. Nevertheless, even relatively brief interglacial humid periods are enough to degrade feldspar completely, necessitating other explanations for their presence in Amazonian marine fan deposits (Irion et al. 1995). Alternative sources for these minerals include rapid erosion and transport of feldspar-rich rock that formed the young Andes, and subsequent erosion of deep sediment layers by the Amazon River during periods of lowered sea level (Irion et al. 1995).

Layers of stones ('stone lines') within sediments allude to buried desert pavements, but also have several alternative explanations. These include fluvial gravel deposits and mineral concretions formed along water penetration lines, both explanations being compatible with current Amazonian geomorphological processes (Colinvaux et al. 2001). Finally, white sands in the Rio Negro basin have been attributed to wind-born deposits associated with arid conditions (Ab'Saber 1982), but it is more likely that they formed as podzols under a humid climate (Colinvaux et al. 2001).

In short, the contraction of Amazonian forests to refugia in the Pleistocene is no longer a tenable theory.

---

[3] 'Science is organized common sense where many a beautiful theory was killed by an ugly fact' (Thomas Henry Huxley).

[4] Coincidence of high richness and endemism for multiple taxa within regions of current high rainfall cannot be construed as biogeographic evidence for refugia, as this would imply circular reasoning—refugia have been proposed as the explanation of such patterns.

[5] Although with some tendency to more seasonal cooler forests (Morley 2000).

peaks of others (e.g. Neotropical forest butterflies) do not even correspond to proposed forest refugia (Brown 2005). Some of the distribution patterns may simply be artefacts of collecting bias (Nelson et al. 1990).

Refugia-based explanations in other regions

Pleistocene refuge theory proposes that past climatic change and forest fragmentation led to the isolation of taxa and influenced modern distributions. Associated isolation has also been identified as a possible cause of genetic divergence and potential speciation. While refuge theory is being abandoned in Amazonia (see previous) it remains a focus of fruitful interest in areas where aridity has had a substantial impact. These include Brazil's Atlantic forests (Carnaval et al. 2009), Madagascar (Goodman and Benstead 2003; Vieites et al. 2006), the Australian wet tropics, and above all in Africa. We shall briefly explore these last two locations.

The importance of Pleistocene refugia may be the result of sifting of species by means of local preservations and extinctions, rather than speciation (Williams 1997; Haffer and Prance 2001). Support comes from the distribution of forest taxa mammals, flightless insects, and reptiles in the Australian wet tropics, which appear to be primarily the result of extinctions in non-refuge areas during the Pleistocene (Williams 1997; Williams and Pearson 1997; Schneider et al. 1998; Schneider and Moritz 1999; Yeates et al. 2002).

Evidence for influential refugia is best established for Africa, where refugia have been used to explain various distributions at various scales. Climatic models and palynological studies infer major refugia in the lowlands of Upper Guinea, Cameroon-Gabon, in West and Central Africa, and in the highlands of the Albertine Rift-Ruwenzori, along with minor refugia in the coastal forests and mountains of East Africa. Such refugia are consistent with many disjunct distributions and various patterns of genetic differentiation. For example, the distribution of gorilla taxa—absent through the central Congo—as well as their genetic distances, is consistent with glacial separations (Anthony et al. 2007).

Large-scale patterns of African plant diversity and endemism also correspond to refugial interpretations (Hall and Swaine 1981; Hamilton and Taylor 1991; Lovett and Wasser 1993). The refugia as museum concept helps explain the high comparative richness of older mountains in East Africa, where the rich flora and fauna of the geologically ancient Eastern Arc Mountains contrast with the much younger and poorer volcanic mountains (Hamilton and Bensted-Smith 1989). Indeed there is increasing genetic evidence that the older forests retained populations of relict taxa that have acted as a source for dispersal to younger forests (Fjeldsa and Bowie 2008; Blackburn and Measey 2009).

Regional patterns can often be interpreted as post-glacial range expansions centred on refugia (Hamilton 1974). For example the co-occurrence of the localized large-seeded caesalpinoid legume tree in Cameroon is believed to reflect proximity to Pleistocene refugia (Tchouto et al. 2009). While the composition of rain forest snail assemblages in Uganda cannot be explained by current environment, it is in accord with ex-refugial spread, with diversity declining with distance from the refugial origin (Wronski and Hausdorf 2008). Interestingly such expansions may also be relevant on the edge of the Amazon region: one proposed example is the steady and otherwise inexplicable advance of the poorly dispersing understorey palm *Astrocaryum sciophilum* in French Guyana (Charles-Dominique et al. 2003). When populations from different refugia expand and meet, the long isolated populations interact in areas called 'suture zones'—in some cases hybridizing and in others maintaining **parapatric** boundaries. Such sites, though long discussed (Remington 1968), have recently become a focus of renewed attention through genetic investigations (Moritz et al. 2009).

Geological instability: resurrection of the species-pump model

Regions that have been subject to recent geological instability, such as mountain building, are often associated with high levels of endemism and species diversity, notably among birds and butterflies (Fjeldsa and Lovett 1997a, b; Hall 2005). Geologically active or topographically complex regions promote speciation through the creation of new habitats and steep environmental gradients. In tropical Africa, for example, altitudinal heterogeneity predicts areas of high endemism (Jetz et al. 2004). Repeated shifts in species distributions driven by climate and tectonic uplift also lead to habitat contraction and expansion, and therefore population isolation, divergence, and subsequent mixing which may contribute to speciation (Graham et al. 2006). Thus montane bird assemblages (as compared to lowland birds) tend to include more recently diverged species groups including greenbuls (*Andropadus*; Pycnonotidae) and rockfowl (*Picathartes*; Picathartidae) in Guinean and Cameroonian highlands in Africa, and spinetails (*Cranioleuca*; Furnariidae) in the Andes (Fjeldsa 1994; Fjeldsa and Lovett 1997b; Roy 1997; Garcia-Moreno et al. 1999; Smith et al. 2005). Speciation history differences along montane gradients have also been shown for the Neotropical butterfly genus *Ithomiola* (Riodininae), with the most recently evolved species occurring in montane regions while older basal species occupy the more geologically stable lowland rainforests (Hall 2005).

These studies have thus given rise to alternative species-pump models, whereby speciation is driven by mountain building events (such as the Andes) or, in Africa, dynamic savannah–forest transitions at the periphery of the main lowland rainforest blocks, such as the mosaics of savannah and gallery forest in southern Zaire (Fjeldsa 1994). According to this model, extensive lowland rain forest regions acted as recipients and repositories (or 'museums') of accumulating species arising primarily from peripheral habitats rather than themselves acting as centres of diversification (Fjeldsa 1994; Fjeldsa and Lovett 1997b; Roy 1997).

Other paleogeographic hypotheses

Isolation of forest blocks following repeated marine incursions (resulting from tectonic activity and fluctuating sea levels) of continental South America during the last 60 Ma may have promoted population isolation and subsequent differentiation (Haffer 1997a). Within the Amazon Basin itself, however, this hypothesis alone appears insufficient to provide the necessary complexity required for intensive speciation (Haffer 1997a).

Riverine barrier hypothesis

The earliest evolutionary hypothesis attempting to explain tropical diversity is from Alfred Russel Wallace (Wallace 1852)[8] who argued that the major Amazonian rivers allowed speciation by population isolation. Distributional patterns of marmosets and tamarins (Hershkovitz 1977), various birds (Haffer 1974, 1997b; Cracraft 1985), and lizards (Avila-Pires 1995) provide support for this theory. For example, similarity of primate faunas on opposite banks of 12 major Amazonian tributaries declines with increasing width and annual discharge—barriers to gene flow diminish as rivers become smaller in their upper reaches (Ayres and Clutton-Brock 1992) (Figure 8.5). Genetic studies, however, provide mixed support for the riverine barrier hypothesis: the strongest support is derived from the largest rivers, but otherwise differentiation of populations across river barriers is limited (Colwell 2000). The riverine barrier hypothesis can at best provide only partial explanation for diversity distributions within Amazonia. While the Amazon River, especially at its widest part, is a significant barrier to dispersal of many forest-dependent birds, it does not represent a significant barrier to many other species, especially in the upper reaches[9] (Hayes and Sewlal 2004).

Environmental heterogeneity

Another set of speciation models propose that strong environmental gradients cause adaptive divergence and speciation. Notwithstanding the abundance of generalist widespread species (Pitman et al. 1999), taxa do often differentiate along topographical, edaphic, or geological gradients (ter Steege et al. 1993; Tuomisto et al. 1995; Tuomisto and Ruokolainen 1997). The frequent occurrence of hybrid zones of sister species within ecotones that straddle habitat boundaries lends further support to

---

[8] It is not clear whether Wallace actually considered rivers as barriers promoting speciation, or whether he merely noted that they were locations where species boundaries meet. In his later work (Wallace 1876), however, he includes Amazonian rivers, along with other examples, as a barrier to dispersal in the context of speciation (Colwell 2000).

[9] Note too that an actively meandering river is not an effective barrier on long timescales as significant tracts of forest habitat are regularly transferred from one bank and to the other as the meanders intersect and rivers change course.

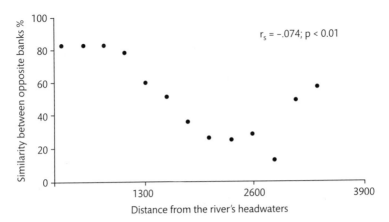

Figure 8.5 Similarity of primate faunas on opposite banks of the Amazon River, plotted as a function of distance from its headwaters. Similarity increases at the upper reaches of the river as the magnitude of the riverine barrier declines. The secondary increase in similarity, near the river's mouth, is probably a consequence of the historical instability of its course and of the existence of islands which provide stepping stones or meander-driven conveyors (see Chapter 11) for dispersal.

Redrawn from Ayres and Clutton-Brock 1992.

the theory of divergence due to habitat heterogeneity (Whinnett et al. 2005).

Satellite images and field analyses of Peruvian rain forests have revealed considerable heterogeneity based on plant associations and geomorphological states, with patch sizes ranging from a few hectares to several kilometres and even regional patterns hundreds of kilometres across (Tuomisto et al. 1995). Widespread rain forest habitat heterogeneity (see also Chapter 7) undoubtedly provides opportunities for diversification, but, as with the refugia and riverine models, there is little agreement about the temporal and spatial structure and location of different forest formations in relation to historical and contemporary habitat heterogeneity, or the degree to which habitat boundaries represent sufficient barriers to dispersal and gene flow to allow diversification (Haffer 1997a).

8.3.2 Ecological mechanisms for the maintenance of species diversity

Tropical tree species show different spatial distributions from fine to broad scales (Figure 8.6). Some appear to be restricted to slopes (*Memora cladotricha* in Figure 8.6), others are aggregated (*Rinorea viridifolia*), widespread (*Iriartea deltoidea*), or both aggregated and

scattered (*Cecropia sciadophylla*). What causes these markedly different distributional patterns? Dispersal and establishment processes, as well as site suitability and availability, undoubtedly play a part, but are these also sufficient to explain the maintenance of the diverse mix of species? When organisms such as trees compete, conventional wisdom suggests that superior competitors ultimately drive inferior competitors to extinction, and so species number declines—a process called **competitive exclusion**. Low diversity tropical forests do exist and may result from such competitive exclusion but they are an exception (Box 8.2). So why does competitive exclusion not lead to low diversity forest communities in the wet tropics more generally? Several processes can prevent competitive exclusion and so allow for species coexistence (Table 8.4). Other theories propose negative density-dependence mechanisms that favour rare species or individuals. Another model implies that limited competition among understorey plants encourages coexistence by rendering competitive exclusion less relevant. These hypotheses are not mutually exclusive, as infrequent competition, niche differences, and density-dependent effects facilitate coexistence of rare species, while negative density-dependence on larger scales regulates recruitment of the few common species (Wright 2002).

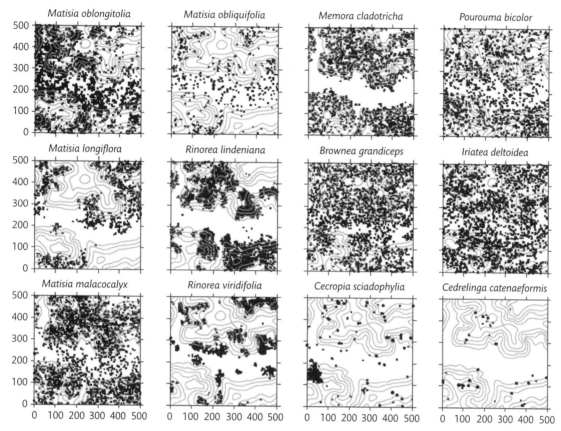

Figure 8.6 Contrasting distribution patterns of 12 tree species in Amazonian Ecuador.

From Valencia et al. 2004.

Table 8.4 Ecological hypotheses seeking to explain species coexistence among tropical forest plants. Each hypothesis violates one (or more) of six conditions required to realize competitive exclusion. The hypotheses are associated with the principal condition violated. Citations are to studies that applied the hypotheses to tropical forest plants. From Wright (2002); note that Wright largely neglects disturbance as a structuring factor, a topic that is discussed in more detail below.

Conditions of competitive exclusion	Hypotheses that violate the condition
(1) Rare species are not favoured demographically	Janzen–Connell hypothesis (Janzen 1970; Connell 1971); Compensatory mortality (Connell et al. 1984)
(2) Species have the opportunity to compete	Recruitment limitation (Hubbell et al. 1999); Low understory densities (this review)
(3) (a) The environment is temporally constant	Regeneration niche, gap dynamics (Grubb 1977; Denslow 1987)
(b) The environment has no spatial variation	Many authors (reviewed by Sollins 1998; Svenning 2001)
(4) Time has been sufficient to allow exclusion	Intermediate disturbance (Connell 1978) and Dynamic equilibrium (Huston 1994); Chance population fluctuations (Hubbell 1979)
(5) Growth is limited by one resource	Nutrient resource ratios (Ashton 1993; Tilman and Pacala 1993)
(6) There is no immigration	Mass effects (Stevens 1992)

Corollary: the greater the degree to which these conditions are broken, the greater the number of species that can coexist.

Niche assembly models

If species specialize in the conditions to which they are best suited, and if the environment presents a range of conditions, then species co-existence would merely reflect the nature of environmental variation. This is the *niche model* of species coexistence.

There is considerable evidence to indicate that species are separated along several biological trait axes: behaving differently along a gradient in resource availability, environmental conditions, or temporal division. In the context of moist tropical forests, three ideas stand out as being particularly important and common: (1) trade-offs between growth rate in high light and survival in low light conditions are well known among many tropical trees, often as the pioneer-climax dichotomy (see Chapter 10) (Poorter and Arets 2003); the related idea of (2) trade-offs among competitive and colonization abilities allow inferior competitors to coexist with superior competitors because they are better colonizers (Levins and Culver 1971), most obviously encountered in the trade-off between seed size and seed number (Coomes and Grubb 2003); and (3) species segregate along environmental gradients such as soil, moisture, nutrients etc. (Harms et al. 2001; Pyke et al. 2001; Pelissier et al. 2002; Phillips et al. 2003). Species might also coexist along a trade-off gradient of vegetative growth and reproduction (Kohyama et al. 2003) without recourse to variation in light or nutrient conditions. Under this model species coexist where an individual's likelihood of reaching reproductive size is negatively related to the species adult size, and per species reproductive output is related to basal area (as a measure of access to canopy light).

One classical view is that gap dynamics (Chapter 11) shapes species diversity: species with different regeneration niches respond differentially to canopy gap formation, driven by varying degrees of disturbance, which is the critical phase that determines the structure and composition of tropical forest communities (Pickett and Kempf 1980; Denslow 1987). According to this 'niche assembly' hypothesis, tropical forest trees are specialists that respond differentially to variation in abiotic conditions that occur within and amongst forest gaps. Gap formation alters light levels, soil environment, litter depth and quality, microclimate and the biotic environment. Each new gap includes a range of microhabitats ranging from its centre to its edge, with larger gaps harbouring a wider range of such conditions. Regeneration niches and gap dynamics theory predicts that species recruit differentially along gap-associated environmental gradients.

A major problem for gap dynamic hypotheses is that, aside from a handful of pioneer species, forest tree species recruitment often appears relatively independent of gap formation. Empirical research on the responses of tropical trees to irradiance have failed to find general support for a simple gap partitioning hypothesis (Brown and Jennings 1998; Brokaw and Busing 2000), notwithstanding some evidence for light partitioning (Montgomery and Chazdon 2002). Over 80% of shade-tolerant saplings in the understorey survive gap formation (Uhl et al. 1988; Fraver et al. 1998) and are then in a prime position to occupy the gap to the exclusion of new recruits regardless of localized site conditions. The spatial heterogeneity created by gaps does not, therefore, create predictable regeneration spaces for most tropical plants (Lieberman et al. 1995) (see also Chapter 11).

Another hypothesis emphasizes specialization to particular soil (edaphic) environments (Terborgh 1992; Sollins 1998) and thus how variations in these properties contributes to beta-diversity (species turnover in space) (Swaine 1996; Clark et al. 1999; Newbery et al. 1999; Potts et al. 2002; DeWalt et al. 2006). Species distributions are often well predicted by soils and habitat characteristics at landscape and regional scales (> 1 km²), but associating seedlings of particular tree species to specific soil properties has been difficult owing to the difficulty of disentangling niche and dispersal processes (habitat conditions and dispersal patterns aggregate at similar scales) and many studies have found it hard to detect significant soil-related associations (Webb and Peart 2000; Harms et al. 2001; Baraloto and Goldberg 2004). Nevertheless, at least one recent large-scale study provides strong support for niche partitioning along soil nutrient gradients: the spatial distribution of up to 51% of tree species in three Neotropical forest plots are associated with local differences in soil nutrient distributions (John et al. 2007).

More complex niche differentiation models partition species' resource use along several resource gradients, including nutrients, water, light, resistance to pests and pathogens. Support for such complex niche differentiation remains in its infancy, but the widespread persistence of generalist species coupled with immense tropical tree alpha-diversity suggest that niche differentiation based on relatively few resource gradients (e.g. light and soil) can only be a partial explanation of how diversity is maintained. Indeed, it is likely that environmental and historical factors, together with dispersal processes, each operating at different scales, combine to structure plant community compositions (Svenning et al. 2004).

A key refinement to niche theory notes that differentiation and specialization may occur at any stages in an organism's life history, and may indeed be brief and transitory. For plants this idea is sometimes termed the 'regeneration niche', defined as the species requirements (both biotic and abiotic) to replace a mature individual (Grubb 1977). The significance of seed and seedling adaptations in establishment often appears key—as in pioneer establishment in early succession. Evidence in favour of this idea operating within communities comes from physiological studies of Bolivian forest trees, suggesting much stronger light partitioning amongst seedlings than amongst adults (Poorter 2007). Similarly the variation in cues required for seed germination reveals various differences that can influence the resulting communities (Pearson et al. 2003). Differentiation amongst adult plants is not excluded by these results. Indeed leaf physiological traits in Malaysian trees have been seen to vary with their adult stature, and imply that variation in juvenile and adult characteristics may not be readily separable (Thomas and Bazzaz 1999).

Spatial and temporal resource variability

Differences among trees in how they trade-off growth under high light with survivorship under shade may contribute to segregation among sites based on gap size (large gaps favour rapidly growing seedlings that have low shade tolerance while the opposite is true of small gaps) or with succession (see *Intermediate disturbance hypothesis*, below). Growth–survival trade-offs are well established among coexisting tropical trees (Figure 8.7) (Hubbell and Foster 1992) and may allow for species coexistence based on temporal changes in resource availability: when resources such as light are low, a shade-tolerant strategy (i.e. slow growth–high survival) is favoured; the converse light-demanding pioneer strategy (fast growth–low shade survival) being favoured during periods of resource abundance (i.e. high light) (Chapter 10).

Other trade-offs in species characteristics may also contribute to species coexistence. A number of these are discussed in Chapter 10. One of special note, already mentioned above, suggests that tree species may be able to partition the vertical light axis in the forest with small understorey species well able to coexist with taller canopy trees. The circumstances in which such trade-offs can stabilize diversity have been a theme in theoretical studies, and have led to predictions that can be examined by field studies (Kohyama et al. 2003; Poorter et al. 2006).

The role of disturbance and productivity

No one disputes that disturbance is needed to maintain pioneer species in the wider landscape. But there is less consensus on the larger role of disturbance in contribut-

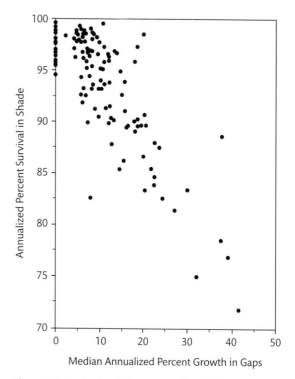

Figure 8.7 The trade-off between survivorship and growth for trees from Barro Colorado Island, Panama. The annual survival rate (vertical axis) is for shaded understorey saplings. The annual growth rate (horizontal axis) is for saplings in treefall gaps. Growth increments of some slow-growing species were undetectable.

From Hubbell and Foster 1992.

ing to diversity maintenance. Is high species richness a stable state, or is it instead a transient successional property that can only be maintained by suitable patterns of disturbance (Connell 1978; Huston 1979)? According to some, competitive exclusion does not seem to operate in tropical forests, and richness is a characteristic of late successional 'equilibrium' communities (Phillips et al. 1997). Certainly it cannot be the only factor at work, as most hurricane impacted forests regrow with little change in diversity (Lugo et al. 1993; Frangi and Lugo 1998). Some consider the population dynamics of many tropical forest plants to be too rapid to be influenced by relatively infrequent disturbances (Condit et al. 1996, 1999; Silva Matos et al. 1999).

The *intermediate disturbance hypothesis* (IDH) generalizes a sucessional model of diversity as an explanation of how species avoid competitive exclusion and therefore enhance alpha-diversity and beta-diversity (Connell 1978; Molino and Sabatier 2001; Sheil and Burslem 2003). Forest communities developing on vacant sites are viewed as progressing through a series of distinct communities—a succession (Figure 8.8). Initially, the vegetation is dominated by colonizing species adapted to disperse well and can make use of abundant light and other resources, but these soon fail to establish in their own shade, and are replaced by more tolerant species. The series ends when the most

effective competitors exclude the rest. In the absence of disturbance, species initially accumulate through succession (by dispersal and establishment) and are later lost through competitive exclusion, leading to low diversity communities dominated by a few extremely shade-tolerant species. Disturbance causes reversion of the community to younger successional states in which species, previously excluded by competition, are once again able to establish and grow (Connell 1978; Huston 1979). Too much disturbance leads to the loss of late-successional species, whereas too little disturbance leads to exclusion of species adapted to colonizing younger sites. An intermediate disturbance regime enables coexistence.

Disturbance can facilitate species coexistence in various distinct ways. 'Within-patch' and 'between-patch' models illustrate the mechanisms involved. Within-patch models consider a homogeneous closed system: competitive exclusion is prevented when the disturbance regime provides intermittent opportunities for the recruitment of species at all points along the life-history spectrum, from early-successional colonizers to late-successional competitors. Species with low tolerance of competition, which would be eliminated without disturbance (**fugitive species**), can endure periods of severe competition in a dormant (seeds) or durable (seedling bank or long-lived adults) life stage, a

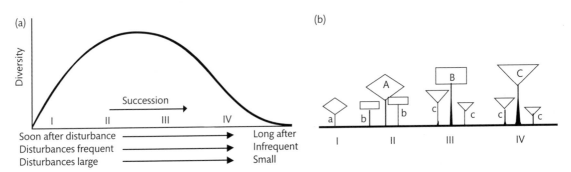

Figure 8.8 The intermediate disturbance hypothesis. One diagram links diversity with 'time since', 'frequency of', and 'size of' disturbances. The diagrammatic sequence of trees in (b) indicates the pattern of species replacement occurring in a successional interpretation of a Ugandan tropical forest (Eggeling 1947). According to this interpretation in colonizing stands the canopy was dominated by a few species (class A), but the juveniles (class B) were of different species. Adults of the class B species occur elsewhere as adults in mixed stands of many species. In these mixed stands, the juveniles were also mainly of different species (class C), those with even greater shade tolerance. Adults of class C occurred in climax stands in which the understorey was composed mainly of juveniles of the same class C canopy species. Roman numbering denotes four stages in the sequence.

From Sheil and Burslem 2003.

mechanism that is sometimes referred to as 'temporal niche sharing' or the 'storage effect'.

Between-patch models include space and dispersal. Pioneers and other fugitive species might be eliminated in any given patch by local competition, but following local disturbances, recolonization can occur from patches that are at earlier successional stages. It should be noted that poor dispersal by dominant species can itself slow exclusion, as it might take many generations to reach and win all the potential sites, even in the absence of disturbance. If fugitive species are highly mobile by comparison, they can take advantage of very limited opportunities in both space and time.

So what is the evidence so far? Certainly, aside from the obvious persistence of pioneers in virtually all forest regions, there are forests where the IDH helps explain local species richness patterns and where a complete absence of extrinsic disturbance would be likely to lead to a major loss of species (Molino and Sabatier 2001). Nonetheless, the contribution of disturbance to overall richness patterns remains contentious: disturbance regimes are hard to measure, and local disturbance-dependent patterns cannot always be readily detected. Evaluation of a large census of sites across Ghana suggests that the IDH may play a much lesser role in wet as opposed to drier forest regions (where perhaps other factors such as density dependence may slow or prevent competitive exclusion). Estimates suggest that variation attributable to disturbance contributes 36% in drier forests as opposed to only 6% and 2% in moist forest and in wet forest respectively (Bongers et al. 2009).

effective dispersal, and shade-tolerant seedlings (De Steven and Wright 2002). Other mechanisms must be inferred to explain why such common species do not begin to dominate (Wright 2002).

It has long been assumed that seedlings and saplings are engaged in a life and death struggle to reach the canopy. In some canopy gaps this is probably the case, but what about the rest of the forest? Low light in the understorey and root competition with canopy trees limit all understorey plants (Canham et al. 1990) (Chapter 10), and herbivory further limits growth (Fine et al. 2004) (Chapter 13). This combination places a heavy burden on understorey plants, which are often highly spaced. It has indeed been suggested that any direct resource competition amongst these understorey plants may be for all intents and purposes insignificant. This implication has been tested twice by experimental removal of some understorey plants and observing the performance of those that remain. Under conditions of competitive interaction, competitor removal should lead to positive performance responses, but in both studies the growth and survival of seedlings and saplings did not increase detectably (Marquis et al. 1986; Paine et al. 2008; Svenning et al. 2008). If this was the only competition that mattered, as might sometimes be assumed, then the apparent lack of such competition necessarily implies little competitive exclusion, allowing species to coexist. No one doubts the reality of competition amongst adult trees, between adults and juveniles, and between seeds for suitable germination sites—thus an absence of direct plant–plant competition in the understorey only reduces one avenue to exclusion.

Coexistence through lack of competition

Limited dispersal of seed that results in recruitment limitation (i.e. failure to reach suitable regeneration sites) may allow coexistence by allowing inferior competitors to occupy sites where the superior competitor is absent (Hurtt and Pacala 1995). Dispersal limitation may affect most tropical trees (Hubbell et al. 1999; Harms et al. 2000; Dalling et al. 2002), and so has the potential to prevent competition among many species and thereby allows coexistence. Nevertheless, abundant and widespread species, such as *Trichilia tuberculata* (Meliaceae), the most common tree on Barro Colorado Island, may escape recruitment limitation due to heavy seed set,

Negative density dependence

'It will be seen that this invasiveness of taxa, due to pest pressure, applies as well to the spread of species from one plant association to another…When we see that by moving from one formation to the other it escapes to a great extent from its own pests, while entering into competition with plants which are still burdened down by theirs, the phenomenon is easy to understand' (Gillett 1962).

Negative density dependence is expressed as an increase in plant performance with increasing isolation from conspecifics. Host specific pests and pathogens or intraspecific competition may result in negative density dependence, favouring recruitment of species other than locally abundant species.

The so-called 'Janzen–Connell hypothesis' (Connell 1970; Janzen 1970; Peters 2003) expands upon Gillet's idea (see above quote) to develop several clear predictions: (1) the incidence/impacts of pests and diseases should be greater at higher host density, and among juveniles close to the mother tree; (2) due to non-random density dependent mortality, the spatial distribution of the host species will become less aggregated through time; (3) there should be greater recruitment of non-susceptible species within the host plant's neighbourhood; and (4) over time plant diversity should be greater than expected compared to random establishment and survival. The hypothesis assumes the existence of species-specific pests and/or pathogens, without which generalist mortality agents would respond to overall plant density irrespective of the identity of species (see Chapter 13).

Studies across the tropics provide considerable evidence to support the first of these predictions. Damping-off and canker diseases (such as *Phytophthera*, see Chapter 3) are density and/or distance dependent for seedlings of many diverse species (Augspurger 1983, 1984; Augspurger and Kelly 1984; Gilbert et al. 1994; Augspurger and Wilkinson 2007). These diseases can mediate shifts in the spatial patterns of surviving seedlings, resulting in a more dispersed distribution over time (Augspurger 1983). Insect predators of seed and/or seedlings can have a similar effect, although vertebrate predators are less likely to have such impacts (Hammond and Brown 1998; Salm 2006). Insects and pathogens tend to act in a more host-specific manner than vertebrates, which are less specialized and far-ranging, hence vertebrates violate the assumption of host specificity that the hypothesis requires.

Disease or pest-related density-dependent mortality of one species, resulting in conditions that favour survival of other non-susceptible plant species (predictions 3 and 4), has been convincingly demonstrated from New and Old World sites (Wills et al. 2006). Other studies, on Barro Colorado Island, provide additional support: growth performance of saplings from 9 out of 11 tree species was lower beneath conspecific adults (Hubbell and Foster 1990), and density-dependent seedling establishment was noted for 53 species (Harms et al. 2000). It is also likely that density-dependent effects have been underestimated, as negative effects on growth and survival may unfold over time intervals longer than, and spatial scales larger than, those of scientific studies. Hence density dependence is generally accepted to be an important mechanism for species coexistence.

Neutral theory and ecological drift

Niche-based theories have been challenged by neutral models that assume ecological equivalence of species and rely on random (neutral) drift processes to slow competitive exclusion and maintain community structure (Caswell 1976; Hubbell 2001; Volkov et al. 2007). Such models have been shown to be capable of generating species rank abundance curves almost identical to those encountered in real life[10] (Hubbell 2001). They have also proved provocative by implying that niche-based explanations are not required for the maintenance of species richness in tropical forests, and so challenge ecologists to demonstrate that differences among species are ecologically important (Hubbell 2001). Proponents of the neutral theory do not deny that species differ, but instead emphasize that the explanatory power of the neutral model, which excludes species competitive differences, points to an underlying simplicity of ecological communities (Volkov et al. 2007).

It is, of course, premature to reject niche partitioning processes for species coexistence. Neutral theory cannot be a sufficient explanation for tropical tree diversity simply because a period of time pre-dating the origin of flowering plants would be required to explain the observed diversity of tropical forests (Leigh et al. 2004). Observations of niche differentiation relating to climate,

---

[10] Note too that while niche-based models predict communities in which composition and resulting diversity patterns are somewhat stabilized and resilient, neutral models make no such claim, with the abundance of each species fluctuating at random over time.

topography, and edaphic and successional requirements also clearly show that there is some niche differentiation at work—and these differences are clearly influencing the local communities. Additionally, there remains insufficient knowledge of the potential role that multiple biotic interactions and trade-offs might contribute to species differentiation (Baltzer et al. 2005; Burslem et al. 2005).

Niche structure can be introduced into neutral models without affecting its main prediction of distribution of species abundances (Purves and Pacala 2008). In other words, real patterns are consistent with both niche and neutral models, and advances in theory show that these patterns could in fact arise from many different multi-species systems and assumptions (Pueyo et al. 2007). Ecological equivalence of species cannot therefore be inferred from such models but must be determined directly (Purves and Pacala 2008). Some niche-structuring of communities cannot be doubted: there are many examples of predictable community responses to environmental characteristics or disturbances, but these factors determine the relative abundance of guilds. For example, rain forests in Borneo that have higher fire frequency also have a higher proportion of (niche-determined) pioneers (Slik and Eichhorn 2003), but the pattern of species abundances within these sub-communities may still match that predicted by ecologically neutral drift (Purves and Pacala 2008).

The richness of tropical rain forests has been used as a testing ground for neutral theory. Its real value is now seen as a baseline. Using this model we can begin to ask whether species distributions, patterns of relative abundance, and diversity are adequately described by neutral processes of population dynamics, dispersal, speciation, and extinction, or whether other structuring factors must be invoked.

8.4   Conclusion

Making sense of the accumulating and now extensive information on the abundance of species and their distribution in the tropics remains challenging, and has been hampered by confusion over scale and different definitions of diversity, as well as confounded and correlated proposed explanatory variables. With the development of null models we can at least now evaluate the extent to which distributional patterns of species diversity and turnover are determined by non-stochastic ecological or biogeographic mechanisms.

Over 100 proposed mechanisms to explain high species coexistence can be condensed to a few main themes. Some of these have accumulated considerable support, including niche segregation associated with micro-topography, niche differences associated with trade-offs between growth and survival, Janzen–Connell effects of low recruitment near conspecifics, and negative density dependence at larger scales. Neutral drift is likely to contribute to the maintenance of species richness within functional groups that shape the broader parameters of community structure. All these mechanisms are likely to operate together to contribute to the persistence of rare species and the limitation of common species, alongside other drivers of high species richness such as productivity and disturbance-mediated turnover. Of course we should not lose sight of the historical environmental changes that have influenced dispersal, speciation, and extinction processes, and it is important to recognize that current patterns of alpha- and beta-diversity reflect the broader legacy of environmental history.

CHAPTER 9

Processes and cycles

'Nowhere does she [nature] penetrate us more deeply with the feeling of her grandeur, nowhere does she speak to us with a more powerful voice than in the tropical world...'

Alexander von Humboldt (1849)

Rain forests are much more than a collection of trees—they are a collection of interacting processes and parts that draw upon and impact on their environment. These interactions range in scale from that of molecules to global atmospheric stability. Defining these interactions, and even clarifying their realms, is often challenging as much remains uncertain. In this chapter we examine a selection of the processes which link forests to their environment. We know enough to show why it is important that we should know more.

9.1   Climate and weather

Equatorial weather is relatively stable, without the large-scale 'fronts' that drive temperatures up or down, or the revolving cyclonic and anti-cyclonic patterns or seasonal temperature swings that dominate at higher latitudes. The length of the equatorial day varies little. Despite shifts in solar angle and air movements there is limited seasonal variation in temperature. Weather is deter-

mined instead more by local convection, atmospheric moisture, and diurnal patterns, with daily temperature changes often exceeding seasonal variations.

This pattern changes at higher tropical latitudes (between 17.5 and 23.5 degrees) where, during the austral or boreal summer, near vertical sunlight occurs daily for several months[1]. This marked seasonal heating is usually coupled with clear skies associated with high pressure belts and continental weather patterns[2].

As hot wet equatorial air rises, it cools and forms clouds. A distinct daily cycle is common in the tropics whereby cumulonimbus clouds gradually build during the day before heavy rain falls in the afternoon and evening. At larger scales, the low pressure generated below rising tropical air masses creates the **inter-tropical convergence zone** (ITCZ) where the northern and southern hemisphere air currents converge. This feature and associated cycling of air to the north and south dominates tropical weather (Figure 9.1).

The location of rising air in the ITCZ shifts to higher latitudes in each hemisphere's summer, and this shift in

[1] The Earth's axis is inclined at 23.5° to the orbit around the sun. During each year the locations where sunlight shines directly overhead move between 23.5° N and S. Within these limits lie the tropics of Cancer (north of the equator) and Capricorn (south of the equator). Indeed the word 'tropics' comes from *tropos*, the Greek for 'turn', referring to the sun's yearly oscillations between the tropics.

[2] Continentality refers to the rapid heating and cooling that occurs over land compared with oceans, and therefore greater extremes over larger land areas further from the sea. The northern hemisphere is warmer than the southern hemisphere as there is more land north of the equator, and the world's 'thermal equator' occurs at about 5° N.

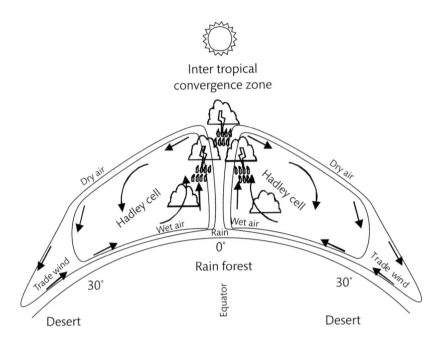

Figure 9.1 Illustration of the inter-tropical convergence zone. Warmed by the sun the equatorial air rises, cools, loses its moisture, and finally falls back to earth between about 30° to 35° to create the 'subtropical high'. The world's principal desert regions lie at these latitudes. From these subtropical highs the trade winds blow westwards and back towards the equator. These winds start out dry but regain moisture as they proceed.

Drawn by Douglas Sheil.

wind generates the monsoons[3]. Tropical monsoon climates are warm throughout the year, but rain is typically concentrated a month or two after the summer radiation peak. Wet monsoon climates dominate the coastal regions of India, Sri Lanka, Indochina, northern Australia, West Africa, Mexico, Brazil, and the Caribbean.

Rain patterns vary and can be localized, reflecting shifts of the ITCZ, the direction of the monsoon winds, ocean temperatures, and terrain. As moist winds rise over high terrain the air expands and cools, promoting cloud formation and rain—and drier winds on the other side. Prevailing winds mean tropical mountains often possess wetter and drier sides, and downwind 'rain shadows'; winds also create localized temperature effects. Shifting monsoon winds give such local patterns a seasonal dimension.

Prevailing winds can thus determine regional patterns: for example, while windward rain forest climates are quite comparable among Pacific islands, seasonal aridity increases markedly on the leeward side from wet Melanesia to arid Hawaii (Mueller-Dombois 2002). In addition, certain low latitude sites are dry, as in East Africa, West Africa[4], the Andean deserts, parts of eastern Indonesia and Papua New Guinea, while some sites at higher 'desert latitudes' are wet (coastal mountains).

Global weather patterns include supra-annual fluctuations. One is the **El Niño** Southern Oscillation. This occurs when the 'normal' high pressure region in the eastern Pacific dissipates, causing significant shifts in global rainfall with droughts in some places and floods in others. Smaller-scale oceanic switches have also been

---

[3] Monsoons are driven by seasonal heat differences between land and adjacent oceans and the resultant transport of moisture between them.

[4] The Dahomey Gap is an arid corridor (rainfall c.700–800 mm yr^{-1}) in Ghana, Benin, and Togo, splitting West Africa's wet forest zone in two: the western 'Upper Guinean' forests and the eastern 'Lower Guinean' forests. In the northern summer the sea winds that carry a wet season inland instead flow parallel to the coast, bringing much less rain inland.

detected in other oceans. Such events lead to unusual weather and have a variety of impacts on rain forests (Chapter 14). Multi-decadal temperature oscillations have also been recognized for western Amazonia (Botta et al. 2002), and drying trends in the northern African tropics may be partially driven by oscillations in Atlantic Ocean temperatures (Malhi and Wright 2004). These natural oscillations should not be confused with anthropogenically-driven climatic changes, although separating the contributions of different drivers is not straightforward.

9.1.1 The climate of tropical rain forest regions

Mean annual climatic values for tropical rain forest are 2,178 mm precipitation, a temperature of 25.4°C, and insolation of 16.5 GJ per year (Malhi and Wright 2005). There is, however, considerable variation around these mean values, with some regions being considerably wetter (such as east Malesia and north-west Amazonia where annual precipitation exceeds 3,000 mm) while Africa is much drier (Table 9.1). Trees in tropical forests can be vulnerable to water stress (Chapter 10).

Consistent directional climatic changes have been recorded in tropical rain forest regions, associated with global warming trends, and climate models suggest that a warming of between 2° and 5°C can be expected in rain forests by the end of the 21st century (Cramer et al. 2004). These trends, and their implications, are explored in more detail in Chapter 14.

9.2   Atmospheric chemistry

Vegetation influences atmospheric chemistry and composition beyond simple carbon metabolism, and rain forests are important sources and sinks of various trace gases. Notably, tropical rain forests are significant global sinks for methane (Wuebbles and Hayhoe 2002), and

rain forest soils absorb 10–20% of the global methane budget[5] (Eggleton et al. 1999; MacDonald et al. 1999). Forests also emit various **volatile organic compounds** (VOCs; Box 9.1), including more than 80% of global isoprene (Potter et al. 2001; Lerdau and Slobodkin 2002; Guenther 2008). These react with (natural) hydroxyl radicals which help cleanse the atmosphere of various trace (greenhouse) gases (e.g. methane, ozone, and nitrogen oxides) (Guenther 2008).

The role of aerosols in forest–climate interactions remains uncertain. These aerosols derive from VOH and particles such as pollen, spores, bacteria, and smoke (Lerdau and Slobodkin 2002; Andreae et al. 2004; IPCC 2007; Möhler et al. 2007). At night, during the Amazonian wet season, fungal spores contribute perhaps 50% of coarse particulate matter in the lower atmosphere (Elbert et al. 2006) but the climatic influence is unknown. Forest soils also release various gases after forest loss (e.g. Veldkamp et al. 2008), and replacing tropical rain forest with other land cover is likely to add to greenhouse gases and increase air pollution (Lelieveld et al. 2008).

9.2.1   Trace gases

Tropical forests are likely to contribute over half the global emissions of biologically sourced global volatile organic compounds[6] (Lathiere et al. 2006). Why forests produce VOCs is poorly understood (Box 9.1), but the implied carbon losses are of similar magnitude to net biome productivity (Kesselmeier et al. 2002). VOCs have significant influence on local climates and global atmospheric chemistry (e.g. governing the dynamics of ozone and nitrogen oxides) and include powerful greenhouse gases (Lerdau and Slobodkin 2002). They also influence atmospheric aerosols and cloud formation.

Halocarbons (organic compounds containing halogens such as chlorine) affect the ozone layer and act as greenhouse gases. Asian rain forests contain major chloromethane (CH_3Cl, 'methyl chloride') emitters,

---

[5] Methane, a more potent greenhouse gas than CO_2, is sequestered in rain forest soils via oxidation by methanotrophic bacteria.

[6] Terrestrial vegetation produces perhaps three-quarters of the mean global emission of all biogenic carbon-based compounds (not including CO_2) estimated at around 750 Tg C yr^{-1}, the main components being isoprene (460–560 Tg C yr^{-1}), monoterpenes (117 Tg C yr^{-1}), methanol (106 Tg C yr^{-1}), and acetone (42 Tg C yr^{-1}).

Table 9.1 Mean climate of tropical rain forest regions over the period 1960–98. P = precipitation, S = solar radiation, T = temperature. Seasonal variation in precipitation and temperature is normalized by dividing by annual mean values.

Region	Mean			Seasonal variation			Dry season
	P (mm)	S (GJ yr⁻¹)	T (°C)	P	S	T (°C)	Length (months)
Central America	2206	17.3	24.0	2.4	0.42	4.4	4.7
North-west Amazonia	2962	15.6	25.9	1.2	0.13	2.3	0.6
South-west Amazonia	2194	16.3	25.7	1.8	0.17	3.2	3.3
Central Amazonia	2420	15.6	26.2	1.6	0.23	2.4	2.3
North-east Amazonia	2260	16.6	26.0	2.0	0.26	2.2	3.7
South-east Amazonia	2103	16.1	25.5	2.3	0.31	2.9	4.5
West Africa	1601	16.6	26.3	2.7	0.28	4.6	5.8
Cameroon	1782	15.2	24.6	2.1	0.18	3.0	4.3
North Congo	1689	17.6	24.5	1.7	0.14	2.5	3.6
South Congo	1530	16.5	23.8	1.8	0.26	2.7	4.3
South-west India	1993	18.3	25.9	3.2	0.32	4.0	5.9
West Malesia	2584	17.1	25.3	2.1	0.31	3.2	2.9
East Malesia	3094	16.7	25.4	1.4	0.18	1.6	1.2
Australia	1702	20.1	24.2	3.5	0.22	6.7	7.0
Pantropical mean	2178	16.5	25.4	2.0	0.24	3.2	3.7

From Malhi and Wright 2005.

Box 9.1   Volatile organic compounds

Some plants emit organic compounds to attract natural enemies of herbivores and to combat pathogens (Holopainen 2004). Emissions may also be accidental 'side-effects' of certain metabolic processes. Isoprene release is often a response to heat stress (Penuelas and Munne-Bosch 2005; Sharkey 2005) and emissions are highly seasonal (Kuhn et al. 2004; Rottenberger et al. 2004). Emissions increase linearly with illumination and temperature (Lerdau and Slobodkin 2002). At La Selva, Costa Rica, isoprene emissions peaked in the dry season (Karl et al. 2004). Emission levels are highly species dependent, varying by over four orders of magnitude across taxa.

Around one-third of the tropical tree taxa examined, including most palms, emit abundant isoprene. In contrast no tropical C4 grasses and only a few C3 grasses (including some bamboo) are big emitters (Lerdau and Slobodkin 2002). This explains why isoprene levels are much higher over African rain forests than over neighbouring savannah woodlands (Klinger et al. 1998; Greenberg et al. 1999). Forest loss can involve major changes—few annual crops are big emitters[1] (Geron et al. 2006).

The full range of volatile compounds produced in rain forests remains unknown. Surprises still occur, such as the recent implication of rain forests as natural sources of various aromatic hydrocarbons[2]. Naphthalene concentrations can reach 3–4 mg kg^{-1} in Amazonian wood samples and in termite nest soil. While the origin of these hazardous compounds remains uncertain, they appear defensive for the termites (Wilcke et al. 2000, 2002).

[1] Some plantation genera are, these include *Bambusa*, *Elaeis*, *Eucalyptus*, *Hevea*, and *Pinus*.
[2] Notably naphthalene, perylene, and phenanthrene.

---

mainly ferns and dipterocarp trees, contributing maybe 910,000 tonnes year^{-1} from this region (Yokouchi et al. 2002)[7]. Wood-rotting fungi in tropical forest may produce a further 80,000 tonnes year^{-1} (Watling and Harper 1998).

9.2.2 Lungs of the world?

Contrary to popular perceptions, rain forests are not 'the lungs of the world'. Under 'stable and natural' conditions forests are in an approximate balance with the atmosphere, neither adding nor removing much oxygen. During the glacial periods rain forests were much reduced (see Chapter 6) and the world still had oxygen. Nonetheless the carbon locked up in tropical forests plays an important role in the global carbon cycle, as do other gases produced and captured by rain forests, and rain forests interact with weather and climate in various ways. It is hard to prove these relationships empirically, and much remains anecdotal and open to alternative interpretations.

Forest loss affects air movements, atmospheric moisture, and cloud formation. Convection-related weather patterns are especially sensitive to forest loss. For example, the daily pattern of land–sea breezes has intensified with forest loss in south-eastern Kalimantan (Nooteboom 1987 quoted in Bruijnzeel 2004). After hurricanes in Puerto Rico, cloud formation rose to higher altitudes, exposing cloud forests. This may reflect reduced transpiration and warmer air following defoliation (Scatena and Larsen 1991). Reduced cloud formation following forest loss also results in reduced

---

[7] Chloromethane is responsible for around 17% of chlorine-catalysed ozone destruction (Harper et al. 2003). Note these natural sources of ozone-depleting chemicals are in a near equilibrium in the atmosphere. Anthropogenic sources of halocarbons remain a much greater threat to the Earth's protective ozone layer.

rain (Eltahir and Bras 1994), and clouds develop later each day over recently deforested regions of southwest Amazonia (Cutrim et al. 1995). Dry season cloud formation in northern Costa Rica has also declined along with forest cover (Lawton et al. 2001). Such changes have impacts on local and regional climates. Daytime clouds reflect heat back into space, but they also trap and absorb outgoing long-wave radiation which leads to night-time warming.

9.3 Soils

General texts continue to imply incorrectly that tropical rain forest soils are uniformly dominated by 'red laterites'. Tropical forest soils are, in fact, diverse, and the greatest diversity is found in the wettest lowlands due to the continuous action of rainfall which determines nutrient inputs, organic matter decomposition and distribution, erosion, and chemical and physical weathering of rocks, all of which lead to soil formation.

These processes are variously shaped by climate and geology and are often mediated by the disturbance regime. In volcanic sites, for example, rock and mineral ash can be major nutrient sources. Soils on steep terrain can be rejuvenated by weathering and landslips that enrich lower landscape positions. Fresh flood sediments can also rejuvenate soils. Over the longer term, inputs from rainfall, anthropogenic atmospheric deposition (Chapter 14), and even remote desert dust storms can all be significant sources of nutrients (Schroth and Burkhardt 2003). Nutrients are lost through erosion, leaching and volatilization, and ash convection by fire.

The variety of processes contributing to soil formation results in a bewildering array of soil classification standards that were not devised for tropical ecologists, and are bewildering even for many soil scientists.

Ageing is the key to soil nutrient availability. In young soils, many nutrients are often predominantly derived from the parent rock. Volcanic ash soils of different ages in Ketambe, North Sumatra, are increasingly more acid and less fertile the older they are (Van Schaik and Mirmanto 1985). Very old rain forest soils no longer gain inputs of local rock-derived nutrients as the weathering rock is too deep. In peat, too, an age-depth phenomenon governs the availability of mineral nutrients. When peat is shallow, roots can access nutrients from the underlying mineral subsoils, but this access declines on older deeper peat (FAO 2001).

Forest organisms influence soil properties and weathering, though the processes are only partially understood. For example, roots affect soil carbonic acid levels (Grimaldi and Pedro 1996) and concentrate some compounds at the root–soil boundary, and phytoliths affect soil silicon inputs (Alexandre et al. 1994). Termites, too, with their amazingly alkaline digestion, may significantly influence soil mineral formation (Chapter 5).

9.3.1 Principal soil types

Tropical soils can be classified into four soil orders: ultisol, oxisol, alfisol, and aridisol, according to their level of development and gradient of rainfall seasonality (Ashton 2004). Alfisols are rare in tropical rain forests, and aridisols occur in regions that are too dry to support tropical rain forests, so we do not concern ourselves further with these (Tables 9.2 and 9.3).

Ultisols

Ultisols occur in the areas of highest rainfall and least seasonality on all three tropical continents. They are generally deep well-drained soils rich in iron oxides which give a red or yellow colouration. They tend to be highly leached and have low cation exchange capacity[8], and therefore have low fertility. There is a considerable variety of ultisols, but this variety falls into two main groups: udult ultisols, with high clay[9] content and rapid decomposition owing to surface pH exceeding 4.2; and humult ultisols, produced by the accumulation of acidic raw humus where clay content is low or rainfall is exceptionally high and leaching intense (Ashton 2004).

---

[8] Negatively charged sites that attract positively charged cations, a major reserve and source of plant nutrients.
[9] Clays are an important mineral component of many tropical soils. Clays are fine-grained minerals that usually possess large, negatively charged surfaces that adsorb cations.

Table 9.2 A broad classification of major soil orders in the tropics based on seasonality and rainfall.

Ultisols		Oxisols	
High rainfall, no seasonality	Weakly seasonal climate		Strongly seasonal climate
Consistent downward water flow		Alternate seasonal downward and upward flow of water	
Recognizable horizons: increased colour intensity with depth	Deep soils: uniform yellow to red brown colour		Accumulation of sesquioxides at certain depths can form an impervious horizon
Leaching and accumulation at depth of clay minerals and sesquioxides*	Paler surface due to leaching of sesquioxides		Pale in the anoxic zone
	Rapid litter decomposition		Shallow or deep rooting
Accumulation of surface humus	Shallow rooting		
Variable rooting depth			

Adapted from Ashton 2004.
* iron and aluminium oxides

Table 9.3 Main soil types of the humid tropics and their extent; grouped by main features.

Soil	Main features	% rain forest area	Example locations
Oxisols and ultisols (see Table 9.2)	Old, infertile, loamy, and clayey	63	Widespread
Alfisols and vertisols	Comparatively fertile, less weathered and locally less leached mineral soils relatively low in organic matter, with relatively high base saturation (alfisols), or with deep wide cracks at some time in most years (vertisols)	4	Many parts of South America, Solomon Islands
Andisols	Very fertile, less weathered soils on volcanic ash	1	Volcanoes of Ecuador and Colombia, Hawaii, Java
Free draining entisols (fluvents)	Comparatively fertile, less weathered, mineral soils lacking developed soil horizons; mainly on alluvial lowlands	12	Amazonian white-water systems
Wet entisols (aquepts)			
Tropepts and lithic soils	Comparatively fertile, less weathered soils on steep slopes	11	
Spodosols and psamments	Infertile often unconsolidated sands; usually moist or well leached	7	White sand formations in Amazonia; South American *campinas* forests; Bornean *kerangas* forests
Histosols	Infertile peats in poorly drained basins	2	Sumatra, Borneo, and New Guinea peat forests

Adapted from Ashton 2004.

Oxisols

The classic tropical soils referred to by general tropical textbooks are the oxisols, highly weathered, mature, red soils up to 20 m depth. The soils are highly weathered, but surface pH usually exceeds 4.5 and litter decomposition rates are high. These soils occur in regions of marked seasonality where in at least four contiguous months evapotranspiration exceeds precipitation (Ashton 2004). Oxisols are widespread across geologically ancient surfaces of all major tropical regions.

9.3.2 Other soil types

Most of the wet tropics, including the Amazon and Congo basins, lie over a gentle terrain of ancient, well-weathered, silica-rich rocks and sediments. Underlying geology no longer has much direct influence. The dominant soils are old, deep, heavily leached, uniform, and low in mineral nutrients. In these environments most elements, even silica, leach from rocks, while less soluble aluminium, iron and manganese oxides and hydroxides accumulate[10]. Typically, clays develop. Fine textured clays resist water infiltration in many wetter parts of the Amazon Basin, much of the Guyana highlands, and also in Malesia. Rooting is shallow, and these soils are especially nutrient-poor, supporting stunted formations rather than tall forests. Kaolin-rich clays are especially widespread in the wet tropics, yielding acidic 'kaolisols': soils with a relatively low capacity to retain nutrients, but good drainage.

In contrast, significant areas of Southeast Asia, New Guinea, Central America, and the Andean foothills possess more varied younger geologies and rugged terrain. Rich soils are associated with volcanic inputs, areas of tectonic up-thrust, or derived alluvial plains. Base-rich younger volcanic geologies often yield clay-rich soils with better nutrient-holding properties, such as vertisols. These fertile soils support tall forests such as Kolombangara in the Solomon Islands. Due to competition with agriculture in these regions, rain forests on rich soils are increasingly scarce.

Spodosols occur in wet climates, and are soils in which all aluminium and iron oxides have been leached. Acid humus may accumulate on the surface, and almost all roots are concentrated in this region. The only source of nutrient input is from rainfall. These soils are widespread on silicaceous substrates and support the *kerangas* formations in Borneo (section 7.1.6 Heath forests).

Sandy soils tend to occur on alluvial coasts, associated with weathered sandstones and other quartz-rich geologies. Nutrient and moisture storage are extremely limited, and organic compounds washing through the profile give streams their dark colour. Rooting depth is typically shallow and plants are prone to drought. Conditions can, however, change as organic compounds slowly aggregate above the water table, forming a tough, cemented layer (or 'pan') that ultimately impedes drainage and saturates the upper layers. Such mature podzols (with a dark upper horizon) are relatively widespread in wetter climates of Asia and South America.

When poor soils, impeded drainage, and aseasonal climates occur together, plant debris can accumulate forming peat. Lignin is the primary material and decomposition is inhibited by anoxic—often acidic—conditions. Soil water is rich with organic acids, leading to dark 'tea-like' streams.

In seasonal forests where air regularly permeates the soil, oxide-rich (red hematite) ferricrete kaolisols—often referred to as 'lateritic'—can develop, although such soils are much more localized than once thought. When dried these harden irreversibly to become 'plinthite'[11].

9.4 Nutrient cycling

The nutrient capture efficiency of rain forests is famous. In one study on infertile sands (spodosols and oxisols) in Amazonian Venezuela, over 99.9% of experimentally applied Ca and P was captured in the soil surface root mat (Stark and Jordan 1978). Yet even the most efficient nutrient cycling is imperfect, and the rain forest nutrient cycle is not a closed system: inputs, outputs, stocks, and flows all contribute.

[10] The bauxite ores used as the raw material in aluminium production are mainly old tropical rain forest soils.
[11] The Khmers built the ancient city of Angkor Wat, Cambodia, from plinthite, making good use of its 'cut and dry' properties.

9.4.1 Nutrient-holding capacity and stocks

In many tropical regions soil carbon is the main factor determining nutrient-holding capacity of the soil, and thus provides a useful index of general fertility. Typically more than 95% of soil N and S, and at least half of soil P, is incorporated in organic matter, and many soil micronutrients (e.g. Zn, Mn, Fe, Cu) are also positively correlated with soil organic matter. Fine-grained organic matter additionally improves soil structure and pH buffering (Zech et al. 1997). The principal reason why rain forest soils soon become less favourable to plant growth after forest clearance is that soil organic matter declines (Tiessen et al. 1994).

Forest organisms (plants and microbes) compete for nutrients—and this competition is fierce when nutrients are scarce, ensuring that few nutrients escape: efficient nutrient capture indicates relative scarcity. As a result, the majority of scarce nutrients are recycled and retained within the top soil, leading to top soil enrichment over time (Jobbagy and Jackson 2001), and nutrients such as P, K, and Ca typically decrease rapidly with depth (see Figure 9.2). Consequently, in most rain forests, but especially in infertile soils, the majority of fine roots, mycorrhiza, soil microbes, and organic matter are in the top 10–20 cm, ready to intercept nutrients that arrive at the soil surface.

Most nutrient uptake occurs in the top few centimetres of soil, but from how deep are nutrients extracted? Deep soils lack the abundant fine roots needed to exploit large volumes of soil, and mycorrhizal fungi are scarce—but deep tap roots which often descend several metres do access soil moisture with some associated nutrients.

9.4.2 Flows

Water can carry significant nutrients to and from forests. For example, below-ground flows carry biologically significant phosphorus between catchments affecting stream side vegetation at La Selva, Costa Rica (Genereux and Jordan 2006).

Organisms also transfer nutrients, especially over small scales. Termite mounds and anthills are typically enriched soil. Places where animal droppings collect (e.g. African tree-hyrax dung heaps) are enriched in P and other nutrients. Fungi capture and translocate nutrients over distances

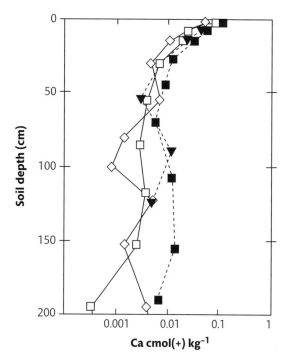

Figure 9.2 Exchangeable Ca^{2+} concentrations versus depth, in four soil profiles from a deeply weathered schist and pegmatite geology near Petit Saut dam, 30 km inland from the Atlantic coast of French Guyana. Sample key: a thick plateau top sandy clay loam (◊); a deep sandy clay hillslope soil (□); a sandy loam over a red loamy saprolite (■) and a loamy sand covering a white loamy saprolite (▼)

From Poszwa et al. 2002.

of several metres (Boddy 1993; Yavitt et al. 2004). Bryophytes too capture and modify nutrients. In Monteverde, Costa Rica, epiphytic bryophytes sequester inorganic N from atmospheric deposition and transform significant quantities to organically bound forms (Clark et al. 1998).

Nutrients are lost from vegetation in various ways, including leaching, herbivory, emissions, and secretions, but mainly through loss of plant parts which constitute fine litter, consisting mainly of leaves, and some flowers, fruits, and fine twigs, and coarse litter, comprising larger woody debris. This litter is in turn a principal source of nutrient inputs and organic matter to soils. In dryland forests, local nutrient inputs tend to vary with litter inputs. Studies in Borneo show that local-scale variations in nutrient inputs (typically 30–70%) are largely explained by local litterfall (Burghouts et al. 1998). Litter inputs and quality decline with reduced soil nutrients, due to plant nutrient conservation strategies. These include increased longevity of leaves, better defended tissues, and more comprehensive withdrawal of nutrients before leaves are shed. Low litter nutrient concentrations also result in slower decomposition (Wood et al. 2009).

Below-ground inputs from roots and mycorrhiza are important, although often overlooked. For example, in three nutrient-poor 'premontane' sites in Venezuela, total above-ground litter production was relatively low (5.6 tonnes ha^{-1} yr^{-1}) while fine root production contributed twice as much (11.14 tonnes ha^{-1} yr^{-1}) (Priess et al. 1999).

9.4.3 Decomposition

Nutrients can be passed from plant to herbivore, and onward through the food chain, but sooner or later nutrients are released through decomposition. Decomposition dynamics represents the intersection of stand carbon and nutrient cycles.

Decomposition processes are both biological (respiration, transformation) and physical (fragmentation, weathering and leaching). In rain forest environments decomposition is often rapid—but not always. Some components are lost rapidly (days to months) while some may linger for centuries, although most litter breaks down within a year. Long-lived items include large pieces of dense lignified wood (Mackensen et al. 2003). Most long-lived soil carbon is processed into complex 'humic compounds' which resist further decomposition (Allison 2006), but a substantial amount is reprocessed into other materials, including glomalin, a fungal glycoprotein that accounts for over 5% of nitrogen in the surface soil at La Selva, Costa Rica (Lovelock et al. 2004).

Decomposers need nutrients too. Decomposer soil microbes require a C:N ration of around 8:1 to effect decomposition, but most plant material, especially wood, has too much carbon. Good nitrogen availability and warmer conditions in most lowland rain forests help explain why decomposition is usually more rapid than in montane forests where nitrogen is scarce (Vitousek et al. 2002). When metabolizing carbon-rich substrates, some microbes secure additional nutrients by fixing atmospheric nitrogen, and fungi translocate significant reserves from elsewhere. Competition among growing microbial populations means that nutrient availability often declines following carbon-rich inputs. Nutrient additions suggest that decomposition rates in many older lowland rain forest soils are limited primarily by P availability (Hobbie and Vitousek 2000; Cleveland et al. 2002). Even in wet sites with rapid decomposition, added P further accelerates the process (Cleveland et al. 2006).

Litter 'quality' too can control decomposition and thus determine nutrient release dynamics. Such limits are most obvious on podzolic soils, deep peats, and in montane forests (Kraus et al. 2003). In nitrogen-limited montane rain forests such as in Hawaii, adding nitrogen-rich nutrients does little to improve decomposition because the tannin-rich litter inhibits decomposition (Hattenschwiler and Vitousek 2000; Vitousek et al. 2002). Short-lived leaves generally possess fewer defensive compounds (such as tannins)—thus litter quality in deciduous rain forests and in young secondary forests often improves decomposition processes.

After decomposition, nutrients may remain sequestered by soil microbes, be bound once again to organic matter or clay, or be leached out into streams or ground water, or may be accessed by plant roots and fungal hyphae.

9.4.4 Soil biota

Rain forest soils are home to an extraordinary biota. There may be more than 10^{16} individual living organisms in a tonne of soil (Curtis and Sloan 2005; Fitter et al. 2005). Live soil biota includes about 10% of soil organic carbon, of which less than half is microbial. Most of these organisms are involved in decomposition. Some processing occurs within organisms and some involves free enzymes released by them.

The soil biota act as both sinks and sources of nutrients. In moist carbon-rich soils most dissolved nutrients are rapidly immobilized by microorganisms. When microbial biomass declines, as happens during drier periods, nutrients become available to plants (Lodge et al. 1994).

Mycorrhizal fungi aid decomposition. Some ectomycorrhizal fungi access phosphate from incompletely decomposed material (Newbery et al. 1997). Plant–mycorrhiza associations have adapted to utilize the tannin-rich organic soils that can accumulate in some rain forests, and specialized mycorrhiza appear able to access N associated with polyphenol-rich environments (Hattenschwiler and Vitousek 2000; Hattenschwiler et al. 2003).

Termites and worms influence nutrient dynamics in rain forests (Chapter 5). Worms require a moist but not waterlogged environment (Fragoso et al. 2003) and can be abundant in montane forests. High earthworm populations promote rich 'mull' soils by allowing air into the soil profile. The Macrotermitinae termites of seasonal Paleotropical forests do not even wait for litter to fall: they cultivate and feed on subterranean fungus 'gardens' that break down cellulose and lignin even from freshly picked material. Neotropical leafcutter ants similarly cultivate fungi on freshly collected vegetation.

9.4.5 Nutrient losses

Phosphorus is rapidly immobilized in rain forest environments due to both chemical[12] and microbial activity. Potassium ions in contrast are highly soluble and would,

in the absence of 'biology', be flushed away by high rainfall. Yet the tightest cycling (lowest losses) of potassium found anywhere occurs in wet rain forests (McDowell 1998; Jobbagy and Jackson 2001).

Nevertheless, while nutrient inputs and outputs are in approximate balance the lower nutrient stocks of older soils (Table 9.4) show that in the long term nutrient losses exceed gains. Nutrient losses include solids (particulate), solutions, and gases, and although most nutrients are carried away in water, losses of Mg and Ca in particulate matter can exceed those in solution (Yusop et al. 2006).

On a catchment scale, nutrient losses typically increase with soil fertility. This is particularly striking when comparing the nutrient-rich white-water rivers that flow from beneath montane Andean rain forests with the nutrient-poor black-water rivers derived from the ancient and infertile Guiana Shield (Table 9.4). There are four non-exclusive reasons for these differences: (a) richer soils tend to occur on younger landscapes where natural weathering losses are high; (b) any soil (and litter) lost to rivers possess higher nutrient concentrations than from poorer sites; (c) root mats are less dense, providing less soil protection; and (d) there is less competition for nutrients, and less investment by plants and microorganisms in nutrient retention.

Table 9.4 Water chemistry of white- and black-water tributaries to the Orinoco and Amazon Rivers.

		Amazon		Orinoco	
Discharge area km^2		6,150,000		1,080,000	
Total sediment t yr^{-1}		1,200,000,000		150,000,000	
Floodplains by main channels km^2		92,000		7,000	
Tributary	Name	Solimoes	Negro	Meta	Atabapo
	Waters:	White	Black	White	Black
	pH	7.0	5.2	7.4	4.4
Conductivity (µS cm^{-1})		58.0	9.0	72.9	14.6
Calcium (mg l^{-1})		8.83	0.39	8.95	0.07
Magnesium (mg l^{-1})		1.26	0.13	1.67	0.03
Potassium (mg l^{-1})		1.40	0.46	0.93	0.15

From Godoy et al. 1999.

---

[12] In older acidic soils, aluminium and iron hydroxides can scavenge and bind inorganic phosphate, reducing availability.

For most nutrients, atmospheric flows out of forests are typically low, but nitrogen is an exception. Denitrification[13] often exceeds fixation (Vitousek et al. 2002) and may account for 24–53% of N loss from tropical forest soils (Houlton et al. 2006). Studies in Guyana suggest floodplains are important for nitrogen loss (Perreijn 2002). Tropical forests are also significant N_2O sources[14] (estimated as 2.2–3.7 Tg of N y^{-1}) (Prather et al. 2001).

9.5 Productivity, biomass, and carbon

'Tropical forests produce around 49 billion tonnes of biomass each year' (Leigh 1999).

Rain forests possess considerable reserves of carbon in living biomass, dead wood (necromass), as well as in litter and soils. Lowland rain forests typically possess an above-ground biomass of 200–600 tonnes ha^{-1}, the higher values being associated with more fertile soils. Estimated soil fertility accounted for a third of the above-ground biomass variation sampled in 65 central Amazonian dryland forests (range 231–492 tonnes ha^{-1}) (Laurance et al. 1999). Necromass is less intensively studied, but above-ground necromass appears to be an order of magnitude less than above-ground biomass (Maass et al. 2002). Two-thirds of necromass consists of branches and logs, while the rest is composed of fine litter.

9.5.1 Total carbon stores

About half (45–50%) of forest biomass is carbon, which moves from biomass to necromass (litter) to soil. Carbon stocks in tropical forest soils are considerable, but poorly known. Tropical peatlands are a major carbon store. Most of Indonesia's 200,000 km^2 of peat forests are deeper than 1 m, and many are more than 15 m deep (Rieley et al. 1997), and in total may store 5 x 10^9 tonnes of soil carbon (Page et al. 2002).

Trends in total forest carbon storage with increasing global temperature remain unclear: an increase in biomass with temperature may be balanced by a decrease of similar magnitude in soil carbon (Raich et al. 2006). **Net primary productivity** (NPP), below-ground carbon allocation, litter production, and surface litter turnover rates generally increase with mean temperature, while soil organic matter decreases (Raich et al. 2006).

Carbon is lost primarily by respiration (and rarely through fire). Terrestrial plants process around 12% of atmospheric CO_2 into carbohydrates each year, and tropical forests are likely to account for about a third of this. Most is eventually returned to the atmosphere or ocean. Rain forests process so much carbon that climatically determined variations in Amazonian productivity and decomposition lead to detectable annual variations in atmospheric CO^2 (Botta et al. 2002).

9.5.2 How productive are tropical rain forests?

In mature forest, the growth of biomass is usually more or less balanced by death and decay. Lowland rain forests typically add 1–5 tonnes (dry weight) per hectare of stem timber each year. Stem timber, though, is only a portion of overall biomass, and a compilation of 39 estimates concluded that NPP[15] varies from as low as 1.7 to as much as 21.7 tonnes C ha^{-1} yr^{-1} (i.e., Mg of carbon per ha per year). Two-thirds of the estimates were between 4 and 8 tonnes C $ha^{-1}y^{-1}$ (Clark et al. 2001b).

Forests with higher standing biomass tend to be more productive (Figure 9.3). Below-ground productivity is much less well characterized than above-ground processes, but is estimated to vary from 20% to 120% of above-ground productivity, with the higher percentages occurring on the least fertile soils (Clark et al. 2001b). Major uncertainties remain, including losses to herbivory

---

[13] There are several denitrification pathways, but a common one involves the reduction and loss of nitrate or nitrite to gaseous nitrogen, either as N_2 or oxides of nitrogen. Microbial decomposition also produces ammonia, which may vaporize into the atmosphere, leach with water, be microbially converted back into nitrates and reassimilated into living organisms, or be dissociated back into nitrogen gas.

[14] Which, like CO_2 and water vapour, is a greenhouse gas.

[15] This is defined as the total new organic (carbon-based) matter produced during an interval of time (usually a year). It is equivalent to the difference between total photosynthesis and total respiration.

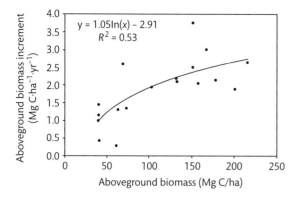

Figure 9.3 The relation between annual above-ground biomass increment and above-ground biomass for 18 tropical forest sites.

From Clark et al. 2001b.

and fungi, and as gaseous or dissolved compounds (Clark et al. 2001a) (Table 9.5).

Rain forests usually have decreasing productivity at higher altitudes. Reasons may include nutrient limitations, waterlogged soils, persistent cloud cover (low sunlight), and wetness (limited transpiration) (Burghouts et al. 1998). However, among wet forests in Asia, warmth of the growing season is the best simple predictor and accounts for the higher productivity of lowland tropical rain forests versus both temperate and montane forests (Takyu et al. 2005). Models suggest that under most circumstances NPP is primarily dependent on temperature and solar radiation, while variation in gross primary productivity (GPP) is dominated by solar radiation (Ichii et al. 2005). Nonetheless, relationships are not linear and different factors can gain significance under different circumstances. In Hawaii, amongst forests matched for elevation and substrate age, NPP increases over a rainfall gradient from 500 to 2,000 mm per year but declines at higher rainfall (Austin and Vitousek 1998).

Soil fertility too is influential. Amongst 104 Neotropical plots, above-ground production varied from 1.5 to 5.5 tonnes C ha^{-1}yr^{-1} increasing primarily in response to soil fertility (Malhi et al. 2004). However, as forests in less fertile regions invest a greater proportion of energy in below-ground biomass, the overall total biomass relationship remains uncertain.

Table 9.5 Estimated ranges (likely upper and lower bounds) of the components of NPP for old-growth tropical forests and the basis for these estimates.

NPP* component	Estimated range (Mg C ha^{-1} yr^{-1})	Basis
Fine litterfall (above-ground)	0.9–6.0	Measured values. Likely to be underestimates as not corrected for pre-collection decomposition.
Above-ground losses to consumers	0.1–0.7	Leaf herbivory as [0.136 x (0.75 x Litterfall)]; then increased by 20% to account for precollection seed and fruit consumption and feeding by sapsucking insects and nectar-feeders.
Tree biomass increment	0.3–3.8	Measured values
Other above-ground biomass increment (understorey, vines, palms)	0.03–0.4	Estimated as 10% of the tree biomass increment, based on relative biomass of these groups
Biogenic volatile organic compounds (isoprenoids, terpenes, etc.)	0.15–0.31 (0.93)	Isoprenes + monoterpenes + other reactive VOCs; model estimates, with high uncertainty
Above-ground organic leachates	Unknown	(No usable data)
Coarse root increment	Unknown	(No usable data)
Fine root increment	Unknown	(No usable data)
Fine root mortality	Unknown	(No usable direct measurements)
Root losses to below-ground consumers	Unknown	(No usable direct measurements)
Root exudates + CHO export to symbionts (mycorrhizae, nodules)	Unknown	(No usable direct measurements)

From Clark et al. 2001b.

9.6 Hydrology and land stability

How do rain forests affect hydrology? In 1931 a study was established in Java to resolve an already long-running dispute concerning whether geology or vegetation was the principal determinant of baseflow from tropical forest catchments[16] (Holscher et al. 2004). Arguments continue to this day.

Forests provide various protective functions, for example by reducing peak flow and protecting water quality. These roles are important but often exaggerated. The popular media describe forests as acting 'like a sponge' by absorbing water in wet periods and releasing it gradually, thereby maintaining streams in drier periods: a view long discredited. Other properties ascribed to rain forests are also questionable or can be substituted by other land cover. For example, rain forests are often claimed to improve catchment level water yields. In fact forest loss will generally increase sometimes decrease yields as forests actually lose more water back to the atmosphere (transpiration and evaporation) than many alternative land covers. Evidence for the true hydrological significance of rain forest cover is drawn mainly from long-term studies of changes resulting from forest loss. To understand rain forest hydrology we must look at inputs, outputs, flows, and controls. The general patterns are illustrated in Figure 9.4.

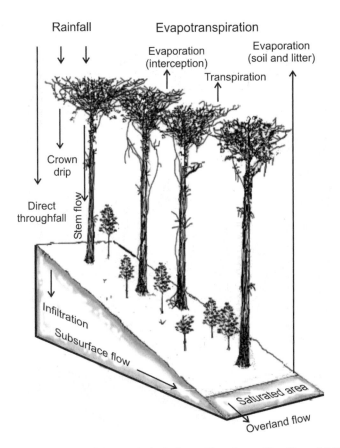

Figure 9.4 General patterns of water flow within a tropical rain forest. There is considerable variability among forest types and regions as to the magnitude of water flow within the different elements.

Figure supplied by Douglas Sheil.

[16] Unfortunately, due to World War II the study was abandoned while still incomplete.

9.6.1 Outputs: why forests evaporate so much water

We can distinguish two types of evaporation: transpiration, which is the flux of water vapour from within plants and is controlled by stomatal opening; and evaporation from wet surfaces, soils, and open water (including ephemeral pools).

Forests evaporate more moisture than other vegetation: large tropical trees can transpire several hundred litres of water each day (Goldstein et al. 1998), and closed tropical forests typically evaporate more than a metre of water per year (Gordon et al. 2005). If moisture is sufficient, forest evaporation is constrained principally by solar radiation and weather (Savenije 2004) and typically exceeds flux from short herbaceous cover by a factor of ten (Calder 2005).

Forests can evaporate water effectively because their height and canopy roughness lead to turbulent airflows. This has been termed the 'clothesline effect' as it echoes the reason why washing dries quicker on a line than when laid flat on the ground (Calder 2005). Access to water reserves is important. High stem volumes allow transpiration to outstrip root uptake, as stem water reserves are depleted by day and replenished at night (Goldstein et al. 1998; Sheil 2003). Trees typically have deeper roots than other vegetation and can access subterranean moisture reserves during dry periods (Calder et al. 1986; Nepstad et al. 1994). Vertical translocation of soil water through the soil profile at night by tree roots may also be important (Lee et al. 2005). In some sites, notably cloud forests and forests subjected to coastal fogs, high leaf areas and bryophytes contribute to efficient mists and dew interception (Dietz et al. 2007).

9.6.2 Inputs

One characteristic of all rain forests is high rainfall. But precipitation data only tell part of the 'inputs' story. Average measures of rain reaching the ground within a forest (**net precipitation**) are usually less than recorded in the open. Rain is interrupted by vegetation, and some evaporates or is absorbed before reaching the ground. Such 'interception loss' accounted for between 25 and 52% (591 and 1321 mm) of the incident annual mean rainfall at five lower montane sites in south Ecuador (Fleischbein et al. 2005).

In the upper montane forests of Cordillera de Talamanca, Costa Rica, epiphytic mosses store moisture equivalent to about 1 mm of rain (Holscher et al. 2004). Excess moisture arrives below only when this canopy storage is exceeded.

Though typically lower in lowland forests, interception remains significant. In a Sarawak (Borneo) forest, interception losses accounted for 14.5% of annual rainfall (2,292 mm) (Manfroi et al. 2004).

Sometimes, however, forest rain gauges receive more water than those placed in the open. 'Fog drip' occurs when fine cloud droplets held in the air coalesce on surfaces, or when water condenses directly from vapour[17]. Such capture is especially obvious in cloud forests where epiphytes, especially pendant mosses with finely feathered forms, play a significant role in stripping water from clouds. Due to evaporation, absorption, and epiphytic storage, the contribution of mist capture is generally greatest *within* the canopy. For epiphytic environments, fog moisture is often significant for local organisms even if it contributes little to measured precipitation (Table 9.6).

Cloud interception typically contributes 5–20% of moisture inputs in montane rain forests. In exposed montane locations, such inputs can exceed 1,000 mm per year (Stadtmüller and Agudelo 1990), varying with altitude and exposure (Cavelier et al. 1996). Even where the annual contribution is lower it can reduce seasonal water stress, especially as the seasonal contribution is often highest when rain is lowest. In Xishuangbanna, south-west China, fog contributes only about 5% of annual rainfall, but over 80% of this occurs in the dry season (Liu et al. 2004).

Moisture returned to the atmosphere contributes to further rain. The same water molecules may fall several times over larger forest regions before being lost. In the Amazon Basin such recycled rain may contribute a quar-

[17] The contribution of this second mechanism is poorly characterized but may be significant. For example, in a Venezuelan cloud forest night-time humidity reached 100% (implying potential condensation as temperature declines) on about half of all nights (Leon-Vargas et al. 2006).

Table 9.6 Measurement of fog precipitation in cloud forest environments.

Location	Elevation (m)	Annual rainfall (mm)	Fog precipitation (mm)	Fog precipitation (% of total water input)
Panama	500–1270	1495–6763	135–2299	2.3–60.6
Puerto Rico	930–1015	3204–4001	0–436	0–26.2
Costa Rica	1500	3191	886	21.7
Colombia/Venezuela	815–3100	450–1125	72–796	3.5–48.3
Guatemala	2100	2559	23	<1
Guatemala	2550	2559	203	7.4%
Hawaii	981–3397	300–2449	134–832	2.6–61.2
Mexico	1330–2425	215–1082	0+339	0–50.7
Venezuela	1750–2150	828–1009	354–592	26–41.7

Modified from Holder 2003.

ter or even a third of all rainfall (Eltahir and Bras 1994). Some have warned that forests may be much more critical in generating the terrestrial water and climate cycle than recycled rain data alone imply, and that large-scale forest cover loss can switch wet climates to arid ones (Makarieva and Gorshkov 2007).

9.6.3 Protection roles

Floods and overland flow

'…everything suggests that logging and forest clearing is less important in large and severe floods than in minor ones, but much is still not understood. Meanwhile politicians and the news media keep blaming most big floods on logging' (Kaimowitz 2004).

Forest loss changes hydrology. Specific impacts are dependent on soil, terrain, geology, vegetation, climate, and the nature of any new land-cover. So generalizations must be treated cautiously.

Rain forests have higher evaporation rates and interception losses than most non-forest vegetation. When montane forests which lack appreciable cloud water inputs are cleared, water yields typically increase by 10–40 cm per year. Despite the higher annual yields, forest clearance often results in reduced dry season stream flows in seasonal regions (Calder 2005). Most dryland rain forest soils possess an open structure allowing rainfall to infiltrate. This reduces run-off and increases belowground moisture reserves. Most streams are fed primarily by sub-surface flows, and soil moisture reserves contrib-

ute to stream flow smoothing out peaks and troughs in local availability. Overland flows occur only when the soil becomes saturated.

Soil infiltration often declines after forest clearing (due to compaction, loss of organic materials, and decreased worm activity). This reduces soil water accumulation and storage, and explains accounts of greater fluctuations in flow and of springs drying up after tropical forest clearance. Soil infiltration generally recovers with forest regrowth. Only five years of forest regeneration was enough to recover good soil infiltration after cultivation in Nigeria (Lal 1996). Though storm flows had increased by around 30% following cultivation in French Guyana, they returned to normal with rapid secondary forest regrowth (Fritsch 1993).

Floods are an increasing problem in many humid lowland tropical regions, most notably in cities, and deforestation is commonly blamed. Forests can reduce the severity of local floods caused by local rain as good soil infiltration (and to a minor extent interception losses and transpiration) reduce run-off. But at larger spatial scales, the protective functions of forests are reduced. Downstream impacts depend on wider rainfall patterns and the sum contributions from other catchments. Nonetheless forest clearance can exacerbate floods, as increased sediment loads fill watercourses.

Sediments and water quality

One result of efficient nutrient capture is water purity. Early attempts to assess forest stream impurities in the Usambara Mountains, Tanzania, failed—the streams were cleaner than the distilled water comparisons (Hamilton and Bensted-Smith 1989). On a local scale, outputs more or less match inputs. Nutrient concentrations in stream waters in Ecuadorian montane forest, for example, differ little from rainfall (Goller et al. 2006). On richer soils, however, water purity declines, as highly efficient nutrient capture is less necessary.

Low rates of soil and sediment loss under rain forest result from the ground protection provided by vegetation and litter, stabilization by roots, good soil infiltration, and the stable watercourse margins. Sediment yields can be as low as 0.1 to 0.5 tonnes ha^{-1} year^{-1} in forested catchments on stable ancient landforms, as in Central Africa, eastern Amazonia, the Guyana Shield, and the granite regions of Thailand–Malaysia (Douglas et al. 1999). In contrast, on the unstable volcanic mountain slopes of Java losses may reach 65 tonnes ha^{-1} year^{-1} (Holscher et al. 2004) and on the geologically dynamic New Guinea slopes 100 tonnes ha^{-1} year^{-1} (Douglas et al. 1999; Holscher et al. 2004).

Sediment flows vary considerably with specific location and moment in time. Heavy rains may saturate soils, after which overland flows may carry much more material into streams and rivers. Scale effects are important too. In small headwater catchments (< 50 ha), yields of suspended sediment are typically below 1 tonne ha^{-1} year^{-1} regardless of geology. Larger catchments can yield considerably higher figures due to cutting of river banks (Douglas et al. 1999).

Sediments and debris may take months to leave larger catchments. Some material moves only during short periods of highest flows (Douglas et al. 1999). Even in a level undisturbed rain forest in central Amazon, storm flows carry 25% of the total annual material export in only 5% of annual water outflow (Lesack 1993). Pools and boulders provide temporary 'storage' locations, and larger pieces of debris are often stranded high on river sides or accumulate in 'debris dams' as waters recede. Such raised 'stores' explain why high river levels move such quantities of loose material (Yusop et al. 2006).

9.7 Conclusion

Tropical rain forests are open systems with significance beyond their extent. They depend on a variety of inputs, and many other systems in turn are dependent on their outputs. Some of these interactions are commonly mischaracterized; other potentially important processes are neglected and may have global significance. Given the rapidly changing nature of many tropical forest regions and the planet in general, and our ignorance of many, and perhaps most, local, regional, and global connections and consequences, there are good reasons to be concerned that we may be taking risks of which we are unaware. Discoveries in these areas must continue for some years before we can assess the consequences of current changes.

CHAPTER 10

Plant form and function: what it takes to survive

'Few there are […] who seem to clearly realize how broad a lesson on the life-history of plants is written in the trees that make the great forest regions of the world.'

H. L. Clarke (1894)

'Even where the trees were largest the sunshine penetrated, subdued by the foliage to exquisite greenish-golden tints, filling the wide lower spaces with tender half-lights, and faint blue-and-gray shadows…Far above me, but not nearly so far as it seemed, the tender gloom of one such chamber or space is traversed now by a golden shaft of light falling through some break in the upper foliage, giving a strange glory to everything it touches.'

W. H. Hudson (1904)

All green plants need, and compete for, the same things: light, water, and nutrients. These demands shape ecophysiological adaptations in tropical rain forests. In this chapter we explore the range of plant adaptations and strategies for growth and survival in rain forest environments.

A number of plant adaptations have been introduced in Chapter 2 and environments in Chapter 7. Here we consider the challenge of effective light capture and utilization, moisture, nutrients, and problem soils. We then consider forest dynamics and the variations inherent amongst rain forest tree species.

10.1 Energy capture: light and shade

Tropical rain forests are solar powered. Sunlight, via photosynthesis (see Box 10.1), ultimately feeds the entire ecosystem. Light availability is fundamental in explaining variation amongst rain forest plants. While full sun bathes exposed canopies, deep shade prevails below. Illumination levels can vary by two orders of magnitude, between the darkest understorey shade and treefall gaps (see Chapter 6), and for most of their life most rain forest plants are limited by light[1]. The vertical stratification of plants responds to the predictable decline in light from the canopy to the shade beneath; but at the same time there are plants that depend on the short-term opportunities provided by increased light nearer the forest floor.

In the darkest understorey few plants survive long without access to alternative forms of energy. In such conditions the seedlings of large-seeded plants are able to draw, for a time, on the seed's resources; in other clonal plants, such as bamboo and some understorey palms, the shoots may be subsidized by neighbouring stems.

[1] Even canopy trees will grow more rapidly if additional light is available (Graham et al. 2003).

Box 10.1 C3, C4, and CAM

All three principal forms of photosynthesis known in higher plants—**C3, C4, and crassulacean acid metabolism (CAM)**—occur in rain forests. Most rainforest plant species, as elsewhere, are of the C3 type. C3 is the most efficient in low light (i.e. fixes more carbon per unit of light energy) but saturation reduces efficiency in strong light. C3 plants also have high water requirements.

C4 photosynthesis[1] is more efficient than C3 in strong light conditions and is common amongst weedy 'forest edge' species, especially grasses and sedges. Only two C4 rain forest trees are known, *Euphorbia forbesii* and *E. rockii* (Euphorbiaceae), both Hawaiian understorey trees (Pearcy 1983).

CAM occurs in perhaps 6–7% of all vascular plant species, including over 30 plant families (Cushman 2005). Though associated in many textbooks with arid climates, CAM is now known to be common in rain forests, notably occurring in about a third of vascular epiphyte species (Luttge 2004). CAM allows plants to store CO_2 allowing daytime stomatal closure

which reduces water loss. The principal benefits are high water use efficiency and drought tolerance, which is important for epiphytes. CAM also allows high photosynthetic capacity in strong light, the ability to photosynthesize when wet, and possibly reduced nitrogen demand (Benzig 1995; Skillman et al. 1999). CAM occurs in various orchids (Silvera et al. 2005) and epiphytic cacti (Andrade and Nobel 1996), and has evolved multiple times in bromeliads (Crayn et al. 2004). In equivalent low light environments, when water is not limited, approximately 10% less carbon is fixed by CAM than C3 (Krause and Winter 1996). There are also understorey CAM herbs and climbers (e.g. *Cissus* spp.). CAM is uncommon in montane forests where drought stress is infrequent.

Many CAM species can switch to C3 when it is advantageous (Borland et al. 1992; Cushman and Borland 2002). Some woody hemiepiphytic *Clusia* spp. (Clusiaceae) switch from C3 to CAM when water is limited (Borland et al. 1992; Wanek et al. 2002).

[1] Most of the 7,500 or so known C4 species (from over 19 plant families) are herbs of open habitats. Various rain forest habitat crops are C4, including maize *Zea mays* and sugar cane *Saccharum officinarum* (both Poaceae).

As illumination increases, plants harness surplus energy to allow growth, repair, and reproduction. The relationships are non-linear. At low light, productivity (fixation of carbon from CO_2) increases with illumination, but the relationship gradually levels off or 'saturates'. Saturation typically occurs well below (i.e. 20–30%) full sunlight (1,600–2,000 µmol m^{-2} s^{-1}) even in exposed canopy tree leaves, whereas saturation occurs at markedly lower intensities for most understorey species[2] (Niinemets and Valladares 2004). Another generalization is that shade-intolerant light-demanding species such as pio-

neers (see below) have higher photosynthetic potentials than shade-tolerant species under similar conditions.

10.1.1 Catching sunshine

Herbs and other understorey plants

Forest plants develop various forms and adopt various strategies that influence their ability to capture light (Boardman 1989). Understorey plants maximize light capture in low light. Despite many differing architectures,

[2] Maximum photosynthetic rates are, however, surprisingly variable and fourfold differences amongst co-occurring canopy tree species may occur (Zotz and Winter 1993; Niinemets and Valladares 2004). Light use efficiency can even vary considerably among otherwise similar individuals of a species exposed to different prevailing conditions.

most understorey plants possess very similar light capture efficiencies (Valladares et al. 2002). But some possess striking adaptations. Lens-like epidermal-leaf cells as in some Begoniaceae may focus light onto chloroplasts or reduce reflectance (Hebant and Lee 1984; Vogelmann et al. 1996). Many taxa both dicot (Begoniaceae, Gesneriaceae, Melastomataceae) and moncot (Commelinaceae, Zingerberaceae) possess red or purple anthocyanin pigments on the underside of foliage and are believed to increase light capture by reflecting unabsorbed light back through the leaf tissue (Lee et al. 1979). A smaller number possess blue iridescence—a physical property resulting from microscopic anatomical features that interfere with light and increase capture of the red wavelengths which prevail in the understorey (Chapter 8). Such iridescence has arisen independently in at least four taxa: *Selaginella* (spikemosses), ferns, Melastomataceae, and Begoniaceae (Hebant and Lee 1984; Graham et al. 1993; Gould and Lee 1996).

Tree structure

Understorey trees possess horizontally inclined (plagiotropic) shoots that produce relatively large and well-spaced leaves that minimize self-shading (Leakey et al. 2005; Pearcy et al. 2005). In such environments multi-layered canopies are uncommon (Figure 10.1). Self-shading is also reduced by leaf shape, size, arrangement, and orientation.

A growing tree must compromise between current and future needs for energy. As the environment changes through their lives, from understorey to canopy, flexible growth forms and turnover of structural elements allow trees to adapt (Kitajima et al. 2005). For example, understorey saplings of the Asian tree *Elateriospermum tapos* (Euphorbiaceae) possess large leaves and low self-shading, while canopy trees of the same species produce smaller leaves within a deep crown (Osada et al. 2002). Some species shift from broad plagiotropic forms with low self-shading to more orthotropic growth on attaining the canopy (King and Maindonald 1999). Only once they are tall enough do saplings start to build crowns with longer lived branches—this height varies from 0.1 to 16 m or more depending on the species (King 1998a).

Leaf placement

Leaves tend to be oriented further from the horizontal the brighter the conditions under which they grow. Upper canopy leaves are often held at steep angles to reduce excess exposure and heat stress, and allow light to enter the lower canopy (He et al. 1996; Falster and Westoby 2003). Trees adapted to high light typically branch with vertically oriented (orthotropic) shoots that successively produce small or very small leaves, which by attenuating[3] but not blocking sunlight provide moderate self-shading, reducing the hazards of **photoinhibition** or **photodamage**. In addition steep leaf angles may increase day-long carbon gain by improving light interception from low angles for emergent stems and upper canopy foliage. The fan-shaped laminae of *Licuala arbuscula* (Arecaceae), an Asian understorey palm, are supported by long petioles in a near vertical position when young to avoid self-shading, but as the leaf ages the angle to the horizontal declines improving light capture (Takenaka et al. 2001). Some plants, and particularly many climbers, adjust leaf positions to track radiation or, as in the understorey treelet *Psychotria limonensis* (Rubiaceae), to avoid shading (Galvez and Pearcy 2003).

Leaf properties and placement change with tree age. For example, saplings of the Asian pioneer *Macaranga gigantea* (Euphorbiaceae) produce their large lobed leaves (up to 60 cm long and wide) near to their stem; expansion takes 21 days, but the petioles continue to extend for 91 days, permitting the ageing leaf to extend beyond the new leaves being produced above. These leaves are smallest and thinnest in young fast growing plants and become longer lived and better defended in adult trees (Ishida et al. 2005).

Leaves in sun and shade

Within each species there is typically marked phenotypic variation between plants growing in high light and low light. Considerable phenotypic variation can even be found within a plant when comparing exposed canopy leaves and those growing in deep

---

[3] As the sun presents a disk, not a point, shadows are fuzzy-edged. With sufficient distance, when the angles subtended by each leaf are smaller than the angle subtended by the sun (about half a degree), sunlight is not blocked but is reduced.

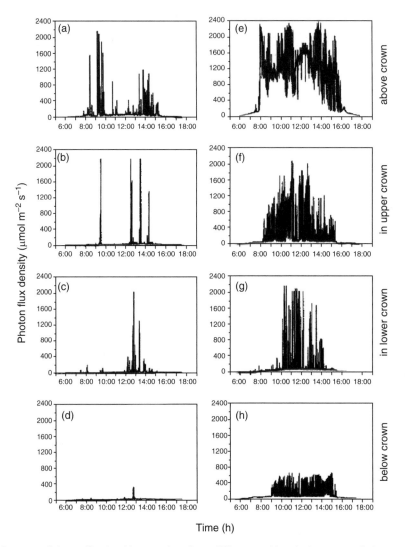

Figure 10.1 Diurnal courses of photon flux densities on a clear day at different positions in the crowns of a late-successional species, *Salacia petenensis* (Celastraceae), in closed forest understorey (A to D); and a pioneer species, *Heliocarpus appendiculatus* (Tiliaceae Malvaceae), growing in a gap (E to H). Note how much more light is available within and beneath the canopy of the pioneer than in the shade-tolerant species. All PFDs were measured with the sensors arranged horizontally relative to the plant's crown height.

Measurements made at Reserva Forestal de San Ramon, Costa Rica (Kuppers et al. 1996).

shade. The main differences relate to light acquisition and photosynthetic efficiency under the differing light regimes (Table 10.1). These traits are expressed mainly in the form and function of their leaves, although some are only partially understood (the large range in leaf sizes found amongst species often dominates inter-specific comparisons [Poorter and Rozendaal 2008]).

'Sun leaves' (leaves growing in direct sunlight) generally have higher maximum rates of photosynthesis than 'shade leaves'. Sun leaves are typically thick and rich in photosynthetic tissues with high nitrogen content per unit leaf area. Leaf rigidity tends to increase with irradiance (Oberbauer et al. 1987), and this stiffness reduces wilting and susceptibility to drought cavitation and can thus increase water flow to leaves (Salleo and Nardini

Table 10.1 How leaves and plants vary between sun and shade conditions–focusing on intraspecific patterns.

Trait	Sun (in contrast to shade)
Leaf physiology (per unit leaf area)	
Light saturation point	Higher
Light saturated photosynthetic rate (maximum energy gain)	Higher
Dark respiration	Higher
Light compensation point	Higher
Water demand at photosynthetic saturation	Higher
Illumination threshold for photoinhibition	Higher
Illumination threshold for photosynthetic induction	Higher
Carbon invested per unit leaf area per unit time	Higher
Lamina morphology (per unit leaf area)	
Stomatal density	Higher
Lamina thickness	Higher
Area to volume ratios	Lower
Specific leaf area (area per unit mass)	Lower
Leaf toughness	Mixed patterns
Individual lamina size	Often smaller
Venation density	Higher
Chlorophyll (per unit mass)	Lower
Chlorophyll (per unit area)	Similar
Leaf behaviour	
Leaf lifespan	Shorter
Leaf turnover	Faster
Biochemistry (per unit leaf area)	
Nitrogen content	Higher
Chlorophyll a to b ratio	Higher
Xanthophyll cycle pigments and/or UV-B-absorbing substances	Higher
Whole plant form	
Canopy layers	More
Leaf inclination	(Often) more vertical
Self shading	Higher
Leaf area index	Higher

From various sources: Givnish 1988; Osmond 1994; Kuppers et al. 1996; Zipperlen and Press 1996; Reich et al. 1998; Roth-Nebelsick et al. 2001; Reich et al. 2003; Niinemets and Valladares 2004; Ishida et al. 2005; Oguchi et al. 2005. There are numerous exceptions and complexities.

2000; Read and Sanson 2003). The thicker cuticle of sun leaves contributes to this rigidity but also acts to reduce moisture loss—though such leaves possess more and/or larger stomata per unit area allowing rapid CO_2 exchange (Sack et al. 2005). Such leaves are usually small and/or narrow and have high transpiration rates to avoid over-heating. They are often lighter-coloured with a preponderance of a form of chlorophyll called 'chlorophyll a' (Kitajima and Hogan 2003) and an abundance of sun-screen compounds that defend against damage caused by UV light (see Box 10.2).

Desiccation stress and wind are lower in the understorey, so shade leaves are typically built to enhance light capture: they are large and thin with a high area to biomass ratio (**specific leaf area** or SLA). They achieve a lower respiration rate in the dark than sun leaves. Shade leaves also tend to be darker, and although they can maintain photosynthetic gain at very low light levels, saturation generally occurs sooner than in sun leaves (Figure 10.2).

Temperature and humidity
Photosynthetic productivity is sensitive to temperature, usually peaking at 15–27°C for C3 plants[4] (see Figure 10.3) (Oberhuber and Edwards 1993; Niinemets and Valladares 2004). This optimum shifts a degree or two higher as illumination increases. In direct sunlight leaf temperatures can easily become too hot for optimal photosynthesis: there is a 60% decline in carbon gain at 38°C as compared to 28°C of *Shorea leprosula* (Dipterocarpaceae) seedlings (Leakey et al. 2005). In strong direct midday sunshine, when canopy temperatures can surpass 30°C, many tropical plants reduce or cease photosynthesis—a topic we discuss further below in relation to moisture stress. Respiration costs also rise with temperature.

10.1.2 Changing irradiance

The rain forest understorey is dim, but not uniformly so. Sunflecks create a patchwork of light across the forest floor which shifts with the position of the sun in the sky and varies with cloud cover and the geometry of the vegetation above. Occasionally full sunlight may reach

---

[4] 30–45°C for C4 and 35°C for CAM plants.

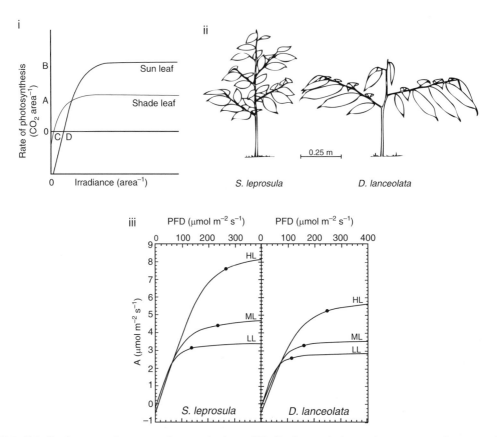

Figure 10.2 (i) Stylized account of apparent photosynthetic rate (CO_2 fixed per unit time and area, commonly expressed in micromol $m^{-2}s^{-1}$) for sun and shade leaves versus irradiance (expressed also as micromol of photons $m^{-2}s^{-1}$). The shade leaf saturation point (A) is lower than that for the sun leaf (B), but the compensation point (C) is also lower than that for the sun leaf (D) and the shade leaves are more efficient at low irradiance. Drawn by DS. (ii) Plant architecture reflects photosynthetic rates among dipterocarps: the shade-tolerant *Dryobalanops lanceolata* (right) avoids self-shading and puts more growth into lateral branches than the light-demanding *Shorea leprosula* (left) which bears the cost of self-shading in exchange for rapid height gain towards the canopy. (iii) Mean fitted photosynthetic response curves for *D. lanceolata* and *S. leprosula* seedlings in high (HL) medium (ML) and low light (LL) (Zipperlen and Press 1996). Note that *Dryobalanops lanceolata* has a higher carbon fixation capacity (and lower compensation point) at low light than *S. leprosula* which does much better in high light conditions.

the understorey. These sunflecks are typically localized, sometimes illuminating only a small area on a single leaf (Figure 10.4). On a daily basis understorey plants experience great fluctuations in light intensity, with sudden changes from 1% or 2% light irradiance in the deep understorey shade to near full sunlight within seconds (Chazdon and Pearcy 1991). To transmit full sunlight, a canopy gap must exceed an angular size of 0.5°[5]. Gaps smaller than this transmit a high proportion of weaker diffuse radiation. Longer-term changes occur following branch or tree fall, resulting in the exposure of the understorey to prolonged full sunlight, although the light regime still varies over the course of a day (Chapter 7). Seedlings, saplings, and understorey plants acclimatized to the deep shade conditions may suddenly, therefore, find themselves bathed in full sunlight, sometimes for as little as a few seconds, sometimes longer for up to an hour or more depending on the

---

[5] Equivalent to the apparent diameter of the sun viewed from the Earth.

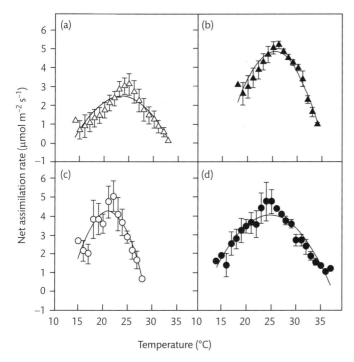

Figure 10.3 Response of apparent photosynthesis with temperature (at 2000 μmol m^{-2}s^{-1} PPFD and 350 μPa Pa^{-1}ambient CO_2) for C3 tree seedlings: *Castanospermum australe* (a, b) and *Flindersia brayleyana* (c, d) at upland (a, c) and lowland (b, d) sites in north Queensland (Swanborough et al. 1998). The assimilation peak occurs at a higher temperature in the warmer lowland than the cooler upland environment.

Figure 10.4 A shaft of light strikes the leaf of a *Licuala* palm in the otherwise shady understorey of the Daintree rain forest in Australia.

Photo by Robin Chazdon.

structure and movement of the overlying canopy and occasionally much longer following gap creation.

Illumination changes represent both opportunities and threats. Sunflecks can contribute 10–85% of total daily light exposure (Chazdon and Pearcy 1991) and can enhance carbon gain by as much as 60% (Pearcy et al. 1994; Leakey et al. 2005). They may also damage the delicate photosynthetic apparatus of shade-adapted plants, depressing photosynthetic rates and causing lasting harm. Canopy gap creation releases seedlings from severe light limitation, which, for many canopy species, is a necessary condition for recruitment into the canopy, and yet sudden and prolonged exposure to full sunlight brings significant stress from which plants may take time to recover.

Plants respond to changes in illumination in various ways. Photosynthetic processes can change within minutes. Prolonged exposure stimulates changes in leaf morphology or leaf replacement which can take days or weeks; and at longer time scales plant architectures reflect months (in the fastest growing species) or even many years of adaptation (Bongers and Sterck 1998).

Short-term adjustments: sunflecks and inductions

In closed canopy rain forest sunflecks are discrete and isolated, although they often come in clusters separated by extended periods of shade (Pearcy et al. 1994). Shade-adapted understorey plants capitalize on sunflecks by adjusting their photosynthetic behaviour through a process called induction: this involves biochemical changes and stomatal opening that allow a more effective photosynthetic gain (Paul and Pellny 2003; Niinemets and Valladares 2004; Houter and Pons 2005). Achieving full induction in light (and deactivation in shade) takes time, although 90% of maximum photosynthetic rates are reached within 3–6 minutes in most understorey species (Kursar and Coley 1993). Rapid induction allows understorey plants to make best use of sunflecks (Rijkers et al. 2000).

Induction responses to brief sunflecks are dependent upon prior irradiance history (Figure 10.5). Long delays between short bursts of illumination, as is characteristic of sunflecks, result in loss of induction state and reduce

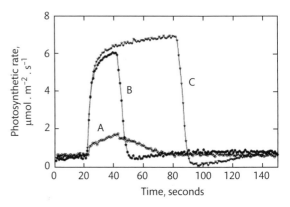

Figure 10.5 Photosynthetic responses of *Alocasia macrorrhiza*, an Australian rain forest understorey plant, to artificial sunflecks of 20 and 60 seconds at different states of induction. Response curve A belongs to a leaf exposed to a 20-second sunfleck following more than 2 hours in low light conditions, and therefore lacking prior induction; response B is also to a 20-second sunfleck but for a fully induced leaf; and C is a fully induced leaf responding to a 60-second sunfleck. In contrast to the situation in A, where the plant's stomata are closed and the photosynthetic machinery 'turned off', in B and C the stomata of the plant are already open and the biochemical pathways are already primed and ready for photosynthesis, thus allowing a fast response. Note in C the post-illumination reduced photosynthetic rate, reflecting inefficient photosynthesis during deactivation of prolonged induction.

From Chazdon and Pearcy 1991.

efficiency (Timm et al. 2002). More consistent illumination, provided by few long-duration sunflecks, therefore allows higher growth rates than a dynamic light regime characterized by many short-duration sunflecks (Tang et al. 1999; Leakey et al. 2005). Very high leaf temperature, which is more likely during long-duration sunflecks, reduces photosynthetic efficiency and may offset some of the benefit in natural conditions (Leakey et al. 2005).

Why plants lose their induction state is unknown, but some cost (in terms of water or energy use) of maintaining induction is assumed (Rijkers et al. 2000; Montgomery and Givnish 2008). Studies of closely related Hawaiian herb species that occur in rain forests and more open habitats indicate that dynamic photosynthetic light responses vary in a manner that allows plants to maximize daily leaf carbon gain from their own typical light environment if a cost for induction can be assumed (Montgomery and Givnish 2008).

Box 10.2 Avoiding sunburn

Rain forest plants possess various means to avoid sunburn—bleaching, damage of photosynthetic apparatus, and loss of photosynthetic performance[2]. Leaves in the upper canopy reduce irradiance per unit leaf area by oblique orientation to the sun and through self-shading amongst foliage. Sun-exposed rain forest leaves also often possess large quantities of coloured carotenoid '**xanthophyll**-cycle' pigments (violaxanthin, antheraxanthin, zeaxanthin) which appear to improve light and heat tolerance by enhancing the dissipation of light energy in the form of heat (Koniger et al. 1995; Dominy et al. 2003).

The epiphyte *Polypodium polypodioides* (Polypodiaceae) rolls its leaves on drying to expose a pale reflective lower leaf surface (Muslin and Homann 1992). Similarly dessicated leaves on the pioneer tree *Cecropia* (Cecropiaceae) twist so as to show their bright abaxial underside to the sun. *Arisaema heterophyllum* (Araceae), a perennial herb with a single palmate leaf, holds its leaf flat in shaded conditions, but folds it defensively under strong sunshine (Muraoka et al. 1998).

Many plants release **isoprene** in response to heat stress—this evaporative process provides some cooling, though the overall process is not well understood physiologically. Although isoprene emission is energetically expensive, it is thought to enhance photosynthesis recovery from short high-temperature episodes (Sharkey and Yeh 2001). Tropical forests are a major source of isoprene and other volatile organic compounds, that are believed to play a significant role in local and global climates (see Chapter 9).

Understorey plants are particularly vulnerable to **photoinhibition**. Although photoinhibition is often tolerable by unstressed seedlings and other understorey plants for extended periods (Lovelock et al. 1994), they minimize vulnerability by altering leaf angles (Table 10.2).

[2] Many of these responses are not specifically associated with tropical plants, although the dynamic light regimes of tropical rain forests make them particularly relevant to plant growth and performance in this biome.

Table 10.2 Leaf angles (± standard error) for a range of rain forest understorey species measured under canopy shade or in full sunlight (horizontal is zero). Plant groups are arranged into three shade-related categories following Whitmore (1984).

Species	Leaf angle	
	Sun	Shade
Group 1 Shade tolerant		
Pandanus krauelianus	14±13	23±13
Licuala sp.	60±6	10±8
Group 2 Shade tolerant but benefit from gaps		
Pometia pinnata	46±9	29±19
Maniltoa sp.	40±25	6±3
Group 3 Shade intolerant, gap specialists		
Macaranga sp.	69±15	50±19
M. quadriglandulosa	75±12	40±8
M. triloba	86±5	74±10
Endospermum sp.	75±10	16±8
Piper aduncum	64±13	11±6
Omalanthus novo-guineensis	88±4	42±3

Data from Lovelock et al. 1994; Watling et al. 1997.

Excess radiation

Too much light can harm plants. Full sunlight brings high temperatures and rapid water loss, and can lead to photorespiration, photoinhibition, and photodamage, resulting in reduced productivity and increased chance of mortality (Box 10.2). Transplant studies (e.g. Kursar and Coley 1999) and observations in natural vegetation (e.g. Krause and Winter 1996) show that shade-acclimatized understorey plants are vulnerable to damage from excess light. Understorey plants exposed to several minutes of direct sunlight show a decline in photosynthetic productivity, and may require several hours to recover (Legouallec et al. 1990; Watling et al. 1997).

Longer-term responses to light regimes

Rain forest plants respond to longer-term changes in light in various ways. The maximum photosynthetic rates of most shade-grown plants exposed to full sunlight rise over a period of several days or even weeks (Krause et al. 2004). Two days after shade-grown understorey shrubs *Hybanthus prunifolius* and *Ouratea lucens* were planted out in a Central American light gap, both showed substantial photoinhibition; seventeen days later both had higher photosynthesis without evident inhibition (Kursar and Coley 1999).

Not all rain forest plants adjust equally well[6]. Panamanian *Psychotria*[7] spp. (Rubiaceae) associated with high-light habitats ('gap species') are more flexible in their response

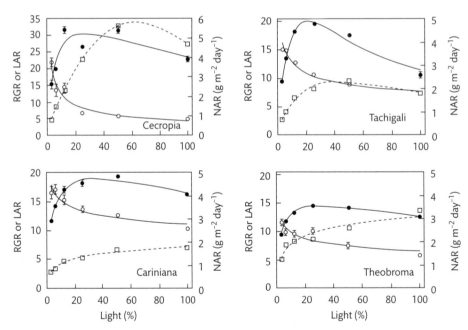

Figure 10.6 Responses of leaf area ratio (LAR, open circles, continuous line), net assimilation rate (NAR, open squares, broken line), and relative growth rate (RGR, filled circles, continuous line) to light for *Cecropia*, a pioneer species, *Tachigali*, an intermediate species, and *Cariniana* and *Theobroma*, two shade-tolerant species. Note that the scaling differs between graphs.

From Poorter 1999.

---

[6] Apart from species identity and ecological character, plants' capacities to adjust to new conditions depend on properties including growth history, foliage age and prior light regime, and water, nutrient, and mycorrhizal status (Lovelock et al. 1997; Kitajima and Hogan 2003).

[7] Mostly understorey treelets and shrubs.

to illumination (i.e. they are more 'plastic') than *Psychotria* species associated with shade, because they more rapidly replace their leaves and develop a new canopy (Valladares et al. 2000). In high light, light-demanding species are generally quicker and more flexible in adapting to changing conditions, but this partly reflects higher leaf turnover and higher intrinsic growth rates. Some general overviews, and a few rain forest studies, suggest that seedlings of 'mid-successional' species are most adaptable to a wide range of light regimes, as these species tend to occur naturally in both bright light and deep shade (Bloor and Grubb 2004; Walters 2005).

A comparison of seedlings of 15 tree species grown under controlled light conditions revealed strong developmental responses (Figure 10.6). While species that were fast growing under low illumination were also fast growing in well-lit conditions, patterns varied as species growth rates peaked, reaching a maximum at intermediate levels of illumination (25–50%). These comparisons indicated that most interspecific variation in relative growth rate was due to a measure of morphology (i.e. the leaf area ratio), whereas at high light it was mainly determined by physiological traits (i.e. the net assimilation rate: NAR), when illumination achieved more than 10–15% daylight (Poorter 1999).

The relative growth rates and shade tolerance of species under different light regimes is a topic to which we return later in this chapter.

Leaf lifespan and habitat
Leaf lifespan varies considerably amongst rain forest trees. Species can choose to allocate resources to short-lived leaves with high photosynthetic assimilation rate or to long-lived well defended leaves. In the Venezuelan Amazon mean species leaf lifespans varied 30-fold (1.5 months to 5 years) amongst 23 tree species (Reich et al. 1991). Longer lived leaves are associated with lower net photosynthesis and leaf stomatal water conductance, but increased leaf toughness and specific leaf area. To be worth the carbohydrate investment, leaves must repay more than their own

construction costs, thus few understorey plants can afford short-lived leaves (Poorter et al. 2006a). Similarly, in very nutrient poor environments, nutrient resources are insufficient to rapidly replace plant parts—thus leaves are long lived[8].

10.2 Water

Plants need water, and the availability of moisture is the single most important factor determining the distribution of major biomes in the tropics including the rain forests. Water shortages prevent transpiration, photosynthesis, the uptake of nutrients, and growth. Like light, moisture availability varies in both time and space over various scales (flood tolerance too is a significant factor in some tropical forest environments [Baraloto et al. 2007]). In many drier tropical forests, tree growth varies positively with annual rainfall (Lugo et al. 1978; Whigham et al. 1991; Bullock, 1997). In seasonal forests too experiments show that dry season irrigation boosts seedling growth (Bunker and Carson 2005). Even in less seasonal well drained sites with high rainfall (>3000 mm yr^{-1}), mean tree and sapling growth rates are often lower in drier than wetter years (Hazlett 1987; Breitsprecher and Bethel 1990; Da Silva et al. 2002; Newbery and Lingenfelder 2004). Patterns are not consistent, as sunlight and rain, both needed for growth, tend to co-vary negatively due to cloud cover: increased cloud cover can itself reduce growth, while short cloud-free droughts can stimulate growth (Clark and Clark 1994; Sheil 2003; Wright and Calderon 2006). In fact, larger trees when able to access deep water often grow fastest in the dry, sunny part of the year (Huete et al. 2006).

10.2.1 Transpiration: water and CO_2

Rain forest trees face a problem common to all trees—that of transporting water from the soil to the canopy.

---

[8] Some theories suggest a trade-off between longevity and photosynthetic capacity, such that total per leaf life-time assimilation is relatively constant (Mediavilla and Escudero 2003) and leaves are shed when the net gain by a leaf per unit time becomes maximal with respect to the leaves' entire lifespan (Hikosaka 2005).

High water demand within a plant can lead to high tensions in the water column. If the water column breaks—an event known as **cavitation**—transpiration stops. Although the water column can recover, too much cavitation can kill the plant. Higher transpiration rates required by faster growth are facilitated by wider diameter vessels, but it is believed that the risk of cavitation then also increases (Tyree 2003).

Rain forest plants show a range of vulnerabilities to cavitation related to the maximum tensions that are normally experienced within the plant (Lopez et al. 2005). Taller plants experience higher maximum tensions than shorter plants, and species with lower wood density appear more susceptible to cavitation than dense timbered species (Lopez et al. 2005). Even for seedlings drought tolerance seems to be a major determinant of inter-specific variation in growth and mortality (Engelbrecht and Kursar 2003).

Rooting depth can play a key role. In canopy gaps, deeper rooted species can maintain higher transpiration rates (Jackson et al. 1995). Nonetheless, shallow-rooted species may have some advantages. Observations in Borneo heath forests show that while shallow-rooted species tend to suffer higher mortality during drought, they also respond more rapidly to light rains that wet only the upper parts of the soil profile (Cao 2000).

Plants lose most water through their stomata, and the simplest means for plants to limit excessive water loss is simply to close their stomata. As this prevents water uptake (and stops photosynthesis), plants employ various additional mechanisms to maintain transpiration while reducing cavitation risks. These include stem water reserves, inclining foliage to reduce heating, the adoption of CAM, and greater investment in roots. When extended droughts are predictable, deciduous behaviour (leaf loss) is favoured (Tyree et al. 1998).

Stomatal opening controls both carbon acquisition and water loss, and plants are forced to balance trade-offs[9]. Thus despite bright illumination, sun-exposed seedlings of the Brazilian *Eperua grandiflora* (Ceasalpinoidae) have low carbon fixation at midday; this is thought to reflect the stomatal closure needed to limit short-term water stress (Pons and Welschen 2004). Emergent trees similarly show declining midday carbon gains (Roy and Salager 1992; Ishida et al. 1999). Very wet conditions also interfere with transpiration by blocking stomata: observations in Colombian montane forest revealed that photosynthetic declines associated with wet leaves were common (Letts and Mulligan 2005).

Air 'clings' to leaf surfaces forming a **boundary layer** that acts as a buffer between the atmosphere and internal leaf humidity. This reduces moisture gradients across the stomata and lowers transpiration (Schuepp 1993). Smaller leaves generally possess a smaller boundary layer, thus canopy leaves (and dry forest leaves) tend to be smaller than in the wet understorey (Monteith and Unsworth 1990; Schreuder et al. 2001).

Seedlings' drought tolerance can influence species distributions (Engelbrecht et al. 2007; Baltzer et al. 2008) especially amongst evergreen species (Poorter and Markesteijn 2008). Species with drought-intolerant seedlings are positively associated with wetter sites such as moist slopes. At Sepilok, Borneo, distinct tree communities grow on sandstone ridges, alluvial valleys, and mudstone slopes that lie between them; apparently the trees of sandstone ridges have higher stomatal closure in drier conditions which reduces their susceptibility to water stress, though at the cost of higher dark respiration rates, lower photosynthetic nitrogen-use efficiencies, and, presumably, reduced growth (Baltzer et al. 2005).

10.3 Nutrients and soil chemistry

The tropical rain forest's luxuriant vegetation is not proof of fertile lands, as often assumed by early Western explorers and colonists. Early disappointments soon generated a different view emphasizing the apparent paradox of abundant vegetation on poor land: the greatest above-ground biomass does not necessarily occur on the most fertile soil, and root biomass is generally higher where soil fertility is lower (Table 10.3).

---

[9] The gradients driving water loss are much greater than those governing CO_2 gain, which is why stomatal controls on productivity, even in wet rain forests, are sensitive not only to available water but to humidity.

Table 10.3 Some stocks and flows in tropical rain forests.

Site	Above-ground biomass (tons/ha)	Root biomass (tons/ha)	Total soil nitrogen (kg/ha)	Total soil phosphorus (kg/ha)	Leaf turnover time (years)
Caatinga, Venezuela	268	132	785	36	2.2
Evergreen forest (oxisol), Venezuela	264	56	1697	243	1.7
Lower montane rain forest, Puerto Rico	228	72.3	–	–	2
Evergreen forest, Ivory Coast	513	49	6500	600	–
Evergreen dipterocarp forest, Malaysia	475	20.5	6752	44	1.3
Lowland rain forest, Costa Rica	382	14.4	20,000	7000	–
Seasonal forest, Panama	326	11.2	–	23	0.9

Modified from Terborgh 1992.

Early research emphasized that most rain forest nutrients are locked up in above-ground vegetation, but in fact a significant fraction typically occurs elsewhere, especially in the top few centimetres of the soil (Chapter 9). Soil nutrients occur in various chemical and physical states and contexts that influence their biological availability. The top 20 cm of tropical peat soils typically contain over 2 tonnes of nitrogen (N) ha^{-1} but little is available to plants without the action of oxidative processes (Fao 2001).

Nutrients influence plant metabolism and productivity. Phosphorous (P) in particular often limits the productivity of forests on old and weathered soils (Vitousek 2004), although nitrogen has also been found to be strongly limiting in some tropical (mainly secondary) forests (LeBauer and Treseder 2008) (see also Chapter 9). Soil variation[10] influences vegetation through the availability of phosphorus and other essential nutrients, and aluminium toxicity (a potential problem when pH drops below 5.5) (Sollins 1998). Thus nutrient-rich white-water flooded Amazonian forests are characterized by higher growth and productivity than nutrient-poor black-water systems (Wittmann et al. 2006). Nitrogen is often scarce in montane forests, and adding nitrogen-rich fertilizer increases tree growth (Tanner et al. 1992; Cavelier et al. 2000). In lowland forests these patterns are not always so clear, and observations in wet lowland forests often fail to detect any direct influence from soils on tree growth or litter production (Ashton and Hall 1992; Clark et al. 1998; Mirmanto et al. 1999; Newbery et al. 2002). This is presumably due to masking by other factors (Coomes and Grubb 2000), insufficiently long term monitoring, and/or limited local variation. Nonetheless, phosphorus availability has been shown to explain differences in growth rates, particularly of large trees—and of overall above-ground net primary productivity[11] in a lowland forest in West Kalimantan, Borneo (Paoli and Curran 2007; Paoli et al. 2008) (Figure 10.7).

High leaf nitrogen and phosphorus are both positively associated with high photosynthetic efficiencies[12] (Reich

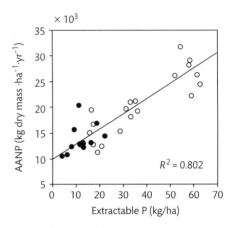

Figure 10.7 Variation in above-ground net primary productivity along a soil P gradient at Gunung Palung National Park, Indonesia.

From Paoli et al. 2005.

et al. 1994; Raaimakers et al. 1995a). The nutrient status of individual plant parts is, however, under complex feedback control, as demand for carbon is, in turn, related to nutrient status. Plants also need various other macronutrients (calcium, magnesium, potassium, sulphur) and micronutrients (boron, copper, iron, manganese, molybdenum, nickel, zinc) to grow effectively. Although sites are often described in terms of 'fertility', nutrient limitations are not equivalent. For example in Hawaiian montane forests low nitrogen availability has a less significant effect on root dynamics than low phosphorus (Ostertag 2001).

The magnitude of nutrient limitation decreases with shade. Differences are easiest to detect among trees growing in high light conditions (Ashton and Hall 1992; Bungard et al. 2000). Light-demanding species are also more responsive to fertilization than shade-tolerant species, although both groups do show clear positive responses under relatively high light (Lawrence 2003). Nutrient addition experiments with shade-tolerant seedlings in only 1% daylight in Singapore indicate small but significant positive effects that will influence competition

[10] Other soil properties apart from chemical composition are also important in influencing vegetation, including drainage and water-holding capacity, which affect rooting depth and tree stability.

[11] The two are linked—as growth of trees over 60 cm diameter contributed 38–82% of plot-wide biomass increment and explained 92% of variation among plots (Paoli and Curran 2007).

[12] Such patterns are complicated by non-photosynthetic nitrogen associated with other functions, such as leaf defence.

and survival[13] (Burslem et al. 1995, 1996). Plants of nutrient-poor sites invest more in nutrient capture, defence, and use-efficiency (Coomes and Grubb 2000; Paoli et al. 2005). This means they tend to have long lived leaves, even in seasonal environments (see Box 10.3).

Rooting reflects both moisture and nutrients. A study in four contrasting Neotropical forests implied that seedlings developed denser (longer and finer) and shallower root systems per unit of photosynthetic area in sites with lower fertility. Wet nutrient poor soils impose a selective pressure to maximize the surface of root absorption over the depth to forage (Paz 2003). Some nutrients, including nitrogen, phosphorus, and potassium, remain partially

mobile within a plant[14], moving to new leaves if needed. Extraction of leaf nutrients prior to leaf fall can be considerable, especially for phosphorus (average of 62%) and nitrogen (43%) (Reich et al. 1995). Minerals from soils are only part of the story. Nutrients are often derived directly from the litter layer, and seedling growth has been demonstrated to decline when leaf litter is removed (Brearley et al. 2003).

Below-ground competition impacts plant growth, especially when soil fertility is low or water is limited (Lawrence 2003; Barberis and Tanner 2005). Trenching (creating a barrier between the roots of neighbouring plants) typically increases tree seedling growth in wet

---

Box 10.3   Deciduous in evergreen rain forests

Contrary to many accounts, wholly evergreen rain forests are rare and canopy phenology can vary markedly. Deciduous species occur as emergents, canopy species, and pioneers in most formations, especially in well drained or shallow soils and seasonal environments. Deciduousness confers drought avoidance advantages, but is more common in adult plants than in seedlings, presumably because seedlings often lack sufficient carbohydrate reserves to replace their leaves annually (Poorter 2008). Moisture stress in the understorey is also typically lower than for more exposed taller plants. Thus many tropical deciduous species possess evergreen seedlings (e.g. *Tooona ciliate*, an Asian Meliaceae, and *Tabebuia rosea*, a Neotropical Bignoniaceae). Contrarian phenologies occur: the Costa Rican shrub *Jacquinia pungens* is leafless through the rainy season when it is in deep understorey shade (Janzen 1983).

Being leafless may have other benefits. Some tree species drop their leaves for only a few days a year—which may reflect a means to shrug off herbivores. Many deciduous species flower and/or fruit during

their leafless phase when visibility to animals is increased (Van Schaik et al. 1993).

How do tropical deciduous and evergreen species differ? Most differences follow general patterns of leaf longevity. Evergreen leaves tend to be tough and well defended while deciduous species possess higher leaf nutrient concentrations, higher photosynthetic efficiencies, and fewer defences (Wright et al. 2004). Deciduous seedlings often have a greater total leaf area (relative to plant mass) and higher overall growth rates than evergreen species (Bowman and Prior 2005). Similar patterns are believed to occur in adult trees (Reich et al. 1997). Deciduous plants usually possess higher hydraulic conductivity than comparable evergreens in the same habitat (Eamus and Prior 2001) and may often achieve higher peak transpiration rates (e.g. in Amazonian Meliaceae) (Dunisch and Morais 2002). Fast growing deciduous species also have an advantage in early succession because evergreens require more time to develop full foliage. Hence secondary forest tree communities are typically more deciduous than older forests.

---

[13] Patterns are complex and therefore not readily summarized. Though most plants growing in full sunlight respond to increased nutrients with more rapid growth, at low irradiance some respond with reduced growth. There is no certain explanation, but it may relate to associated respiration costs incurred by the plant.

[14] Retranslocation tends not to occur for boron, calcium, copper, iron, manganese, nickel, sulphur (in part), and zinc.

lowland forests on nutrient-poor soils (Whitmore 1966; Fox 1973; Coomes and Grubb 1998; Lewis and Tanner 2000), in forest gaps (Lewis and Tanner 2000; Barberis and Tanner 2005), and in seasonal forests (Gerhardt 1996), but not detectably in shaded wet forests on fertile soils (Denslow et al. 1991; Ostertag 1998).

An area drawing recent attention is the proposal that juvenile plants may sometimes gain nutrition (minerals and/or carbohydrates) from adults via below-ground connections. This is of course the nature of various clonal plants such as bamboo—but what about genetically distinct trees? Transfers might be via fused roots—a phenomenon specifically noted in some rain forest trees. Alternatively it could occur via mycorrhizal networks. In either case the ability of an adult to be selective in assisting genetically similar juveniles would appear to make biological sense. There is in any case likely to be some benefit to juveniles from improved mycorrhizal colonization, even if there were no net transfer of nutrients from adult to juvenile (Onguene and Kuyper 2002). A species-specific ectomycorrhizal network leading to positive seedling selection near to adults is currently one of the leading theories to explain monodominance in several Caesalpinoid tree species (Mcguire 2007) (see Box 3.2). There is also evidence that soil specialization reflects indirect effects, related to plant defence and herbivory (Fine et al. 2006) (see Chapter 8).

Problem soils

Tropical soils can inhibit plant growth, not only through nutrient scarcity and poor physical properties, but also by elements in excess (Vonuexkull and Mutert 1995; Iturralde 2001; Proctor 2003; Brady et al. 2005). Aluminium (Al), for example, limits crop productivity on many acid tropical forest soils and is almost certainly an important—if poorly understood—factor for tropical forest plants. Toxicity tends to occur at pH values below 5.5 and includes a large (but uncertain) proportion of tropical rain forest dominated by oxisols and ultisols. The toxicity arises from the creation of specific soluble forms of aluminium. Rain forest plants occurring on aluminium-rich soils typically employ one of two basic strategies: 'excluders' prevent aluminium from entering their tissues (mycorrhiza may be involved, especially in the Ericales); and 'accumulators' take up and tolerate aluminium (Box 10.4). Similar strategies occur on soils rich in other toxic metals such as the nickel-rich ultramafic soils of Cuba and New Caledonia, and the cobalt- and copper-rich soils of Congo.

Box 10.4   Hyper-accumulators

Hyper-accumulators are plants defined by their ability to accumulate high concentrations of specific elements (including Mn, Zn, Ni, Cu, Se, Cd, Cr, Pb, Co, Mg, Al, and As) without incurring any ill effects. Most hyper-accumulators probably occur in tropical rain forests. Aluminium hyper-accumulators occur in over 45 mostly woody plant families[3] from the wet tropics (Jansen et al. 2002). Some plants accumulate aluminium above 1,000 parts per million (ppm), the record being 72,240 ppm (i.e. 7.2% of plant mass) for an Indonesian *Symplocos spicata* (Symplocaceae) (Jansen et al. 2002).

Reasons proposed for accumulating such elements include: (1) inadvertent uptake, (2) tolerance via internal sequestration, (3) improved drought resistance, (4) competitive **allelopathy**, by enriching litter with compounds that have a greater negative impact on competitors (see Box 10.5), and (5) anti-herbivore or anti-pathogen effects (Boyd 2004). Different advantages may operate in different cases, but the first three ideas have gained little support. Allelopathy may explain nickel and lead enrichment of soils from the litter of the rain forest tree *Sebertia acuminata* (Sapotaceae) of New Caledonia (Boyd and Jaffre 2001). 'Elemental defences' have been documented in several contexts, but not yet in rain forests (Coleman et al. 2005). By passing concentrated and often toxic material into food webs, accumulators may have significant local impacts on tropical forest food chains (Boyd 2004).

[3] *Lycopodium* spp. are an exception in being non-woody.

10.4 Growth and survival in the space race

Forest changes unfold at multiple temporal and spatial scales, and taken together define what is meant by forest dynamics. Components of change include disturbance and recovery from disturbance, including succession (topics covered in Chapter 11), but patterns of change are also shaped by smaller-scale processes acting at individual or species levels, foremost of which are plant growth and survival.

'Trees a crowd'

With enough time, tiny seedlings can become towering forest trees—it can take centuries. Some rain forest trees can grow nearly 10 m height in a single year under ideal conditions[15], but within forests conditions are less benign.

There is limited space in a closed canopy forest: for some trees to grow large others must die. Few trees survive to adulthood—but those that do are standing in the space of many fallen adversaries. Competition for sunlight mediates tree growth and death, which together drive the continuous state of change that all tropical forests experience. Such turnover is part of the normal inherent dynamics that maintain the forest. Senescence and size probably play a role, thus even the biggest trees cannot live forever.

Tropical tree growth rates are often expressed as changes in stem diameter, and diameter growth in tropical trees (over 10 cm dbh) is usually around 1–10 mm year^{-1} (Worbes 1999). When the canopy opens and illumination increases, plant growth surges as seedlings compete for light and space (Figure 10.8)—but light and space are limited, competition is intense, and most plants die young. In the deep understorey many seedlings barely grow at all: some *Chrysophyllum* seedlings in Australian rain forest had virtually no growth for 27 years (see Figure 10.12) (Connell and Green 2000). Typical estimates suggest only one seed in 100,000 may become a 1-cm-diameter sapling (Harms et al. 2000).

Figure 10.8 These 5 m tall *Shorea leprosula* seedlings in Sabah, Borneo, were planted in a canopy gap as young seedlings of 20 cm only 18 months before this photograph was taken.
Photo by Alex Wild.

---

[15] *Paraserianthes falcataria* trees in plantations have achieved 9.91 m annual height growth.

Much local variation in tree growth rates is explained by competition for light (Clark and Clark 1992; King et al. 2005), though other life-forms, especially lianas can also play a significant role (Box 10.5). Seedlings and saplings are much more impacted by adult trees than adult trees are by seedlings and saplings—a form of competition that is termed 'asymmetric'. Other plant life forms, including herbs, shrubs, and palms, can also impact on seedlings, and tree seedling densities are commonly inversely related to the abundance of these life-forms (Harms et al. 2004).

Although herbivores and falling debris are major causes of mortality of herbs, seedlings and other understorey plants (Clark and Clark 1989), mortality is principally a function of tree growth and competition. Amongst individuals, as an indication of health and a good competitive status, the most rapidly growing stems are more likely to survive—while the slowest growing stems more often die. This dependence explains why, when populations are observed over long periods, the average growth rates of surviving stems are typically higher than that observed over short intervals.

Taller plants cast shade on shorter neighbours. Even among closely planted seedlings, height differences of 1 cm can distinguish likely winners and losers (Tanner et al. 2005). Although size and light are important, the seedlings of some species have intrinsically higher growth rates than others and can catch up. In artificial canopy gaps in Bornean dipterocarp forest, the tallest seedlings maintained their height advantage for 3–4 years but were finally overtopped by late starting but faster growing *Shorea johorensis* (Whitmore and Brown 1996). Smaller stems endure more shade and larger stems gain more light—thus mean diameter growth rates typically increase around tenfold as stems move from 1 to 10 cm diameter (Figure 10.9). Mortality rates tend to decrease with increasing size in the understorey until plants reach their maximum size. Patterns among canopy trees are less size-consistent: plant growth slows as trees mature and invest energy reserves in flower, fruits, and seeds (Thomas 1996b). The very largest trees of any given species often have lower growth rates and higher mortality indicative of senescence (Vanclay 1991; Clark and Clark 1999)[16].

Box 10.5   Competition between trees and lianas

Lianas occur in the canopies of many larger tropical rain forest trees. They reduce tree growth and fecundity, and increase the likelihood of, and susceptibility to, physical damage (Grauel and Putz 2004; Kainer et al. 2006; Schnitzer and Bongers 2002). Clearing away lianas has long been recognized as a silvicultural management intervention that improves tree growth (Dawkins 1958; Pena-Claros et al. 2008). Shade is only part of the competition story—much of the battle takes place below ground (Schnitzer et al. 2005). Unlike trees, lianas need not invest in massive structural support—instead they are believed to develop more extensive, deeper and more dynamic root systems, and are thus able to exploit soil nutrients and water reserves more effectively (Andrade et al. 2005; Schnitzer 2005).

Trees may fight back. Architecture and behaviour may reflect the evolutionary consequences of trees' ancient battle with lianas. There is some evidence that dropping leaves, especially compound leaves, helps to shed lianas. Other potentially beneficial characteristics include rapid growth, self-pruning branches, narrow crowns, flexible main stems, and hosting protective ants (Putz and Mooney 1991).

---

[16] Senescence is poorly studied in rain forest trees, and the physiological basis remains uncertain. Size rather than age may be key. As an organism gets bigger the energy needed to maintain it also increases. As crown size (and thus energy capture) is related to area (dimensions squared), while some physiological costs relate to volume (dimensions cubed), there is likely to be a maximum size after which maintenance costs begin to outweigh energy gains.

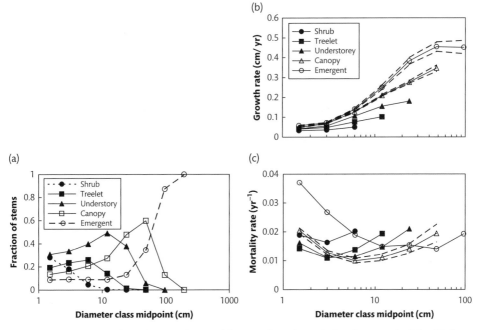

Figure 10.9 Population dynamics of woody plants by size. (a) Proportion of trees belonging to each of five life forms in each octave-wide (two-fold) stem diameter class in Pasoh Forest Reserve, Peninsular Malaysia. Life forms are 'shrubs' (95% of stems over 1 cm dbh are less than 5 cm dbh), treelets (95% are less than 12 cm dbh excluding shrubs), understorey trees (95% < 25 cm dbh excluding treelets), canopy trees (95% < 60 cm dbh), and emergents (95% ≥ 60 cm dbh). (b) Mean diameter growth and (c) mortality rates vs. octave-wide diameter class for trees of each life form.

From King et al. 2006.

The lifespans of tropical trees remain poorly known (Box 10.6). Background mortality rates among larger stems typically average around 1–2% per year[17] across a wide range of tropical rain forest sites and rarely exceed 3% per year (Lugo and Scatena 1996; Muller-Landau et al. 2006). Competition ensures mortality in the absence of exogenous causes, but often these factors operate in concert, and in Chapter 11 we examine tree turnover as a disturbance process. Any resources or well-lit space will provoke competition from plants in any position to take advantage (Box 10.7). Thus it is not simple to allocate mortality to endogenous or exogenous causes.

What of growth differences among species? Generally, when grown under shared conditions, species characteristic of productive environments (viewed in terms of light, nutrients, or water) have higher growth rates than species associated with less productive conditions—the actual conditions of the trial have little impact (Chapin et al. 1986; Grubb 1998; Baker et al. 2003; Russo et al. 2005). It appears that 'slow species' invest more in defence while 'fast species' take greater risks.

Differences in species-specific demographic rates are generally much greater at seedling stages and are extenuated by the contrasting light environments within which young plants find themselves (deep shade of the understorey versus the bright light of a canopy gap). Tall adult trees, by comparison, tend to experience much less variation in light regime even in the subcanopy.

10.5 Sun trees, shade trees: pioneer and 'climax' species

Even the least observant rain forest visitor should notice a difference in vegetation growing along forest edges and abandoned fields on the one hand, and within more

---

[17] Similar to temperate and boreal forests.

Box 10.6   How old are rain forest trees?

Wild papayas, *Carica papaya*, in Mexican rain forests can live out their lives, from seedling to senescence, in only eight years (Martinez-Ramos and Alvarez-Buylla 1998). Most rain forest trees live much longer. The oldest reasonably reliable date for a planted tropical forest tree is 288 BC for a *Ficus religiosa*, planted by King Tissa at Anuradhapura in Sri Lanka, and (we believe) still alive as we write (Lewington and Parker 1999 cited in Harrison 2005). Such records are exceptional, and, aside from papaya and

a handful of other short-lived pioneer species, direct observation of trees' lifespans takes too long to be practical. Various indirect tree-ageing approaches have been developed. These range from approximations based on size and extrapolations from growth rates to sophisticated isotopic methods.

Size alone is only a rough guide to age: while larger trees tend to be older, there can be considerable variation both between and within species (Figure 10.10). For example, Amazonian trees of 60 cm diameter

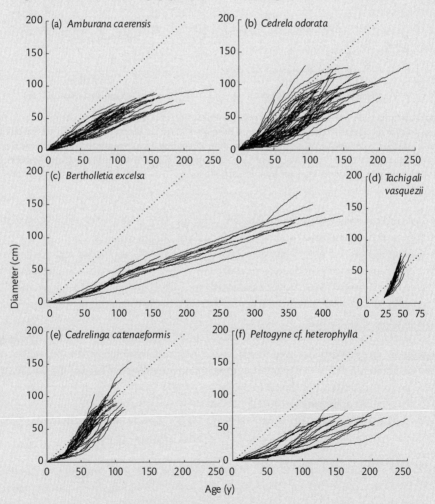

Figure 10.10 Age and diameter of six different rain forest tree species as determined by growth rings in the Bolivian Amazon. The dotted line is equivalent in all the figures to help comparison (Brienen and Zuidema 2006).

Figure 10.11 Contrary to common beliefs, many rain forest timbers, like this specimen of *Sclerolobium setiferum* (Fabaceae: Caesalpinoidae), from Amazonia, Brazil, possess distinctive growth rings. This specimen shows two dated scars that have been used to judge the relationship between time, season, and ring formation.

Photo by Douglas Sheil.

vary two- or threefold in age even within a given species at a single site (Brienen and Zuidema 2006).

Tree rings—a well-established means to date temperate trees—have less often been exploited in the wet tropics where many (but not all; see Figure 10.11) timbers lack clear annual rings. Even in seasonal locations, where rings are more common, they are often irregular and/or hard to distinguish. Thanks to improved methods tree rings are now increasingly being used to evaluate growth and age (Worbes 1999; Worbes et al. 2003; Parolin et al. 2004; Brienen and Zuidema 2005).

Radiocarbon (carbon-14 isotope) methods provide relatively reliable ages for wood samples taken from older stems (more than 500 years old). A distinct radioisotope approach uses traces of past nuclear bomb tests—a dated event that can be detected in wood—as markers to determine subsequent growth which can be used to estimate age (Worbes and Junk 1989).

Table 10.4 Maximum age estimates of large broad-leaved tropical rain forest trees estimated or measured at particular diameters or at their maximum observed diameter (max).

Location	Maximum age (y)	Diameter (cm)	Species	Source
Projections (based on short term dynamics)				
Costa Rica	608	100	*Lecythis ampla*	(Clark and Clark 1992)
Central Amazon	981	Max	*Pouteria manaosensis*	(Laurance et al. 2004)
Mexico	22–1030	Max	*Brosimum alicastrum*	(Martinez-Ramos and Alvarez-Buylla 1998)
Ring analysis				
Costa Rica	650 (3)	156	*Hymenolobium mesoamericanum*	(Fichtler et al. 2003)
Cameroon	220		*Celtis zenkeri*	(Worbes et al. 2002)
emergents	124	–	*Celtis zenkeri*	(Worbes et al. 2002)
canopy	220	–	*Celtis zenkeri*	(Worbes et al. 2002)
understorey	146	–	*Celtis zenkeri*	(Worbes et al. 2002)
Thailand	257	–	*Afzelia xylocarpa*	Baker pers. comm.
Bolivia	427	180	*Bertholletia excelsa*	Brienen
Brazil (varzea)	183	104	*Macrolobium acaciifolium*	(Wittmann et al. 2006)
Brazil (igapo)	502	108	*Macrolobium acaciifolium*	(Wittmann et al. 2006)
Radiocarbon dating (only >350 yr)				
Central Amazon	1370 ± 80	180	*Cariniana micrantha*	(Chambers et al. 2001)
Sarawak	1207	121	*Eusideroxylon zwageri*	(Kurokawa et al. 2003)
Brazil	440± 60	233	*Bertholletia excelsa*	(Camargo et al. 1994)
Guyana	350	110	*Chlorocardium rodiei*	(Zagt 1997)

Adapted from Brienen 2005.

All these methods have weaknesses. For example, none account for the sometimes considerable time spent as a seedling (see later). Ageing seedlings indirectly is also problematic, although leaf spacing can be used in some (King and Clark 2004).

Leaf-scar counting can be used to age palms. In drawing these estimates together, it becomes clear that most rain forest tree species can readily live for centuries and some for over a millennium (Table 10.4).

Box 10.7   Spiteful behaviour

The life and death struggle of competition—only winners survive to reproduce—means that plants may be spiteful. Growing faster than your neighbours can be achieved by speed, and also by spite. Thus, plants can benefit from inhibiting the growth of their neighbours and from reducing the establishment of competitors.

Plants may seek to shade their neighbours. Models suggest that competing trees should produce more foliage overall, and more horizontally distributed leaves, than might be predicted by optimization of light capture alone—the reason is to minimize useful light reaching and helping neighbours (Hikosaka 2005). Such predictions remain unverified.

Competing directly for resources is one form of competition. There may be others. Plants may produce chemical compounds primarily to interfere with the germination, growth, or viability of other plants: such interactions are called 'allelopathic'. Though often considered, clear evidence of allelopathic interactions among rain forest plants has proved elusive due to the difficulties of distinguishing possible effects from other competitive processes[4] (Inderjit

and Weiner 2001). The presence of adult *Grevillea robusta*, a non-gregarious proteaceous tree, has been inferred to be toxic to conspecific seedlings in the Australian (sub)tropical forest (Webb et al. 1967). Tropical examples are scarce, though they have been implied for specific observations such as the scarcity of strangler figs growing on other strangles (Titus et al. 1990).

An instructive case is the Amazonian tree *Duroia hirsuta* (Rubiaceae), which forms monodominant stands lacking undergrowth (Pfannes and Baier 2002)—this according to local beliefs represents the work of evil spirits. Early research suggested the soil itself somehow inhibited seed germination and seedling development (Campbell et al. 1989). However, the clear area is now known to be due, at least in part, to the effect of symbiotic ants (*Myrmelachista schumanni*) that kill other plants with formic acid (Chapter 13). Whether or not this is considered 'allelopathic' becomes a matter of semantics (Frederickson et al. 2005). Such 'allelopathic symbiosis' may be favoured in rain forest climates where costly and soluble allelopathic compounds are soon washed away or broken down.

[4] For example allelopathic interaction has been suggested: to reduce adhesive climbers on rain forest trees in Queensland Australia (Talley et al. 1996); to assist the invasion of exotic *Lantana camara* into forest margins; and to affect plant spacing patterns on the Caribbean island of Guadeloupe (Prugnolle et al. 2001). Sometimes impacts are hard to classify—for example fast growing bamboo can acidify soil, thereby increasing aluminium toxicity for neighbouring plants—but given that this is a side effect of growth, it is not considered allelopathic, though it may play that role.

Table 10.5 Some typical relative characteristics of pioneer and non-pioneer tree species.

	Pioneer species	Climax species
Alternative names	*Early successional, secondary, pioneer, nomads,* (light-demanders)	*Late successional, primary, non-pioneer, dryads,* (shade-bearers)
Habitat association	Disturbed areas, more common in fertile areas	Undisturbed areas
Population structure	Often in even-aged cohorts	Abundance declines with size, more juveniles than larger trees
Longevity	Often short	Long, sometimes very long
Germination	Gap-dependent	Not gap-dependent, though may benefit from gaps
Seedlings	Persist only in open well-lit conditions (rare in understorey)	May persist in closed forest, and often abundant in understorey
Seed/fruit production begins	Often early in life	Usually late in life
Seed size	Usually small and plentiful, often many per fruit, and often or continuously	Often large and produced annually (or less frequently)
Seed dormancy	Often long-lived (orthodox), and many can be dormant in seed bank	Seldom long-lived-most germinate rapidly (recalcitrant)
Dispersal	Often long distances (wind or animal)	Usually short-range (diverse mechanisms)
Growth rate (diameter height and biomass)	High (under good illumination)	Low
Compensation point	High	Low
Branching	Sparse–often few orders	Denser–often multiple orders
Leaves	Usually thin (low mass per unit area), large sized, often with long petioles or sometimes very small	Relatively small-sized, often with drip-tips
Leaf longevity	Short, high turnover, often deciduous	Long, overlapping leaf generations, slow turnover
Leaf defence	Often low (soft, few chemicals)	Often high (tough, chemically defended)
Defensive ant mutualisms	Common	Rare
Potential photosynthetic rates of seedlings	High	Low
Photosynthetic water use efficiency	Low	Variable
Carbohydrate 'energy' reserves	Usually low	Often higher
Roots	Highly branched (sometimes deep) mycorrhizal associations may be absent	Mycorrhizal (length variable)
Root turnover[*]	Rapid	Slow
Nutrient absorption per unit root mass[*]	High	Low
Principal nitrogen source[*]	Often nitrate	Often ammonium
Wood density	Usually low	High
Wood properties	Usually pale, not durable	Pale to dark, often durable, sometimes siliceous
Ability for adults to re-sprout from broken stem?	Generally poor	Often good
Mortality rate	Higher	Lower
Sensitivity to pathogens	Higher	Lower
Acceptability to generalist herbivores	Higher	Lower

From Whitmore 1998; Turner 2001; Pons et al. 2005 and other sources.

or less continuous unbroken forest on the other. Trees differ in ability to establish and survive in shaded environments, and differences in physiology and life form of many plants can be attributed to membership of two ecological groups: 'pioneers' that, to a rough approximation, depend on high light environments throughout their existence[18]; and 'climax' species, that germinate in the deep shade of the understorey which they tolerate for many years before finally reaching the canopy (Swaine and Whitmore 1988; Turner 2001). Here we summarize a range of traits associated with the pioneer–climax dichotomy (Table 10.5).

10.5.1 Pioneers

Pioneers have high photosynthetic and respiration rates and low wood density[19] allowing rapid growth under high light conditions (Raaimakers et al. 1995). In open areas on good soil pioneer tree species typically establish and grow rapidly while light, nutrients, and water are available. Pioneer species in genera such as *Cecropia* and *Vismia* in the Amazon, *Macaranga* in Asia, and *Musanga* in Africa have net photosynthetic capacities (both per unit leaf mass and per unit leaf area) several times greater than many non-pioneer species. Their crowns are usually open-branched to occupy a large space for maximal light capture. Pioneers possess markedly higher hydraulic conductance per unit leaf area in their root and shoot than climax species, allowing them to transpire more rapidly and maintain productive photosynthesis under strong illumination (Tyree 2003). While biomass accumulation rates within the first few years of growth can be rapid, they often decline with age to rates comparable to many climax species. Most pioneers are, however, comparatively short-lived[20], and early rapid growth is sufficient to maintain their position in a high light environment and to maintain the carbon assimilation rates necessary for near continuous and prodigious reproductive output.

Pioneer leaves (both large and small) tend to have high metabolic rates and high turnover. By preferentially investing in rapid growth, pioneers sacrifice chemical (secondary compounds) and structural (fibre content) leaf defences, and consequently their leaves tend to be palatable to herbivores. Many pioneers (e.g. species belonging to *Cecropia*, *Piper*, *Ochroma*, and *Macaranga*) instead rely on defence by ant-guard mutualists (Chapter 13). Low wood density associated with rapid growth may have facilitated the evolution of ant defences by providing ants with suitable nesting locations (some pioneers even have hollow stems), while high photosynthetic rates in high light conditions allow inexpensive production of sugar-based foods for ants.

Pioneer tree species generally achieve high individual growth rates, even as shaded juveniles. Due to high nutrient requirements to sustain rapid growth, pioneers differ more markedly in growth rates when compared across sites of variable nutrient abundance than do slower-growing species (Huante et al. 1995; Fetcher et al. 1996; Raaimakers and Lambers 1996). Indeed, most likely due to their greater competitive advantage in the most productive environments, pioneers appear more common on more fertile soils (Ashton and Hall 1992; Hawthorne 1996; ter Steege and Hammond 2001).

Pioneers typically produce numerous seeds but provide few resources for each[21]. A single *Cecropia obtusifolia* tree, for example, may produce several million seeds a year (Alvarez-Buylla and Martinez-Ramos 1990) each with an air-dried mass of approximately 0.68 mg (Dalling and Hubbell 2002). The light seeds of pioneers are dispersed far, often by wind, which may explain why pioneer genera are widely distributed and species-poor—efficient dispersal reduces the development of distinct populations[22].

---

[18] That is, germination, seedling establishment, and subsequent growth.

[19] Exceptions include *Milicia excelsa* and *M. regia* (Moraceae), African pioneers with dark, dense, durable wood; and *Casuarina* (Casuarinaceae) in the Asian tropics, which are similarly heavy-timbered pioneers.

[20] Although the pioneers *Gustavia superba* (Lecythidaceae), *Macrocnemum glabrescens* (Rubiaceae), and *Jacaranda copaia* (Bignoniaceae) have low mortality rates at Barro Colorado Island (Condit et al. 1995).

[21] Exceptions include *Aleurites* spp. (Euphorbiaceae), which have about 8 g (walnut-sized) seeds, and more familiarly, the coconut *Cocos nucifera* (Arecaceae).

[22] The Asian tropics are richest in pioneer species—a pattern most likely due to the fragmented populations created by the Melanesian Islands.

Seeds may lie dormant until a tree fall or similar event presents favourable conditions for germination (Agyeman et al. 1999). Tropical forest soils are often rich in a range of dormant pioneer seeds (Chapter 12).

Most pioneer trees belong to just a few families: Euphorbiaceae, Malvaceae, Sterculiaceae, Tiliaceae, Ulmaceae, and Cecropiaceae. Major genera include *Cecropia* (Cecropiaceae, Neotropics), *Macaranga* (Euphorbiaceae, Paleotropics), *Musanga* (Cecropiaceae, Africa), *Trema* (Ulmaceae, Pantropical), and *Ceiba* (Malvaceae/Bombacaceae, Neotropics and Africa, though widely naturalized in Asia).

10.5.2 Climax species

Climax[23] species generally germinate, establish, and persist below canopy shade (Swaine and Whitmore 1988b), though many (including some large light-demanding canopy trees such as several *Parashorea* and *Shorea* of Southeast Asia) do not last long as shaded seedlings and

die within a year unless exposed to full light. Other climax tree seedlings and saplings persist for many years in the shaded understorey, often hardly growing in that time (Figure 10.12). In Guyana it has been estimated that some species take 12–126 years to reach 3 m in height (Zagt 1997). These slow growing strongly shade-tolerant species possess dark, dense, and often silicaceous timber and dense crowns with long-lived well defended leaves. Dense wood provides strong stems that are resistant to damage and attack, and contribute to longevity.

The seeds of climax species are often large and rich in resources, allowing young seedlings to establish even in deep shade, and to develop roots that can penetrate thick litter. Seedlings persisting in the understorey form 'seedling banks'—a mass of small plants waiting for the opportunity afforded by gap formation. The seedlings respond rapidly to increased light and have a headstart over pioneer species that germinate from newly dispersed or dormant seed (see Chapter 12).

10.5.3 Shade tolerance

Concepts of seedling shade tolerance encompass the illumination required for establishment, survival, and growth. Complexity arises from the recognition that these are not necessarily the same thing. A broad compilation of seedling growth and survival studies suggests that light-demanding species actually grow faster than shade-tolerant seedlings in both high and relatively low light conditions (Poorter 2005). Light demanders have larger specific leaf area and higher photosynthetic potential, resulting in a carbon gain that exceeds that of shade-tolerant species regardless of the light environment. By definition, though, shade-tolerant species have an advantage over more light-demanding species in low light. What is this advantage? Shade-tolerants are frugal: they require less energy to maintain themselves, and so it is often assumed that they can continue to grow in low light levels when less tolerant species cannot (Agyeman et al. 1999; Baker et al. 2003). In fact, many controlled studies find that pioneers (and other light-demanding species) maintain higher relative

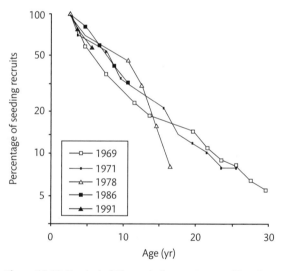

Figure 10.12 Survival of *Chrysophyllum sp. nov.* seedlings in Queensland, Australia. The years indicate when the recruits were initially mapped; the survival curves all begin with the number of seedlings aged 2.5 years. Only 6% of seedlings recruited in 1969 were still alive 27 years later and on average had only doubled their height. Note log scale on y-axis.

From Connell and Green 2000.

---

[23] The term 'climax' has fallen out of favour as it implies that a climax exists, which is debatable. A better term would be shade-tolerant canopy species, but as climax is simpler and widely understood we continue using this term.

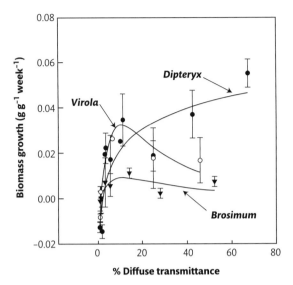

Figure 10.13 Seedling mean relative growth rate (± 1 SE) to percent diffuse transmittance (i.e. proportion full sunlight) for seedlings of *Dipteryx panamensis* (Fabaceae/Papilionoidae) (●), *Virola koschnyii* (Myristicaceae) (○), and *Brosimum alicastrum* (Moraceae) (▼), growing across a forest–pasture edge at La Selva Biological Station, Costa Rica. *Dipteryx panamensis* grows fastest of the three at high light and *V. koschnyii* better at low light, while *B. alicastrum* grows slower than both under all illuminations.

From Montgomery 2004.

growth rates than shade-tolerant species even down to 3% sunlight (Kitajima 1994; Veneklaas and Poorter 1998). These observations suggest that shade tolerance may be largely about survival (Bazzaz and Pickett 1980; Kitajima 1994; Veneklaas and Poorter 1998).

Field studies are few, but a comparison of three Costa Rican tree species found patterns suggesting that seedling responses among species may be very sensitive to slight differences in light regime. Thus variation in the range of only 1–7%, and by implication even very small gaps, elicits differential responses among species (see Figure 10.13) (Montgomery 2004). In any case, at very low light, growth is slow for all species and survival—waiting for improved conditions—is crucial. Rapidly growing species die sooner in shade than do most slow growing species. This likely reflects that the fastest growing species are also the least well defended (Kitajima 1994; Veneklaas and Poorter 1998), while shade-tolerant seedlings better repel herbivores and pathogens

(Mccarthy-Neumann and Kobe 2008) and can better survive repeated mechanical damage (Clark and Clark 1991; Coley and Barone 1996; Montgomery 2004; Alvarez-Clare and Kitajima 2007). In the shaded under-storey, biomass loss cannot be readily replaced, so shade-tolerant species reduce mortality through higher investment in long-lived well-defended leaves (Veneklaas and Poorter 1998; Poorter 2005). In high light pioneers dominate by growing fast, while in low light shade-tolerators dominate by simply surviving.

Patterns of plant adaptation to shade may differ with tree age, due to the necessity of time to build up sufficiently deep crowns and leaf areas (Sack and Grubb 2003), and also between seasonal and less seasonal environments (Poorter 2009). Multi-species comparisons in Bolivian wet forest species suggest that more shade-tolerant species tend to possess tougher and more persistent leaves, while in the drier more seasonal (and deciduous) forests shade tolerance is more clearly associated with high light interception and high water use. This suggests that the relative mechanisms that determine light partitioning vary over the climate gradient: in moist evergreen forest a growth–survival trade-off dominates while relative growth dominates in seasonal forests (Poorter 2009).

While some studies and theories suggest that there is likely to be a trade-off between shade and drought tolerance (Smith and Huston 1989), detailed evaluations in Bolivia suggest that in fact the traits associated with drought and shade tolerance are largely independent (Markesteijn and Poorter 2009). This offers potential for species to specialize separately for light and moisture, thus allowing a range of possible strategies and perhaps for increased species richness.

10.6   Functional traits as predictors of demographic rates

Functional traits are plant species attributes that influence survival, growth, and reproduction, which collectively determine plant fitness (Ackerly 2003). Trade-offs among these traits influence the outcome of interactions among plant types. Beyond the simple pioneer–climax distinctions discussed in Box 10.8, various functional traits appear to have explanatory power. For example an

Box 10.8   The pioneer–climax classification or continuum

The criterion of 'gap-dependent germination' as being sufficient to define pioneer tree species (Swaine and Whitmore 1988a) has been found to be inadequate (Whitmore 1996). Definitions of 'gap' and diverse seed germination behaviours (Chapter 12), confound simple dichotomous classifications. We now know that many species with 'pioneer' traits (Table 10.6) readily germinate in deep shade, even if they do not survive long (Alvarez-Buylla and Martinez- Ramos 1992; Kennedy and Swaine 1992; Kyereh et al. 1999; Novick et al. 2003).

Rather than a dichotomy based on germination alone, a broader multi-factor approach recognizing a pioneer–climax continuum, along which species are spread, may often be more insightful. Combined measures of leaf lifespan, plant growth, and population turnover could, for example, form the basis of a more refined classification system. Yet still there are complications. Growth rates change over the course of a plant's life, and mortality within a species is often a function of tree size, and can also be highly variable among otherwise similar species (Condit et al. 1995).

As long as pioneer and climax are recognized to be ends of a continuum, this simple categorization for all its faults helps considerably in making sense of tree survival strategies in tropical rain forests. Some species, such as South American mahogany *Swietenia macrophylla* (Meliaceae), have been classed as both a pioneer and climax species: germination of *S. macrophylla* seed in seasonal forest in Quintana Roo, Mexico, is higher under shade (a climax trait), but strong light is required for subsequent growth without which the seedlings quickly die (a pioneer trait) (Morris et al. 2000). Similar species elsewhere include African *Entandrophragma* and *Khaya*, and Asian *Toona* (all Meliaceae), and some *Shorea* and *Parashorea* (both Dipterocarpaceae). Such species are common amongst canopy species and have been termed 'non-pioneer light demanders' (Jones 1956; Clark and Clark 1992; Hawthorne 1995). Other cases include the 'cryptic pioneers' which establish in open areas but can later persist in forest shade, examples being the Panamanian *Alseis blackiana* (Rubiaceae) (Dalling et al. 2001), and African *Irvingia gabonensis* and *Klainadoxa gabonensis* (both Irvingiaceae) (Sheil et al. 2006). A classification of differences between seedling and adult light requirements has been described (Table 10.6).

Table 10.6 Species groupings based on perceived juvenile and sub-canopy shade-tolerance.

| | | Seedlings (<5cm diameter) primarily found in | |
		Well-lit habitat	**Shade**
Adults (> 20 cm diameter) primarily found in	Well lit habitat	*Pioneers* Seedlings develop most profusely in open areas and canopy gaps. Seedlings may be found in shade but seldom survive long in such locations. Includes many low light-timber species.	*Cryptic pioneers* Seedlings found in open areas and canopy gaps, but trees persist and reproduce even when the forest canopy has closed above them. Few species, but appear to be a widespread group.
	Shade	*Non-Pioneer Light Demanders (NPLDS)* Seedlings may be found under closed canopy, but illumination is needed for further development. Many species. Includes many high value (medium density) timber species.	*Shade bearers* All sizes >5cm dbh can be found under closed canopy, and will persist in these conditions. Many understorey species. Generally very high density timbers.

From Hawthorne 1995.

Another useful classification is based on combining the pioneer–climax dichotomy with an understorey–canopy (short–tall) dichotomy to make four classes (Turner 2001) or two axes of variation (Falster and Westoby 2005; Poorter et al. 2006). Further divisions can be useful. Applying these characters in the central Amazon identifies four relatively discrete tree guilds: fast-growing pioneer species, shade-tolerant subcanopy species, canopy trees, and emergent species (Nascimento et al. 2005). Of course all such classifications involve relatively arbitrary divisions along what are, in reality, continuous axes of variation in shade tolerance, height, and other characters.

exploration of demographic performance among 240 large tree species at five Neotropical rain forest sites found that all four of the traits they considered—seed size, specific leaf area, wood density, and plant height at maturity—were significantly correlated with relative growth rate and/or mortality rates across all five sites, indicating that species within these rain forests face the same trade-offs and adopt a similar set of life-history strategies (Poorter et al. 2008). Let us look at these traits in greater detail.

Seed size is a key predictor of reproduction and establishment. Large seeds provide plenty of resources for seedling establishment, even under low resource conditions. Seed size is associated with shade-tolerance related traits, including the slower growth and lower mortality of adult trees (Hammond and Brown 1995; Osunkoya 1996).

Specific leaf area (SLA) is a measure of the light-capturing foliar area per unit of leaf biomass invested. Species with high SLA tend to have high nutrient concentrations and high photosynthetic and respiration rates (Wright et al. 2004; Poorter and Bongers 2006). These are the 'pioneer-type' sun leaves associated with rapid growth, high turnover, and are highly palatable. Low SLA leaves are thick and dense, physically robust, and well chemically defended. They are therefore more expensive to produce and tend to be longer lived: the leaves of shade-tolerant 'climax' species. Specific leaf area is a major determinant of variation between species in seedling growth rates (Wright and Westoby 1999), but surprisingly it is not clearly linked to demographic rates among adult trees (Poorter et al. 2008). Leaf defence and removal are perhaps expected to have the greatest impact on young understorey plants, as they have fewer stored resources from which to replace losses. Large trees with substantial stored carbon reserves (Box 10.9) are better positioned to replace losses, and can therefore be more flexible in their leaf use strategies.

Wood density is biomass invested per unit wood volume. High growth rates are achieved by sacrificing structural robustness in the production of low density wood (King et al. 2005). High density wood, by contrast, is composed of small and densely packed sells with thick cell walls that build stems resistant to breakage (Van Gelder et al. 2006) and pathogen attack (Augspurger 1984).

Plant height at maturity determines access to light for adult trees. There is a tendency for taller tree species to be less shade-tolerant than species of smaller stature, even when observed as juveniles (Givnish 1988; Sheil et al. 2006) and tree growth rates are generally higher among the tallest species (King et al. 2006; Poorter et al. 2008). Shorter species possess inherently lower leaf photosynthetic capacity, and again this relationship appears to hold at sapling and adult tree stages (Thomas 1996a). Tall species hold their leaves in the brightest region of the forest canopy, and they also tend to have larger crown areas (Poorter et al. 2006). Hence their faster growth rates may reflect more effective light capture relative to smaller species. Tall species tend to have higher survival per unit time, but as they can take longer to reach adult sizes than short species, they may nonetheless suffer greater losses before gaining sexual maturity. These costs, and the investment taken in reaching large sizes, seem best explained by the high reproductive output that can be achieved by a well illuminated canopy tree. Trade-offs in size and reproductive output have been postulated as contributing to the maintenance of multiple tree

Box 10.9   Energy stores

Plants produce, store, use, and lose carbon compounds. The size of the mobile fraction of carbon, that is the non-structural carbon that can be made available on demand, reflects the balance between photosynthetic carbohydrate **(NSC)** uptake and the investments in plant structures. Except for defensive compounds, the mobile C-pool reflects a plant's overall carbon supply status, with the greatest fraction of this pool commonly present as non-structural carbohydrates (NSC, largely starch and sugars) (Figure 10.14) (Wurth et al. 2005). These reserves can be exploited for growth, repair, leaf flushing, or reproduction. Alternatively they can be invested, for example in structural and chemical defences, or stored in anticipation of future needs.

Carbon reserves vary with conditions, phenological cycles, and other demands. NSC-pools of tropical forest species have been observed to increase in drought periods even when coincident with major carbon-demands of flowering, fruiting, or leaf flushing. The relatively low NSC-concentrations in wet seasons may reflect plant investment in structural growth which represents a major C-sink, resulting in a smaller contribution to the NSC-pool. Less available

light due to wet season cloud formation may also limit photosynthetic carbon capture.

Despite some seasonal fluctuations, NSC-pools of canopy trees are sufficiently high to be able to provide the carbon necessary for a complete canopy leaf-replacement independently of any additional carbon inputs through photosynthesis during regrowth (Table 10.7). In some tree species high NSC concentrations are even found deep in the heart of the stem, which suggests that far from being 'dead' heartwood, there is instead active storage in living ray tissues in this region.

Photosynthesis is more limited in the understorey or subcanopy, so trees in these habitats are expected to be particularly dependent on carbon reserves for leaf replacement, and during periods of negative carbon balance imposed by shade or damage, as has been shown for *Piper arieianum* in the Neotropics (Marquis et al. 1997), and for shade tolerant Neotropical tree seedlings (Myers and Kitajima 2007). Carbohydrate reserves are strongly related to juvenile tree survival in wet forests, where more shade-tolerant species typically possess higher reserves (Poorter and Kitajima 2007).

Table 10.7 Forest NSC-pools as estimated from forest biomass and mean tissue type specific NSC, averaged across 17 species. All dry mass related data in t ha-[1].

Organs	Biomass (t ha[-1])	NSC (d.m %)	NSC (t ha[-1])	Starch (t ha[-1])	Sugar (t ha[-1])
Leaves	8	6.3	0.50	0.18	0.32
Branches	38	7.0	2.66	1.14	1.48
Stems	112	9.1	10.2	7.28	2.91
Coarse roots	34	7.4	2.52	1.73	0.78
Fine roots	8	3.0	0.24	0.12	0.13
Total	200	8.0	16.1	10.5	5.6

From Wurth et al. 2005.

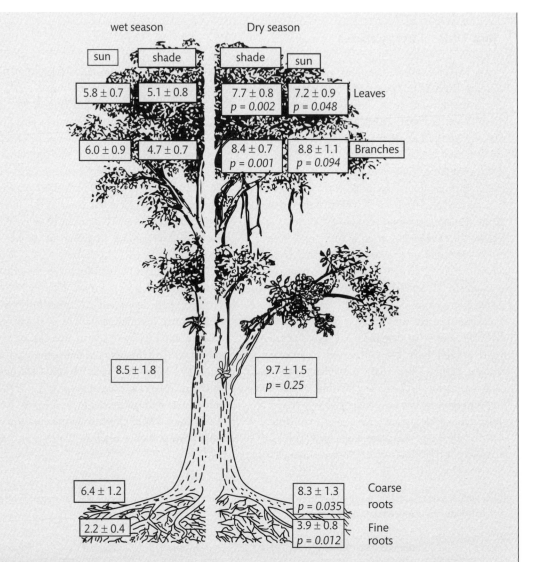

wet season

Dry season

| sun | shade | shade | sun |

| 5.8 ± 0.7 | 5.1 ± 0.8 | 7.7 ± 0.8 *p = 0.002* | 7.2 ± 0.9 *p = 0.048* | Leaves |

| 6.0 ± 0.9 | 4.7 ± 0.7 | 8.4 ± 0.7 *p = 0.001* | 8.8 ± 1.1 *p = 0.094* | Branches |

8.5 ± 1.8

9.7 ± 1.5
p = 0.25

| 6.4 ± 1.2 | 8.3 ± 1.3 *p = 0.035* | Coarse roots |

| 2.2 ± 0.4 | 3.9 ± 0.8 *p = 0.012* | Fine roots |

Figure 10.14 Wet and dry season means of non-structural carbohydrate concentrations (dry matter % ± SE, in large starch) averaged across nine tree species (P-values for season effects); from (Wurth et al. 2005). Note that these concentrations do not represent absolute reserves available to the plant, but provide some upper limit—even under situations of energy starvation values do not drop to zero.

strategies, thus maintaining species diversity (Kohyama et al. 2003).

10.7   Conclusion

The challenges of being a successful rain forest plant are to live, grow, and reproduce in an environment where many other individuals of many other species are trying to do the same. Though the wet, sunny and frost-free environments are generally well suited to plant growth, space and resources are finite, and competition is intense.

Light is especially important in the dim understoreys of tropical rain forests, and plants have evolved several strategies to aquire it. Light is probably the main factor explaining variation in plant form and dynamics within any given forest site, but other factors including available moisture and nutrients and soil properties often play a major role in explaining variation between sites. Nutrient and moisture acquisition has also led plant species to evolve various specializations and adaptations. While much of this variation between plant individuals and species remains only partly understood, it is clear that many characteristics and behaviours co-vary in a con-sistent and non-random manner. There are encouraging signs that the diversity of tree species strategies might be understood in terms of a limited number of key charac-teristics, that include juvenile shade tolerance, adult plant height, leaf-longevity, and drought tolerance.

CHAPTER 11

The ever changing forest: disturbance and dynamics

'Large masses of forest, including trees of colossal size, probably 200 feet in height, were rocking to and fro, and falling headlong one after the other into the water. After each avalanche the wave which it caused returned on the crumbly bank with tremendous force, and caused the fall of other masses by undermining them.'

H. W. Bates (1863)

The forests on the coast of Ujong Kulon National Park in western Java reverberate with the noise of hornbills and monkeys squabbling over figs amongst massive trees. But this is no ancient rain forest: in 1883 the Krakatau volcanic eruption and tsunami obliterated any previous vegetation; this is regrowth.

Old-growth rain forests have often inspired poetic visions of a changeless, primal nature. Biologists, too, often assume that these landscapes, if left alone, attain and remain in a state of approximate equilibrium. Such inspirations and assumptions are often misleading. The devastation wrought by the Krakatau tsunami seems extreme and unusual, but forest landscapes are subject to constant change. In this chapter, we consider natural processes that build, sustain, and transform tropical rain forests, with a focus on trees and the stands they comprise.

Forest dynamics describes the interaction of physical and biological forces that determine ongoing changes in forest structure, composition, and function. Forests are subject to continuous change, driven by interactions among individuals and species within the community as well as by external perturbations. To a great extent, forest dynamics is also a function of chance events, and some effort in tropical forest research has focused on determin-ing the relative importance of stochastic versus deterministic factors in shaping forest dynamics and the resulting forests. This chapter emphasizes two principal elements of forest dynamics: disturbance and succession.

11.1 Disturbance

High species diversity of tropical forests has been associated with their long-term stability, allowing for niche diversification and low extinction rates (Stebbins 1974). But stability may be illusory as rain forests are now known to be subject to diverse disturbances at various spatial and temporal scales (Huston 1994; Whitmore and Burslem 1998). Small events, such as falling branches, are frequent, but have only minor local consequences. Large and heavily destructive events are rare, but leave a long-lasting imprint. Definitions of disturbance vary but most emphasize a reduction in population density and/or a rapid release of resources for organisms to exploit (Sousa 1984; Rykiel 1985; Vandermaarel 1993; Sheil 1999b; White and Jentsch 2001).

Different types of disturbance have differing impacts and influence on the nature and character of rain forests, their ecological functioning, and their diversity.

Disturbances span a range of scales of severity, spatial extent, frequency, and duration which vary greatly within and among tropical regions, reflecting broad differences in climate, geology, biogeography, and local factors such as tree growth and turnover, soil structure and fertility, and species composition, as well as human interventions. Disturbances can be endogeneous[1], being a function of intraspecific competition or plant senescence, or exogenous, including pathogen outbreaks, landslides, hurricanes and fire, which tend to be less frequent and larger scale, with qualitatively different but nevertheless important impacts on community dynamics and diversity.

Disturbance is a central theme of research relating to the process of succession and the maintenance of tropical forest species diversity. Forest dynamics, in this context, refers to the study of changes over time in the composition of the regenerating community (Burslem and Swaine 2002). Ecologists have sought to understand the role of disturbance in contributing to the maintenance of species diversity (Chapter 8), and the implications of disturbance for successional processes and the equilibrium status of tropical forests.

11.1.1 Gaps and tree falls

Gaps are a natural part of the growth cycle in almost all tree-dominated forests (Figure 11.1). Many authors have found it useful to adopt Aubréville's gap-mosaic model of forest dynamics, at least as a basis for simple description. In this model forest turnover cycles through three phases: first a 'gap-phase' immediately following a tree fall or other canopy opening event; then a 'building phase' characterized by regeneration and growth; and finally a 'mature phase' where the high forest canopy is re-established (Aubréville 1938b; Whitmore 1978). The potential for forest structural heterogeneity created by this model of the forest growth cycle, summarized as 'gap-phase dynamics' has formed a major theme of tropical ecology research.

Gap formation

Recent tree fall canopy gaps provide a striking break in the gloom of understorey shade (Figure 11.2). Yet within a few weeks a near-impenetrable thicket of dense vegetation may develop; in a few years it may be hard to tell the gap ever existed. Tree fall gaps are typically tens to hundreds of square metres in area (Table 11.1), and a bigger tree crown leaves a bigger space. The size, form, and character of gaps also reflect whether trees fall against each other, catch in each other's branches, or are pulled down by lianas.

Gaps can also be minimal: trees may die standing and drop their branches before the stem collapses, while the surrounding canopy slowly closes around the space. A standing death is more likely with buttressed, sound[2], and dense-timbered stems growing on stable ground without wind storms (Gale and Barfod 1999; Hall et al. 2004). Gaps tend to be larger when stems snap, or are wholly uprooted—especially when the falling debris brings down neighbouring vegetation. Large, multiple tree fall gaps are common in cyclone-prone regions, on steep slopes, and among the slender 'top heavy' trees that develop on loose fertile soils found in some volcanic regions (Hall et al. 2004). Older taller forests reportedly have larger gaps (Nicotra et al. 1999). Gap-forming processes are intimately linked to the mortality of canopy trees. Though competition and senescence usually predominate, other processes also operate. Aerial images in some of Borneo's peat forests reveal a smooth canopy dominated by *Shorea albida*, pockmarked by the cumulative destructive impacts of lightning strikes. Trees can also be extensively defoliated and ultimately killed by the predations of termites, ants, or caterpillars.

Due to different gap definitions and measurement procedures, it is difficult to summarize patterns among studies, but it appears that gaps occupy a smaller proportion of stand area in a typical rain forest than in most other major forest types.

Gap dynamics are more complex than a sudden opening and gradual closing of canopy spaces. Not all gaps are

---

[1] Endogenous refers to causes that arise within individuals of the same population; exogenous causes are attributed to external factors.

[2] In many tropical forests a significant proportion of larger stems are infected by fungi, leading to pockets of rot or hollows, and the resulting loss of mechanical integrity may increase the likelihood of snapping stems or dropping branches.

A

B

Figure 11.1 Hemispherical photos taken in a non-gap heavily shaded environment (A) and within a small (*c*.40 m²) gap environment (B) at Sepilok, Sabah, Borneo. The rapidly growing seedlings in the gap environment can be seen crowding the edge of the field of view.

Photo by Julia Born.

Figure 11.2 Tree fall in Kibali Forest, Uganda, East Africa.

Photo by Rhett A. Butler.

in the canopy, especially in secondary forests where cohorts of early successional small trees are subject to high mortality when overtopped, leaving sub-canopy gaps where they have died. Trees and branches also fall in the sub-canopy and understorey, varying the light environment beneath. Not all gaps result from tree falls: in Zagne, Ivory Coast, many small canopy gaps result from branches dropped by emergent trees (Jans et al. 1993). Similarly, lianas die, lightening the shade beneath.

When larger gaps occur, neighbouring tree crowns often extend rapidly into the vacant space, making the unbalanced trees more likely to fall into than away from the opening (Young and Hubbell 1991; Ackerly and Bazzaz 1995; Muth and Bazzaz 2003). Such factors mean that larger gaps can possess complex dynamics: fluctuating in size and even moving about (Jansen et al. 2008).

Gap consequences

Gaps provide opportunities for new recruitment and accelerated growth, but their creation inflicts considerable damage to seedlings, saplings, and young trees (Aide 1987). Woody plants often suffer breaks from falling branches or other plant material, such as palm fronds, and the danger to understorey plants can be substantial (Peters et al. 2004). In *Prestoea montana* palm forest in Puerto Rico, 70% of total tree mortality was caused by the crushing of smaller trees by a few falling canopy trees (Frangi and Lugo 1991). In Crater Mountain, Papua New Guinea, a seedling has a 35% chance of suffering serious damage each year (Mack 1998); the risk is even higher on slopes. More than 80% of seedlings were damaged in a similar study in Costa Rica (Clark and Clark 1989) (Table 11.2). Rates in Amazonian forest were 20–24% (Scariot 2000).

Gap regeneration comes from pre-existing vegetation (seedlings, saplings, suckers, resprouts from broken stems), from seeds, and even occasionally from epiphytic seedlings in the crowns of fallen trees (Lawton and Putz 1988; Brown and Jennings 1998; Busing and Brokaw 2002). In small and cluttered gaps regrowth usually depends on pre-existing vegetation, while seeds play a more significant role in larger, emptier gaps.

Different conditions result, depending on whether trees die standing, snap, or by uprooting. Uprooted stems create a hollow and a mound of mineral soil from the root plate. At La Selva, Costa Rica, and Los Tuxtlas, Mexico, different tree species establish preferentially in distinct parts of such gaps (Brandani et al. 1988). For example, *Cecropia obtusa* in French Guyana and *Trema tomentosa* in Malaysia establish preferentially on root plates (Charles-Dominique et al. 1998; Whitmore 1998).

In larger gaps, light levels are high, and many seedlings and saplings display a surge of growth, rapidly competing for space. Towards the shadier edges growth can be slower. In such large gaps the different sizes and often short lives of the initial colonizing vegetation lead to localized succession, as various pioneers jostle for space and are ultimately outlived or replaced by more shade-tolerant species.

Gaps may also fill with liana thickets that suppress tree seedlings (Schnitzer et al. 2000). Palms, heliconias, bamboos, gingers, ferns, wild bananas, and other large herbs can also thrive and shade out tree seedlings. Animals also

Table 11.1 Gap characteristics of different tropical rain forest sites.

Location	Forest type	Gap fraction (percent)	Average gap size (m²)	Gap range (m²)	Annual gap formation rate (percent)	Turnover rate (year)
Ivory Coast	Tropical moist	0.84	41	11–244	–	244
India	Wet evergreen	3.6	–	≤500	1.13	83
Ivory Coast	Lowland evergreen	0.84	41	–	–	–
Panama	Tropical moist	4.3	79	8–604	0.45–6.5	–
Panama	Lowland tropical	2.0	64	22–232	0.63	159
Panama	Lowland tropical	2.8	86	26–342	0.88	114
Christmas Island	Tropical	–	45.9	17–700	–	–
French Guiana	Lowland rain forest	1.1	120	–	0.96	–
French Guiana	Lowland rain forest	–	–	–	1.3–1.5	–
Costa Rica	Tropical wet	3.6–7.5	54–120	–	0.72–1.3	80–138
Panama	Tropical moist	–	–	–	0.73	137
Mexico	Tropical cloud	–	–	–	0.5	158
Costa Rica	Tropical cloud	–	–	> 4–135+	0.8–1.4	95
Ecuador	Lowland rain forest	1.4	10	–	–	–
		5.1	15	–	–	–
Boreal and subalpine		6–36 (21)	41–141 (78)	15–1245	0.6–2.4 (1.0)	87–303 (147)
Temperate hardwoods		2–20 (10)	28–239 (79)	8–2009	0.4–1.3 (0.8)	45–240 (134)
Temperate coniferous		11–18 (14)	77–131 (85)	5–734	0.2	280–1000 (650)
Tropical		0.8–8 (4)	10–120 (50)	4–700	0.5–6.5 (1.0)	80–244 (137)
Southern hemisphere		3–35 (8)	40–143 (93)	24–1476	0.25–0.28 (0.3)	320–794 (408)

Adapted from McCarthy 2001.

Table 11.2 Causes of damage to single artificial seedlings at two tropical rain forest sites: Crater Mountain, Papua New Guinea, and La Selva, Costa Rica.

Cause	Crater Mountain (%)	La Selva (%)
Unknown	11.0	42.2
Litterfall	13.8	19.2
Animal-digging	7.0	21.0
Water wash	2.9	–
Undamaged	65.3	17.6

From Mack 1998.

play a role. Many herbivorous animals, large and small, feed preferentially on lush low growth in canopy gaps and thereby suppress regrowth. Indeed elephants and other large mammals can ensure that gaps are long-lived.

Lianas influence gap dynamics, generally increasing the prevalence of tree falls and the size of the resulting gaps. Climbers are more abundant in areas where larger gaps have previously occurred, so positive feedbacks can occur. In Peru some forests are dominated by climbing bamboos (*Guadua weberbaueri* and *G. sarcocarpa*), which colonize saplings and small trees in disturbed forest. As disturbance favours the bamboo, this cycle appears self-reinforcing (Griscom and Ashton 2003, 2006).

11.1.2 Landslides

Landslides are caused by weathering processes on steep slopes and typically occur after heavy rain or in conjunction with earthquakes. They are frequent in young tropical mountains including the Andes, Hawaii, New Guinea, Sumatra, and Sulawesi. For instance 120 major landslide scars were recorded in the up-thrust area of Gunung Mulu National Park in Sarawak, Borneo (Gyasi et al. 1995). In Brunei, Borneo, slopes of 40 degrees or more are estimated to collapse at intervals of less than 10,000 years[3] (Dykes 2002). In the montane forests of Central America and the Caribbean, earthquake-induced landslides disturb about 1-2% of the canopy per century (Garwood et al. 1979; Restrepo and Alvarez 2006), and 8-16% in northern Papua New Guinea (Garwood et al. 1979). Erosional landslides, by comparison, are estimated

to affect around 3% of the canopy per century in the same region.

Landslides are more frequent in areas cleared of forest than in intact forest areas. Forest vegetation slows erosion, and deep roots may help secure the upper soil mantle against deeper layers reducing the probability of shallow landslides. Usually, however, the falling land mass shears at the weathering front where the looser saprolite contacts the rock beneath. Often this is deeper than most tree roots—nonetheless the role of deeper roots in facilitating water percolation and drainage is not yet well understood and such hydrological influences may also be important (Sidle et al. 2006a). In any case roads and tracks (even paths) increase landslide risk. Landslides are typically more than ten times more frequent along roads than in similar undisturbed rain forest terrain (Sidle et al. 2006b).

Landslides result in the movement and loss of vegetation, soil, and seed banks, yet impoverished seed banks and accumulated topsoil can remain at the depositional zone creating foci of rapid regrowth by pioneer trees and herbs (Guariguata 1990). Exposed subsoils and underlying rock create distinctive edaphic conditions. Nutrient availability is often low (Fetcher et al. 1996) and soil structure poor, ongoing erosion can be rapid, and few plants can establish. A topsoil rich in organic matter can take many years to develop (Larsen et al. 1999; Wilcke et al. 2003). Such sites are often colonized by specialized ferns (typically *Dicranopteris linearis* from Africa to Polynesia; *D. pectinata* in the Caribbean) that appear to retard establishment of woody plants (Walker 1994; Russell et al. 1998).

11.1.3 Floods and rivers

In 2003, in the forested hills of Gunung Leuser National Park in North Sumatra, the rain was unusually heavy. A small landslide had formed a debris dam in a narrow valley. The dam collapsed. Waters, sediments, debris, and fallen trees were catastrophically released. More than 100 people died downstream and a significant swathe of forest was washed away. Floods are not just about water. Soil, rock fragments, and vegetation can be

---

[3] Thus even in these rugged old-growth rain forests about 1% of the steeper vegetation is naturally less than 100 years old.

picked up by strong water flows to form highly destructive debris flows. Large tree trunks can be especially destructive, battering down vegetation as the current sweeps them past. A localized, species-poor forest in lowland Ecuador has been attributed to a catastrophic flood 500 years previously; the low diversity is thought to result from slow recolonization by poorly dispersing understorey species (Pitman et al. 2005). Floods can, however, also carry fresh fertile sediments, and sometimes seeds or living organisms, over long distances[4].

Meandering rivers are less violent, but their ever-changing pattern of channels has profound significance. Fast flow on the outside of each bend erodes sediment, while slower flow on the inside allows sediments to fall from suspension. Over time, each bend grows and migrates until the sinuous channels intersect, usually during a flood, cutting a channel across the inside of the bend, and often isolating the former meander as an oxbow lake[5]. As rivers meander over the landscape they intersect old lakes, channels, and other relic drainage features, adding complexity to this dynamic landscape (Chapter 7, Figure 7.10). The upper tributaries of the western Amazon carry abundant silt from the Andes. The tributaries meander through the soft alluvial plains, replacing around 0.2% of the floodplain each year (Puhakka et al. 1992). More than 10% of Peru's Amazonian forests appear banded when viewed from above, reflecting different ages and stages of compositional succession (Puhakka et al. 1992; Terborgh et al. 1996). Similar processes occur in other alluvial river plains.

In seasonally flooded forests, shifting sediments and vegetation succession act in concert (Wittmann et al. 2004). Once established, vegetation usually increases deposition and stabilizes the alluvium, gradually raising the terrain. Thus, mature flooded forests are typically shallower (less standing water) than younger stands

(Godoy et al. 1999; Wittmann and Junk 2003; Wittmann et al. 2004). As drainage patterns shift, some areas may be isolated from external flows—these sites gradually lose nutrients resulting in localized 'black-water' (nutrient poor) channels and lakes even within rich 'white-water' systems.

11.1.4 Wind

In the early 1990s satellite images revealed numerous mysterious fan-shaped secondary forests, each 30–100 ha in extent, in a north–south belt passing through the Central Amazon and covering a total area of about 900 km^2. The cause was identified as convectional storms with violent down-drafts (Nelson et al. 1994). Large destructive windstorms occasionally occur in most rain forest regions, and localized blowdowns similar to those in the Amazon have been observed in the equatorial forests of Borneo (Proctor et al. 2001). In Africa 'corridors' of storm-felled trees occur even deep in the Congo. Destructive wind events are often accompanied by heavy rainfall that exacerbates the damage through flash floods and landslides.

Cyclonic wind storms—cyclones, hurricanes, and typhoons—are the most violent and regular, occurring seasonally in two belts 10°–20° north and south of the equator. Islands and coastal forests in the Caribbean, the Western Pacific including Australia's wet tropics, and the southern Indian Ocean are the most frequently and severely affected[6] (Figure 11.3). Plants growing in these windy sites tend to develop shorter, stronger stems; a review of 106 studies describing lowland rain forest found a significant correlation between shorter rain forests and frequent windstorms[7], and the low average canopy height of Madagascar's lowland rain forests has been explained on this basis (de Gouvenain and Silander

---

[4] Rafts of debris are a significant means by which organisms may disperse across rivers, and even along coasts or across seas (see Chapter 6).

[5] Note that meandering rivers pose no long-term barrier to species, as chunks of habitat are frequently transferred from one side of the river channel to the other. Riverine barriers are thus less important to biogeographical patterns in floodplains than in less dynamic settings.

[6] The main discrepancy in the global pattern is the absence (with rare, recent exceptions) of such winds in the southern tropical Atlantic.

[7] Exceptions include the 48 m tall forests of Kolombangara, Solomon Islands, which are regularly struck by hurricanes, although these forests are typically young, so height may not be a fixed characteristic.

Figure 11.3 A forest blowdown at Bisley, Luquillo Experimental Forest in Puerto Rico, following the passage of Hurricane Hugo in 1989.

Photo by Ariel Lugo.

2003). Strong winds are largely absent from the low understorey where the largest leaved plants typically occur. Taller monocots with large leaves have either strong divided leaves (most tall palms) or leaves that can tear in a manner allowing continued functioning, such as Musaceae.

The intensity and frequency of windstorms defines the structure and dynamics of these forests. Windstorms cause considerable structural damage and increase mortality, but they also release nutrients to the forest floor through defoliation and increase light reaching the understorey, both of which provide opportunities for regeneration and changes in successional pathways (Lugo et al. 1983; Guzman-Grajales and Walker 1991; Walker 1991; Whigham et al. 1991; Everham and Brokaw 1996; Lugo and Scatena 1996; Ostertag et al. 2003).

The resistance of forests to severe winds is a function of both individuals and communities. Stand properties—stem sizes and densities, canopy structure, and rooting depth—as well as the direction and intensity of the wind in conjunction with local terrain and soil properties, are important factors determining impact (Everham and

Brokaw 1996), but so is past disturbance history (Ostertag et al. 2005). Species differ in their resistance to wind, although patterns can be difficult to disentangle as many factors affect resistance. Species with low wood density appear to suffer more damage (Zimmerman et al. 1994; Curran et al. 2008), although this pattern is not always apparent (Ostertag et al. 2005). Large size also increases likelihood of damage (Lugo et al. 1983; Walker 1991; Ostertag et al. 2005), but again these responses are not universal (Curran et al. 2008). In the Australian wet tropics, species with dense wood and low specific leaf area (SLA: area per unit mass; Chapter 10) were most resistant to cyclone damage, while species with low wood density and high SLA (traits associated with rapid growth) suffered damage but recovered rapidly. This combination of trade-offs across the community suggests both high resistance and resilience[8] to cyclones at the community level (Curran et al. 2008).

Wind damage is highly heterogeneous. Some areas of forest may be devastated while adjacent patches are practically unaffected (Lugo et al. 1983; Brokaw and Walker 1991; Walker 1991; Bellingham et al. 1992). In forests

[8] 'Resistance' is the ability to avoid damage; 'resilience' is the ability to recover.

subjected to frequent windstorms, regrowth is primarily from resprouting trees and surviving understorey plants (Frangi and Lugo 1991; Whigham et al. 1991; Zimmerman et al. 1994). Apart from a brief pulse of pioneers, there is little evidence that tree composition changes much; apparently these communities are well adapted to the conditions. For example, Kolombangara in the Solomon Islands was struck by four cyclones between 1967 and 1970, and most broken trees simply resprouted, resulting in little change in composition (Burslem et al. 2000). Hurricane Joan, which struck Nicaragua in 1988, caused extensive damage to the Atlantic coast forests: 27% of trees had fallen and 53% had snapped (Yih et al. 1991). Nevertheless, four months after the hurricane almost all (95%) individuals still standing had resprouted, as had the majority of the snapped trees (76%) and fallen trees (66%). Thus after catastrophic wind damage forests recover mainly through 'direct regeneration', such that species diversity and composition remain similar to those immediately preceding the disturbance event (Yih et al. 1991; Burslem et al. 2000).

11.1.5 Drought

Wet rain forests are not always wet. An explorer who visited Kutai, East Kalimantan (Borneo) in 1879 reported a drought[9] that lasted more than eight months and killed a third of the forest canopy (Bock 1881 quoted in Brookfield et al. 1995). The next similarly severe drought occurred 119 years later, in 1998. This sounds like a long time, but when we know that many trees are centuries old it seems clear that even such 'rare' droughts can have lasting impacts.

Severe droughts are often associated with the climatic phenomenon known as the **El Niño Southern Oscillation** (ENSO), although during ENSOs some regions receive more rainfall while others receive less, and each event varies in regional severity. ENSO events greatly affect forest dynamics and tree mortality through drought stress (Barber and Chavez 1983; Siegert et al. 2001;

Charrette et al. 2006). Their frequency and severity have increased in recent decades, and are likely to continue to do so (Collins 2005; Hansen et al. 2006).

Not all droughts can be blamed on El Niño. In 2005 the severe drought affecting the south-western Amazon reduced river water flows considerably, dried up lakes, and facilitated the propagation of fires (Aragao et al. 2007). Fish died and their decomposing bodies clogged the depleted rivers and poisoned the waters. On this occasion, the drought, estimated to be the worst in 40 years, was driven by anomalous warm North Atlantic Ocean surface temperatures which weakened the transport of moisture into Amazonia by the trade winds (Marengo et al. 2008).

Droughts kill seedlings, saplings, and even adult trees (Delissio and Primack 2003; Van Nieuwstadt and Sheil 2005). Short-rooted plants and seedlings are especially vulnerable in exposed or well-drained locations (Dalling and Hubbell 2002) (see Chapter 10 for additional factors that influence plant water demand and drought tolerance). Though many larger trees can initially exploit a larger and deeper soil volume than smaller stems, in prolonged droughts exposure to sun and wind can ultimately lead to greater desiccation. Severe droughts can cause canopy unevenness by killing many larger trees.

Plants under, or recovering from, drought stress are more susceptible to insects and pathogens. In central Panama an outbreak of forest caterpillars occurred after the 1997–8 droughts with foliage damage more than double normal incidence (Van Bael et al. 2004). Similar outbreaks occurred in Borneo (Itioka and Yamauti 2004)[10].

Trees killed by drought initially remain standing, though begin to fragment by shedding branches. Such open and weakened forests are susceptible to wind throw. Many surviving trees are later damaged or knocked down by collapsing neighbours, thus tree mortality rates tend to remain high for several years after a severe drought.

Some plants may benefit from droughts. Large woody lianas often possess deep roots and may be favoured

---

[9] Droughts can be viewed from two perspectives. From a climatic perspective drought in rain forest might be defined as, for example, less than 100 mm monthly rainfall for three consecutive months. Plants, however, differ in their vulnerability to relatively dry periods.

[10] An alternative explanation lies in the simultaneous flush of young undefended leaves at the end of drought providing a vast food resource for rapidly breeding and opportunistic herbivores.

by droughts and their aftermath, and some trees only regenerate effectively following intense droughts, presumably due to the space and light made available. Stands of *Microberlinia bisulcata* (Fabaceae, Caesalpinioideae) in Korup, Cameroon, for example, appear to have had three previous bursts of regeneration that coincided with severe droughts in 1740–50, 1820–30 and 1870–95 (Newbery et al. 2004).

11.1.6 Fire

Until recently fire was considered irrelevant to intact rain forests. After all, as many campers can testify, dry fuel is scarce. For most of the time rain forests are too damp to burn, but during droughts dry air penetrates the understorey and dries the fuel that has accumulated from drought-induced leaf fall and tree mortality. The forest, consequently, becomes flammable. It is worth emphasizing that rain forests never burn without a preceding drought—the resulting destruction is the effect of both combined. Together, drought and subsequent fires are highly destructive as they preferentially kill larger and smaller stems respectively (Figure 11.4). A forest burnt once can recover, but remains vulnerable to repeated burning due to a more open canopy and generally more flammable vegetation (Cochrane et al. 1999; Barlow and Peres 2008). Indeed disturbed forests and secondary forests are more vulnerable to fire than are more pristine rain forests (Cochrane et al. 1999).

Fire directly affects the structure and composition of forests by killing trees and saplings. Severe fires, particularly during periods of drought, can kill as much as 94% of trees in logged forest and 71% of trees in unlogged forest, although mortalities of around 40–50% are more common (Woods 1989). Nevertheless, plants differ in their vulnerability to fires. When a rain forest burns, dry litter is the main fuel[11]—low flames move slowly over the ground (Laurance 2003). Small plants, seedlings, and thin-barked saplings usually die, while larger trees survive better. Palms also often survive relatively well. Few rain forest species appear well-adapted to fire, although some bamboos in drought-prone forests take advantage of occasional fires or other canopy-opening events to

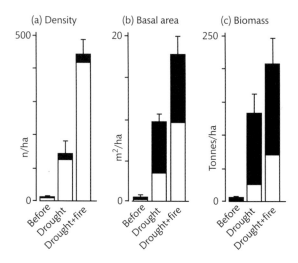

Figure 11.4 Density (a), basal area (b), and above-ground biomass (c) of dead stems before and 21 months after drought, and drought and fire at Sungai Wain, East Kalimantan in 1998. The contribution of stems between 10 and 40 cm dbh are coloured white, and those larger than 40 cm are black. Error bars are 95% confidence intervals. Fire contributed more to stem deaths than drought alone, but most of these stems were small and so the loss of basal area and biomass was lower.

From Van Nieuwstadt and Sheil 2005.

spread rapidly into open areas. The thicker bark found in most seasonal rain forest, when compared to wetter and less seasonal locations, may reflect selective pressures due to greater fire exposure. Most dead trees remain standing for months.

An accumulation of dry fuel, particularly following drought, can promote hotter and longer burns, which are more deadly. Plants occupying litter-free slopes and ridges and moist valley bottoms are more likely to survive fires—the fire may not even reach them. Unburned forest patches (6% by area in one eastern Amazon study [Cochrane and Schulze 1999] and 2% in another in Kalimantan [Van Nieuwstadt and Sheil 2005]) may lack fuel, or be somehow protected by prevailing winds, residual moisture, and/or natural firebreaks. These unburned patches are important refuges from which vulnerable forest species can later recolonize surrounding areas.

---

[11] Peat forests are an exception—after drought, dried peat itself can burn (Page et al. 2002).

Selective logging, fragmentation, and a drying climate increase forest vulnerability to fire. Long-term habitat changes resulting from repeated burning cycles may be more important for biodiversity than the immediate short-term effects of burning. Even so, short-term effects shape subsequent habitat–fire interactions, particularly with regard to rate of recovery from fire and therefore subsequent vulnerability to it. Cyclones or hurricanes also increase forest vulnerability to fire. Rain forest fires in 1989 in Yucatan (Mexico) were exacerbated by prior opening of the closed forests by Hurricane Gilbert in 1987, an event that increased fuel loads. These woody fuels were desiccated by the subsequent drought of 1988–9, and the forest was ignited by land-clearing fires, ultimately affecting an area of 90,000 hectares (Goldammer 1992).

In East Kalimantan, and in parts of Sumatra, fires burn within thin coal seams that are widespread in the subsoil. These may burn slowly for years beneath the ground surface, occasionally giving rise to new ignition points. A total of 76 coal fires were found over a two-year period in the vicinity of the small (10,000 ha) Sungai Wain reserve in East Kalimantan—some had spread over 300 m from their ignition sites. The shallow burning killed the forest, leaving behind areas of dead trees up to 1,000 m² (Fredriksson 2001). Estimates suggest that East Kalimantan alone may have many thousand such fires. Efforts have now been expended to detect these fires, drill down, and extinguish them (Whitehouse and Mulyana 2004).

Recovery from fires may be slow: density and species richness of seedlings and saplings in a burnt Bornean forest remained lower than unburned plots for several years after the fire event, although pioneers had higher growth and recruitment (Cleary and Priadjati 2005). Prolonged changes in vegetation communities in lightly burnt forest may alter ecosystem functioning, and, by changing litter fall dynamics and microclimate, may affect the frequency and severity of subsequent fire events. Thus a positive feedback of increasing vulnerability to fire is established which, if continued unchecked, inexorably converts forests to scrub or savannah (Cochrane et al. 1999). Fires may also lead to the depletion of soil cover, resulting in increased run-off and erosion, with downstream consequences in the form of mudflows, landslides, flooding, or siltation of reservoirs.

Much of the soil seed bank remains viable after fire, as heat seldom penetrates deeply. Initial regrowth is usually by pioneers and ferns which benefit from the open space and ash-enriched soils. Some woody species resprout from basal shoots and root suckers, after fire kills their above-ground tissues: the longer-term patterns of survival and the relative contribution of different forms of regrowth remain poorly known.

11.2   Patterns in space and time

The concept of a 'disturbance regime' encompasses patterns of disturbance events in space and time (Baker et al. 2005). These patterns determine the mix and spatial arrangement of different successional stages within the forest.

Local disturbances can result in or derive from larger regional patterns (ter Steege and Hammond 2001). In Guyana, for example, the more diverse forest formations possess a higher proportion of pioneer and early successional species, whereas the low-diversity communities are characterized by dense-timbered, slow-growing, shade-tolerant species—implying an absence of regular disturbance. What is cause and what is effect remains uncertain.

It is instructive to compare modes of mortality in terms of their prevalence, so as to understand better how they shape community composition and structure. Large-scale sudden disturbances catch our attention. Certainly some regions are especially disturbance prone. New Guinea, for example, is a geologically active region: mountains rise and fold as the Australian plate subducts below the Pacific plate. The foothills of West Papua's Foja Mountains reverberate at night with the sound of landslides. Rivers frequently change course. Drought and fires also have a long history in shaping the vegetation. Occasional volcanic eruptions, tsunamis, and storm-force winds leave their mark, as does the extensive shifting cultivation long practised by the Papuan peoples. In some sites, palynology suggests continual disturbance, maintaining pioneer-dominated vegetation for at least 3 millennia (Jago and Boyd 2005). Thus it seems unsurprising that New Guinea possesses an unusually high proportion of aggressive pioneer species such as *Paraserianthes* spp. and *Eucalyptus deglupta* (Johns 1986, 1989; Whitmore 1998).

Aside from possibly unusual cases like New Guinea, do dramatic perturbations have as great an impact on cumulative tree mortality and turnover as less dramatic but comparatively frequent tree falls resulting from endogenous processes? In practice it is not simple to allocate mortality to endogenous or exogenous causes.

Tree dynamics weaves together the influence of endogenous and exogenous processes. Large trees often have greater risk of windthrow, drought, flood, or pathogenic attack—though typically smaller stems have a lower overall rate of survival (Van Nieuwstadt and Sheil 2005; Muller-Landau et al. 2006). A burst of short-lived pioneers that often follows major disturbances will raise turnover rates for several years.

Exogenous disturbances meanwhile have impacts that unfold long after the event. Hurricanes, for example, may increase vulnerability of surviving trees to fire (Whigham et al. 1991), and drought or fragmentation may increase plant stress and thus vulnerability to pathogens and other causes of mortality (Lugo and Scatena 1996; Williamson et al. 2000; Laurance and Williamson 2001).

The natural processes of tree death and forest turnover averaged over a large area allow predictable summary statistics, but at a local scale such processes are uneven and stochastic—between 1939 and 1993 in 1.8ha of a Ugandan forest 60% of canopy turnover was the result of a loss of only seven stems over a metre in diameter (Sheil et al. 2000). Notwithstanding these complications, the intensity of mortality by various causes can be compared in terms of percentage forest affected annually. Background tree mortality is that attributed to mortality in the absence of exogenous disturbances, in other words competition, turnover, and senescence. Background mortality rates amongst larger stems typically average around 1–2% per year[12] across a wide range of tropical rain forest sites, and very rarely exceed 3% per year (Muller-Landau et al. 2006).

In contrast, exogenous mortality caused by hurricanes, landslides, and fires can destroy millions of trees within hours. In such circumstances tree mortality exceeds background rates by an order of magnitude (Lugo et al. 1983; Van Nieuwstadt and Sheil 2005). However, observations over long-time scales show that, even in forests prone to catastrophic disturbances, the impacts of these disturbances on tree turnover are typically low compared to background rates. Thus over a 100-year time period at Luquillo Experimental Forest (Puerto Rico) in the hurricane zone, background mortality accounted for between 1,973 and 2,650 trees (over 10 cm dbh) ha^{-1} while hurricanes killed 472, and landslides a mere 30–64 (Lugo and Scatena 1996). Nevertheless, catastrophic events remain a major influence on forest composition and structure.

11.2.1 Succession—forest colonization and recovery

Succession is defined as a long-term directional change in community composition, structure, and function following a (usually large) disturbance event. Successional theory dates back to the early 20th century where succession was considered to be a deterministic process with a largely predictable outcome, notably a stable **climax community** (Clements 1916). This concept, developed from a temperate perspective, has been elaborated and developed over the decades to incorporate more stochasticity (chance events) and less determinism (Gleason 1926; Tansley 1935). In tropical forests the 'climax' forest concept was replaced by gap-phase dynamics concepts (Aubréville 1938a). By the 1970s most ecologists accepted that stable equilibrium conditions were no longer tenable. Succession was considered to be driven mainly by interactions among component species and their responses to changing environmental conditions. The theoretical development of these ideas has been very influential in developing frameworks for understanding tropical forest dynamics.

Successional pathways following catastrophic disturbances differ according to whether the environment is radically altered as a result of disturbance. Landslides, volcanic eruptions, debris flows, and massive floods can completely remove vegetation, destroy seed banks, and even remove or sterilize the soil itself. Alternatively, fires, hurricanes, and cyclones do not usually severely impact soils or microclimate, and their effects on plants are also muted by comparison.

---

[12] Similar to temperate and boreal forests.

The more complete and extensive the loss of vegetation incurred, the slower the recovery (Box 11.1). Distance from remnant forest patches or seed trees can slow recovery due to poor dispersal. Small disturbances may have little effect on composition. Small gaps may merely release saplings already present and just waiting for their day in the sun.

Even areas cleared of rain forest seldom remain free of vegetation for long. Typically the first arrivals include ferns, herbs, shrubs, and climbers (Finegan 1996a). Initial ground cover by herbs and shrubby vegetation can inhibit recruitment by woody colonists (Ferguson et al. 2003), although this stage is usually short-lived and gives way to the building phase, when pioneer trees establish and become dominant (Chazdon 2008). This early woody regeneration is derived primarily from seed rain and the seed bank, and from resprouts (Box 11.2) (Uhl et al. 1981; Kammesheidt 1998).

In hurricane-prone areas resprouting is common among tree species—even after logging or fire, the early regenerating vegetation is often dominated by resprouts—and the stand initiation phase is often bypassed (Lugo et al. 1983; Vandermeer et al. 1990, 1995, 1996; Yih et al. 1991, Boucher et al. 2001). Intense competition often leads to rapid thinning and mortality of these tree thickets (Vandermeer 1996). Details vary with location, but short-lived pioneer tree species will typically establish a closed canopy on formerly cleared land within 5–10 years, and often dominate for up to 30 years (Finegan 1996a). They are gradually replaced by a taller canopy of longer-lived, but still light-demanding species ('secondary forest species' or 'long-lived pioneers') which may dominate for more than a century (Table 11.3) (Finegan 1996; Denslow and Guzman 2000; Poorter et al. 2005; Chazdon 2008). During this period the density of seedlings in the understorey decreases, and there is high mortality of light-demanding herbs and treelets as well as seedlings and saplings of long-lived pioneers (Capers et al. 2005; Chazdon et al. 2005). Shade-tolerant species establish at various stages during this process and will eventually begin to reach the canopy.

Box 11.1   The role of soil

Soils and vegetation interact to influence the consequences of disturbance, subsequent recovery, and forest succession. Processes that kill forest vegetation and open up the canopy can impact soils directly. If the ground is exposed to sunlight, surface desiccation and high temperatures will impact soil biota. Trees that fall without snapping result in lifted root plates and hollows. Sudden influx of fine and coarse plant debris affects impacts microbial behaviours and nutrient processes, and decomposition provides a nutrient bonus to surviving and colonizing vegetation (Denslow et al. 1998).

Much depends on the presence or absence of vegetation cover. If plants are wholly absent there is no nutrient cycling, organic materials will decompose, and soluble nutrients may be lost quite rapidly. When vegetation is present, fast growth and nutrient acquisition generally ensure capture of free nutrients (even on richer soils). Sudden changes in vegetation cover cause rapid changes in soil properties which may drive divergent successional patterns (Guariguata and Ostertag 2001).

Landslides, floods, and volcanic deposits expose new surfaces for weathering. Most forest communities have a handful of species adapted to colonizing such environments as soils develop. On nutrient-poor substrates, succession is slower as initial colonists need time to develop soils and accumulate nutrients. Such areas are often dominated by low herbaceous vegetation.

Soil disturbance also unfolds at smaller scales. Organisms such as termites and worms turn over the soil, and larger fauna also play a role. In Borneo, bearded pigs (*Sus barbatus*) often dig under ridge-top tree roots, which influences erosion in these rugged landscapes and probably contributes to sediment flows, tree falls, and even to landslides.

Box 11.2 Tolerating damage by resprouting

Physical damage to seedlings and saplings is a frequent hazard caused by falling woody debris (Martinez-Ramos et al. 1988; Gartner 1989; Chazdon 1992). Understorey shrubs and small trees, exposed to such hazards their entire life, often respond by spreading vegetatively and resprouting after damage (Gartner 1989; Kinsman 1990) (Figure 11.5). Clonal palms reduce mortality risk from falling debris by being multi-stemmed, as it is unlikely that all stems would be killed by any single debris fall event (De Steven 1989; Chazdon 1992; Svenning 2000). Juvenile canopy tree species can minimize their exposure by growing rapidly and escaping the 'hazard zone', but seedlings and saplings remain vulnerable, and this is particularly a concern for the slower growing shade-tolerant species that may have to persist in the understorey for many years. Many canopy tree saplings that have been bent or snapped also respond by resprouting from adventitious shoots (Greig 1993; Guariguata 1998). There is considerable variability among species in their resprouting capacities.

Many adult trees in lowland (Putz and Brokaw 1989; Guariguata 1998; Paciorek et al. 2000) and montane rain forest (Kinsman 1990; Matelson et al. 1995) are able to respond through resprouting after damage: more than half of 165 trees recorded as snapped in Barro Colorado Island, Panama, resprouted (Paciorek et al. 2000) (see also responses to hurricanes in main text). Resprouting allows fast-growing, structurally-weak trees to compensate for their fragility (Putz et al. 1983), but it appears to be equally common in long-lived, slow growing, and structurally robust species (Pauw et al. 2004). For example in Borneo large trees of *Eusideroxylon zwageri* (Lauraceae), cut more than half a century earlier by shifting cultivators, survive as coppice within old secondary growth (DS, pers. obs.).

Figure 11.5 Resprouting *Cinnomomum*, here in the Seychelles, where it has become invasive. Cutting back the tree has little impact owing to its ability to regrow through resprouting. The young leaves are rich in anthocyanins, probably to protect against herbivores (Chapter 10).

Photo by Jaboury Ghazoul.

Resprouting can also be a means to capture space. Many bamboo and palm species, and some trees, develop dense suckers or runners that rapidly fill any well-lit space and smother competing seeds. This site occupancy can be pre-emptive. In Uganda the exotic Asian tree *Brousonettia papyifera* (Moraceae) produces a surrounding thicket of shoots in the dark forest understorey that will grow rapidly whenever a gap occurs (DS, pers. obs.). A dramatic example is *Tetramerista glabra* (Tetrameristicaceae), a canopy tree of Melanesian peat forests, which grows rapidly under high light, but its weak stems often collapse. These fallen stems produce vigorous, fast-growing sprouts, which may collapse in turn, and so on, filling space (Gavin and Peart 1999) (Chapter 12).

Table 11.3 A framework for vegetation dynamics processes across successional phases in tropical forests (from Chazdon 2008). The time periods for each phase are indicative rather than absolute.

Phase 1: stand initiation phase (0–10 years)
- Germination of seed bank and newly dispersed seed
- Resprouting of remnant trees
- Colonization by shade-intolerant and shade-tolerant pioneer trees
- Rapid height and diameter growth of woody species
- High mortality of herbaceous old-field colonizing species
- High rates of seed predation
- Seedling establishment of bird- and bat-dispersed, shade-tolerant tree species

Phase 2: stem exclusion phase (10–20 years)
- Canopy closure
- High mortality of lianas and shrubs
- Recruitment of shade-tolerant seedling, saplings, and trees
- Growth suppression of shade-intolerant trees in the understorey and sub-canopy
- High mortality of short-lived shade-intolerant pioneer trees
- Development of understorey and canopy tree strata
- Seedling establishment of bird- and bat-dispersed, shade-tolerant tree species
- Recruitment of early colonizing, shade-tolerant tree and palm species into the sub-canopy

Phase 3: understorey reinitiating stage (25–200 years)
- Mortality of long-lived, shade-intolerant pioneer trees
- Formation of canopy gaps
- Canopy recruitment and reproductive maturity of shade-tolerant canopy and sub-canopy tree and palm species
- Increased heterogeneity in understorey light availability
- Development of spatial aggregations of tree seedlings

Turnover in species composition of the canopy and sub-canopy may continue indefinitely as canopy trees die and create gaps and opportunities for new recruits. Turnover in species in the understorey is also apparent as the advance regeneration shifts from pioneer-dominated species to shade-tolerant canopy species that are more characteristic of old growth forests (Guariguata et al. 1997; Denslow and Guzman 2000; Chazdon 2008).

Pioneers are vulnerable 'fugitive species' that must stay ahead of their more shade-tolerant competitors (Chapter 10). Good dispersal, high seed output, and rapid maturity allow pioneers to spread quickly and occupy suitable sites. Large areas long free of significant disturbances possess few pioneers trees or seeds. At a pristine site only recently accessible by road at Danum Valley in Sabah, Borneo, the seed bank was depleted and contained few typical pioneers (e.g. *Macaranga, Trema* [Whitmore 1998]). When such areas are disturbed few pioneers are present to colonize any vacant area, which may slow initial recovery. In contrast, increased disturbance in neighbouring vegetation boosts the flow and stocks of pioneer seeds and increases rapid colonization by pioneers (Janzen 1983; Sheil and Burslem 2003). This is likely to be the case in most circumstances as the impacts of fire, wind, and landslides at the landscape scale are highly heterogeneous (Garwood et al. 1979; Lugo and Scatena 1996).

The soil seed bank may contribute to forest recovery, and indeed dormant seeds await opportunities afforded by disturbances, to the extent that the composition of the seed bank can influence the course of forest succession (Swaine and Hall 1983; Hopkins and Graham 1984; Young et al. 1987; Saulei and Swaine 1988; Quintana-Ascencio et al. 1996). Dormant seeds allow pioneer species to be the first to access short-lived colonization opportunities, and tree seeds in rain forest soil seed banks are dominated by pioneers (Saulei and Swaine 1988; Garwood 1989; Kennedy and Swaine 1992; Vazquez-Yanes and Orozco-Segovia 1993) (see Chapter 12). Their longevity is, however, often limited to little more than a year (Dalling et al. 1997; Guariguata 2000). Additionally, seeds of any trees constitute a small proportion (sometimes less than 5%; Dupuy and Chazdon 1998) of seeds in the soil as the seed bank is mainly comprised of early successional herbaceous species (Young et al. 1987; Quintana-Ascencio et al. 1996; Dupuy and Chazdon 1998). For this reason the contribution of the soil seed bank to the recovery of long-lived canopy trees is limited (Butler and Chazdon 1998).

Given adequate time, rain forests will recover from even severe disturbances. Resilience depends on maintaining or regaining suitable species and the ecological processes needed for them to recolonize. Such development processes and the interplay of dispersal and chance establishment are well illustrated by observations of succession on tropical islands such as the Krakatau group (Box 11.3).

11.2.2 Turnover

Average tree replacement, or turnover, provides a summary of forest change. Canopy tree turnover is typically higher in tropical than in temperate forests, but rates vary with conditions, increasing on more fertile soils and declining with elevation (Stephenson and van Mantgem 2005) (Figure 11.6). History also plays a role: rapid

---

Box 11.3   Krakatau—long-distance rain forest assembly

Krakatau volcano exploded on 27 May 1883. No animals or plants survived on the island's fragmented remains. Although it remains biologically impoverished compared to nearby mainland Java and Sumatra (41 and 32 km away respectively), forest has returned.

Various spores and seeds arrived, carried by wind, sea, bats, birds, and humans. The first animal-dispersed plants (six species) were recorded in 1897, a mere 14 years after the eruption; but after 36 years the scrubby vegetation was still noted for its monotonous nature. 'One could wander for hours through ravines and over the ridges without finding a new species' (Doctors van Leeuwen 1936 in Whittaker et al. 1989). Closed forest formed about 50 years after the eruption (Whittaker et al. 1998). Frugivorous bats and birds are now resident and regularly cross to the mainland; at least 100 animal-dispersed plant species now occur. No gravity-dispersed Dipterocarpaceae—the dominant tree family in the wider region—have yet reached the islands. Long-distance dispersal and establishment is a matter of chance. *Neonauclea calycina* (Rubiaceae) was established early on one island where it is now common, but it remains scarce elsewhere. Other species, including *Ficus pubernis* (Moraceae) and a climber *Smilax zeylanica* (Smilacaceae), have each become established on only one island where they have become locally common (Whittaker et al. 2000).

The area remains volcanically active: eruptions, ash falls, powerful electric storms, and poisonous gas have intermittent impact—adding further complexity to these rapidly changing forests.

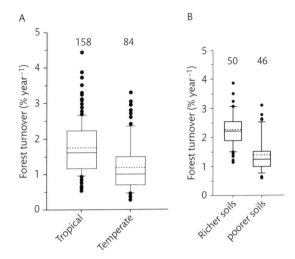

Figure 11.6 (A) Tropical and temperate forest turnover based on a compilation of published data. Turnover rates in Amazonia according to soil fertility. (B) Each box encompasses the 25th through 75th percentiles; the other solid horizontal lines indicate the 10th, 50th (median), and 90th percentiles. Dotted horizontal lines indicate the mean. The number above each box plot indicates the number of sites in the sample.

From Stephenson and van Mantgem 2005.

turnover occurs in young or disturbed forests due to the preponderance of short-lived pioneers (Burslem and Whitmore 1999; Sheil et al. 2000).

Data from long-term censuses of permanent forest plots in the Neotropics and Paleotropics show that tree turnover, tree biomass, and large liana densities have increased in mature tropical rain forests in the late 20th century (Phillips and Gentry 1994; Phillips et al. 2002, 2007, 2008; Laurance et al. 2004). Compositional changes have also been detected. Permanent plot surveys spanning 15 years (1984–99) in Amazonia reveal that population density and basal area of large fast-growing canopy trees have increased at the expense of small slower-growing sub-canopy genera, while pioneer genera show no apparent change in density or basal area (Laurance et al. 2004). These changes signify a global shift in the ecology of tropical rain forests, which is not fully understood. It may be a result of atmospheric CO_2 increases, changes in cloud cover, and climate change. Successional changes from hypothetical historical large-scale disturbances are often dismissed as the main driver of these changes, owing to the global extent of recorded changes in turnover, and because mean mortality rates

lag behind growth and recruitment rates, the opposite of what would be expected under successional change scenarios (Phillips et al. 2005). Long-term increases in atmospheric CO_2, and perhaps increased insolation in western Amazonia, remain the most promising explanations for these regional and global patterns, and interactions among CO_2 and insolation could be synergistic with respect to growth responses. Differences in the scales of effect (global versus regional) among these two variables allow for future investigation of their relative importance by comparing growth and turnover rates across regions (Phillips et al. 2005).

11.2.3 Resilience and climax communities

The term 'climax community' is sometimes used to describe a stable old-growth forest in which there are no large-scale shifts in the distribution of community and species traits, such as biomass, shade tolerance, and wood density (Table 11.4). Thus in modern usage the 'climax' state does not necessarily refer to stable relative abundance of species in the local communities, and indeed such changes will happen naturally due to the vagaries of reproduction, dispersal, establishment, and growth. The concept implies that succession has a natural end point and that, if left undisturbed, vegetation tends towards a similar predictable climax state. Such concepts, once influential, appear to have fallen from fashion. Nonetheless many ecologists still believe or assume that old-growth rain forests are close to some kind of equilibrium. So is the climax concept still valid?

There is general acceptance that a truly stable successional end point is an abstract simplification never encountered in the real world. In real forests, disturbance and recovery are never wholly excluded and environmental instability ensures constant change. Spatial heterogeneity is maintained through canopy gap formation and small-scale patch dynamics. Nonetheless, as forests recover following major canopy loss—assuming fire and other repeat disturbances are excluded—tree densities and stand basal area usually return to near pre-disturbance states. Species composition is less predictable. We can be confident that shade-tolerant species replace light-demanding species—but species-specific trends are less certain. It remains unclear the degree to which forests under similar conditions and drawing from the same species pool do or

Table 11.4 Structural characteristics of lowland tropical rain forest as a function of age since disturbance various sources.

Characteristic	Young secondary forest	Old secondary forest	Old-lgrowth 'climax' forest
Stand basal area	Lowest	Intermediate	Highest
Distribution of tree stem diameters	Lowest coefficient of variation (CV)	Intermediate (CV)	Highest (CV)
Seeds	Small, light, wind-dispersed	Mainly large, animal-dispersed	Mainly large, animal- and gravity-dispersed
Canopy structure	Even canopy, few gaps	Even canopy, small gaps common	Variable canopy height, gaps frequent and mixed size
Lianas/epiphytes	Absent	Rare	Common, diverse, and large
Large logs	Present or absent	Usually absent	Always present
Shade tolerance of young regeneration	Low	Mixed	Highest
Tree wood density	Low	Mixed	High
Very large trees	Usually absent except as obvious remnants	Often present	Occasional

Adapted from Clark, 1996; Guariguata and Ostertag 2001.

do not tend towards a similar forest community (Terborgh et al. 1996; Sheil 1999a; Vandermeer et al. 2004).

In many respects our knowledge of tropical forest dynamics remains too limited to make conclusive statements about equilibrium states. Most studies are short: a few years or less (Chazdon et al. 2007). Very few have spanned decades, while most trees can live for centuries (Sheil et al. 2000). At larger scales climate is always changing: a recent drying trend in West Africa appears to be influencing forest composition (Wittmann et al. 2006). Our understanding of longer-term processes is mainly indirect, drawn from comparisons among sites, models calibrated with a few years of data, or just credible supposition. We need more long-term studies, as doubtless such studies will surprise us.

11.3 Conclusion

This chapter has considered the processes that destroy, but also build, sustain, and modify tropical rain forests.

Tropical rain forests are highly dynamic and are impacted by disturbances at every scale. Rare and catastrophic disturbance events are difficult to research, but are dramatic when they occur, and are increasingly recognized as having had a considerable influence in shaping current forest structure. Nevertheless, the cumulative impacts of many small-scale disturbances also play a major, if variable, role in shaping every forest. Both endogenous and exogenous sources of disturbance are clearly major drivers of rain forest ecology and forest dynamics, but the importance of these drivers may vary greatly within and among tropical regions, and even over time, reflecting broad-scale differences in climate, geology, and biogeography, as well as local factors such as soil and topography that affect gap formation, and the processes of environmental change that have never allowed contexts to remain static.

Despite remarkable resilience in the face of most natural disturbances, forest ecologies are vulnerable to many modern-day threats (as described further in Chapters 14 and 16). Rain forests are, paradoxically, both resilient and vulnerable.

CHAPTER 12

The bloomin' rain forests: how flowering plants reproduce

'I see trees of green, red roses too. I see them bloom for me and you. And I think to myself what a wonderful world.'

***What a Wonderful World*, Louis Armstrong (1901–70)**

The high diversity of tropical forest plants ensures that individuals of the same species are widely spaced. A perennial question, then, is how do widely separated plants reproduce through the sexual exchange of pollen? How too do the resulting progeny leave the mother plant to arrive in suitable sites? The answer is that tropical flowering plants rely overwhelmingly on comparatively efficient animal vectors for pollen and seed movement, which has led to a bewildering array of interactions and floral forms driven by the coevolution of plants with their pollinators and seed dispersers. By contrast, wind pollination and dispersal is rare in the tropics, even though wind currents above the forest canopy can be strong. Neither is wind targeted, and it is therefore inefficient in an environment where pollen donors and recipients are few and far between. The evolution and ecology of plant–pollinator and plant–seed disperser interactions is, of course, only one aspect of plant sexual reproduction. Successful plants need to ensure that developing embryos are protected from pathogens and pests, and that seedlings are appropriately equipped to survive in a hostile environment of herbivores and competitors.

While flowering plants (angiosperms) dominate the vegetation of tropical rain forests, sexual reproduction does not have a monopoly on procreation. Many flowering plants reproduce by other means, including asexual seed set and

vegetative spread. Non-flowering plants, such as ferns, bryophytes, and gymnosperms, reproduce sexually but rely on wind and water as vectors of gamete and spore dispersal, though in this respect they differ little between tropical and temperate regions. In this chapter, we focus on the reproductive processes of the flowering plants that are so immensely diverse in wet tropical forests.

12.1 Flowering

Variation in the flowering of angiosperms is reflected by the variety of floral forms, as well as the timing of flowering, the number of flowers, the manner in which they are produced, the breeding systems that exist to prevent (or promote) self-pollination, and the floral colours, rewards, and volatiles used to attract pollinators. Certain combinations of associated traits that attract particular pollinator groups have been recognized as 'pollination syndromes' (Faegri and van der Pijl 1966), although the variety of pollination systems belies any attempt to derive neat categorical patterns. While some groups, such as the Orchidaceae, Moraceae, and Clusiaceae, are renowned for highly derived and complex floral structures evolved in association with specialized pollinators, there are many generalist floral forms visited and polli-

---

Box 12.1   Specialization and generalization of plant-pollinator mutualisms

Are plant–pollinator interactions in the tropics more specialized than those in temperate regions? It has long been thought that this is the case, figs and orchids serving as prime examples, and that specialization maintains high species diversity by providing mechanisms for reproductive isolation among congeners. For example, pollinator relationships for 11 species of *Costus* (Costaceae), a Neotropical understorey herb, revealed that each species was primarily pollinated by either euglossine bees or hummingbirds, and that pollinator specificity contributed strongly to reproductive isolation (Kay and Schemske 2003).

Recent work has, however, questioned this idea. Taxonomic analyses suggest that prevalence of specialist pollination mutualisms are broadly similar in temperate and tropical systems (Olesen and Jordano 2002; Ollerton and Cranmer 2002; Armbruster 2006; Ollerton et al. 2006). It also seems that many

tropical specialist plant–pollinator mutualisms are somewhat less specialist than first thought (Waser et al. 1996). More revealing from an evolutionary perspective is variation in pollinator specialization across a plant species' range. For example, the pollinators of the morphologically specialized resin-rewarding *Dalechampia* (Euphorbiaceae) in the Neotropics vary across the geographic distribution of the species' populations, and in relation to the presence of other *Dalechampia* species (Armbruster 1985). Thus species taken to be pollination specialists at specific locations may be considered generalist across their wider range. Such geographic variation in plant–pollinator interactions has given rise to new 'mosaic' concepts of coevolution (Thompson 2005). Such concepts better describe the structure of plant–animal interactions, instead of the previous simple dichotomous generalist vs. specialist categorization.

---

nated by many different insects and animals. We recognize a continuum between extreme specialization and generalization (Box 12.1). Unsurprisingly it is the highly specialized floral forms that often attract most interest, due to the often remarkable structures involved and curiosity concerning specialized relationships.

12.1.1 Phenology

In temperate regions, annual flowering, determined by seasonal cues, is the norm. In the aseasonal tropics, the environmental cues that trigger the onset of flowering are less apparent, resulting in greater variability of flowering phenologies. Reproductive phenological variability is expressed by differences among species in the frequency of flowering events, their duration, the proportion, and timing of flowering. Species show tremendous variation even at a single location, from near continuous flower production to supra-annual events.

Such variation in flowering, and subsequent fruiting, times has implications for the availability of resources for

pollinators and seed and fruit consumers. Supra-annual 'masting' of Southeast Asian dipterocarps, involving near-synchronous flowering, and even more synchronous fruiting, of many canopy species, leads first to a massive but short-lived increase in floral resources and, subsequently, a glut of seed availability that is thought to be an adaptive strategy to ensure effective regeneration by satiating seed predators (Janzen 1974; Sakai 2002). Masting events often include both dipterocarp and non-dipterocarp species, over thousands of square kilometres (Sakai 2002). Flowering among the main canopy dipterocarp species may be staggered sequentially during such events, which may reduce competition for shared generalist pollinators (Appanah 1985). Staggered flowering and/or fruiting may also occur among other taxa that share pollinator or seed disperser groups, such as *Miconia* in Trinidad (Snow 1965), and hummingbird-pollinated *Heliconias* in Costa Rica (Stiles 1977). Sequential flowering is not, however, an obvious feature of forests, for example in Côte d'Ivoire (Anderson et al. 2005), Costa Rica (Bawa et al. 2003) and Australia (Boulter et al. 2006).

Other trees, notably figs, and various understorey shrubs (Rubiaceae etc.) produce flowers and fruit at irregular non-synchronous intervals, such that fruit are patchily distributed but continuously available, providing important resources for frugivores throughout the year. The onset of flowering among individuals of *Terminalia lucida* (Combretaceae) and *Dipteryx panamensis* (Fabaceae) in Costa Rica is also non-synchronous, which is possibly a response to avoid pollinator limitation by bees in these large canopy trees (Bawa 1979).

Monocarpy—success through suicide

Monocarpic (or semelparous) plants reproduce only once in their lifetime, often in synchronous 'big bang' events, and then die. Iteroparous plants, by contrast, have several reproductive events during their lifetime. Aside from various bamboo species, monocarpy is rare among long-lived plants within stable environments, involving as few as 4 genera and around 30 species of monocarpic rain forest trees (Whitmore 1996). Nevertheless, some are very successful: *Tachigalia vasquezii* (Fabaceae) in the Bolivian forests is the tenth most abundant tree species (Poorter et al. 2005), while the monocarpic sago palms (*Metroxylon sago*) of New Guinea and neighbouring islands dominate vast swamps. Most bamboos (Chapter 2) also flower once, synchronously, over wide areas at multi-decade intervals[1]. By investing only in vegetative tissue, juveniles of monocarpic species attain rapid growth compared to iteroparous neighbours. At the time of reproduction, massive translocation of stored resources to reproductive tissue results in offspring production that far exceeds that of closely related iteroparous species during any reproductive episode.

Long-lived monocarpic species may avoid specialist seed predators and satiate generalist predators during their synchronous and massive fruiting episodes. Amongst synchronized species, the resulting period without adult plants (or living roots) may help shrug off any accumulating burden of parasites and herbivores. The mass 'suicide' that follows synchronous reproduc-tion creates gaps in the forest canopy, which provide favourable conditions for the establishment and growth of their locally abundant offspring (Foster 1977). Despite these apparent advantages, monocarpy among long-lived plants remains an oddity in tropical forests, not only because it is rare among species, but also because the few monocarpic species are, paradoxically, so successful.

What triggers flowering?

In seasonal rain forests, as in Panama, many plants respond to seasonal cues, such as chilling of the developing flower buds, for the initiation of flowering. Climatic variations associated with the periodic El Niño events may provide the stimulus for mass flowering of dipterocarps in Malesia (Sakai et al. 2006), and may enhance reproduction of trees generally in Borneo (Curran et al. 1999; Sakai 2002) and Panama (Wright and Calderon 2006). Thus periods of low temperature (Ashton et al. 1988) and El Niño-associated drought have been proposed as flowering triggers (Sakai et al. 2006).

Photoperiodic induction has also been proposed as a mechanism to explain synchronous flowering of rain forest trees (Borchert et al. 2005). In the tropics changes in day length large enough to affect reproductive development occur only at the equinoxes, and it is at these times that flowering peaks. Flowering synchrony is less precise closer to or at the equator, presumably because the variations in day length have less magnitude. Thus vegetative growth of *Bombax malabaricum* is synchronous at 5° N and S, but asynchronous at 1°N in Singapore (Borchest et al. 2005).

12.1.2 Breeding systems

The low densities coupled with presumed poor flight capacities of many tropical pollinators led many early ecologists to suppose that tropical trees were primarily self-pollinated (Baker 1959; Fedorov 1966). Studies of **outcrossing rates** show this was generally incorrect (Hamrick and Murawski 1990; Murawski et al. 1994; Doligez and Joly 1997), and most tropical plants came

---

[1] This monocarpic habit of bamboos makes species identification very difficult, as taxonomy of many co-occurring and morphologically near-identical species is based on floral structure, forcing the botanist to wait several decades before confirmation of species identities can be achieved. The application of genetic methods should soon address this challange.

to be considered primarily outcrossing, and therefore dependent on the pollen transfer between individuals to effect seed set (Table 12.1) (Ashton 1969; Bawa 1974; Ward et al. 2005; Dick et al. 2008). More recently the apparent prevalence of outcrossing has been tempered by the recognition that outcrossing is variable, and that several breeding systems that exclude outcrossing may be more widespread than previously thought. Nevertheless, the consensus remains that outcrossing is common among rain forest plants, and large distances between individuals is not necessarily a barrier to pollen exchange.

Self-incompatibility mechanisms to prevent self-fertilization are widespread among tropical plants, and include physiological barriers to the germination and growth of pollen on the stigmatic surface, floral morphologies that ensure that pollen-bearing stamens are far removed from pollen-receiving stigmas, and temporal separation of pollen production and stigma receptivity. For example, the Neotropical liana *Drymonia serrulata* (Gesneriaceae) is completely self-compatible but minimizes self-pollination by producing only a few flowers daily over several months. Each of these flowers has a male phase that lasts one day, and this is followed by a female phase the next day (Steiner 1985).

Even so, many self-compatible species do self-fertilize, and in other species the self-incompatibility barrier is imperfect (Bawa 1974; Bawa et al. 1985). Having moved away from the idea of predominant outcrossing among tropical trees, ecologists are now once again re-evaluating the extent of outcrossing. Recent studies show that many canopy trees produce many selfed fruit, although selfing rates can be highly variable within a population (see also *Apomixis* below). Thus parental analysis of seeds from only eight mother trees of *Shorea acuminata* (Dipterocarpaceae) revealed selfing rates from 7.6% to 88.4% (Naito et al. 2008) and in *Ceiba pentandra* (Malvaceae) a range of 0% to 97.8% selfing has been

Table 12.1 Breeding systems among tropical rain forest trees. Outcrossing rates range from 0 (complete selfing) to 1 (complete outcrossing).

Tree species	Country	Pollinator	Outcrossing rate
Carapa guianensis (Meliaceae)	Brazil	Small insect	0.93
Carapa guianensis (Meliaceae)	Costa Rica	Small insect	0.99
Carapa procera (Meliaceae)	French Guiana	Small insect	0.85
Caryocar brasiliense (Caryocaraceae)	Brazil	Bats	0.84
Cavanillesia platanifolia (Malvaceae)	Panama	Hawk moths, monkeys, bats	0.57
Ceiba aesculifolia (Malvaceae)	Mexico	Bats	0.96
Ceiba pentandra (Malvaceae)	Mexico	Bats	0.90
Dinizia excelsa (Fabaceae)	Brazil	Small insects	0.90
Dryobalanops aromatica (Dipterocarpaceae)	Malaysia	Large bees	0.92
Entandrophragma cylindricum (Meliaceae)	Cameroon	Bees and moths	0.98
Enterolobium cyclocarpum (Fabaceae)	Costa Rica	Insects (moths, bees, beetles)	1.00
Helicteres brevispira (Sterculiaceae)	Brazil	Hummingbirds	0.66
Leucaena esculenta (Fabaceae)	Mexico	Small insects	0.64
Pachira quinata (Malvaceae)	Costa Rica	Bats and sphingid moths	0.91
Balizia elegans (Fabaceae)	Costa Rica	Sphingid moths	0.99
Pterocarpus macrocarpus (Fabaceae)	Thailand	Insects	0.95
Samanea saman (Fabaceae)	Costa Rica	Sphingid moths	0.99
Sextonia rubra (Lauraceae)	French, Guiana	Small insects	0.90
Shorea curtisii (Dipterocarpaceae)	Malaysia	Small insects	0.93
Shorea megistophylla (Dipterocarpaceae)	Sri Lanka	Large bees (Apis)	0.87
Symphonia globulifera (Clusiaceae)	Costa Rica	Hummingbirds	0.90

Adapted from Dick et al. 2008.

reported among individual trees (see Ward et al. 2005). The development of selfed fruit which is aborted only at late stages of development is common among dipterocarps, and is curious as it appears to waste significant resources. Such developing fruits may serve as sacrificial seed predator sinks, improving the probability of outcrossed seed escaping predation (Ghazoul and Satake 2009).

Spatial separation of male and female functions: monoecy and dioecy

Another way of minimizing or avoiding selfing is the physical separation of male and female functions. Spatial segregation of the sexes may occur among flowers on a single plant (monoecy) or among plants (dioecy). Compared to temperate regions, a relatively high proportion of plants in the tropics are dioecious, including many figs (Box 12.2), and pioneers such as *Cecropia* (Bawa 1980) (Table 12.2). Dioecious plants often have small open flowers, a feature usually considered typical of generalist pollination syndromes (Bawa and Opler 1975). In fact, high pollinator fidelity has been noted among dioecious species that bear such flowers[2]. Other tropical dioecious species[3] are wind pollinated (Bullock 1994; Carpenter et al. 2003; Ibarra-Manriquez & Oyama 1992; Renner & Ricklefs 1995), though the relative importance of wind and animal pollination has not been well documented.

Some plants undergo sex change, which may be a function of age, or a response to changing environmental variables. Several tropical orchids when exposed to sunlight change sex from male to monoecious to female, the sequence being reversed if plants are returned to shade, and in natural conditions males are most abundant in shady areas, while females are found mainly in sunlight (Dodson 1962).

Cleistogamy

At the opposite extreme to dioecy, the anthers and styles of cleistogamous flowers are entirely enclosed within a non-opening flower, thereby ensuring selfing. Although **cleistogamy** occurs within 50 plant families, it is rare outside the grass family Poaceae[4]. In the tropics, cleistogamy is rare and mainly found among the herbs and epiphytes of the Bromeliaceae and Orchidaceae (Gilmartin and Brown 1985; Culley and Klooster 2007), and may allow plants growing in relatively stressful conditions, such as nutrient or water stress, some assurance of seed set (Lecorff 1993). *Plowmanianthus*, a recently described Neotropical herb genus consisting of five species within the Commelinaceae, is unusual in having a particularly high frequency of cleistogamy, with nothing but cleistogamous flowers known for some species. The shallow roots of these plants do not extend beyond the rain forest leaf or humus layers, suggesting that they too may be under water and nutrient stress (Hardy and Faden 2004).

Apomixis

Seed production without fertilization (Box 12.3) can occur through apomixis by which embryos are derived from the maternal tissues of the ovule, and are therefore genetically identical to the mother plant[5]. Apomixis appears to have evolved many times within the angiosperms. It has been recognized in over 400 species from 86 families, including both monocotyledonous and eudicotyledonous plants (Allem 2003). Plants that regularly reproduce through apomixis are often found in frequently disturbed habitats, or where successful outcrossing may be limited by low population density, as may be the case in tropical rain forests (Asker and Jerling 1992).

Asian rain forest trees that can reproduce apomictically include species of *Shorea* and *Hopea*

---

[2] In one study 25% of dioecious species were visited by moths, 22.5% by bees, 20% by beetles, 12.5% by flies, and only 10% by generalist diverse insects (Renner and Feil 1993).

[3] Including gymnosperms like *Podocarpus*.

[4] Excluding the Poaceae, cleistogamy has been documented in only 370 mostly temperate species from 140 genera. It occurs among a further 371 species in 140 Poaceae genera (Culley and Klooster 2007).

[5] The initial discovery of apomixis is attributed to a solitary female plant of Australian holly *Alchornea ilicifolia* (Euphorbiaceae), native to the subtropical rainforests of eastern Australia, which continued to develop seeds when planted at Kew Gardens in England.

Box 12.2   Dioecy in figs: an enigma

The persistence of dioecy in many species of figs is enigmatic because it represents an intense conflict between the interests of the pollinating wasps and those of the plant (Weiblen 2002). The maturation of pollinators and seeds is segregated in two types of figs on separate plants (Weiblen et al. 1995). Short-styled flowers in male 'gall' figs are consumed by pollinator larvae which complete development to fully functional adults. In female 'seed' figs the ovules of the long-styled flowers are unharmed as wasp ovipositors are too short to reach the ovules. It seems odd that this mutualism should remain stable, because pollinators would be expected to evolve a preference for gall figs or evolve longer ovipositors, yet neither appears to have occurred (Grafen and Godfray 1991; Janzen 1979; Patel et al. 1995). Experiments have shown fig wasps to be incapable of distinguishing between fig types despite their assured reproductive failure in seed figs (Patel et al. 1995; Weiblen et al. 2001). This failure is particularly enigmatic considering that fig wasps are capable of identifying unique fig hosts from amongst a range of co-occurring and closely related species. Thus, the persistence of dioecious fig pollination may be based on pollinator deception, with selection favouring seed figs that mimic gall figs in attractiveness (Grafen and Godfray 1991). Dioecy may also benefit both pollinator and plant by reducing the incidence of parasitism of the mutualism. Parasites of fig wasps waste time probing seed figs thereby increasing the plant and pollinator fitness through increased pollinator production. Parasitism, therefore, may be crucial in stabilizing the dioecious fig pollination mutualism.

Table 12.2 Floral sexuality of trees in tropical lowland rainforests. Typical values of dioecious plants from temperate regions are 3–4%.

Breeding system	% Species				
	Lowland forest		Montane forest		Island forest
	C. America	SE Asia	Jamaica	Venezuela	New Caledonia
Hermaphroditic	65	60			
Monoecious	11	14			
Dioecious	23	26	21	31	16

From Bawa 1992.

(Dipterocarpaceae), *Garcinia*[6], *Kayea* and *Calophyllum* (Clusiaceae), *Diospyros* (Ebenaceae), *Memecylon* (Melastomaceae), *Syzigium* (Myrtaceae), and *Mangifera* (Anacardiaceae) (Allem 2003; Chan 1981; Kaur et al. 1978; Lee et al. 2000). In the Neotropics apomixis appears less common, but has been reported in species of *Clusia*, *Erythroxylum* and in *Bombacopsis glabra* (Allem 2003).

12.1.3 Pollination

Tropical flowering plants depend overwhelmingly on animals, particularly insects, as agents of pollen transfer: over 80% of the woody plants in Neotropical forests do so (Gentry 1982). Animal pollinators that actively seek out flowers are a far more efficient means of securing pollen transfer than reliance on passive agents of

---

[6] The commercially important mangosteen *Garcinia mangostana* reproduces almost exclusively in this way (Lasso and Ackerman 2004), and widespread apomictic reproduction among *Garcinia* species may favour the female bias frequently observed within this entirely dioecious genus (Thomas 1997).

pollen transfer such as wind or water. The vast majority of pollinators are drawn from the ranks of bees and wasps, beetles, butterflies and moths, thrips, and flies, as well as some vertebrate groups, mainly birds and bats, though occasionally also non-flying mammals and even lizards. Despite the great variety of plant-pollinator systems, it is possible to associate floral traits with particular pollinator groups as a series of 'syndromes' (Table 12.3).

Plants might obtain the services of pollinators by luring them with exotic odours, by tricking them, or even by trapping them. Mostly, however, flowers bribe pollinators with a variety of floral resources, mainly pollen and nectar, but occasionally other rewards. Nectar is the principal reward for most flower visitors, but some, including bees, beetles, heliconid butterflies, and thrips, specialize on the collection of pollen for their own consumption or for that of their larvae (bees). Even glossophagid bats derive important proteins from pollen (Herrera and Del Rio 1998; Mancina et al. 2005).

Rather more unusual rewards have evolved with certain pollinator groups. Oils from oil-secreting organs called elaiophores are collected by some anthophorid and melittid bees, which pollinate primarily Neotropical plants from families as diverse as Gesneriaceae,

Table 12.3 Pollination syndromes of tropical rain forest trees. Adapted from various sources.

Pollinator	Floral characteristics	Typically-pollinated plants families and genera
Bats Glossophaginae (Neotropics),Megachiroptera (Old World)	Large, pale flowers or inflorescences with musky odour, held away from the foliage. Abundant nectar rich in hexose (Neotropics) or sucrose (Paleotropics)	Bignoniaceae, Malvaceae, Myrtaceae, Caesalpinioideae, Caryocaraceae
Birds Hummingbirds (New World), sunbirds and honeyeaters (Old World), honeycreepers (Hawaii)	Diurnal red or orange flowers, lacking scent. Often tubular corollas or open flowers with an array of protruding brush-like stamens. Abundant dilute nectar (hummingbird flowers may be sucrose rich).	*Erythrina, Spathodea, Symphonia globulifera, Musa salaccensis,* a variety of herbaceous understorey plants.
Bees Apidae, Halictidae, Megacilidae, Anthophoridae, Andrenidae	Extremely varied in structure. Specialist bee-flowers tend to be brightly coloured, with nectar guides on the petals. Both pollen and nectar offered as a reward.	Fabaceae, Bignoniaceae, Melastomataceae, Burseraceae, Euphorbiaceae. Oils offered as a reward by some Malpighiaceae and Orchidaceae, and resin by species of *Clusia* and *Dalechampia.*
Moths Sphingidae, and some Noctuideae, Pyralidae and Geometridae	Nocturnal heavily sweet-scented flowers with long tubular corollas. Plentiful sucrose-rich nectar.	Rubiaceae, Apocynaceae, Meiaceae, Mimosoidea.
Butterflies Piendae, Papihonidae, Nymphalidae, Danaidae	Long tubular corollas, brightly coloured flowers, often aggregated into loose inflorescences.	*Delonix, Caesalpinia, Ixora, Cordia* and *Mussaenda*
Beetles Very diverse range of beetle pollinators, including weevils, dung beetles, chrysomelid beetles	Highly variable, but mostly open flower structure. Flowers can be relatively large (Annonaceae) or small and aggregated into huge inflorescences (palms)	Annonaceae, Lauraceae, Myristicaceae, Palmae, Cyclanthaceae, Dipterocarpaceae
Thrips and other small insects	Small unspecialized flowers, often pale or white and faintly scented.	Anacardiaceae, Annonaceae, Dipterocarpaceae Euphorbiaceae, Guttiferae, Moraceae
Flies Hoverflies, carrion and dung flies, midges	Small pale flowers with small quantities of nectar.	Anacardiaceae, *Neolitsia dealbata* and *Litsea leefeana* (Lauraceae, Australia), some palms (*Asterogyne martiana* in New Guinea and *Prestoea schulzeana* in Ecuador).

Iridaceae, Krameriaceae, Malpighiaceae, Orchidaceae, Scrophulariaceae, and Solanaceae (Simpson and Neff 1983; Buchmann 1987). Fragrances from Neotropical orchids are collected by male euglossine bees, apparently to attract females (Dressler 1968, 1982; Chase and Hills 1992). Several bees and wasps use floral resins for nest construction (Armbruster 1984). Some plants offer no rewards at all and rely on deceiving pollinators by presenting showy but empty flowers (*Plumeria rubra*; Apocynaceae), or even temporarily trapping them (*Amorphophallus johnsonii* and other Araceae) (Haber 1984; Beath 1996).

Among the most highly evolved plant-pollinator systems are those that involve brood rearing within the flowers where the plants' ovules themselves comprise the reward. While this is also well known for examples outside the tropics, it is among the tropical fig species where it has been most developed (Box 2.4, Chapter 2).

The variety of pollinators

Bees are overwhelmingly important in all tropical locations, and visit such a wide variety of floral forms, with varying degrees of specialization, that classification of a typical bee-pollination 'syndrome' is difficult. As many as 60% of canopy and sub-canopy tree species at La Selva, Costa Rica, are bee-pollinated, predominantly by medium- to large-sized, long-tongued bees (Kress and

---

Box 12.3   Space invaders: asexual reproduction to capitalize on empty space

Asexual reproduction is a means by which space can be rapidly colonized. A dramatic example is *Tetramerista glabra* (Tetrameristaceae) where vegetative reproduction is taken to an extreme. *Tetramerista glabra* is a large common tree in primary peat swamp rain forest in Sumatra, Borneo and peninsular Malaysia (Gavin and Peart 1999). In this species, regeneration by seed is only effective in high light conditions, and even then germination rates rarely exceed 10%. Recruitment is more commonly achieved by vegetative propagation of **ramets** (stem sprouts). Young ramets grow rapidly, particularly in the high light conditions of a treefall gap, but shoot elongation soon exceeds the stem's ability to support the sapling, so the shoot bends and collapses. Fallen stems produce numerous vegetative sprouts which grow and, in turn, collapse to repeat the cycle (Figure 12.1). Occasionally a ramet gains support from surrounding vegetation and grows without collapsing. Stilt roots develop when stem diameter reaches 4–8 cm (by which time the sapling has attained 5 m height) allowing the sapling to support itself independently. Stilt roots thicken as the tree continues to grow and eventually fuse forming a cylindrical trunk of up to 150 cm in diameter.

Other examples of vegetative reproduction are just as effective in occupying space. Many bamboo and palm species, and some trees, develop dense suckers or runners that rapidly fill any well-lit space and smother competing seeds. This site occupancy can be pre-emptive. In Uganda the exotic Asian tree *Brousonetia papyifera* (Moraceae) produces a surrounding thicket of shoots in the dark forest understorey that will grow rapidly whenever a gap occurs (DS, pers. obs.).

The water hyacinth *Eichhornia crassipes* (Ponte-deriaceae), native to tropical South America, has been widely introduced to tropical waterways in Africa, Asia and Australia where it has become a serious invasive pest. Its rapid population growth and spread to new areas is greatly facilitated by vegetative reproduction through runners that establish large numbers of independent daughter plants. Daughter plants are carried by water currents to colonize new locations.

Many tropical forest ferns also reproduce asexually by ramets, fragmentation, and regrowth, and by producing plantlets. These genetically identical copies of the parent plant appear to be particularly common in montane rain forests where they are able quickly to colonize new space created by frequent landslides and tree falls (Koptur and Lee 1993).

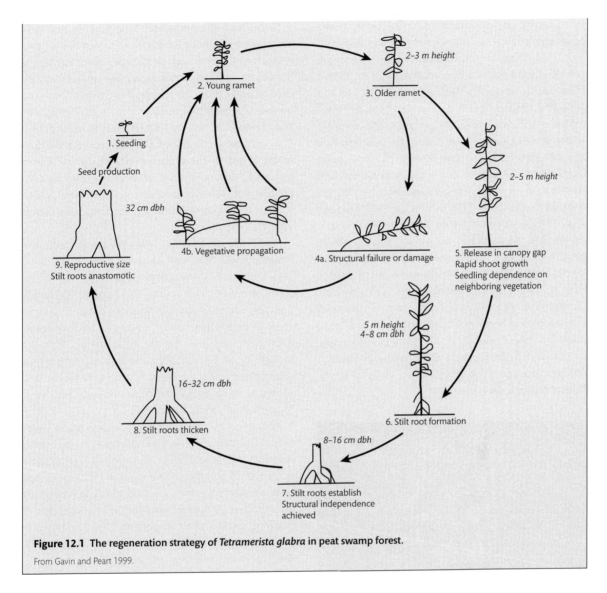

Figure 12.1 The regeneration strategy of *Tetramerista glabra* in peat swamp forest.

From Gavin and Peart 1999.

Beach 1994). There is a tendency for 'small bee' flowers to be relatively undifferentiated generalists that may be visited by other pollinator groups (e.g. flies and wasps), while flowers pollinated by large bees are more structurally complex and specialized to host specific bees, such as the many Neotropical orchids pollinated by euglossine bees[7]. Bee pollination also dominates in the Asian rain forests, although here the larger, long-tongued bee species are not as important as pollinators (Momose et al. 1998b). In Southeast Asian dipterocarp forests, the breeding cycles of many social bees, and particularly the giant Asian honeybee *Apis dorsata*, may be synchronized to seasonal flowering peaks. In Sarawak, Borneo, *Apis dorsata* bees respond to supra-annual mass flowering

---

[7] Euglossine bees are the exclusive pollinators of around 2,000 of the 7,000 Neotropical orchid species, and are hence commonly called orchid bees. They are most diverse in cloud forests, matching the diversity of tropical orchids they pollinate.

events, typical of the lowland equatorial forests in this region, by migrating to flowering forests areas[8] (Momose et al. 1998b). Long-tongued bees are poorly represented in the Australian tropics, where many flowers are instead visited by small beetles, flies, and thrips (Irvine and Armstrong 1990).

Beetles pollinate many canopy trees, and scramble among dense clumps of small flowers typically with little apparent discrimination. The flowers of many Asian Dipterocarpaceae, as well as Araceae, Annonaceae, and Arecaceae in all major tropical rain forest biomes, are visited by beetles (Beath 1996; Gottsberger 1990; Momose et al. 1998b; Sakai et al. 1999). Beetles are particularly important in the Australian wet tropics, where, perhaps due to fewer bees, up to one quarter of all flowering plant species are beetle-pollinated (Irvine and Armstrong 1990). Most pollinating beetles belong to the Chrysomelidae, Curculionidae (Figure 12.2), and Nitidulidae, but examples of more unusual beetle-pollinated systems include pollination of eleven species of *Orchidantha* (Lowiaceae, Zingiberales) by *Onthophagus* dung-beetles in Sarawak, Borneo (Sakai and Inoue 1999) that are attracted by the flowers' distinctly dung-like odour.

Figure 12.2 A curculionid beetle visiting the flowers of *Mikania micrantha*.

Photo by Alan Timmermann.

Nocturnal hawkmoths have long tongues that seek nectar from within the long narrow corollas of the flowers with which they have clearly coevolved. The tips of hawkmoth tongues may support fine spines that are used to break open the tissues of nectarless flowers to release sap. Neotropical *Heliconius* butterflies gather pollen into a ball held by the proboscis to which a drop of clear liquid (probably nectar) secreted from the proboscis is added. The proboscis is then used to knead this mix for several hours after which the resulting amino-acid rich solution is drawn in—a foodstuff which sustains these butterflies for a much longer life than is usual (Gilbert 1972).

Tiny thrips (up to 2 mm in length) are ubiquitous in flowers, where they feed on pollen, nectar, and floral tissues (Lewis 1973). Weak flight suggests that they mostly promote self-pollination (Baker and Cruden 1991), but their tiny size facilitates wind-assisted dispersal, and, as aerial plankton, they have the potential to carry pollen over great distances (as for fig wasps). Thrips are thought to be the primary pollinators of some *Shorea* dipterocarps[9], and members of the Annonaceae and Moraceae in Southeast Asia (Appanah and Chan 1981; Momose et al. 1998a; Sakai 2001).

Pollination by flies (Diptera) is known to be widespread in Sterculiaceae and climbers in the genus *Aristolochia* (Prance 1985), and, famously, flies pollinate many species of *Rafflesia* (Beaman et al. 1988), the world's largest single flower (Box 2.7, Chapter 2), to which they are attracted by specific volatile scents reminiscent of decaying carrion. More important from a commercial perspective, tiny midges are the pollinators of cocoa *Theobroma cacao* (Young 1982).

Bats pollinate many tropical forest trees and woody shrubs. Nectar-feeding bats are particularly important in the Neotropics where the specialist nectar-feeding subfamily Glossophaginae (family Phyllostomidae) have hairs at the tips of their tongues to extract nectar and pollen. At 6–25 g, these are also among the smallest of

---

[8] Interestingly local people in Borneo recognize this phenomenon and claim that pigs follow the bats which follow the bees.
[9] Pollination of dipterocarps by thrips was widely accepted until recent studies by Momose et al. (1998) and Sakai et al. (1999) failed to attribute much significance to thrips in Bornean rain forests, where dipterocarps were predominantly pollinated by bees and beetles.

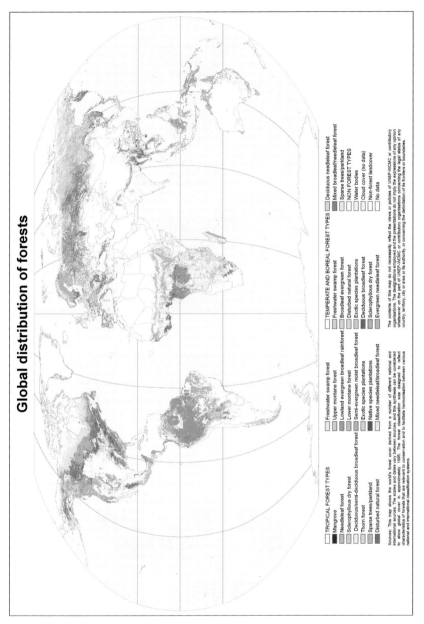

Plate 1 Map of forest distributions worldwide. (See Figure 1.3, page 4.) Map courtesy of WCMC-UNEP. http://www.unep-wcmc.org/forest/global_map.htm

Plate 2 A drawer of butterflies collected and arranged by Alfred Russel Wallace to illustrate Batesian mimicry, these specimens being from India, Africa, Southeast Asia, and Central America. The inedible model species are arranged in the far right column and the presumed edible mimics are to the left of them. Note that the males of two of the mimic species are not mimetic. All the writing is Wallace's. (See Figure 5.3, page 78.) Photo by George Beccaloni.

Plate 3 Müllerian mimicry complexes form where co-occurring noxious species evolve to resemble one another to reinforce visual unpalatibility cues. Each pair of pictures shows a separate 'sub-ring' with the models (each a different species of *Melinaea*, Ithomiinae) above, and the mimics being different *Heliconius numata* (Heliconiinae) morphs, shown below. All of these species come from the same site in Ecuador. (See Figure 5.4, page 79.) Photo by George Beccaloni.

Plate 4 Many tropical forest insects have become masters of camouflage, both for the purposes of protection and/or predation. Insects commonly emulate flowers. (See Figure 5.12 C, page 89.) Photo by Carsten Bruehl.

Plate 5 Lesser bird of paradise (*Paradisaea minor*) from the lowland rainforests of New Guinea. (See Figure 4.8, page 58.) Photo by Rhett A. Butler.

Plates 6 and 7 The variety of invertebrates of tropical rain forests. Bugs (right) are important herbivores, but some groups are also predatory. Butterflies (center) are probably the most familiar and best known. (See Figures 5.1 D and 5.1 H, pages 72–3.) Photos by Karin Beer.

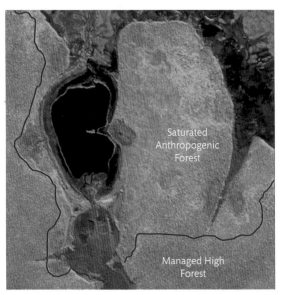

Plate 8 Black seeds within a bright red fruit, as shown here by the fruit of *Sterculia* sp. (Sterculiaceae) at Sepilok, Sabah, Malaysian Borneo; this is a common visual display to potential seed dispersers. (See Figure 12.11, page 266.) Photo by Julia Born.

Plate 9 An anthropogenic landscape in southern Amazonia (from Heckenberger et al. 2003). Areas to the south and west of the red line are relatively undisturbed forest, while regions of former cultivation are revealed in lighter shades of green. (See Figure 14.3, page 300.)

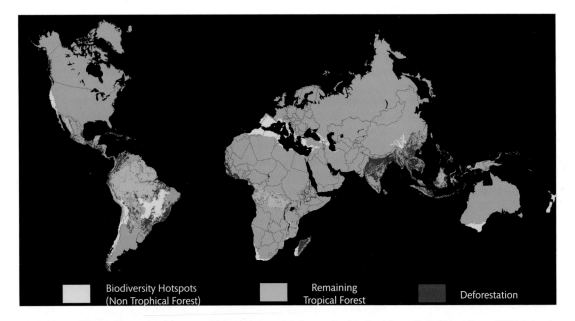

Plate 10 Modified Fuller projection map of forest cover and deforestation, derived from Global Land Cover 2000 dataset. This equal area projection shows the true extent of tropical terrestrial regions relative to global land area. Map generated and provided by Clinton Jenkins. (See Figure 14.5, page 309.)

bats. Small size reduces the energetic costs of hovering over flowers (Voigt 2004). Other bats, including larger members of the Phyllostomidae in the Neotropics, and flying foxes (Pteropodidae) in the Oriental tropics, visit similar flowers of the Bignoniaceae, Malvaceae, and Fabaceae.

Neotropical hummingbirds (Trochilidae) are common pollinators of understorey and forest gap plants (Figure 12.3). Sunbirds perform a similar function in the Asian and African tropics. Bird-pollinated plants differ considerably in floral morphologies, with some differentiating among birds based on tongue length and bill curvature. Bat-pollinated plants show no such specialization, possibly owing to constraints in the morphological structure of bat mouth-parts. One remarkable exception is a recently discovered bat, *Anoura fistulata*, from the cloud forests of the Ecuadorian Andes, which can extend its tongue twice as far as closely related species, and is apparently the specialist pollinator of *Centropogon nigricans* (Campanulaceae), which has a corolla length to match (Muchhala 2006).

Although a few insect and vertebrate groups dominate the spectrum of pollinators, new and unusual interactions continue to be uncovered. Cockroaches, for example, have been discovered to pollinate annonaceaeous lianas (Nagamitsu and Inoue 1997) and *Artrocarpus* trees from Borneo (Momose et al. 1998b). It is curious to note that

Figure 12.3 A *Eulampis holosericeus* hummingbird visiting *Charianthus grenadensis*.

Photo by Alan Timmermann.

pollination by ants is absent in tropical rain forests given their ubiquity in this habitat (Ghazoul 2001).

Pollination by non-flying mammals is rare. Pollination of a small number of plants is known by lemurs in Madagascar (Sussman and Raven 1978; Kress 1993; Nilsson et al. 1993; Kress et al. 1994; Birkinshaw and Colquhoun 1998), prosimians and rodents in tropical Africa (Grünmeier 1990), rodents (pollinating *Blakea*, Melastomataceae)and marsupials in Australia and New Guinea (Goldingay et al. 1991; Carthew and Goldingay 1997), and marsupials and primates in the Neotropics (Prance 1980; Janson et al. 1981; Lumer and Schoer 1986; Tschapka and von Helversen 1999). Flowers pollinated by non-flying mammals are often strongly attached close to the ground on the stems of tree, a trait known as cauliflory (Figure 12.4).

Lizard pollination can be important on some tropical island communities (NyHagen et al. 2001; Olesen and Valido 2003; Godinez-Alvarez 2004) (Figure 12.5). Examples include Mauritius where flowers of endemic *Trochetia* spp. produce coloured nectar as an apparent signal to their pollinator the endemic *Phelsuma* geckos (Hansen et al. 2006).

Wind pollination

Wind is an inefficient agent of pollen transport when compared with living vectors, particularly in the tropics where forest structure limits wind below the canopy (Whitehead 1983) and frequent rain washes pollen out of the air (Regal 1982). Consequently, pollination by wind is relatively rare in the tropical forests, comprising 2–3% of lowland Neotropical rain forest floras (Bawa and Opler 1975) and 7–9% of plants in the Australian wet tropics (Irvine and Armstrong 1990), compared to 70–100% of temperate arboreal flora. Even grasses, almost all of which are wind-pollinated in other habitats, have insect-pollinated species in lowland rainforests (Soderstrom and Calderon 1971).

Despite this, wind-pollinated species belonging to several families occur in all tropical regions (e.g. *Chamaedorea alternans*, Arecaceae, in Mexico; *Trophis involucrata*, Moraceae, in Costa Rica; and *Artocarpus rigidus*, Moraceae, in Malaysia) among both understorey and canopy species, and particularly among dioecious plants (Bawa and Crisp 1980; Listabarth 1993; Bullock 1994; Otero-Arnaiz and Oyama 2001). In lowland African

Figure 12.4 (continues).

Figure 12.4 Examples of cauliflory where flowers are produced directly from the stem: *Coffea macrocarpa, Syzygium mamiluatum, Diospyros revaughanii, Tambourissa peltata,* and *Sideroxylon puberulum.*

Photos by Christopher Kaiser-Bunbury.

Figure 12.5 *Anolis richardii* visiting the flowers of *Marcgravia umbellata.*

Photo by Alan Timmermann.

The apparent low prevalence of wind pollination may need to be re-examined. The syndrome itself may have been underestimated, as differentiating between biotic and abiotic pollination is difficult when these typically small inconspicuously coloured flowers are also visited by various insects (Bullock 1994). Many rain forest plants are spatially aggregated; wind in the upper canopy can be substantial, and updrafts of air currents in the evening can carry pollen-bearing insects (and presumably pollen too) high above the canopy (Compton et al. 2000; Compton 2002). Finally, rainfall is frequently seasonal and need not limit wind pollination by washing pollen out of the air.

12.2 Seed set

Plants often set only a tiny proportion of their ovules as seed, and tropical plants are no exception. Insufficient or inappropriate pollination is a primary reason, but seed losses also arise from selective abortion or pre-dispersal predation. Pollination limitation due to non-visitation is widespread (Burd 1994), although studies on tropical plants are relatively rare. Even visited flowers may not be effectively pollinated: only 54% of the flowers visited by bats in the bat-pollinated Costa Rican palm *Calyptrogyne ghiesbreghtiana* were actually pollinated (Cunningham 1996). Reduced pollination may be particularly pronounced for plants that are relatively isolated from

rain forests pollination by wind is reported to be widespread among the Moraceae[10].

Pollination by very small insects (e.g. thrips and fig wasps) is at the very least wind-assisted: tiny fig wasps can be dispersed by wind for tens of kilometres above the forest canopy (Harrison 2003; Harrison and Rasplus 2006). Insect-assisted wind pollination also occurs in the understorey palm *Chamaedorea pinnatifrons* (Arecaceae) from the Peruvian Amazon, where the activity of pollen-feeding thrips and ptiliid beetles[11] within inflorescences triggers the release of pollen which becomes airborne and is then carried to female plants by wind (Listabarth 1993).

---

[10] Based on a personal communication with D. Leston quoted in Bawa and Crisp (1980).
[11] 'Feather-winged' beetles of the Ptiliidae family that are among the smallest of all beetles

conspecifics, which might result naturally or from human activities such as forest fragmentation or logging (House 1993; Ghazoul and McLeish 2001; Ghazoul 2005).

Even if pollinated successfully, ovule abortion due to genetic incompatibility may result from selfing or mating with neighbours or others that have similar genotypes. The Neotropical herb *Costus alleni* (Costaceae), for example, achieves better seed set when it receives pollen from a larger number of neighbours, which reduces the likelihood of inbreeding (Schemske and Pautler 1984). Selective abortion of inbred progeny may be adaptive by favouring the best quality progeny under conditions of resource limitation.

12.2.1 Seed predation

Successful plant recruitment and persistence requires negotiating a series of steps from pollination to establishment and growth until maturity. Each step has associated hazards for individuals within a cohort, and the most critical of these in terms of proportional mortality is often predation of seed by invertebrates and vertebrates, and their loss to fungal pathogens. Seeds are vulnerable to predation at all stages of development and dispersal[12]. Seed predation contributes to the structure and composition of tropical rain forests. Indeed, the rich diversity of rain forest plants is maintained, at least in part, by differential predation of dispersed seeds (Chapter 14).

Pre-dispersal seed predation
Flowers and seed represent accessible and nutritious packages that many animals readily exploit. Consequently, and despite massive seed production, seed predators can cause near complete destruction of a seed crop before maturation. Loss of propagules occurs through opportunistic as well as targeted seed and flower 'predation'. Flowers can form as much as 38% of the dry season food of Brazilian scaly-headed parrots (Galetti 1993). Often more serious is loss to insect seed-predators. Many of these often specialized

insects, which include pyralid and tortricid moths and bruchid and curculionid beetles, lay their eggs in flower buds or young fruit and consume the developing seed as larvae. While large flowers and inflorescences are effective in increasing pollination success, the resulting large seeds, and/or infructescences, appear to be more susceptible to infestation, as has been shown for the genus *Piper* (Greig 1993a). In some cases, the proportion of seed lost to such insects can exceed 90%, and may even cause complete loss of seed, as for some dipterocarp trees (Toy et al. 1992) and the understorey shrub *Hybanthus prunifolius* (Augspurger 1981). Even during synchronous fruiting (or masting) events, losses to pre-dispersal insect seed predators can be very high, with no stage of seed development being invulnerable (Nakagawa et al. 2005).

Post-dispersal seed and seedling predation
Survival to maturity and subsequent dispersal of seed from the mother tree do not guarantee protection, as a wide variety of post-dispersal seed predators, mostly vertebrates, value such seeds for their own reproductive success. Migratory and fecund populations of Bearded Pigs *Sus barbatus* destroy large quantities of dipterocarp seed during mast seeding events in Borneo, and the landscape-scale distribution of dipterocarp fruiting events determines migratory patterns of these seed consumers (Curran and Leighton 2000). While masting contributes to seed survival through predator satiation, seed survival may also be due to escape from groups of nomadic pigs that arrive late in the fruiting period after establishment of these rapidly germinating seedlings (Curran and Webb 2000).

Seed predators can play central roles in regulating the demography of tropical trees, and their local extinction can have cascading effects in tropical ecosystems. This has been demonstrated through a fortuitous local extinction and recolonization of white-lipped peccaries, a major consumer of fallen palm fruits, at Cocha Cashu in south-eastern Peru: during the 12-year absence of peccaries, seedling density of the common palm *Astrocaryum*

---

[12] Animal seed dispersal agents can also incur a cost by destroying a portion of the seed produced, but this is often deemed a subsidiary cost of dispersal.

murumuru increased by 70%, but dropped to its former level after peccaries returned (Silman et al. 2003).

Vulnerability to predators and pathogens

Heavy seed predation and pathogen attack should favour selection for defence strategies, but these may be constrained by trade-offs acting on offspring traits such as dispersal and resource provision. Thus large seeds usually have thicker seed coats, which protect against predator and pathogen attack, and plentiful reserves that allow for rapid germination. However, plants with large seed produce fewer of them. Large-seeded species may also be more susceptible to host-specific predators and pathogens (Pringle et al. 2007), although this is by no means universal (Augspurger and Kelly 1984).

12.2.2 Seed size, production, and dispersal

Seed dispersal exemplifies trade-offs in ecology through the balance of seed size and seed number, and has been central to theories concerning species coexistence (see Chapter 14). The competition–colonization trade-off hypothesis predicts that species abundances are limited by dispersal, and this limitation is greater for superior competitors. Thus competitively inferior pioneers should produce abundant, well-dispersed small seed (due to seed abundance/size trade-offs). Although the pattern is considered to be true in general, seed sizes of pioneer trees from the rain forests in Panama vary in size over four orders of magnitude, and other co-existing species that share similar adult ecologies also differ widely in seed size and dispersal (Dalling and Hubbell 2002).

Across all tropical rain forest plants, seed weights span six orders of magnitude between 0.001 mg in some Orchidaceae, Piperaceae, and Melastomaceae to almost 1 kg in trees within the families Arecaceae, Fabaceae, and Lecythidaceae and to over 10 kg if we include the coco-de-mer palm *Lodoicea maldivica* (Figure 12.6) (Moles et al. 2005). Orchids are exceptional in that their tiny dust-like propagules depend on fungal associations for embryonic nutrition, and are therefore provided with almost no reserves (other taxa must provide enough resources to sustain the seed until it can initiate photosynthesis as a seedling). Such variation in seed size may be driven by another trade-off, that of dispersal versus establishment success (small-seeded plants have lower germination success as they are susceptible to drought and inhibition of germination by litter). Thus evaluating the ecological functions and evolutionary derivation of seed traits must consider size, production, and dispersal together.

Importance of dispersal

Seed dispersal allows propagules to escape from seed predators or pathogens associated with the parent plant, to spread the risks encountered by seeds in a heterogeneous environment, to avoid competition between parent and offspring, and to locate 'safe sites', such as forest gaps, where seeds can establish. The recruitment of fast growing and relatively short-lived pioneer trees depends on their seeds finding ephemeral forest gaps or disturbed ground for establishment, and they maximize their chances by producing vast numbers of highly dispersed and often long-lived seed. Dependency on ephemeral habitats is therefore associated with the evolution of effective seed dispersal mechanisms (or longevity, i.e. dispersal in space *and* time).

Dispersal also offers dramatic benefits to non-pioneers through avoidance of seed predation. Seeds of *Virola surinamensis* falling beneath the parent tree were 44 times more likely to be killed by a curculionid weevil

Figure 12.6 The giant seed of coco-de-mer *Lodoicea maldivica*. It is not obvious why this seed has evolved such a large size, but one possibility is that the very nutrient poor soils of the granitic Seychelles, which is its native range, have selected for increasing investment in seed resources to allow for effective seedling establishment.

Photo by Jaboury Ghazoul.

seed predator than seeds from the same tree dispersed 45 m away (Howe et al. 1985). Similarly, seed survival of the palm *Welfia georgii* was lowest within 10 m of adult fruiting trees, owing to high predation (Schupp and Frost 1989). Primate- and bird-dispersed seeds may also benefit from protection against fungal pathogens through removal of the sticky fleshy pulp that otherwise remains attached to non-dispersed seeds falling directly beneath the parent tree (Lambert 2001).

Each of these explanations for the evolution of dispersal mechanisms has received empirical support, but it should also be noted that several tropical plants (e.g. some unwinged dipterocarps such as *Parashorea densiflora* and *Vatica* spp.) lack traits to promote seed dispersal, and that this is not necessarily disadvantageous. Indeed, seedling recruitment in an absolute sense may be much higher near to parent trees, owing to satiation of seed predators (Burkey 1994) or more favourable growing conditions, as for *Ocotea whitei* in Panama (Gilbert et al. 2001). Even so, dispersal has been shown to improve the survival of seeds and seedlings for a great variety of plants throughout the tropics, indicating that dispersal offers substantial benefits to the probability of recruitment success, notwithstanding apparent exceptions to this general rule.

Many tropical seeds are large

There is a striking pattern of increasing seed size with decreasing latitude, such that mean seed mass increases 320-fold from 60° to the equator; this may be explained by several factors, including changes in seed dispersal syndrome, vegetation type, net primary productivity, and plant growth form (Moles et al. 2007). Large seed size is thought to be favoured in the tropical understoreys because large seeds offer resources for developing seedlings (Lusk and Piper 2007), allowing them to perform better in response to deep shade, deep litter, defoliation, mineral deficiency, drought, and competition from established vegetation (Westoby et al. 1996). The abundance of vertebrate seed dispersers in the tropics may also have allowed plants to evolve larger seed sizes, and

the proportion of vertebrate-dispersed seeds does indeed increase towards the equator (Willson et al. 1989; Lord et al. 1997). Across the tropics, Paleotropical fruit tend to be larger than Neotropical fruit (Mack 1993), reflecting the paucity of large frugivores in the Neotropics[13] (see Figure 12.7). Higher net primary productivity of tropical ecosystems also permits the production of more propagules, leading to more intense competition and therefore selection for large seed size, which provides a competitve advantage to young seedlings (Moles and Westoby 2003).

12.2.3 Agents of dispersal

In tropical rain forests, seeds are primarily dispersed by biotic agents that are attracted to the nutritious fruit (although some seeds are passively dispersed by attachment to the fur or feathers of dispersal agents). Fruits are often brightly coloured for this purpose, although odour may play an important role in attracting dispersers (Figure 12.8). Vertebrates are particularly important, with

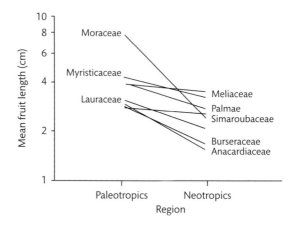

Figure 12.7 Paleotropical fruits tend to be larger than Neotropical fruits, represented here by comparisons across eight plant families. This pattern may reflect the relative scarcity of large-sized frugivores in the New World.

From Herrera 2002.

---

[13] Those that do exist, such as papaya and pineapple, appear to have developed their large size by recent coevolution with humans.

Figure 12.8 (A) While some plants advertise themselves to animal seed dispersers using bright fruit colours, others such as the Southeast Asian durians (*Durio* spp.) use odour. (B) The durian fruit are much valued for their flavour, but they smell so strongly that they have been banned from many hotels, busses, and airplanes through the region.

Photos by Douglas Sheil (A) and Indah Susilanasari (B).

DEAR GUEST,

Please be informed that Durian is strickly not allowed to be brought into the hotel premises.

Thank you for your cooperation.

The management
Gloria Hotel

as many as 85% of woody rain forest species depending on birds and mammals for dispersal (Terborgh et al. 1990). There are also examples of dispersal by terrapins (Moll and Jansen 1995), tree frogs (Da Silva et al. 1989), lizards (Hansen and Muller 2009) (Figure 12.9), and fish (Reys et al. 2009). Other than ants (Pizo and Oliveira 2000; Pizo 2008), invertebrates play a subsidiary role, and even ants are relegated to the (sometimes important) role of secondary dispersal agents, i.e. dispersing seed that has been deposited by vertebrates (Passos and Oliveira 2002) (Figure 12.10).

Seed dispersal by animals may be a consequence of deliberate seed harvesting by frugivores or granivores, or an incidental transport of seed either externally or internally (see also Box 12.2). Dispersal by animals requires a nutritious reward, usually the fruit, the seed being an indigestible and unwanted portion that is discarded either before or after ingestion. Over 90% of tropical rain forest woody species enclose their seeds within fleshy fruit, compared to around 45% in tropical dry forests and 35% in temperate deciduous forests (Jordano 1992). Many birds and mammals are, however, attracted by the seed itself (i.e. granivores), which is consumed and destroyed, although a small proportion may be dropped or hoarded and not recovered. This dispersal system is

Figure 12.9 Seed dispersal by lizards is unusual in most rain forests. This *Phelsuma cepediana* gecko is visiting the fruit of *Roussea simplex* on Mauritius. The seeds are swallowed and pass through the gut unharmed.

Photo by Dennis Hansen.

well known from temperate forests, and may be much more common in tropical forests than is generally acknowledged, with rodents fulfilling this function (Forget 1990, 1993; Forget et al. 1999; Ratiarison and

Forget 2005). Thus seed predators can also be effective dispersers, and a distinction between predators and dispersers may not exist.

This potential source of conflict between plants and their animal dispersers or seed predators lies at the centre of the debate about the extent of coevolution between them. In the tropics, possibilities for coevolution are limited by the many different types of dispersers in different parts and elevations of a plant species' range (Murray 1988). Evidence for coevolution among plants and their dispersers is further undermined by a phylogenetic survey of 910 fruit species, which showed that almost all fruit traits could be accounted for by phylogeny: only fruit diameter could be interpreted as an adaptation to seed disperser type, with larger diameter fruit associated with an increasing participation of mammalian dispersal agents (Jordano 1995). Early examples of coevolved plant-disperser mutualisms have, on closer examination, turned out to be myths, the most celebrated being the dodo and the tambalacoque tree (*Sideroxylon grandiflorum*; Sapotaceae) of Mauritius (Witmer and Cheke 1991) (see Box 14.1, Chapter 14).

Figure 12.10 Ants are among the few invertebrates that do disperse seeds, as shown here for *Ectatomma ruidum* dispersing seeds of *Renealmia alpinia* (Zingiberaceae) at La Selva Biological Station, Costa Rica. Their role is usually as secondary seed dispersers, which can nevertheless be important for increasing the probability of seed survival and its location to a favourable germination site.

Photo by Carlos Garcia-Robledo.

Indeed, previous notions that specialized plant-disperser interactions are a characteristically tropical phenomenon have been challenged, as many 'specialist' frugivores (e.g. toucans, hornbills, trogons) have been discovered to include significant amounts of animals and insects in their diets (Remsen et al. 1993; Sun and Moermond 1997).

Investment in fruit pulp to attract dispersers represents a cost to plants, and is the source of another potential conflict among plants and their dispersal agents. Plants should seek to minimize their costs while retaining the benefits of dispersal, and some plants with brightly coloured seeds achieve this by deceiving dispersers by posing as 'fruit' without offering any nutritional reward (Box 12.4).

Dispersal by birds

Many frugivorous birds species (e.g. hornbills, barbets, bulbuls, and pigeons in the Old World tropics; toucans [Figure 12.12], trogons, manikins, and touracos in the Neotropics) disperse a wide variety of tropical rain forest tree seeds. Hornbills are important seed dispersers in the African and Asian tropics, and in Cameroon they account for the dispersal of seed from around 22% of forest species (Whitney et al. 1998). Bellbirds (*Procnias tricaruncu-lata*, Cotingidae) tend to deposit the seeds of the Neotropical tree *Ocotea endresiana* (Lauraceae) to microsites that have a lower incidence of fungal pathogens, and dispersal by any other species may not be beneficial to the tree (Wenny and Levey 1998). Among the largest frugivores are the imperial pigeons *Ducula* spp. from

---

Box 12.4 Cheating plants and deceived dispersers

Producing nutritional fruits is a costly way to secure the services of dispersal agents. Some plants dispense with this cost by favouring passive dispersal agents such as wind and water that do not require nutritious rewards. In tropical rain forests such dispersal strategies are, however, less efficient than active dispersal by animals. Consequently, some plants have evolved strategies by which animal dispersal is retained but the reward offered is minimized or even eliminated entirely. One such strategy is to hide small inconspicuous fruit among leaves that may be consumed by large folivores, which thereby inadvertently disperse the seed. No fruit reward needs to be provided and dispersal costs become largely limited to the loss of leaf material, which would probably be lost to herbivores anyway. This 'foliage is the fruit' hypothesis (Janzen 1984) has not received much attention in tropical rain forest biomes, but is in any case unlikley to be efficient as a dispersal strategy owing to its unreliability.

Another strategy is to display colourful seeds that 'mimic' the rewarding fruit of other plants (van der Pijl 1982). Most such seeds offer no nutritive reward, but attract birds by being bright (e.g. red seeds in *Sophora secundiflora* and *Erythrina* spp.; Fabaceae) or contrastingly coloured (black seeds within a red pod in *Parachidendron pruinosum*, *Pithecellobium* spp. and *Archidendron* spp.; Fabaceae) (Figure 12.11). Mimetic fruit occur in at least seven families, but are particularly common in the Fabaceae (Galetti 2001).

Other hypotheses advanced to explain brightly coloured non-rewarding seeds have received little empirical support in rain forests (Galetti 2001). One posits that galliform birds actively seek the hard seeds of such species to use as grit in their gizzards, and that no mimetic function is implied (Peres and van Roosmalen 1996). The aposematic hypothesis contends that the bright colours signal toxicity to mammals and parrots, and indeed the seeds of some species (*Erythrina* and *Sophora*) are known to be highly toxic to vertebrates and insects. However, Galetti (2001) concludes that mimetic seeds are simply parasitic on mutualistic plant–seed disperser interactions by suggesting a nutritious reward but offering none.

Figure 12.11 Black seeds within a bright red fruit, as shown here by the fruit of *Sterculia* sp. (Sterculiaceae) at Sepilok, Sabah, Malaysian Borneo; this is a common visual display to potential seed dispersers. (See colour plate 8.)

Photo by Julia Born.

Figure 12.12 *Toco toucan* feeding on manduvi fruit (*Sterculia apetala*) in the Pantanal, Brazil. Although this image is not from the rain forest, toucans are important seed dispersers of many Neotropical rain forest plants.

Photo by Ellen Wang.

Southeast Asia and tropical Pacific Islands, which take large fruits such as nutmeg that other birds cannot handle, and cassowaries *Casuarius* spp. (Casuariidae) (Figure 12.13) of Australia and New Guinea. In Australia cassowaries[14] (*Casuarius* spp.) disperse the fruits of over 70 rain forest trees, the seeds of which are deposited in a fecal mass which may provide protection from seed predators as well as moist and nutrient-rich conditions for germination and growth (Stocker and Irvine 1983).

Birds are important seed dispersers, but efficiency varies. The pantropical pigeons (Columbidae) and parrots (Psittaciformes) destroy most seeds they consume, mashing them in their gizzards, although the fruit pigeons of Asia, New Guinea, and Australia have much thinner walled gizzards through which seeds pass unharmed (Corlett 1998). Parrots crush seeds before swallowing them, but drop some. The seed of *Cabralea canjerana*

[14] In New Guinea and the Australian wet tropics there are no large mammalian frugivores, and dispersal of large seeds is dependent on the cassowary.

fruits, rich in fats and proteins, versus opportunistic frugivores that feed on less nutritious carbohydrate-rich fruit (Snow 1981). Frugivory by specialist birds is particularly well represented among the Annonaceae, Myristicaceae, Burseraceae, Lauraceae, and Arecaceae, with unspecialized birds favouring plants within the families Rubiaceae, Ulmaceae (*Trema* spp.), Moraceae (*Ficus* spp.), Zingeberaceae (*Heliconia* spp.), and Melastomataceae (*Miconia* spp.) (Snow 1981). Regional differences are apparent, with leguminous fruits being important for birds in Africa and Southeast Asia but not in the Neotropics. Arecaceae and Lauraceae are favoured by specialized frugivorous birds in the Neotropics, Southeast Asia, and Australasia, but not in Africa where species from these families are less common. Perhaps because of this there is no group of frugivorous birds in Africa that has radiated to the same extent as the cotingas of the Neotropics or the birds of paradise of New Guinea.

Mammals as dispersers

Seed-dispersing mammals include a diverse group of animals ranging from bats, rodents, lemurs, monkeys, tapirs, gorillas, rhinoceros, and elephants. Many of these animals feed on fruit during periods of abundance, switching to worms (peccaries, bearded pig), insects (spiny rats and some bats), leaves and stems (primates and elephants) or grasses (rhinoceros) when fruits are less abundant.

The agouti (*Dasyprocta* spp.; Figure 12.14) and acouchi (*Myaprocta* spp.) are frugivorous throughout the year (Henry 1999). During fruit abundance these animals feed mainly on the fruit pulp and not on the seeds themselves, and their impact as seed predators is likely to be minimal. In lean periods they survive by consuming cached seeds and therefore act as seed predators, although many seeds fail to be relocated and subsequently establish as seedlings. The apparently antagonistic behaviour of agoutis as seed predators needs to be placed in the context of the fate of seeds that are not cached by these rodents, the very large majority of which are destroyed by bruchid or curculionid beetles, peccaries, ants, are infected by fungal pathogens, or simply dry out (Forget 1990; Hammond et al. 1999; Harms and Dalling 2000).

Figure 12.13 Cassowaries are the only large animal, and only large seed disperser, found in the forests of New Guinea and surrounding islands. They can be dangerous.

Photo by Douglas Sheil.

(Meliaceae) in Brazilian Atlantic forests are dispersed by a diverse assemblage of 33 specialist and opportunistic frugivorous birds that vary qualitatively and quantitatively in dispersal function (Pizo 1997). Some of these birds consume and transport seeds far from the source, while others deposit seeds beneath the branches of the parent tree on which they are perched, rendering seeds vulnerable to predation by rodents and insects. Additionally, birds differ in the amount of fleshy material they remove from the seeds, the presence of which reduces the likelihood of successful germination (Pizo 1997). The quality of dispersal agents may even vary within a species: male manakins (Pipridae) disseminate seed to a restricted area within the lek display arenas, while females disseminate seeds more widely (Krijger et al. 1997).

A distinction can be made between relatively specialized frugivores that feed on large and high-quality

Figure 12.14 An agouti *Dasyprocta* sp., a common seed disperser in South America, here in Guiana.

Photo by Douglas Sheil.

Scatter-hoarding by agoutis and acouchis has long been known to play an important role in the secondary dispersal of relatively small seeds in Neotropical forests, but it is now apparent that other animals perform a similar function in African, Asian, and Australian tropical rain forests. This should not be surprising given the many species of nut-producing Fagaceae (*Quercus*, *Lithocarpus* and *Castanopsis*) in Southeast Asian forests (Forget and Wall 2001). Large ground squirrels in Makokou, Gabon, scatter-hoard nuts of *Panda oleosa* (Pandaceae) trees, while in Malaysia nocturnal long-tailed giant rats (*Leopoldamys sabanus*) and diurnal three-striped ground squirrel (*Lariscus insignis*) hide fruits in soil and plant litter (Forget and Wall 2001). In Australian rain forests, seeds of *Beilschmedia bancroftii* (Lauraceae) are cached by white-tailed rats (*Uromys caudimaculatus*) in a multi-step process (Harrington et al. 2001), and fruits and seeds are scatter-hoarded by the musky rat-kangaroo (*Hypsiprymnodon moschatus*). Indeed, recent work suggests that scatter-hoarding rodents and marsupials are critical to the survival of Australian rain forest trees (Forget and Wall 2001).

Primates consume and disperse seeds of many Neotropical, African, and Asian plants. In the Guianas and Amazonia, approximately 40–50% of woody plants are dispersed by primates (Peres and van Roosmalen 2001), of which woolly monkeys (*Lagothrix* spp.) and spider monkeys (*Ateles* spp.) provide particularly high-quality seed-dispersal services for large seeded plants (Chapman 1989, Stevenson 2000). African guenons (*Cercopithecus* monkeys) consume fruit of many tree species, and appear to be the only dispersers of several African rain forest trees (Kaplin and Lambert 2002). Guenons differ from Asian and Neotropical primates in having check pouches in which they temporarily store fruit and from which the seeds are spat out as the monkeys move through the forest (Kaplin and Lambert 2002), although they also defecate seeds intact. Defecation of seeds by other primates and apes results in a more clumped seed distribution (although secondary dispersal may reduce the degree of aggregation of resulting seedlings).

Long-distance seed dispersal by tapirs in Asian and American forests reduces attack by bruchid beetles by covering the seeds in protective fecal material (as for cassowaries) and by moving them to sites far from parent tree aggregations where seed predators are likely to be

SEED SET **269**

less abundant (Fragoso et al. 2003). Aggregated patterns of *Maximiliana maripa* palms associated with tapir latrine sites in the Maraca Island Ecological Reserve, Roraima, Brazil, suggest that tapirs are responsible for the generation of palm clumps (Fragoso et al. 2003).

Bat-dispersed seeds are common among Fabaceae, Chrysobalanaceae, Moaraceae, Humiriaceae, and Anacardiaceae. Asian and Australian flying foxes (Pteropididae) carry ingested seeds over many kilometres, although they do not ingest all seeds and spit out many beneath the parent tree. They seem to be efficient dispersers of figs: around a third of fig species in the Philippines are dispersed by pteropid bats (Kalko et al. 1996), and fig seeds that pass through the digestive tracts of flying foxes are almost twice as likely to germinate (Utzurrum and Heideman 1991). In Australia *Pteropus conspicillatus* bats[15] disperse seeds of at least 23 rain forest plants (Hall and Richards 2000).

As seed size increases, the number of animals that can process them decreases. Thus the largest seeds are exclusively dispersed by the largest dispersers, often primates and ungulates. These animals are usually amongst the most vulnerable to hunting (Corlett 1998; Peres and van Roosmalen 2001). The conjecture that hunting of seed-dispersing mammals in tropical forests impacts on seed dispersal and seedling recruitment is supported by studies from several tropical regions (Nunez-Iturri and Howe 2007; Peres and Palacios 2007; Stoner et al. 2007; Wang et al. 2007). Mammals typically exploit a different set of fruits from birds[16] and the limited overlap in dietary preference suggests that the effects of hunting of mammals on plant recruitment are unlikely to be compensated by birds (Poulsen et al. 2002).

Dispersal by fish (ichthyochory) and water (hydrochory)
Consumption of fruits and seeds has been documented in approximately 182 species belonging to 32 families

of freshwater fishes (Figure 12.15) (Correa et al. 2007). Up to 30% of fleshy fruits in the riverine forests in South America have their fruits dispersed by fishes (Galetti et al. 2008; Reys et al. 2009).

In Amazonian floodplain forests, peak fruiting coincides with maximum inundation, when marginal forests can be flooded annually to a depth of over 10 m (Junk 1989). Many fish migrate into the flooded forests to feed on buoyant fruits and seeds (Gottsberger 1978; Kubitzki and Ziburski 1994; Williamson and Costa 2000). Many seeds remain buoyant for prolonged periods (Goulding 1980) and release chemical cues to attract fish (Araujo-Lima and Goulding 1997). Many tree species occur in both floodplain and non-inundated forests, and have different types of seeds in each of the different locations, with adaptations for animal or wind dispersal in the **terra firma** habitat and water or fish dispersal in the floodplain (Kubitzki and Ziburski 1994).

Myrmecochory: dispersal by ants
Unlike temperate or arid zones, ants are rarely primary mechanisms of dispersal in tropical rain forests. Exceptions include *Calathea micans* (Marantaceae), an evergreen herb commonly found in lowland rain forests from Mexico to Peru, which adds oily arils (elaiosomes) to seeds to attract ants dispersers (Lecorff and Horvitz 1995; Le Corff and Horvitz 2005). Ant-dispersed plants in the Amazon are mostly epiphytes and often occur in 'ant-gardens' (see Chapter 13). Ants are also attracted to the sweet fruit of epiphytic *Codonanthe* (Gesneriaceae), the seeds of which are sticky and adhere to the ants and are consequently dispersed along tree branches. The germinating seedlings of *Codonanthe* produce adventitious[17] roots which form cavities within which the ants live (Macedo and Prance 1978).

Ants are more significant as secondary seed dispersers, and increase germination success and performance

---

[15] Bats may be especially important as seed dispersers in Australia, which lacks the larger vertebrate dispersal agents common in other tropical regions.
[16] Birds tend to favour purple-black and red fruits, while mammals are more catholic in their choices but show a preference for green or brown fruits.
[17] 'Adventitious' refers to root or shoot buds formed from largely differentiated tissue.

Figure 12.15 School of piraputangas *Brycon hilarii* (Characidae) foraging at Formoso River, Bonito, western Brazil. Note one fish jumping out of the water to feed on *Gomphrena elegans* fruits (Amaranthaceae).

Photo by J. Sabino.

of seedlings as a result. In the Neotropics many seeds expelled in the faeces of birds or mammals, or dropped with bits of attached pulp, are rapidly located by ants and removed to the nest where they are cleaned before being deposited within or just outside the nest. Germination success and seedling performance of these seeds is improved owing to reduced infection by fungal pathogens and relocation to favourable ant nest micro-sites (Levey and Byrne 1993). In the sandy rain forests in south-east Brazil, over 90% of the seeds of *Guapira opposita* (Nyctaginaceae) and *Clusia criuva* (Clusiaceae) dropped or defecated by birds are relocated to more fer-tile ant nests within 12 hours (mainly by *Odontomachus* spp. and *Pachycondyla* spp.) (Passos and Oliveira 2002, 2004). Many similar examples from the Neotropics (Pizo and Oliveira 2000, 2001) contrast with the absence of such observations from the Paleotropics, though this may simply reflect a lack of research effort in this direction.

Wind dispersal (anemechory)

Dispersal of propagules by wind is a relatively minor strat-egy in lowland rain forests (Irvine and Armstrong 1990), presumably for the same reasons that wind pollination is relatively scarce. Nonetheless it plays a significant role in many forests: 16% of the large trees of Barro Colorado Island (Gentry 1982) and 21% of tree species in New Caledonia (Carpenter et al. 2003) possess wind-dispersed seeds. Wind dispersal tends to be more common on tropical islands such as New Caledonia, perhaps due to nutrient-deficient soils (light wind-dispersed propagules are less costly to produce) and low abundance and diver-sity of vertebrate dispersers[18]. Emergent trees, including Neotropical *Vochsia* and mahogany *Swietenia macro-phylla*, and the more widely distributed tropical *Pterocarpus*, *Cariniana*, and *Ceiba* (Figure 12.16) have seed that benefit from higher wind speeds above the canopy. Similarly, canopy lianas from the Bignoniaceae, Sapindaceae, and Malpighiaceae are often also wind-

---

[18] This may also reflect phylogenetic history in terms of higher likelihood of successful dispersal to remote islands by light-weight dry seeds that are resistant to desiccation.

dispersed (see also gliding seeds below) and possess light seeds supported by wings or a pappus to increase aerial support.

Dispersal by gravity, gyration, and gliding

The large fruit of several rain forest trees have no obvious means of primary dispersal and simply fall to the ground, although they may then be secondarily dispersed by animals or water. The heavy fruit of the Brazil nut tree *Bertholletia excelsa* (Lecythidaceae) drops directly beneath the canopy where it is chewed open and the seed dispersed by scatter-hoarding agoutis, with such dispersal restricted to a few tens of metres (Peres and Baider 1997).

The seeds of other gravity-dispersed rain forest trees bear morphological structures to impede downward movement and promote lateral dispersal. Dispersal of seeds of the Asian Dipterocarpaceae is facilitated by two, three, or five wings that cause the fruit to whirl, allowing for moderate dispersal by wind. Experimental work has shown that the wings of dipterocarp fruits allow dispersal by up to 30 m (Osada et al. 2001), but this is a low estimate given that most dispersal occurs during high winds associated with thunderstorms in Asian tropical forests, and it is not uncommon for open areas hundreds of metres from forest to gain a significant seed rain during such storms (DS and JG, pers. obs.). The fruit of dipterocarps belonging to the genera *Vatica* and *Anisoptera* lack wings, and so dispersal is presumably very limited.

Other gyrating fruit include the South American tipu tree (*Tipuana tipu*), unusual among the Fabaceae in that its seeds bear a single wing causing them to spin as they fall. Species of *Gyrocarpus* (Hernandiaceae), a pantropical genus of tropical trees, shrubs, and lianas, have winged fruit that closely resemble those of the more famous Dipterocarpaceae.

Gliding seeds are flattened and flanked by membranous wings allowing the seed to sail through the air, often for considerable distances. *Alsomitra* (Cucurbitaceae), a genus of tropical vine native to Southeast Asia and New Guinea, releases hundreds of winged seeds from large gourds that hang high in the forest canopy. The membranous wings of *Alsomitra macrocarpa* seeds have wingspans of up to 15 cm, matched in size only by the Central America quipo tree (*Cavanillesia platanifolia*; Malvaceae) (Figure 12.17).

Figure 12.16 The seeds of kapok trees (*Ceiba*) are enveloped by cotton allowing dispersal by wind, as shown here in Manu National Park, Peru. Kapok cotton was once used to stuff life preservers and pillows.
Photo by Rhett A. Butler.

12.2.4 Evolutionary implications of clumped versus scattered seed dispersal

Regardless of the relative costs and benefits of dispersal, differences in dispersal modes of different tree species determine the long-term spatial structure of the forest community (Seidler and Plotkin 2006). Trees with small-animal dispersed fruit have more clumped distributions than those with large fruits dispersed by larger birds and mammals which have larger home ranges (Figures 12.18 and 12.19) (Janson 1983; Wheelwright 1985; Kelt and Van Vuren 2001). Trees with gravity- and wind-dispersed seeds also tend to be aggregated, and generally more so than animal-dispersed species, because the tropical forest canopy impedes strong winds in the understorey (Nathan and Katul 2005). In the Neotropics, however, the seeds from many gravity-dispersed fruits are secondarily dispersed by small mammals, such as seed-caching rodents (Brewer and Rejmanek 1999; Yasuda et al. 2000; Vander Wall et al. 2005). Rapid germination of gravity-dispersed dipterocarp fruit in Asian forests limits the likelihood of secondary dispersal, and consequently these species tend to be more highly clumped (Seidler and Plotkin 2006) than trees in Neotropical forests (Hubbell 1979). Dispersal by birds and bats produces the most diffuse spatial distribution of trees in both Neotropical and Asian sites (Hubbell 1979; Seidler and Plotkin 2006).

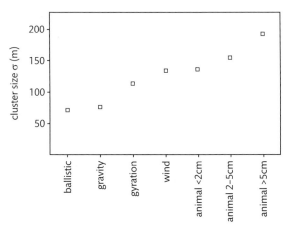

Figure 12.17 Seed of the quipo tree *Cavanillesia platanifolia*; Malvaceae.

Photo by Steven Paton, reproduced with permission of Smithsonian Tropical Research Institute.

12.3   Genetic neighbourhoods

Gene flow by pollen and seed was once thought to be very restricted in tropical forests (Baker 1959; Fedorov 1966), but while most pollination and seed dispersal events are indeed limited to the vicinity of the parent tree, it is now clear that pollinators of all shapes and sizes are far more effective in dispersing pollen over

Figure 12.18 The relationship between dispersal syndrome and spatial aggregation among 425 tropical tree species from Pasoh, Peninsular Malaysia. Dispersal syndromes are significantly associated with spatial aggregation of mature trees.

From Seidler and Plotkin 2006.

much larger distances than was previously thought. Fig wasps, for example, have been shown to move pollen between flowering fig trees that are separated by many kilometres (Nason et al. 1996; Harrison and Rasplus

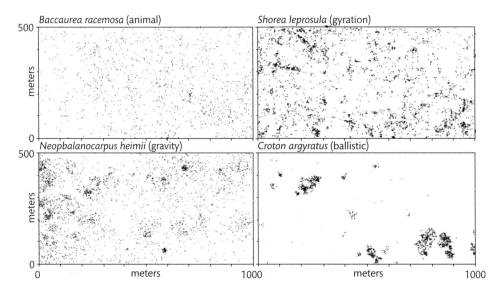

Figure 12.19 Examples of mapped distributions of adult trees of four species at Pasoh 50-ha plot, Peninsular Malaysia. The seeds of each tree species are dispersed by different mechanisms, with the uniform distribution of the animal-dispersed *Baccaurea racemosa* at one extreme contrasting with the ballistic seed dispersal of *Croton argyratus* at the other.

From Seidler and Plotkin 2006.

2006). Bats also move pollen over similar distances (Ward et al. 2005). Other pollinators regularly move pollen between trees separated by several hundreds of metres (Table 12.4).

Gene flow by seed dispersal is thought to be less efficient than pollen-mediated gene flow in tropical rain forest trees, but seed dispersal distances are characterized by long-tailed (sometimes called 'fat-tailed') distributions, such that a small but ecologically significant proportion of seed is transported over great distances (Portnoy and Willson 1993). Many species (e.g. *Anacardium* spp.; Anacardiaceae) have small flowers, and by implication small and poorly dispersing pollinators, but their seeds are dispersed by bats and birds that have the potential to carry seeds over great distances. Bat-mediated seed dispersal from Java and Sumatra to the Krakatau islands some 30 km distant illustrates the potential for long distance gene flow by seed (Chapter 11, Box 11.3).

Such patterns of long-distance pollen and seed dispersal may limit the emergence of fine-scale spatial genetic structure (i.e. a rapid decrease in relatedness with distance) within tropical rain forest tree populations (Hardy et al. 2006; Dick et al. 2008). Nevertheless, differences in local genetic structure are apparent among tropical species whose seeds are dispersed by birds, bats, or monkeys (which have lower genetic differentiation with distance), versus those dispersed by gravity, wind, or scatter-hoarding rodents (Dick et al. 2008). At larger regional scales (distances exceeding 50 km) populations of tropical trees do appear to be more genetically differentiated than temperate tree species (Dick et al. 2008), although this difference disappears upon exclusion of wind pollinated species (comprising almost half of the temperate species tested).

Although most tropical trees appear mostly highly outcrossed (Hamrick and Murawski 1990; Ward et al. 2005; Cloutier et al. 2007; Dick et al. 2008), they also, paradoxically, appear to have lower **heterozygosity** than trees in temperate zones (Hamrick et al. 1993). This may be due to lower densities and therefore smaller effective population sizes (fewer potential pollen donors). It may also be due to an underestimation of selfing, and

large variations in outcrossing rates among individuals within tropical tree populations suggest that the apparent predominance of outcrossing may not reflect the true complexity of breeding patterns (Lee et al. 2000; Naito et al. 2005; Naito et al. 2008).

12.4 Seed dormancy and germination

12.4.1 Seed dormancy

Many rain forest species have seeds which can persist in the soil for months before germinating. Seed dormancy is the ability of seeds to avoid germination when conditions for germination are suitable, but the probability of subsequent seedling survival and growth is low (Fenner and Thompson 2005). Dormant seeds therefore respond to specific environmental cues to trigger germination to maximize seedling establishment and success. 'Delayed germination', on the other hand, is distinct from dormancy in that germination is simply postponed for some period independent of any external environmental cues. Many tropical seeds, for example, are enclosed within a hard impermeable coat which prevents germination until the coat becomes permeable through weathering and microbial degradation. Delayed germination is common among rain forest trees, particularly Fabaceae and Malvaceae (Ng 1980). Delayed germination usually varies within a cohort of seed and is presumably a bet-hedging mechanism to spread risk in a temporally heterogeneous environment.

True seed dormancy is a strategy to ensure that all seeds postpone germination until conditions are in some manner indicated to be suitable for germination. Dormancy is therefore likely to improve initial seedling survival in comparison to delayed germination—though suitable comparisons are not readily found. There is, however, a cost to dormancy, and in rain forests dormant seeds rarely survive in the soil for long[19] (and often less than a year; Guariguata 2000), as high respiration rates (a function of high temperature and moisture) and plenty of seed-eaters and fungal pathogens conspire to

---

[19] There is some evidence that tree seeds become more eager to germinate as time goes on...which means that disturbing the litter can trigger a local eruption of seedlings (DS, pers. obs. in Uganda).

Table 12.4 Estimates of pollen dispersal distances for tropical forest trees.

Tree species	Country	d.b.h. (cm)	Density (ha^{-1})	Pollinator	Pollen flow (m and km)
Dinizia excelsa (Fabaceae)	Brazil	≥40	0.017	Small insects	188 and 212 m (TG)
Sextonia rubra (Lauraceae)	French Guiana	≥10	2.4	Small insects	65 and 80 m (TG)
Calophyllum longifolium (Clusiaceae)	Panama	≥18	0.028	Small insects	≥210 m (62% of pollen)
Spondias mombin (Anacardiaceae)	Panama	≥20	0.33	Small insects	≥300 m (5% of pollen)
Turpinia occidentalis (Staphyleaceae)	Panama	≥20	1.27	Small insects	≥130 m (1% of pollen)
Carapa guianensis (Meliaceae)	Brazil	≥30	2.5	Small insects (butterfly)	75 and 265 m (TG)
Neobalanocarpus heimii (Dipterocarpaceae)	Malaysia	≥30	0.83	Large and mid-sized bees	191.2 m (mean ±104 SD); max ≥ 663.6
Dipterocarpus tempehes (Dipterocarpaceae)	Sarawak	≥50	3.97	Large bees (year 1) and moths (year 2)	192 m (moths) and 222 (bees); ≥600
Dicorynia guianensis (Fabaceae)	French Guiana	≥20	3.925	Large bees	≥700 m
Platypodium elegans (Fabaceae)	Panama	–	0.78	Large bees	≥750 m (25% of pollen)
Tachigali versicolor (Fabaceae)	Panama	–	–	Large bees	³750 m (21% ≥ 500 m)
Glyricidia sepium (Fabaceae)	Guatemala	≥10	0.37	Large bees (Xylocopa)	275 m max
Symphonia globulifera (Clusiaceae)	French Guiana	≥10	10.9	Perching birds, hummingbirds	27 and 53 m (TG) FG
Ceiba pentandra (Malvaceae)	Brazil	≥90	0.001	Bats (Phyllostomus)	≥18 km
Ficus dugandii (Moraceae)	Panama	–	0.004	1–2 mm wasps	14.2 km (10.9–16.9)
Ficus obtusifolia (Moraceae)	Panama	–	0.072	1–2 mm wasps	5.8 km (5.2–6.4)
Ficus popenoei (Moraceae)	Panama	–	0.013	1–2 mm wasps	9.7 km (7.7–11.4)

Adapted from Dick et al. 2008.

Maximum reported pollen dispersal distances (i.e. outside of the study plot) are preceded by ≥ or indicated by 'max' if it is the maximum value for precisely measured gene flow events; mean values are indicated for studies employing the TwoGener method (TG).

consume seed resources (Vazquez-Yanes and Orozco-Segovia 1993). Nevertheless, hard-coated seeds of *Acacia, Cecropia, Piper,* and *Trema*[20] can remain viable after more than a year in forest soil (Hopkins and Graham 1987; Vazquez-Yanes and Orozco-Segovia 1993), and at a Panamanian site where an isolated pioneer (*Trema*) had died more than nine years previously, a quarter of all buried seeds could still be germinated (Dalling et al. 1997).

Banking for trees

Among trees, pioneer and other gap-colonizing species (Chapter 7) that depend on unpredictable and ephemeral habitats are most likely to occur in the soil seed bank, while large primary forest species tend to have little or no seed dormancy (Garwood 1989; Kennedy and Swaine 1992; Vazquez-Yanes and Orozco-Segovia 1993). Usually the most common members of the soil seed bank are herbs, shrubs, and vines, which in some forests account for more than 75% of the species richness and abundance in the soil seed bank (Dupuy and Chazdon 1998). Altogether, the accumulation of viable seeds in the soil can reach very high densities of up to 15,000 seeds m^{-1} in secondary forests, although lower densities of up to 3,000 seeds m^{-1} are more typical of old-growth primary forests (Garwood 1989; Dupuy and Chazdon 1998).

Many canopy species, such as Asia's dipterocarps, possess recalcitrant[21] seeds incapable of dormancy. Seeds that germinate rapidly form a blanket of relatively persistent seedlings that have been termed the 'seedling bank'. Functionally the consequences are similar to a seed bank with the seedlings barely growing while waiting for suitable growth opportunities. Such seedling banks are especially characteristic of the forests of Southeast Asia.

12.4.2 Germination

Cues for breaking seed dormancy include wider soil temperature variation, increased soil moisture, and a change in the intensity and spectral quality of light (Chapter 10), often as a consequence of gap formation or soil and litter disturbance (Vazquez-Yanes et al. 1990). Gap formation also causes changes in soil nitrate concentrations, temperature, and water potential, that may also initiate germination. Differential responses to combinations of these variables by *Piper* species facilitate germination in only the most suitable microsite conditions (Daws et al. 2002).

Sunflecks also influence germination of seedlings and herbs in forest understoreys. Sunfleck activity in a Mexican rain forest promoted germination in the pioneer species *Piper auritum* and *P. umbellatum*, and percentage germination is higher in response to longer sunflecks that have higher R:FR ratios (Chazdon and Pearcy 1991). Germination responses to irradiance and temperature among Neotropical pioneers are differentiated by seed size, which is positively correlated with the R:FR germination threshold (Figure 12.20) (Daws et al. 2002; Pearson et al. 2002). Small-seeded species respond to an irradiance cue for germination, while larger-seeded pioneers respond to increasing daily temperature fluctuation (Pearson et al. 2002). The scarce resources of small seeds limit their ability to delay germination, and they are consequently less discriminatory in response to changes in light quality, while large seeds generally persist until more favourable light conditions are encountered (Pearson et al. 2003).

Germination behaviour is one of several traits used to classify tropical rain forest trees as 'pioneers' that require canopy gaps for germination and establishment, and 'non-pioneers' that are able to establish and grow in shade (Chapter 10). Tree seeds that persist in the seed bank have often been considered pioneers, awaiting the formation of a canopy gap to break seed dormancy and initiate germination (Swaine and Whitmore 1988). This classification system is overly simplistic, as many species with seeds in the soil seed bank have considerable shade tolerance and germinate under canopy, while many

---

[20] The seeds of the balsa tree *Ochroma lagopus* from tropical America remained viable after 44 years of herbarium internment, demonstrating the potential for long-term dormancy in favourable storage conditions (Vazquez-Yanes and Orozco-Segovia 1993).

[21] Recalcitrant seeds do not survive drying and must germinate almost immediately after dispersal. Many dipterocarp seeds germinate even before dispersal (JG, pers. obs.). Orthodox seeds are able to withstand desiccation for variable periods, and germination occurs upon rehydration.

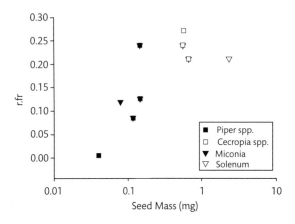

Figure 12.20 Threshold R:FR for germination for pioneer species spanning a range of seed sizes.

From Pearson et al. 2003.

light-demanding pioneers also germinate and grow rapidly (though soon die) in shade (de Souza and Valio 2001). In African rain forest light-mediated germination was found to be relatively rare even among several pioneers (Kyereh et al. 1999). Thus seed dormancy and germination are not necessarily useful traits by which to separate pioneers and non-pioneers (see Chapter 10 for further consideration of this issue).

12.5 Vegetative regeneration

The absence of gene exchange in vegetative reproduction limits genetic variability and dispersal opportunities[22] (but see Box 12.3). Nevertheless, an ability to resprout following damage or tree fall is an important survival strategy for many plants (Chapter 11, Box 11.2). For some it is a primary means of regeneration.

Many tropical rain forest plants regenerate, propagate, and disperse vegetatively. Clonal propagation is now considered to be relatively common in the tropics, particularly among understorey plants (Kinsman 1990; Sagers 1993), and many tropical lianas propagate vegetatively

through the production of adventitious shoots (Penalosa 1984; Nabe-Nielsen and Hall 2002).

Several means of vegetative propagation can be illustrated by the single genus *Piper*[23] (Greig 1993b). Many *Piper* species produce multiple stems, usually following damage to the main stem, that originate from stem suckers or replacement shoots. This ability is widespread among *Piper* species (Greig 1993b). Alternatively, adventitious roots and new stems may grow from decumbent stems or branches that contact the ground, with the original stem eventually rotting to leave independent vegetatively propagated stems. Other *Piper* species shed twigs or small branches, or even just a leaf, which produce adventitious roots on the moist forest floor and, eventually, new shoots and leaves (Kinsman 1990; Greig 1993b). *Piper auritum* and *P. pseudobumbratum* produce root sprouts that spread laterally just below the ground, producing shoots at intervals and sometimes several metres away from the parent plant (Greig and Mauseth 1991; Greig 1993b). Vegetative regeneration in *Piper* is more frequently associated with the shade-tolerant mid- or late-successional species, with early successional species reproducing mainly by seed. This is expected for pioneer species that depend on dispersal into ephemeral light gap habitats, although the thin and flimsy structure of fast growing early successional plants may also constrain regeneration from cuttings or fragments.

Regeneration of broken and shed fragments has been found among Piperaceae, Melastomaceae, Acanthaceae, Solanaceae, and Rubiaceae at Monteverde in Costa Rica (Kinsman 1990), Lauraceae and Piperaceae in Panama (Paciorek et al. 2000), and in *Tetramerista glabra* (Tetrameristaceae) in Borneo swamp forests (Gavin and Peart 1997, 1999) (Box 12.3). Such forms of regeneration may be especially important to long-lived understorey species that are exposed to frequent damage by falling debris over their lifetime (Kinsman 1990; Paciorek et al. 2000) (Chapter 11).

A number of traits of shed plant fragments, including high nitrogen concentration, high photosynthetic capabilities, and high tannin content, may enhance the capa-

---

[22] Dispersal may be less limited for vegetatively reproducing water weeds such as water hyacinth *Eichhornia crassipes* (Pontederiaceae) (Box 12.3).

[23] *Piper* is one of the most diverse plant genera in the tropics with over 1,000 species worldwide, and forms an important element of the understorey community in wet neotropical forests (Gentry 1990).

bility of these fragments to develop into fully functional independent plants (Sagers 1993). *Psychotria horizontalis* (Rubiaceae), an abundant Neotropical shrub, reproduces vegetatively from shed leaves and rooting stems. The plant withdraws only 7% of nitrogen from its leaves prior to their abscission, unlike most plants that recover 50–90% of leaf nitrogen (Chapin 1980), which allows sufficient photosynthetic activity to persist and grow[24]. High leaf tannin content may protect leaves from fungal attack (Sagers 1993).

Vegetative reproduction allows rapid colonization and dominance of space in the local vicinity (Box 12.3), but also reduces genetic heterogeneity of the population. In some species (including *Psychotria horizontalis*) individuals may differ in their propensity to reproduce vegetatively, suggesting the presence of trade-offs between sexual and vegetative reproduction. It is widely believed that under stressful conditions, such as might be produced by drought, sexually produced propagules are favoured. Parent plants may also incur high costs of vegetative production by the loss of nutrient-rich leaves, and this may limit subsequent sexual production either through shorter plant lifespan or lower subsequent seed output.

Vegetative reproduction appears particularly prevalent among understorey herbs[25], shrubs, and climbers, and comparatively rare among trees, although no broad-scale surveys have been undertaken. In one study only one out of seven tree species expressed a capacity for vegetative reproduction, compared to five out of six shrubs (although three belonged to the genus *Psychotria* and cannot be considered independent). The association of vegetative reproduction with the lower understorey, may be related to a lack of pollinators or seed dispersers in this region, although evaluation of such ideas is needed.

12.6   Conclusion

Animal dispersal agents are particularly important for tropical plant species, the majority of which are animal pollinated and dispersed. Yet tropical plant populations are increasingly becoming reduced, fragmented, and isolated through human activities (Chapters 15 and 17), and these environmental impacts often cause pollinators and seed dispersers to decline in abundance or change their foraging behaviours such that dispersal is curtailed. This, in turn, may detrimentally affect reproductive output and the extent of gene exchange and therefore the capacity to respond to environmental change. Unfortunately, the long-term implications of environmental change for ecosystem processes that underlie tropical rain forest plant reproductive viability remain largely unknown.

---

[24] Presumably these abscised leaves are very sensitive to dessication through transpiration during this early stage prior to root development. It is not clear whether these leaves are specialized for this purpose.

[25] Horticulturalists will be familiar with the ease of propagation from leaf stem cuttings of African violets *Saintpaulia* spp. (Gesneriaceae), native to East Africa. This has made African violets probably the world's most popular house plant.

CHAPTER 13

Nature's society: life's interactions

'As in music the person who understands every note, will, if he also has true taste, more thoroughly enjoy the whole, so he who examines each part of a fine view, may also thoroughly comprehend the full and combined effect.'

Charles Darwin (1845)

'When one tugs at a single thing in nature, he finds it attached to the rest of the world.'

John Muir (1911)

A memorable image of Neotropical rain forests is a stream of fresh green leaf fragments meandering along the forest floor. Closer inspection reveals that each fragment is borne aloft by an ant that is often much smaller that the fragment itself. Yet further investigation shows that these leafcutter ants harvest leaves to cultivate fungi on which the ants feed. Pests and parasites afflict these fungus gardens, to which ants respond by way of bacterial 'pesticides'. This complexity of antagonistic and mutually beneficial (mutualistic) interactions among ants, plants, fungi, and bacteria is but the centrepiece of a network of interactions that unfold within the leafcutter ant colony.

From an evolutionary perspective, mutualisms often emerge from initially antagonistic interactions. Conversely, established mutualisms can be undermined by the evolution of antagonists, the parasites of cooperative partnerships. Thus interactions can themselves drive adaptation and variation among species as they become refined, undermined, and defined by local conditions.

The time over which ancient tropical rain forests have existed has allowed the evolution of a myriad of increasingly elaborate interactions which have wider importance, in that together they shape community structure and stability.

Describing the full range of tropical rain forest interactions is clearly a heroic and (for these authors at least) an impossible task. In this chapter we focus instead on just a few examples of antagonistic and mutualistic[1] interactions, namely herbivory, a widespread ecological process that has considerable bearing on plant competition and growth, and a suite of ant–plant interactions (which can be both mutualistic and antagonistic). Ants and plants are both ubiquitous, and the breadth of interactions among them is sufficient to fill a whole book, as indeed it has (Rico-Gray and Oliveira 2007). Both herbivory and the variety of ant–plant interactions encompass ecological and evolutionary lessons from which an understanding of important elements of tropical forest functioning can be gleaned.

---

[1] Plant reproductive interactions, both mutualistic and antagonistic, are covered in Chapter 12. Plant competitive interactions (a particular type of antagonism) are discussed in Chapter 10.

13.1 Herbivory (and plant responses to it)

A typical rain forest is, to a first approximation, green and lush. Given an opportunity to comment, herbivores would almost certainly disagree. Plants have evolved a variety of biotic, chemical, mechanical, and phenological means by which to attack, poison, starve, or simply avoid herbivores, and the ability of herbivores to overcome these defences governs the functioning of tropical rain forest food webs, as well as nutrient recycling and community diversity (Coley and Barone 1996). Insects are usually the most important consumers of leaf material in tropical rain forests, although fungi and other pathogens also have an impact. On Barro Colorado Island, Panama, between two-thirds and three-quarters of annual leaf consumption is due to insects[2], with almost all of the rest attributed to pathogens (Coley and Barone 1996; Leigh 1997). Insect herbivores are very diverse, with 171 phytophagous insect families at La Selva, Costa Rica (Marquis and Braker 1994). Their impacts on plants are not always obvious. Holes in leaves betray the activity of leaf chew-

ers (Figure 13.1 and 13.2), but there is little to indicate feeding by phloem feeders, even though their impacts on plants may be just as great (Leigh 1997).

According to one summary, annual herbivory rates in temperate forests average 7% of leaf area, and in tropical

Figure 13.1 Unidentified caterpillars eating their way through the leaf of a *Shorea parvifolia* seedling in Sabah, Borneo.

Photo by Arthur Chung.

Figure 13.2 The results of herbivory by a caterpillar on a *Shorea argentifolia* seedling.

Photo by Julia Born.

[2] With about 90% of this damage attributed to specialist insect herbivores (feeding on plants from a single family) (Coley and Barone 1996).

humid forest 11% for shade-tolerant species and 48% among gap-specialists[3] (Coley and Barone 1996). Such damage influences plant survival and growth. For example, seedling densities were 230% greater in forests where mammal herbivores were almost absent (Dirzo and Miranda 1990), and experimental exclusion of insect and vertebrate herbivores increased the growth rate of an understorey shrub by an order of magnitude (Sagers and Coley 1995). Several studies have demonstrated the impact of herbivory on seedling growth and survival, which, particularly among shade-tolerant species, are mediated by investment in defences (Marquis 1984; Clark and Clark 1985; Terborgh and Wright 1994).

13.1.1 Plant defences

To a human, the most obvious form of plant defence is physical, such as the spines that many stems of rattans, bamboos, and acacias bear to ward off vertebrate herbivores (Figure 13.3). Prickly leaves are occasionally encountered in the understorey (e.g. *Rinorea ilicifolia* Violaceae, and various *Acanthus* spp. Acanthaceae, in forest–grassland edges in Africa), especially when herbiv-

ores are common, but are usually absent in the canopy. Stinging plants occur throughout the tropics including climbers like *Tragia/Cnesmone* (Euphorbiaceae). The seed pods of *Mucuna* (Fabaceae) climbers are covered in stinging hairs. The leaves of various Urticaceae are armed with fine silica-based tubules that inject formic acid, histamine, and other irritant chemicals—some species carry quite a punch (JG, personal testimony, Kalimantan 1999; DS, personal testimony, Halmahera 2005). A report from New Guinea in the 1920s reports a human death following contact with a *Dendrocnide* tree (Urticaceae). For palms, especially single-stemmed palms, serious damage to the growing bud means death. This may explain why palms such as *Oncospermum horridum* (Southeast Asia), many rattan climbers, and some bamboos (Figure 13.3) possess such horridly spiny stems.

Most herbivory in most tropical forests, however, is due to invertebrates, to which some plants have responded by investing in tougher leaves or specialized hairs (trichomes) on the surface of leaves (Letourneau and Barbosa 1999; Molina-Montenegro et al. 2006). Leaf toughness in tropical lowland forests is particularly associated with monocots, which have tougher leaves than

Figure 13.3 Sharp thorns around the stem of a rattan in Sepilok, Sabah, Malaysia.
Photo by Julia Born.

---

[3] But these constitute a small proportion of individual trees in a rain forest.

dicots at both immature and mature leaf stages (Dominy et al. 2008). Indeed, unlike dicots, monocot leaves can remain tough throughout the expansion phase of development (Dominy et al. 2008). Immature monocot leaves also expose less of their surface to herbivores by remaining tightly folded or rolled until they reach more than half of their final length. Presumably due to such leaf traits, leaf area loss to herbivory on monocots is substantially less than it is on dicots (Grubb et al. 2008).

Almost all forest plants contain chemical compounds to deter or poison prospective herbivores. Observations in a Ugandan forest frequented by bark-gouging elephants found that trees known to have very toxic bark (*Antiaris toxicaria*, *Strychnos mitis* and *Erythrophleum suaveolens*) remained relatively unscarred compared to most others (Sheil and Salim 2004).

Many woody plants, and some herbs, exude fluids such as latex, resins, gums, or sap when they are damaged. These can interfere with feeding, be repellent, toxic, and/or wound sealing, and prevent infections. Insects show various strategies to reduce getting gummed up, for example destroying the main leaf veins (Dussourd and Eisner 1987). Some of these exudations are hazardous to humans. The irritating resin of the Asian *Gluta renghas* (Anacardiaceae) can provoke skin rashes in sensitive people even without them touching it. Even minor amounts of the bark of some *Strychnos* (Loganiaceae) species if ingested will provoke severe vomiting. Latex from *Antiaris toxicaria* is the main ingredient of the poison applied to blow-pipe darts used by many indigenous groups in Borneo and the wider region—the poison has to be prepared with great care as even a tiny amount entering the body can kill in minutes. *Hura crepitans* Euphorbiaceae in Central and South America combines conical spines on the trunk with toxic latex to deter vertebrate and invertebrate herbivores, and it is said that people who fall asleep in the shade of this tree wake up blind on account of the dripping toxic latex (JG, pers. comm. with farmers in Costa Rica). The famous British Admiral Lord Horatio Nelson's poor health was ascribed to the time he drank from a spring contaminated by a mancanilla tree (*Hippomane mancinella*) during his stay in the Caribbean.

Some defences involve extraordinary biochemistry. For example the cocoa tree *Theobroma cacao* (Sterculiaceae) produces elemental sulphur in response

to pathogens (Williams and Cooper 2003); some plants like the blue-sapped *Nauclea* spp. accumulate, and somehow tolerate, toxic metal salts that make them inedible to most herbivores (Chapter 10). The development of new plant defence systems appears to spur speciation and diversification (Ehrlich and Raven 1964; Cornell and Hawkins 2003): for example plant families with latex are more speciose than those that have none (Farrell et al. 1991).

Young leaves have higher nutritional quality and are preferred by herbivores, such that almost 70% of the lifetime damage incurred by a leaf occurs while it is expanding (Coley and Barone 1996; Kursar and Coley 2003). Leaf toughness, high fibre content, and other physical defences are effective against herbivores in mature leaves, but such traits are incompatible with leaf expansion (Kursar and Coley 2003). Herbivory on young leaves is minimized instead by investing in abundant and diverse chemical defences such as alkaloids, tannins, and phenols (Coley and Barone 1996; Coley and Kursar 1996). Higher herbivore pressure and longer leaf longevity in the tropics are presumed to have resulted in tropical leaves being generally better defended than temperate leaves (Levin 1976; Levin and York 1978; Basset 1994). Shade-tolerant tree species dominate most old-growth tropical rain forests in terms of number of species, individuals, and biomass, and their well-defended seedlings persist in the shade of the forest understorey for considerable time (Chapter 10) (Welden et al. 1991; Lieberman et al. 1995). The interaction between herbivores and the young leaves of tree seedlings has considerable importance in regulating trophic dynamics, by determining herbivore population sizes and/or seedling persistence and recruitment among species (Kursar and Coley 2003; Fine et al. 2004).

Reducing leaf expansion time can also minimize herbivory by limiting the period of vulnerability. Thus slowly-expanding leaves of *Pentagonia* (Rubiaceae) suffer twice as much damage as rapidly expanding leaves of the same species (Ernest 1989). Energy and nitrogen losses can be reduced by delaying the development of photosynthetic apparatus (chloroplasts) of young tropical leaves during their expansion (Kursar and Coley 1992), which is why many young leaves lack green chlorophyll and are pale coloured. The cost of reducing herbivory by delayed greening is forfeited photosynthate, thus delayed greening is particularly prevalent among understorey plants where photosynthesis is anyway limited by shade, and among plants that expand leaves rapidly (see also Chapter 2).

Investment in chemical defences can be expensive in terms of energy and nutrients, and often has a trade-off with growth rate. Heavy investment in toxins to protect long-lived leaves from herbivory is associated with relatively shade-tolerant slow-growing trees, while the most palatable leaves belong to fast-growing shade-intolerant trees, including many pioneers. A continuum of leaf properties representing strategies from herbivore escape to active defence has been suggested. Defence-strategy species are characterized by slow leaf expansion, normal greening, effective chemical defence, and low rates of damage. Escape-strategy species lack chemical defence, but employ delayed greening, and rapid development (leaf area may double each day), and sometimes synchronous flushing to satiate herbivores (Aide 1993; Kursar and Coley 2003; Coley et al. 2005) (Table 13.1).

13.1.2 Managing mercenaries

Herbivory represents a major cost to plants, to which many respond by developing chemical or physical defences. Such defences are, however, expensive to produce and maintain, and some plants instead bring in paid defenders or mercenaries. In exchange for a home and sustenance, ants will aggressively deter potential herbivores and liana seedlings (Schupp and Feener 1991). Plants pay for such services with extra-floral nectar, food bodies, and nesting sites (domatia). Ant-plants (myrmecophytes) range from trees and climbers to herbs and epiphytes. Although such interactions are found among many different plant communities across the world, there is a marked increase in their variety, abundance, and complexity in tropical forests[4]. At least 45 clear-cut myrmecophytic species belonging to 20 genera and 14 families in Peninsular Malaysia alone (Moog et al. 2003).

Domatia, the cavities within which ants are housed, are structures evolved specifically for ant occupation. These can be relatively simple hollow swellings along the stem of a plant, leaf pouches (*Clidemia tococoidea*, Melastomataceae), or within thorns, but can also include remarkably complex structures that comprise the bulk of the host plant. The epiphytic *Myrmecodia* (Rubiaceae) in Southeast Asia and New Guinea is little more than a swollen tuber within which a complex arrangement of cavities and chambers is designed to accommodate ants, maintain a favourable microclimate for them, and draw nutrients from their deposits through the cavity walls[5].

Plant domatia often house aggressive ants, which defend the plants from both invertebrate and vertebrate browsers. Even elephants avoid the African *Barteria* trees (Passifloraceae, the 'adulteress tree' in Gabon), which host aggressive *Tetraponera* ants. Many *Cecropia* in the Neotropics and *Macaranga* in Asia are well defended by *Azteca* and *Crematogaster* ants respectively, which swarm out to sting and bite any intruder (Itioka et al. 2000). Many ants also remove the seeds of epiphytes and chew through encroaching liana stems, thereby maintaining a relatively competitor-free environment for the host plant (Davidson et al. 1988). Ant activity on some *Piper* ant-plants re sults in an increase in fitness for the plant through the removal of stem borers and fungal spores (Letourneau 1998).

Table 13.1 Characteristics of young tropical leaves with 'escape' and 'defence' strategies.

	Escape	Defence
Herbivory	high	low
Toughness	low	low
Leaf expansion rate	fast	slow
Nitrogen for growth	high	low
Chemical defences	low	high
Chloroplast development	delayed	normal
Nitrogen for greening	low	high
Synchrony of leaf production	high	low

From Kursar and Coley 2003.

---

[4] Ant-defence systems are found within 17 plant families in the Neotropics, 26 in the Asian tropics and 10 in the African tropics.

[5] Myrmecophyte epiphytes are especially common on nutrient poor soils such as podsols.

Some members of the Melanesian rattan genera *Calamus* and *Daemonorops* possess interlocking combs of spines forming well defended horizontal galleries around their leaf sheath. These are often occupied by ant colonies (mainly from the genus *Camponotus*) (Moog et al. 2003). Getting too close can be unnerving—the dry sheaths hiss with the percussive drumming of the ants.

Although ant-plants save resources by removing the need to invest in secondary defence compounds, housing and feeding ants also incur a cost of resources diverted from growth and reproduction. Thus the extent to which plants invest in ant defences may depend on evolutionary history as well as local environmental conditions (Itioka et al. 2000). Plants specializing in the high light conditions of forest gaps, for example, have high carbon accumulation rates making sugar and lipid rewards relatively cheap to produce, and consequently gap species are more likely to have ant bodyguards (Schupp and Feener 1991). Some plants that house ants continue to produce plant defence compounds. This apparent redundancy makes sense when ants, such as *Pheidole bicornis* on *Piper cenocladum*, are only effective in deterring specialist herbivores but not at deterring more generalist herbivores (Dyer et al. 2001).

Ants are not the only mercenaries. Leaf domatia, little more than tufts of hairs and other minute structures in the vein axils, occur widely in tropical flora, and provide nurseries, protection, and moulting sites for mites. A survey of the Australian rain forests revealed that 15% of trees had domatia, and 50% of domatia contained mites (O'Dowd and Willson 1989, 1991). These mites may benefit the plants by removing fungi and/or herbivorous arthropods such as phytophagous mites (Norton et al. 2001). When the leaf domatia of *Cupania vernalis* (Sapindaceae) saplings in Brazil were blocked with resin, the plants consequently possessed fewer defensive mites and suffered more chlorosis than untreated plants (O'Dowd and Willson 1989, 1991; Romero and Benson 2004).

13.1.3 Diet specialization in insects

Evolution of host specialization is favoured among herbivores faced with a diverse array of toxic leaves.

Specialists that have been able to overcome or tolerate the defences of particular plant species will not only grow and reproduce more quickly on those plants, but also are freed from competition from non-specialized herbivores. On the other hand, the costs of locating rare host plants may favour a generalist herbivore strategy (Beaver 1979; Jaenike 1990; Basset 1994).

Much recent effort has been expended to determine the degree of host specificity of herbivores on their hosts (Barone 1998; Novotny and Basset 2005; Novotny et al. 2006, 2007; Dyer et al. 2007). These studies provide information on the number of species that can be packed together in a habitat and, when undertaken at many sites spanning large spatial scales, the extent of species turnover (beta-diversity; Chapter 8) (Dyer et al. 2007; Novotny et al. 2007). Recent work from Papua New Guinea has shown that only a minority of herbivores are host-specific (Novotny and Basset 2005), and host plant range may be more closely associated with genera, and other groupings of related taxa, rather than species. Specialization of herbivores on plants differs by herbivore type, and granivores and leaf-miners seem to be more specialized than leaf- or wood-chewers or root-feeders (Novotny and Basset 2005). Other studies on other trophic groups have also shown more generality than has been previously assumed. Parasitoids in Belize are mostly generalists, although here too the leaf-miners they parasitize are relatively specialist on their host plants (Lewis et al. 2002). Although comparative data are limited, tropical plant–herbivore webs do not, overall, seem to be more or less specialized than those of temperate communities[6].

13.1.4 Ecological impacts of herbivory

The trade-off between growth and anti-herbivore defence can contribute to habitat specialization by intensifying differences in species responses to abiotic conditions such as soil type (Fine et al. 2004). When protected from herbivores, lightly defended species outperform species that invest heavily in leaf defences,

---

[6] This has implications for estimates of tropical species richness which have generally assumed high proportion of host-specific insects in tropical rain forest communities (Erwin 1982; Chapter 5).

owing to the energetic trade-off between growth and defence, regardless of soil type. When exposed to herbivory, well-protected plants on poor soils perform better than poorly-defended species, due to the impacts of herbivory, but on rich soils fast-growing species outcompete others in spite of herbivory (Coley et al. 1985). Further interactions with light regime, water availability, mycorrhiza associations, and seed predators among others is also likely to create opportunities for niche differentiation required to support high species richness (see Chapter 8).

A continuum of leaf properties representing strategies from herbivore escape to active defence has been suggested. A leaf intended for longer use would typically be better defended than short-term leaves. Defence-strategy species are characterized by slow leaf expansion, normal greening, effective chemical defence, and low rates of damage. Escape-strategy species lack much defence, but employ delayed greening and rapid (leaf area may double each day) and sometimes synchronous leaf flushing to satiate herbivores (Kursar and Coley 2003).

The type of defence will reflect the relative cost of N versus C. Leaf nitrogen is generally related to photosynthetic capacity, but tropical rain forest canopy leaves typically possess relatively low nitrogen levels, probably reflecting both limited soil nitrogen and high risk of attack by herbivores (Coley and Barone 1996; Turner 2001). When soils are less fertile, carbon-based defences (tannins and toughness) are favoured. This also reflects leaf longevity: defences such as the nitrogen-based alkaloids and cyanogenic glycosides are relatively cheap for plants to produce but have maintenance costs making them expensive for longer lived tissues (Coley et al. 1985).

13.2   Ant-mediated interactions

Mutually beneficial interactions between plants and other organisms extend beyond the pollination and seed dispersal partnerships described in Chapter 12 and the plant–mycorrhiza interactions described in Chapter 3. Reflecting the complexity of interactions, mutualistic associations occupy a whole range of functions, from highly integrated mutually dependent symbioses to more loosely mutualistic and even on occasion parasitic interactions. For example, seedlings generally benefit from mycorrhiza, but these associations have a cost under low light and limited soil moisture—carbon assimilation rates of seedlings of *Diconyina guianensis* (Fabaceae, Caesalpiniodeae) in French Guiana are actually reduced by the carbohydrate demands of the mycorrhizal fungus (Bereau et al. 2005).

Hemiptera[7] exchange protection by ants for plant-derived sugars, the consequences of which may be negative for host plants (see Christmas Island example in Chapter 16), due both to consumption of sap and exposure to plant pathogens by hemipteran vectors, or positive when the attraction of ants serves to limit losses due to other non-hemipteran herbivores (Rico-Gray and Oliveira 2007). By indirectly drawing on plant resources through tending homopterans, ants may even be the most abundant 'herbivores' in the canopy arthropod community (Davidson et al. 2003).

13.2.1 Ant-fed plants, ant gardens, and devil's gardens

Ants are ubiquitous in continental tropical rain forests, so it is perhaps not surprising that they have established a range of intimate interactions with a variety of plants. While plants often feed ants in return for protection (see above), sometimes plants are fed by ants. **Myrmecotrophy** refers to the ability of plants to draw nutrients from ant-derived debris (Beattie 1985; Benzing and Clements 1991) and has been demonstrated among many epiphytes, usually from the families Rubiaceae, Orchidaceae, Polypodiaceae, and Asclepediaceae, and particularly in nitrogen-limited conditions. Myrmecotrophic plants (Figure 13.4) harbour ants within cavities or chambered tubers where the ants deposit organic materials (debris piles) which as they decompose release nutrients that are absorbed by the

---

[7] These are sap-sucking insects that include aphids, scale insects, coccids, whiteflies, leafhoppers, and treehoppers.

Figure 13.4 *Mymercoedia* sp. (Rubiaceae), a genus of small mostly epiphytic shrubs, restricted to tropical Asia and New Guinea, that develop a symbiosis with ants (including *Iridomyrmex* spp.). The ants inhabit the swollen stem base/root. Such plants are especially common in swamp environments where ants have difficulty nesting on the ground, as here in a mangrove formation (Lei-Lei Island, Northern Moluccas Islands, Indonesia).

Photo by Douglas Sheil.

plants. Such plants were first described ecologically in Southeast Asia but have since been reported among many plants including ferns. Neither are ant-fed plants restricted to epiphytes: *Cecropia peltata* (Cecropiaceae) in Trinidad may obtain 98% of its nitrogen from ant deposits, and its *Azteca* ants derive 18% of their carbon from the tree (Sagers et al. 2000); also the shrubs *Piper fimbiulatum* and *P. obliquum* (Piperaceae) have been shown to derive nitrogen naturally from the detritus generated by *Pheidole bicornis* ants (Fischer et al. 2003). Many other examples of ant-fed plants have been found across South America among both epiphytes and **geophytes** (Rico-Gray and Oliveira 2007).

Among the most complex of all plant–ant mutualisms are 'ant gardens' that comprise an aggregation of epiphytes assembled through the action of ants. These ant gardens are common and diverse throughout the Neotropics and the Asian tropics, and yet are poorly understood. An ant garden is established when ants bring the seeds of specialized epiphytes into their fragile carton nests. The plants flourish on accumulated detritus and the plant roots become part of the nest, increasing its robustness and resistance to damage from heavy rain. It is also possible that structural support by epiphytes allows ants to build larger nests in more open and light-rich areas, which would increase the availability of food derived from homopteran-tending ants, and allow the ants to dominate such resource rich areas (Yu 1994). The ants feed on the plants fruits, food bodies, and elaiosomes, and from extra-floral nectaries (Rico-Gray and Oliveira 2007). The ants may also disperse the seeds and provide protection against herbivores. Specialist ant-garden epiphytes are drawn from at least 16 genera (including ferns), and while they sometimes grow untended they are clearly adapted to this symbiosis in that their seeds produce substances that are particularly attractive to ants.

Evil forest spirits have been invoked to explain many mysterious natural phenomena, and this applies too to Amazonian 'devil's gardens', the strange patches dominated by a single tree species *Duroia hirsuta* appearing in the midst of an otherwise dramatically contrasting species-rich forest (see also Box 10.7). The truth is almost as strange, as these patches are maintained by an ant, *Myrmelachista schumanni*, that creates these clearings by poisoning all plants in the vicinity except *D. hirsuta*, the tree within which it nests (Figure 13.5) (Frederickson et al. 2005). The ants poison the plants by injecting formic acid into the base of the leaves, and in so doing they favour the spread of their host tree, allowing for larger colony sizes. The ants attack herbivores, but the gardens attract higher rates of herbivory as they grow larger, which ultimately limits the expansion of the gardens and constrains the devil (Frederickson and Gordon 2007). Both ant gardens and devil's gardens provide remarkable examples of how mutualisms can influence local plant dominance and shape community structure.

Figure 13.5 A 'devil's garden' of forest clearing created by *Myrmelachista* ants which kill undesirable trees to maintain the open space for the growth of its preferred host tree.

Photo by Douglas Yu.

13.2.2   Leafcutter ants

Leafcutters are members of the tribe Attini (subfamily Myrmicinae) and comprise approximately 200 species distributed throughout the Neotropics. Their colonies can grow large, with nests excavated in the ground to a depth of 6 m and containing several million ants. They are successful in several habitats, including savannahs, but have considerable impact as herbivores of a wide range of plants in rain forests (Herz et al. 2007). *Atta cephalotes*, for example, harvests leaf material from 30–50% of plants in a Guyanese rain forest (Cherrett 1968). By selectively harvesting leaves from certain plants, and particularly seedlings, leafcutter ants affect the plant composition within forests as well as patterns of succession and even soil processes.

Collected leaves are composted for the benefit of a fungus *Leucoagaricus* (formerly *Leucocoprinus*) *gongylophorus* which naturally occurs only within leafcutter ant nests[8]. The ant larvae feed upon the fungus, which is tended and fertilized by tiny 'minima' workers (Figure 13.6). The fungus, far from being a passive partner in this relationship, chemically influences ant foraging to ensure that ants avoid plants with strong fungicidal properties (Ridley et al. 1996)—thus the fungus avoids plant antifungal defences, and the ants circumvent plants's secondary defence compounds by feeding on the fungus instead of directly on the leaves.

The fungus benefits from this arrangement by having a habitat that is virtually free of competitors by the constant tending of their ant partners. Even so, fungal parasites of the genus *Escovopsis* occur at low levels in virtually all fungal gardens and can destroy a colony if not kept in check. These parasites are maintained at low levels due to the contribution of a third partner in this interaction, a *Streptomyces* bacterium which lives on the bodies of the ants (Currie et al. 1999) and produces specific antibiotics against *Escovopsis*. This complex suite of antagonistic and mutualistic interactions represents an evolutionary arms race between the fungal parasite on

---

[8] Although the point is debated, this single fungal species is probably common to all leafcutter ants.

Figure 13.6 Leafcutter ants are expert farmers. These *Atta cephalotes* leafcutter ants are exclusively fungivores, cultivating a specialized fungus (the white matter in this image) in extensive underground galleries. The cut leaves harvested by these ants are not consumed directly but are used to grow the fungus on which the ants feed.

Photo by Alex Wild.

the one hand and the ant–fungus–bacterium alliance on the other.

13.3   Interactions across multiple trophic levels

13.3.1. Food webs

Tropical rain forest community behaviour cannot be understood solely by their species richness, but rather by the nature and strength of interactions among the component species. The feeding interactions among species within a community constitute the food web, and food webs (Box 13.1) underlie many ecosystem properties, including species population dynamics, energy flows, nutrient cycling, primary production, ecosystem stability, and biodiversity. For example, knowledge of whether herbivores are limited by predators (top-down control) or the availability of palatable plant biomass (bottom-up control) contributes to our understanding of how changes in predator populations or plant biomass (perhaps resulting from anthropogenic interventions) might affect community organization and stability. Quantified

food webs that map energy flows across biological communities are one approach by which ecosystem-level implications of changes in particular feeding groups (trophic guilds) might be determined. We remain a long way from a comprehensive understanding of such interactions within tropical forests, although some headway is being made in this direction (Box 13.1).

Recent work in the tropics and elsewhere has shown that indirect interactions among species can have far-reaching impacts on community dynamics and structure. Species may interact indirectly through food webs even though their individuals may never actually meet: a common species that hosts a parasite might impact on another species that shares the same parasite by supporting a large parasite population. Such 'apparent competition' has rarely been identified, but convincing evidence for it has been obtained from the Belizean rain forest, where leaf-mining beetles and flies were removed from the system by removing their host plants (Morris et al. 2004). The result was to reduce parasitism and increase abundance of other insect species that share parasites with the leaf miners. Thus removal or addition of species may have consequences that are expressed

---

Box 13.1   A tropical rain forest food web

Quantifying food webs is complex, not just because of the sheer number of organisms that feed on each other, but also because many consumers are inconspicuous parasites or parasitoids that are nevertheless important. Some tropical forests have lost certain trophic groups, particularly top predators, providing an opportunity to investigate the impacts of such losses (Terborgh et al. 2001). Others have experimentally manipulated web structures by removing specific elements such as herbivores or ants (Letourneau and Dyer 1998).

Work in the Luquillo Experimental Forest on the island of Puerto Rico has revealed some of the dynamics and structure of its food web (Reagan and Waide 1996). A hurricane-driven disturbance regime, absence of large herbivores and predators, low faunal richness but an abundance of frogs and lizards[1] does not make the site necessarily representative of continental tropical forests. Even so, Luquillo's food web is far more diverse than most published (temperate) food webs and reveals some unexpected properties.

Theory predicts that trophic loops, in which species A eats B, which eats C, which eats A, would be rare because predators generally feed on prey smaller than themselves, and therefore body size limits a prey's ability to consume its predator (Pimm and Lawton 1978). In fact, trophic loops are relatively common at Luquillo, with loops often involving reciprocal predation at different life history stages. For example, frogs often predate a host of juvenile arthropods, which as adults themselves consume tadpoles of the frog (Reagan et al. 1996). The relatively large size of many predatory invertebrates at Luquillo, coupled with their use of venoms, also allows organisms such as *Scolopendra* centipedes (Chapter 5) and *Phrynus* whip scorpions to prey on frogs, toads, and lizards larger than themselves (Reagan et al. 1996).

Another surprise is that the Luquillo Forest food web, far from being limited by efficiency of energy transfer to 3–5 trophic links as predicted and observed in temperate systems (Pimm and Lawton 1977; Pimm 1991), is long: possessing a mean food chain length of 8.6 links and a maximum of 19 (Reagan et al. 1996). The abundance of ectothermic frogs and lizards, which have high energy conversion efficiencies and small body sizes, allows the establishment of longer food chains and trophic loops (Reagan et al. 1996).

[1] High numbers of amphibians and lizards are characteristic of island communities due to a lack of large predators, but Luquillo Experimental Forest has some of the highest population densities ever recorded (Waide and Reagan 1996).

---

through the rich tropical forest community network (Lewis et al. 2002).

Quantitative food webs describe community structure in terms of the abundance of different species and the number and strength of interactions among them. The complex food webs of diverse tropical rain forests have been described as both highly fragile and extremely stable (Pimm 1986). The frequency of links within a food web allows for alternative pathways providing redundancy within and among trophic levels, thereby ensuring, theoretically at least, that the impact of species losses, for whatever reason, will be buffered. Alternatively, strong interdependence among species and trophic levels can make them vulnerable to perturbation. We do not yet know enough about tropical forest food webs to resolve these issues, but progress is being made by studies that explore the degree of specialization among interacting species, and the extent to which changes at one trophic level cascade through the web to affect other trophic levels.

Keystone species are those that are widely and strongly connected within the network of community interactions, and by definition have profound and disproportionate impacts on the flow of energy and resources (Terborgh 1986). They have been the subject of considerable interest, particularly with respect to how community processes are affected by the loss of these species. Fig trees (*Ficus* spp., Chapter 2) that provide resources to a wide variety of organisms during times of scarcity have been touted as keystone species in Barro Colorado Island, Panama, and Manu National Park, Peru (Terborgh 1986), and in Borneo (Harrison 2003, 2005). Fig trees fruit asynchronously and the fig tree population therefore provides food to frugivores throughout the year. Other contenders for keystone status include nitrogen-fixing plants, decomposers such as termites and dung beetles, and dispersal agents of important forest tree seeds, such as agoutis. The keystone concept is not, however, well defined: in the case of fruiting trees, the ideal keystone species would be 'all things to all frugivores in all places and at all times' (Westcott et al. 2002). No such species exists, and keystone status is often a matter of context (Dew and Boubli 2002). Thus figs are not considered keystone species in Gabon, where monkeys and birds instead depend on the fruit of other species[9] during lean times (Gautier-Hion and Michaloud 1989).

The transfer of energy and nutrients from one trophic level to the next can also represent a transfer of energy and nutrients from one location to another, and in some systems this can be substantial and ecologically important. Within the Amazon Basin migratory fish transfer nutrients from the nutrient-rich white-water river systems to the nutrient-poor blackwater systems, characterized by low productivity but high piscivore abundance (Winemiller and Jepsen 2004). During the wet season fish migrate into the productive Orinoco and Amazon floodplains to feed and spawn, and with the onset of the following dry season massive schools of juveniles migrate back to blackwater rivers where many are consumed by piscivores. This represents a 'food web spatial subsidy' that supports the high diversity and abundance of the piscivore community in the nutrient-poor regions of the Amazonian Basin[10]. A similar transfer of nutrients occurs on a daily basis in terrestrial systems, albeit on smaller spatial scales, from the forest canopy to the soil through the feeding and excretions of countless her-

bivores, and from the soil via mycorrhizal fungi and roots back to the canopy.

13.3.2 Trophic cascades

Interaction between plants and their herbivores are mediated by predators, parasitoids, and pathogens of herbivores. For example, plant chemical and mechanical defences slow the growth rates of herbivores and therefore expose them to higher predation pressure. This might dramatically reduce herbivory because per capita herbivory rate is at its highest in the final instars of insect larval development. Conversely, predators of plant mutualists, such as beetles predating ant-guards on *Piper*, allow herbivores access to leaves previously protected by the ants, resulting in reduced vegetative cover (Letourneau et al. 2004).

Although these examples are instructive, studies of knock-on effects across several trophic levels (trophic cascades) in the tropics are few owing to the difficulty in large-scale experimental manipulation. We can, however, draw inferences about multi-trophic interactions from observations of the impacts of hunting by humans, which has greatly reduced the populations of many vertebrates in tropical rain forests. The consequences of large vertebrate losses, be they predators, herbivores, or frugivores, have been expressed through reduced seed predation and dispersal, and seedling survival (Terborgh et al. 2001). Hunting of vertebrate seed-dispersers reduces plant reproduction by intensifying competition and vulnerability to natural enemies of poorly dispersed seedlings (Beckman and Muller-Landau 2007; Muller-Landau 2007; Stoner et al. 2007; Wright et al. 2007). In Venezuela, forest islands were created by the inundation of a watershed following dam construction. Islands which lacked large predators had densities of rodents, howler monkeys, and iguanas which were 10 to 100 times greater than nearby mainland forest, indicating that predation and not food availability limits populations of these groups. Densities of canopy tree seedlings and saplings on such islands were severely reduced, providing evidence of trophic

---

[9] Notably *Pycnanthus angolensis* and *Coelocaryon preussii* (both Myristicaceae) and *Polyalthia suaveolens* (Anonnaceae).
[10] Dams may create barriers to these flows that have potential implications beyond the migratory fish populations.

cascades following loss of predators[11] (Terborgh et al. 2001).

Both top-down and bottom-up effects have been shown to act through leafcutter ants along the edges of Neotropical forest fragments. Leafcutter ant colonies are far more abundant along forest edges, in forest fragments, and within secondary forests than in the interior of large undisturbed forest tracts (Wirth et al. 2003, 2007; Urbas et al. 2007). This has been attributed to bottom-up effects from the increased availability of fast-growing and poorly-defended pioneer species for which the leafcutter ant species show a clear preference. Release from top-down forces may also play a role. Armadillos and army ants are suspected of limiting colony establishment and growth by feeding on the brood of young colonies, while parasitoid phorid flies attack and kill individual workers and can impose substantial costs on colonies (Rao 2000; de Almeida et al. 2008). It seems that both top-down and bottom-up forces shape the interactions among leafcutter ants, habitat type, and plant species composition.

Such effects can even begin to alter ecosystem processes. High densities of red howler monkeys *Alouatta seniculus* on small isolated predator-free islands disproportionately impacted on palatable tree species. Chemically well-protected tree species with low nutrient concentrations (high C:N ratio) became more abundant as a result, but this reduced leaf litter quality and slowed decomposition rates and nutrient mineralization (Feeley and Terborgh 2005). African elephants *Loxodonta africana* in Kenya's Shimba Hills coastal rain forests impede recovery of disturbed forests by uprooting trees and preferentially foraging on pioneers. Counterintuitively, this favours the persistence of the preferred pioneers (such as *Leptonychia usambarensis*, Sterculiaceae) that are able to tolerate and recover from high levels of browsing, while seedlings and saplings of other species, although not consumed by elephants, are crushed and trampled (Hoft and Hoft 1995).

The relative importance of top-down versus bottom-up regulation of animal populations is an issue of some complexity, because herbivore populations are influenced by food availability *and* predator pressure, the relative importance of which is expected to vary across a productivity gradient (Richards and Coley 2007, 2008). Treefall gaps in tropical forests are patches of high productivity in otherwise less productive forest understorey. Light availability in gaps increases plant resources, and therefore herbivores should be limited more by predators in gaps, and by resource availability in understorey. Tests of these ideas in Panama indeed found that, as the theory predicts, plant resources, herbivores, and predators were more abundant in gaps than in the understorey, and herbivory rates as well as predation were higher in gaps (Richards and Coley 2007, 2008).

13.4   Animal diseases and parasitism

Animal diseases make headlines when they concern large charismatic animals of conservation significance, and even more so when the diseases also threaten human populations in rain forest regions. Ebola, a fearsome haemorrhagic fever that affects apes and occasionally humans, has wiped out many wild apes in Gabon and Congo in recent years. As many as 5,000 gorillas (and up to 95% of some populations) and around 83% of chimpanzees are thought to have perished within the Lossi Sanctuary in Republic of Congo between 2002 and 2003 (Bermejo et al. 2005). The Lossi outbreak was just one of several gorilla and chimpanzee die-offs caused by Ebola since the early 1990s, and altogether Ebola may be responsible for killing as much as 25% of the world gorilla population during this time.

From an ecological perspective, infectious diseases are a special form of host–parasite relationship, and are as much a part of ecosystems as are predator–prey or plant–herbivore relationships. Most parasites and pathogens co-exist with the host with little or even no ill-effect; because killing your host is rarely beneficial, highly virulent genotypes and susceptible hosts are selected against. Outbreaks of virulent diseases, such as Ebola among gorilla populations, occur when a naïve population is exposed to a novel pathogen due to contact with the native host, which in the case of Ebola is probably bats.

---

[11] This conclusion has been challenged on the grounds that a more parsimonious explanation for widespread mortality of trees is the effects of flooding itself, which exposes tree roots to long periods of inundation (White 2007).

The global decline of amphibians, including rapid declines and extinctions in pristine habitats, has been attributed to outbreaks of chytridiomycosis, a potentially lethal infection of the amphibian epithelium by the fungal pathogen *Batrachochytrium dendrobatidis* (Berger et al. 1998; Lips et al. 2006; Pounds et al. 2006; Laurance 2008). Growth and pathogenicity of *B. dendrobatidis* are inhibited by warm temperatures (Kriger and Hero 2007), which presumably explains why most observed amphibian declines have occurred in montane areas.

Parasites are ubiquitous in vertebrates and invertebrates—even parasites have parasites—and while the degree to which they impact on individuals and populations in tropical rain forests is not well known, it is generally accepted that parasites do have some role in regulating wild animal populations. Despite this, at least one study on a number of rodents and marsupials in the Brazilian Atlantic forest found no correlation between body condition and parasite load (Puttker et al. 2008). Nevertheless, gastrointestinal nematodes and protozoan parasites are common in all rain forest mammals, although they may be less prevalent among arboreal species due to lower infection probability (Puttker et al. 2008). Forest disturbance and fragmentation has been hypothesized to increase mammal gastrointestinal parasite burdens in response to increased stress. This idea has received support from work on primates in Kibale, Uganda, where logged forest increased primate infection risk compared to unlogged forest (Gillespie et al. 2005), but has not been found to be the case among rodents and marsupials in the Atlantic forests of South America (Puttker et al. 2008).

Parasites improve their dispersal in various ways. One dramatic means can be by transforming the behaviour or morphology of their host. One parasitic nematode that infects ants in the canopies of Central American and Amazonian tropical forests dramatically transforms the ants to resemble bright red berries, presumably to trick birds into eating infected ants and thereby spreading the parasites in its faeces (Yanoviak et al. 2008). Some fungal pathogens of ants and other invertebrates cause the host to climb an exposed twig or leaf, whereupon the ant dies and the fungal fruit body promptly develops and releases its spores in the air current (see also Chapter 3).

13.5 Conclusion

The web of life's interactions is the fabric that binds species. The full complexity of the web still largely defies current understanding. Yet the fabric is slowly unravelling as populations and species are extirpated, and we have little ability to predict specific outcomes or the wider implications.

Section III

Our future legacy

CHAPTER 14

Forests in the Anthropocene

'Much fruit of all kinds.'

Friar Gaspar de Carvajal (1541–2)[1]

'I was rather disappointed in these forests as we were led to understand by the guides that they were extensive and practically virgin in character. This we found to be very far from the case, and the whole tract of country showed unmistakeable signs of villages, having been once pretty well inhabited. Large tracts of forest were found to be of secondary origin, and signs of villages having once existed here were also not wanting.'

H. Thompson (1910)[2]

The current geological age (epoch) can be called the 'Anthropocene' in recognition of humanity's impact on the planet (Steffen et al. 2007). While these impacts have increased dramatically with the Industrial Revolution starting in the 18th century, significant impacts date back over 40,000 years. Today every corner of the globe, even the remotest tropical forest, is affected by people[3].

An enduring myth that dates back to the earlier history of Western exploration of the tropics is that of the 'noble savage', indigenous people living in harmony with the environment. Native tribal people occupied lush and productive environments, so it seemed obvious that they had used and managed their environments sustainably (Borgerhoff Mulder and Coppolillo 2005). The noble savage concept has, however, long been criticized, debated, and ultimately rejected, as tribal peoples' relationship with their environment has been more carefully scrutinized (Hames 2007). Examples of both good and bad indigenous environmental management are now well established (Diamond 2005). In many cases, evidence suggests that the ancient human legacy may be much greater than is often recognized. In these last chapters we consider the interactions between people and the world's tropical rain forests, from prehistory through to future challenges.

---

[1] From the Jesuit friar's account, first published in 1910, of the first European exploration of the Amazon, led by Francisco Orellana, from the Napo River in Peru down the Amazon to its mouth: Carvajal G.d. (1970) *The Discovery of the Amazon*. AMS Press, New York, USA.

[2] Thompson, H. (1910) *Gold Coast: Report on Forests*. Colonial Reports-Miscellaneous No. 66. HMSO, London, UK, p. 46.

[3] Those who contest this claim should note the changes in atmospheric composition, depositions, climate, and pollutants recorded even in the remotest uninhabited locations.

14.1 Humans arrive: Pleistocene and Holocene extinctions

Over the last 50,000 years, extinctions among large terrestrial vertebrates have been common, with as many as 90 genera of larger mammals extirpated. Humans are implicated, though alternative explanations are hotly debated (see Burney and Flannery 2005; Koch and Barnosky 2006). The earliest extinctions attributed to humans in tropical rain forests extend back at least to the late Pleistocene around 11,000 years ago, a period marked by the extinction of many large mammals, such as the giant ground sloth *Megatherium*, gomphotheres and horses[4], in the Americas (Cione et al. 2001). Hunting of the naive fauna, along with more or less extensive habitat modification (by fire), appear the most favoured explanations for these events, though other possibilities such as the introduction of diseases have also been advanced. Debate also surrounds the significance and role of climatic changes which—with the end of the last glacial—were marked at that period (Koch and Barnosky 2006).

During the Holocene humans certainly contributed to many tropical extinctions. Increased human activity on Java, Indonesia, during and since the Neolithic (2,600 years ago), for example, has contributed to the extinction of Javan tapir, elephant, and Malayan bear, and widespread declines in rhino and tiger[5] (Figure 14.1) (Morwood et al. 2008). Across Asia, probable human-aided extinctions included the giant pangolin, *Manis palaeojavanica*, and possibly the giant tapir *Megatapirus augustus*, Stegodonts (dwarf elephants), and the ape *Gigantopithecus* (Corlett 2007). Cambodia's herds of wild cattle, once described as resembling East Africa's ungulate herds, now fail to impress (Corlett 2007). However, people have lived and hunted in tropical Asian forests for

Figure 14.1 Last known photo of the Javan tiger, which likely became extinct in the 1990s. The original photo hangs in a Ministry of Forestry building in Bogor, Indonesia.

This copy provided by Erik Meijaard.

---

[4] These are believed primarily to have been species of dry forests and forest edges rather than wet forest.
[5] Rhino still persist in a very limited area. Even in the early Dutch colonial period they were shot as vermin around Jakarta. Javan tigers may still persist—sightings are still occasionally claimed—but the last accepted records suggest extinction in the 1990s. Tigers still occur in Sumatra. Teeth excavated in the Niah caves in Sarawak imply that tigers once occurred in Borneo, while other remains have been found on the south-western Philippine island of Palawan.

at least 40,000 years and perhaps considerably longer[6], and perhaps because of the long history of cohabitation relatively few extinctions occurred in Asia (at least compared with the Americas) during the Pleistocene and Holocene. In Africa, where the fauna evolved alongside our human ancestors, there were few extinctions (only 10 mammalian genera were lost in the last 100,000 years), presumably owing to less naivety towards human hunters, resistance to human-associated diseases, and adaptation over a longer period to human habitat modifications (Koch and Barnosky 2006).

Oceanic islands, large and small, experienced numerous human-driven extinctions. An estimated 2,000 birds and many other species became extinct as people moved eastwards across the Pacific, starting some 30,000 years ago in the Bismarck Archipelago and Solomon Islands (Steadman 1995; Steadman and Martin 2003). An extinction wave occurred in Madagascar shortly after the estimated arrival of the first human colonists around 2,000 years ago. Madagascar's losses included the elephant bird, the leopard-like giant fossa, two hippopotamus and 17 large lemur species—butchered bones put humans at the scene for at least some of these deaths (Perez et al.

2005). More recently, starting around 500 years ago, many other Indian Ocean island species began to disappear, again following human settlement. These included several species of giant tortoise on the Seychelles and Mascarene Islands, as well as a host of birds, including the dodo, on Mauritius[7].

Inevitably forests that lose species are ecologically altered as a result. Large herbivores influence forest structure, while lemurs, birds, and tortoises dispersed seeds. Nonetheless, these changes are hard to clarify, despite provocative proposals that various trees and lianas—including many large seeded legumes such as *Enterolobium cyclocarpum* and *Dioclea* spp.—are adapted to dispersal by extinct megaherbivores (Janzen and Martin 1982). As many of these plant species remain, common evidence of coevolved dependence is weak. Even in the modern case of elephant-dispersed species in Africa, observations suggest most species can persist in the absence of these animals (Hawthorne and Parren 2000).

A notorious and erroneous claim of a disrupted mutualism is that of the dodo and the tambalacoque tree in Mauritius (Box 14.1). Whatever the details, our

---

Box 14.1   Resurrection of the living dead: the myth of the dodo and the tambalacoque tree

The extinction of the dodo *Raphus cucullatus* on Mauritius in the late 18th century was said to doom another Mauritian endemic, the tambalacoque tree *Sideroxylon grandiflorum* (previously *Calvaria major*), to a similar fate. In a widely read essay it was suggested that germination of tambalacoque seeds was dependent on passage through the digestive tract of the dodo, the extinction of which explained the absence of trees younger than 300 years (Temple 1977). Thus the tambalacoque represented a species of the 'living dead'—an extant species doomed to extinction.

This attractively tragic story of mutual dependence and doom was widely accepted. But it is wrong. Tambalacoque seeds do not require any abrasion in dodo gizzards in order to geminate. Wild tambalacoque trees younger than 300 years old do occur, showing that dodos are not essential for their recruitment (Witmer and Cheke 1991). The decline of the tambalacoque is now interpreted to be a result of deforestation, seedling destruction by introduced monkeys and pigs, and competition from aggressive exotic plants.

---

[6] Bones of *Homo erectus*, a tool-using precursor of modern humans, found in Java have been dated to over one and a half million years old. These dates are disputed and 500,000 years is now considered more likely. In either case 'people' have been a long time in the region.

[7] A combination of hunting, introduced species, and vegetation alteration caused these extinctions—specific contributions are hard to assess based on current evidence (Grayson 2001).

point is to highlight how these extinctions affect our conceptions of 'pristine' forests.

14.1.1 Prehistoric legacies of human occupation

Signs of past human occupation and disturbance have been found in many tropical rain forests, including many remote and currently uninhabited sites (Figure 14.2). Archaeologists continue to record new and significant finds in virtually all regions. Mysterious megaliths and earthworks occur in forested regions scattered through much of the tropics—a testimony to sophisticated but unknown ancient peoples. Only the Australian rain forest seems to have been largely uninhabited at the time of their discovery by Western explorers, although there had

been a long history of fire use, with forest converted to savannah (Bowman 1998). In Madagascar and many Pacific Islands habitation was relatively recent (2 millennia or less) but impacts on tropical forests have been severe nonetheless.

Agriculture has a long history in the wet tropics and probably developed independently in Asia–Africa, New Guinea, and the Americas. The adoption and selection of plants from this early period still provide the main genetic stock upon which many tropical farmers depend. Domestications included maize and seedless bananas, which have been dependent on humans for their persistence for many thousands of years. The original domestication sites are long lost in the subsequent transfers that occurred within and

Figure 14.2 Forest soon reclaimed the areas cleared by the Maya, as shown here on the steps of a temple complex in Mexico (A) and in among the Tikal ruins in Guatamala (B).

Photos by Douglas Sheil (A) and Andrew Macgregor (B).

between continents—no one know for certain how cassava came to Central America, teak to Java, maize to the Philippines, sweet potatoes to New Guinea, or bananas to Africa (Box 14.2).

As people moved they introduced livestock such as pigs and chickens. These sometimes turned feral—as has been the case with pigs on many islands. Stowaways, such as rats, were widely albeit inadvertently spread, while less expected species such as agoutis, cassowaries, and cuscus were introduced to new locations, probably as pets.

Many tropical regions, especially the more fertile areas, were well populated and supported heavily cultivated landscapes long before European contact[8]. Rich civilizations such as those of the Olmec, Maya, and Inca in America, the Javanese and Khmer in Asia, included significant rain forest territories. These peoples were well

able to engineer terraces, canals, dams, drainage systems, reservoirs, and even to shift rivers (Lentz 2000).

Cultures across the tropics developed rules, taboos, and social norms which, seen through modern eyes, can be seen as having conservation significance. Even today many areas of forest that persist in heavily deforested regions possess grave sites or other cultural significance to the surrounding peoples[9].

14.1.2 Paleohuman settlements

Until recently the conventional view of indigenous tropical forest communities was that they were small, ephemeral, and had very low population densities (0.01–0.3 persons km^{-2}). Consequently, it was assumed that they had had little impact on tropical forests. Such assumptions are now being overturned by detailed historical,

---

Box 14.2 The remarkable history of the domestic banana

The centres of wild banana (*Musa* spp.) diversity are in Melanesia, including New Guinea. Archaeological evidence suggests that bananas were already deliberately planted in the highlands of New Guinea by 5000 BC (Denham et al. 2004), while genetic evidence indicates that *Musa acuminata* ssp. *Banksii*, an ancestor of many modern-day varieties, was also domesticated in New Guinea (Lebot 1999). Domestication involved the selection of seedless varieties which then became wholly dependent on human agency. Intensive vegetative propagation and selection created many derived cultivars.

The existence of bananas in South America in pre-Columbian times indicates early Polynesian transfers

(Langhe and Maret 1999). Evidence for the early cultivation of (Asian) banana[1] in Africa has recently been reported following the discovery of *Musa* phytoliths dated 840–350 BC in Cameroon (Mbida et al. 2001) and (more tentatively) from the late 4th millennium BC in Uganda (Lejju et al. 2006). Plantains (also *Musa*) are rare in Asia (outside India), and perhaps the same Polynesian groups who carried bananas to the Pacific Islands also carried them to Africa where they were later diversified through local selection and adaptation (Langhe and Maret 1999). Certainly the dry suckers of many *Musa* spp. are able to survive storage for several months, allowing transport by way of long sea voyages.

[1] The African false banana, *Ensete ventricosum*, which once covered a much larger area than today, has also been domesticated and remains an important staple in the Ethiopian highlands.

---

[8] A modern example is the Dani people of the Baliem Valley in Papua: even with Stone Age implements the valley was almost completely deforested and supported high population densities when it was first discovered by the Western observers in the 1930s. Remarkably when they were first encountered the main crop of the Dani was the sweet potato—a plant of Neotropical origin that had crossed the Pacific, counter to the direction of explorations, via Polynesian seafarers.

[9] Sacred forests, often small groves, occur in Western, Southern and Eastern Africa as well as in India and scattered through Southeast Asia (Wadley and Colfer 2004; Khumbongmayum et al. 2005; Barre et al. 2009).

archaeological, and anthropological studies. It now appears that tropical forest communities in America and Asia were, prior to European interference, not only substantial, but had considerable impact on the forested environment. With satellite imagery and other advances, the 'footprint' of these communities is now being revealed in the structure and composition of apparently 'pristine' forests and in the findings of large-scale archaeological features (see Box 14.3) (Heckenberger et al. 2003, 2007; Mann 2008). This raises many questions, including the degree to which ancient cultural forces are responsible

for patterns of biodiversity that are now the subject of conservation interest.

Archaeological evidence, mainly in the form of ceramics and human-modified soils, indicates that the first neotropical hunter-gatherers (i.e. pre-ceramic foraging populations) were living at localized sites dispersed across Amazonia between 11,000 and 10,000 BP (Roosevelt et al. 1996). Horticulture using native tubers and seed plants, and probably also the tending and deliberate planting of trees, increased between 10,000 and 8,600 BP, and by 7,000 BP large-scale food

---

Box 14.3   Cultural legacy in the Brazilian Amazon

Large-scale landscape transformation in the Amazon prior to European influence has been particularly well documented in the Upper Xingu region in the southern Amazon (Heckenberger et al. 1999, 2003, 2007). Long-term presence of Xinguano peoples in this region over more than 1,000 years indicated by archaeological work on several interconnected settlements from around AD 1250–1650 and from substantially altered forests and wetlands in the region.

These settlements covered a total combined residential area estimated at 40–80 ha in an area of 400km², with an estimated population of 2,500–5,000 people (6–12.5 persons per square km). These areas were intensively managed, and included a formally arranged transportation network and agricultural areas, as well as patches of secondary regrowth and managed forests. These archaeologically mapped areas correspond to a distinctly different forest cover type (Figure 14.3) that contain tree species associated with *terra preta* anthosols (Box 14.4). For example, the moriche or buriti palm, *Mauritia flexuosa* (Arecaceae), widely used by the Xinguanos, is closely associated with ancient settlement areas, as is *Acrocomia aculeata* (Arecaceae), which, based on extremely tall size and number of whorls, may date to the time of ancient communities. Fruit trees are also particularly common near abandoned Xinguano settlements. Any understanding of the local ecology

and biodiversity in these areas requires an appreciation of their cultural history and recognition of habitat alteration by populations spanning many generations.

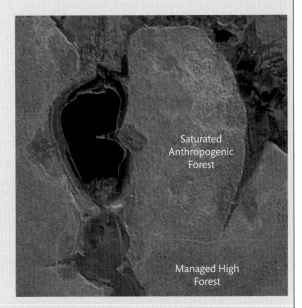

Figure 14.3 An anthropogenic landscape in southern Amazonia (from Heckenberger et al. 2003). Areas to the south and west of the red line are relatively undisturbed forest, while regions of former cultivation are revealed in lighter shades of green. (See colour plate 9.)

production in Central America involved the cultivation of fields away from houses (Piperno and Pearsall 1998). In Rondônia, Brazil, evidence of occupation by hunter-gatherers dates to 9,000 BP, and agricultural activity appears around 4,500 BP (Miller and Nair 2006). Native fruit trees were domesticated by prehistoric people within early agricultural systems, a process that may have started by the incidental discarding of edible fruit seeds near settlements, and developed into home gardens of fruit trees and other useful plants (Lentz 2000).

Population densities in the rural Mayan area today are only about 5 people km^{-2} compared with perhaps 400–500 people km^{-2} during the height of the Maya civilizations (Turner 1976). Combined, the Olmec and Maya civilizations of south-eastern Mexico persisted for about 3,000 years (Turner 1976). The conversion of forests into a mosaic of fields and settlements left its mark on modern forested landscapes. For example, local topography of the southern Yucatán Peninsula is shaped by former house mounds, boundary walls, and agricultural terraces of the Mayan civilization (Turner 1974; Beach 1998; Lentz 2000). Extensive soil erosion impacted the soils in Yucatán at larger scales, and is associated with deforestation in AD 700–900, an event detected in distinct 'Maya clay' sediment layers in lakes and wetlands (Foster et al. 2003).

Evidence suggests that typical population densities of Amazonian pre-Columbian communities[10] were at least an order of magnitude greater than current indigenous communities. Recent studies have documented areas of dense, interconnected settlements representing complex societies which converted forests into patchy, managed landscape mosaics (Heckenberger et al. 2003; Mann 2008). This management, which dates from the late Pleistocene and often lasted several centuries, changed subsequent forest composition (Box 14.3) and transformed soils (Box 14.4).

Fire is probably the most extensively used tool for land management. Forest fires produce charcoal which is gradually buried within the soil profile, and examination of nearly 300 soil profiles across 60,000 ha in

---

Box 14.4 *Terra preta*: the rich forest soils of the Amazon

Pre-Columbian peoples left another lasting legacy: fertile 'dark earth' soils. At the edge of the dryland forests along some Amazonian rivers, patches of nutrient-rich black soils overly deep strongly weathered infertile soils. These are known as *Terra Preta de Indio* (Indian black earth), often abbreviated to *terra preta*, and can be as much as 2 m deep and cover areas of less than 1 ha to several hectares. *Terra preta* soils have been improved by adding large quantities of crumbled charcoal, as well as ash, green manure, or in some cases fish meal (Lehmann et al. 2003). They often contain broken ceramics and bone fragments attesting to their human origins (da Costa and Kern 1999). These sites are often associated with fruit trees and other useful species (Clement et al. 2003). A key ingredient in these soils is the presence of long-lived organic matter that originates from charred woody material, which is also responsible for good nutrient-holding characteristics. There is debate as to whether the soil improvement was intended or inadvertent—resulting from land clearing practices and waste—or a combination of both (German 2003).

The elevated nutrient-holding properties and richness (particularly of phosphorus, nitrogen, potassium, and calcium) of *terra preta* allowed near-permanent cultivation and supported high population densities (Glaser et al. 2001). The oldest known *terra preta* sites in Rondônia, Brazil date from about 4,500 years ago, and are still preferably used for agriculture by modern farmers (Lima et al. 2002).

---

[10] Estimates of human population in Amazonia at the time of European arrival range from 1 to 11 million (Bush et al. 2007).

Guyana recorded charcoal in all of them (Hammond et al. 2006). Radiocarbon dating of charcoal layers from across the Amazon Basin, especially in more seasonal areas, shows an expansion of agricultural activity at around AD 250, then a series of peaks between AD 700 and 1550, with little activity after AD 1600, the time of European contact (Figure 14.4). This pattern matches archaeological estimates of settlement expansion in the Amazon Basin (Bush and Silman 2007).

Similar charcoal-based evidence along with ceramic artefacts in the Las Cruces reserve, Costa Rica, shows that this region has been inhabited for at least 3,000 years (Clement and Horn 2001). Sedimentation of charcoal and maize pollen in the nearby Laguna Zoncho indicates that the intensity of forest clearing and burning increased over a 2,500-year period leading up to around AD 1500. Thereafter, cultivation was abandoned and the forest recovered to form Costa Rica's largest remaining tract of

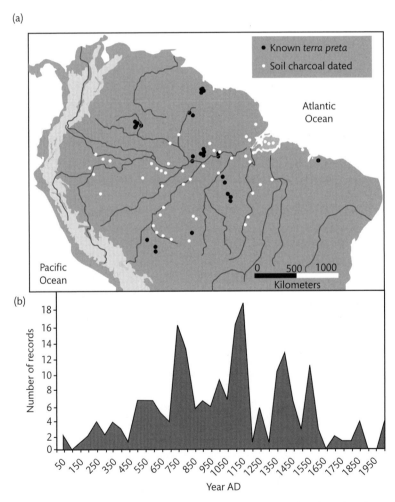

Figure 14.4 Soil carbon data from Amazonia. (a) Sketch map showing major Amazonian rivers, locations of radiocarbon-dated soil charcoal, and *terra preta*. (b) Ages of 228 carbonized layers in Amazonian soils. All ages are expressed in calibrated calendar-years BP.

From Bush and Silman 2007.

lower montane rain forest (Clement and Horn 2001; Anchukaitis and Horn 2005).

Combining soil charcoal with other archaeological evidence suggests that Pre-Columbian Amazonian settlements were relatively widespread and sophisticated. The collapse of indigenous American populations (largely through disease, migration, conflict, and slavery) following European contact (1492) resulted in the abandonment of indigenous managed landscape mosaics across Amazonia and Central America. These regenerated into dense forests[11], by which the vision of indigenous communities as small, primitive, and dispersed with only marginal impact on the land was introduced, despite some earlier accounts to the contrary (Lovell 1992; Newson 1996). Adjectives such as 'untouched', 'virgin', and 'pristine' became incorporated into scientific thinking, but with new archaeological and paleoecological evidence these views have been increasingly challenged. Some anthropologists are now even referring to the forests of Amazonia as 'cultural parkland' and 'human created' (see Heckenberger et al. 2003; Bush et al. 2007). Some studies suggest an enrichment of local flora in areas used by pre-Columbian Amerindians (ter Steege and Hammond 2000).

Referring to the Amazon forest as a manufactured landscape may, however, be only locally true (Bush et al. 2007). Pre-Columbian settlements, although widespread, were scattered and concentrated at the main river channels, and human impact on the forest hinterland was probably low (Heckenberger et al. 2003). Concentric zones of anthropogenically-altered environments around village sites are a more likely model and are consistent with the distribution of anthropogenic soils (Lehmann et al. 2003). On the other hand, recent recognition of the abundance and wide distribution of rock paintings, circular or square areas ranging in size from 100 to 350 m across and outlined by trenches 1–7 m deep, and other major earthworks spread across the south-west Amazon, indicate that a sophisticated human agrarian society may have extended over a much larger area than has previously been suspected (Mann 2008).

Pre-colonial forest impacts in the Solomon Islands

The New Georgia Group of the western Solomon Islands in the south-west Pacific are among the few places on Earth where there remain large tracts of coastal rain forest, extending from sea level to 1,000 m high mountain peaks. These forests have sometimes been interpreted as being 'natural' and 'virgin' (Lees et al. 1991, cited in Bayliss-Smith et al. 2003), but are now known to have supported large agricultural communities (Bayliss-Smith et al. 2003). Up until about 1800 and the arrival of Europeans these communities practised agroforestry and enriched secondary forest with *Canarium* nut trees and other useful species. Evidence for these agroforestry systems exists in archaeological remains, oral histories, and in vegetation patterns. Overgrown remains of settlements and ceremonial grounds with associated stone-walled terraces occur frequently in these forests. Extensive stands of even-canopied forest dominated by *Campnosperma brevipetiolata*, a light-demanding tree typical of severely disturbed forest sites, occur in many locations that are also associated with signs of former habitation[12] (Bayliss-Smith et al. 2003). Early human occupation may also have favoured hard-wooded species such as *Calophyllum neoebudicum*, which are difficult to fell with stone axes (Whitmore 1966).

Historical human impacts on African rain forests

Africa has sustained human populations far longer than any other tropical region, and human impacts on African tropical forests certainly have an ancient history. Unfortunately, research on such impacts is sparse, though it has often been suggested that a signature of human activity dating back several centuries can be detected in current tree species composition, including groves of oil

---

[11] As evidenced by Mayan ruins deep in the forests of Central America, though the Maya collapse appears to have been well underway prior to European arrivals.

[12] Alternatively, gaps produced by tropical cyclones might account for stands of light-demanding trees such as *Campnosperma*, but in the Solomon Islands these natural disturbances have not led to an increase of these trees at the expense of the shade tolerant species (Burslem and Whitmore 1989).

palm that still occur in many parts of the Congo Basin, as well as secondary regrowth following abandonment. Some regions that are now totally uninhabited, like the Nouabale-Ndoki National Park in the Republic of Congo, reveal evidence of human occupation with pottery, furnaces, and charcoal below the dense forest (Brncic et al. 2007). Evidence of human activities extending as far back as 3,000 years ago has also been found in the form of charcoal and pottery in forest soils in Nigeria, Zaire, and Cameroon (Hart et al. 1994; White and Oates 1999; Mindzie et al. 2001).

14.1.3 Forest recovery and concepts of 'naturalness'

In view of widespread evidence of fire, agriculture, and even urbanization in forested areas that were until relatively recently thought to be pristine, concepts of 'pristine' or 'virgin' forests are now considered untenable and ecologists are increasingly using alternatives such as 'old growth'. Yet concepts of pristine and ancient rain forests undisturbed by humans retain a hold over ecologists' thinking, as reflected in the confusion of terminology concerning primary old growth and secondary forests (Clark 1996).

It allows optimism that forests can be perceived as 'pristine' within a few centuries or even decades following the abandonment of cultivation. Needless to say, the regrowth forests that result do nonetheless bear the ecological signs of past human activity, if only they can be read. In some cases, as in locally enriched sites in the Amazon and Central America or the Solomon Islands, tree species richness appears to have been enhanced. In other locations, perhaps in some African forests, or in heavily utilized Pre-Columbian Amazonian settlements or intensively cultivated sites in the Solomons, species richness is depressed compared to non-settled sites or recovery is still underway—though this may simply reflect successional dynamics. Although human population densities within tropical rain forests were often much higher than was previously thought, they were probably concentrated in favourable settlement sites, and the geographic extent of their impact was therefore somewhat limited.

14.2 European expansion

'... The bark trees [Cinchona] are rapidly verging on extinction, and it is therefore of the first importance that steps should be taken for their cultivation on a large scale, in some country where there is no lack of industrious hands' (Richard Spruce's letter to the British Foreign Office, 1858, quoted in Honigsbaum 2001).

During the first account of travelling the Amazon from Andes to coast (1541-2)—a search stimulated by stories of El Dorado and imaginary cinnamon trees—Francisco de Orellana observed a heavily populated landscape[13]. But these dense populations did not persist. European penetration of tropical regions, especially those in the Neotropics, led to the calamitous collapse of many indigenous societies newly exposed to European diseases to which they had little resistance[14]. Between 1520 and 1600 diseases such as smallpox, measles, diphtheria, and influenza may have reduced Amazonian populations by 95% (Denevan 2003).

Initially European settlements were mostly restricted to coastal areas and navigable rivers. Often explorations were undertaken by missionaries, explorers, and fortune hunters. In many lowland areas permanent outposts became possible only when malaria was tamed—a story itself dependent on exploring for the Cinchona tree (see below).

Europeans saw tropical rain forests as a source of various high-value, non-perishable commodities: gold ('El Dorado'), slaves, spices (which were often believed to have magical powers), and exotic luxuries such as bird of paradise feathers. It was these treasures that spurred many of the great exploratory voyages, including those of Magellan, Columbus, and Raleigh. Vast private ventures,

---

[13] Friar Gasper de Carvajal, also a member of this expedition, wrote of a tribe of female warriors. These captured the popular imagination, hence the river of Amazons, after the Ancient Greek name for a tribe of female fighters. This story was repeated in Acuña's account of a 1639 expedition. Even today the men of some ethnic groups hunt and forage for extended periods leaving the women to defend their village.

[14] Columbus's voyages of discovery unleashed a plague on Neotropical peoples, but also in Europe as syphilis, the 'great pox', broke out in 1495—a phenomenon that impacted on society in a manner similar to HIV today (Harper et al. 2008).

often a prelude to imperial enterprise, followed. These ventures were largely commercial in aim. Some sponsored the transfer of high-value crops from one part of the tropics to another (Miller and Nair 2006). In Brazil the Portuguese introduced mango and jackfruit in the 16th and 17th centuries, and nutmeg, breadfruit, and sugarcane followed in the 19th century. Travelling on the Amazon in 1849, the British naturalist Henry Walter Bates described home gardens with banana, papaya, mango, orange, lemon, guava, avocado, as well as coffee shrubs growing under the shade of the fruit trees (Bates 1863). Cassava and maize were similarly spread from the Neotropics to other tropical regions.

In Latin America the Spanish and Portuguese administrations encouraged forest clearance for crop plantations in the 17th and 18th centuries (Hecht and Cockburn 1989; Bryant 1996). Sugar plantations in the Caribbean were developed by various European powers, leading to high demand for labour and driving the slave trade— which in turn fuelled considerable conflict and upheaval in West and Central Africa.

In tropical Africa and Asia, European enterprises arrived in force later than in the Americas, and it was not until the 19th century that political and economic changes began to have major impacts on how the forests of Asia and Africa were exploited (Tucker and Richards 1983;

Box 14.5   Ancient trades

It is easy to adopt the Western-centric view that puts Europeans at the centre of world history. But tropical forests had not been immune to outside influences prior to European expansion. The Chinese were trading for high-value forest products and tropical crops in Southeast Asia for many centuries[2]. Arab and Indian trade along the East African and Asian coast is millennia old. In Asia, Hindu kingdoms and later Islamic sultanates established near river mouths controlled a valuable trade from the interior with the outside world. The first inhabitants of Madagascar were seafaring people from Asia. The Polynesians successfully explored the Pacific, and perhaps the Chinese had circumnavigated the globe before Europeans had even reached the Azores.

The international trade in tropical forest products, especially in spices, is ancient. Excavations in the Mesopotamian site of Terqa (modern Syria) found a clove dating back to 1721 BC—especially remarkable considering that before modern times the clove tree was restricted to only five small islands (in the Moluccas, Indonesia), each less than 20 km across. The Roman Empire imported spices from far and wide, especially black pepper, and clove and nut-

meg from the Moluccas too, by the 4th century AD. Spices were used for more than just flavour. Black peppercorns have been found in the mummified remains of Egyptian Pharaoh Ramesses II (dated to 1224 BC). The Romans used cinnamon primarily to burn in funerals, while the Maya used allspice in embalming. Cinnamon and other spices appear in Old Testament accounts of holy anointing oils. Spices were attributed to have diverse medical, mystical, and aphrodisiac properties. Many have mildly disinfecting properties, reducing the likelihood of food poisoning from spoiled meat. Early traders were happy to exaggerate stories concerning the exotic origins of spices—some of these inspired stories such as Sinbad's voyages.

The urge to control the vast wealth represented by the spice trade was the main spur to much of the West's explorations of both the East and the West. Columbus was seeking the Spice Islands when he set sail for what we now call the Americas. Magellan too was seeking a route to the Indies—and indeed managed to find it—even though they were already well known to Arab traders. It is considered likely that the ships that carried bubonic plague to Europe were part of the Indian spice trade (Turner 2004).

[2] Some plantation developments, such as *Uncaria gambir* in Sumatra and Malaysia, served a regional trade. In Borneo legacies persist in the jealously guarded bird's nest and Eaglewood (*Aquilaria* spp.) trades—for many Punan their most prized heirlooms are still ancient Chinese vases and bronzes traded for forest products.

Dargavel et al. 1988). Although the history of trade in forest products is long (Box 14.5, 14.6), demand for tropical products increased during the Industrial Revolution, and by the second half of the 19th century rubber plantations had been established in Asia using *Hevea brasiliensis*, an Amazonian species. Plantations of coffee, tea, cocoa, and many spices were also established. Deforestation in Madagascar was increased by the introduction of coffee, which displaced the original cultivators from the most fertile land, forcing them to clear forested land for their own subsistence. The colonial enterprise was often precarious, with ill health, especially malaria, ravaging settlers in the wet lowlands. The strategic significance of *Cinchona* (Rubiaceae), an Andean–Amazonian tree and the sole source of quinine and quinidine, a cure for malaria, led to a rush by the global powers to develop their own plantations.

Exploitation of forest products often followed a pattern of boom and bust through overexploitation. At the start of the 20th century the search for electrical insula-

---

Box 14.6 The history, and prehistory, of timber management

Prehistory

Trade in timber has a long history. Teak *Tectona grandis*, a durable timber from monsoonal Asia (Burma and Indochina), has been excavated from the ancient cites of Ur (3rd millennium BC) and also from Babylon, Nineveh, and Ctesiphon in modern-day Iraq. Such timber probably followed the ancient maritime spice routes. Teak was introduced and actively managed in Java by the 10th century (Dawkins and Philip 1998).

Imperial ambitions

The early history of European forest exploitation in the wet tropics was linked to the twin forces of commercial opportunism and imperial ambition. Initially, tropical timber was mined with often little regard to impacts. The Caribbean mahoganies (*Swietenia macrophylla* and *S. mahogany*) were exploited for the luxury timber and shipbuilding markets of Europe from as early as the 17th century, and forest depletion led to establishment of forest reserves in Tobago, and St Vincent in the Caribbean by 1764 (Grove 1996). Similar trade involved various species from Western Africa. Damage from timber cutting (and introduced animals) was especially devastating on islands such as Mauritius, the Seychelles, and St Helena (Grove 1996).

Naval timber demand became severe during the Napoleonic Wars (Grove 1996), and the world's

superpowers acted to secure their timber supplies. Professional forestry was developed by British, French, and Dutch interests in India, Burma, and Java in the 18th century. The Dutch 'Javanese cutting rules' of 1770 defined a minimum teak beam as 25 cm by 7 m, implying a lower diameter limit of approximately 35 cm (Dawkins and Philip 1998). The need for scientific monitoring was formally recognized—plots were established in India as early as 1884 and efforts by the British Ecological Society in the 1920s and 1930s emphasized the careful evaluation of tropical vegetation (Sheil 1998).

Industrialization

Management systems that emphasized the need for careful harvesting to protect future crop trees were in place in India, Malaysia, and Philippines as well as in various African and Caribbean countries by the 1930s. However, it was only after World War II, with the rapid post-war economic expansion and improvements in heavy vehicles (notably 'caterpillar tracks') and chainsaws, that large-scale industrial exploitation started. Significant markets rapidly developed in Japan, Korea, Taiwan, and elsewhere. Some commentators began to question whether mechanized selective harvesting of moist tropical forest was truly compatible with sustained timber management. This led to systems to control harvesting intensities and related practices (e.g. Dawkins 1958). Such systems were widely adopted but seldom implemented for long.

During this same period political independence was sweeping through the tropics. Environmental damage, illegal logging, and governance became hot political issues. While earlier management had focused on sustaining timber outputs and controlling access, new demands highlighted conservation impacts and the rights of local people.

The term 'reduced-impact logging' (RIL) arrived in the 1990s. The concept resonated not only with timber managers but also with influential environmental organizations. In consequence, RIL gained legitimacy (Gustafsson 2007). During the same period considerable efforts were invested in assessing and improving tropical forest management more generally. For more than a decade, guidelines, codes of practice, and 'crite-

ria and indicator' systems have proliferated with many national and international systems vying for legitimacy and influence (Gustafsson 2007).

By 1996, FAO had published a 'model' code of forest harvesting practice, a guideline that includes RIL along with more general aspects of good management (Dykstra and Heinrich 1996). This spurred various tropical countries to develop their own codes and practices (Gustafsson 2007). Various codes and guidelines have been developed and promoted by international bodies such as the International Tropical Timber Organization (ITTO). The latest ITTO guidelines for maintaining biodiversity in production forests were written collaboratively with **IUCN** (IUCN/ITTO 2009). Consumer-based initiatives, most notably the **Forest Stewardship Council (FSC)** (Chapter 17) has sought to provide incentives for implementing good practices.

tors suitable for undersea telephone lines led to high demand for gutta-percha[15], the coagulated sap of *Palaquium gutta* (Sapotaceae), and the near complete depletion of this species[16] from Singapore and later Borneo's forests (Potter 1997). Similarly European demand for resin (used for high-value varnish) from *Agathis* trees led to intense tapping and subsequent death of many *Agathis* trees in Kalimantan and Sulawesi (Indonesia) (Whitmore 1998).

Eager to secure export revenues from timber and cash crops, colonial administrations enforced new forest laws that codified state ownership of forests, regulated resource exploitation, and preserved forests for these purposes. Forest management as advanced by colonial forest departments was driven by the demand for timber and by Western concepts of technical progress (Bryant 1996). Colonial control was particularly widespread in South and Southeast Asia in the 19th century. In Dutch-controlled Java, and in British India and Burma, extensive areas of state-reserved forests were created and commercially important species were protected

(e.g. teak *Tectona grandis* in Java, Burma, and South India, and sal *Shorea robusta* in India) (Guha 1989; Bryant 1992; Peluso 1993). Many forest reserves established by colonial administrations remain forested even today, as forest policies changed little with post-colonial governments.

Diverse forms of cultivation and agroforestry (Chapter 15) had been practised since the Neolithic, but the arrival of Europeans ushered in new systems of 'scientific' land management and resource exploitation for the economic benefit of colonial administrations and companies. Cultural prejudices often prevented European foresters from understanding or even recognizing the benefits and subtleties of indigenous management systems, to the extent that the activities of local indigenous communities were often viewed as inherently destructive. Such prejudices were by no means universal, but have often persisted among post-colonial administrations.

Local forest uses and indigenous forest management systems were consequently often ignored and marginalized. The imposition of colonial administrative systems

---

[15] Still used in dentistry due to its biological inertness and thermal (rather than its electrical) insulation properties.

[16] The tree was felled to extract the latex. Luckily, this species resprouts after cutting and so has been able to persist.

often led to the exclusion of native people from forests, which prompted the growth of local resistance to forest access restrictions that had become integral to colonial forest management. Illegal burning became a destructive symbol of protest against state authority. While colonial interventions sometimes caused hardship for the indigenous populations, experiences differed across territories. In some cases, such as the early days of Brazil, much of the Caribbean, the Dutch control of the Banda Islands (modern-day Indonesia), and the Belgian Congo, slavery and disregard for indigenous lives were the norm. Other regions had a more nuanced experience. Rajah Brooke, the British adventurer King of Sarawak, included local leaders as advisers in his government. More usually European administrations had scarcely any form of local participation in decision making.

Over time the excesses gave way to more constructive and often paternalistic colonial philosophies. For example in the Ugandan Protectorate, traditional forest users gained controlled access to specific resources, and improved health care and elimination of forest-associated disease vectors (such as the simulian flies which carry the disease Onchocerciasis) brought larger populations into contact with some forested regions. Yet colonial administrative legacies shaped post-colonial forest management institutions, and conflict often remains a central feature of forestry management[17] (Peluso 1993;

Gadgil and Guha 1994), despite recent moves towards devolution participation and community management (Poffenberger and McGean 1996).

14.3 Tropical rain forest degradation and loss in the industrial era

The politics of deforestation have often sought scapegoats. Smallholders and shifting cultivators have often shouldered the blame for clearing and burning forests, but often more powerful interests lie behind deforestation (Table 14.1). Certainly the causes of deforestation cannot be easily disaggregated and assigned. Drivers of deforestation encompass many actors making decisions across multiple scales, and transparency is often lacking. The history of tropical deforestation is often a story of power and opportunism.

The pattern, causes, and extent (Figure 14.5) of deforestation differ among the three main tropical regions. In broad terms deforestation has been driven by agricultural expansion, both in response to planned development and settlement programmes as well as more spontaneous colonization of forest frontiers. Ranching has played an important role in Central and South America, while plantation developments have been more important in South and Southeast Asia. Demand

Table 14.1 Actors in Amazonian deforestation

Actor category	Principal affected areas	Activities
Landless migrants	Southern Pará	Annual crops, pasture
Colonists	Transamazon highway, Rondônia	Annual crops (rice, maize, manioc), pasture
Ranchers	Northern Mato Grosso, southern Pará	Pasture
Gold miners	Rondônia, Pará	Mining
Labourers, debt slaves	All ranching areas	Labour for deforestation
Capitalized farmers	Mato Grosse, Santarém, eastern Rondônia	Soybeans
Landgrabbers	Terra do Meio, southern Amazonas, Santarém	Pasture
Sawmill operators, loggers	BR-163 highway, Tailândia (eastern Pará)	Logging, building access roads

From Fearnside 2008.

[17] It can be hard to judge the past without a biased perspective. Some academics have sought to develop 'alternative narratives' which oppose conventional accounts (Leach and Mearns 1996). These typically highlight Western failures to understand and engage with local cultures and institutions, as a cause and legacy of both major injustices and environmental problems in tropical countries.

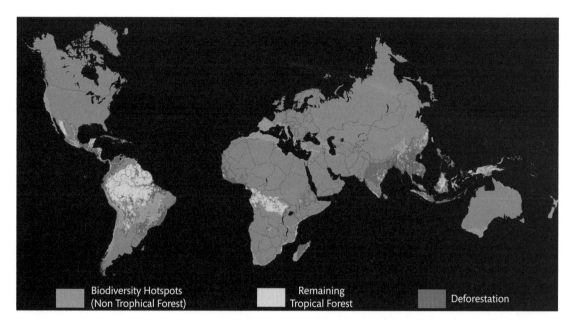

Figure 14.5 Modified Fuller projection map of forest cover and deforestation, derived from Global Land Cover 2000 dataset. (See colour plate 10.) This equal area projection shows the true extent of tropical terrestrial regions relative to global land area.

Map generated and provided by Clinton Jenkins.

for charcoal has meanwhile been a significant problem around various urbanized regions in Africa. Population growth and swidden cultivation are a major problem in Madagascar.

Deforestation trends

Historical patterns of forest loss have been more severe in some regions than in others. Tropical rain forests of the Amazon, much of Southeast Asia and Africa were still largely intact until the 1970s, although this was not the case in Mesoamerica and South Asia, which had already experienced widespread deforestation and forest fragmentation. Since the 1970s there has been a dramatic increase in the rates of deforestation, resulting in large swathes of tropical forest conversion and degradation. In the Brazilian Amazon an area roughly equivalent to the size of Portugal had been deforested in the five centuries of colonial presence up to 1970, but by 2008 the deforested area amounted to 731,000 km², an area greater than that of France, representing around 18.2% of the original forested area of the Brazilian Amazon (data from Fearnside 2005, updated using information derived from

Figure 14.6). Currently it is estimated that some 130,000 km² of the world's forests are lost each year, including 55,000 km² of primary tropical forests, equivalent to almost 0.3% of the biome (Table 14.2) (FAO 2007; Asner et al. 2009), although one should note that the accuracy of forest cover data is debatable (Box 14.7) (Grainger 2008).

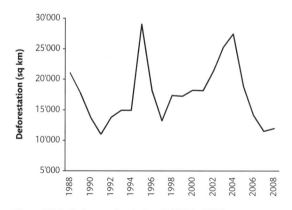

Figure 14.6 Deforestation in Brazil (1988–2008).

Data from Brazilian National Institute of Space Research (INPE).

Table 14.2 Approximate geographic extent of contemporary forest cover, deforestation, and selective logging by region in the humid tropical forest biome. Values are in km², with percentage of biome extent also given.

Region	Total biome extent	Area with 0–50% forest cover, 2005	Area with 50–100% forest cover, 2005	Forest area cleared 2000–2005	Selective logging (2000s)
Africa	2,918,511	1,085,941 (37.2%)	1,832,569 (62.8%)	14,972 (0.5%)	561,153 (19.2%)
Asia/Oceania	7,191,529	5,234,293 (72.8%)	1,957,236 (27.2%)	93,955 (1.3%)	1,777,963 (27.2%)
Central America/ Caribbean	685,840	501,415 (73.1%)	184,425 (26.9%)	9,687 (1.4%)	36,097 (5.3%)
South America	8,826,966	3,194,632 (36.2%)	5,632,334 (63.8%)	156,001 (1.8%)	1,603,166 (18.2%)
TOTAL	**19,622,846**	**10,016,282 (51.0%)**	**9,606,564 (49.0%)**	**274,615 (1.4%)**	**3,978,379 (20.3%)**

From Asner et al. 2009.

Box 14.7   Quantifying forest cover and forest quality, deforestation and degradation

Ambiguities regarding definitions of land cover and reliability of data have made the gathering and comparison of forest area and area change statistics notoriously controversial. Basic figures vary greatly depending on the source, leaving considerable scope for differing interpretations. Key factors in any forest cover classification include extent of canopy cover, spatial resolution, how plantations are addressed, and how temporary changes are considered.

While remote sensing technologies are improving, assessments of forest change remain dependent on existing data which can seldom be used to map reliably past canopy cover, vegetation height, or biomass across regions. Indeed, deforestation was omitted from the Kyoto treaty in 2001, due in part to concerns over how to define and monitor compliance. Definitions will once again be a central practical and political challenge in implementing **REDD** mechanisms (Chapter 17). In the future, as technologies improve, land cover definitions may better reflect satellite-based monitoring abilities (Olander et al. 2008).

The United Nations Food and Agriculture Organization (FAO) has been the main agency responsible for global forest monitoring since 1948, and regularly publishes forest cover statistics in its Global Forest Resources Assessments (latest version FAO 2006). Problems remain as FAO, to a great extent, relies on national reporting of forest statistics. Yet tropical forest countries have seldom been monitored comprehensively or frequently enough to track change with meaningful confidence. Differing definitions of forests, limited capacity to undertake comprehensive forest assessments, and difficulties in differentiating between secondary and primary forest formations add to the complexity of producing meaningful forest cover and deforestation statistics (Grainger 2008).

The largest forest losses over the past decade have been in South America, particularly in the Amazon Basin, where large tracts of rain forest have been cleared for cattle ranches and soybean plantations. In absolute terms, there has been more forest clearance in Brazil than in any other nation. Droughts, such as the severe Amazonian drought in 2005, can greatly exacerbate forest degradation through the spread of anthropogenic fires. Across South America around 31,000 km² of forest are estimated to have been lost each year from 2000 to 2005 (Asner et al. 2009).

Losses in Africa are largely due to subsistence-based agriculture. Nigeria has the dubious honour of having

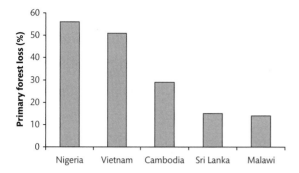

Figure 14.7 Greatest proportional losses of primary forest cover by nation (2000–05).

Adapted from information provided by Rhett Butler, Mongabay.com.

the highest annual deforestation rate of 11.1%, and on this trajectory will soon lose virtually all its remaining forest (FAO 2007) (Figure 14.7). Losses elsewhere have been almost as dramatic. Vietnam lost over 50% of its forests between 2000 and 2005, though losses over tropical Asia as a whole stand at around 1% per year.

Why have some countries maintained a high proportion of forest cover when others have not? There are different answers for each country, but some generalizations too are possible. Some, such as Brunei, Gabon, and Venezuela, are wealthy through oil—labour costs are consequently high and there is little need or opportunity to export timber or cash crops (Wunder and Sunderlin 2004). In other regions, such as Guyana, Surinam, and large areas of Congo and Western New Guinea, inaccessibility due to flooding or terrain, poor soils, or limited security have provided few commercially attractive opportunities for forest exploitation and development.

14.4 Drivers of deforestation

The drivers of deforestation are multiple and intertwined. Among the most important direct drivers are industrial logging, forest conversion to agriculture and plantations (Figure 14.8), and forest fires. Industrial logging is often at the forefront of deforestation, as logging roads increase access for conversion to agriculture. However, it is the underlying social, economic, and political drivers of deforestation activities that need to be understood to gain a proper appreciation of land cover change in the tropics. The basic issue is the opportunity cost of retaining forest. At the national level major influences include resource privatization and other tenurial changes, inflation-driven land speculation, tax breaks, subsidies and other financial incentives, and long-term government agendas striving for rapid economic growth, industrialization, and development (Chapter 17). At supra-national scales they include the creation of markets through globalization and fluctuations in regional and global economic health.

Across tropical regions most modern deforestation involves conversion to agriculture (Box 14.8). Macroeconomics play a significant role. As prices of agricultural commodities increase, so does deforestation as land is cleared to plant more crops (Kaimowitz and Angelsen 1998; Fearnside 2008). The increases in agricultural prices worldwide in 2008 have driven sharp increases in deforestation in the Brazilian Amazon, especially in Mato Grosso where soybean crops have been expanding. Deforestation rates in the Brazilian Amazon have also been positively correlated with beef prices[18], to which can be attributed the Amazonian deforestation peak in 1995 (Fearnside 2008; Karsenty 2008). Recent global demand for biofuels, including palm oil (Figure 14.9), soya, and sugarcane, has also encouraged corporations to exploit large blocks of forest for logging and conversion to plantation agriculture (Koh 2007; Fargione et al. 2008; Koh and Ghazoul 2008).

Economic recessions generally lead to a slow-down in deforestation rates as global demand for products wanes, and individuals, corporations, and governments lack the funds to expand agricultural investments or infrastructure. Declining deforestation rates in Brazil from 1987 to 1991 corresponded to national economic recession (Fearnside 2005). The 2005–7 decline in deforestation for pasture and soybeans coincided with

[18] Cattle, unlike other agricultural commodities in the Amazon, are both a product and the means of production. When prices rise in response to market demand, ranchers tend to retain their stocks to breed more cattle, and so supply fails to keep pace with demand. The result is more exaggerated and prolonged price rises that drive deforestation for ranching over subsequent years (Fearnside 2008).

A

B

Figure 14.8 Amazonian deforestation for pasture (A), and a smaller swidden for cash crops in an Arawak Amerindian village in Surinam (B).

Photos by Douglas Sheil.

Box 14.8   Forest transitions

Large-scale trends in a region's forest cover are addressed by the theories of 'forest transition' (Rudel et al. 2005). Forest transitions occur when prolonged episodes of decline in forest cover are replaced by increases in forest cover. The theory links such changes to consistent stages in economic development. A forested nation generally loses forest in the early stages of development, as more and more land is converted to more intensive uses.

Eventually when higher income levels are achieved, forest cover is protected, restored (some as forest plantations), or regrows on abandoned land following urbanization, and cover increases once more. The two key driving forces for the transition are increased urban employment opportunities and the scarcity of forest goods and services, which encourages investment in forest planting, protection, and management.

Figure 14.9 An oil palm plantation established on recently forested land near Pembuang in Central Kalimantan. This estate is around 1400 km².

Photo by Matthew Struebig.

declining international prices for these commodities and a worsening exchange rate of the Brazilian real against the US dollar which impacted on Brazilian exports (Fearnside 2008). However, the crash of the world financial markets in 2008 drove gold[19] prices to record levels, which may place renewed pressure on Amazonian gold-mining activities.

Population growth continues to be an underlying driver that interacts with rural economic policies, poverty, and tenure insecurity, while population and economic growth in urban areas drive continuing demands for for-

---

[19] Gold is a typical 'safe haven' for investors during times of financial crisis.

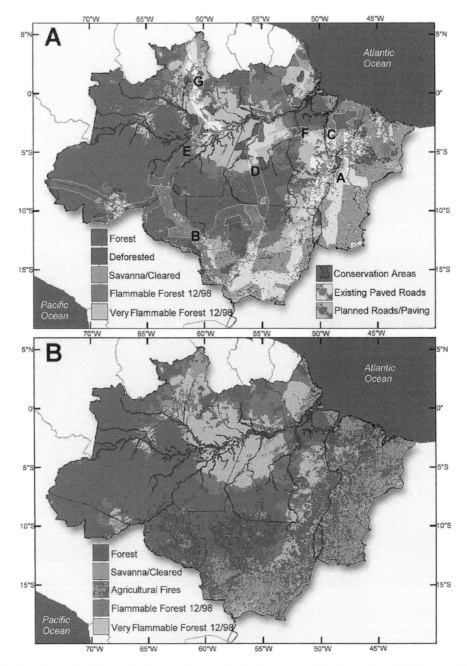

Figure 14.10 An estimated 1 million km² of Amazonian forest in Brazil was classed by Nepstad et al. 2001 as 'very flammable' and a further 0.5 million km² as 'flammable' owing to soil moisture depletion following the 1997/1998 El Niño event. Existing or planned roads (Panel A) increase the likely penetration of people, and therefore ignition sources, which are currently restricted to the southern and eastern boundaries of the Amazon (Panel B).

est resources. Increasing demand for energy by a growing population and a developing economy impacts on forests by the construction of dams for hydropower: Brazil's Altamira dam will flood 6,140 km², and is only one of five planned dams along the Xingu River (Fearnside 2006).

Government-sponsored road building, settlements, and large-scale agrobusinesses acting in synergy have been behind most recent deforestation. Forests that remain inaccessible (e.g. the central Congo Basin, much of Papua, and north-western Amazon) still contain extensive forests. As new roads are built (by logging companies or through government development policies), access to migrants and markets accelerates the clearing of forests in remote areas (Nepstad et al. 2001) (Figure 14.10). Indeed, road density is considered to be the best predictor of settlement and encroachment at the forest frontier (Kaimowitz and Angelsen 1998), and more than two-thirds of Amazonian deforestation has occurred within 50 km of paved highways (Alves 2000; Nepstad et al. 2001). Illegal unpaved roads, or *viscinais*,

built by illegal loggers also provide pathways for the expansion of deforestation. These are particularly problematic as they are not under government control, and often penetrate through indigenous territories, government land, and ecological reserves. It is reported that there are more than 170,000 km miles of *viscinais* in the Amazon region (Phillips 2007).

14.4.1 Deforestation history in Brazil

The history of deforestation in Brazil is instructive as it is broadly representative of other Central and South American tropical regions, but given that two-thirds of the Amazon and all the tropical Atlantic forest lie within Brazil, it is also the main story.

In Brazil, migration has been the important driver of deforestation, not just in terms of numbers of colonists, but also in terms of the movement of investments (Fearnside 2008). Deforestation of the Atlantic forests was a direct result of settlement since the mid 16th century, and the

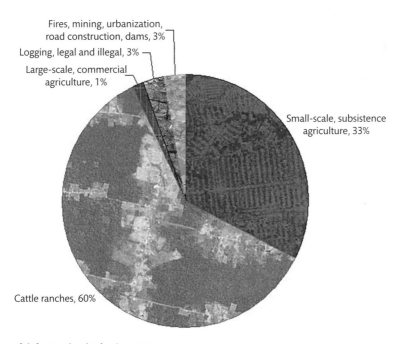

Causes of Deforestation in the Amazon, 2000–2005

Fires, mining, urbanization, road construction, dams, 3%

Logging, legal and illegal, 3%

Large-scale, commercial agriculture, 1%

Small-scale, subsistence agriculture, 33%

Cattle ranches, 60%

Figure 14.11 Causes of deforestation in the Amazon.

Image from Rhett A. Butler.

expansion of sugar, cotton, and coffee plantations as well as cattle ranching. This was accelerated with Brazil's post World War II development objectives of rapid industrialization and economic growth. In the 1950s and 1960s development was focused in the hinterland of the big cities of Rio de Janeiro, Sao Paulo, and Belo Horizonte, causing further losses and fragmentation of the remaining Atlantic forest.

The assault on the Amazon (Figure 14.11) began later, and arose out of similar objectives of national economic development, but was further driven by a political undercurrent to secure government control over land, resources, and political boundaries against what were perceived to be internal and external threats. In a pattern that was to be repeated under different programmes in successive decades, the Brazilian government implemented a strategy (Operation Amazonia 1964–7) to develop the Brazilian Amazon region through a road-building programme to promote colonization (particularly of the states of Rondônia and Acre) and to facilitate exploitation of mineral resources. Road building opened up the forest and connected it to urban markets. Small holdings and cattle ranches became established along the new highways, often supported by an array of financial incentives including generous tax credits, highly favourable loan arrangements and subsidies. Consequently land was cleared primarily for speculation, which gave considerable return over the short term for both smallholders and larger enterprises alike (Mahar 1989). Without these credits it is unlikely that cattle ranching, the main land use, would have been profitable (Downing et al. 1992). While immigrant smallholders flooded into the region, they were quickly displaced by large cattle ranches[20] and soybean plantations with wealthy owners who readily exploited the financial incentives being offered (Hecht et al. 1988; Hecht 1993).

Government programmes have not been effectively supported by coherent and organized land management and planning authorities. Such authorities are decentralized, resulting in confusion arising from overlapping land claims that cannot be easily verified owing to distributed and incomplete document repositories. This situation makes it easier for illegal 'landgrabbers' to clear forest for timber and other resources before selling the land to others either with or without legal tenure (Fearnside 2007).

14.4.2   Deforestation history in Asia

Government-sponsored schemes to promote development of forested areas have been prominent in Southeast Asia, particularly in Indonesia and Malaysia. Transmigration of people from densely populated inner islands such as Java, Bali, and Lombok to the outlying regions of Sumatra, Kalimantan, and Irian Jaya (now Indonesian Papua) has been underway since the beginning of the 20th century, although this became a pillar of government policy from the mid 1960s. From the early 1970s through to 2000, the Indonesian Transmigration Project resettled around 5 million people (Kartawinata et al. 1981; Secrett 1986; Fearnside 1997). This programme was accompanied by additional spontaneous and unrecorded migrations of an estimated 10–15 million. It is very difficult to estimate the overall impacts of these migrations on forests, particularly as many migrants settled on already drained wetland areas (Fearnside 1997).

Transmigration project planners had intended that households would be supported through farming (agroforestry and annual crops), but yields were often too low. A lack of farming experience on such relatively poor soils, inefficient supporting services, and labour shortages limited productivity. Soil erosion often resulted from inappropriate land clearing for transmigration development. The few transmigration projects that have been successful have attracted more migrants, resulting in more deforestation.

Transmigration has not been uniform in scale, location, or objectives. Some schemes have sought to promote self-sufficient independent farming by small holders, while others provide houses and other infrastructure for transmigrants who are then employed on private forestry plantations. Many transmigration projects associated with plantation development were supposed to be developed on *Imperata cylindrica* grassland areas, but plantations that exceed 200,000 ha often involved

---

[20]   The average size of which was 24,000 ha. A small proportion of people own a high proportion of the land through much of the Amazon region.

forest clearing as few grassland areas were of this size. As in Brazil, Indonesian government policies contributed to deforestation through generous loan agreements for plantation establishment, and transmigration schemes were viewed as an opportunity for securing cheap labour (Fearnside 1997).

Rates of forest clearing in Indonesia at the peak of transmigration during the 1980s have been estimated at around 500,000 ha yr^{-1} by smallholders, including transmigrants, and 250,000 ha yr^{-1} by development projects that include plantation projects using transmigrant labour (World Bank 1990). It remains uncertain how much old-growth forest was lost in this period.

Aside from transmigrations, Indonesia, Malaysia, and other Southeast Asian countries have sought rapid economic development. Malaysia particularly has pursued a policy of expanding plantations of valuable commodities, including rubber and oil palm. In consequence of this, forest cover on the Malaysian peninsula dropped from 73% in the early 1950s to 51% by 1982, and stands at around 44% in 2005 (FAO 2007). Vietnam's forests became subject to heavy degradation by chemical agents during the protracted war, and now much of this land too has been converted to coffee and other cash crops.

14.4.3 Direct causes of forest loss and degradation

Timber extraction and its consequences

The large-scale commercial timber industry is a major income producer and employer in several tropical countries, particularly in Asia (Box 14.9). In Brazil, small-scale sawmill operators accounted for much of the

---

Box 14.9   The production and trade of tropical hardwood

In 2006 the international tropical timber trade from producer countries had a value of US$ 9.7 billion, of which US$ 2.1 billion was from unprocessed logs and US$ 7.6 billion from plywood, veneer, and sawnwood. Tropical timber processing and subsequent use support an estimated 200 million people (ITTO 2009).

Production of tropical hardwood logs in ITTO[3] producer countries totalled 125 million m^3 in 2007, which is just under 10% of the global total production (ITTO 2009). Of this, 59% came from the Asia–Pacific region, 26% from the American Tropics, and 14% from Africa. The top four tropical log-producing countries (Brazil, Malaysia, India, and Indonesia, ranked by 2007 production) together comprised 68% of total production. Production of tropical timber declined 4% from 2003 to 2007, but imports worldwide declined by 13% over the same period. Europe, Japan, Korea, and Taiwan are major importers of tropical hardwood, but since 2003 all have greatly reduced their dependency. Europe reduced imports of tropical hardwood logs from 2003 to 2007 by 25%. Japan reduced imports over the same period by 35%[4], Korea by 45%, and Taiwan by 36% (source data from ITTO 2009). China's imports also declined over this period by almost 6%, but this has to be set against an 800% increase in imports between 1993 and 2003 due to China's growing economy, a continuing ban on domestic harvesting, and a zero tariff on log imports. Due to the disproportionate weight of China's imports, the decline in imports for the Asia–Pacific region as a whole is only 15%. Imports have fallen in recent years, due to supply constraints and increasing reliance on Russian timber. Despite this decrease, China's imports dominate the tropical log market, constituting in 2007 over 53% of all ITTO imports[5]. Note that China re-exports much of this wood in processed form.

The great majority (90% in 2007) of tropical hardwood logs are, however, consumed within producer countries (source data from ITTO 2009) where there is limited demand for certified timber (Ghazoul 2001).

[3] International Tropical Timber Organization.

[4] Japan's imports have been declining for the past 15 years and are around 85% less than in 1994.

[5] And 70% of consumer country imports.

destruction of the Brazilian Atlantic forests, and have since moved on to the Amazon where logging affects an area as large as that which is deforested (Asner et al. 2009). Globally it is estimated that more than 3.9 million km^2 of humid tropical forests (or about 20.3%) have been allocated to selective timber harvesting (Asner et al. 2009).

Timber extraction does not cause deforestation on its own, as it is a selective process in which only larger stems of a few species have commercial interest—a logged forest is still a forest. Nevertheless, logging causes damage and increases vulnerability to fire and subsequent encroachment, and government policies often encourage the further conversion of logged forest to other land uses. In Indonesia, if a production forest's standing volume of timber falls below a threshold of 20 m^3 ha^{-1} it is reclassified as conversion forest that can be converted to plantations. This, together with government subsidies for plantation establishment, has encouraged companies first to log forests heavily and then apply for contracts to establish plantations on the same land (A. Hadi Pramono, quoted by Fearnside et al.

2007). In the Amazon, logged forest is four times more likely to be cleared than unlogged forest (Asner et al. 2006).

Careless timber extraction causes considerable damage to non-commercial species as well as advance regeneration and soil structure. Commercial species typically comprise 2–10% of the standing volume in Amazonia but poor logging practice destroys as much as 60%, and the forest soils are additionally compacted and eroded (Uhl and Vieira 1989). In most cases forests would eventually recover from even intensive log removal were it not for logging roads that provide access to others who engage in land clearance. Large logging operations have declining significance in much of Southeast Asia owing to the ever diminishing extent of forest, although former natural forest areas are being replaced by large-scale oil palm and pulp wood plantations (Koh 2007).

Illegal logging refers to activities related to timber harvesting that transgress international, national, and subnational laws[21] (see also Box 14.10). Logging within strictly protected areas is clearly illegal, but logging within

---

Box 14.10   Conflict timber

Timber sales have supported, and in some cases continue to support, armed conflicts in Asia and Africa. 'Conflict timber' refers to the timber used for financing these conflicts.

There is a strong association between conflict timber and poor and inequitable governance (Glew and Hudson 2007). Corruption makes it difficult to regulate forest resource use, leading to unsustainable exploitation[6]. Where government control is strong, as in Burma, logging firms collaborate with governments

to harvest timber equipment, and transport infrastructure can often only be guaranteed by the state military. Thus in many nations sections of the military are implicated in control of illegal harvesting.

In some cases rebel movements such as the Khmer Rouge insurgents in Cambodia are largely sustained through conflict timber. The Khmer Rouge were forced to enter peace negotiations in 1996 when the Thai and Cambodian governments cooperated to prohibit log exports (Global Witness 1996).

[6] Heightened insecurity can also reduce timber harvesting (as during the civil war in the Democratic Republic of Congo ongoing from 1997).

---

[21] Note that these statutory laws often neglect traditional local 'ownership' and use rights, or may themselves have unclear 'legality' owing to contradictions among laws, as in Indonesia where the Soeharto military government claimed all forested land as belonging to the state in a manner which was not only unjust but illegal by its own laws (Lynch and Harwell 2002).

Table 14.3 Summary of existing estimates for some ITTO producer countries.

Country	% wood harvested illegally	Source
Bolivia	80	Contreras-Hermosilla (2001)
Brazil (Amazon)	85	Greenpeace (2001)
Cambodia	90	World Rain Forest Movement and Forest Monitor (1998)
Cameroon	50	Global Forest Watch Cameroon (2000)
Colombia	42	Contreras-Hermosilla (2001)
Ghana	34	Glastra (1995)
Indonesia	51*	Scotland (2000)
Myanmar	80	Brunner (1998)

Exploitation of forests can contribute to and even support conflicts including civil wars (Box 15.4). Recent examples include civil wars in Cambodia, Sierra Leone, Ivory Coast, Democratic Republic of Congo (DRC), Burma, and Liberia.

allocated concessions could also be considered illegal if it targets restricted species or trees below the allowable size limits, or if it exceeds allowable quotas. High market demand, corruption at many levels, and difficulties in law enforcement allow illegal logging-related activities to proceed. This issue has become a major concern in recent years, with the realization that in some regions it accounts for most of the marketed timber, although exact figures are very difficult to generate (Table 14.3).

Mining

Gold, silver, copper, aluminium, and other precious metals as well as coal for fuel, are mined from rain forests around the world, often by way of large-scale open-pit mining operations (Figure 14.12) which are extremely destructive of forests and underlying soils. Damage can also extend much further through the effects of sedimentation and release of pollutants. Mineral deposits are mined by large-scale enterprises as well as small-scale miners, but in both cases miners destroy river banks (using explosives and high-power water cannons), clear floodplain forests, and use heavy machinery to unearth deposits. Mercury is used in the extraction of gold, and much of what is used contaminates rivers and streams, and ends up in the food chain where it can have toxic effects particularly on top predators and humans (Fostier et al. 2000). Cyanide is also often used to separate gold from gravel deposits, and spills from waste-holding pools can cause widespread poisoning.

The construction of roads associated with mining operations makes remote forest areas accessible to settlers, land speculators, and small-scale miners. One consequence of this is that miners and settlers often come into conflict with indigenous populations, agriculturalists, or other land users (Hammond et al. 2007).

Fire

Natural rain forest fire cycles may be in the order of hundreds or thousands of years, but the combination of widespread forest fragmentation, associated edge effects, and the penetration of forested areas by humans has increased the exposure and vulnerability of rain forests to fires (Cochrane et al. 1999). This is exacerbated by climatic changes, notably drought-causing events (Kinnaird and O'Brien 1998; Taylor et al. 1999). Rain forest fires have in recent decades caused severe pollution problems as well as biodiversity losses and led to permanent forest conversion. Large smoke episodes became international news during the El Niño drought years of 1997/1998 when fires raged across rain forests in Brazil and Indonesia. In Brazil during this period at least 20,000 km^2 of forest were affected by understorey fires (Nepstad et al. 2001). In Borneo, Indonesia, an estimated 50,000 km^2 of forest was burned during the same period, leading to severe smog as far away as Kuala Lumpur and Singapore (about 1,000 km away). Associated short-term human health impacts included irritations of the respiratory tract, skin and eyes, bronchitis, conjunctivitis, and increased asthma attacks. Air traffic in the region was cancelled for days on end, and ships collided at sea.

Figure 14.12 A gold mine in the Amazon rain forest.

Photo by Rhett A. Butler.

In modern times human impact has increased the frequency and extent of rain forest fires, but there is much evidence that forest fires were also associated with earlier human occupation of rain forests. The cause, frequency, and extent of pre-human natural fires remain uncertain; intense droughts provided fuel, and fires were then ignited, perhaps by lightning, volcanic activity, swamp gas, microbial fermentation, or any sparks or friction. Fires, man-made or natural, are now seen as a plausible explanation for vegetation patterns in apparently pristine rain forests. For example in French Guyana, soil charcoal and pulses of pioneer pollen in cores suggest

that past fires may explain the modern distribution of some peculiar plant formations such as liana forests, palm forests, and bamboo thickets (Charles-Dominique et al. 1998).

The scale and extent of forest fires have increased over the last century as people and policies have generated increasing exploitation and penetration of primary forest. Forests far from human activities seldom burn even if they are highly susceptible to fires, while close to settlements, roads, and railway lines, fire frequency is often high. Forests can recover after a single burn (it is a mistake to equate an area of tropical forest burned with an area of 'deforestation'), but repeated burning, without decades of recovery time, can ultimately convert forest to grassland.

The underlying causes for forest fires are diverse and often contested (Dennis et al. 2005). Many blame the deliberate actions of small-scale farmers and large-scale companies seeking to open up land for agriculture or plantations. In 1994, the Indonesian government accused smallholder farmers of being responsible for 85% of the 5 million hectares burned in Borneo. In the same year environmental organizations blamed forest concessionaires and plantation owners as being primarily responsible. Information from satellite imagery indicates that large-scale land clearing for pulpwood and oil palm plantations was the major cause of Indonesian fires in 1997 and 1998.

In view of their carbon storage and hydrological functions, peatland forest fires raise particular concerns. Peatland forest fires have become increasingly widespread, with perhaps about 2.1 million ha burning in the 1997/1998 El Niño event in Indonesia. They generate thick smoke and haze and result in major carbon emissions (Page et al. 2002). Estimates suggest that Indonesian forest and peat fires in 1997 released as much as 0.81 and 2.57 Gt of carbon to the atmosphere, or 13–40% of the annual contribution attributed to global use of fossil fuels (Page et al. 2002).

Rain forest fires may have wider biodiversity impacts extending even to coral reefs. Oceanic primary productivity is limited by the availability of iron, and atmospheric fallout of iron from the 1997 Indonesian wildfires[22] may have been the main cause of exceptionally dense and extensive oceanic algal blooms in the region, which in turn killed or damaged 59% of the corals in the western and central Indian Ocean (Abram et al. 2003).

14.5 Regional and global atmospheric change

The world is undergoing climate change at unprecedented rates (Solomona et al. 2009). Temperatures are rising and global weather patterns are changing. Patterns of seasonality and the intensity and frequency of more extreme

---

Box 14.11 Future rainfall scenarios in the tropics

Rainfall is critical to rain forests, which are generally very susceptible to drying (Chapter 10). Future changes in rainfall may therefore have profound implications for rain forest cover, composition, and distribution. Climate models agree that temperatures will increase over most areas of the globe in the coming decades, but there is considerable uncertainty and regional variability associated with projected precipitation patterns. Here we summarize regional IPCC precipitation projections (Christensen et al. 2007) for which there is 66–90% confidence (*likely*) or more than 90% confidence (*very likely*).

Africa: Rainfall in tropical West Africa is *likely* to decrease, but is *likely* to increase in East Africa. Future rainfall patterns along the Guinean Coast remain very uncertain.

Asia: Precipitation is *likely* to increase in South Asia and Southeast Asia. The frequency of intense precipitation events in South Asia and Southeast Asia is *very likely* to increase, as are tropical cyclones

America: Annual precipitation is *likely* to decrease in most of Central America, although large local variability is expected in mountainous areas. Annual and seasonal rainfall patterns remain uncertain over the Amazon forest, though marked regional variability is expected with rainfall projected to increase in Ecuador and northern Peru, and decrease in the eastern Amazon.

Australia: Changes in rainfall in northern and central Australia are too uncertain to make meaningful projections.

---

[22] An estimated 11,000 metric tonnes of iron were released from Sumatran wildfires in 1997 (Abram et al. 2003).

events, such as hurricanes and droughts, are being altered. There has been a 10% increase in atmospheric CO_2 concentration over the past four decades (Prentice 1998). Mean global temperature has increased by about 0.5°C over the same period (Malhi and Wright 2004). Temperatures in tropical rain forest regions are projected to increase by between 2° and 5°C by the end of the 21st century (Cramer et al. 2004). Rainfall projections are less certain and more variable geographically (Box 14.11), and yet probably most crucial for anticipating climate change impacts on rain forests (Christensen et al. 2007).

At smaller regional scales, changes in atmospheric conditions and climate are affected by regional land-use patterns. Nutrient inputs from atmospheric deposition of nitrogen pollution have increased within tropical regions, though the impact on growth is complicated and remains uncertain (Box 14.12). Deforestation and forest burning change the regional and landscape climatic conditions, which may exacerbate global climatic trends. Thus recent temperature increases since 1960 over tropical rain forest regions are particularly marked (Figure 14.13), as are precipitation declines (Figure 14.14), and these can probably be attributed to a combination of regional and global scale effects brought about by land cover and global atmospheric changes (Malhi and Wright 2004).

Box 14.12 Nitrogen deposition and carbon balance in tropical forests

Through the 20th century, terrestrial (and oceanic) ecosystems have experienced a three- to five-fold increase in nitrogen inputs, mainly in the form of nitrogen oxide (NO) derived from industrialization and fossil-fuel burning, and ammonia (NH_3) from the intensification of agriculture (IPCC 2007). These emissions and their subsequent deposition affect plant growth, and therefore carbon sequestration by forest ecosystems. Projections are, however, complicated by uncertainties that include a limited understanding of the interaction between atmospheric nitrogen deposition and ecosystem carbon dynamics (Reich et al. 2006; Reay et al. 2008).

Nitrogen deposition is expected to have a positive impact on forest growth and biomass where deposition is high[7] and where nitrogen limits growth. Yet it remains uncertain as to which tropical forests are limited primarily by nitrogen rather than phosphorus. Warmer and wetter tropical climates enhance nitrogen mineralization and plants' nitrogen use efficiency (Lloyd and Taylor 1994; Schlesinger and Andrews 2000), and ancient tropical soils have become phosphorus-depleted by age (Chapter 9) (Walker and Syers 1976). For these reasons phosphorus is expected to exert more significant limits

on growth in equatorial regions than nitrogen. Supporting evidence is, however, scarce. Nitrogen limitation of net primary production has been reported in montane and secondary tropical forests (it is well established that nitrogen availability declines after forest loss), but information from primary lowland forests is limited and equivocal (LeBauer and Treseder 2008).

The soil carbon sink depends on the balance between N-induced increases in carbon inputs (through increased plant growth) to the soil and the influence of increased nitrogen on carbon losses through soil decomposition, respiration, and mineralization processes. Higher C/N ratios tend to be associated with lower decomposition rates, so decomposition rates might be expected to increase if C/N ratios decrease with increasing nitrogen deposition.

The world's oceans, forests, and soils play a crucial role in buffering anthropogenic CO_2 emissions. While N-deposition inputs have potential for enhancing such carbon sinks, it is the protection of existing tropical forests from deforestation that is likely to provide more reliable and substantial climate mitigation.

[7] Mainly in the intensively farmed regions of South and Southeast Asia where nitrogen deposition is much higher than in South America or Africa.

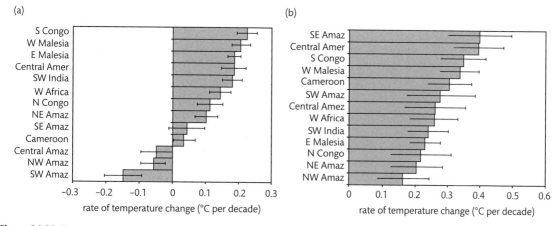

Figure 14.13 Temperature trends in tropical rain forest sub-regions 1960–98 (a) and 1976–98 (b).

From Malhi and Wright 2004.

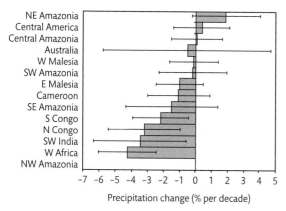

Figure 14.14 Precipitation trends in tropical rain forest regions 1960–98.

From Malhi and Wright 2004.

What are the implications for rain forests? Temperature increases will be greatest at high latitudes and high elevations. This is scant comfort given the overall changes currently predicted. Even moderate scenarios imply that by 2100 three-quarters of current tropical forest regions will be subject to mean annual temperatures greater than found in any region supporting closed-canopy forest in the middle of the last century (Wright et al. 2009). While most models agree such basic global patterns and trends, predictions for specific sites remain uncertain. In any case local climate changes reflect not just global trends but local influences such as land-cover changes and atmospheric aerosols (Bonan 2008) (Chapter 9).

The effects of climate change on tropical rain forests and their species will reflect the physiological ability of organisms to survive and reproduce under the new climatic and atmospheric conditions, to adapt to any changes in the ecological context, and to withstand additional impacts. We have little knowledge of the dependencies of most species, so wider ecological assessments remain largely speculative. Nonetheless, they may be severe (Tewksbury et al. 2008). The cues and requirements for plant phenology patterns remain poorly understood but appear vulnerable to multiple factors including temperature, moisture, sunshine, and seasonal cycles (Guenter et al. 2008). Already there are claims of forest plants now failing to set viable seed (Chapman et al. 2005). The current decline of tropical amphibians has also been attributed, at least in part, to climate change (Chapter 16).

The forest habitat itself is vulnerable. Increased temperature interferes with effective photosynthesis and increases respiration costs, and there is some evidence that during unusually hot weather tropical forest trees fail to grow (Clark 2004). As weather changes so will disturbance regimes (Chapter 11). In continental interiors, disruptions of forest-dependent processes that carry moisture inland can cause regional droughts (Makarieva and Gorshkov 2007; Sheil and Murdiyarso 2009). High temperatures, droughts, and fire will benefit C4 grasses (Sage and McKown 2006) (Chapter 10) and in some circumstances may even allow a switch to non-forest vegetation.

Migrations

To persist, species must either adapt or move if they are to stay within tolerable conditions. Cooler refuges will be more readily accessible to species that occur near or on mountains (Colwell et al. 2008), and there is already evidence of species moving higher on tropical mountains. It is good news that many species with small distributions are associated with mountains (Ohlemüller et al. 2008), but these species remain vulnerable as habitat generally shrinks with altitude—and those on mountain tops have nowhere higher to go (Colwell et al. 2008).

In the flatter lowlands, it means larger moves to reach higher elevations or latitudes. The distances between current locations and the nearest temperature refuges in a moderate 2100 scenario projection suggest that over half the equatorial areas are more than 400 km^2 from a potential refuge, and significant areas of the Amazon and Congo Basins and West Africa are over 1,000 km from a refuge (Wright et al. 2009).

In the longer term forests will bounce back as they have always done in the past (Chapter 6). Global forecasts imply expansion of the climatically tropical zone, suggesting that the extent of future tropical forests may even be greater than at present (Seidel et al. 2008).

Carbon sequestration trends

Tropical forest responses to climatic and atmospheric changes are critical to the outcome of future atmospheric CO_2 levels and climate change. Conventional wisdom suggests that mature forests are neither sources nor sinks for atmospheric carbon—sequestration from growth is balanced by emissions from decomposition. Long-term measurements (spanning 40 years) of Amazonian and African tropical forest net biomass accumulation[23] undermine this view (Lewis 2006; Phillips et al. 2008; Lewis et al. 2009). African rain forests are estimated to have been absorbing 340 million tonnes of carbon per year in recent decades, an amount that is roughly equivalent to the total emissions from

deforestation in Africa (Lewis et al. 2009). What has caused this increase in biomass, and is this trend likely to continue?

An increase in net primary productivity (NPP) is expected in response to higher temperature, which increases soil nutrient availability[24], and higher atmospheric CO_2, which reduces photorespiration by increasing CO_2:O_2 ratios (Lewis et al. 2004). Some Amazon Basin models suggest that increasing atmospheric CO_2 will initially drive increasing plant biomass accumulation, signifying an increased forest carbon sink[25], but as growth rates saturate, this sink will switch to a carbon source due to higher respiration rates and mortalities at higher temperatures (Phillips et al. 2008). More frequent and severe periodic droughts and a general drying of the basin will exacerbate this shift (Phillips et al. 2009). Tree species will differ in the degree of their responses to climatic changes: fast growing, low wood density species are likely to capitalize quickly on opportunities afforded by temperature-induced nutrient release and higher atmospheric CO_2, but conversely will often be more vulnerable to drought.

In addition to these gradual changes, climate change is also predicted to increase the frequency and severity of rare but large-scale disturbance events such as windstorms, heavy rainfall events, droughts, and associated fires. Such events can cause considerable mortality, which could reduce biomass accumulation and carbon storage functions of tropical forests, but their relative impact compared to frequent but individual tree mortality events over a decadal period is, at least under current regimes relatively insignificant (Nelson et al. 1994).

Predictions of rain forest die-back

Although projections remain very uncertain, some scenarios of global climate anticipate alarming collapse of the Amazonian forest and its transformation to savannah or even bare soil (Cowling et al. 2004; Cox et al. 2004).

---

[23] A trend punctuated by short-term declines in NPP due to droughts (Phillips et al. 2009).

[24] On the other hand, higher temperature is also expected to increase photorespiration and respiration costs, thereby reducing NPP.

[25] Estimated to be at least 0.49±0.18 Pg C per year (Phillips et al. 2008), rising to 0.79±0.29 Pg C per year with the inclusion of other biomass and necromass components. This is approximately equal to the carbon emissions to the atmosphere by Amazon deforestation.

Table 14.4 Mean climate and land storage from an interactive climate–carbon cycle projection for the decades of the 1990s and 2090s.

	1990s		2090s	
	Globe	Amazonia	Globe	Amazonia
Screen Temperature (K)	288.0	301.4	292.1	310.6
Precipitation (mm day^{-1})	2.96	4.56	3.05	1.64
Evaporation (mm day^{-1})	2.96	3.14	3.05	1.20
Vegetation Carbon (GtC)	539.5	45.6	549.7	10.1
Soil Carbon (GtC)	1204.5	19.8	1049.2	5.5

From Cox et al. 2004.

While these are not mainstream views, they do warrant attention in that the consequences are catastrophic, and the scenarios are at least plausible. These fully interactive climate-carbon cycle projections predict severe warming, by almost 10°C, and a 64% decline in rainfall in western Amazonia by 2100 (Table 14.4). Vegetation feedbacks promote such warming, principally through CO_2-induced stomatal closure which reduces transpiration and amplifies surface warming (Betts et al. 2004). Under this scenario, climatic changes are predicted to reduce net primary productivity, due to soil moisture constraining photosynthesis and increased respiration costs, which in turn cause tree die-back leading to a 78% loss in plant carbon and a 72% soil carbon loss by 2100[26] (Cox et al. 2004).

Evidence for Amazonian drying is, however, mixed: precipitation has declined in some Amazonian regions but not others, where if anything there has been a slight increase in rainfall (Figure 14.13 above). On the other hand, these projections do not include forest fragmentation and fire impacts, which may accelerate Amazonian drying and dieback.

African rain forests, already the driest of rain forest regions, have become progressively drier in recent decades. Their spatial extent may therefore be particularly vulnerable to additional drying, a view that is supported by paleo-records which show that large areas of African tropical rain forests were covered by savannah in the last ice age (Morley 2000) (see also Chapter 7).

Many uncertainties remain, particularly with respect to vegetation-climate feedbacks, and the probabilities of catastrophic change in the Amazon, and extensive contraction of African rain forests, remain matters of conjecture. Nevertheless, the scale and potential rapidity[27] of change, and its immense implications for biodiversity human welfare, surely warrant serious attention and response.

14.6 Conclusion

We shall never be able to experience directly what forests were like before the advent of humans. This has implications for how we perceive nature and for what purpose we will manage and protect tropical forests in the future. Accumulating evidence for prehistoric and historic deforestation and forest degradation does not justify more recent episodes of forest clearing, but it does clarify that even significantly modified forests can maintain significant conservation and other values.

Humans have marked tropical rain forests in many ways. Some, such as megafaunal extinctions, are permanent, while others, such as timber harvesting, despite their apparent violence to the vegetation, may, given

---

[26] Although CO_2-fertilization of photosynthesis is expected to maintain rain forest cover in the first half of the 21st century, with warming and drying impacts leading to rapid biomass reductions thereafter (Cox et al. 2004).
[27] Tropical forests can switch from carbon source to sinks within one year under relatively minor changes in climate (Tian et al. 1998; Clark et al. 2003).

time, leave little lasting impact if only they are permitted to recover. Many common processes that have eaten away at tropical forests, such as agriculture, plantations, and fire, are now joined by new concerns, including climate and atmospheric change, invasive species, and new crops driven by new economic imperatives. Another major legacy of our own generation will be considered further in Chapter 17—our ability (or inability) to achieve effective conservation of remaining tropical rain forests.

CHAPTER 15

People of the forest: livelihoods and welfare

'The forest was our source of wealth.'

Translated from Ziterera Raphael, Nyamabare Parish, Uganda (Namara et al. 2000)

Our closest relatives—the great apes—are tropical forest creatures. Humans, in contrast, now occupy virtually all the Earth's biomes. Regardless of where we now live, tropical forest can still rouse strong feelings. It is our ancestral home and refuge, but also a current home and refuge for many. It is a place of danger and disease, but also of cures and remedies. It is a source of ancient wisdom and knowledge, but also of marginalized peoples. It offers land, wealth, and an escape from poverty, but is also home to many of the world's poorest people.

People's relationship with rain forests depends on context and experience. Many live in rain forests, and many more farm their periphery. Many depend on forest products, some of which have been widely traded for millennia. Many may never see a rain forest, and yet daily consume rain forest products (coffee, chocolate, cola, soya, and palm oil).

Anybody addressing the fate of tropical rain forests must confront people's needs and perceptions. Needs and perceptions reflect concrete concerns such as food, health, and income as well as less tangible factors such as histories, cultures, and individual hopes and fears.

People are integral to many rain forest landscapes. If we are to achieve equitable and acceptable conservation and land use outcomes, then we must appreciate and understand the cultural richness, value systems, and resource management strategies of indigenous forest people and immigrants, and acknowledge different perspectives among cultures and genders (Chapter 17). It is with this vision that we provide an overview of people and peoples' livelihood strategies in rain forests.

15.1 Indigenous people

Richard Archibald flew over the interior of Western New Guinea in 1938, and discovered the 60 km long Baliem Valley. He was surprised to find farmland divided up in tidy parcels as if he was in Europe. Far from being densely forested as he had expected, this was the home to a long-established indigenous farming system which, without the use of any metal or other imports, supported some 50,000 people.

A staple of Western mythology is the meeting of explorers with previously 'uncontacted' peoples. Such incidents are increasingly scarce, as few people remain untouched by the modern world. Nonetheless, the **NGO** Survival International estimates that over 100 uncontacted tribes still remain worldwide. Such groups are likely to be small and living in remote forested locations (New Guinea and the Amazon). Recognized examples include some (or all) of Bolivia's Yanaigua, Peru's Kirineri,

Guyana's Wapishana (Arawak), Ecuador's Tagaeri, and Brazil's Karafawyana and Papavo (in 2007 the Brazilian government officially recognized the existence of 67 different uncontacted tribal groups). New Guinea alone is estimated to be home to at least 44 groups. Responses to contact vary. The archers of the North Sentinel (Andaman) Islands discourage contact, even firing arrows on approaching helicopters. Of two previously unknown groups encountered in West Papua in 1996, one soon began trading for steel axes and other goods, while the other immediately disappeared.

Many other groups have long had interactions with communities outside rain forests, but have themselves continued to live as hunter-gatherers within the rain forest (e.g. the Punan and Penan in Borneo, or the Aka, Baka, and Batwa pygmies of the Congo Basin, and various groups of Amerindians in Amazonia). As hunter-gatherers they predominantly live off the forest with little if any formal cultivation[1], but most have long had trading relationships with farming communities for food and implements. Indeed it has been suggested that a purely self-reliant hunter-gatherer livelihood without any culti-

---

Box 15.1 The hunter-gatherer dilemma: is there enough food?

Exclusively hunter-gatherer societies rely wholly on wild food. Yet in rain forests fat-rich animals, oil-rich seeds, and carbohydrate-rich tubers and roots are often scarce, toxic, or otherwise well defended, unpredictable in supply, or demanding to extract and process. Is such a hunter-gatherer existence possible? Perhaps hunter-gatherers always depend on subsidies from cultivation or trade: the 'cultivated calories hypothesis' (Headland and Bailey 1991). It appears that all contemporary hunter-gatherer societies indeed cultivate or trade for food, but the idea that independent hunter-gatherer subsistence has never been possible remains debated (Bahuchet et al. 1991; Brosius 1991).

The original debate centred on the availability of carbohydrate-rich tubers. Proponents of the cultivated calories hypothesis argue that such plants are principally adapted to seasonal environments and generally restricted to secondary-forest habitats (Hart and Hart 1986). However, studies of rain forests inhabited by the Babinga (Aka) pygmies of the Central African Republic suggest that densities of edible tubers could support far greater use than they are subjected to by the Aka today (Bahuchet et al. 1991). Similarly in Asian forests wild palms, including *Eugeissona utilis*, are an important source of sago

for Penan and Punan people in Borneo (Brosius 1991).

Rain forests provide other nutritionally important foods, including nuts, fruit, and honey. Fat and fat-soluble vitamins, rather than starch, appear to be limiting in parts of Southeast Asia. The fat provided by the bearded pig *Sus barbatus* plays a key role in the diets of virtually all Bornean hunter-gatherers and accounts for a high percentage of calories consumed; it can also be stored for several months (Brosius 1991).

Foraging people often influence and manipulate wild plants and their environments, which blurs definitions of the hunter-gatherer concept (see also Chapter 14). For example the Aka of western Congo, like other pygmies, replant the parts of tubers that are attached to the rest of the plant, allowing the plant to survive (Bahuchet et al. 1991).

Contemporary hunter-gatherer societies do, however, trade and/or cultivate themselves. Trade also provides iron tools, pottery, salt, and other essentials. Whether or not a pure hunter-gather existence is possible remains uncertain. We now recognize that clear answers will depend on location and a more complete nutrition than simply calories.

---

[1] Most cultivate to some degree, perhaps inadvertently, by planting fruit trees around regularly used camping sites, and will tend desirable wild plants.

vation or trade may be impossible due to the limited food resources in rain forests (Headland and Bailey 1991) (Box 15.1). Agriculture was adopted centuries earlier in the wet alluvial plains of Asia where cleared forest was permanently replaced with irrigated rice, in the highlands of New Guinea, and in dryland forests where cultivation was developed in all main rain forest regions.

15.1.1 Diversity of rain forest peoples

As with many biological taxa, human cultural and linguistic diversity (the latter often considered a proxy for cultural diversity) is greatest in tropical forest regions. Ten of the top 12 countries for biodiversity are also among the 25 linguistically richest (Harmon D. 1996 in Maffi 2005). Hotspots occur in New Guinea (Australasia) and the Gulf of Guinea (Africa). High rainfall, low seasonality, topography, and habitat diversity all correlate positively with linguistic diversity (Nettle 1998; Cashdan 2001). In the Neotropics modern and pre-Columbian cultural patterns differ, and correspondence between biophysical and cultural diversity is weaker, suggesting that such patterns also reflect historical stability (Manne 2003).

As with declining biological diversity, there is concern over the loss of cultural diversity. Estimates suggest that 50–90% of over 6,000 languages spoken worldwide may be lost by 2100 with a concurrent loss of cultures and knowledge (Krauss 1992). Various international initiatives seek to document and maintain biocultural diversity (Maffi 2005). An ethical and practical challenge is to achieve this while encouraging development and integration with modern societies, making local aspirations a central concern.

When we speak of 'forest-dependent peoples' we are talking about very diverse groups with different concerns and aspirations. Nevertheless, we can generalize about three principal types of forest-dwelling peoples: hunter-gatherers, forest cultivators, and immigrants. Hunter-gatherer communities are typically small and mobile. Their cultures lack all but the most portable, or rapidly made, cultural items. Mutual dependence creates a culture of shared responsibility and leadership. With little need of possessions and wealth, and a strong bond of collaboration, these people generally have (or had) a reasonably easy life and healthy diet (see *Human Health* below).

Cultivator societies often settled in one site for many years, living in large communities with food surpluses and accumulated wealth. The requirement of defence and leadership often led to a strong hierarchy—sometimes with slaves. The violence of some of these communities is legendary as it was advantageous to scare and intimidate potential adversaries (head-hunters and similar practices once occurred in all the main tropical regions).

Hunter-gatherers and cultivators tend to have specific knowledge and cultural affinity with their surroundings, which may give the forest various intangible values. Settlers lack such knowledge, culture, and perceptions. Without access to the many positive values, and still subject to the negative (diseases, animals etc.), they may view the forest simply as a place to avoid or convert to more desirable land cover.

15.2 Farming the forests

The earliest forms of forest farming most likely included the tending and enrichment of valued food trees and the opening and maintenance of clearings for hunting. Before the advent of metal tools, clearing old-growth forest would have been a formidable task. Prehistoric cultivation may have initially invested in locally intensive systems that did not necessitate regular forest clearing, as with the Amerindians who developed *terra preta* (see Box 14.4) and the Baliem Valley cultivators mentioned above. It is, therefore, plausible that regular forest clearance for agriculture followed the introduction of steel through recent millennia. In recent decades the widespread availability of chainsaws has further alleviated restraints on forest clearance.

A wide range of forest-based agricultural systems have been identified and classified around the world (ICRAF 2000). These range from the unplanned enrichment of forests with favoured edible species, through the casual collection of edible fruits and dispersal of their seeds, to more planned selection of species, and to complex agroforests and intensively managed home gardens, and ultimately plantations (Table 15.1). Each of these systems is typically associated with different institutional and land tenure arrangements.

Table 15.1 Forest and tree utilization and management practices and associated institutional arrangements.

Plant-exploiting practice	Ecological effects	Socio-economic conditions	Institutional structure
Casual and opportunistic gathering and collecting of wild tree products	Incidental dispersal of propagules, no transformation of natural vegetation composition and structure	Segmented societies, low population density, subsistence economy	Open access
More or less systematic gathering and collecting of wild tree products	Incidental dispersal of propagules, no transformation of natural vegetation composition and structure	Low population density, incipient social stratification at the community level	Common property rights, sometimes priority rights to valuable tree species
Systematic collection of wild tree products with tending of preferred species	Reduction of competition, limited transformation of forest structure and composition	Increased social stratification and incipient local commercialization	Combined common property rights on forests and private property rights on claimed trees
Selective cultivation of wild trees by transplanting, and dispersal of seeds in forest environment	Purposeful dispersal of propagules to new habitats, partial transformation of forest structure and composition	Increased population density and socio-economic stratification	Priority rights to forest plots for tree planters
Tree crop cultivation (possibly in combination with annual crops)	Land clearance, total or almost total transformations of forest structure and composition	Medium to high population density, increased incorporation in market economy	Private land and tree rights
Cultivation of domesticated trees in plantations	Propagation of improved varieties, land clearance and soil modification, with inputs of fertilizers and pesticides	High population density, fully commercialized resource use	Private land and tree rights

Adapted from Wiersum 1997.

15.2.1 Shifting cultivation

Shifting cultivation, also called swidden agriculture and more pejoratively 'slash and burn' agriculture, has been defined as 'a continuous system of cultivation in which temporary fields are cleared, usually burned, and subsequently cropped for fewer years than they are fallowed' (Peters and Neuenschwander 1988). Swidden agriculture can be viewed as a form of agroforestry in which crops and tree cover are combined in time rather than in space.

There are many forms of shifting cultivation, a system that is still widespread in Southeast Asia, tropical Africa, and tropical America, and is estimated to support some 40–500 million people worldwide (Mertz et al. 2008) at densities of 10–20 people per km², although exceptionally up to 50 people or more per km² (Table 15.2).

Traditional shifting cultivators manage land on a more or less regular cycle with fallows. By contrast, forest pioneers are migrants who clear forests to establish permanent or semi-permanent fields, often for cash crops as well as subsistence crops. If the land becomes unproductive they may abandon it (or sell it as pasture) and move on to clear more forest. Both groups are referred to as shifting cultivators, although the term should strictly be applied to the former: forest pioneers are sometimes called 'shifted cultivators' as they are often displaced people or transmigrants (Myers 2000) (Box 15.2).

Usually only a small forested area of up to a few hectares is cleared and burned by each household. Large canopy trees may be left uncut if the wood is too hard or if they provide useful products such as fruit or nuts[2]. Annual staples, typically upland rice in Asia and maize in Africa and the Neotropics, are planted, sometimes with (or followed by) longer-lived species such as chilli, fruit trees, or root crops such as cassava. After 1–3 years yields begin to decline and the area is left fallow for 4–60 years[3]. Longer-lived crops continue to provide produce in the first years of fallow, and other plants may be deliberately nurtured and planted (Finegan and Nasi 2004) (Table 15.3). Shifting cultivators also often leave patches of forest from which they obtain timber or medicinal plants, or use them for hunting.

Land is selected based on various criteria, typically including location, vegetation, and soil characteristics that indicate perceived productivity, ease of clearance, and possible risks. In many cultures areas are cleared communally, as is common with many of the tribes of the Borneo interior, leading to extensive openings that are easier to protect from pigs and other pests (Basuki and Sheil 2005). The Makushi people of Central Guyana tend to cultivate near their relatives, close to water, and they avoid places where aggressive ants occur. In contrast to tribes in Borneo, these people prefer primary forests which they consider to be easier to clear and have fewer aggressive ants (Forte 1996).

Table 15.2 Population densities of shifting cultivators in Asian rain forests.

Location	People	Population density (km⁻²)	Fallow period (years)
Kalimantan, Borneo	Kantu	16	7
Sarawak, Borneo	Kenyah	11–18	12–20
Sarawak, Borneo	Iban	18	12
Philippines	Hanunoo	48	12
Chittagong Hills, Bangaldesh	Unspecified	1961: 29 2003: 96*	1961: 15–20 2003: 3–5
Democratic Republic of Congo	Unspecified	11–20	15
New Guinea	Tsembaga	34	15–25

Based on Whitmore 1998 and supplemented with information from Zhang et al. 2002 and Borggaard et al. 2003.

* This population density is unlikely to be sustainable without incurring land degradation.

[2] Examples are the tree *Bertholletia excelsa* which provides valued Brazil nuts in the Amazon, and the Southeast Asian *Koompasia* trees which are retained because they are a favoured nesting site for honey-providing bees.

[3] Often a combination of fallow lengths is found within one community and even one household. For example, some households may cultivate a subset of their land more frequently simply because they prefer to be closer to their village or to other fields, even though this may lower productivity per unit area (Brown 2008).

Box 15.2 **Shifting vs. shifted cultivators**

A distinction should be made among two types of agriculturalists of forest margins. 'Shifting' cultivators have used forest ecosystems sustainably through a swidden system with managed forest fallows. This practice has served people for centuries, though these shifting cultivators have probably never exceeded more than a few tens of millions of people worldwide.

The negative stereotype of shifting cultivation should really only be attributed to 'shifted' cultivators, small-scale farmers who have been displaced from their traditional farmlands, or who have voluntarily emigrated, often spurred by government incentives (Brown and Schreckenberg 1998). Examples include government transmigra-

tion schemes that attract opportunistic poor migrants for whom forest cultivation may be a strategy borne out of desperation—and possibly a temporary one at that. Shifted cultivators generally have little comprehension of tropical forest ecosystems, and consequently practise unsustainable forms of exploitation. Much forest destruction may be accounted for by such migrants who clear and exploit land until it is exhausted, then may sell it to accumulate the capital to buy better quality land with better access to social infrastructure (Brown and Schreckenberg 1998). These shifted cultivators may number in the hundreds of millions and, by some accounts, their numbers are increasing (Myers 2000).

Table 15.3 Useful plants in the fields and forest fallows of Bora Indian shifting agriculture, Amazonian Peru.

Stage	Time	Planted	Spontaneous
High forest	—	—	Numerous species for construction, medicine, handicrafts, and food
Newly planted field	0–3 months	—	Dry firewood from unburnt trees
New field	3–9 months	Corn, rice, cowpeas	Various early successional species
Mature field	9 months–2 years	Manioc, tubers, banana, cocona (*Solanum sessiliflorum*), and other quick maturing crops	Vines and herbs of forest edges
Transitional field; seedlings of useful trees appear	1–5 years	Replanted manioc, pineapples, peanuts, coca, guava, caimito (*Poumeria caimito*), uvilla (*Pourouma cecropiifolia*), avocado, cashew, peppers; trapped wildlife	Medicinal plants within field and on edges; abandoned edges yield straight tall saplings of *Cecropia* and balsa *Ochroma lagopus*
Transitional fruit field, with abundant forest regrowth	4–6 years	Peach palm, banana, uvilla, guava, annatto (*Bixa orellana*), coca, some tubers; propagules of pineapple and other crops; hunted and trapped wildlife	Many useful soft construction woods and firewoods; palms, including *Astrocaryum* (for oil); many vines
Orchard fallow	6–12 years	Peach palm, some uvilla, macambo (*Theobroma bicolor*); hunted wildlife	Useful plants as above; self-seeding *Inga*; probably most productive fallow stage
Forest fallow	12–30 years	Macambo, umari (*Poraqueiba sericea*), breadfruit, copal (*Dacryodes* sp.)	Self-seeding macambo and umari; some hardwoods becoming harvestable (cumala *Virola sebifera*); many large palms (*Astrocaryum* spp., *Euterpe* spp.)
Old fallow, high forest	Over 30 years	Umari, macambo	A few residual planted and managed trees

From Whitmore 1998.

Yields decline because crops rapidly exhaust the nutrients released from burning, and because of the accumulation of pests, diseases, and weeds. Weeds include *alang alang Imperata cylindrica* which are difficult to clear once established. The fallow period restores soil fertility, provides weed control, and provides a variety of plant resources, both planted or naturally established, such as medicines, firewood, fruit, and building materials (Colfer et al. 1997) (Table 15.3). Much of the land area within a shifting cultivation region is likely to be forested fallow land in various stages of regeneration and development (Vieira et al. 2003).

For a period towards the end of the last millennium, shifting and particularly 'shifted' cultivators (see Box 15.2) were often held responsible for tropical deforestation (Lanly 1982; Allen and Barnes 1985; Peters and Neuenschwander 1988). Estimates of their share ranged as high as 60% (Myers 1992). Most attempts to control and integrate these cultivators into national development schemes, and to replace the practice with more intensive sedentary systems, have not been very successful. The reasons for this failure include an inadequate understanding of the logic of shifting cultivation and factors influencing farmers' decision-making. In reality, deforestation has been driven by several factors. Shifting cultivation is undoubtedly one of the more visible causes, but it is driven by many factors and is probably overshadowed by burning, cattle ranching, unregulated logging, economic development projects, road building, and mining (Chapter 14). Yet the emphasis on shifting cultivation coupled with population growth was, until recently, entrenched in explanations of forest destruction, exemplified by a World Bank overview of deforestation in West and Central Africa which argued that although shifting cultivation was sustainable at low population densities, 'with the shock of extremely rapid population growth…these practices could not evolve fast enough. Thus they became the major source of forest destruction and degradation of the rural environment' (Cleaver 1992).

This view is now more often challenged (Brown and Schreckenberg 1998). Shifting cultivation, rather than being a wasteful and destructive practice as has often been portrayed, is often an appropriate, sustainable, and sophisticated land-use system. Changing socioeconomic conditions, growing populations, and reduced land availability (with the best land often appropriated for other uses) have, however, led to a shortening of fallow periods (de Jong et al. 2001). Over recent decades the typical fallow period in lowland Bolivia has declined from 6–12 years to around 4 years; in north-east India fallows of 3–10 years are now considerably shorter than the more traditional 60-year fallows of the past (Finegan and Nasi 2004), and in Bangladesh fallow periods of 15–20 years have been reduced to 3–5 years (Borggaard et al. 2003). Shorter fallows reduce crop yields and increase weeds,[4] undermining sustainability and encouraging additional forest clearance (de Rouw 1995; Becker and Johnson 2001; Finegan and Nasi 2004).

Recently, shifting cultivation is receiving renewed attention, not as a problem but as a rational land use. Land-use degradation and collapse at higher population densities have been predicted, but the evidence for such collapse, even at short fallow periods, is not generally evident (Bruun et al. 2006; Mertz et al. 2008)—although evidence of severe land degradation in Madagascar suggests exceptions where land shortages occur (Pfund 2000; Styger et al. 2007).

15.2.2 Agroforestry

Agroforestry is the integrated management of trees with crops. The practice of integrating trees in farming systems is ancient, but is now being widely promoted as a more sustainable land use that can also support biodiversity. There are a great many different agroforestry systems, some little more than monospecific plantations with an understorey crop, while others are complex and diverse. Diverse agroforestry systems are characterized by a complex vegetation structure that includes trees, lianas, seedlings, shrubs, and herbs.

Agroforests are often small (1–2 hectares) but complex individually-owned and managed units that in aggregation form a forested mosaic. Apart from home gardens[5], they are rarely intensively managed. They may originate

---

[4] Short fallows in Côte d'Ivoire led to a 72% increase in weed biomass (Becker and Johnson 2001).
[5] Examples include Mesoamerican *solar* and Javanese *pekarangan*. More extensive home gardens are found on richer soils, as on volcanic Java, or shambas plots of the Chagga people on Mt Kilimanjaro.

by progressive transformation of the original forest eco-system through tree planting, thinning, and selection of naturally occurring useful species, or are established after complete removal of the original vegetation by shifting cultivation and the planting of desired species. Some modern agroforestry systems developed from shifting agriculture. For example, in Sumatra and Kalimantan (Indonesia) shifting cultivators sometimes introduce rubber trees *Hevea* spp. into their otherwise short-lived cropping systems, and following decades of fallow these develop into 'jungle rubber' agroforestry systems (Joshi et al. 2002). Shifting cultivation is, however, usually differentiated from agroforestry, as the agroforestry farming system is spatially and temporally integrated.

Farmers usually plant a staple food crop such as upland rice, or maize, which is intercropped with several timber or fruit tree species, or other useful long-lived species such as rattan. The staple crop may subsequently be replaced with other perennial woody plants to create a complex and relatively diverse structure that is not dissimilar in appearance to natural forest. Examples include the damar *Shorea robusta* and durian *Durio zibethinus* agroforests of Sumatra (Michon and de Foresta 1999) and coffee or cocoa agroforests established with canopy trees retained from the natural forest, a system that is

found in the Neotropics, West Africa (de Rouw 1987; Johns 1999), and southern India (Bhagwat et al. 2008).

15.3 Forest products

Rain forests provide a multitude of products besides timber, many of which have important local value while a few are traded on global markets. Although tropical forest plants are often well defended (being tough, astringent, toxic, or thorny), edible nuts, berries, roots, and animal foods occur in most regions. Foods and construction are only a subset of uses to which indigenous people put forest products. A study with hunter-gatherer Punan and swidden farming Merap in East Kalimantan underlines the huge diversity of plant species with local uses. From over 18,000 plant records over a range of habitats, 2,141 separate distinct uses or values were reported from 1,457 species spread across a range of uses (Figure 15.1) (Sheil et al. 2009).

15.3.1 Animal resources

Animals hunted for subsistence use and for sale represent a valuable forest product ('**bushmeat**') for approximately

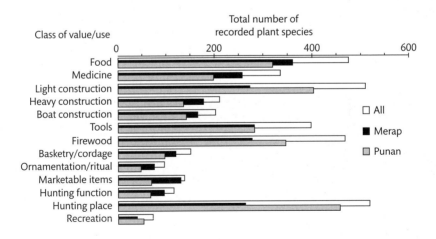

Figure 15.1 Useful plant species (including trees, herbs, climbers, and others) recorded by value category overall and for Punan and Merap informants separately in Malinau, East Kalimantan. Overall, 119 species were recorded as having uses that were in some way viewed by the informant as exclusive to that plant alone—most of these are ritual/ornamental, medicinal, or tools (Sheil et al. 2009).

Figure 15.2 In Cambodia some larger forest spiders are collected to be eaten as a delicacy.

Photo by Manuel Boissiere.

60 million across the tropics (Pimentel et al. 1997; DFID 2002; Robinson and Bennett 2004). About 30 million people living in or near the forested regions of the Congo Basin regularly eat bushmeat, which provides up to 80% of the protein and animal fat in rural diets. Over 2 million tonnes of wild animals are consumed annually from this region alone (Nasi et al. 2008). In the Amazon the value of wild meat harvested exceeds US$ 175 million per year, and overall the global trade in wild animal products is estimated at US$ 3.9 billion (Nasi et al. 2008). Bushmeat provides a 'safety net' during periods of seasonal shortfalls: in eastern Democratic Republic of Congo consumption during the hungry season rose by 75% (de Merode et al. 2004). Thus many people, and particularly those living in extreme poverty, depend on bushmeat and other animal resources (Figure 15.2) from local markets or forests to meet their most basic needs.

Figure 15.3 Forest-derived crafts in the market in western Uganda. The baskets are made from the palm *Phoenix reclinata*, while the winnowing tray is made from leaves and small branches of *Smilax anceps*, *Draceana laxissima*, and *Securidaca welchii*.

Photo by Robert Bitariho.

Box 15.3 'Fruits' of the forest: commercialization of some non-timber forest products

Açaí

Açaí palms (*Euterpe* spp.) grow in large numbers in low-lying areas of the Amazon estuary, and for at least 1,000 years the fruits and palm hearts have been used as a food, though it is not clear how it was managed in antiquity (Miller and Nair 2006). Açaí

was reported as a staple for the population of Belem in 1859 (Miller and Nair 2006). Today it continues to have great economic importance (Figure 15.4), and stands of the palm are actively managed for the fruit and palm hearts (Anderson 1988).

Figure 15.4 An açaí farm. *Euterpe* trees from which açaí is derived surround the homestead, here near Belem, Brazil.
Photo by Douglas Sheil.

Brazil nuts

Brazil nuts are almost entirely sourced from natural forests. The trade accounts for the protection of millions of hectares of forests in Brazil, Bolivia, and Peru, and supports the livelihoods of thousands of Amazonian residents (Ortiz 2002). The Brazil nut tree (*Bertholletia excelsa*, Lecythidaceae) has never been viably domesticated owing to its slow growth and the dependency of Brazil nut product on pollination by large-bodied solitary bees that are associated with natural forest. Unlike social bees, they are not easily manipulated by humans.

Cacao

Cacao (*Theobroma cacao*) is an understorey forest tree that was widely used, and perhaps managed or cultivated, in Amazonia and Mesoamerica prior to European arrival. Europeans first became aware of cacao following the Spanish conquest of Mesoamerica (Miller and Nair 2006). It was one of the first Amazonian products exported by the Portuguese (Miller and Nair 2006) and was the dominant regional export by the 1730s (Alden 1976). By the 19th century cacao was being cultivated in agroforests along the Amazon. By the end of the

19th century, cacao production was displaced to southern Bahia following the rubber extraction boom (see below). Cacao is now cultivated in an area of 7.4 million hectares worldwide by both smallholders and in plantations (FAOSTAT 2008).

Rattan

Rattans (mainly *Calamus* spp., Arecaceae; Chapter 2) are Paleotropical climbing palms with strong flexible stems; they form the basis of a regional and international craft and product trade, primarily in cane furniture and baskets. Indigenous uses are diverse and include fish traps and bridges. During the last two decades of the 20th century the rattan trade exploded as export values increased 250-fold in Indonesia and 75-fold in the Philippines (Sunderland and Dransfield 2002). No quantitative estimates of the true value of rattan are available, although domestic trade and use of rattan create annual benefits estimated at US $3 billion, and another estimated US $4 billion are generated through global exports (Sastry 2002). Indonesia is the supplier of about 90% of the world's demand.

Rattan is an open access resource through much of its range, and is vulnerable to over-exploitation. Enrichment planting has been successfully adopted in logged forest in East Kalimantan, Indonesia, and involves the creation of artificial gaps to favour rattan growth. Rattan has also long been integrated in shifting cultivation and agroforestry systems in Southeast Asia, and more recently has been incorporated into rubber plantations in Peninsular Malaysia.

Rubber

Rubber, derived from the latex of the rubber tree *Hevea brasiliensis* (Euphorbiaceae), has been used by indigenous rain forest people in South America for centuries[1] (Figure 15.5). Demand for rubber soared with the invention of the automobile in the late

Figure 15.5 Children gather latex to make toys from a wild tapped *Hevea* rubber tree near Belem.

Photo by Douglas Sheil.

19th century, transforming the city of Manaus in the heart of the Amazon into a major commercial centre. The tragedy of rubber is that the 'rubber barons' of Manaus built their fortunes on the backs of native Indian slave labour[2] at the cost of incredible brutality and thousands of deaths (Williams 2003).

[1] Rubber was used for the large heavy balls used in the sacred Maya, Olmec, and Aztec ball games. The games are still played, but since the losers are no longer sacrificed the game has somewhat lost its appeal.

[2] Exploitation of people with scant regard to human rights continues today in the form of slave labour within the Amazonian charcoal-making industry and in the clearance of forests for the benefit of Amazonian cattle ranchers. New immigrants employed as labourers are caught in debt traps by which they become bound to unscrupulous intermediaries who supply labour to ranchers (Fearnside 2008). The number of people working in such conditions in the Brazilian Amazon is not certain, but is reported to number 25,000 in Pará alone (quoted by Fearnside 2008).

This Amazonian tragedy came to an end with the establishment of large-scale Southeast Asian rubber plantations[3] in the early 20th century. By 1910 Brazilian production had fallen to 50% of the market share, and by 1940 accounted for only 1.3%. Today, around 10 million tonnes of natural rubber is produced annually on around 9 million hectares, with Asia (particularly Indonesia, Malaysia, and Thailand) accounting for 94% of output (2007 values) (FAOSTAT 2008). Much of this is in large commercial plantations, but smallholder 'jungle-rubber' agroforests cover 3 million ha in Indonesia, mostly in Sumatra and West Kalimantan. In the Amazon Basin, rubber tapping continues to play a minor role within extractive reserves, though even here its importance is declining (Chapter 17).

[3] Attempts at establishing *Hevea* plantations within its native range proved unsuccessful owing to leaf blight (Hecht and Cockburn 1989). Leaf blight remains absent from Asia.

15.3.2 Plant resources

Rain forest plants include the ancestors and relatives of many cultivated species, including cocoa, rubber, banana, citrus fruits, coconut, maize, avocado, sugarcane, and oil palm, and the conservation of these wild progenitors is essential as a source of genetic diversity for plant breeding (Table 15.4). Some wild plants provide significant economic benefits to rain forest dwellers: Brazil nuts and açaí fruits together account for annual revenue of over US$ 50 million in the Brazilian Amazon alone, and when other non-timber forest products (NTFPs) are included this amounts to revenues of around US$ 70 million. The combined value of these products is, however, dwarfed by revenues from timber (US$ 2,311 million), which itself is only the third most important economic activity in the Brazilian Amazon after industrial mining and cattle ranching (Veríssimo and Lentini 2007).

Nevertheless, across the tropics hundreds of millions of rural people living in forested areas depend directly on NTFPs for income or subsistence (Figure 15.3) (de Beer and McDermott 1989; Falconer 1990), and their variety and commercial and non-commercial value to tropical poor communities worldwide is immense (Table 15.4). Poor households tend to be disproportionately dependent on forest resources and the contributions to livelihoods can be substantial (Vedeld et al. 2004). For example, NTFPs contribute as much as 60% of the cash income of Soligas tribal communities in the Western Ghats forests of southern India (Uma Shaanker et al. 2004). Some NTFPs are collected infrequently, but are important in filling seasonal gaps in food or income derived from other activities, and provide emergency sustenance during times of hardship (Neumann and Hirsch 2000; de Merode et al. 2004). In such cases forest resources act as a form of 'natural insurance', this role being of particular importance to those with few assets and little land (McSweeney 2005). However, in many forest areas access restrictions, and forest loss, have impacted on the availability of NTFPs to local communities, as has over-harvesting of some commodities (Hegde and Enters 2000; Larsen and Olsen 2007).

Through the 1990s there was significant investment by development agencies in the commercialization of NFTPs as a basis for livelihood enhancement through integrated conservation and development initiatives (Salafsky and Wollenberg 2000; Arnold and Ruiz-Pérez 2001). Although some NTFPs have been successfully commercialized to the benefit of harvesting communities (Box 15.3), for the large majority of forest communities NTFPs remain a subsistence commodity.

15.4 Human health

Horrific new diseases and wonder-cures awaiting discovery are two popular perspectives concerning rain forests and human health[6]. Indeed, the rain forest can be

[6] Hollywood portrayals of these perspectives include the films *Outbreak* (1995) and *Medicine Man* (1992).

Table 15.4 Some commercially important products derived from rain forest plants (compiled from multiple sources).

Product/Substance	Taxonomic name	Organism/part	Likely origin	Use and notes
Açaí	*Euterpe oleracea*	Palm tree, fruit (and bud)	South America	Food
Allspice	*Pimenta dioica*	Tree, fruit	Caribbean	Spice/flavouring
Avocado	*Persea americana*	Tree, fruit	South America	Food
Balatá	*Manilkara bidentata*	Tree, latex	South America	Rubber (outer part of golf balls)
Banana	*Musa* spp.	Herb, fruit	Asia	Food (Several domesticates occur. All are dependent on cultivation as they are seedless.)
Benzoin	*Styrax* spp.	Tree, resin	Southeast Asia	Incense, perfume, medicine, flavouring
Betel	*Areca catechu*	Palm, seed	Malaysia or the Philippines	Stimulant (mixed with Piper and lime)
Black pepper (white pepper, green pepper)	*Piper nigrum*	Climber, fruit	Asia (probably India)	Spice/flavouring
Brazil nut	*Bertholletia excelsa*	Canopy tree, seed	South America, Amazon region	Food
Breadfruit	*Artocarpus altilis*	Tree, fruit	Southeast Asia and Pacific New Guinea	Food
Breadnut (Ramón)	*Brosimum alicastrum*	Tree, fruit	Mesoamerica	Food
Cardamom	*Elettaria cardamomum* (some *Amomum* spp.)	Herb, fruit	Asia	Spice/flavouring
Chicle	*Manilkara zapota*	Tree, latex	Central America	Chewing gum
Chilli (red pepper)	*Capsicum* (5 species)	Shrub, treelet, fruits	South and Central America	Spice/flavouring/food
Cinnamon	*Cinnamomum zeylanicum* *Cinnamomum burmani*	Tree, bark, and leaves	Sri Lanka SE Asia	Spice/flavouring
Clove	*Syzygium aromaticum*	Tree, flower buds, leaves	Asia (Moluccas Islands)	Spice/flavouring, cigarettes
Cocaine	*Erythroxylon coca*	Tree, leaves	South and Central America	Stimulant/Narcotic
Cocoa	*Theobroma cacao*	Tree, seeds (fruit)	South America	Food
Coconut	*Cocos nucifera*	Palm, seeds	Origins uncertain: possibly Philippines	Food, oil
Coffee	*Coffea arabica C. canephora*	Tree, shrub, seeds	North-east Africa	Drink, stimulant
Cola nut	*Cola nitida* and *Garcinia cola*	Tree, seeds	West Africa	Spice/flavouring/tonic
Damar	*Shorea javanica, Shorea robusta* *Agathis* spp.	Canopy trees, resin	Asia	Varnish
Eaglewood (gaharu)	*Aquilaria* spp.	Tree, resin (produced as reaction to fungal infections)	Southeast Asia	Incense and perfume (some other genera may also produce suitable resins)
Ginger	*Zingiber officinale*	Herb, rhizome	South Asia (China, Indochina)	Spice/flavouring, medicinal
Grains of paradise	*Afromomum melagueta*	Herb, seeds	West African (coastal swamp forest)	Spice/flavouring,

(Continued)

Table 15.4 (Continued)

Product/Substance	Taxonomic name	Organism/part	Likely origin	Use and notes
Guava	*Psidium guijava*	Tree, fruit	America	Food
Guiana chestnut	*Pachira aquatica*	Tree, seed	Central and South America (freshwater swamps)	Food
Gutta percha	*Palaquium gutta*	Tree, latex	Southeast Asia	Rubber (electrical insulator, dentistry, conveyor belts)
Illipe nuts	*Shorea* spp.	Tree, seed	Southeast Asia	Cocoa butter substitute
Jackfruit	*Artocarpus heterophyllus*	Tree, fruit	India, Sri Lanka	Food
Jelutong	*Dyera costulata*	Tree, latex	Melanesia	Chewing gum
Macadamia/Macadamia nut	*Macadamia integrifolia* and *M. tetraphylla*	Tree, seed	Australia	Food (not eaten by aboriginals), cosmetics
Maize	*Zea mays*	Cereal grain	Mesoamerica	Food (not found in wild but believed to be a mutant and hybrid form of teosinte)
Mango	*Mangifera*	Tree, fruit	India	Food
Mangosteen	*Garcinia mangostana*	Tree, fruit	Sunda Islands and Moluccas	Food
Nutmeg, mace	*Myristica fragrans*	Tree, fruit	Asia (Moluccas Islands)	Spice/flavouring/mild narcotic
Oil palm	*Elaeis guineensis, Elaeis oleifera*	Palm tree, fruit	West/central Africa	Food, oil, chemical feedstock
			South America	
Palm heart	*Euterpe* spp.	Palm tree, fruit, heart, leaves	Amazonia	Food, thatching, weaving
Papaya	*Carica papaya*	Tree, fruit	Brazil and Paraguay	Food
Pepper (black)	*Piper nigrum*	Vine, fruit	South India	Spice
Pineapple	*Ananas comosus*	Herb, fruit	South or Central America	Food
Rambutan	*Nephelium* spp.	Tree, fruit	Southeast Asia	Food
Rattan	*Calamus* spp.	Climbing palm, (fruit) and stem	Africa, Asia	(Food and fruits) furniture, resin
Rubber 'Para rubber'	*Hevea brasiliensis*	Tree, latex	South America	Material
Sago	*Metroxylon sago, Pigafetta, Raphia*	Tree, pith	New Guinea and islands	Food
Soursop	*Annona muricata*	Tree, fruit	America	Food
Star anise	*Illicium verum,*	Tree, fruit	Asia (Southern China)	Spice/flavouring
Starfruit	*Averrhoa carambola*	Tree, fruit	India and Southeast Asia	Food
Sugar cane	*Saccharum* spp.	Perennial grasses, sap	South and Southeast Asia, New Guinea	Sugar, alcohol, fuel
Sweet potato	*Ipomoea batatas*	Herbaceous perennial vine, tuber	South America Caribbean	Food
Sweetsop	*Annona squamosa*	Tree, fruit	America	Food
Turmeric	*Curcuma longa*	Herb, rhizome	Asia	Spice/flavouring/food colour
Vanilla	*Vanilla* spp.	Climber (twining orchid), fruit	Mesoamerica	Flavouring (Tahiti, has its own special hybrid form presumably derived from early introductions)
Zedoary	*Curcuma zedoaria*	Herb, rhizome	Asia (possibly Melanesia)	Spice/flavouring

viewed as both a source of major human health threats or as a vast repository of future cures. Both perspectives have some basis, but warrant further scrutiny for a realistic assessment. Additionally, human health is about more than just the avoidance of physical ailments, and should be more broadly defined to encompass cultural[7], mental, and nutritional health[8]. Many forest hunter-gatherer communities suffer significant health costs as they attempt to adapt to new lifestyles (Box 15.4).

15.4.1 Infectious diseases in the rain forest

Many tropical infectious diseases are associated with moist tropical regions (Table 15.5). Malaria affects 350–500 million people annually, causes 1–3 million deaths (70% in Africa), and is the fourth leading cause of death in children

under five years across the tropics (WHO 2005). Incidence of malaria infection has been linked to heavily forested regions, and some mosquito vectors do indeed occur at high concentrations in tropical rain forests (Butler 2008). Increased incidence of malaria in some areas of Africa, South America, and Southeast Asia is also linked to deforestation, probably due to the increased availability of suitable mosquito breeding sites (Vittor et al. 2006; Pattanayak and Yasuoka 2008).

Plasmodium, the malaria parasite, is a single-celled protozoan that is transmitted to humans by the bloodsucking female anopheline mosquito. Only 60 or so of the approximately 380 anopheline species are believed to be effective vectors. Disease transmission is aided by high human densities and the arrival and movement of infected carriers (Patz et al. 2004). Parasite development

---

Box 15.4 Health and welfare when forest nomads settle

Nomadic lifestyles of forest hunter-gatherers such as the African pygmies and the Punan of Borneo[4] were, for many thousands of years, well adjusted to avoiding diseases and parasites prevalent in the wet tropics. High mobility, with groups moving every few weeks to new encampments, was facilitated by light burdens and small communities. Extensive spatial distribution of resources encouraged periodic movement, and hence diseases such as malaria were avoided as people moved beyond the flight range of mosquitoes before the parasites could reproduce (Dounias and Froment 2006). Following settlement of hunter-gatherer communities, malaria outbreaks become more common—a result of increased access to more severe

forms of malaria through temporary employment in logging camps or plantations or following visits to towns, and through the creation of favourable mosquito breeding sites such as stagnant water under houses and in discarded materials. Increased human density accompanies the shift from nomadic to sedentary lifestyles, which then enhances propagation of transmissible diseases (Dounias and Froment 2006). Permanent settlement also facilitates contacts with outsiders, and consequently exposure to diseases.

Among the Punan, such changes in lifestyle are dramatically reflected in the decline of their health (Dounias et al. 2007). In 2005–6 several medical treatment campaigns were carried out among the

[4] 'Pygmies' include various ethnic groups, like the Baka and Batwa, associated with the rain forests of Central Africa. 'Punan' and 'Penan' are generic terms for the hunter-gatherers of Borneo. Most of these formerly nomadic peoples have largely been settled in permanent villages but still depend on hunting and gathering for their livelihood.

---

[7] Many traditional rain forest societies, for example, suffer from alcoholism and violence due to erosion of their traditional laws and customs following the encroachment of logging and mining activities and the introduction of the cash economy (see Box 15.4).
[8] The World Health Organization defined human health as a state of 'complete physical, mental and social wellbeing and not merely the absence of disease or infirmity' (WHO 1978).

Punan. These showed that more than 50% of Punan health problems are now due to communicable diseases (malaria, measles, tuberculosis, acute respiratory infections, amibiase, and diarrhoea among others), which rarely occurred in the past when the Punan were nomadic. For instance, post-measles infections and complications commonly affect adults, demonstrating that they had not experienced this infectious disease when they were young.

Settlement is often also associated with food insecurity (as wild meat is locally exhausted, farming skills are limited and land is marginal). Excessive consumption of previously limited sugar and fat-rich foods leads to nutritional disorders such as diabetes and obesity.

Social support networks, such as reciprocal aid (Figure 15.6), collective activities, and food sharing—a critical part of the lightweight nomadic life—are replaced by individualism. Traditional healers and leaders are unable to address the new problems and lose their influence, while the younger generation loses respect for its cultural heritage and knowledge is lost. Conflict increases as does domestic violence and sexually transmitted disease. Oppor-

tunities for cash income are often squandered to poor time-keeping and drunkenness. Heavy alcoholism, smoking, and drug addiction are increasing health concerns (Levang et al. 2007). Stress and depression are a common part of a downward spiral.

There have been gains too. Despite their good diet and health, nomadic lifespan was typically short due to the prevalence of cultural forms of demography regulations (infanticides), accidents, and war. Such wars are largely a thing of the past, and access to modern medicines and education brings genuine opportunities. Child and infant mortality rates are much lower among settled groups (Dounias and Froment 2006; Levang et al. 2007).

The healthy future of formerly nomadic groups depends on many factors, including access to education and recognition of their basic rights. Medical assistance will help address symptoms, but other interventions are needed, considering the ecological, social, political, and economic changes that indirectly affect the health of forest people. Improving health of formerly nomadic forest-dwelling people is not simply a medical question.

Figure 15.6 Delousing is common social behaviour among the Baka pygmies of southern Cameroon, who temporarily abandon their permanent villages for seasonal camps in the forest. Excessive parasites in the camp often provide motivation to move to another place.

Photo by Edmond Dounias.

Table 15.5 Examples of rain forest-associated emerging infectious diseases.

Disease-causing agents	Distribution	Hosts and/or reservoirs	Exposure	Possible emergence mechanisms
Viruses				
Yellow fever	Africa South America	Primates	Vector	Deforestation and expansion of settlements along forest edges Hunting Water and wood collection Domestication of vectors and pathogen
Dengue	Pantropical	Primates	Vector	Mosquito vector and pathogen adaptation Urbanization and ineffective vector control programmes
Chikungunya	Africa Indian Ocean Southeast Asia	Primates	Vector	Pathogen and vector domestication
Oropouche	South America	Primates	Vector	Vector composition changes
SIV	Pantropical	Primates	Direct	Human expansion into forest Hunting of forest wildlife Pathogen adaptation
Ebola	Africa	Primates Bats	Direct	Hunting and butchering Logging and agricultural encroachment into forest margins Alteration of natural fauna
Nipah virus	South Asia	Bats Pigs	Direct	Pig and fruit production on forest margins
SARS	Southeast Asia	Bats Civets	Direct	Harvesting, marketing, and mixing of bats and civet cats Wildlife trade for human consumption
Rabies	Worldwide	Canines Bats Other wildlife	Direct	Human expansion into forest
Protozoa				
Malaria	Africa Southeast Asia South America	Primates	Vector	Deforestation, habitat alteration beneficial for mosquito breeding Human expansion into forest, non-human primate malaria among humans
Leishmaniasis	South America	Numerous mammals	Vector	Human expansion into forest and settlement of forest margins Domestication of animal vectors Deforestation and habitat alteration
Sleeping sickness (Trypanasomiasis)	West and Central Africa	Humans	Vector	Human expansion into forest (disease incidence associated with forest edge)

Adapted from Wilcox and Ellis 2006.

is temperature-sensitive, and malaria is expected to expand into tropical highlands due to climate change (Zhou et al. 2004).

Over the last century the land area supporting endemic malaria has been approximately halved, but nonetheless demographic changes mean that 2 billion additional people are at daily risk of malaria infection (Hay et al. 2004). The success of malaria control and treatment pro-grammes relies on the recognition that effective disease management is dependent on many factors beyond ecological and ecosystem variables, including demography, human migration, and efficiency of institutional organizations. Often weak institutions, high rates of human migration, and little community cohesion undermine efforts to control malaria in forested and recently deforested regions (de Castro et al. 2006).

The concept of emerging infectious diseases has been prompted by pathogens such as human immunodeficiency virus (HIV), severe acute respiratory syndrome (SARS), and Ebola virus, as well as the spread of well-known diseases such as malaria and dengue. Tropical deforestation and forest fragmentation, as well as bush-meat hunting, are thought to contribute to the emergence of diseases (Wolfe et al. 2005). Some of the first plague-causing pathogens, such as smallpox, may have originated in tropical Asia when humans first domesticated animals and cleared forests for cropland. Understanding host reservoirs (Box 15.5), pathogens, and transmission mechanisms within and between species is necessary, but for many diseases knowledge remains limited.

Nipah virus emerged in Southeast Asia in 1998 among people working with pigs: it appears that the pigs were infected with the virus by bats that had been displaced from forests during El Niño-associated droughts and fires in that year. Bats are the reservoirs for a number of tropical forest diseases, probably including Ebola in Africa and Asia (Leroy et al. 2005). Vampire bats appear to be the vector for rabies in the Neotropics, where one species now regularly feeds on cattle and dogs—which can infect people (people too are occasionally fed on by bats).

15.4.2 Medicinal species

Rain forests are recognized as important sources of medicinal compounds—especially from plants. Plant families that are rich in pharmacologically active ingredients, e.g. Apocynaceae, Rubiaceae, and Menispermaceae, are widespread and species-rich in rain forests, and extracts of bark, leaves, and roots from innumerable plants[9] have been used to treat illnesses, control parasites,

---

Box 15.5 Sloths as reservoirs of pathogens

Human diseases associated with tropical forests are often transmitted between a variety of potential hosts. These interactions remain poorly appreciated, and their ecological implications are mostly unknown. Sloths alone are a reservoir of a bewildering array of parasites and pathogens (Gilmore et al. 2001). Most seriously for humans, sloths (both *Choloepus* spp. and *Bradypus variegatus*) are reservoirs of various *Leishmania* species. Infection results from the bite of a phlebotomine sandfly vector, and leads to human cutaneous and/or mucosal *Leishmaniasis*, an often disfiguring and sometimes life-threatening disease. Sloths survive infection without evidence of pathology.

Sloths are also implicated as hosts to a number of arthropod-borne viruses. The Mayaro virus, carried by mosquitoes, causes occasional outbreaks of human febrile illness in forested areas of South America and Trinidad. In Panama, a high prevalence of St Louis encephalitis virus has been found in sloths, particularly *Choloepus* species. Whether sloths play an important role in the transmission of this disease to humans remains uncertain. In Pará, Brazil, *Bradypus tridactylus* has been found to harbour Oropouche virus, which is transmitted by culicoides midges, and causes periodic epidemic febrile illness in humans. Utinga virus (midge- and mosquito-carried) has also been isolated from Brazilian and Panamanian sloths. While sloths can apparently be infected with yellow fever, they are not considered an important host.

A considerable number of other viral infections have also been detected in both two- and three-toed sloths, and while few of these currently pose a major threat to humans, the sheer number and variety of pathogens harboured by sloths suggests that many other diseases remain to be discovered among other animals that are pathologically less well known.

---

[9] Forest animals too possess medicinal values. In 1993, 23% of the 150 most commonly prescribed drugs in America were based on active ingredients derived from animals (Grifo et al. 1997). Some poisonous compounds, including polypeptide toxins and dendrotoxins, are purified from venomous animals such as the poison arrow frogs of Central and South America (Wilson 2002).

Table 15.6 Common indigenous medicinal plants from tropical rain forests.

Species, Family	Common name	Life form, part used	Region	Source	Afflictions treated
Ampelocera edentula, Ulmaceae		Small tree, bark extract	Bolivia	Natural forest	Cutaneous leishmaniasis
Carapa guianensis, Meliaceae	Andiroba	Tree, seed oil	Brazil	Natural forest	Insect repellent, sprains, arthritis
Copaifera reticulata, Fabaceae	Capaiba	Tree, oleoresin	Brazil	Natural forest	Wounds, sore throat
Himatanthus sucuuba, Apocynaceae	Sucuúba	Tree, bark exudates	Brazil	Natural forest	Worms, herpes, uterine inflammation
Parahancornia fasciculata, Apocynaceae	Amapa	Tree, bark latex	Brazil	Home gardens and managed forests	Respiratory diseases and recovery from malaria
Piper methysticum, Piperaceae	Kava	Shrub, root	Polynesia	Forest understorey	Stimulant/Narcotic
Prunus africana, Rosaceae	Red stinkwood, pygeum	Tree, bark extracts	Africa	Natural forest and cultivated	Chest pains, benign prostate hypertrophy
Stryphnodendron barbatima, Fabaceae	Barbatimão	Tree, bark	Brazil	Natural forest	Haemorrhage, uterine and vaginal infections
Tabebuia impetiginosa, Bignoniaceae	Pink ipê or Pink lapacho	Tree, bark	Brazil	Natural forest	Inflammations, ulcers, skin ailments
Uncaria tomentosa, Rubiaceae	Cat's claw, vilcacora	Vine, bark, root and leaves	Neotropics	Natural forest and cultivated	Treatment of cancer, HIV, general health tonic, contraceptive, anti-inflammatory, diarrhoea, rheumatic and urinary tract disorders
Croton lechleri, Euphorbiaceae	Dragon's blood	Pioneer tree, latex	Neotropics	Natural forest and cultivated	Wounds, fractures, and haemorrhoids, intestinal and stomach ulcers
Garcinia lucida, Clusiaceae	Essock	Tree, bark, and seeds	West Africa	Natural forests	Food poisoning, stomach and gynaecological pains, snake bites

Information sourced from Cunningham et al. 2008.

and repel insects (Table 15.6); perhaps the best known of these treatments is the use of quinine (an alkaloid) from the bark of South American *Cinchona* spp. (Rubiaceae) to treat malaria. In Brazil and Africa, oil from *Carapa guianensis* (Meliaceae) is used to keep malarial mosquitoes and jiggers at bay. Other extracts include familiar alkaloids like caffeine, nicotine, and cocaine, as well as a host of other compounds with biologically significant activities. Even more than a decade ago, more than 10,000 alkaloids had been identified in 300 plant families (Raffauf 1996).

Indeed, for most of human history doctors were basically herbalists, and most rain forest communities still use a variety of plants and animals for medicinal use (Figure 15.7) (Cunningham et al. 2008). Many forest-dwelling peoples have their own health traditions and folklore, often intertwined with spiritual beliefs and practices, with knowledge passed from generation to

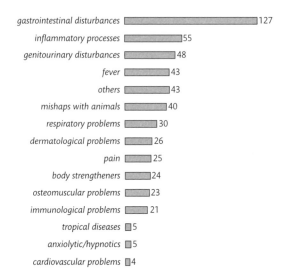

gastrointestinal disturbances 127
inflammatory processes 55
genitourinary disturbances 48
fever 43
others 43
mishaps with animals 40
respiratory problems 30
dermatological problems 26
pain 25
body strengtheners 24
osteomuscular problems 23
immunological problems 21
tropical diseases 5
anxiolytic/hypnotics 5
cardiovascular problems 4

Figure 15.7 The 519 distinct medicinal uses of plants and animals recorded amongst the inhabitants of Jaú National Park, Amazonas, Brazil, summarized into 15 categories. A total of 120 plants and 29 animals were used (Rodrigues 2006).

generation. For the Tabwa and Kongo of Zaire the capacity to heal is associated with those in positions of leadership and power (Janzen 1978; Roberts 1997). Sometimes women and men hold very distinct knowledge, e.g. relating to childbirth. Such issues make it complex to assess traditional knowledge fully[10] (Box 15.6), and recognizing the intellectual property rights of forest-dwelling people has been an issue of considerable sensitivity (Box 15.7).

Traditional healers remain the dominant medical care providers in many forested areas. Indeed traditional practitioners can be preferred over 'modern' alternatives—in Cameroon, they demand and gain higher prices, though the belief that healers can cure otherwise incurable diseases, and also harm patients with their magic, encourages such payment (Leonard 2003).

Health care systems are beginning to place more value on traditional systems of health care, particularly in remote regions that lack access to pharmaceutical medicines. In Brazil, traditional medicines are being incorporated into local community health care, and in the Peruvian Amazon an indigenous federation (FENAMAD) is also addressing local health care needs through traditional medical knowledge (Alexiades and Lacaze 1996). Such systems of health care have long been known in South and Southeast Asia and have been developed into a highly sophisticated industry to provide herbal treatments for many common ailments (Wahlberg 2006). In view of the variety and apparent effectiveness of many traditional remedies, it is often claimed that many new medicines from rain forests await discovery and refinement by Western pharmaceutical industries, though the realization of such rewards has proved difficult (Pearce and Puroshothaman 1995) (Chapter 17). Nevertheless, rain forest products have considerable importance, both to traditional forest-margin communities and as the basis of a highly profitable international trade in forest-derived traditional medicines that encompasses urban dwellers worldwide.

[10] While many traditional medicines have genuine pharmaceutical values, the use of a medicinal product, however widespread, does not prove that it works. Tiger bone and rhino horn are highly valued in Chinese medicine. Nutmeg and clove were sold as a cure for the bubonic plague in Europe in the 16th and 17th centuries. In both examples there is no scientific basis for the curative benefits of such substances. 'Traditional' cultures are not invulnerable to quack remedies.

Box 15.6   Tapping traditional knowledge and local know-how

Local knowledge helps. In 1818 the then British Governor of the East Indies, Thomas Stamford Raffles, and his botanist friend Joseph Arnold asked locals to guide them to see interesting plants in Bengkulu, Sumatra; a giant flower was 'discovered'—we know it as *Rafflesia arnoldi*. 'New' species continue to be discovered with local help, but now there is more recognition of local knowledge.

Traditional knowledge and scientific perspectives can show strong inherent similarities (Rist et al. 2010). Examples include the correspondence between Mayan categorization of local vegetation and the result of multivariate species analysis in Yucatan, Mexico (Hernandez-Stefanoni et al. 2006).

The role of local knowledge in drug discovery and other 'bioprospecting' is well established, but there are other roles too. Local people, sometimes after training as parabiologists, have already helped biologists to address goals ranging from taxonomic collection and developing estimates of local or even global species numbers (Novotny et al. 2002), distinguishing habitat types in the Peruvian Amazon (Shephard et al. 2001), explaining fire regimes and consequences in local contexts (McDaniel et al. 2005), and detailing environmental trends (Hellier et al. 1999). They can also be vital in more applied topics such as raising environmental awareness (Basset et al. 2000), improving timber harvesting approaches (Sheil et al. 2006), monitoring environmental impacts (Basset et al. 2000), as well as in planning, implementing and evaluating conservation interventions (Danielsen et al. 2000; Sheil and Lawrence 2004).

Local knowledge is not necessarily accurate or unambiguous (Rist et al. 2010). For example, traditional people often deny that malaria is transmitted by mosquitoes (Colfer et al. 2006). The Makushi people of Guyana recognize various 'local fauna' that do not correspond to animals recognized by biologists—examples included several additional monkeys and cats. Some of these may be new species, but local people also note that four of the undescribed 'cat' species live entirely underwater, along with an 'aquatic anteater', suggesting that mythical creatures and/or unorthodox categories are also in use (Forte 1996).

Knowledge varies among and within communities. While non-indigenous groups often possess significant knowledge, indigenous groups typically have a richer insight. In Acre, Brazil, both the indigenous Yawanawa and Kaxinawa communities and non-indigenous rubber tapper communities knew many uses for several common palm species, but the indigenous communities know significantly more (Campos and Ehringhaus 2003). Typically knowledge varies amongst individuals, and to some degree with gender, age, and experience (Case et al. 2005).

Given the potential commercial benefits, many countries now have strict regulations concerning local knowledge. The ethics of benefit-sharing remain a hotly contested topic—though some basic principles were agreed in the **Convention on Biological Diversity**.

15.4.3 Psychoactive plants

Cultural richness and well-being of tropical rain forest communities are often intimately associated with the ritual use of plants and animals, for cultural and religious purposes. In some regions and cultures, some of these plants are psychoactive (Shepard 2005) (Table 15.7), and while some are effectively recreational drugs, many play a specific cultural role. The indigenous people of Papua New Guinea, for example, make a drink by boiling the leaves and bark of the tall forest tree *Galbulimima belgraveana* with leaves from *Homalomena* spp. which, when imbibed, causes them to become violently intoxicated before falling into a deep sleep during which they experience visions and fantastic dreams.

Many mildly addictive forest-derived substances are very familiar: caffeine from coffee (*Coffea* is a genus of African understorey tree), and chocolate containing caffeine-like theobromine from the American cacao tree

---

Box 15.7   Intellectual property rights

Capitalizing upon forest people's knowledge through, for example, bioprospecting (Chapter 18) raises difficulties associated with intellectual property rights (Greene 2004; Hamilton 2004). Intellectual property rights are a specific form of property law related to the protection of the products of human creativity by which owners of information can secure rewards associated with its profitable use (Phillips and Firth 1990). It has been estimated that much less than 0.001% of the market value of medicines derived from plants used by indigenous communities has ever been returned to the source communities (Posey et al. 1995). In the context of bioprospecting based on ethnobotanical knowledge of forest communities, the intellectual property rights of the communities must be recognized, allowing them to secure appropriate financial rewards stemming from the use of their knowledge.

Yet intellectual property rights have been established to protect individual inventors and their inventions rather than the collective knowledge of indigenous communities. The Convention on Biological Diversity aims to promote, among other things, 'fair and equitable sharing of the benefits arising out of the utilization of genetic resources', but achieving this objective through the provision of intellectual property rights remains subject to continuing debate (Monagle 2001). Even if rights were to be awarded to communities, exerting them is difficult, expensive, and time-consuming. Without a workable system of intellectual property rights for indigenous and forest margin communities, the benefits of bioprospecting for these communities are unlikely to be realized, and hence they will have little incentive from bioprospecting to conserve forests. Control over information has become one of the main struggles of the indigenous movement, who are now demanding intellectual property rights over research information and just compensation for economic benefits that may eventually accrue (Posey 2002; Greene 2004).

(*Theobroma cacao*). Cola[11] drinks are based on caffeine-rich *Cola* nuts (a genus native to African rain forests). These nuts are widely used by truck drivers in Central and West Africa to inhibit fatigue. While some plants such as cocaine[12] are already illegal, some others such as kava and kratom enjoy a growing international trade.

15.4.4 Growing trade and declining supply

There is increasing concern for the continued provision of medicines from rain forest plants, as a result of forest degradation and contraction and over-exploitation of medicinal plants to meet a growing trade (Hamilton 2004). Many examples of declining availability of medicinal plants have been demonstrated (Shanley and Luz 2003; Cunningham et al. 2008). In some cases logging has resulted in the loss

of community access to medicinal resources. This is particularly problematic for 'dual-purpose' species that have medicinal importance but also provide valuable timber, such as *Prunus africana*, *Tabebuia impetiginosa*, and *Hymenaea courbaril*. Integration of medicinal plants within agroforestry systems has taken place, but often this is not sufficient to prevent over-exploitation of wild populations, as has been the case with *Prunus africana* in Cameroon (Cunningham et al. 2008).

15.5   Forests, subsistence, and poverty

It is hard to find a widely acceptable definition of poverty. Economists tend to use measurable financial or material criteria, and perhaps include access to services

---

[11] The ingredients supposedly used in one famous cola drink include numerous forest-derived products: cinnamon, cola nut, nutmeg, vanilla, and processed coca leaves (now with the cocaine removed).

[12] Many farmers in the foothills of the Colombian Andes still find this a highly profitable crop.

Table 15.7 Common psychoactive plants from tropical rainforests.

Species, Family	Common name	Life form, part used	Region	Source	Uses and trade
Areca catechu, Arecaceae	Betel nut	Palm, fruit	South and Southeast Asia	Agroforestry systems	Ritual and recreational use, widely traded
Banisteriopsis caapi, Malpighiaceae	Ayahuasca	Vine, stem	South America	Home gardens and managed forests	Ritual, small-scale trade
Erythroxylum coca, Erythroxylaceae	Coca	Shrub, leaves	South America	Cultivated in the Amazon	Prevents fatigue and altitude sickness
Mitragyna speciosa, Rubiaceae	Kratom	Tree, leaves	Southeast Asia, New Guinea	Home gardens and natural forest	Prevents fatigue, treatment of morphine addiction
Piper methysticum, Piperaceae	Kava	Shrub, root	Western Pacific	Agroforestry systems	Ceremonial and recreational use, widely traded
Psychotria viridis, Rubiaceae	Chacruna	Shrub, leaves	South America	Cultivated in South and Central America	Hallucinogen, eye drops and treatment of migraine, widely traded
Tabernanthe iboga, Apocynaceae	Iboga	Shrub or small tree, roots	Central Africa	Natural forest	Hallucinogen, widely traded
Virola spp., Myristicaceae		Tree, resin	South America	Managed forests	Hallucinogen, barter

Adapted from Cunningham et al. 2008.

such as health and education. It is also possible to emphasize a more comprehensive view of well-being that may include power and status. It is also reasonable to ask whose criteria count. Many in New Guinea, and other regions still possessing strong traditional societies, lack significant financial income, education and modern health care, but are insulted to be called 'poor'—they gain most of their material needs from their community, lands, and forests. Nonetheless, it is also obvious that many millions of people are undoubtedly poor regardless of definitions. According to World Bank figures for 2005, about half of all humans are poor (living on less than US $2 a day), and 25% live in extreme poverty (less than US $1.25 a day).

What relevance do forests have to poverty? Over 90% of the 1.4 billion people living in extreme poverty depend on forests for some part of their livelihoods (World Bank 2002). Forests are often essential to the needs of the poorest who rely on them for a wide spectrum of products and goods (Ogden 1990). Often trading in forest products also provides an important source of cash income. Forests can also provide an important security function as a safety-net during times of hardship, as when crops fail (Ogden 1990; Takasaki et al. 2004).

Areas of rural poverty and natural forest tend to overlap in the tropics where the poorest people tend to live in remote regions (Sunderlin et al. 2005). Many forested areas have low population densities and are simply neglected. Lack of infrastructure, poor communication, isolation from markets, and limited health and education services constrain livelihood options. Often the inhabitants of forested areas are minorities, with their own languages and cultures, and such groups are often marginalized or indeed persecuted due to ethnicity, their perceived primitive state, and their lack of political voice and representation.

Can forests help people escape poverty? In some cases conversion of forests to alternative land uses has led to significant livelihood enhancement for those who subsequently use the land. But there are also examples where deforestation has increased local hardship, with reduced food security and well-being. Usually some gain while others lose, and it is often the poorest who are most vulnerable.

Other development initiatives over the past two decades have emphasized the potential for non-timber forest products to help lift people out of poverty (Belcher 2005). The notion that NTFPs are available for helping poor people implies that timber, often the most valued trade commodity, is unavailable (Dove 1993). Many countries possess regulations that ensure forest trees are owned by the state, which effectively outlaws a commodity that could best help local people attain wealth. A recent study in Kalimantan (Indonesian Borneo) found that 7 out of 10 of the most valued forest species from the perspective of indigenous forest-dependent people were actually timber species (Sheil et al. 2006). In Mexico, where many forests are owned by communities who can legally sell timber, there are noted benefits for both the communities and the protection of forests (Bray et al. 2003).

Among the many challenges confronting those who seek to improve the flow of wealth from forests to the poor is ensuring that the poor genuinely benefit. Valuable resources are frequently captured by powerful elites, often by displacing the intended beneficiaries (Capistrano and Colfer 2005). Another challenge is to maintain the forest as incomes rise, as this often stimulates forest loss by raising demand for land (Wunder 2001).

15.6 Conclusion

While humans have long derived many free goods and benefits from rain forest lands, others have found ways to cultivate the land and to increase productivity. Competition has often led to depletion and scarcity—but this has also led to a growing awareness that sustainability requires restraints. The erosion of local cultures and knowledge is a significant concern with many parallels to biodiversity conservation. Living in or near to forests is not wholly beneficial—costs include various associated health risks. Even those living far from forests are threatened by emerging diseases that originate in forest regions. Many poor people continue to depend on rain forests, this claim along with those of the global community and of future generations ensuring that the future of rain forests and rain forest lands is as much an ethical issue as a biological or political concern.

CHAPTER 16

Biodiversity in a changing world

'…A tropical forest which has once suffered by the hand of man never recovers its original splendour, even were it left to itself for a century. Some will say that this indelible mark is the seal with which man, as king of creation, impresses his conquest; others will be inclined to think that this miserable biped has, like the fabled harpies, the sad faculty of soiling and withering whatever he touches.'

Paul Marcoy (1872)

Human impacts on tropical forest cover[1] (Chapter 15) threaten tropical rain forests and their biodiversity. Indeed, the very concept of biodiversity has grown alongside concern for tropical rain forest loss, and the two issues have been closely intertwined through their conceptual development (Box 16.1). The scale and rapidity of deforestation have been suspected to be driving a contemporary mass extinction event (Myers 1993; and see Box 16.1) comparable to the mass extinctions of the geological past. While global in extent, the current extinction event is considered most prominent in tropical rain forests where most of the world's species reside. Catalogues of threatened taxa compiled by the IUCN suggest that around 32% of amphibians, 20% of mammals, 12% of birds, and 3% of flowering plants are considered to be threatened with extinction (http://www.iucnredlist.org).

Many of these species occur in 'biodiversity hotspots' (Myers et al. 2000), vulnerable regions of high biological uniqueness. Many of these hotspots are centred on tropical rain forest biomes (Box 16.2), where dramatic species losses have been predicted. Widespread declines of taxa have been attributed to habitat clearance, but also to disease, climate change, and invasive species (Box 16.3). Dramatic amphibian declines have made headline news, particularly when well-known species such as the golden toad *Bufo periglenes* of Monteverde, Costa Rica, are declared extinct[2]. What seldom make the news are the many invertebrates that may also be heading for extinction without even the dignity of documentation (Dunn 2005).

The consensus among scientists is that anthropogenically-driven environmental change will, over the coming decades and centuries, increase extinction rates by

---

[1] The principal driver is deforestation through clearance for agriculture, with secondary importance being allotted to forest degradation by logging or other activities, hunting, invasive species, fragmentation, fire, and climate change, although interactions among these drivers have considerable significance. An IUCN report on threatened mammals, birds, and plants concluded that habitat loss or degradation was the principal cause of vulnerability to 89%, 83%, and 91% of the species in these groups respectively (Hilton-Taylor 2002).

[2] The golden toad was declared extinct by IUCN in 2004.

Box 16.1   Biodiversity and deforestation concepts

Tropical rain forest destruction and biodiversity loss have been inextricably linked in the public conscience with respect to rising environmental awareness since the 1960s. The term 'biodiversity' itself cannot be easily defined, and has different meanings depending on the context in which it is used. The term is perhaps best presented as encapsulating the intersection between measurable species, habitat, and ecosystem properties on the one hand, and more abstract social, ethical, and political environmental concerns on the other. It is, perhaps, 'a tool for a zealous defence of a particular social construction of nature.' (Takacs 1996).

The development of biodiversity as a concept paralleled a rising concern for loss of rain forest. Paul Richards in his classic 1952 text *The Tropical Rain Forest* was among the first to raise concerns relating to the widespread and rapid clearance of rain forests. At this time his views were not widely shared by other writers on the environment, though H. C. Dawkins suggested even then that a third of every area exploited should be protected from any damage to ensure the maintenance of ecological processes and services (Dawkins 1958).

It was not until the late 1960s and early 1970s that it dawned on Western minds that tropical landscapes were not indestructible and resilient but were inexorably being transformed by bulldozers, chainsaws, and insatiable demand for resources. In 1972 Arturo Gomez-Pompa and colleagues expressed their concerns about the long-term viability of tropical rain forests (Gomez-Pompa et al. 1972). The following year Paul Richards reiterated that rain forest destruction was directly responsible for the loss of 'genetic diversity' and 'species richness', introducing these terms perhaps for the first time to a wider audience (Richards 1973).

The publication, some years later, of the first global assessment of the status of tropical moist forests suggested that 11 million hectares were being lost annually (Sommer 1976).

By the end of the 1970s tropical deforestation had become accepted as an environmental problem of international significance. This was largely due to its association with the economic and, especially, ecological value of biological diversity. Deforestation and species diversity became symbols of the environmental movement, supported and further popularized by a number of books by leading environmentalists, including Norman Myers' *The Sinking Ark: a New Look at the Problem of Disappearing Species* (1976) and Paul and Anne Ehrlich's *Extinction: the Causes and Consequences of the Disappearance of Species* (1981). The idea that tropical deforestation was leading directly to catastrophic species extinctions became entrenched in environmental thinking.

The term 'biodiversity' was first introduced to the world in 1986 by Walter G. Rosen within the title of the National Forum on BioDiversity, a meeting sponsored by the Smithsonian Institution and the US National Academy of Sciences. The edited proceedings of this meeting were subsequently collated in 1988 by E. O. Wilson under the singular title *Biodiversity* (Wilson 1988). In these contexts biodiversity was shorthand for concepts such as 'species richness', 'variety of life', and 'genetic diversity', but it quickly became entrenched in popular literature and culture as encapsulating various aspects of natural heritage, the environmental movement, and, especially, the deforestation discourse. Within the space of a few years, 'biodiversity' as a concept was officially enshrined within the Convention on Biological Diversity, as declared at the Rio de Janeiro Earth Summit in 1992.

Box 16.2   Biodiversity hotspots

Many terrestrial species have small geographic ranges, and the distribution of these species is overwhelmingly concentrated in just a few locations. These regions have been mapped as 34 'biodiversity hotspots', representing 2.3% of the world's land surface but contain, as endemics, around 50% of all plant and terrestrial vertebrate species (Myers et al. 2000, 2004). To qualify as a hotspot, plants endemic to the area must comprise at least 0.5% of all the world's plants (based on a global estimate of 300,000 species) and must have lost at least 70% of their primary vegetation. Hence the Amazon Basin, the most species rich region in the world, is excluded.

Of the 34 hotspots, 20 are wholly or partly tropical rain forest areas[1] (Table 16.1). These collectively represent the sole remaining habitats of an estimated 34% of the Earth's plant species and 32% of its terrestrial vertebrates. In this subset of rain forest hotspots only 14% of the original primary habitat remains (compared to about 50% for tropical rain forest regions as a whole).

About 10% of the hotspot area total is under some form of protected area designation, although many designated protected areas continue to be impacted by human activities. But even if protection afforded to these areas were sufficient to prevent further human encroachment, the lag time to extinction associated with habitat area losses is likely to result in many species extinctions over the coming decades. Species-area relationships suggest that even if all remaining habitats in tropical forest hotspots are saved, around 18% of their species will still be lost over 100 years (Figure 16.1). If only the currently protected areas remain, the extinction curve is correspondingly worse.

[1] A further two include tropical dry or highly seasonal forests.

Table 16.1 The twenty tropical wet forest biodiversity hotspots, as defined by Myers et al. (2004).

Hotspot	---Habitat---		---Plants---		---Vertebrates*---		Predominant biomes
	Original (km²)	Remaining %	Species	Endemics (%)	Species	Endemics (%)	
Tropical Andes	1,542,644	25	30,000	50	4,442	39	Tropical and Subtropical Moist Broadleaf Forests; Montane Grasslands and Shrublands
Tumbes-Chocó-Magdalena	274,597	24	11,000	25	1,955	19	Tropical and Subtropical Moist Broadleaf Forests
Atlantic Forest	1,233,875	8	20,000	40	2,330	31	Tropical and Subtropical Moist Broadleaf Forests
Mesoamerica	1,130,019	20	17,000	17	3,334	36	Tropical and Subtropical Moist Broadleaf Forests
Madrean Pine-Oak Woodlands	461,265	20	5,300	75	1,539	9	Tropical and Subtropical Coniferous Forests
Caribbean Islands	229,549	10	13,000	50	1,521	60	Tropical and Subtropical Dry Broadleaf Forests

(continued)

Table 16.1 (continued)

Hotspot	---Habitat---		---Plants---		---Vertebrates*---		Predominant biomes
	Original (km²)	Remaining %	Species	Endemics (%)	Species	Endemics (%)	
Guinean Forests of West Africa	620,314	15	9,000	20	2,077	20	Tropical and Subtropical Moist Broadleaf Forests
Maputaland-Pondoland-Albany	274,136	25	8,100	23	1,092	7	Tropical and Subtropical Moist Broadleaf Forests; Montane Grasslands and Shrublands
Coastal Forests of Eastern Africa	291,250	10	4,000	44	1,405	8	Tropical and Subtropical Moist Broadleaf Forests
Eastern Afromontane	1,017,806	11	7,598	31	3,340	30	Tropical and Subtropical Moist Broadleaf Forests; Montane Grasslands and Shrublands
Madagascar and the Indian Ocean Islands	600,461	10	13,000	89	1,241	82	Tropical and Subtropical Moist Broadleaf Forests
Western Ghats and Sri Lanka	189,611	23	5,916	52	1,232	41	Tropical and Subtropical Moist Broadleaf Forests
Himalaya	741,706	25	10,000	32	1,849	8	Tropical and Subtropical Coniferous Forests; Montane Grasslands and Shrublands
Indo-Burma	2,373,057	5	13,500	52	3,801	27	Tropical and Subtropical Moist Broadleaf Forests
Sundaland	1,501,063	7	25,000	60	2,793	39	Tropical and Subtropical Moist Broadleaf Forests
Wallacea	338,494	15	10,000	15	1,402	41	Tropical and Subtropical Moist Broadleaf Forests
Philippines	297,179	7	9,253	66	1,317	45	Tropical and Subtropical Moist Broadleaf Forests
East Melanesian Islands	99,384	30	8,000	38	661	44	Tropical and Subtropical Moist Broadleaf Forests
New Caledonia	18,972	5	3,270	74	269	37	Tropical and Subtropical Moist Broadleaf Forests
Polynesia-Micronesia	47,239	21	5,330	58	475	49	Tropical and Subtropical Moist Broadleaf Forests
	Total	%	Total	Total	Total	%	
Summary	13,282,621	14	228,267	911	38,075	32	

* includes freshwater fishes.

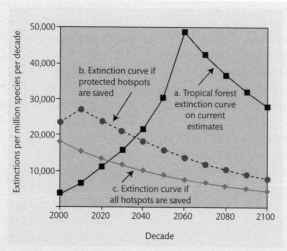

Figure 16.1 Projected extinction in tropical forests, 2000–2100 (Pimm and Raven 2000). Three projections show extinction rates (millions of species per decade) under three scenarios: a) extinction curve on current estimates of tropical forest loss; b) extinction curve assuming all currently remaining habitat within hotspots are saved; and c) extinction curve if only currently protected habitat within tropical forest hotspots are saved. The curves are derived from species area calculations detailed in Pimm and Raven (2000) and are based on 17 hotspots that include all types of tropical forest. Extinction peaks are apparent because of the non-linear relationship between species and area, and the time lags before extinctions unfold.

Box 16.3   Amphibians are croaking

Since the 1980s amphibians have undergone a dramatic global decline, and 32% of the 5,743 known amphibians are now considered threatened with extinction—a disproportionate number of these occurring in Neotropical montane tropical forests (Stuart et al. 2004). Most vulnerable are amphibians that have the combination of low fecundity, high specificity to rain forest habitats, and larval development in streams (Lips et al. 2003, 2005b).

Proposed causes include habitat loss, over-exploitation, acid precipitation, increased UV-radiation following depletion of the ozone layer, increased air pollution, disease, and changes in global climate (Young et al. 2001; Stuart et al. 2004; Lips et al. 2005b, a) although none of these explanations is wholly convincing on its own. Nevertheless, the emerging consensus is that the disease chytridiomycosis, caused by the chytrid fungus, *Batrachochytrium dendrobatidis*, is the main cause of amphibian decline

(Berger et al. 1998; Daszak et al. 2003; Lips et al. 2005a, 2006). Studies suggest that the fungus performs poorly at temperatures above 22°C, which may explain why chytridiomycosis-induced amphibian declines occur primarily in cool regions, like mountain chains (Kriger et al. 2007; Andre et al. 2008). Global warming may nevertheless contribute to the vulnerability of amphibians to disease by shifting local temperatures closer to the optimum for the *B. dendrobatidis* pathogen (Pounds et al. 2006).

Some researchers contend that the focus on chytridiomycosis has made amphibian conservation efforts dangerously myopic. Increasing climatic variability, also associated with global warming, may cause developmental problems and increase stress among amphibians, which increases their vulnerability not only to chytridiomycosis, but also to other diseases (Alford et al. 2007; Di Rosa et al. 2007) and other causes (Kiesecker et al. 2001).

three to four orders of magnitude over background rates (as determined by the fossil record). There is a glimmer of hope that a time lag between environmental modification and extinction allows opportunities for conservation to save species that would otherwise be destined for extinction. We have many challenges to overcome, many of them in tropical rain forests. The task encompasses a broad range of issues: effective conservation of tropical forests and their myriad species is generally more about politics, economics and people than it is about biology. In this chapter we outline some of the patterns and causes of biodiversity loss in tropical rain forests, leaving a description of conservation responses to the next chapter.

16.1   How many species are going extinct?

Birds are the most intensively recorded group of species worldwide. Based on known bird extinctions since 1500, species losses have occurred at the rate of around 26 extinctions per million species per year (E/MSY) over this period (Pimm et al. 2006). This figure does not account for species that became extinct before their scientific description (most bird species were described after 1850), and neither does it include species that are probably extinct but have not yet been formally declared so. Revised estimates accounting for these sources of error predict bird extinction rates[3] of around 100 E/MSY, two orders of magnitude greater than the fossil background rate of 1 E/MSY, a benchmark against which to measure human-driven extinction rates (May et al. 1995).

Projecting extinction rates into the future has relied on the well-established relationship between species number (S) and area (A) where $S = cA^z$, where c is a case-specific constant and z varies but is often around 0.25. Species loss can thus be estimated based on the difference between original and current habitat cover. Using the species–area relationship, estimates of bird species extinctions appear to correspond well with the number of threatened[4] birds in the Brazilian Atlantic forest and in insular Southeast Asia (Brooks and Balmford 1996; Brooks et al. 1997). Threatened status is not, however, the same as extinction, and although South America's Atlantic forest has suffered more than 90% clearance, with a corresponding prediction of over 40% extinction of species from this region, only one bird, the Alagoas curassow *Mitu mitu*, is currently believed extinct in the wild[5]. It is presumed that time lags explain this discrepancy, and that threatened species have a very high probability of becoming extinct within 100 years (Pimm et al. 1995).

Accepting the validity of the species–area relationship, its extrapolation to the 25 biodiversity hotspots (Box 16.2) suggests that around 1,700 endemic bird species will be lost out of a total of 2,821 hotspot endemic birds under current scenario projections (Pimm et al. 2006). This translates to extinction rates of around 1,000 bird E/MSY resulting from land transformations that have already taken place. In other words, we should expect around 10 bird extinctions each year over the coming decades.

Are these values credible? Species–area relationships have formed the basis of most estimates of species extinctions[6], and so it is not altogether surprising that independent estimates mostly correspond (Figure 16.2), but they incorporate various assumptions. For example, the application of species–area relations to fragmented habitat is questionable as species may persist in the intervening matrix or as resistant metapopulations spread across patches (Koh and Ghazoul 2010).

Nevertheless, current estimates based on species–area curves take no account of other potential causes of extinction, such as hunting, invasive species, and climate change, or forest regrowth following human demographic change (Box 16.4). Projections to 2050 estimate 10% extinction (i.e. species committed to extinction) of all tropical rain forest species based on habitat loss alone[7], but a far greater extinction of 24% under projected

---

[3] Or at least species that considered to be committed to extinction over the coming decades.

[4] As designated by IUCN Redlist criteria.

[5] Although *Mitu mitu* seems to have been rather rare even before extensive deforestation, as since its discovery in 1648 no further sightings were recorded until 1951.

[6] Other estimates rely on extrapolations of the rate at which species move up the ranks of IUCN threat categories.

[7] Using projected annual rain forest conversion values of 0.43%.

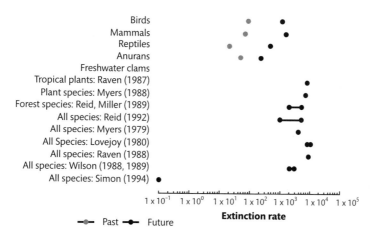

Figure 16.2 Estimates of extinction rates expressed as extinctions per million species per year. For birds through to clams, past extinction estimates are derived from known extinctions in the past 100 years, and future extinctions on the assumption that currently threatened species will be extinct within 100 years.

From Pimm et al. 1995.

mid-range climate change scenarios[8] (Thomas et al. 2004). Regional differences relating to different climate change projections are expected: under minimum expected climate change Mexico is expected to lose 2–4% of its species, while species loss for Queensland is expected to be 7–13% (Thomas et al. 2004).

16.1.1 Extinction proneness and coextinctions

Species are clearly not equally prone to extinction following habitat clearance, hunting, or other anthropogenic pressures. If characteristics associated with vulnerability could be identified, then this would be a useful tool in identifying species at greatest risk.

The most vulnerable butterflies in Singapore[9] are those that have high larval host-plant specificity and adult habitat specialization (Koh et al. 2004c). Species dependent on particular host plants or habitats are obviously highly sensitive to the loss of these hosts or habitats. Habitat or host specialization appears to be a feature

of species that have relatively narrow geographic ranges, and restricted-range butterflies (in this case confined to the Oriental region) were three times more likely to become extinct than more cosmopolitan species. The tendency of cosmopolitan taxa to be more adaptable in their ability to exploit a wider range of habitats and resources allows them to persist in a changing environment. The correlation between local vulnerability and species range has also been found for butterflies elsewhere (Hamer et al. 1997, 2003; Ghazoul 2002) and other taxa such as primates (Harcourt et al. 2002; Harcourt 2006), birds (Jones et al. 2001), and mammals in general (Meijaard et al. 2005). In Koh et al.'s study, more mobile butterfly species were also more resistant to extinction than less mobile species, presumably because they are less likely to be isolated through fragmentation (see below).

Extinctions of plants and animals can have cascading effects, leading to associated extinctions among mutualists, predators, and parasites (Koh et al. 2004a, b). Such

---

[8] Note that these projected extinction rate values are also very controversial.
[9] Singapore provides a natural laboratory for such studies, as its fauna has been species-rich (almost 400 butterfly species in 600 km²) and has been studied in detail. It has also undergone substantial habitat change, and at least 38% of its butterfly species have become extinct in the past century (Brook et al. 2003).

Box 16.4 Is the future of tropical forest biodiversity dependent on human demography?

The rates of tropical deforestation, the principal driver of biodiversity loss, show few signs of diminishing, and many consequently argue that there is little chance of averting an extinction crisis among tropical forest species (Bradshaw et al. 2009). Recently, a more optimistic outlook argued that because humans living in rural settings are the main causes of deforestation, and because human demographic trends suggest slowing population growth rates and rural to urban population shifts, therefore deforestation will slow and natural forest cover will increase through regeneration (Wright and Muller-Landau 2006). In consequence, the anticipated forest losses and mass extinctions will be far less severe than currently projected. The pessimists object to this argument noting that even if the demographic assumptions are correct, most essential habitats for biodiversity would be lost during the time taken for deforestation rates to decline and eventually reverse (Brook et al. 2006). The optimist's response to these criticisms is that many extant tropical forest species have survived much longer periods of forest reductions during glacial cycles and prehistoric human impacts, and are therefore likely to persist within the landscape matrix over the next few decades until such time as they can expand once more into regenerating areas (Wright and Muller-Landau 2006).

A crucial factor that will shape the outcome for biodiversity of these scenarios is the conservation value of secondary forests. Secondary and degraded forests account for around 45% of the remaining 11 million km^2 of remaining tropical forest (Wright 2005), and although conservationists generally agree that they are relatively depauperate compared to primary forests, there is little agreement on just how depauperate they are (Chapter 17). Similarly, whether landscape mosaics (comprising primary forest, regenerating secondary forest, agricultural land, and urban areas) can sustain biodiversity remains uncertain, although recent studies suggest that an optimistic outcome is not unwarranted (Ricketts 2001; Horner-Devine et al. 2003; Chazdon et al. 2009a).

To evaluate the probability of the Wright and Muller-Landau scenario we need more information on the social elements of the thesis, notably a better understanding of how changing demographics, and increasing wealth, and expanding markets are likely to impact on rural land use in the tropics. We also need to know more about the ecological elements, notably the extent to which tropical species are dependent on pristine old-growth forests, as opposed to being able to persist in degraded landscape mosaics.

---

effects will magnify extinction events. While many of the vulnerable taxa will be various ecto- or endo-parasites (Figure 16.3), many others will be more glamorous species that garner more conservation concern. For example, butterfly extinctions have been found to be coupled to extinctions of their host plants in Singapore (Koh et al. 2004b), and the same may apply to over 100 vertebrate and invertebrate species affiliated with army ants which often disappear from fragments that are less than 50 hectares.

Ultimately, it may be the ability of taxa to use and colonize secondary forest and regrowth areas that will

provide the best protection from forest losses, and taxa differ substantially in such abilities (Barlow et al. 2007) (Figure 16.4A). The proportion of old-growth species increases with secondary forest age (Figure 16.4B), and the value of secondary forest for biodiversity generally, initially very low, steadily increases following a few decades of recovery (Chazdon et al. 2009b). As old-growth forest continues to diminish in extent, secondary forests will inevitably play a more significant role for rain forest biodiversity. The ultimate value of these secondary forests will depend on whether and where they persist.

A

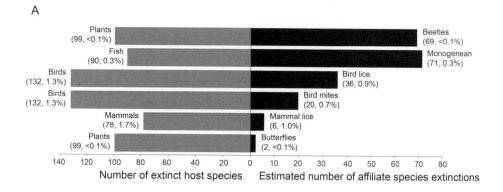

B

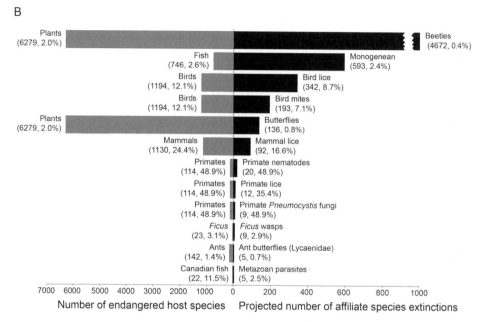

Figure 16.3 Number of coextinctions anticipated from historically recorded extinctions of host species (A) and expected if all currently endangered species were to become extinct (B). The figures in brackets are the absolute numbers and the percentage values.

From Koh et al. 2004a.

16.2   Causes of biodiversity decline

The Food and Agriculture Organization of the United Nations (FAO) defines deforestation as 'the conversion of forest to another land use *or* the long-term reduction of tree canopy cover below the 10% threshold' (FAO 2000). Changes within the forest class (e.g. from closed to open forest) that negatively affect the stand or site and, in particular, lower the production capacity are termed forest degradation. Forest degradation, on the other hand, is defined as 'a reduction of the canopy cover or stocking within a forest' (FAO 2000). Both impact on biodiversity, but the rate of decline of species richness with increasing degradation remains unknown for most taxa.

16.2.1 Logging

Selective timber harvesting reduces the abundance of large mature trees but need not extensively fragment the forest beyond the road networks, skid-trails, and log

A

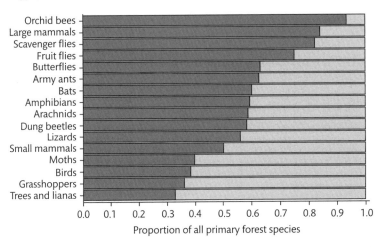

B

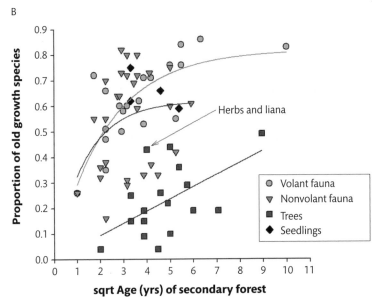

Figure 16.4 (A) Proportion of primary forest species recorded in 14–19-yr-old secondary forests across 16 taxonomic groups of plants, invertebrates, and vertebrates sampled throughout the Jari forest landscape of north-eastern Brazilian Amazonia. (From Barlow et al. 2007.) **(B)** Successional influx of different taxa based on a pantropical compilation of studies (see Chazdon et al. 2009b). The proportion of old-growth species that occurred in each second-growth area sampled is estimated. Taxa are divided into four categories: volant animals, non-volant animals, trees, and seedlings. No trend line was fitted for seedlings, for which there are only four data points.

bays, especially if heavy machinery is avoided[10]. Because large trees contribute more to pollination than small ones, it is frequently assumed that their selective removal may disproportionately affect pollination success of the affected species. In fact there is little evidence from rain forests that either supports or refutes such a suggestion[11], though long-distance pollinations may be sufficient to overcome reduced tree density. Dipterocarps in equato-

[10] Unfortunately the need to keep earth roads dry in the wet tropics means that it is general practice to clear forest on either side of logging roads.
[11] Although reduced pollination and seed set due to logging has been documented at a dry forest site in Thailand (Ghazoul et al. 1998).

rial Southeast Asian forests may be more vulnerable to logging, as their successful recruitment is dependent on massive synchronous fruiting to overcome seed predation by both invertebrates and vertebrates (Curran et al. 1999). A reduction in the number of fruiting trees through logging activities may mean that fruit production across the community is no longer sufficient to satiate seed predators.

Unfortunately timber harvesting is often poorly implemented, leading to damage of most residual trees in harvested blocks. Trees are often indiscriminately removed with little attention being given to forestry regulations. At sites in West Kalimantan, Borneo, logging operations removed 15 mature dipterocarp trees per hectare, damaged or killed an additional 7.5 trees, and left only 0.8 trees in an undamaged state (representing 3% of the original number) (Curran et al. 1999). Eight years later there was some recovery to 5.8 reproductive dipterocarps per hectare, but this still represented an 83% reduction from pre-harvest levels of 25.5 trees per hectare.

The responses of animals to logging, as with fragmentation, are highly variable (Box 16.5). Butterflies have higher species richness and abundance in unlogged forests (Hill et al. 1995), but some butterflies are equally abundant in both forest types (Hill 1999). Bird species richness shows no consistent trend: logging reduced bird species richness in Indonesia (Marsden 1998) and French Guiana (Thiollay 1992, 1997), increased richness in Sabah, Malaysia (Johns 1997), and Sumatra (Danielsen and Heegaard 1994), and had no significant effect on species richness in Uganda (Owiunji 2000; Ngabo and Dranzoa 2001) and Belize (Whitman et al. 1998). Nonetheless community composition changes greatly, with various forest specialists, especially understorey insectivores, often declining or temporarily disappearing from logged sites (Lambert and Collar 2002).

Less is known about mammal responses to logging, but as with other groups, responses are taxon-specific. Logging certainly changes the abundance and type of available food—nutritious fruits are often more readily available in unlogged forest—affecting the distributions of many primates (Johns 1986; Johns and Skorupa 1987), ungulates, particularly mouse deer (Heydon and Bulloh 1997), and some civets (Heydon and Bulloh 1996), but

---

Box 16.5   Sensitivity to logging

Tropical forest vertebrates respond differently to selective timber extraction. Some benefit, such as deer that can feed on the resulting herbaceous growth, while others decline.

Why do some species decline with logging? There are various reasons and these have been well enumerated in Borneo. The drier microclimate of logged forests may impact on humidity-dependent species such as land-breeding amphibians and agamid lizards. Logging creates a disjointed canopy that can impede movement and access for canopy species such as gibbons, (some) squirrels, and binturong that avoid the ground. Reduced numbers of larger trees may deplete nesting sites and hollows required by species such as hornbills, woodpeckers, civets, and squirrels. The increased vegetation density in the understorey may impede species adapted to

more open conditions such as owls and frogmouths. Loss of specific tree species impacts on any dependent species. Fruit species are believed to exert strong control over densities of frugivorous animals, so any change in their abundance following logging can have a direct impact. Damage to special sites, such as salt and clay licks and fruit tree groves, is also likely to impact on animal populations.

Some impacts are not wholly related to timber extraction but rather result from the larger operations including road building and increased access. Thus forest fragmentation, siltation of water bodies, and increased hunting and collection can all have a marked impact. Some silvicultural treatments, such as understorey cutting, reduce habitat quality for many species, although steps can be taken to reduce such negative impacts (Meijaard et al. 2005).

interestingly not Asian fruit bats, most of which do well in human-dominated landscapes on account of their high mobility[12]. Insectivorous bats do not do so well in logged forest, as invertebrates are less abundant and the efficiency of echolocation declines in more open areas (Kingston et al. 2003). Primate vulnerability to logging is thought to be related to their 'arborealness' (Marsh et al. 1987): semi-terrestrial species such as orang-utan generally fare better in secondary forest, although this often exposes them to hunting, particularly when they raid crops adjacent to logged forest. Another theory is that tolerance to logging is related to phylogenetic age: data from Borneo show that species that evolved earlier are more vulnerable to selective logging (Meijaard et al. 2008). Most of these species are endemic to insular Southeast Asia, and have more specialized diets. Species that evolved more recently are more tolerant to logging: this group tends to include generalist omnivores or herbivores, tends to use all vegetation strata, and its species are regionally widespread. The possible generality of such relationships offers the ability to predict the sensitivities of lesser-known species, and thus has important conservation implications (Meijaard and Sheil 2008).

Frogs and toads (anurans) in both larval (tadpole) and adult life-forms are sensitive to desiccation. Logging causes the loss of water bodies (e.g. within leaf axils, bromeliads, tree holes, and puddles) and increases stream temperature and sedimentation[13], to which aquatic tadpoles are very sensitive (Feder and Burggren 1992). Anurans that complete their entire life cycle in leaf litter are likely to be sensitive to the desiccating effects of canopy opening (Zou et al. 1995). It is therefore surprising that few studies have actually demonstrated declines in anuran abundance and diversity in logged areas. Although logged forest in North Sumatra, Indonesia, contained only 20% of the abundance of anurans of unlogged mature forest (Iskander 1999), in East Kalimantan (also Indonesia) anuran abundance was actually higher in logged forest[14] (Meijaard et al. 2005). Increases in abundance of taxa in logged forests can often also be attributed to an increase in generalist species, but another study in Sabah found that anuran species richness was higher in 8–10-year-old selectively logged areas (albeit under reduced impact logging regimes; Chapter 17) than in unlogged forest, a result that was attributed to the availability of more microhabitats in logged forest (Meijaard et al. 2005). Similar results have been obtained in African and Neotropical sites. In Bolivia there was little difference in frog densities among 1-year-old logged and unlogged forests (Fredericksen and Fredericksen 2004), and in Kibale National Park, Uganda, anurans (and reptiles) were more common and species-rich in logged forest (Vonesh 2001). In Madagascar neither amphibian diversity nor abundance were significantly affected by low-level selective logging (Vallan et al. 2004). It appears that anurans, or at least some common species, are more tolerant of logging impacts than might be anticipated.

Fish within tropical forest streams depend on inputs from the surrounding forest, including animals and plant material, for food. Logging can reduce these inputs by changing the structure of the surrounding habitat, particularly by reducing the amount of overhanging vegetation. Reduced shading increases stream temperature, which lowers dissolved oxygen levels while increasing fish metabolic rates. Under these conditions fish mortality can increase from oxygen depletion. Increased turbidity can kill fish directly by clogging gills, and smothers food resources (e.g. algae on rocky beds) and spawning grounds (Kottelat et al. 1993). Despite these apparently negative impacts, fish appear quite resistant to selective logging. At sites in Sabah, selective logging did cause a few species to decline initially, but had little longer-term impact on fish faunas overall (Martin-Smith 1998).

If left alone, logged forests recover much of their original composition and structure within a few decades, and

---

[12] Large Asian fruit bats, the 'flying foxes' or *Pteropus* spp., can cover more than 70 km each night (A.N. Start, quoted in Meijaard et al. 2005).

[13] One study in Sabah recorded an 18-fold increase in stream sediment loads five months after logging, and a 4-fold increase a year after logging (Douglas et al. 1993).

[14] The contrast between the Sumatra and Kalimantan sites may be due to the intensity of logging: in Sumatra every large tree had been extracted, while in East Kalimantan lower logging intensity maintained a more or less closed logged forest canopy.

forest species are therefore expected to recover provided source populations are available. Logging, however, is rarely an isolated disturbance agent, and is often associated with deforestation, fragmentation, and fire, the interaction with which can prevent forest recovery in the long term.

16.2.2 Fragmentation and edge effects

Fragmentation

Fragmentation is caused by the irregular clearance of forest across a landscape, resulting in the subdivision of continuous forest into blocks, separated perhaps only by a road, or in other cases separated by extensive areas under alternative land uses. Forest fragmentation encompasses three main impacts: a reduction in forest habitat; isolation of remaining forest blocks; and edge effects where forest abuts another habitat type. The study of forest biodiversity and ecology must consider each of these elements, but this is difficult as the factors are confounded. Additionally, the surrounding habitat matrix influences the severity of edge effects and responses of plant and animal groups to forest fragmentation (Box 16.6).

The responses of birds to forest fragmentation are particularly well studied. Isolated Amazonian forest fragments smaller than 100 hectares lose half of their forest-dependent birds within 15 years—larger isolated fragments also lose species but at a slower rate (Ferraz et al. 2003). Birds deemed forest specialists on account of their foraging over large forest areas, such as those that follow army ants, are most sensitive to habitat

---

Box 16.6   The landscape matrix: putting fragmentation into context

The matrix, as used in the context of forest fragmentation, is the intervening land (and land uses) surrounding fragmented forest blocks, and the nature of this habitat matrix affects the severity of edge effects. Tree mortality along Amazonian edges bordering cattle pasture, for example, is higher than that bordering pioneer-dominated secondary regrowth areas (Mesquita et al. 1999). Highly contrasting microclimatic and hydrological conditions across the forest–matrix boundary may therefore be transient as secondary regeneration expands and begins to protect the forest edge from desiccating conditions (Camargo and Kapos 1995).

The habitat matrix also affects possibilities for dispersal between fragments. A 'favourable' matrix, perhaps a patchwork of agroforests, may provide ample possibilities for resource use by animal species outside the forest environment, which may further promote plant gene exchange and dispersal (through pollen and seed movement) between forest fragments. Less favourable habitat matrices, such as pasture or extensive monoculture crops, which provide few resources to animals, may prove substantial barriers to animal or plant movement and exacerbate isolation impacts.

As the matrix changes, so too can communities within forest fragments. A marked reduction in euglossine bees (Powell and Powell 1987) and dung beetles (Klein 1989) at sites in Amazonia in the immediate aftermath of forest fragmentation for pasture could no longer be detected when the same sites were re-sampled 6 and 15 years later respectively—previously impacted communities had largely recovered their original populations and compositions as more favourable secondary regrowth encroached onto previously cleared forests (Becker et al. 1991; Quintero and Roslin 2005).

Indeed, one of the criticisms of using species-area curves to predict extinctions from diminishing habitat area is the failure to consider that species may persist or utilize the intervening habitat matrix (Koh and Ghazoul 2010). Increasing recognition of the capacity for secondary or predominantly agricultural landscapes to support a variety of forest species has raised awareness of the potential conservation value of some forms of human-dominated landscapes (Harvey et al. 2008; Chazdon et al. 2009a).

fragmentation (Van Houtan et al. 2007) (Figure 16.5). Sedentary birds appear least vulnerable to fragmentation as their requirements can be met within a relatively small area. For other less forest-dependent birds the severity of forest fragmentation and isolation is largely a function of the properties of the habitat surrounding forest fragments over which they might disperse (Gascon et al. 1999; Daily et al. 2003; Sekercioglu et al. 2007).

Insectivorous birds are especially sensitive to fragmentation, as they tend to be sedentary understorey specialists with specialized foraging strategies (Stouffer and Bierregaard 1995; Canaday 1996; Sekercioglu et al. 2002; Sodhi et al. 2004). Their mixed-species flocks often disintegrate following isolation, but can recover if the birds are able to move across the matrix between forest fragments (Stouffer and Bierregaard 1995). Generally, however, insectivorous species have poor dispersal ability and limited adaptability to modified habitat (Canaday 1996; Develey and Stouffer 2001; Sekercioglu et al. 2002).

Many invertebrates, including some beetles, butterflies, and bees are also sensitive to forest fragmentation (Brown and Hutchings 1997; Didham et al. 1998). Beetles that are common in non-fragmented primary forest become depleted in 100 ha fragments, provoking suggestions that competitively dominant (but poorly dispersing) species are highly sensitive to habitat fragmentation (Didham et al. 1998). Bee taxa also differ in their responses to fragmentation. Social stingless meliponine bees in Costa Rica are associated with larger fragments, while introduced honeybees are more commonly found in open less-forested landscapes (Brosi et al. 2008). The responses of other groups are less well demarcated, and pronounced annual variability in population abundances confounds short-term studies (Roubik 2001). Butterfly species richness and abundance correlate more strongly to habitat heterogeneity than to fragment size *per se*—a moderate amount of forest fragmentation increases heterogeneity and elevates butterfly richness and abundance at local and landscape scales (Brown and Hutchings 1997).

Edge effects

Cleared lands have lower humidity and lower water tables, higher temperatures, and considerably more light than adjacent forest (Wright et al. 1996; Gash and Nobre 1997), and these conditions often penetrate forest interiors causing a number of changes to affect forest communities, collectively termed 'edge effects'[15].

Changes in moisture, temperature, and light increase **evapotranspiration** by understorey plants along forest edges, and deplete soil moisture. This can result in increased mortality of drought-sensitive trees, an impact that can extend up to 300 m into the forest (Kapos 1989; Didham and Lawton 1999; Sizer and Tanner 1999) (Figure 16.6). Edge-related microclimatic and soil moisture changes also cause many trees to lose their leaves and die standing (Sizer and Tanner 1999). The resulting accu-

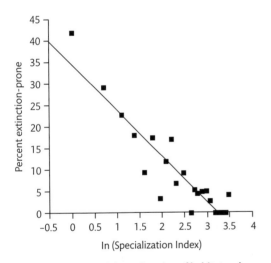

Figure 16.5 Extinction risk as a function of habitat and dietary specialization among birds. Specialization decreases with an increase in the value of the specialization index: thus more specialized birds, based on habitats used and food types consumed, are more prone to extinction. Extinction risks derived from simulations based on IUCN extinction possibilities for threatened species.

From Sekercioglu et al. 2004.

---

[15] Such edge effects become increasingly important with continuing fragmentation and consequently increasing edge to area ratio. Around 17,800 km^2 of Amazonian forest lie within 300 m[15] of a forest edge (Jenkins and Pimm 2003), which amounts to around 0.3% of Amazonian forest area. The far more fragmented Brazilian Atlantic forest has around 29,400 km^2 within 300 m of an edge, and this is equivalent to 29% of the remaining forest.

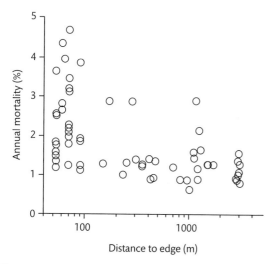

Figure 16.6 Tree mortality rises sharply near forest edges in central Amazonian experimental forest fragments.

From Laurance 2005.

mulation of litter and coarse woody debris reduces seedling establishment and lowers established seedling survival (Bruna 1999; Scariot 2000). Accumulating dead plant material, lower humidity, and higher temperatures, as well as proximity to sources of ignition, also greatly increase vulnerability to fire in periods of drought[16]. Fire further increases canopy openness, creating positive feedbacks that extend the penetration of edge effects (Cochrane and Schulze 1999; Nascimento and Laurance 2004; Vasconcelos and Luizao 2004).

Trees at forest edges are also more exposed to wind—the risk of being snapped or uprooted increases with proximity to the forest edge (D'Angelo et al. 2004). In central Amazonia large trees (exceeding 60 cm dbh) at forest margins appear particularly vulnerable, succumbing to wind and drought-induced mortality nearly three times faster than equivalent trees in the forest interior (Laurance et al. 2000). Vulnerability is also correlated with wood density: denser shade-tolerant species decline while light-timbered pioneer species increase[17] (Laurance et al. 2006b). Thus pioneer species and lianas are favoured, both of which may additionally impact on large trees by exacerbating evapotranspira-

tion effects and reducing their recruitment opportunities (Viana et al. 1997; Laurance et al. 2001). The direct consequences of these effects are reflected in changing forest composition through the loss of species and higher species turnover (Figure 16.7) (Laurance et al. 2006a, b).

Edge effects also affect animal communities, albeit not necessarily in obvious ways. For example, beetle abundance and richness in Amazonia often peak at mid-distances (ranging from 26 to 105 m depending on local

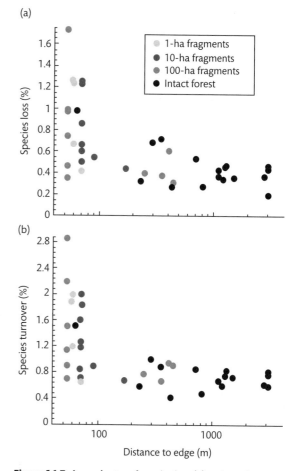

Figure 16.7 Annual rates of species loss (a) and species turnover (b) in Amazonian tree communities as a function of the distance from the forest edge.

From Laurance et al. 2006a.

[16] Rain forest trees are not adapted to fire, and exposure to even low-intensity ground fires causes the mortality of up to half the trees (see Chapter 11) (Cochrane and Schulze 1999; Barlow et al. 2003).

[17] Which has implications for carbon storage in fragmented forests (Laurance et al. 2006a).

site conditions) along the edge–interior gradient, perhaps representing a zone of overlap of edge- and interior-associated species (Didham 1997). The abundance and diversity of butterflies also generally increases at forest edges, with light-preferring butterflies penetrating to at least 300 m into forest interiors (Brown and Hutchings 1997).

Roads, power lines, and other narrow linear features that subdivide forest may clear only a tiny fraction of the forest they penetrate, but the area of forest they affect can be considerably more substantial, due both to edge effects and as barriers to species movement (Goosem 1997). Animals often avoid artificial linear features, particularly when these are roads, trails, or tracks along which people move. In the rain forests of southern Gabon, Central Africa, species richness and often the abundance of nocturnal primates, smaller ungulates, and carnivores was depressed within 30 m of roads independently of hunting pressure (Laurance et al. 2008). In Sumatra it has been shown that forest paths may affect the behaviour of mammals if the trails are regularly used (Griffiths and van Schaik 1993).

Effects on ecosystem processes

Forest fragmentation can alter a number of ecosystem processes by changing community composition or species behaviour. The various stages of plant recruitment, pollination, seed dispersal, germination, and seedling survival have all been demonstrated to be negatively impacted by fragmentation (Ghazoul 2005).

Pollination processes are affected when fragmentation reduces pollinator abundance and subdivides formerly continuous plant populations to the extent that pollinators are no longer able to bridge the gaps between isolated subpopulations (Ghazoul and McLeish 2001; Ghazoul 2005; Aguirre and Dirzo 2008). Nevertheless, it is now apparent that for many species long-distance pollen movement, previously greatly underestimated, allows for gene flow between fragments separated by a few to many kilometres (Mawdsley et al. 1998; Hanson et al. 2008). Isolated trees within the intervening habitat matrix may greatly facilitate pollen

movement between fragments (Dick et al. 2003). Despite these caveats, pollination among plants in forest fragments may become restricted to a smaller pool of individuals, potentially leading to inbreeding and reduced seed set, reduced seed viability, or both (Chapter 12).

The decline of animal seed-dispersers such as primates and birds in small fragments in Africa, Asia, and South America has been linked to lower recruitment of animal-dispersed seedlings and saplings; recruitment of abiotically dispersed propagules is unaffected (Cordeiro and Howe 2001; Chapman et al. 2003; Cramer et al. 2007).

Seed germination may also be impacted as higher temperatures, drier conditions, and increased light penetration, all associated with forest edges and small fragments, can affect germination cues. For example, seeds of *Heliconia acuminata* in continuous forest are between three and seven times more likely to germinate than those in 1 and 10 ha forest fragments (Bruna 1999).

Seedlings are vulnerable to changes in animal communities (Chapter 13). Vertebrate herbivore densities increase 10- to 100-fold following the loss of large predators on small forest islands, with consequential negative impacts on seedlings and saplings of canopy trees (Terborgh et al. 2001). The abundance of leafcutter ants in small Neotropical forest fragments also increases by an order of magnitude[18] which greatly reduces tree recruitment by killing seedlings (Rao et al. 2001; Terborgh et al. 2006; Lopez and Terborgh 2007).

16.2.3 Hunting

There is typically little management of bushmeat resources (Chapter 15), and consequently wildlife is being widely depleted (Table 16.2), especially where high populations of local consumers rely on bushmeat for protein, and commercial hunters supply urban and international markets (Bennett and Robinson 2000; Nasi et al. 2008). As human populations grow, the

---

[18] Though the reasons are unclear.

Table 16.2 Decrease in population densities in hunted areas compared to non-hunted areas.

Location	Country	Reduction in density of target species by hunting (%)	Reference
101 Amazonian sites	Brazil	>90	(Peres 2000; Peres and Palacios 2007)
Quehueiri-ono	Ecuador	35.3	(Mena et al. 2000)
Mbaracayu	Paraguay	53.0	(Hill and Padwe 2000)
Ituri I	D.R. of Congo	42.1	(Hart 2000)
Ituri II	D.R. of Congo	12.9	(Hart 2000)
Mossapoula	C. African Republic	43.9	(Noss 2000)
7 sites in Sarawak and Sabah	Malaysia	62.4	(E.L.Bennett, unpublished data)
Nagarahole	India	75.0	(Madhusudan and Karanth 2000)
Makokou	Gabon	43.0 to 100	(Lahm 2001)
Mbaracayu	Paraguay	0 to 40	(Hill et al. 2003)
Mata de Planalto	Brazil	27 to 69	(Cullen et al. 2000)

From Nasi et al. 2008.

demand for bushmeat across tropical forested areas is likely to increase, and solutions for both wildlife-dependent human populations and exploited species are urgently needed (Bennett et al. 2007). Site-specific integrated solutions that regulate sustained off-take of certain species with protection of more vulnerable species will be needed (Hackel 1999; Wilkie and Carpenter 1999; Davies 2002; Robinson and Bennett 2002). Success is unlikely without parallel development of appropriate land management and tenure, the strengthening of community participation in resource management, and improved governance and institutions (Bennett et al. 2007). Solutions to the bushmeat issue will require an interdisciplinary approach that incorporates biological, economic, social, and political components (Nasi et al. 2008).

Although bushmeat is a primary driver of rain forest animal depopulation, regional and international markets for live birds and other animals exist to trade goods valued as traditional medicines (e.g. tigers, bears) or pets, and which can severely impact on some species (Corlett 2007).

16.2.4 Species introductions

Human-assisted dispersal of flora and fauna has a long history. The early Polynesians, for example, wandering among Pacific oceanic islands, took with them a variety of plants and animals including the cuscus *Phalanger orientalis* which they introduced to New Ireland from New Guinea some 19,000 years ago (Hurles et al. 2003); and much later, at around 3500 BP, pigs *Sus scrofa*, dogs *Canis familiaris*, jungle fowl *Gallus gallus*, and the Pacific rat *Rattus exulans*, along with various plants, were also transported to remote Pacific Islands. While many introductions have been intended, some—like rats and mosquitoes—were not. Some of these species have spread greatly, at the expense of the native biodiversity, and also at considerable expense to the livelihoods and economies of people, communities, and nations.

Most alien species fail to establish in the new habitat to which they have been introduced, but a small minority thrive. The extent of habitat alteration and native species displacement accountable to these invasive aliens can be dramatic. The woody shrubs *Chromolaena Eupatorium odorata* and *Lantana camara* both originate from the Neotropics, but now dominate extensive areas including disturbed forest margins in Southeast Asia, Australia, and parts of Madagascar and Africa (Jenkins and Pimm 2003). The impact of invasive species extends beyond their direct influence on the natural communities. By displacing farmers from productive land they may drive the conversion of other forested areas to farmland.

Oceanic islands are particularly vulnerable to invasive species. Their isolation, low species richness, and simpler community structure are thought to offer less resistance. Rich and interactively complex communities, such as continental tropical rain forests, are widely considered more resistant to invasion, although this view is not universally shared. Rain forest communities that have been substantially altered by alien species occur predominantly on heavily disturbed islands such as Guam, Hawaii, Christmas Island, Mauritius, Vanuatu, and the Seychelles. These islands have been impacted by alien plant and animals including invertebrates, with the spread of these species often facilitated by forest degradation. Thus Mauritius is almost entirely devoid of native forest, having been replaced by dense alien woody shrubland following extensive forest clearance. The lowland forests of the granitic Seychelles are now dominated by *Paraserianthes*, *Cinnamomum*, *Chrysobalanops*, and other alien plants. Common mynas *Acridotheres tristis* and red-vented bulbuls *Pycnonotus cafer* now occur widely in degraded and fragmented forests in South and Southeast Asia.

Intact continental rain forest appears more resistant to invasion, even when surrounded by considerable environmental change. The small remaining area of natural rain forest in Singapore has been infiltrated by only a Neotropical herb, *Clidemia hirta* (Melastomataceae) (Teo et al. 2000). *Clidemia hirta* has also invaded Malaysian rain forests, but is restricted to canopy gaps and degraded areas (Peters 2001). It also occurs in African forests and Pacific islands where it appears to be more shade-tolerant than in its native range, suggesting that the absence of natural enemies reduces light requirements of these species (DeWalt et al. 2004). Other invasive plants of tropical rain forest have similar ecologies in their new ranges. *Piper aduncum* (Piperaceae), introduced from the Neotropics to New Guinea, dominates early successional communities and suppresses native pioneer species. It has, however, never been found in closed primary forests, although it is beginning to spread into naturally disturbed forest habitats such as treefall gaps (Leps et al. 2002).

A classic example of the development of an invasive plant and its consequences is provided by the fast-growing pioneer tree *Maesopsis eminii* (Rhamnaceae). *M. eminii* was introduced to the submontane forests of the East Usambara Mountains, Tanzania, from Central Africa early in the last century. The tree, the seed of which is dispersed by hornbills and monkeys, now dominates large areas of the forest, despite initial observations suggesting that it would not regenerate naturally. This species has been the focus of several studies (Cordeiro et al. 2004).

The invasion remains unchallenged, owing to the massive intervention that would be required for successful control, and as a pioneer it is not considered a threat to pristine forest.

While invasion is often facilitated by human alteration of natural habitats, it need not be: in Hawaii mosquitoes introduced in the early 19th century thrived in native forests and spread avian malaria, decimating native bird species that lack natural resistance to the disease (Atkinson et al. 2005). On Guam the brown tree snake *Boiga irregularis* is probably responsible for the extinction of 12 of the 18 native birds (Fritts and Rodda 1998; Wiles et al. 2003). Brown rats *Rattus rattus* have destroyed island bird population throughout Polynesia, through a series of introduction events that accompanied the spread of Polynesians across the Pacific over the last 2,000 years. They are also held to be partly responsible for the destruction of palm seeds and the loss of palm forests on the formerly forested Easter Island (Hunt 2007).

The spread of invasive species might be facilitated by interactions with native species that act as pollinators or seed dispersers, or indeed the lack of enemies in the novel environments may release invasive species from pressures they encountered in their native range (often referred to as 'enemy release'). Little about such interactions is known in tropical rain forests, although the variety and abundance of caterpillars on the leaves of the invasive Neotropical plants *Piper aduncum* and *P. umbellatum* in New Guinea were just as high if not more so than herbivore pressure on comparable native species, in contrast to enemy release expectations (Novotny et al. 2003).

Positive interactions between introduced species may also facilitate their spread. On the Indian Ocean island of Reunion, the seeds of four fleshy-fruited invasive plants, *Clidemia hirta*, *Rubus alceifolius*, *Lantana camara*, *Schinus terebinthifolius*, are predominantly dispersed by another invasive species, the red-whiskered bulbul *Pycnonotus jocosus* (Mandon-Dalger et al. 2004). Invasion in this case is accelerated through the positive feedback engendered through this alien–alien mutualism.

There is widespread concern that invasive species may cause dramatic changes and disruption of community interactions may lead to an 'invasion meltdown' (O'Dowd et al. 2003). Evidence for such catastrophic change within tropical forest systems remains limited, but examples do exist. Pollination of native species on Guam has declined to almost nothing following the extirpation of the native bird pollinators by the introduced brown tree snake (Mortensen et al. 2008). On Christmas Island the alien crazy ant *Anoplolepis gracilipes* has achieved huge population densities of around 1,000 ants per square metre, and in the process has nearly extirpated the native red land crab, the dominant consumer on the forest floor. This has released seedling recruitment and slowed litter breakdown, resulting in a much denser forest understorey. In the canopy, native host-generalist scale insects have benefited from the attentions of these ants and their populations have increased to an extent where canopy die-back and even the death of canopy trees is now evident (O'Dowd et al. 2003).

In summary, invasive species are rarely problematic in intact closed canopy rain forests, but they do come to dominate disturbed or open forest areas, very likely impacting the extent and rate of forest recovery. Island rain forest habitats appear more vulnerable to invasive species, although successful invasion often accompanies and is facilitated by other damaging activities. Furthermore the low intrinsic rates of population growth of many rain forest trees, slow stand turnover, and the potential for strong selective shifts during naturalization of introduced species, suggest that invasive species will become a greater problem in most if not all tropical forest regions over the next century.

16.3   Conclusion

The impacts of land use on biodiversity are many and interacting. Forest clearance reduces forest area, but also impacts on biodiversity through fragmentation of habitats and the degradation of habitats by edge effects, fire, and invasive species. Forest clearance is often accelerated by other anthropogenic impacts, such as road construction by logging and mining operations, or infrastructure development projects, while logged forest is more vulnerable to fire, the occurrence of which renders forest increasingly susceptible to subsequent fires.

As yet, we know little about the responses of different taxa to forest degradation and deforestation, or the extent to which taxa can persist in a landscape mosaic comprising several habitat types, including forest patches. The shape of response curves to increasing deforestation will determine the rate of biodiversity loss and the differential impact on different components of biodiversity.

While biodiversity is undoubtedly declining, we may be surprised to discover that many forest species are able to persist in a patchwork landscape that includes remnant and regenerating forest patches. Nevertheless, measuring population declines and determining extinction rates are both problematic, as impacts often take years to take effect. These time lags mean that the long-term repercussions of decisions made on short-time horizons are underemphasized.

CHAPTER 17

A matter for scientists and society: conserving forested landscapes

'Nearly the whole of Sarawak is smothered in dense and luxuriant jungles.'

Robert W. Shelford (1916)[1]

'There have been rumours for a long time of the existence of a magnificent forest in Brunei.'

C. Hummel (1914)[2]

Much has changed since Shelford and Hummel expressed wonder about the vastness of the unexplored tropical forests of Borneo in the early 20th century. Less than 100 years later, 'Save the Rain Forests' has become a familiar rallying cry of the environmental movement. But how do we save rain forests? Despite the best efforts of an international environmental movement and the growing indignation of individuals, institutions, and many governments, the area of old-growth tropical forests continues to shrink as people, commercial enterprises, and governments seek to cash in on rain forest land and resources. It is estimated that in 2005 about half of the tropical humid forest contained 50% or less tree cover, and that at least 20% of this biome was subject to timber extraction between 2000 and 2005 (Asner et al. 2009) (Chapter 14). The 'magnificent forest in Brunei' remains magnificent, due to Brunei's oil wealth which alleviates national demands on forest goods or land. Few other regions have escaped with such little loss, and while some extensive rain forest tracts remain intact due to their inaccessibility (such as the north-west of the Brazilian Amazon, eastern Congo, and much of New Guinea) the expectation is that these too will become threatened.

One obvious solution is to create large 'forest protected' areas. To some extent this has worked, in that almost a quarter of the world's rain forests are at least nominally protected by some sort of conservation legislation. On the other hand, protected area systems could be seen as only for those rich enough not to worry about where their next meal is coming from, a colonial relic often synonymous with the exclusion of local communities. Global conservation efforts will continue to rely on protected area networks as a primary strategy for the foreseeable future, and so ensuring their effectiveness and their acceptability remains a serious challenge.

Much conservation effort focuses on implementing protected areas, and indeed we begin this chapter by considering this approach. However, many other activities can contribute to forest and species conservation. A recent classification by IUCN identifies the basic classes: area-based protection, area-based management, species-centred management, education and awareness, law and policy, livelihoods and incentives, and capacity building (Salafsky et al. 2008). This classification implies that an awareness of human livelihood issues, particularly those relating to local people, is also central to conservation efforts (Box 17.1).

---

[1] Referring to the period of 1897–1905.
[2] Deputy Conservator of Forests of the Federated Malay States.

Box 17.1   Conservation and poverty: a story of winners and losers

For hundreds of millions of people, biodiversity has been about eating, staying healthy, and finding shelter. Wild and semi-wild plants and animals contribute significantly to nutrition, health care, income, and culture in developing countries, and the poorest and most vulnerable people often rely on those resources most. Depleting those resources or making them inaccessible can impoverish these people even further. There is a clear ethical argument for 'pro-poor conservation', that is, conservation that explicitly seeks to address basic human needs (Kaimowitz and Sheil 2007).

Many people depend on forests for a substantial proportion of their subsistence needs, including for food, fibre, medicines, and other uses (Neumann and Hirsch 2000). Others perceive forest exploitation as a means by which to escape poverty (Belcher 2005; Sunderlin et al. 2005). Forest conservation is likely to be low on these peoples' priorities if it limits their possibilities for livelihood support, and such costs are seldom effectively compensated or mitigated (Balmford and Whitten 2003). For example, when the mountain gorilla national parks of Uganda, Rwanda and DR Congo were established, a significant population of Batwa 'pygmies' suddenly found themselves without access to forest resources or land. The Batwa people were largely illiterate, marginalized by other ethnic groups, and their

ability to respond to their changed predicament was further undermined by poor health, low status, and cultural collapse (Box 16.4, Chapter 16). What are the rights of the younger generation of Batwa to the forests, to compensation, to local jobs, or to other preferential treatments? There is no unique right answer, but similar tragedies are being played out again and again across the tropics.

Pro-poor conservation objectives might also conflict with powerful commercial interests. In some cases action by local communities, often supported by NGOs, has linked local livelihoods with environmental concerns, and has delivered some notable successes. Most famously, Francisco Alves 'Chico' Mendes, a rubber tapper in the Brazilian Amazon, devoted his life to the defence of the workers and people of the forest. Chico Mendes combined the protection of labour rights with the defence of the forest and was instrumental in forming the National Rubber Tapper Council, which lobbied for rubber tapper livelihoods through forest protection and agrarian reform against the powerful interests of large-scale cattle ranching. Chico Mendes was murdered for his efforts on 22 December 1988, but his legacy is the creation of a network of extractive reserves across Brazil which place the economic interests of local people before those of large commercial enterprises.

This theme underpins various other approaches to conservation which we describe further below.

17.1   Defined conservation areas

17.1.1 Protected area systems

Worldwide there are over 100,000 terrestrial protected areas, covering 12% of land area[3] (Chape et al. 2003).

Coverage is greatest in the tropics (Figure 17.1). Given current interests in protecting tropical forests for mitigation of climate change, coverage of tropical forest is expected to grow.

Tropical forest protected areas (PAs or 'parks'), of which there are many different kinds, are designated to prevent deforestation and thereby secure conservation of biodiversity and natural resources. In tropical moist forest zones 23.3%[4] (2003 value) of the land surface is

---

[3] This coverage varies widely by IUCN categories (whereby, broadly speaking, Categories I–IV are strict nature reserves while Categories V and VI are managed to preserve cultural features and ecosystem services respectively).
[4] Equal to 2,450,344 km^2 within 3,422 sites. The extent of protected rain forest area varies depending on sources and definitions.

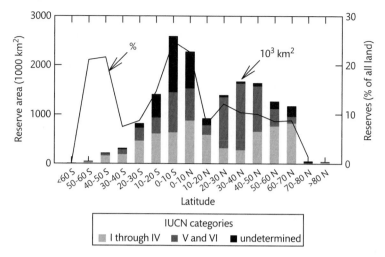

Figure 17.1 The latitudinal distribution of protected areas as a percentage of all land area (connecting line) and as absolute area (histogram). From Brooks et al. 2009.

All data from the World Database on Protected Areas (2007).

under some sort of national conservation designation, a considerable increase from around 8.8% in 1997[5] (Chape et al. 2003). Most of this is located in the Neotropics, where the largest rain forest blocks remain (Table 17.1).

Table 17.1 Proportion of land area by tropical rain forest region protected for conservation. Globally, 23.3% of tropical moist forest lies within a protected area (see main text). Values for Europe and North America are also given by way of comparison.

Region	Proportion (%)
Caribbean	11.7
Central America	24.8
Pacific	2.1
South America	24.9
Brazil	18.3
South Asia	6.8
Southeast Asia	14.8
Western and Central Africa	8.7
Europe	13.1
North America	18.2

Adapted from Chape et al. 2003.

Effectiveness of protected areas

The effectiveness of protected areas can be undermined by limited capacity to withstand pressures for forest conversion and degradation (Bruner et al. 2001, 2004). For example, the protected lowland forests of West Kalimantan have suffered a 56% decline in forest cover from 1985 to 2001 due to clearance and conversion to plantations (Curran et al. 2004).

So is there evidence that protected areas can be effective in preventing deforestation and conserving biodiversity? Measuring park effectiveness is not straightforward, as it is not possible to measure 'avoided deforestation' directly. Simply comparing deforestation rates in protected and non-protected areas is biased, in that protected areas are often located in inaccessible or low-population density regions where deforestation is already low. Additionally, forest protection can displace deforestation activities to other regions[6] including the surrounding areas, a problem often referred to as 'leakage' (Armsworth et al. 2006; Oliveira et al. 2007; Andam

[5] The 1997 analysis under-represents the protection of biomes at that time by about 30%, because only 70% of the global protected areas network was classified (Chape et al. 2003).
[6] Though note that if protected areas have been established in the most valuable conservation areas, leakage to lower priority sites cannot be called a 'failure'. In any case leakage will decline to zero if ever all the remaining tropical forests are within effective PAs (Brooks et al. 2009).

et al. 2008; Ewers and Rodrigues 2008). Failing to account for such biases may overestimate the effectiveness of protected areas (e.g. Andam et al. 2008). Nonetheless at least nine large-scale studies, including two that allow for leakage (Oliviera et al. 2007 in the Peruvian Amazon, and Andam et al. 2008 for Costa Rica) and two further case study based meta-analyses (Naughton-Treves et al. 2005; Nagendra 2008) suggest that PAs do slow deforestation in tropical regions (Brooks et al. 2009), although the conservation benefits may be considerably less than previously calculated after accounting for biases and leakage.

A pantropical satellite-based assessment found that the density of detected fire events in and around 823 tropical and subtropical moist forest PAs was significantly lower inside reserves than in contiguous buffer areas, but varied by five orders of magnitude among reserves.

Much of the variation in relative protection is correlated with national wealth (positive) and perceived corruption (negative) (Wright et al. 2007).

Protected areas are most vulnerable to degradation and encroachment at their accessible edges, where biological edge effects and human activities interact. In some contexts it has been possible to develop buffer zones (usually external but sometimes internal) around park boundaries. These are intended to protect against encroachment, often by allowing local people limited low-impact resource uses such as the collection of fruits and nuts, medicinal plants, weaving materials, or animal fodder (Hansen and Rotella 2002). External buffer zones also augment the area available to species of conservation interest (Hansen and Rotella 2002). Forests within external buffer zones have, unfortunately, often been degraded or even lost: from the early 1980s to 2001, 69% of 163 moist tropical forest reserves[7]

Figure 17.2 The boundary of Bwindi Impenetrable National Park, south-western Uganda, East Africa. Intensive cultivation abuts the protected area, and the lack of an external buffer zone results in a well defined and stable edge that has remained intact for several decades.

Photo by Robert Bitariho.

[7] These reserves are from IUCN protected area management categories I and II. Category I refers to strictly protected nature reserves managed for nature protection and science. Category II are national parks managed for ecosystem protection and recreation. For further information about these categories see: http://www.unep-wcmc.org/protected_areas/categories/index.html.

lost forest area within their buffer zones, and 25% of these sites experienced forest loss within the boundaries of the reserves themselves (DeFries et al. 2005). Incursions into buffer zones, and indeed all loss of native habitat in the surrounding landscape, increase reserve isolation and work against efforts to maintain a favourable landscape mosaic to facilitate movement of animals and plant propagules across a wider area (Laidlaw 2000). Of course, many protected areas lack external buffer zones entirely (Figure 17.2).

How much is enough? Priorities and priority setting

Setting aside debates about the effectiveness of protected areas, we also need to identify their appropriate extent in order to achieve conservation objectives. Arbitrary targets of 10% of each biome have been widely used in conservation planning (McNeely 1993). For the tropical rain forest biome the 10% target has already been exceeded (Table 17.1). But is this enough?

There is no objective scientific answer to the 'how much is enough' question—this is a matter of societal choice (Wilhere 2008). Society is undoubtedly concerned about biodiversity loss and its conservation, but society also has other resource-demanding concerns, among which are economic growth, social security, education, poverty alleviation, and health. The amount of resources and land allocated to protected areas is therefore not, ultimately, a scientific issue, but a societal value judgement. What science can offer is tools to evaluate whether protected areas are appropriately distributed, given resources made available by societal decisions, to encompass maximum biodiversity and habitat variety along with any other conservation goals.

'Gap analysis' assesses the extent to which current protected areas encompass biodiversity[8], and identifies priority regions for expansion of the protected area network (Brooks et al. 2004). These studies conclude that the existing global protected area system is inadequate, and priority non-protected areas are over-

whelmingly (65%) located in tropical and subtropical moist forests (Rodrigues et al. 2004). For example, of the 595 global sites identified as possessing the only remaining population of one or more highly threatened species of terrestrial vertebrates or conifers (the majority of which include tropical forest), fewer than half (290) have any protected status even in part (Ricketts 2005).

To decide how these gaps might best be filled there are a range of approaches that can help set priorities (Margules and Pressey 2000). Specific tropical forest priority regions that are commonly identified in such studies include the Western Ghats (India) and Sri Lanka in South Asia, Vietnam, peninsular Malaysia and the Philippine and western Indonesian islands in Southeast Asia, and New Guinea and Australia's wet tropics (Rodrigues et al. 2004). Many Pacific and Indian Ocean islands such as the Solomon Islands, Polynesia, Seychelles, Mauritius, Reunion, and Comoros also emerge as important on account of their threatened birds. In Africa the biodiversity of Madagascar[9] is the least well represented within parks, although the coastal forests of East Africa are also under-represented. In the Neotropics it is the Brazilian Atlantic forests, the Caribbean, and Central American regions that emerge as priority areas. The Congo Basin in Africa, and Amazon Basin and Guiana Shield in South America, are not priority areas on account of the relatively broad distributions of most of their species and existing protected area coverage (Rodrigues et al. 2004).

Other global evaluations of priority conservation areas exist (Brooks et al. 2006), including biodiversity hotspots (Myers et al. 2000; Chapter 16), Endemic Bird Areas (Stattersfield et al. 1998), and Global 200 Ecoregions (Olson and Dinerstein 2000), but these do not consider existing protected area networks in their identification of priority regions. Other priority setting efforts emphasize 'wilderness' areas (Mittermeier et al. 1998) while yet others base priority-setting on maximizing conservation

---

[8] In practice, gap analysis usually equates 'biodiversity' with terrestrial vertebrate richness and endemism, for which distributional and threat data are readily available. These data are compiled by various initiatives, including the IUCN Global Mammal Assessment, Birdlife International Threatened Birds of the World, EMYSystem World Turtle Database, and the Global Amphibian Assessment, among others.

[9] Although in 2003 the president of Madagascar committed to a tripling in protected area coverage over a 5-year period.

return for available financial resources (Balmford et al. 2003; Naidoo et al. 2006)[10].

Moving beyond protected areas

Protected areas are usually established in recognition of the uniqueness of particular habitats, often based on species richness or endemism, or with a view to conserving particular flagship species, all for the public good. Many biologically rich tropical forest regions are also areas of considerable human poverty. Recognition that many poor people are highly dependent on forest resources[11] implies that strict forest protection objectives might conflict with livelihood needs in remote areas (Box 17.1). Since the 1980s, 'Integrated Conservation and Development Projects' (ICDPs) sought to integrate biodiversity conservation with human livelihood objectives to secure win–win solutions. These projects arose out of the perception that 'sustainable use' promises to reconcile people, environment, and economies. Yet ICDPs have had limited success in either halting forest destruction or alleviating poverty (McShane and Newby 2004; Wells and McShane 2004; Linkie et al. 2008), largely due to unrealistic goals, misplaced assumptions about incentives and compensation to gain local support, complexities of project implementation, failure to appreciate market realities, and inadequate accounting of the temporal and spatial dynamics of ecosystems (Wilshusen et al. 2002; Wells and McShane 2004; Spiteri and Nepal 2006).

The adoption of the Millennium Development Goals (United Nations 2008) has served to emphasize that forest loss is only one of several global challenges that include climate change and poverty, each of which is linked to the others in terms of both impacts and solutions. In the last 10 to 15 years, new approaches to forest conservation have been advanced by an environmental movement that has increasingly acknowledged the growing diversity and immensity of the challenges we face in the 21st century, and the potential power of regional and global markets to resolve such challenges.

This remains an area of considerable research interest and debate.

17.1.2 Indigenous and extractive reserves

The Amazonian region harbours an extensive network of indigenous and extractive reserves. Indigenous reserves are officially recognized lands allocated for the use and habitation by native Amerindian people (Peres 1994). Several indigenous reserves are very substantial, exceeding a million hectares and dwarfing other types of reserve[12], and in the Brazilian Amazon they are in total five times more extensive than areas protected for biodiversity (Nepstad et al. 2006). Indigenous reserves ostensibly offer benefits for conservation (by maintaining large forest tracts that are subject only to relatively low impact management systems practised by indigenous people) against the depredations of logging, ranching, and mining interests. Whether indigenous communities will continue to engage in non-destructive management activities is uncertain (Peres 1994). However, analysis suggests that indigenous reserves are the most important barrier to deforestation in the Brazilian Amazon (Nepstad et al. 2006).

In the Brazilian Amazon, conservation and development objectives converged in the context of extractive reserves, which were established through the activities of rubber tappers' associations seeking to secure their rights in the face of pressure from large-scale cattle-ranching interests (Box 17.1). Extractive reserves promote the concept of guaranteeing local peoples' rights while conserving the environment (Ruiz-Perez et al. 2005) (Box 17.2). Since the establishment of the first extractive reserve in 1990, Alto Juruá in the western Brazilian Amazon, a further 37 reserves had been created by 2006 covering more than 10 million hectares of the Brazilian Amazon (Salisbury and Schmink 2007).

Extractive reserves recognize the rights of exclusive use by cooperatives of rubber tappers, yet it remains

[10] All these methods are top-down. There are in addition a growing number of conservation planning tools that can be useful for engaging with local interests (Enters 2000; Lynam et al. 2007).

[11] Over 90% of the 1.2 billion people living in extreme poverty have been said to depend on forests for some part of their livelihoods (World Bank 2004).

[12] The 13 million ha reserve of the Kayapó and Upper Xingu peoples across Pará and Mato Grosso states is larger than any tropical forest park anywhere in the world (Nepstad et al. 2006).

Box 17.2   Safe sex saves rain forest

Make love to save the Amazon rain forest?

In September 2007 the Brazilian government opened a condom factory in the Amazon, which may contribute to the preservation of forest cover. The factory, located in the remote town of Xapuri[1], in the state of Acre, western Brazil, uses latex extracted from local rain forests to make around 100 million condoms a year. Around 500 local rubber tappers are provided with a secure livelihood and an incentive to preserve the rain forest. So, to ensure the long-term viability of this project, use Natex condoms.

[1] Xapuri became infamous in 1988 when cattle ranchers assassinated environmental activist and rubber extractor Chico Mendes.

challenging to secure economically, environmentally, and socially viable forest-based production. The creation of the first extractive reserves coincided with a decline in rubber prices which, despite additional payments by the Brazilian government to rubber tappers for environmental services (the Chico Mendes Law introduced in 1999), has forced extractivists to diversify their livelihood portfolios. Many extractivists have reduced considerably their former dependency on rubber, and some have abandoned rubber taping entirely. Pig and cattle farming, annual cash crops such as beans, and timber harvesting have largely replaced rubber as the main sources of income in several extractive reserves in Acre (Almeida 1996; Ruiz-Perez et al. 2005; Salisbury and Schmink 2007). This has led to deforestation within extractive reserves but, probably owing to low population densities that are concentrated along rivers, such deforestation remains modest: it is only marginally higher than in national parks and indigenous reserves, and considerably lower than that associated with rural development projects of the Brazilian National Land Reform Institute (Ruiz-Perez et al. 2005).

Given the dynamic changes in extractivist livelihoods in the past decade, the future trajectory of deforestation in extractive reserves remains uncertain. It is particularly uncertain as younger households tend to be more entrepreneurial and more likely to engage in activities that encourage deforestation such as cattle ranching (Salisbury and Schmink 2007), a trend that is set to increase as more than half of the population of extractive reserves in 1999 was under 18 years (Câmara 2002, quoted in Salisbury and Schmink 2007). On the other hand, young people tend to be more mobile and are more likely to seek employment opportunities in urban centres than remain bound to a forest-dependent existence.

17.1.3 Conservation concessions

A conservation concession is a voluntary agreement whereby a government and/or other affected stakeholders are compensated for foregoing economic development on public lands. Rather than paying for the right to exploit resources, as would be the case in a logging concession, the buyer pays to preserve the forest concession from other uses (Ferraro and Kiss 2002; Niesten and Rice 2006; Karsenty 2007b). The main elements of conservation concessions include the determination of appropriate payments, the duration of the contract, and the development of a management plan. In theory, payment is determined by the opportunity costs of foregone employment and taxes incurred as a result of conservation, but one of the barriers is the country's capacity to value resources and stakeholder dependencies on the resources[13]. The duration of concession contracts is variable but is often similar to that for alternative land-use contracts such as a logging concession. Payment is also conditional on a management plan that includes measurable performance criteria to demonstrate that conservation objectives are actually being met.

Few conservation concession applications have been completed at the time of writing, although there are several under negotiation. Among the first are the 81,000 ha concession in Upper Essequibo, Guyana, granted to

---

[13] A full economic analysis would also include foregone export earnings in foreign currency and employment opportunities. On the other hand, the economic value of the services provided by an intact ecosystem (e.g. soil protection, flood mitigation), if considered, would offset some of the opportunity cost (Fearnside 1997).

Conservation International (CI), and a 135,000 ha concession along the Madre de Dios river in Peru, granted to a Peruvian NGO[14] in collaboration with CI (Hardner and Rice 2002). In both these cases the sites were remote and alternative demands on the land, and thus opportunity costs to be compensated, were low. Indeed projected costs for conservation concessions that were in various stages of negotiation in 2006 were estimated at US $0.5 to US $15 ha^{-1} year^{-1} (Niesten and Rice 2006). The limitation of the approach is the cost where forests have high commercial value[15]. Thus regions that are rich in resources, and close to markets or heavily populated areas, are not conducive for conservation concession establishment.

The advantage of conservation concessions is that they transfer the cost of conservation to those stakeholders who are able and willing to pay. Conservation organizations can also pay local people to manage and protect the forests, thereby providing local income. Politically, conservation concessions may be more palatable than nature reserves, as they are valid only for the duration of the contract, raise few 'sovereignty' concerns, and the costs are borne by the investor. The income stream may be more dependable than timber revenues. From the perspective of conservation, the temporary arrangement means that there are no guarantees beyond the contract end date. Another concern is that governments may be less willing to create new permanent protected areas if instead investors can be attracted to finance time-limited conservation concessions (Karsenty 2007b).

17.1.4 Debt-for-nature swaps

A debt-for-nature swap involves developed-country NGOs purchasing from banks or governments the debt that is owed by developing countries at greatly discounted 'secondary market value'[16], converting the debt into local currency, and using the proceeds to fund conservation activities managed by local NGOs (Visser and Mendoza 1994). The great advantage is that the debt is obtained at the secondary market value and exchanged for an environmental benefit in the indebted country that is valued at a higher price.

Another version is bilateral debt-for-nature swaps, where one country forgives a portion of the public debt of a debtor nation in exchange for environmental commitments from that country. The concept of debt-for-nature swaps was first conceived by **WWF** in 1984 to address the problems of developing-nation indebtedness and their consequential reduced capacity to address environment issues[17].

Debt-for-nature swaps can leverage substantial funds for conservation. Recent examples include a 2006 bilateral debt-for-nature swap agreement between the USA and Guatemala worth $24 million over 15 years; an even larger USA–Costa Rica 2007 agreement which absolves $26 million of Costa Rica's foreign debt in exchange for the country spending the same amount on tropical forest conservation over 16 years; and a $25 million agreement in 2008 to support conservation of Peru's rain forests. Both these arrangements were facilitated by conservation NGOs who contributed to the debt purchase. The majority of agreements are between the USA[18]

[14] Amazon Conservation Association (Asociación para la Conservación de la Cuenca Amazónica).

[15] The cost of the Guyana concession amounted to 37 cents ha^{-1}, but another potential concession, Ngoyla-Minton, offered by the Government of Cameroon, was deemed by Conservation International to be too expensive at $2 ha^{-1} (Anonymous 2008).

[16] Lenders are willing to sell debt at lower than the face value because of the uncertain ability of developing countries to make repayments in the future.

[17] This was particularly targeted to the Latin American debt crisis in the 1980s, which contributed to poverty-driven forest clearance of marginal lands, exploitation of natural resources to generate revenue, and reduced governmental capacity and willingness towards investing in environmental conservation (Gullison and Losos 1993; Thapa 1999).

[18] The centrality of the United States government is due to its role as a major creditor, and to the enactment of the Foreign Assistance Act of 1989 which authorized government agencies to 'furnish assistance, in the form of grants on such terms and conditions as may be necessary, to nongovernmental organizations for the purchase on the open market of discounted commercial debt of a foreign government of an eligible country which will be cancelled or redeemed under the terms of an agreement with that government as part of a debt-for-nature exchange', and the Tropical Forest Conservation Act (TFCA) of 1998 by which bilateral debt-for-nature agreements were also made possible by the US government.

and tropical American nations, including Belize, Peru, Colombia, Jamaica, and Panama, although the USA also has agreements with Bangladesh and the Philippines, and the French government concluded a $25 million arrangement with Cameroon[19] in 2006 and a $20 million deal with Madagascar in 2008, both facilitated by WWF.

In addition to forest conservation, debt-for-nature agreements can include provision for strengthening civil society by supporting national and local NGOs and local communities, and the potential to create jobs in remote areas also exists. In terms of debt relief, the amounts forgiven are miniscule compared to the total indebtedness of most countries[20] (Gullison and Losos 1993). Nevertheless, the contribution to a country's conservation budget can be substantial. For example, two debt-for-nature swaps facilitated by WWF, in Ecuador in 1987 and 1989, established a US $10 million fund for conservation that was twice the size of the parks and reserves budget (Thapa 1999), and for Costa Rica the interest alone from conservation funds worth US $53.7 million raised by 1990 was several times greater than the annual budget allocated to the country's Park Service (Reilly 1990).

Problems with debt-for-nature swaps include the inflationary risks of increased domestic spending, though this is resolved by limiting the size of swaps and by issuing long-term interest-bearing bonds that spread out the payments (Gullison and Losos 1993). Some countries perceive these transactions to infringe their sovereignty[21], although the funds generated are usually used to fund programmes already approved by their governments, or requiring governmental consent (Ayres 1989; Gullison and Losos 1993; Thapa 1998).

17.2 Conservation and livelihoods: biodiversity as business?

'Biodiversity is a commodity that can be bought and sold; conservation is business' (Nicholls 2004). The establishment and enforcement of protected areas often create conflicts between the proponents of conservation and commercial enterprises, such as mining and logging companies, as well as local communities who might also be excluded from land and resources. Reserves have therefore often been perceived as being for the benefit of a wealthy elite, to the detriment of the national economy and the rural poor (Bawa et al. 2004; Sodhi et al. 2006). Recognition of the legitimate claims of a diverse set of stakeholders leads to more considered land-use designation, whereby the economic interests of rural communities and commercial enterprises are addressed alongside forest management and conservation objectives (Wollenberg et al. 2009). Various market-based

Table 17.2 Potential investments for biodiversity conservation, from the least direct form of investment downwards to the most direct.

Investment	Examples
Support for the use/marketing of extracted forest products	Logging, non-timber forest products, hunting
Subsidies for reduced-impact land and resource use	Sustainable agriculture on already cultivated lands
Support for the use or marketing of biodiversity within intact forest	Ecotourism, sport hunting, bioprospecting
Payment for environmental services	Watershed protection, carbon sequestration
Payment for conservation land or retirement of biodiversity use rights	Conservation concessions
Performance-based payment for biodiversity conservation	Paying for bird or turtle breeding success

Adapted from Ferraro and Kiss 2002.

[19] The agreement requires Cameroon to earmark funds for education, health, and infrastructure as well as natural resources.

[20] The above-mentioned debt-for-nature swaps for Guatemala, Cost Rica, and Peru account for 0.25%, 0.21% and 0.08% respectively of the foreign debt of these countries (calculations based on data from the CIA 2008 Factbook).

[21] The first debt-for-nature swap in 1987 led to the establishment of the Beni Biosphere Reserve in the Bolivian Amazon, but was delayed by 21 months partly due to concerns over sovereignty.

approaches have been developed to make conservation activities more financially viable, and also to compensate local people for their opportunity costs. In some cases direct compensation is paid to stakeholders, be they large commercial entities, governments, or local communities, to overcome the opportunity costs of conservation (Table 17.2) (see 17.1.3 *Conservation concessions* above).

A variety of alternative approaches seek to create livelihood opportunities that depend on healthy forest cover (Table 17.2). Thus many conservation agencies have promoted ecotourism, the harvest and commercialization of non-timber forest products (NTFPs; Chapter 15), and bioprospecting, to deliver the multiple objectives of development, poverty alleviation, and conservation (Kiss 2004). Other approaches offer payments for ecosystem services generated by forests.

Some people doubt whether stand-alone economic viability from extractive forest uses can be compatible with the continued delivery of environmental benefits and values, and instead, in a long-running and recurring debate, advocate strictly protected areas for conservation (Bowles et al. 1998; Terborgh 1999). Ongoing deforestation, degradation, and biodiversity loss attest to the difficulties of conserving tropical forests by any of these means.

17.2.1 Sustainable forest management and Reduced Impact Logging

Can logging and conservation coexist? Concepts of 'sustainable forest management' differ, but generally they emphasize the continued existence of the forest and the viability of multiple forest products and services, including biodiversity, over the long term (Pearce et al. 2003). In 2005 estimates suggested that perhaps 4–5% of the world's tropical forests (not including strictly protected areas) were being managed in a manner that might be broadly considered 'sustainable' (ITTO 2005).

A major barrier to the realization of sustainable forest management is that the costs of management and the combined growth and appreciation of the timber stock are not sufficient to make it a competitive and viable

financial investment. Annual price-appreciation of timber, estimated at 3%, together with an annual timber stock growth of 4%, give an overall annual increase in stand value of around 7% (Rice et al. 2001). This is less than typical discount rates of 8–15% for developing countries, so it is financially rational to liquidate the forest and invest the profits elsewhere.

Thus it costs more to manage a forest properly than simply to cut and market the timber—any producer wishing to cover management costs must charge higher prices and consequently loses market competitiveness against 'timber miners'. Sustainable forest management has, nevertheless, been widely promoted by governments, conservation NGOs, and forest enterprises, and efforts to increase its profitability through price premiums by certification are discussed below (Nepstad et al. 2002; Fearnside 2003; Kainer et al. 2003).

Case studies from Southeast Asia, Africa, and South and Central America convincingly demonstrate that the detrimental effects of selective logging on stand structure, biodiversity, and forest soils can be reduced substantially through the adoption of Reduced Impact Logging (RIL) guidelines[22]. Reduced Impact Logging seeks to reduce the impact of timber harvesting through a series of improved management interventions which include a well-trained workforce, careful harvest planning, cutting of climbers and lianas before felling, directional tree felling to minimize impact to the surrounding forest, the establishment of buffer zones along water courses, and using improved technologies to reduce soil damage and compaction during log extraction (Putz et al. 2008). Nevertheless, while RIL reduces forest damage, it is not clear whether the long-term recovery of forests, their associated biodiversity, or their potential to sustain economically competitive future harvests can be achieved (Sist and Ferreira 2007; Zarin et al. 2007; Peña-Claros et al. 2008). While RIL has traditionally focused on timber production, even here there are notable omissions, such as the absence of explicit measures to ensure seed production (Sist et al. 2003).

Although RIL is widely accepted to be effective in reducing logging impacts, and is an essential compo-

---

[22] RIL guidelines are a necessary component of sustainable forest management but are not sufficient to ensure sustainable forest management, as it will also require monitoring and management of biodiversity, ecosystem functions and services, and disturbance among others.

nent of sustainable forest management, a number of constraints result in infrequent uptake of RIL, and sustainable forest management generally (de Blas and Perez 2008; Putz et al. 2008). These constraints include high initial costs (especially in training) and insufficient awareness of the benefits; lack of interest in forest management stemming from a dearth of appropriate policies and incentives and a lucrative illegal timber market; and absence of the longer-term tenure necessary to encourage investment. Additionally, it is not always the case that RIL is more profitable than conventional methods of timber extraction. A review of 23 studies indicated that adoption of RIL resulted in an average increase in total costs of around 43%, although different techniques and harvesting intensities may undermine meaningful comparisons (Killmann et al. 2001). Certainly there is considerable variation in the range of values (Holmes et al. 2000): case studies in the eastern Amazon show that estimates of net revenue from RIL range from 18% to 35% higher than for conventional logging, but in Sabah, Malaysia, estimates were 62% lower due to reduced timber extraction (Forest Trends undated). In view of the often uncertain economic profitability of RIL, coupled with not inconsiderable up-front costs and effort for training and planning, it is unsurprising that uptake has been limited, but there are potential incentives that are changing this.

17.2.2 Forest certification

One way to encourage good management is to promote a market for its products. Certifying products as sourced from well-managed forests provides a mechanism by which discerning consumers can reward such management by paying a premium. Buyers must be able to discriminate these products from others through a 'label' issued by an independent certifying authority. Certification is most prominent in the timber industry, where the concept, which includes legal and social concepts along with environmental concerns, has been spearheaded by the Forest Stewardship Council (Box 17.3).

Although forest certification has grown since its inception in the mid 1990s, growth in tropical forest certification has been relatively slow. As of April 2008, less than 0.7% of the remaining tropical and subtropical forests have been certified by FSC, and less than 1.5% has been certified at all by any accreditation agency (Bennett 2008). The impact of certification in promoting sustainable practices has, therefore, been limited. Part of the problem is that sustainability through certification relies on an environmentally aware market that does not yet appear to be sufficiently large to ensure its success. The majority of tropical timber is illegally harvested and traded to less discriminating consumers. Much tropical timber is also harvested to meet the demands of domestic markets where external environmental concerns have less weight (Ghazoul 2001). Current market-based incentives for forest certification are therefore not sufficient to encourage wide uptake in tropical forest regions, because the costs of implementing certification standards often exceed the market benefits (Gullison 2003). Nonetheless there are some positive signs: in Mexico several community-based projects and nearly 7,000 km^2 have achieved certification (FSC 2002), while in Indonesia public companies achieving FSC certification have observed considerable increases in their stock prices (Meijaard and Sheil 2007).

The efficacy of the certification concept relies on trust in the label and what it represents. In response to the potential market demand for certified products, and to counter the negative public image of logging, a wide variety of regional certifiers and certification methods have become established, including the Pan African Forest Certification, Lembaga Ekolabel Indonesia, and the Brazilian CERFLOR Forest Certification Programme. Some certifications assess compliance to legal standards and may be self-evaluated by the commercial companies themselves. Different standards and methods may undermine the objectives of certification by confusing consumers and reducing credibility (de Blas and Perez 2008). Credibility may also be undermined if non-compliance to standards by certified companies is demonstrated, and this has influenced trust in even the FSC label (Schulze et al. 2008a, b).

Smallholders are often excluded from participation in certification schemes by high transaction costs (Grieg-Gran et al. 2005). RIL was originally developed for relatively large industrial forestry operations, and as such is often unsuitable for smallholder and community-based forest management (Rockwell et al. 2007). Yet about 25% of forests in developing countries are managed by communities, and this proportion is expected to rise to 50%

Box 17.3 Forest Stewardship Council

Encouraged by the UN Conference on Sustainable Development in 1992 (the Rio Summit), representatives from business, social groups, and environmental organizations joined forces in 1994 to establish the Forest Stewardship Council (FSC), an independent, non-governmental, not-for-profit organization with a purpose to improve forest management worldwide. FSC sets standards for good management, and provides formal recognition of certifiers in accordance with agreed standards and requirements (i.e. accreditation). Producers must meet FSC standards. Certified products can then bear the FSC label (Figure 17.3) by which buyers can identify products that come from a well managed source.

The FSC prescribe how forests should be managed to meet the social, economic, and ecological needs of present and future generations. The ten FSC Principles are:

1. Compliance with all applicable laws and international treaties
2. Demonstrated and uncontested, clearly defined, long-term land tenure and use rights
3. Recognition and respect of indigenous peoples' rights

A

Figure 17.3 (continues).

Figure 17.3 (A) Sawn timber sourced from FSC certified location, sitting on the dockside in Belem, Brazil, ready to be exported. Great care has to be taken throughout the whole transport chain to ensure that only genuinely sourced timber is included, as the higher prices create an incentive not only for good management but also (regrettably) to cheat. (B) FSC certified timber stamped with the FSC label.

Photos by Douglas Sheil (A) and courtesy of FSC (B).

4. Maintenance or enhancement of long-term social and economic well-being of forest workers and local communities and respect of workers' rights
5. Equitable use and sharing of benefits derived from the forest
6. Reduction of environmental impact of logging activities and maintenance of the ecological functions and integrity of the forest
7. Appropriate and continuously updated management plan
8. Appropriate monitoring and assessment activities to assess the condition of the forest, management activities, and their social and environmental impacts
9. Maintenance of **High Conservation Value Forests (HCVF)** defined as environmental and social values of outstanding significance or critical importance
10. In addition to compliance with all of the above, plantations must contribute to reduce the pressures on and promote the restoration and conservation of natural forests.

The FSC standards are applicable to all forestry systems, in the tropical or non-tropical zones. Uptake has, however, been mainly by temperate forestry operations.

By April 2008, 12 million ha of tropical or subtropical forest had been certified by FSC, but this represented only 12.9% of all FSC-certified forests, and even some of these are plantations rather than natural forest. The distribution is also uneven, with tropical America accounting for 10.3% of the world FSC-certified area, tropical Asia less than 1.7%, and tropical Africa a mere 0.9% (FSC 2009).

by 2020 (Molnar 2003; White and Martin 2005). Smallholders and community management are therefore likely to have an increasingly important role to play in tropical forest conservation, and with more limited access to heavy machinery community-managed logging may often be more 'reduced impact' than RIL. In view of this, FSC initiated new certification procedures[23] and associated forest management guidelines specifically for smallholders (Humphries and Kainer 2006).

Other forms of certification seek to address conservation in complex and diverse agroforestry systems where tropical forest elements such as native trees are retained within the landscape matrix. Thus organic, fair trade, and bird-friendly labels attached to internationally traded crops such as coffee and cocoa have been promoted in theory and practice as benefiting conservation and poor farmers (Mas and Dietsch 2004; Ferraro et al. 2005; Perfecto et al. 2005; Taylor 2005). However, experience shows that price premiums are limited, providing few incentives and low rates of seeking and gaining eco-certification in the tropics.

Furthermore, wholesalers and exporters need large supplies of product that is uniform in quality, which further favours large-scale intensive agriculturalists rather than smallholders in rain forest margins and agroforests. New certification approaches are being constantly developed with a view to overcoming barriers and promoting inclusivity. New proposals include certification of landscapes that are conserving biodiversity or providing ecosystem services (Ghazoul et al. 2009), but many of these have yet to be made operational. Eco-certification remains a relatively novel approach and is still evolving, as is the market for certified products. Despite its difficulties and limited success, certification has been very effective in raising awareness regarding conservation as an economic and social iterative, as well as an ecological one (Rametsteiner and Simula 2003; Nebel et al. 2005). Its future success ultimately depends on being able to motivate consumers to give more weight to environmental and social concerns in their consumer behaviour.

Less costly than formal certification schemes are green branding and marketing initiatives, in which an item is sold with the promise, or implication, that some benefit will be given to a good environmental cause. Examples include Body Shop rain forest-sourced cosmetics and Ben & Jerry's Rainforest Crunch ice cream. Such products are marketed with allusions to fair trade and forest protection. These initiatives are unregulated but have been key in promoting public awareness—for example, green cosmetics are now amongst the best selling in Manaus, the capital of Amazonas state. However, delivering real benefits to rain forest communities and conservation is more difficult to demonstrate, and the marketing of some of these products has been criticized for failing to deliver on the promise of supporting livelihoods of rain forest-based people (Corry 1993).

17.2.3 Non-Timber Forest Products (NTFPs)

A concept broadly similar to extractive reserves is the promotion of conservation based on the extraction of non-timber forest products (NTFPs; Chapter 15), the harvesting of which is assumed to incur less environmental damage than timber extraction (Arnold and Ruiz-Pérez 2001). Early studies appeared to show that the economic potential of NTFPs exceeded that from cattle ranching or timber extraction (Peters et al. 1989). These results were highly influential in promoting conservation through the development of NTFP-based livelihoods. There has been modest success, but in most cases inherent limitations such as distances to market, low and unpredictable yields, and unstable market demand limit the viability of this idea. Opportunity costs are, despite the earlier indications to the contrary, often very high, while markets are limited and often difficult to access (Godoy et al. 2000; Paoli et al. 2001; Sheil and Wunder 2002). Most NTFPs have limited market value and as incomes rise are quickly substituted for superior quality goods, even by harvesters themselves (Arnold 2002). Products that do attract high market demand often attract more powerful commercial interests that take control of supply, processing, and marketing, and therefore exclude forest communities from access to the resource, the market, or both (Dove 1993, 1994). Assumptions of low-impact harvesting are quickly undermined as demand increases, and overharvesting and destructive impacts are commonplace for many commercially valuable NTFPs

---

[23] Small and Low Impact Managed Forests Streamlined Certification Procedures (FSC 2002).

(Moegenburg and Levey 2002; Ruiz-Perez et al. 2004). Furthermore, some argue that promoting NTFPs as a livelihood strategy can bind people into a state of persistent poverty (Arnold and Ruiz-Pérez 2001).

17.2.4 Ecotourism

Rather than value forests based on the tangible products they contain, another approach has been to capitalize on tourism. Ecotourists are visitors motivated by an appreciation and observation of nature, although ecotourism can be more narrowly defined as responsible travel to natural areas that conserve the environment and sustain the well-being of local people (International Ecotourism Society 2006). 'Community-based ecotourism' is a form of ecotourism where the local community has substantial control over ecotourism development and management, with a major proportion of the benefits remaining within the community. Since the early 1990s, ecotourism has been growing 20%–34% per year[24]. Most of this growth is occurring in and around the world's remaining natural areas (Christ et al. 2003). The benefits of ecotourism can extend far beyond the local area that tourists visit, as government policies and public opinion towards conservation may be positively influenced by the contribution of foreign revenues derived from ecotourism.

The dramatic growth in ecotourism has heralded new opportunities for conservation in the tropics. But how successful are such ventures in delivering local development and conservation objectives? The assessments that have been conducted suggest that the main economic benefits of ecotourism accrue to outsiders and to relatively few members of local communities, usually not the poorest (Butynski and Kalina 1998). Ecotourism in Madre de Dios[25], Peru, for example, sustains around 70 tourist lodges and is a profitable source of income for lodge owners (Box 17.4), but is not a sufficient source of income to support the families of lodge employees (Nicholls 2004).

Tourism incomes can, however, provide a much-needed source of conservation funds. Tourists visiting Bwindi Impenetrable National Park in Uganda to view gorillas have increased foreign exchange earnings, generated employment, and attracted financial investment to the area: each year several thousand tourists pay US $500 for one hour of gorilla viewing, thereby paying for much of Uganda's conservation. Some of Uganda's national park gate-fees go into community projects, but the issues of overall share, how it is controlled, and who qualifies to benefit are a source of considerable contention (Adams and Infield 2003). In most such cases, less than half of the overall revenues accrue to the host country (Boo 1990) and much less reaches local people (Southgate 1998).

Neither has ecotourism been entirely free of detrimental environmental impacts (Butler 1991; Duffus 1993; Butynski and Kalina 1998). Impacts such as cutting trees and building roads are obvious and can be easily controlled; other impacts are less obvious. Transmission of disease to wildlife, or disturbance by increased stress, may increase mortality and lower breeding success (Griffiths and van Schaik 1993; de la Torre et al. 2000). For example, in Cuyabeno reserve in the Ecuadorian Amazon juvenile hoatzins *Opisthocomus hoazin* visited by tourists had double the levels of stress hormones and survival was around 30% that of chicks in areas from which tourists were excluded (Mullner et al. 2004). Disease transmission from humans to immunologically naive primates might lead to extirpation of local populations (Macdonald 1996; Kondgen et al. 2008). Concern over human to gorilla transmission remains high (Guerrera et al. 2003; Kalema-Zikusoka et al. 2005), but rules that limit proximity between tourists and gorillas have been difficult to implement (Sandbrook and Semple 2006) (Figure 17.4).

Ecotourism is especially vulnerable to political and economic instability. The financial crisis at the end of 2008 is likely to have a severe impact on tourism, and may undermine those conservation efforts where incentives to conserve forest are largely dependent on tourism. Ecotourism is also challenged by the profitability of more exploitative income-generating activities such as logging, cattle ranching, and hunting (Box 17.4). Despite these problems, the dramatic and continuing growth in

---

[24] A rate that is around three times faster than the tourism industry as a whole (International Ecotourism Society 2006).

[25] Heralded by *The Economist* (10 April 2008) as a triumph of the capitalist approach to conservation: http://www.economist.com/world/americas/displaystory.cfm?story_id=11017681.

Box 17.4 Ecotourism in Madre de Dios, Peru

Chris Kirkby and Douglas W. Yu

The province of Madre de Dios in south-east Peru, an Amazonian frontier region bordering Bolivia and Brazil, is renowned for its biologically and culturally rich landscape. One area of this region in particular, known as Tambopata, is now also firmly entrenched in the minds of international travellers as the principal Amazon rain forest destination, attracting 39,565 ecotourists in 2005. Two large protected areas, the Tambopata National Reserve (TNR, 2,747 km²) and the Bahuaja-Sonene National Park (BSNR, 10,914 km²) owe their existence in part to early tourism entrepreneurs who built jungle lodges in the 1970s and lobbied government for the creation of private reserves. These reserves are home to primates, giant otters, macaws, and feature intact oxbow lakes and clay licks that concentrate fauna, allowing for easy observation.

High profitability provides the incentive and the means for lodge-owners to protect their businesses by protecting forest cover: in 2005, ecotourists spent a total of US$ 11.6 million to visit Tambopata, of which 50% was lodge revenues and another 32.5% was spent locally within Tambopata. Many lodges have taken advantage of government legislation, passed in 2002, that allows businesses 40-year renewable leases on public forest outside protected areas. By 2005, lodges had leased 325 km², with 90% of this area acquired by the four most profitable operators which together manage eight lodges. Another 162 km² have been provisionally awarded as of early 2008, totalling 486 km².

An economic analysis of ecotourism profitability shows that it compares favourably with most other forms of land use, and therefore appears to provide a viable strategy for income generation, coupled with forest conservation outside formally protected reserves (Table 17.3).

Lodges acquire forest concessions in order to secure legally recognized land titles to forest in a bid to protect their livelihood base. Lodges have successfully sued and evicted loggers and miners from their concessions and have entered into benefit-sharing agreements with neighbouring communities to cease extraction and hunting. In one notable episode in 2007, the ecotourism industry collaborated with Peru's conservation community to successfully lobby against a government proposal to de-gazette a portion of the BSNP for oil exploration.

Tambopata's ecotourism industry has the incentive and the means to continue protecting and even expanding their concession holdings. It remains to be seen how effective ecotourism will be in protecting Tambopata's biodiversity against new threats, including the paving of the Interoceanica Sur Highway, a westerly extension of the Trans-Amazon Highway, which will connect Brazil to the Pacific Ocean.

Table 17.3 Land use profitability compared across alternative options. Ecoutourism compares favourably against all options except for unsustainable cattle ranching (where number of cattle exceeds 3.5 per hectare).

Land use	Pre-tax profits (US$ ha-¹)
Lodge-controlled land (Ecotourism)	38.9
Specialized land use (> 50% of household revenues)	
Unsustainable cattle ranching (> 3.5 animals ha⁻¹)	40.3
Sustainable cattle ranching (> 3.5 animals ha⁻¹)	35.3
Timber extraction	35.7
Agriculture (rice, maize, manioc, and bananas etc.)	21.3
Land used for multiple purposes	
Agriculture, livestock, and timber	27.1

Based on unpublished data from Kirby and Yu.

Figure 17.4 A healthy mountain gorilla *Gorilla gorilla beringei* in Bwindi Impenetrable National Park, Uganda. Close contact between gorillas and tourists is the probable cause of disease outbreaks among visited populations. In two incidents in 1988 and 1990 in Volcanoes National Park, Rwanda, over three-quarters of mountain gorillas regularly exposed to tourists became infected with a respiratory disease and several individuals died (Butynski and Kalina 1998).

Photo by Damien Caillaud.

ecotourism is likely to ensure that ecotourism will have a major role to play in future conservation programmes.

17.2.5 Payments for Environmental Services (PES)

Direct payments for conserving forest have been promoted through the concept of environmental services that forests are supposed to provide. Forests support biodiversity, but also sequester carbon, provide pollinators of crops, regulate water flows, and contribute to the aesthetic beauty of the landscape and the natural heritage of the region. The beneficiaries of these services and attributes may be far removed from the landscape and therefore incur none of the costs of maintaining these services. Thus under **Payments for Environmental Services (PES)** schemes, landowners who incur the costs of forest maintenance for the delivery of forest services are compensated by direct payments from the beneficiaries.

Such schemes have been implemented with varying degrees of success in tropical forest regions. Among the best publicized is the Costa Rican payment for ecosystem services scheme whereby landowners are paid for the preservation of forest on their land, in recognition of the ostensible provision of four services: biodiversity protection, carbon sequestration, hydrological services, and preservation of scenic beauty. Uptake of this scheme has been modestly successful, although problems remain over transaction and implementation costs, lack of trust among potential service providers in the implementing agencies, and late payments by such agencies to ecosystem service providers (Pagiola 2008). Another limitation of the Costa Rican scheme and other similar systems is that the benefits accrue to the wealthy: they exclude many smallholders[26] for whom the transaction costs may be prohibitive, or they may simply not have enough land to take part in the PES scheme (Wunder 2008).

To overcome some of these problems, alternative PES schemes have been developed that target and reward poor smallholders, or at least encompass their interests (Ghazoul et al. 2009). ICRAF's Rewarding Upland Poor for Environmental Services (RUPES) programme highlights social mobilization, in a community-based action to empower communities socially and politically to engage in PES schemes (ICRAF 2008). RUPES experience has shown that the likelihood of achieving broadly acceptable PES systems for smallholders depends on shared perceptions of environmental services and opportunity costs, representative community institutions that manage the implementation of the PES scheme, and trust between communities. Similar to RUPES is the Mexican Payments for Hydrological Environmental Services Program (Spanish acronym: PSAH) (Munoz-Pina et al. 2008).

Ecosystem service values accrue differently at different scales, and differently to different stakeholders[27]. They may also have a trade-off against each other in the context of different land uses. Consequently ecosystem service values are uncertain and often unrealized, and

---

[26] For example, the Ecuadorian PROFAFOR scheme operates only with landowners who have a minimum of 50 ha (Wunder and Alban 2008).

[27] For example, timber values vary from \$20 to \$4400 ha^{-1}, genetic values from \$0 to \$3,000, and climate values from 360 ha^{-1} to 2200 ha^{-1} (CBD 2001).

few ecosystem services have been the subject of direct market transactions. Their limited inclusion in formal markets undermines their visibility for land-use planning and decision-making as compared to market-valued products. While considerable progress has been made to internalize ecosystem services and recognize their values through markets, this approach remains of limited importance for large-scale conservation. This may change with the establishment of carbon sequestration payments through REDD (see below), and with the growing interest in this field.

17.2.6 Bioprospecting and intellectual property rights

Responding to the multitude of plant pathogens, parasites, and herbivores, tropical rain forest plants have evolved a rich chemistry. Secondary compounds, chemicals that are not essential for plant life, are retained by plants owing to their defensive properties (Chapter 13). They include terpenoids, phenolics[28], bitter-tasting alkaloids[29], and cyanogenic compounds. This natural pharmacopoeia has prompted many to emphasize the value of tropical rain forests as a source of raw materials for drugs of the future. There is some justification to this, as many existing medicines and drugs are originally derived from tropical plants. The most famous examples are the Madagascar periwinkle[30] *Catharanthus roseus*, used for the treatment of various forms of cancer, and the anti-malarial alkaloid quinine derived from Andean *Cinchona* trees (Voeks 2004).

The close correspondence between ethnobotanical knowledge and practice by forest margin communities and the use of plant-derived compounds in Western medicine indicates potential for developing new medicines in collaboration with these communities (Farnsworth 1988; Cox et al. 1989). This view, coupled with the realization that less than 5% of the world's flora had been subject to any kind of chemical investigation (Fellows and Scofield 1995), has prompted demands for conservation justified by medical need and funded by industrial bioprospecting.

Initiatives of this kind have been implemented in Costa Rica, where in 1991 a bioprospecting agreement was reached between Merck Pharmaceutical Ltd and the National Biodiversity Institute of Costa Rica (INBio), a non-profit, public interest organization established by the Costa Rican government (Mateo 1997). Panama pursued a different approach by initiating, in 1998, a programme that relies on national expertise and facilities and seeks to use discoveries from bioprospecting to establish research-based industries within the country (Kursar et al. 2006). No substantive medicinal product has yet emerged from any of these programmes[31], and expectations have declined in the last decade.

Some have questioned the very basis of bioprospecting for justifying rain forest conservation, by claiming that most medicinally useful compounds are derived not from undisturbed rain forests but from weeds within highly modified tropical landscapes (Soejarto and Farnsworth 1989; Stepp 2004; Voeks 2004): the Madagascar periwinkle, the mascot of the bioprospecting movement, is a wayside weed commonly encountered along forest edges and trails, and never found in primary forest. By this view, pristine rain forest as the primary repository of nature's medicinal store is little more than hype that has been marketed by environmental entrepreneurs and optimistic conservationists (Voeks 2004).

Advocates of bioprospecting as a strategy for forest conservation remain vocal (Kursar et al. 2006), and may

---

[28] Including the coloured flavanoids which play a role in pollinator attraction, and astringent-tasting tannins that provide defence against leaf herbivores.

[29] Particularly effective against mammalian herbivores—though some may develop a taste for certain compounds, as with the common addiction for caffeine found in coca and coffee seeds and the quinine in tonic water.

[30] Two of its 76 alkaloids have been used in the treatment of forms of leukaemia, lymphosarcoma, Hodgkin's disease, and other forms of cancer.

[31] Another business, Shaman Pharmaceuticals, invested in drug discovery by focusing on plants with a history of medicinal use among traditional forest communities, intending to share any profits with these same communities. After spending more than $90 million in 10 years, the enterprise went bankrupt as no product of commercial significance was generated, demonstrating the challenges of such ventures.

yet be vindicated by a major discovery. On balance, though, bioprospecting has made limited gains for medical advancement, and perhaps even less for conservation (Gomez-Pompa 2004; Costello and Ward 2006). This is unlikely to change, as advances in biotechnology are improving our ability to develop synthetic biologically-active compounds (Firn 2003).

17.2.7 Land sparing vs. wildlife-friendly farming

Growing populations and increasing affluence create higher food demand. In the tropics this demand has been met by converting forests to agriculture, a process that probably represents the greatest threat to rain forests (Foley et al. 2005). Intensification of temperate-region agriculture has, on the other hand, been immensely successful at raising agricultural yields per unit area, and significant production gains have been achieved without incurring significant expansion of agricultural area. Intensifying tropical agriculture has therefore been proposed as a means to reduce tropical deforestation rates: the delivery of high agricultural yields in a relatively small area alleviates the pressure on land elsewhere. This strategy has been called 'land sparing' (Balmford et al. 2005; Matson and Vitousek 2006; Fischer et al. 2008), and analyses of agricultural yields through the 1980s provide evidence for its viability: countries with higher yield growth had lower rates of land conversion (Barbier and Burgess 1997; Southgate 1998).

Intensive agriculture has other impacts that extend beyond the farm boundaries, such as pollution from fertilizers and pesticides, and diversion of water, which would undermine putative benefits derived from land sparing (Tilman et al. 2002; Matson and Vitousek 2006). Additionally, if intensification reduces labour demands or displaces farmers, then these people might turn to other environmentally-damaging livelihood activities (Angelsen and Kaimowitz 2001; Kaimowitz and Smith 2001). Advocates of land sparing should also acknowledge that continued future growth of productivity with increasing intensification is not assured.

A contrasting approach is agricultural extensification by adopting 'wildlife-friendly' farming practices with lower chemical inputs, higher crop diversity, integrated water and pest management, and incorporating fallow and natural areas within the farm system (Harvey et al. 2008). In the tropics this approach is typified by agroforestry (Chapter 16) and organic farming, where the unmanaged biodiversity is generally higher than that of other tropical agricultural systems (Thiollay 1995; Bhagwat et al. 2008). Recognizing the higher labour demands and lower yields than more intensive farming systems (Green et al. 2005), agroforestry advocates nevertheless argue that agroforestry can be more productive than cattle pasture or monoculture oil palm, and if adopted as an alternative can reduce deforestation pressure and improve regional conservation values (Bhagwat and Willis 2008).

From a conservation perspective, the debate between land sparing and wildlife-friendly farming rests on the response of biodiversity value and crop productivity to different degrees of farming intensities (Figure 17.5) (Green et al. 2005). If biodiversity value follows a convex response to increasing intensification, then species losses may not become apparent until highly intensive monoculture farming systems are approached (Curve A in Figure 17.5). A concave response implies that even the

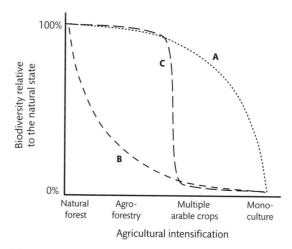

Figure 17.5 Possible responses of biodiversity (value, species, richness, etc) on land subject to increasing agricultural intensification. Along the convex curve (A) biodiversity is resistant to even quite drastic land transformation. Along the alternative concave curve (B) even conversion from natural forest to diverse agroforesty scenarios would lead to considerable biodiversity losses. Discontinuous responses (C) describe sudden and dramatic biodiversity changes at a particular threshold. Functional response curves are not well known for any taxa, but are likely to differ between taxa, due to different life-history requirements.

Drawn by Jaboury Ghazoul.

conversion of natural forest to highly diverse agroforest may significantly impact on biodiversity (Curve B in Figure 17.5). Response curves may be punctuated by thresholds at which dramatic changes in biodiversity occur with only slight increases in intensification (Curve C in Figure 17.5). At present, there is little information on these so-called density-yield functions, but quantifying the nature of the trade-off between food production and biodiversity will be essential for developing policies for biodiversity values in human-dominated rain forest environments.

Wildlife-friendly farming and land sparing should not be considered as mutually exclusive alternative concepts, but each should be pursued in consideration of what is appropriate for the region concerned, and in any one landscape both can be jointly implemented according to diverse human and ecological needs (Koh et al. 2009).

17.3   Governance and land ownership

'If the misery of the poor be caused not by the laws of nature, but by our institutions, great is our sin' (Charles Darwin).

Almost all large-scale conservation and land use initiatives unfold within a policy environment which may be enabling or obfuscatory. Thus debt-for-nature swaps and conservation concessions would not be possible without a facilitatory governance process. Governance is relevant to conservation by shaping the decision-making context for landowners and resource users, which in turn affects land-use outcomes. Issues include capacity, leadership, equity (Box 17.5), awareness, and corruption, each of which influences conservation outcomes.

Box 17.5   Hearing the voice of women

Carol J. Pierce Colfer

If we wish to manage and protect biodiversity effectively, then we must listen to forest people. Some, however, are hard to hear: women's voices particularly are often absent in discussions of the future of rain forests (Wilde and Vainio-Mattila 1995). Reversing this requires overcoming various hurdles, including women's overburdened work schedules, lower levels of literacy and numeracy, limited fluency in national languages, social norms constraining female–male interactions, and gendered organizational preferences, among others. Yet the possibility of doing so exists (Colfer 2005; Vernooy 2006).

Ecologists, conservationists, and forest managers need to listen to women as it is they who are generally more intimately associated with some aspects of forest use and management than men. Women will often make regular use of a wide variety of forest plants for the preparation of food and medicines (Allotey and Gyapong 2005; Colfer 2008; Smith 2008). For these reasons women tend to have a more holistic view of the forest than men (Bolaños

and Schmink 2005). Such views can complement and reinforce ecologists' perspectives in their bid to protect species and habitats. Forest women typically have a different knowledge repertoire than men, but also different perspectives on the social features of their own societies, features that may be important in collaborative work to develop incentives to encourage human behaviour change.

Improving the ability of women to plan family size can contribute both to population stabilization and to women's own prospects for a life characterized by better opportunities (Colfer et al. 2008). Looking to the future, women of all cultures have greater responsibility for raising children. Women therefore represent a potent opportunity for influencing the views of the next generation.

Finally, the inclusion of women is a matter of equity and justice. Excluding half the population from resource management or biodiversity conservation issues that clearly directly affect them and their families is not only inequitable, but also undermines any efforts towards such objectives.

17.3.1 Poor governance and corruption

Any attempts to promote biodiversity conservation must recognize that many tropical forest areas are subject to conflict and poor governance. Existing governance institutions often lack capacity to manage forests, an issue that goes far beyond just law enforcement. In most tropical countries decision-makers have had comparatively little experience or training in conservation and resource management, but even where this is not the case, conservation often remains low on the list of priorities, as poverty, health, and education have greater resonance among the national populations. Management of natural forests for multiple benefits including conservation requires a diverse set of skills and a broad knowledge base, both of which are often insufficiently developed where there is chronic underinvestment in training. Building skills and human resources to improve forest management, be it for conservation or production, is a slow process even if financial constraints were to be alleviated.

Corruption often undermines conservation and management efforts, by ignoring or distorting legislation and undermining the effective functioning of the justice system. Institutional corruption has been demonstrated to encourage widespread illegal logging in Kalimantan (Indonesian Borneo) and the Philippines (Kummer and Turner 1994; Smith et al. 2003). The payment of bribes, or 'sweeteners', to secure contracts for resource exploitation is common practice in many tropical regions. In the Amazon, as elsewhere, the inability to enforce environmental legislation, which would otherwise immensely curtail deforestation, is primarily due to corruption (Soares-Filho et al. 2006). It also accounts for considerable losses to national revenues derived from forest use: illegal logging is estimated to cost the Indonesian government US\$ 4 billion annually (Laurance 2004).

The emergence of more decentralized governance paradigms (Box 17.6) has prompted the emergence of multi-stakeholder and inter-disciplinary partnerships to deliver more equitable systems of resource governance. Inclusivity engenders broad legitimacy and strength through the diverse set of knowledge, skills, and resources of the different stakeholders (Adger et al. 2003). Nonetheless, crafting new institutions and steering organizational change towards collaborative governance

is a rocky political process in which participants seek to safeguard their own interests. Despite the intrinsic appeal of participatory governance arrangements, their ability to deliver upon their expectations remains to be proven. Interactions among institutions and organizations, which may have fundamentally different goals, cultures, perspectives, and power, are not expected to be conflict-free (Young 2002).

17.3.2 Land ownership

Land rights are at the centre of effective decentralization and community management processes, and uncertainty over land tenure remains a fundamental issue in how social–ecological interactions unfold in the tropics (Sayer et al. 2007) (see also Box 17.6). Conservation through sustainable management might be best achieved by granting legally recognized land and user rights to forest margin communities (Sayer et al. 2007). Improvements in forest management following the allocation of secure user rights and tenure have been documented in Mexico and East Africa (Bray et al. 2003; Sayer et al. 2005). In Mexico, for example, effective community-based management is most evident where the *ejido* system gives strong community ownership of land and resources (Dalle et al. 2006).

Even the perception of tenure security, rather than the actual legal status of tenure, can provide sufficient investment security (Kusters et al. 2007). In Indonesia, a decree issued in 1998 enabled communities in Krui, Sumatra, to register for concession rights for their forests. None of the Krui communities ever formally applied for their concession rights, and yet the right to do so provided the legal security for their continued investment in their forest and agroforest systems, and wide recognition of the decree has been instrumental in stopping outsiders' attempts to appropriate these lands (Kusters et al. 2007).

Yet despite the trend towards devolving rights and responsibilities to communities, legal recognition of community ownership rights in the tropics remains a slow bureaucratic process. The reluctance to transfer legal tenure to communities stems from a long history of institutionalization of forest resources and lands, and the inertia inherent in making large changes of this nature. It has also been common for communities to be given user rights over NTFPs, which are perceived to have low

Box 17.6   Who owns, lives in, or controls the world's rain forests

Globally, around 300 million people live within forests, while another 200 million live on forest edges (Pretty 2002). At the end of the last millennium it was estimated that over a billion people living within 25 of Conservation International's global biodiversity 'hotspots' earned less than US$ 1 per day (McNeely 1999 cited in Molnar et al. 2004). While often viewed as a threat, these people can and often do contribute to conservation (Kaimowitz and Sheil 2007).

The effectiveness of forest governance and protection is partly explained by ownership (Agrawal et al. 2008). Forest tenure determines who can own, use, and modify forest lands and under what conditions. Under customary tenure (determined in most cases at the local level) people tend to view themselves as the owners. Under statutory tenure systems (determined by governments) formal laws often designate the government as controlling forest access. As local claims often go unrecognized by governments, customary and statutory tenure claims often conflict. There are signs that this is changing.

An estimated 71% by area of developing country forests are claimed by governments; around 22% are owned or designated for use by indigenous groups or other communities; and the remaining 7% are owned by individuals or firms (White and Martin 2002). Patterns of ownership, however, vary across nations. No forests are held by communities in DR Congo and Venezuela, while in Mexico and Papua New Guinea the figures are 80% and 90% of forest land respectively. Trends from 2002 to 2008 suggest ongoing reduction in government forest ownership (Sunderlin et al. 2008) and, in a handful of countries including Bolivia, Brazil, Cameroon, Colombia, and India, increases in forest lands owned or designated for use by local people. On the other hand, virtually no progress has been made in allocating forest to local people in other regions.

Legal rights, though helpful, cannot alone guarantee people's rights and forest protection (Ghazoul 2009b). In India the 2006 Recognition of Forest Rights Act codifies the rights of forest-dwelling communities to land and other resources which had been denied to them ever since colonial administrations assumed control of forests. While seemingly restoring rightful ownership, this new law has been the subject of considerable debate: opponents claim it will lead to unregulated forest exploitation, while supporters maintain that tribal communities are the best conservators (Sekhsaria 2007). In Liberia, West Africa, some communities possess legal ownership to forested areas but not the forest—the government owns the trees and can legally allocate 35-year concessions without any local involvement or right of veto (Alden Wily 2007). A similar situation in India and various other countries generates perverse incentives for landowners to clear native trees when opportunities arise and replace them with alien species that are excluded from government ownership (Ghazoul 2007).

Indigenous property rights are constitutionally recognized in Papua New Guinea, but powerful companies still open industrial timber concessions without local consent (Bun et al. 2004). The Peruvian government has allocated four-fifths of the country's Amazon forests for oil and gas exploration, and has apparently allowed investors to violate indigenous land titles (Sohn 2007). Brazil's 2006 Law on Public Forest Management recognizes local rights to forest, and allows communities to run forest concessions—but at the same time extractive reserves in Brazil remain vulnerable to various incursions by ranchers and other powerful interests (Sunderlin et al. 2008).

Many NGOs address community rights and empowerment. The right to hold and manage tropical forests legally can bring various benefits to the people and encourages improved forest management and conservation. Either way, Western perceptions of biodiversity values need to be weighed against the rights and needs of forest-dependent people to manage their forests for their benefit. Often we will all have common goals, and sometimes we will not. Who is to decide whose values take precedence?

market value, while timber rights are retained by the state or large commercial enterprises. Clarifying and formally recognizing user- and land-rights, and increasing access to knowledge and justice systems, provide a strong foundation for effective community management and a barrier to external illegal activities that encroach on such rights (Komarudin et al. 2008).

Some conservationists argue that awarding tenure over forest lands to local communities will simply accelerate forest conversion (Oates 1999; Terborgh 1999). Certainly for many local people conversion of forest to agriculture and other land uses (or the sale of rights to corporate entities) offers a far more effective means to escape poverty than environmentally sustainable options (Boedhihartono et al. 2007). Much of the deforestation in the Amazon can be attributed to the expectation that legal land tenure will eventually be granted following clearance of state-owned forest land (Fearnside 2008). A long history of placating landless families through the legal recognition of land claims in Brazil has encouraged forest clearance by colonists and land speculators.

Community control of forests is least likely to lead to conservation when communities have been excluded from forest lands for decades. During the 20th century most forest land in Mexico was controlled by the state, which offered large concessions to industrial timber companies. The resulting alienation of people from their previously community-owned land led to extensive timber smuggling and land conversion (Klooster 1999, 2000). Since the 1980s, community-based management has been promoted. While forest management and benefits to the community improved, some problems remained, including corruption, the disproportionate capture of forestry benefits by local elites, and timber smuggling (Klooster 1999, 2000). Decentralization and local ownership of forests will therefore not necessarily lead to favourable community management. Positive outcomes additionally require appropriate institutional structures and capacity building to promote communal accountability, fair distribution of benefits, and effectively

implemented resource management regulations (Berkes 2004, 2007; Komarudin et al. 2008; Ros-Tonen et al. 2008).

17.3.3 The Kyoto Protocol

The state of tropical forests, the clearance of which represents an important source of carbon dioxide (CO_2) emissions (Box 17.7), should be central to responses to the challenge of climate change and could be meaningfully coupled with conservation efforts. Unfortunately, political squabbling has, to date, prevented an adequate response. Discussions on the inclusion of avoided tropical deforestation in international climate change agreements, notably the Kyoto Protocol, have foundered on political manoeuvring[32].

Wealthier industrial nations (called 'Annex I countries' in the protocol) who signed the Kyoto Protocol committed themselves to reductions in annual greenhouse gas emission relative to baseline emissions in 1990. Forests are specifically included in provisions regarding Land Use, Land Use Change, and Forestry (LULUCF), and Article 12 of the Kyoto Protocol established the Clean Development Mechanism (CDM) whereby reductions in greenhouse gas emissions of Annex I countries can be achieved by investing in environmentally appropriate development in developing countries where the costs of doing so are lower (Hardner et al. 2000; Fearnside 2006).

In 2001, the Parties to the Protocol decided to exclude forest conservation from consideration within the CDM, and forest-related development projects were therefore restricted to afforestation and reforestation. This decision represents a lost opportunity for addressing biodiversity conservation and climate change challenges, and places additional pressure on forest (Niesten et al. 2002; Niesten and Rice 2006). In its current state, the Kyoto Protocol provides perverse incentives to clear natural forests to establish plantations which can be credited for the carbon sequestered without regard for carbon lost through prior deforestation[33]. Another threat to tropical

---

[32] For a description of the political currents that underlay this decision see Fearnside (2006).

[33] Technically, projects that are unable to demonstrate net carbon sequestration would not qualify for CDM credits, and reforestation of forests cleared after 1990 are also excluded from consideration for CDM credits, but this has not prevented some timber companies from seeking to profit from such activities (Niesten and Rice 2006).

Box 17.7 Deforestation and carbon emissions

Carbon dioxide (CO_2) is sequestered by growing plants through photosynthesis, and is released by the respiration and decomposition of dead plant material. Forests become a carbon sink when the sequestered CO_2 exceeds CO_2 losses, or a carbon store when these are balanced. Tropical forests and their soils may store as much as 375 Gt of carbon, more than 50% of which is within the soil (Brown et al. 1993). Deforestation releases CO_2 to the atmosphere through the burning or decomposition of plant biomass, and by disturbing forest soils which are significant repositories of stored carbon. Changed land cover also substantially reduces sequestration capacity. Indeed, around 20% of global emissions are attributed to tropical deforestation, roughly equivalent to 35–50% of fossil fuel CO_2 emissions (Stern 2007). Carbon stocks can recover following forest regrowth, and this forms the conceptual basis for many carbon sequestration schemes.

---

forests emerging from the Kyoto Protocol is the decision to provide credits for management of forests in Annex I countries for carbon storage and sequestration. This gives forests in Annex I countries a carbon value, which represents an opportunity cost for logging. Were logging activities to decline in Annex I countries in response to this, the shortfall in timber on the global market might then be met through increased harvesting in the tropics.

17.3.4 REDD alert

Although credits for avoided deforestation were excluded from Kyoto Protocol arrangements, there is anticipation of their inclusion from 2013[34]. This would take the form of a Reduced Emissions from Deforestation and Degradation (REDD) mechanism. The exact details of such a mechanism remain unclear, but it is likely to include national-level accounting and baselines, with options for project-level implementation. Financing will be in the form of international development funds or market mechanisms based on tradable carbon credits. Debates on the determination of baseline forest carbon values, demonstration of additionality[35], and proving the avoidance of leakage[36] have yet to be resolved. Some of the issues that led to the exclusion of avoided deforestation mechanisms from the Kyoto Protocol in 2001, including the issue of permanence[37], have also yet to be resolved. Concern remains that the adoption of REDD could undermine the current carbon cap-and-trade mechanisms that have been adopted by the Kyoto Protocol (Karsenty 2008).

Determining forest biomass, and therefore carbon value, is a key stumbling block. The difficulty in appraising the accuracy of values, as well as the variability in structure and density of tropical rain forest formations, leads to considerable leeway in the selection of values, with the potential for selection made to suit particular interests (Fearnside 2006). Thus baselines are subject to

---

[34] At the UNFCCC Conference of Parties in Bali in 2007, it was agreed that the post-2012 climate agreement should include a mechanism to provide developing countries with financial incentives for REDD.

[35] Additionality refers to the need to achieve net reduction in greenhouse gas emissions compared to expected emissions in the absence of a carbon mitigation project.

[36] Leakage refers to the displacement of negative impacts from one location to another, resulting in the negation of benefits achieved by a carbon sequestration project.

[37] A non-emitted tonne of carbon is considered to be 'definitively non-emitted', but this leads to problems in the case of forest-related projects, as deforestation, intentional or accidental, may still occur after carbon credits have been awarded.

political influence by governments and companies. This clearly has repercussions for assessing additionality, which is further fraught with complexity owing to the number of variables that affect deforestation trends and outcomes (Karsenty 2008).

An initial proposal[38] to adopt a historic reference, determined as the average of past deforestation converted into carbon emissions, favours countries that have already exploited their forests—so states that have retained large tracts of forest are disadvantaged both in terms of lost opportunities for development and limited capacity to benefit from REDD mechanisms. Disadvantaged countries include those of the Congo Basin and Brazil (although Brazil supports historical baseline determination).

Another approach to the assessment of baselines and additionality is to develop business-as-usual scenarios (Chomitz 2007). In contrast to historical baselines this approach could be more realistic were it not so complex.

Even a single variable, agricultural commodity prices, is largely unpredictable though key to forecasting forest conversion rates. There are many other factors to consider in determining deforestation rates, and therefore in establishing reasonable business-as-usual scenarios.

Displacement of deforestation activities from sites where carbon emission mitigation projects are being implemented to elsewhere constitutes 'leakage', which could negate any avoided emissions benefits. Successful and meaningful carbon crediting through REDD must have assurances against leakage. Complete prevention of leakage may not be possible, but projects should take steps to minimize leakage and invest in verifying its extent.

Africa's Congo Basin countries had made efforts to implement compulsory forest management plans prior to the CoP 13[39] discussions, and in recognition of this the principle of rewarding emissions reductions by reducing forest degradation was accepted. Evaluating carbon emissions from degradation presents an even more

---

| **Box 17.8** | **Can REDD support conservation concessions?** |

Selective extraction of 3 m^3 ha^{-1} timber, which is typical of highly selective logging in the Congo Basin, results in emissions of around 37 tonnes of CO_2 ha^{-1} (equivalent to 10 tonnes of carbon) (Laporte et al. 2007). Assuming REDD carbon credit values of $4.40 per tonne of CO_2 (based on current voluntary market prices; Butler et al. 2009) gives a REDD value of $163 ha^{-1} which on a 30-year rotation is equivalent to $5.40 ha^{-1}. This has to be set against the opportunity costs of conservation of unlogged forest. For Cameroon, this has been estimated at around $21 ha^{-1}, which is among the highest in the Congo Basin (Karsenty 2007a). Thus REDD is unlikely to cover all the opportunity costs of establishing a conservation concession in an area slated for logging, but is likely to offset a substantial proportion (at least 25%) of the required funds.

Much depends on the pricing of future REDD carbon credits compared to values of other commodities derived from tropical lands. Growing demand for palm oil from emerging economies such as China and India, for example, is likely to support high palm oil prices for the next decade (World Bank 2008). Oil palm agriculture is therefore likely to remain an attractive alternative land use to REDD schemes. To improve the long-term economic prospects of REDD, the mechanism will need to be encompassed within a Kyoto-sanctioned compliance market where higher carbon credit prices (in the region of $30–50 t^{-1} C) can be assured (Butler et al. 2009). Trading in compliance carbon markets will ensure a stable supply of funds at higher carbon prices, and allow REDD to be buffered against uncertainties and fluctuations in opportunity costs, such as palm oil price increases.

---

[38] Presented by Papua New Guinea and Costa Rica in 2005: http://unfccc.int/resource/docs/2005/cop11/eng/misc01.pdf.
[39] 13th Session of the Conference of the Parties and 3rd Session of the Meeting of the Parties to the Kyoto Protocol, Nusa Dua, Bali, 3–14 December 2007.

complex problem than the evaluation of emissions from deforestation, because remote sensing currently lacks the resolution to differentiate among forests subject to different logging regimes (Foody 2003; Foody and Cutler 2003; Asner et al. 2009).

The realization of carbon payments through REDD will undoubtedly generate substantial new funds for improved management and conservation. Nevertheless, it remains questionable as to whether economic viability of REDD will be sufficient to exceed the returns from alternative land uses, and some modelling studies suggest that some environmentally destructive land uses, such as oil palm agriculture, will remain more profitable than likely incomes from REDD carbon credits (Butler et al. 2009) (Box 17.8).

17.4 Conclusion

Protected areas probably provide the best general form of protection for rain forest biodiversity conservation in the long term. Protected areas need not always exclude humans, and through ecotourism development and selective resource use in buffer zones some of the conflicts that have arisen between conservation bodies, governments, and forest margin people may be alleviated and eventually overcome. There are, however, other approaches that offer conservation benefits and have fewer costs for local people and other stakeholders. In many cases conservation results have been disappointing, and many obstacles, including weak institutions and limited political commitment, remain. Nonetheless, there is little doubt that the great extent of protected areas and other measures are starting to stem the tide against biodiversity loss, and that species extinctions are being prevented.

Conservation is, however, far broader than the implementation of protected areas. Investing in education and skills, and empowering communities with the knowledge and confidence to engage in political lobbying and addressing issues of tenure security and justice, are several ways in which environmental management might be

---

Box 17.9 Valuing and funding biodiversity

Perhaps the most fundamental barrier to conserving the world's remaining tropical forests is the difficulty in valuing biodiversity, and therefore the artificiality of many arguments purporting to provide economically rational justification for conservation. Biodiversity is a global commons, and as a commons there is little or no incentive for any individual to invest in its protection (Hardin 1968). At a global scale one consequence of this is the paltry financial commitments made to conservation worldwide (Ghazoul 2009a, b). The global community spends around US $1.5 billion each year on biodiversity conservation. A further US $500 million is spent on development projects that include a biodiversity conservation component (Hardner et al. 2000; Niesten and Rice 2006). To date, UNDP has disbursed a total of US $830 million in GEF funding and leveraged a further US $1.3 billion to address biodiversity loss (UNDP-GEF 2008). Collectively,

conservation NGOs have an annual budget of around US $1 billion. Taken together, the global willingness to pay for conservation therefore amounts to around US $5 billion annually, a sum that seems large, but is dwarfed by the commitment of the USA alone of US $700 billion to bail out the American financial system during the 2008 economic crisis. Were conservation and the planet's environmental wellbeing to be taken as seriously, then an equivalent sum would support current conservation activities for the next 140 years, without even making allowances for investments over this time. Conservation is clearly under-funded: even if financial commitment were to double to US $10 billion each year, as has been suggested as being necessary (Balmford and Whitten 2003), then an equivalent sum to the one that saved the bankers from ruin in 2008 would still buy 70 years to save the global environment from ruin.

equitably improved. Many non-governmental organizations (NGOs) at international, national, and local levels are engaged in such initiatives. Others seek to raise conservation awareness among the public, and hold companies and governments accountable for their actions.

Nonetheless, conservation remains predominantly a top-down process in much of the tropical world. It is often viewed as an externally driven, even foreign, intervention, and this can cause resentment and reluctance. Meanwhile, funds allocated to conservation are paltry compared to those invested in other sectors (Box 17.9). This will need to change if it is to gain the support and backing of emerging democratic nations.

Trends are encouraging, even if agencies have been slow to respond: environmental movements have expanded globally, and polls often imply environmental concerns in developing countries similar in magnitude to those found in wealthy Western countries (Steinberg 2005). Even in remote regions, such as the interior of Borneo, consultations with local people reveal a genuine support and desire for effective, democratically accountable conservation—even if this is not the model that they observe (Padmanaba and Sheil 2007). While local people sometimes appear hostile to conservation projects, this reflects distrust and orientation rather than any genuinely anti-conservation perspective (Sharpe 1998).

CHAPTER 18

Requiem or revival

'...but when I first entered on and beheld the luxuriant vegetation in Brazil it was realising the visions in the Arabian nights—the brilliancy of the scenery throws one into a delirium of delight...'

Charles Darwin (1832)

We have sought to provide a scientific exploration of rain forests. We remain aware of how much we have left unsaid. There is no one tropical rain forest, no common biogeographic trajectory, no pre-eminent diversity theory and no single conservation solution. We have provided an overview, with selected emphases, reflecting our own choices and compromises—likely to please no one entirely. Indeed, we have frequently disagreed among ourselves, though gained much through the resolution of our differences.

One point we agree upon wholeheartedly is the urgent need to address rain forest conservation, and we expect this to be uncontroversial for most of our readers. But not everyone agrees. Therefore at the end of this book we need to consider some uncomfortable questions: What is the future for tropical rain forests? Do humans have the willingness and capacity to stop forest loss? Both pessimism and optimism are justified.

18.1 A question of societal will

Can we afford not to protect rain forests? Estimated global values for ecosystem services range into the trillions of dollars (Costanza et al. 1997; Pimentel et al. 1997) of which rain forests are a key aspect, but such 'virtual' val-

ues are meaningless unless they can be integrated into economic assessment and decision making (Chapter 17). Certainly the loss of tropical rain forest will be a grave loss—our previous chapters indicate why.

Can we afford to protect rain forests? Current expenditure on area-based conservation (i.e. not including education and *ex situ* work etc.) is around US$ 6 billion per year globally, only part of which is spent in the tropics. Estimates of what is needed vary, but one figure is US$ 28 billion annually[1] (Balmford et al. 2002). Effective management and protection of existing tropical protected areas has been estimated to cost US$13 billion annually (Bruner et al. 2004) while expenditure to slow deforestation by 95% through direct financial incentives (REDD) has been estimated at US$ 30 billion (Strassburg et al. 2009).

These big numbers are nonetheless affordable: US$ 28 billion is only 5% of the USA's annual defence budget, little more than twice what Europeans spend on ice cream, or half what Americans spend on pet products each year.

The greatest barrier is not financing but societal will. At first sight this is a paradox: rain forests and their plight have become part of popular culture—the subject of kindergarten projects, Hollywood films, and consumer campaigns. But not everyone is worried. Some attach lit-

[1] Note that these estimates do not address global warming.

399

tle concern to forest loss, and place greater weight on other societal issues. In many regions forest conservation is simply not a priority, especially where it comes at substantial local cost. Hunger, land security and ownership, education, and many other issues take precedence. Other people accept the seriousness of the problem but believe that human ingenuity will find solutions. In tropical rain forest countries, the popular rhetoric of conservation is often perceived as foreign and intrusive to many local and national decision-makers. This perception is bolstered by international campaigns that seem, and often are, more concerned with animals than people. To ignore or deny these schools of thought is to fail to engage with the most important obstacles to rain forest conservation.

Concern for the environment must be balanced with concern for poverty and human health, and for energy generation, national security, and economic development. These concerns need not compete, but they often do. It is indeed hard to argue for the environment taking precedence when thousands die every day from malnutrition and easily preventable disease.

In view of these issues, the greatest challenge for conservation is to build societal willingness to invest in conservation actions. Until society in both developed and developing countries prioritizes and politicizes forest protection, then cash and political capital can continue to be preferentially invested in clearing forest land for farms, cattle, or plantations.

A society that values conservation is necessary but not sufficient to deliver conservation: protecting and maintaining tropical rain forests in the long term also requires adequate capacity. Local, regional, and international institutions that develop and implement environmental plans and legislation are necessary, as are skilled labour and relevant knowledge to make these institutions effective. National institutions charged with forest protection and management exist in virtually every tropical country, and while most possess committed and hard-working staff, they often lack the resources required to perform their function. As an example, it has been estimated that between Ugandan independence in 1962 and 1988 the purchasing power of a forest officer's salary (a graduate post) in their forest service fell by an astonishing 99.6% (Howard 1991). This is a concern if we wish for tropical rain forest lands to be in any sense managed by a well-

trained cadre of professionals. Careers that pay so little will seldom gain the best people, particularly when such work takes people to remote regions far from schools and other services.

Capacity includes access to knowledge and good judgement. In many countries, especially poorer nations, decision-makers and their advisers remain poorly informed on environmental issues. This requires stimulating education and building powerful information and networking facilities. It also requires an active research community with a vibrant culture of critical evaluation, discussion, and communication. Scholarship programmes, syllabus development, internet access, and attractive and informative materials in local languages can all help. A free press and platforms to allow for debate and even protest will serve to engage a wider section of society, including those most directly impacted by changes in land use and forest loss. Favourable socio-political environments allow the proliferation of non-governmental and grassroots organizations that play an important role in information dissemination, advocacy, and representation of interest groups that might otherwise have little visibility in social processes.

International aid programmes target capacity building, conservation, and development in tropical forests by various means. International agreements have been negotiated and institutions established that reflect concern for tropical forests (see Table 18.1). This substantial political effort has delivered a number of valuable advances, but many of these initiatives remain hard to assess objectively as tangible outcomes are often elusive.

18.2   What role for scientists?

We know next to nothing about most rain forest species, but ignorance is no excuse for inaction. We do not need to count, let alone assess, every single forest species to recognize the value of forests. We know enough to know that we have to act, and we know enough to act effectively.

We know that tropical rain forests are the Earth's most species-rich terrestrial biomes, that they are being degraded and lost over much of their range, that they have long supported human populations, and that

Table 18.1 Selected international institutions and agreements relevant to tropical forests compiled by Douglas Sheil from various sources.

Organisation or initiative	Abbreviation	Initiated or revised	Summary of purpose, role and relevance
World Bank and affiliates	WB	1944–7	Development funding, lending and policy. Many forest relevant initiatives. Emphasis now on the Millennium Development Goals. The 'Forests Strategy and Operational Policy', approved 2002, links economic development, poverty reduction, and forest-value protection.
Food and Agriculture Organization of the United Nations	FAO	1946	UN agency for food security. Also compiles forest cover statistics. FAO Forestry Department promotes sustainable forestry practice through technical support and information sharing.
World Conservation Union (formally International Union for Conservation of Nature)	IUCN	1948	Membership (government and non-government) organization for conservation. Involved in many initiatives including the assessment of species conservation status via red-lists.
United Nations Environment Programme	UNEP	1972	UN body with the mandate to lead on global environmental concerns and policies.
World Heritage Convention	WHC	1972	UNESCO agreement which oversees selected cultural and natural sites considered of value to the global community. It provides a status to protected areas and other sites (cultural and natural) including several that contain tropical forests. There are 97 official forest sites (July 2008).
Convention on Trade in Endangered Species of Flora and Fauna	CITES	1975	Intergovernmental agreement to regulate international trade in wild species of conservation concern.
International Tropical Timber Organization	ITTO	1983, 1992, 2006	Intergovernmental organization promoting the conservation, sustainable management, use, and trade of tropical forest resources. Develops policy and guidelines and provides project funding.
World Conservation Monitoring Centre	WCMC	1980s	Mandated by UNEP to inform relevant conventions and initiatives.
Tropical Forest Action Plan	TFAP	1985,1991 (ended)	Funding coordination initiative intended to reduce tropical forest loss, and to promote improved management. Outcomes widely criticized.
International Union of Forest Research Organizations	IUFRO	Mid-1080s	Membership organization. Runs a 'developing countries programme' which supports forest research in the tropics.
Bruntland Commission (World Commission on Environment and Development)	WCED	1987	UN commission created to address sustainable development and environmental conditions relevant to human well-being.
Convention on Biological Diversity (see also GEF)	CBD	1991/93	Non-binding agreement on biological conservation. Seeks scientific guidance via the Subsidiary Body on Scientific, Technical, and Technological Advice (SBSTTA).
Global Environment Facility	GEF/CBD	1991	Coordinates funding in support of CBD UNFCCC and other environmental conventions.

(Continued)

Table 18.1 (Continued)

Organisation or initiative	Abbreviation	Initiated or revised	Summary of purpose, role and relevance
United Nations Conference on Environment and Development,	UNCED	1992	The 'Earth Summit' that developed environmental agreements including the CBD, FCCC, Agenda 21, and the Forest Principles.
United Nations Framework Convention on Climate Change	UNFCCC	1992	International treaty guiding 'protocols' on greenhouse gas emissions (e.g. Kyoto Protocol). Regular 'Conferences of the Parties' or COPs. COP-14, Poznan, Poland, agreed to address forest protection.
Centre for International Forestry Research	CIFOR	1993	Research institution largely focused on policy issues, addressing tropical forests and livelihoods.
World Commission on Forests and Sustainable Development	WCFSD	1994 (to 1999)	Advisory body on global policy for promoting sustainable management of forests.
Seven/Eight eight major industrialized democracies	G7/G8	1990s	The 1998 G8 Action Programme on Forests provides a political commitment to international support to forests. It focuses on (1) monitoring and assessment; (2) national forest programmes; (3) protected areas; (4) private sector; and (5) illegal logging.
Interagency Task Force on Forests	ITFF	1995	Informal consortium of eight international organizations to support the IPF/IFF and assist governments in implementing their proposals.
Intergovernmental Panel on Forests	IPF	1995 (to 1997)	Ad hoc panel formed by the UN Commission on Sustainable Development (CSD). Developed proposals for action on forests (see also IFF).
Intergovernmental Forum on Forests	IFF	1997 (to 2000)	Successor to IPF. The main legacy of the IPF/IFF is a set of approximately 270 proposals for action, known collectively as the 'IPF/IFF Proposals for Action' which address policies for sustainable forest management.
Kyoto Protocol		1997, 2005 (to 2012)	Developed under UNFCCC. Binding commitment by richest industrialized nations to limit greenhouse gas emissions (forest-based sequestration is excluded). Ends in 2012.
Millennium Development Goals	MDG	2000	Eight broad goals for 2015 agreed by 192 UN states. They include reducing extreme poverty.
Collaborative Partnership on Forests	CPF	2000	Consortium of 14 international forest-related organizations. Supports the UNFF and international collaboration on forests and forestry.
United Nations Forum on Forests	UNFF	2000	UN policy forum, mandated to meet annually and advance the forest agendas agreed at UNCED.
Global Strategy for Plant Conservation	GSPC	2002 (to 2010)	CBD programme intended to slow plant extinction by 2010.

they continue to deliver a wide range of goods and services upon which hundreds of millions of people depend. We judge that the balance of evidence strongly implies that human-caused climate and land use change will cause irrevocable damage to tropical forests and their species, and that widespread forest loss will exacerbate climate change and leave us with a legacy of societal and environmental problems.

Research is undoubtedly still needed to generate innovative interventions, and to test and adapt them. Interdisciplinary approaches are considered essential to deliver acceptable and viable long-term solutions, but such changes are often part of a gradual and often iterative research process. Even more important is the engagement that can generate real solutions for problems, and the need to ensure that local concerns, values, and aspirations are recognized and integrated into these solutions. Just as knowledge needs to be more widely shared across society (Figure 18.1), we should also be open to the appreciation and understanding of alternative belief and knowledge systems if ultimately we are to succeed in securing cross-societal conservation objectives.

This is a technical book, but caring for forests is not dependent on technical issues. Caring is based upon emotion. Charles Darwin experienced a 'delirium of delight' when he first experienced the tropical rain forest. Such a sense of wonder motivates our interest in and concern for the natural world. Such sympathies are not universal. We have no inherent right to expect, let alone demand, similar reactions from others who may lack our vantage point and privileges. Nonetheless we find hope in the realization that such views *are* common. We believe such empathy is stimulated by knowledge and engagement and can thus be actively promoted. Books like this (we hope), and more popular media, can make a contribution.

On the other hand we cannot expect people to value rain forests if they lack the necessary understanding. There is a danger, even in wealthier societies, that as people become increasingly centred in cities and built environments they become increasingly alienated from direct experience of the natural world and may lose any affinity or concern for its fate.

Yet this is not inevitable. Almost all young people are interested in animals and the natural world. This seems to be near universal and can be nourished through schools, education campaigns, and opportunities. Engagement for adults is also possible. Many people learn a deep appreciation of forests and the natural world only by direct experience in later life, and this is something that can and should be acted upon.

18.3   Societal oversight

To find and implement solutions we require both the skill and will to act, and unless we advocate dictatorship (which we do not) we will need to carry a broad section of society with us. Tropical rain forests are claimed by a diverse range of people with different needs and concerns. Some want these areas to remain (or become)

Figure 18.1 Education about local species can stimulate pride and environmental engagement. Here, Pak Parman from Papasena Village, Papua (Indonesia), uses a book to show a child the bird species that occur in the nearby forests, and explains, as he has recently learned from the visiting researchers, that many of these species are not found elsewhere.

Photo by Douglas Sheil.

untainted wilderness, some want to extract the goods they can find there, others want to clear away the forest and use the land for something else entirely—how can such claims be reconciled?

An era of increasing democracy and decentralization places the future of tropical rain forests in the hands of local people. Conservationists have had a mixed reaction to this trend, but we would urge that it is a trend that conservationists oppose at their peril. After all, in the long term forests will only be protected from clearance if enough people are in a position to do something about it. Local support is invaluable.

Global environmental awareness is growing, as is appreciation that most people are willing to support some form of conservation so long as it respects their needs and concerns. Our main unease is that such changes may not be occurring fast enough, and may not gain the support that they clearly need soon enough to become an effective force for conservation.

Certainly the global environmental movement has become an impressive force, able to leverage substantial funds for various forms of conservation, but even these few billions pale into insignificance alongside other aforementioned funding priorities. Many of the issues mentioned above come back to investing in people and institutions, in keeping people well informed, engaged, and empowered. There is a role for anyone and everyone in advancing education, experience, dialogue, and a vibrant culture of knowledge sharing, open-minded, debate, and applied research. There is a need for consumers, voters, and investors to be well informed, to take an ethical and informed choice, and to encourage others to do so. There is a role for governments to develop the kinds of institutions and support required. There is a role for institutions and non-governmental organizations to be outward looking and open minded, to share information, and encourage critical debate. And there is a role for scientists, not just as researchers contributing new knowledge, but as engaged and credible experts informing and influencing policy. We have our own fora and societies, such as the Association of Tropical Biology and Conservation (http://www.atbio.org/) and the Society for Conservation Biology (http://www.conbio.org/) by which scientists can and do influence policy. Scientists need to be actively engaged in such societies, which have impact by the weight of the collective scientific community they

represent, and by which they can influence debate without undermining individual objectivity.

18.4 Final words

We do not know what the next few decades or centuries hold in store. Will it be a world without rain forests where stories of jungles filled with extraordinary species, of jaguars, gorillas, and flying frogs, recede into the realms of myth? On current trajectories this is a real possibility. Will that world have the technologies needed to survive and flourish without forests? Probably, but will we be the richer for it? We think not.

Pessimists can find solace in the resilience of nature over geological timescales. Even if we indeed precipitate the mass extinction that many currently foretell, we can note that such catastrophes have happened before and can happen again (Chapter 6). We can anticipate that,

Figure 18.2 All may not be lost—grow your own rain forest.
Photo by Douglas Sheil.

long after people are gone, a rich new biota will almost certainly arise within a few million years—and that rich tropical forests will likely exist once more. But this misses the point. Rain forests do not have intrinsic value—their value reflects what is attached to them by society and individuals. Forests will recover, but this is of little consequence to us in a world without humans.

The next half century seems set to be the key era when the survival of tropical rain forests and their species, and our associated cultural and natural heritage, are finally decided. We believe we can, collectively, make a positive and effective contribution by empathizing, educating, and engaging. If not, then let us hope for some effective innovations (Figure 18.2).

Glossary

For brevity we do not include taxonomic labels and categories that refer to modern species. In terms of names of institutions and agreements we include here only those used in the text—a more comprehensive summary is provided in Table 18.1

Abscission	Process by which a plant discards one or more of its parts, such as a leaf, fruit, flower or seed.
Aerial roots	Above-ground roots, found in diverse plant species, notably in hemi-epiphytic species such as strangler figs.
Afrotropical	Tropical region of Africa including Madagascar and neighbouring islands.
Alkaloid	Basic organic nitrogenous compounds, often highly complex, formed by living organisms. Many possess pharmacological effects.
Allelopathy	The release by a plant of toxic chemical substances to suppress the growth of nearby competing or potentially competing plants.
Allopatry	Referring to geographically separated populations or species. Allopatric speciation arises when populations that are physically isolated by an extrinsic barrier adapt to local conditions and thus diverge, the process becoming complete when reproduction among individuals of the two groups is no longer viable.
Alluvial	Environment, processes and products of rivers and streams.
Alpha-diversity	A measure of diversity that equates to species richness within a defined area or habitat.
Anastomozing	Forming a network (e.g. of streams, branches, roots, or leaf veins) that both divides and reconnects, producing a reticulate pattern.
Anecic (earthworms)	Drawing leaf litter from the surface into underground burrows where it is consumed.
Angiosperm	Flowering plants; their ovules are enclosed in an ovary.
Anoxic	Deficient in oxygen.
Anthocyanin	Water-soluble pigments of terrestrial plants that may appear red, purple, yellow, or blue. Anthocyanins can occur in all tissues of higher plants, including leaves, stems, flowers, roots, and fruits.
Anthropoid	'Of human form', monkeys and apes–a grouping of primates that excludes lemurs, lorises, and tarsiers.
Anthrosols	Soils formed, or profoundly modified, through human activities.

Apical	At the apex or tip.
Aposematism	Warning signal to potential predators indicating the organism's defence mechanism (e.g. toxicity).
Arbuscule	A well-branched, microscopic hyphal structure of a fungal mycorrhizal symbiont that forms within living cortical cells of plant roots.
Archaea	Group or kingdom of single-celled microorganisms (i.e. prokaryotes) that lack a nucleus or any other organelles. Previously known as 'archaebacteria', but recognized as evolutionarily distinct from more familiar bacteria.
Archaeopteris	An extinct genus of spore producing tree-like woody plants with fern-like leaves.
Arthropod	Invertebrate with an exoskeleton, a segmented body, and jointed attachments called appendages.
Ascomycetes	Largest phylum (or division) of fungi, containing over 30,000 species, including many plant pathogens.
Autotroph	Organism that produces complex organic compounds from simple inorganic molecules using energy from light (and/or inorganic chemicals).
Avifauna	Assemblage of bird species.
Axil	Of a plant, the angle where a smaller stem joins a larger one, e.g. where a petiole (leaf stalk) meets the twig on which it is born.
—	
Bacteriovore	Consumer of prokaryotic organisms.
Basal area (of woody vegetation)	The sum sectional area of the tree stems per unit of ground area based on measurements of stem circumference/diameter. Normally quoted for measurements at 1.3 m above ground level or above any buttresses.
Basidiomycetes	Phylum of mainly filamentous fungi that includes mushrooms, bracket fungi, and other polypores.
Beta-diversity	A measure of diversity that equates to the rate of change in species composition across space, habitats, or among communities. The turnover in species among habitats.
Biofilm	Structured community of microorganisms encapsulated within a self-developed matrix attached to a surface.
Biomass	Mass of biological material (often measured per unit area). NB clarified as live and/or dead, above-ground and/or below-ground.
Black-water	Waters coloured dark by tannins and other organic matter, associated with drainage from nutrient poor catchments and peats. Also used to refer to the catchment itself.
Bole	Main stem of a tree, up to the branching crown.
Boundary layer	Layer of fluid or gas in the immediate vicinity of a surface. Used in plant physiology to describe the air which clings to a leaf or other surface.
Brachiation	Form of arboreal locomotion in which gibbons and primates swing from tree limb to tree limb using their arms.
Brachidodromous	Pattern of leaf venation occurring in dicotyledonous plants in which secondary veins loop and connect near the leaf margin.
Bract	Modified or specialized leaf often associated with reproductive structures.
Branching, order of	Referring to the number of divisions (branchings) encountered in travelling from the end of a (typical) twig to the base of the main stem. In some plant forms, especially simple forms, the number is fixed. In others with more complex structures it is not.

Bromeliads	Species belonging to the monocot Bromeliaceae family of flowering plants comprising around 2,400 species native to the Neotropics (one in Africa).
Bryophytes	All non-vascular plant species in the Marchantiophyta (liverworts), Anthocerotophyta (hornworts), and Bryophyta (mosses).
Bushmeat	Meat from wild animals.
Buttress	Of a tree bole or root: a broad prop-like or plank-like structure that extends from the lower portions of a tree stem generally reaching its broadest dimensions at or near the ground.
–	
C3 metabolism	Form of photosynthesis by which carbon dioxide and ribulose bisphosphate is converted into 3-carbon phosphoglycerate (a step which actually occurs in all higher plants as the first step of the Calvin cycle). This is the most common form of photosynthesis in tropical forest plants.
C4 metabolism	Form of photosynthesis that overcomes the tendency of RuBisCO (the first enzyme in the Calvin cycle) to photorespire by separating RuBisCO from atmospheric oxygen, and using 4-carbon oxaloacetate and malate to transport fixed carbon to it. This form of photosynthesis is mostly encountered in herbs of open environments.
Caatinga Amazonica	Heath forests occurring in the Amazon region. Note that *caatinga* is also used to refer to a distinct semi-arid bushland in north-east Brazil.
Caecilian	Burrowing legless amphibians of the family Caeciliidae order Gymnophiona. They are associated with wet tropical climates.
Caecum	Gut cavity where the large intestine starts.
CAM	Crassulacean Acid Metabolism, form of photosynthesis that stores carbon dioxide first as the 4-carbon acid malate (typically during the night). During photosyenthesis this stored carbon dioxide is released to the enzyme RuBisCO. The CAM pathway allows stomata to remain shut during the day and thereby to conserve water, and is found mainly in epiphytes and succulent plants.
Cambium	Plant tissue comprising a layer or layers of cells between the xylem and phloem which retain the ability to divide. These cells provide the source of woody growth.
Campina	Low, often open form of Amazonian heath forest; the canopy is typically 8 m or less.
Campinara	Taller closed form of Amazonian heath forest closed forest usually around 20 m tall.
Caratenoid	Organic lipid-soluble pigments (mostly yellow orange) found in plant tissues that include carotenes and xanthophylls.
Catina	The variation or gradient in edaphic properties associated with local scale slopes derived from a common parent material or origin.
Cauliflory	Borne on the trunk of woody plants.
Cavitation	Formation of gas bubbles within the xylem of vascular plants.
Cellulose	Structural polysaccharide carbohydrate found primarily in plant cell walls and composed from polymerized glucose.
Cercopithecine	Subfamily of around 71 Old World monkeys that include baboons, macaques, and vervet monkeys.
Chavascal	*Varzea* forest inundated for more than six months each year.
Chitin	Structural polysaccharide containing nitrogen found in many invertebrates, fungi, and microorganisms, and similar to cellulose.
Chlorophyll	Green pigment found in most plants, algae, and cyanobacteria, which is responsible for capture of light energy during photosynthesis.

Chloroplast	Organelle within eukaryotic plant cells where photosynthesis occurs.
Cirrus	See *flagella*.
Cladoxylopsid	Fossil plants considered likely ancestors of ferns and horsetails.
Cleistogamy	Self-pollination within an unopened flower.
Clear-water	Water low in sediments and organic impurities associated with drainage from certain (usually geologically stable) catchments. Also used to refer to the catchment itself.
Climax community	A community of plants and animals that, following ecological succession, has reached a steady state composed of species best adapted to average conditions in that area.
Colobines	A subfamily of Old World monkeys in 10 genera, including langurs and colobus monkeys. They possess a multi-part stomach, which aids digestion of leaves and other cellulose-rich material.
Commensalism	Relationship between two species where one benefits and the other is unaffected.
Competitive exclusion principle	Assertion that given a suite of competing species within a bounded system, interspecific competition will result in the elimination of all but one species.
Compound leaf/leaves	Leaf composed of two or more leaflets on a common stalk.
Congener (of a species)	Sharing the same genus.
Convention on Biological Diversity	An international treaty intended to help sustain the diversity of life on Earth (see also Table 18.1).
Cord (of fungi)	Thickened string like aggregation of hyphae.
Coriaceous (of leaves)	Thick and leathery.
Cortex	The outermost layer of an organ. In plants, the outer portion of stem or root between epidermis and endodermis.
Crepuscular	Primarily active at dawn and at dusk. Mobile around twilight.
Cryptic species	Species that is not readily distinguished from other species based on readily accessible criteria.
Cryptogams	'Cryptogamae', a now outdated grouping that encompasses all plants and fungi that reproduce through spores.
Cultivar	Plant selected for particular characteristics that persist through propagation.
Cuticle (of plants)	Protective waxy covering produced by the epidermal cells of leaves, young shoots, and other aerial plant organs.
Cyanobacteria	Photosynthetic bacteria in the phylum Cyanophyta.
Cycads	Evergreen seed-bearing plants belonging to the gymnosperm division Cycadophyta. They are characterized by a crown of pinnate compound leaves and a stout trunk.
—	
Dark respiration	Metabolism of photosynthetic plants when devoid of light, and thus dependent on stored energy and the release of carbon dioxide.
Dbh	*Diameter at breast height* (1.30 m) of trees, used as a standard measure of tree stem breadth/circumference. Various conventions are employed to measure above buttresses, and avoid other distortions.
Decadal	Periods of ten years or decades.
Deciduous	(1) Shedding leaves seasonally/annually. (2) Synchronous leaf fall in response to a specific cue such as drought. (3) Also applied to the leaf, or other organs, shed.
Denitrification	Process by which nitrate is chemically transformed to molecular nitrogen (gas or in solution).

Diatom	Diatomaceae, a family of minute unicellular (though sometimes communal) algae possessing a stiff siliceous cell wall.
Dicotyledons	Flowering plants whose seed typically have two embryonic leaves or cotyledons. Also referred to as 'dicots'.
Dimorphism/Dimorphic	Possessing two distinct forms.
Dioecious	Organisms segregating male and female function to separate individuals.
Diploid	Two times the base number of chromosomes.
Dipterocarp	(1) Trees in the family Dipterocarpaceae. (2) Forests (of Asia) that are dominated by this family. NB Despite the name, not all species in this family have a two-winged seed.
Diurnal	(1) Active during the day. (2) Daily.
Domatia	Chambers or other structures produced by plants to shelter arthropods. Domatia differ from galls in that they are produced by the plant rather than being induced by their inhabitants.
Drip-tip	Extended acuminate leaf tips presumed to facilitate the shedding of water from the leaf surface.
—	
Echolocation	Sense based on sonar: a sound is released and the echo provides information about the physical environment. It is used by many bats, as well as some aquatic mammals and cave-dwelling birds.
Ecotone	Transitional zone or community intermediate between two typified communities and possessing intermediate characteristics.
Ectomycorrhiza	Mycorrhizal fungi in which the hyphae seldom penetrate cells in the roots of their host plants but sheath the root within a net of hyphae (these penetrate the tissues by passing between cells). Includes various fungi in the Ascomycota, Basidiomycota, and Zygomycota. Plants such as dipterocarps that associate with such mycorrhiza are termed 'ectomycorrhizal'.
Edaphic	Environmental conditions, habitats, or vegetation communities characterized or distinguished primarily by soil conditions.
El Niño	See ENSO.
Elfin forest	Vegetation characterized by stunted contorted trees—often associated with high mountain conditions nearing the upper tree-line.
Endemic/endemism	Possessing a distribution restricted to a specified geographic region.
Endogeic (earthworms)	Feeding mainly on soil organic matter.
Endophyte	Living within plant tissues.
ENSO	El Niño Southern Oscillation (commonly referred to as 'El Niño'). A global coupled ocean–atmosphere phenomenon driven by temperature fluctuations in surface waters of the tropical Eastern Pacific Ocean. ENSOs are associated with atypical weather, drier in some regions and wetter in others.
Entomopathogenic	Causing disease/pathology in insects.
Epigeic (earthworms)	Living at, or close to, the soil surface and feeding on organic surface debris.
Epilithic	Living on rock surfaces.
Epiphyll	Living on leaf surfaces.
Epiphyte/epiphtic	Growing on living plants.
Eusocial	Living as a structured cooperative community.
Eutrophic	Rich in nutrients. Applied to aquatic systems—associated productivity may lead to anoxia.

Evapotranspiration	The sum of transpiration and other evaporation from the land surface to atmosphere. Evaporation includes water loss to the air from the soil, canopy interception, and surface water.
Evergreen	Maintaining foliage throughout the year.
Exudate	Liquid substances released from cut plant tissues.

—

Faunivorous	Consuming animals.
Fern	Vascular seedless plants in the division Pteridophyta.
Filmy fern	A family, the Hymenophyllaceae (filmy ferns and bristle ferns) of primarily tropical rain forest ferns, generally restricted to very damp places.
Fisher's Alpha	A measure of taxonomic diversity, based on relative abundances, that is considered relatively robust to scale and sample size.
Flagella (of palms)	A specialized hook-spined climbing organ developed from leaves occurring on some climbing rattan palms.
Foliicolous	Growing on leaves.
Folivorous	Consuming leaves.
Forest dynamics	The continuous state of change, imposed by the interaction of biotic and abiotic factors, that determines the composition and structure of a forest.
Frugivorous/Frugivores	Referring to animals that feed primarily on fruit.
FSC (Forest Stewardship Council)	A non-governmental body that sets standards for certifiable forest management. See Box 17.3.
Fugitive species	A species that typically colonizes ephemeral or periodically extreme environments. Often characterized by strong dispersal ability and rapid life cycles allowing rapid population expansion during favourable periods but eventually eliminated by persistent competition with other species. Any species that cannot persist without temporary release from inter-specific competition.
Functional traits	Attributes of organisms (usually plants) believed to influence their life-histories and/or ecological interactions.

—

Gallery forest	Forests that form along rivers or freshwater alluvial wetlands.
Gamma-diversity	The species diversity within all habitat types within a region diversity (or regional diversity).
Geomorphology	The study of landforms and earth surface processes including erosion, deposition, and formation of landforms and sediments.
Geophagy	Consumption of soil.
Geophyte/geophytic	Growing on rock surfaces. See also *epilithic*.
Gigantism	The tendency to achieve large 'giant' size. Achieving distinctly greater dimensions than considered typical (within a taxa or group of taxa).
Gomphotheres	Extinct proboscideans (elephant-like animals) widespread in North America during the Miocene and Pliocene epochs, 12–1.6 Ma, and extending their range into South America around 3 Ma on the formation of the isthmus of Panama. The group survived in Central America until the end of the Pleistocene, and some species possibly persisted in South America into the last millennium.
Gondwana	The ancient (Jurassic-Cretaceous) southern supercontinent derived from Pangea along with Laurasia in the north. Gondwana included continental crust now divided among Australia, Antarctica, South America, Africa (here including Italy), Madagascar, and India.

Granivores	Feeding primarily on seed.
Guild	A non-taxonomic grouping of organisms based on shared behaviours, dependencies and/or other ecologically influential biological properties. Defined by context.
Gymnosperm	Non-flowering seed-bearing plants that have exposed ovules borne on scales that are usually arranged in cone-like structures.

—

Hadley cell	Atmospheric circulation with air rising near the equator and descending in the subtropics (see Figure 9.1). See also ITCZ.
Haplorrhine	Clade of primates that include all modern monkeys and apes as well as tarsiers.
Haustoria	Specialized structure that penetrates host tissues and allows parasitism of higher plans by fungi and parasitic plants.
Heath forest	Short stature dryland forest formations occurring in nutrient-poor settings. Adaptations for nutrient retention capture and recycling are normally highly developed. Though usually well drained, the water table is sometimes close to the surface.
Heliotherm	Requiring warmth from exposure to sunlight.
Hemi-epiphyte	Plants having both epiphytic and grounded life phases.
Hemiparasite	Parasitic plant that is also photosynthetic.
Herb	Generic term referring to any non-woody plant (usually terrestrial). Not a defined technical term.
Herbivore	Feeding primarily on vegetative (non-reproductive) plant tissues.
Herpetofauna	A collective term for reptiles and amphibians.
Heterotroph	Deriving energy from organic substrates. Dependent on forms of carbon produced by other organisms.
Heterozygosity	Possessing more than one allele at a given genetic locus.
High Conservation Value Forest (HCVF)	Forest habitats which have environmental values that are considered to be of outstanding significance or critical importance, as determined by a formal assessment procedure. Various definitions are in use.
Holocene	A geological epoch beginning approximately 11,700 years ago and continuing to the present. The current interglacial.
Holoparasite	A plant completely parasitic on other plants, with no dependence on its own photosynthesis, and drawing its organic nutrient supply from the host.
Home gardens	Intensively managed multi-species tree-dominated gardens adjacent or close to homes, usually small (0.1 to 0.3 ha).
Homoiochlorophyllous	Form of desiccation tolerant, *poikilohydric*, physiology involving maintenance of chlorophyll and associated photosynthetic apparatus during dehydration.
Homozygous (genotype)	Possessing identical alleles on two (or more) matched chromosomes.
Hornworts	Bryophytes, or non-vascular plants, comprising the division Anthocerotophyta. The common name refers to the sporophyte which is elongated in the form of a horn. The flattened, green plant body of a hornwort is the gametophyte plant.
Horsetails	Jointed and hollow stemmed vascular plants in the class Sphenopsida which possess a single genus, *Equisetum*.
Hypha(e)	Branched filaments produced by fungi.

—

IUCN	The World Conservation Union—a global consortium of people, governments, institutions, and non-governmental organizations dedicated to conservation (see also Table 18.1).

Igapó	Relatively nutrient-poor freshwater swamp forests (black-water) of tropical Americas.
Indochina	The region encompassing Laos, Cambodia, and Vietnam.
Induction (of photosynthesis)	Short-term processes by which exposure to light influences(increases) plant's rates of carbon fixation during photosynthesis compared with response after prolonged darkness.
Inflorescence	A cluster or aggregation of flowers arranged on a stem.
Insectivorous	Consuming insects/arthropods.
Inselberg	Solid rock hill or ridge that rises abruptly from a surrounding landscape.
Insolation	Measure of solar radiation arriving at a location.
Inter-tropical convergence zone (ITCZ)	Belt of converging trade winds and low pressure in the tropics, where the surface level winds from the northern and southern Hadley Cells meet. It shifts seasonally and is associated with the monsoons and other seasonal weather patterns.
Irradiance	Radiation (usually light).
Isoprene	Volatile organic unsaturated hydrocarbon, C_5H_8 (2-methylbuta-1,3-diene), that can polymerize to make rubber. Produced naturally by many organisms and believed to influence atmospheric chemistry.
IUCN	The World Conservation Union – a global consortium of people, governments, institutions, and non-governmental organizations dedicated to conservation (see also Table 18.1).
—	
Karst	Topography, shaped by the dissolution of soluble bedrock, usually carbonate rock such as limestone or dolomite. Sometimes of extreme ruggedness. Surface water bodies are often scarce or absent with the principal drainage being subterranean.
Kerangas	Heath forest growing on acidic sandy soils mainly on Borneo and other smaller Indonesian islands. From the Iban language of Borneo, referring to a place where rice cannot be cultivated.
Kerapah	Inland peat formations in Asian wet tropics.
Knee-roots	Roots forming a loop resembling a bent knee above the substrate surface. Often among trees in swamp forests.
—	
Lamina	The flattened body of a leaf or leaf-like organ. Leaf blade.
Laurasia	Ancient (Jurassic-Cretaceous) northern supercontinent derived from Pangea along with Gondwana to the south. Laurasia included continental crust now divided among Eurasia, North America, and Greenland.
Leaching	Transfer, extraction or loss in aqueous solution. Generally used only for the (slow) processing of relatively insoluble compounds.
Leaf Area Index (LAI)	The ratio of total upper leaf surface of vegetation divided by the land area on which the vegetation grows. Values are dimensionless and range from 0 for bare ground to around 8 for dense forest.
Leaf margin	Perimeter of the leaf lamina/blade.
Legume	A plant in the pea family Fabaceae (or Leguminosae).
Lek (cock of the rock)	A localized competitive display in which male animals seek to attract mates.
Lianas	Woody stemmed climbing plants.
Lichen	A symbiotic combination of algae and fungi that in its vegetative biology acts as a coherent photosynthetic organism.

Light compensation point	Irradiance level where carbon gain from photosynthesis equals loss from respiration—the leaf/plant has only enough energy to stay alive.
Light saturated photosynthetic rate	Rate of carbon fixation occurring at the light saturation point (the maximum photosynthetic rate possible by varying irradiance levels).
Light saturation point	Level of irradiance at which photosynthesis fails to fix additional carbon with further increases in irradiance.
Lignin	A complex and rigid organic compound derived from polymerization of phenolic elements (phenylpropanoid). The major component of woody tissues (along with cellulose).
Lithophytes	Growing on rock.
Liverworts	Small bryophyte plants (usually less than 10 cm long) belonging to the division Marchantiophyta.
Living dead	Species found in the wild but which can no longer maintain viable populations and are therefore doomed to extinction.
Lycopods	Pteridophytes of the family Lycopodiaceae. They possess small simple leaves and bifurcating (forking into two) stems. Extant species include herbs and epiphytes. Extinct species included tree forms that dominated the swamp forests of the Carboniferous.

—

Ma	Million years ago.
Macronutrients	Elements required in significant amounts in order for an organism to survive and develop normally. For plants these include nitrogen, phosphorus, potassium, sulphur, calcium, and magnesium. In addition, the material needed to make carbohydrates: carbon, hydrogen, and oxygen, derived from air and water are also necessary (though are often omitted in favour of solely soil derived nutrients). For animals sodium and chlorine too are macronutrients. See also micronutrients.
Makatea	Land formed from uplifted coral reef. A limestone habitat associated with coastal environments.
Malagasy	Of Madagascar.
Malesia	A biogeographical region reaching from the Malay Peninsula to New Guinea and the Australian wet tropics, and including the islands of Indonesia, Sumatra, Java, Bali, and Borneo as well as those of the Phillipines and East Timor. See also Melanesia.
Mangal	Mangrove vegetation.
Mangrove	Trees adapted to waterlogged brackish and marine environments. Also refers to the habitats created by such vegetation when they occur in tidal regions.
Mass elevation effect	See *Massenerhebungseffekt*.
Massenerhebungseffekt	The effect that leads to altitudinal vegetation zones occurring higher on larger than upon (otherwise comparable) smaller mountains/highlands.
Mast fruiting	Patterns of synchronized abundant fruiting events interspersed with extended multi-year periods of absent or low fruiting. Can refer to a species as well as to a vegetation community.
Mata de igapó	See *igapó*.
Mata de várzea	See *varzea*.
Matrix	Surrounding habitat and spatial context.
Megaherbivores	Large-bodied plant-eating animals.
Megathermal	Possessing a frost free climate.

Mekong Basin	Catchment area that feeds the Mekong River of Indochina.
Melanesia	Region of Oceania north and north-east of Australia extending from Fiji in the east to New Guinea in the west. See also Malesia.
Mesoamerica	Central America.
Mesophyll	Leaf having a (one-sided) area of 4,500 to 18,200 mm^2. Small mesophyll of Raunkier (see Table 2.3).
Microclimate	Climatic conditions associated with a specific location (these may differ from neighbouring locations or not).
Micronutrients	Elements required for an organism to survive and develop normally—but needed only in comparatively smaller amounts than macronutrients. For plants these include Boron, Chlorine, Cobalt, Copper, Iron, Manganese, Molybdenum, Nickel, (Sodium), and Zinc. For animals various additional elements are required, e.g. arsenic, chromium, iodine, selenium, tin, vanadium (but not boron). See also Macronutrients.
Microphyll	Leaf having a (one-sided) area of 225 to 2025 mm^2.
Mineralization	Change from organic to inorganic compounds associated with decomposition and/or microbial metabolism.
Moluccas	Islands in eastern Indonesia between Sulawesi and New Guinea. They are the original home of nutmeg and cloves, and hence the archipelago is also known as the Spice Islands.
Monocotyledons	Flowering plants whose seed typically have one embryonic leaf or cotyledon. Also referred to as 'monocots'.
Monodominant	Vegetation predominantly comprising a single (canopy) species.
Monopodial	Plant crowns constructed from a single leading apical shoot.
Monotypic	Higher level taxon possessing a single member species.
Monsoon	(1) Types of seasonal climate occurring in much of the non-equatorial tropics that switch between distinct wet and dry periods. (2) The wet season winds and rainfall associated with such seasonal climates. From *mausim*, the Arabic word for season.
Montane	Associated with mountains.
Moss	Small (1–10 cm) seedless plants with simple leaves within the division Bryophyta.
Mutualism	A symbiotic, or mutually beneficial relationship.
Mycelium/mycelia	Filament(s) of threadlike hyphae constituting the vegetative (non-reproductive) body of a fungus.
Mycoheterotroph	Plant obtaining its carbon, water, and/or nutrient supply through a parasitic association with saprophytic fungi. It is increasingly being recognized that many plant species can be considered mycoheterotrophs for some, often brief, phases in their life cycle.
Mycorrhiza	Fungi closely associated with plant roots and usually involved in a mutually beneficial symbiotic relationship with the plant.
Mycotroph	Plant obtaining some part of its carbon, water, and/or nutrient supply through association with fungi.
Myrmecophyte	Plant possessing specialized structures to benefit the ants, with which it forms a close symbiotic relationship ('ant plant').
Myrmecotrophic	A plant that derives some of its nutrients from waste materials accumulated by resident ant colonies.
—	
Naturalized	Introduced, established, and persisting in a viable state, outside its native range without targeted interventions. Often refers to domesticated species 'gone wild'.

Nectarivorous	Feeding on nectar.
Nematode	Unsegmented roundworms, phylum Nematoda. Many are free-living and many more are parasites.
Net Assimilation Rate	In plant physiology: the overall rate of dry matter production per unit of leaf area.
Net precipitation	The average measures of rain reaching the ground within a forest.
Net primary production	Net production of organic compounds from CO2 through (mainly) photosynthesis.
NGO	Non Governmental Organisation.
Niche	Range of environmental conditions to which an organism is suited.
Nitrogen deposition	Biosphere's gain of reactive nitrogen compounds from the atmosphere.
Non structural carbohydrate (NSC)	Organic (carbon based) material that is mobile within an organism: typically sugars and starch. Plant 'fibre' such as cellulose and lignin is excluded.
Notophyll	Leaf having a (one-sided) area of 2,025–4,500 mm^2. Large mesophyll of Raunkier (see Table 2.3).
NTFP	Non-timber forest product.

–

Obligate	Dependent upon.
Old World	Regions known to Europeans before the discovery of the Americas. Europe, Asia, Africa and associated offshore islands.
Oligotrophic	Low in nutrients. Applied especially to aquatic systems and peats.
Ombrogenous (peatlands)	Hydrological (near) surface flows limited, thus nutrient inputs derived primarily from the atmosphere.
Onchocerciasis	Parasitic disease, also known as 'river blindness', caused by an African worm *Onchocerca volvulus* carried by tiny Simulian flies that breed along fast-moving rivers.
Onychophora	Predatory terrestrial invertebrates, commonly known as velvet worms, with the superficial appearance of short-legged worms. Belong to the ancient superphylum Protostomia.
Oomycete	A group of filamentous, unicellular Heterokonts. Though long considered fungi they lack chitin in their cell walls and are more closely related to some algae. Many are pathogens on plants.
Opportunity cost	Of a specific course of action: the potential benefit from an alternative to be forgone.
Oscines	Singing birds in the suborder Passeri in the Passeriformes.
Outcrossing rate	The proportion of progeny that result from matings between genetically distinct individuals.
Overstorey	The collective vegetation that forms the forest mid and upper canopy.
Ovule	Plant structure containing the female germ cell.
Oxbow lakes	Lakes formed when river meanders are shed from the main channel and no longer carry flowing water.
Oxisol	Old red soils up to 20 m depth formed in the seasonal wet tropics.

–

Paleotropics	Tropical regions of Africa and Asia—tropical portion of the Old World (see above).
Palynology	The science of contemporary and fossil organic particles ranging in size from 5 to 500 micrometres, including pollen.
Pangea	Ancient (but most-recent) supercontinent that included all major continental land masses in the Permian and Triassic. Splits into Laurasia and Gondwana about 200 Ma.
Parapatric	Species or races separated in space but potentially meeting at a boundary.

Paratropical forests	Predominantly evergreen forests occurring in moist generally frost-free environments outside the true tropics.
Pathogen	Disease-causing organism.
Peat	Soils rich in organic matter derived predominantly from vegetable matter.
Peristome (of pitcher plant)	The smooth ridged surface at the rim of the insect trap pitcher of *Nepenthes* pitcher plants.
Petiole	Leaf stalk.
Payment for Environmental Services (PES)	A transaction in which a service provider is (financially) rewarded for maintaining service provision.
PFD	See photon flux density.
Phagotrophic	The physical means by which eukaryotic cells can engulf food particles for processing within a phagocytic 'digestive' vacuole.
Phloem	Plant vascular tissue principally involved in the translocation of sugars and organic nutrients.
Photodamage	The injury and/or impairments caused by exposure to excessive irradiance. The lasting negative effects of excess light upon plant tissues.
Photoinhibition (PFD)	A reduction in photosynthetic productivity due to excess light. Any temporary decrease in carbon fixation due to surplus irradiance.
Photon flux density	A measure of irradiance that summarizes light-quanta per unit time and area.
Photorespiration	A side reaction during normal photosynthesis that involves light-dependent release of carbon dioxide and uptake of oxygen.
Photosynthetic induction	The short-term process by which exposure to photosynthetically active radiation influences the plant's abilities for photosynthesis.
Photosynthetically Active Radiation	Electromagnetic wavelengths able to contribute to photosynthetic carbon fixation.
Phylogeny	Branched pattern of evolutionary relationships usually indicating descent from a shared ancestor.
Physiognomy	Form and structure of vegetation communities.
Phytolith	Microscopic mineral particles derived from plants.
Phytophagous	Consuming plants or plant derived material.
Phytosaurs	An extinct group of large (2–12 m) semi-aquatic predatory Late Triassic reptiles resembling modern crocodiles.
Phytotelmata	Water bodies held by plants including bromeliad and pitcher plant tanks, water-filled tree hollows, *heliconia* inflorescences, bamboo internodes, and water collected at the base of leaves, petals, or bracts.
Pinnate	Having leaves or leaflets arranged on opposite sides of a common axis.
Pioneer	To ecologist pioneers are colonizing species that typically establish early in an ecological succession. In tropical forest ecology the term often refers to species which establish in open well-lit areas that may already be long colonized by other vegetation.
Pleistocene	Period following the Pliocene from 1.81 Ma to the start of the Holocene 11,700 years ago. This period was characterized by multiple glaciations and the emergence of modern humans.
Ploidy	Referring to the number of chromosomes.
Pneumatophores	Peg-like modified roots which allow breathing (gaseous exchange) in waterlogged environments. Associated with mangrove vegetation and swamp forests that do not involve large long-term (months) changes in flood depth.

Podzol	Ancient heavily leached soils developing from sustained long-term weathering and from which virtually all soluble minerals have been lost.
Poikilochlorophyllous	Form of desiccation tolerant, poikilohydric physiology involving loss of chlorophyll and associated cell components during dehydration.
Poikilohydric	Survives desiccation.
Polycotyledonous	Possessing multiple cotyledons (in contrast to most angiosperms which have either one or two).
Polypore	Fungi of the Polyporaceae (the Boletaceae are included by some authors) having the spore-bearing surface within tubes or pores.
Potential evapotranspiration	A representation of the amount of water that would evaporate from a saturated surface, which is a function of temperature and moisture availability. It is directly related to plant production. Evapotranspiration is said to equal potential evapotranspiration when there is ample water.
Primary production/productivity	Rate of creation of organic compounds from carbon dioxide through photosynthesis (also chemosynthesis).
Prokaryote	Single-celled organisms lacking internal membranes or organelles (includes bacteria and archeae).
Prosimian	A polyphyletic grouping of primates that excludes monkeys and apes. It includes the primates suborder Strepsirrhini (lemurs, lorises, galagos, pottos) and the suborder Haplorrhini (tarsiers) and also their common ancestors with monkeys.
Prototaxites	Earliest large-sized terrestrial 'forest forming' organisms. Known from Silurian and Devonian fossils. Long considered plants but now believed to be more similar to fungi or lichen.
Pseudosporochnales	An extinct group of fern like plants.
Pteridophytes	Group of plants comprimising four classes: the Psilosida, including the most primitive vascular plants, found mainly in the tropics; the Lycopsida, including the club mosses; the Sphenopsida, including the horsetails; and the Pteropsida, including the ferns. They do not produce seeds.
—	
Ramiflory	Borne from the surface of mature branches.
Rattans	A paleotropical group of (predominantly) climbing and spiny stemmed palms.
Ramet	Individual member of a clone (usually connected to other members).
REDD	Reducing Emissions from Deforestation and Forest Degradation: one form of payment for ecosystem service where carbon storage value of forests that are threatened by degradation or clearance is financially recognized through payments to forest owners (usually, but not exclusively, nation states) to conserve the forest (see section 17.3.4).
Refugium/refugia	Safe haven(s) for species, typically relating to pockets of persisting or stable habitat.
Relative Growth Rate	Measure of the rate of increase of an organism or population that is determined with reference to the starting size. Proportional rather than absolute measure of increase.
Resilience	The ability to recover after disturbance.
Restinga alta	*Varzea* forest typically inundated for between two and four months each year.
Restinga baixa	*Varzea* forest typically inundated for between four and six months each year.
Reticulate	Network pattern, often with reference to leaf venation.
Rhizomes	Horizontal underground stem of a plant, often sending out roots and shoots from its nodes.

Riverine	Associated with rivers and streams.
Root hairs	Thread-like cells or 'rhizoids', outgrowths of trichoblasts that emerge from the root epidermis of many vascular plants and improve uptake of water and nutrients from the soil.
Root mats	Tightly woven and/or interlinked roots and hyphae that sometimes form on the soil surface.
Ruminants	Hoofed ungulate mammals in the order Artiodactyla that digest cellulose-rich plant-based food. They possess a compartmentalized stomach that allows gastric fermentation—'cud chewers'.
—	
Saprolite	A chemically weathered rock containing predominantly quartz and a high percentage of kaolinite with other clay minerals.
Saprophyte	Living on dead matter of biological origin.
Saprotrophic	An organism that obtains its nutrients from dead and decaying non-living organic matter.
Sclerophyll	Stiff/hard leaved.
Secondary compounds	Organic compounds not directly involved in metabolism required for normal growth and development, but derived from side-branches. Many secondary compounds serve a defensive role.
Secondary consumers	Organisms feeding upon primary consumers.
Secondary forest	Woody regrowth vegetation in areas where forest cover was previously removed, destroyed or absent.
Secondary tissue	Produced by a lateral meristem, e.g. secondary xylem.
Seed bank	Viable seeds present in the soil.
Seedling bank	Juvenile trees which persist but grow little under normal forest understorey shade.
Semi-evergreen/ semi-deciduous	Forest characterized by a mixed deciduous and evergreen forest.
Serpentine	Pertaining to magnesium rich minerals—soils derived in areas with such minerals are often also rich in nickel, chromium, and cobalt. See also ultramafic.
Shrub	Low stature (height usually less than 5 m) usually multiple-stemmed woody plant. Not a strict scientific term but widely used.
Slime moulds	Single-celled bacteriovores that form coordinated communal aggregations of many individuals, which behave as an individual multicellular organism as part of their life cycles. Includes two distinct and only distantly related taxa: the Dictyostelids and the Myxomycetes.
Sorensen's similarity index	A measure of the proportion of species shared between two samples (maximum 1, minimum 0).
Spatulate	Spoon-shaped.
Speciation	Processes by which new species arise from ancestral forms.
Specific humidity	Mass of water vapour per unit mass of moist air.
Specific leaf area (SLA)	The leaf area to dry weight ratio, used as a measure of leaf thickness.
Speciose	Rich in species.
Sphenopsids	Ancient class of plants originating in the Devonian, with common living species including the horsetails.
Spodosol	See podzol.
Stilt roots	Multiple prop-like woods structures that support a plant stem.

Stomata	Pores found mainly in the plant epidermis, especially the leaves and other green tissues, and used for gas exchange.
Strata (of forest)	Layers, or pattern of layers, used to describe the structure of tree-dominated vegetation.
Structure (of forest)	Canopy height and stratification, tree size and stem densities, canopy forms and arrangements, the presence of lianas and epiphytes.
Succession	Sequence of changes in the composition and / or structure of an ecological community following disturbance or environmental change.
Sucker	Shoot or sprout which grows from a bud at the base of a tree or shrub or from its roots. Suckers may also arise from the stumps of cut trees.
Sunfleck	Discrete patch of sunlight reaching the forest understorey.
Syconium	The closed inflorescence of a fig (*Ficus*).
Symbiotic	Pertaining to a mutually beneficial relationship between two species.
Sympoidal	Plant crown constructed from multiple leading shoots.
–	
Tardigrade	Arthropods, usually microscopic, of the division Tardigrada. Sometimes called 'water bears'. Most eat plants or bacteria though some are predatory.
Tendril	Specialized structure, derived from an elongated stem, leaf or petiole, used by climbing plants for attachment, generally by twining around a support.
Tepuis	Table-top mountains in the Guiana highlands of Venezuela, Brazil, and Guyana in South America.
Terra firma	Dry land not subjected to regular flooding.
Terra preta	Dark carbon-rich *anthosols* of the Amazon region associated with pre-Colombian Amerindian societies.
Topogenous (peatlands)	With nutrients arriving through flowing or tidal waters
Transpiration	Evaporation of water from plant organs, notably leaves, but also stems, flowers, and other plant parts.
Treeline	Effective limit of tree occurrence (associated with high altitude).
Trichomes	Any outgrowth from the plant epidermis.
Trophic guild	A classification of an organism's place within a food web: primary producers, primary consumers, secondary consumers, etc.
Turnover	(1) Rate of production and loss with relation to what is present. (2) Change in a community, typically relating to composition, with time or distance.
–	
Ultisol	Mineral soils resulting that lack any calcareous material and have less than 10% weatherable minerals in the top layer of soil.
Ultrabasic	See ultramafic.
Ultramafic	Denotes regions characterized by geology with high concentrations of Mg, Fe, Cr, Co, and Ni, and low concentrations of P, K, and Ca. Roughly synonymous with 'serpentine' and 'ultrabasic'.
Understorey	Forest stratum below the forest canopy. NB Generally this boundary is not well defined.
–	
Urticating (hairs/spines)	Inducing physical irritation.
Varzea	Relatively nutrient-rich freshwater swamp forests (white-water) of tropical Americas.

Vascular (of plants)	Possessing specialized conducting tissues. Includes pteridophytes and all seed-bearing plants.
Vesicular-arbuscular mycorrhizas	Form of mycorrhizal fungi, in the division Glomeromycota, which penetrate the cortical cells in the roots of host plants. Formerly VAM, now known as arbuscular mycorrhizas.
Volant	Able to fly.
Volatile organic compounds	Organic molecules that occur as vapour at ecologically relevant temperatures ($< 45\,^{\circ}C$). Many can diffuse from plants to the atmosphere.
Volatilization	Becoming vapour or aerosol.

—

Wallace's Line	One of several biogeographical boundaries intended to separate the biota of Asia and New Guinea–Australia. The line passes between Lombok and Bali Islands in modern-day Indonesia.
White water	Waters rich in sediments and nutrients, associated with drainage from certain, often geologically young, catchments. Also used to refer to the catchment itself.
WWF	World Wide Fund for Nature and Natural Resources (World Wildlife Fund), a conservation NGO.

—

Xanthophyll	Class of molecules related to carotene and associated with photosynthesis. They generally absorb light at frequencies distinct from chlorophyll, and contribute to photosynthetic efficiency, protection, and colour.
Xeromorphic	Physical attributes associated with low moisture availability.
Xerophytic	Adapted to low (or infrequent) moisture availability.
Xylem	Plant vascular tissue principally involved in the translocation of water and associated solutes. The plant tissue that forms wood.

—

Yeast	Various unicellular fungi that reproduce principally by budding or fission. Not a taxonomically coherent grouping. Most species remain undescribed.

—

Zygopterids	Group of extinct ferns or fern-like plants.

References

Aanen, D. K. and Eggleton, P. (2005) Fungus-growing termites originated in African rain forest. *Current Biology* **15**, 851-855.

Abram, N. J., Gagan, M. K., McCulloch, M. T., Chappell, J. and Hantoro, W. S. (2003) Coral reef death during the 1997 Indian Ocean dipole linked to Indonesian wildfires. *Science* **301**, 952-955.

Ab'Saber, A. N. (1982) The paleoclimate and paleoecology of Brazilian Amazonia. *Biological Diversification in the Tropics.* Prance, G. T. Columbia University Press, New York, USA. Pp. 41-49.

Ackerly, D. D. (2003) Community assembly, niche conservatism, and adaptive evolution in changing environments. *International Journal of Plant Sciences*, **164**, S165-S184.

Ackerly, D. D. and Bazzaz, F. A. (1995) Seedling crown orientation and interception of diffuse-radiation in tropical forest gaps. *Ecology* **76**, 1134-1146.

Adam, P. (1994) *Australian Rainforests.* Clarendon Press, Oxford, UK.

Adamczewska, A. M. and Morris, S. (2001) Ecology and behavior of *Gecarcoidea natalis*, the Christmas Island red crab, during the annual breeding migration. *Biological Bulletin* **200**, 305-320.

Adams, W. M. and Infield, M. (2003) Who is on the gorilla's payroll? Claims on tourist revenue from a Ugandan National Park. *World Development* **31**, 177-190.

Adger, W. N., Brown, K., Fairbrass, J., Jordan, A., Paavola, J., Rosendo, S. and Seyfang, G. (2003) Governance for sustainability: towards a 'thick' analysis of environmental decision making. *Environment and Planning A* **35**, 1095-1110.

Adis, J. (2001) Arthropods (terrestrial), Amazonia. *Encyclopedia of Biodiversity.* Levin, S. A. Academic Press, London, UK. Pp. 249-260.

Adis, J., Harada, A., da Fonseca, C., Paarmann, W. and Rafael, J. (1998) Arthropods obtained from the Amazonian tree species Cupiuba (*Goupia glabra*) by repeated canopy fogging with natural pyrethrum. *Acta Amazonica* **28**, 273-283.

Adis, J. and Harvey, M. S. (2000) How many Arachnida and Myriapoda are there world-wide and in Amazonia? *Studies on Neotropical Fauna and Environment* **35**, 139-141.

Adis, J., Lubin, Y. D. and Montgomery, G. G. (1984) Arthropods from the canopy of inundated and terra firma forests near Manaus, Brazil, with critical considerations on the pyrethrum-fogging technique. *Studies on Neotropical Fauna and Environment* **19**, 223-236.

Adler, G. H. (1994) Avifaunal diversity and endemism on tropical Indian Ocean islands. *Journal of Biogeography* **21**, 85-95.

Adler, G. H. and Kestell, D. W. (1998) Fates of neotropical tree seeds influenced by spiny rats (*Proechimys semispinosus*) *Biotropica* **30**, 677-681.

Agrawal, A., Chhatre, A. and Hardin, R. (2008) Changing Governance of the World's Forests. *Science* **320**, 1460-1462.

Aguirre, A. and Dirzo, R. (2008) Effects of fragmentation on pollinator abundance and fruit set of an abundant understory palm in a Mexican tropical forest. *Biological Conservation* **141**, 375-384.

Agyeman, V.K., Swaine, M.D. and Thompson, J. (1999) Responses of tropical forest tree seedlings to irradiance and the derivation of a light response index. *Journal of Ecology*, **87**, 815-827.

Aide, T. M. (1987) Limbfalls: A major cause of sapling mortality for tropical forest plants. *Biotropica* **19**, 284-285.

Aide, T. M. (1993) Patterns of leaf development and herbivory in a tropical understorey community. *Ecology* **74**, 455-466.

Alden, D. (1976) The significance of cacao production in the Amazon Region during the late Colonial period: an essay in comparative economic history. *Proceedings of the American Philosophical Society* **120**, 103-135.

Alden Wily, L. (2007) *So Who Owns the Forest? An Investigation into Forest Ownership and Customary Land Rights in Liberia.* FERN. Brussels, Belgium, The Sustainable Development Institute. 318 Pp.

Alexandre, A., Colin, F. and Meunier, J. D. (1994) Phytoliths as indicators of the biogeochemical turnover of silicon in equatorial rainforest. *Comptes Rendus de l'Academie des Sciences Serie II* **319**, 453–458.

Alexiades, M. and Lacaze, D. (1996) FENAMAD's program in traditional medicine: An integrated approach to health care in the Peruvian Amazon. *Medicinal Resources of the Tropical Rainforest*. Balick, M. J., Elizabetsky, E. and Laird, S. A. Colombia University Press, New York, USA. Pp. 341–365.

Alford, R. A., Bradfield, K. S. and Richards, S. J. (2007) Global warming and amphibian losses. *Nature* **447**, E3–E4.

Algeo, T. J. and Scheckler, S. E. (1998) Terrestrial-marine teleconnections in the Devonian: links between the evolution of land plants, weathering processes, and marine anoxic events. *Philosophical Transactions of the Royal Society of London Series B* **353**, 113–128.

Allem, A. C. (2003) Optimization theory in plant evolution: An overview of long-term evolutionary prospects in the angiosperms. *Botanical Review* **69**, 225–251.

Allen, G. R., Hortle, K. G. and Renyaan, S. J. (2000) *Freshwater Fishes of the Timika Region, New Guinea*. PT Freeport, Timika, Indonesia.

Allen, J. C. and Barnes, D. F. (1985) The causes of deforestation in developing countries. *Annals of the Association of American Geographers* **75**, 163–184.

Allison, S. D. (2006) Brown ground: A soil carbon analogue for the green world hypothesis? *American Naturalist* **167**, 619–627.

Allotey, P. and Gyapong, M. (2005) *The Gender Agenda in the Control of Tropical Diseases: A Review of Current Evidence*. Special Programme for Research and Training in Tropical Diseases. World Health Organization, Geneva, Switzerland. Pp. 38.

Almeida, M. (1996) The management of conservation areas by traditional populations: the case of the Upper Jurua Extractive Reserve. *Traditional peoples and biodiversity conservation in large tropial landscapes*. Redford, K. H. and Mansour, J. A. Arlington, Nature Conservancy. Pp. 137–157.

Alvarez-Buylla, E. R. and Martinez-Ramos, M. (1990) Seed bank versus seed rain in the regeneration of a tropical pioneer tree. *Oecologia*, **84**, 314–325.

Alvarez-Buylla, E. R. and Martinez- Ramos, M. (1992) Demography and allometry of *Cecropia obtusifolia*, a neotropical pioneer tree: An evaluation of the climax pioneer paradigm for tropical rainforests. *Journal of Ecology*, **80**, 275–290.

Alvarez-Clare, S. and Kitajima, K. (2007) Physical defence traits enhance seedling survival of neotropical tree species. *Functional Ecology*, **21**, 1044–1054.

Alves, D. (2000) An analysis of the geographical patterns of deforestation in Brazilian Amazonia the 1991–1996 period. *Patterns and Process of Land Use and Forest Change in the Amazon*. Wood, C. and Porro, R. University of Florida Press, Gainesville, FL, USA.

Amaral, I. L., Adis, J. and Prance, G. T. (1997) On the vegetation of a seasonal mixedwater inundation forest near Manaus, Brazilian Amazonia. *Amazoniana* **14**, 335–347.

Amedegnato, C. (1997) Diversity in Amazonian canopy grasshoppers community in relation to resource partition and phylogeny. *Canopy Arthropods*. Stork, N. E., Adis, J. and Didham, R. K. Chapman and Hall, London, UK. Pp. 281–319.

Amoroso, V. B., Zamora, P. M. and Rufila, L. V. (2001) Species richness, assesment and distribution of *Lycopodium* in the Philippines. 5th International Flora Malesiana Symposium 9–14 September 2001, Sydney, Australia. From http://www .anbg.gov.au/fm/fms5/fms5_fern_abstracts.html.

Anchukaitis, K. J. and Horn, S. P. (2005) A 2000-year reconstruction of forest disturbance from southern Pacific Costa Rica. *Palaeogeography Palaeoclimatology Palaeoecology* **221**, 35–54.

Andam, K. S., Ferraro, P. J., Pfaff, A., Sanchez-Azofeifa, G. A. and Robalino, J. A. (2008) Measuring the effectiveness of protected area networks in reducing deforestation. *Proceedings of the National Academy of Sciences of the United States of America* **105**, 16089–16094.

Anderson, A. B. (1988) Use and management of native forests dominated by açaí palm (*Euterpe oleracea* Mart.) in the Amazon estuary. *The Palm-Tree of Life: Biology, Utilization and Conservation*. Balick, M. J. New York Botanical Garden, Bronx, NY, USA. Pp. 144–154.

Anderson, C. L., Bremer, K. and Friis, E. M. (2005) Dating phylogenetically basal eudicots using rbcL sequences and multiple fossil reference points. *American Journal of Botany* **92**, 1737–1748.

Anderson, D. P., Nordheim, E. V., Moermond, T. C., Bi, Z. B. G. and Boesch, C. (2005) Factors influencing tree phenology in Tai National Park, Cote d'Ivoire. *Biotropica* **37**, 631–640.

Anderson, J. M., Proctor, J. and Vallack, H. W. (1983) Ecological studies in four contrasting lowland rain forests in Gunung Mulu National Park, Sarawak. 3. Decomposition processes and nutrient losses from leaf litter. *Journal of Ecology* **71**, 503–527.

Andrade, J. L. and Nobel, P. S. (1996) Habitat, CO_2 uptake and growth for the CAM epiphytic cactus *Epiphyllum phyllanthus* in a Panamanian tropical forest. *Journal of Tropical Ecology*, **12**, 291–306.

Andrade, J. L., Meinzer, F. C., Goldstein, G. and Schnitzer, S. A. (2005) Water uptake and transport in lianas and co-occurring trees of a seasonally dry tropical forest. *Trees-Structure and Function* **19**, 282–289.

Andre, S. E., Parker, J. and Briggs, C. J. (2008) Effect of temperature on host response to *Batrachochytrium dendrobatidis* infection in the mountain yellow-legged frog (*Rana muscosa*) *Journal of Wildlife Diseases* **44**, 716–720.

Andreae, M. O., Rosenfeld, D., Artaxo, P., Costa, A. A., Frank, G. P., Longo, K. M. and Silvas-Dias, M. A. F. (2004) Smoking rain clouds over the Amazon. *Science* **303**, 1337–1342.

Andrews, J. H. and Harris, R. F. (2000) The ecology and biogeography of microorganisms of plant surfaces. *Annual Review of Phytopathology* **38**, 145–180.

Angelsen, A. and Kaimowitz, D. (2001) When does technological change in agriculture promote deforestation. *Trade offs or synergies? Agricultural intensification, economic development and the environment.* Lee, D. R. and Barret, C. B. CAB International, Wallingford, Oxon, UK. Pp. 89–114.

Anonymous (2008) The unkindest cut. *The Economist*, 14 February 2008.

Anthony, N. M., Johnson-Bawe, M., Jeffery, K., Clifford, S. L., Abernethy, K. A., Tutin, C. E., Lahm, S. A., White, L. J. T., Utley, J. F., Wickings, E. J. and Bruford, M. W. (2007) The role of Pleistocene refugia and rivers in shaping gorilla genetic diversity in central Africa. *Proceedings of the National Academy of Sciences of the United States of America* **104**, 20432–20436.

Anthony, P. A., Holtum, J. A. M. and Jackes, B. R. (2002) Shade acclimation of rain forest leaves to colonization by lichens. *Functional Ecology* **16**, 808–816.

Appanah, S. (1985) General flowering in the climax rain forests of South East Asia. *Journal of Tropical Ecology* **1**, 225–240.

Appanah, S. and Chan, H. T. (1981) Thrips: the pollinators of some dipterocarps. *The Malaysian Forester* **44**, 234–252.

Aptroot, A. (2001) Lichenized and saprobic fungal biodiversity of a single *Elaeocarpus* tree in Papua New Guinea, with the report of 200 species of ascomycetes associated with one tree. *Fungal Diversity* **6**, 1–11.

Aragao, L.E.O.C., Malhi, Y., Roman-Cuesta, R. M., Saatchi, S., Anderson, L. O. and Shimabukuro, Y. E. (2007) Spatial patterns and fire response of recent Amazonian droughts. *Geophysical Research Letters* **34**, L07701.

Araujo-Lima, C. R. M. and Goulding, M. (1997) *So fruitful a fish: Ecology, conservation, and aquaculture of the Amazon's tambaqui.* Columbia University Press, New York, USA.

Armbruster, W. S. (1984) The role of resin in angiosperm pollination: Ecological and chemical considerations. *American Journal of Botany* **71**, 1149–1160.

Armbruster, W. S. (1985) Patterns of character divergence and the evolution of reproductive ecotypes of *Dalechampia scandens* (Euphorbiaceae) *Evolution* **39**, 733–752.

Armbruster, W. S. (2006) Evolutionary and ecological aspects of specialized pollination: views from the arctic to the troipcs. In: *Plant-pollinator interactions: From specialization to generalization.* Waser, N. M. and Ollerton, J. (eds.). University of Chicago Press, Chicago, IL, USA. Pp. 260–282.

Armsworth, P. R., Daily, G. C., Kareiva, P. and Sanchirico, J. N. (2006) Land market feedbacks can undermine biodiversity conservation. *Proceedings of the National Academy of Sciences of the United States of America* **103**, 5403–5408.

Arnold, A. E., Maynard, Z., Gilbert, G. S., Coley, P. D. and Kursar, T. A. (2000) Are tropical fungal endophytes hyperdiverse? *Ecology Letters* **3**, 267–274.

Arnold, A. E., Mejia, L. C., Kyllo, D., Rojas, E. I., Maynard, Z., Robbins, N. and Herre, E. A. (2003) Fungal endophytes limit pathogen damage in a tropical tree. *Proceedings of the National Academy of Sciences of the United States of America* **100**, 15649–15654.

Arnold, J. E. M. (2002) Clarifying the links between forest and poverty reduction. *International Forestry Review* **4**, 231–233.

Arnold, J. E. M. and Ruiz-Pérez, M. (2001) Can non-timber forest products match tropical forest conservation and development objectives? *Ecological Economics* **39**, 437–447.

Ashton, P. S. (1969) Speciation among tropical forest trees: some deductions in the light of recent evidence. *Biological Journal of the Linnean Society* **1**, 155–196.

Ashton, P. S. (2003) Floristic zonation of tree communities on wet tropical mountains revisited. *Perspectives in Plant Ecology, Evolution and Systematics* **6**, 87–104.

Ashton, P. S. (2004) Soils in the tropics. In: *Tropical Forest Diversity and Dynamism: Findings from a Large-Scale Plot Network.* Losos, E. C. and Leigh, E. G. (eds.). University of Chicago Press, Chicago, IL, USA. Pp. 56–68.

Ashton, P. S., Givnish, T. J. and Appanah, S. (1988) Staggered flowering in the Dipterocarpaceae: new insights into floral induction and the evolution of mast fruiting in the aseasonal tropics. *American Naturalist* **132**, 44–66.

Ashton, P.S. and Hall, P. (1992) Comparisons of structure among mixed dipterocarp forests of north-western Borneo. *Journal of Ecology* **80**, 459–481.

Asker, S. and Jerling, L. (1992) *Apomixis in Plants.* CRC Press, Boca Raton, Florida, USA.

Asner, G. P., Broadbent, E. N., Oliveira, P. J. C., Keller, M., Knapp, D. E. and Silva, J. N. M. (2006) Condition and fate of logged forests in the Brazilian Amazon. *Proceedings of the National Academy of Sciences of the United States of America* **103**, 12947–12950.

Asner, G. P., Knapp, D. E., Broadbent, E. N., Oliveira, P. J. C., Keller, M. and Silva, J. N. (2005) Selective logging in the Brazilian Amazon. *Science* **310**, 480–482.

Asner, G. P., Rudel, T. K., Aide, T. M., DeFries, R. and Emerson, R. (2009) A contemporary assessment of global humid tropical forest change. *Conservation Biology* in press.

Asquith, N. M., Wright, S. J. and Clauss, M. J. (1997) Does mammal community composition control recruitment in neotropical forests? Evidence from Panama. *Ecology* **78**, 941–946.

Atkinson, C. T., Lease, J. K., Dusek, R. J. and Samuel, M. D. (2005) Prevalence of pox-like lesions and malaria in forest bird communities on leeward Mauna Loa Volcano, Hawaii. *Condor* **107**, 537–546.

Aubréville, A. (1938a) La forêt coloniale: les forêts de l'Afrique occidentale française. *Annales, Académie des Sciences Coloniales* **9**, 1–245.

Aubréville, A. (1938b) Regeneration patterns in the closed forest of Ivory Coast. In: *Foundations of Tropical Forest Biology* Chazdon, R. L. and Whitmore, T. (eds.). University of Chicago Press, Chicago, IL, USA. Pp. 523–537.

Augspurger, C. K. (1981) Reproductive synchrony of a tropical shrub: Experimental studies on effects of pollinators and seed predators on *Hybanthus prunifolius* (Violaceae) *Ecology* **62**, 775–788.

Augspurger, C. K. (1983) Seed dispersal of the tropical tree *Platypodium elegans*, and the escape of its seedlings from fungal pathogens. *Journal of Ecology* **71**, 759–771.

Augspurger, C. K. (1984) Seedling survival of tropical tree species: Interactions of dispersal distance, light-gaps, and pathogens. *Ecology* **65**, 1705–1712.

Augspurger, C. K. and Kelly, C. K. (1984) Pathogen mortality of tropical tree seedlings: Experimental studies of the effects of dispersal distance, seedling density, and light conditions. *Oecologia* **61**, 211–217.

Augspurger, C. K. and Wilkinson, H. T. (2007) Host specificity of pathogenic *Pythium* species: Implications for tree species diversity. *Biotropica* **39**, 702–708.

Austin, A. T. and Vitousek, P. M. (1998) Nutrient dynamics on a precipitation gradient in Hawaii. *Oecologia* **113**, 519–529.

Avila-Pires, T. C. S. (1995) *Lizards of Brazilian Amazonia (Reptilia: Squamata)*. Backhuys Publishers, Leiden, The Netherlands.

Ayres, J. M. (1989) Debt-for-equity swaps and the conservation of tropical rain forests. *Trends in Ecology and Evolution* **4**, 331–332.

Ayres, J. M. and Clutton-Brock, T. H. (1992) River boundaries and species range size in Amazonian primates. *American Naturalist* **140**, 531–537.

Azevedo, J. L., Maccheroni Jr, W., Pereira, J. O. and de Araújo, W. L. (2000) Endophytic microorganisms: a review on insect control and recent advances on tropical plants. *Electronic Journal of Biotechnology* **3**, 40–65.

Bachman, S., Baker, W. J., Brummitt, N., Dransfield, J. and Moat, J. (2004) Elevational gradients, area and tropical island diversity: an example from the palms of New Guinea. *Ecography* **27**, 299–310.

Bahuchet, S., McKey, D. and Degarine, I. (1991) Wild yams revisited: is independence from agriculture possible for rain forest hunter gatherers? *Human Ecology* **19**, 213–243.

Bajpai, S., Kay, R. F., Williams, B. A., Das, D. P., Kapur, V. V. and Tiwari, B. N. (2008) The oldest Asian record of Anthropoidea. *Proceedings of the National Academy of Sciences of the United States of America* **105**, 11093–11098.

Baker, H. G. (1959) Reproductive methods as factors in speciation in flowering plants. *Cold Spring Harbour Symposium on Quantitative Biology* **24**, 177–191.

Baker, J. D. and Cruden, R. W. (1991) Thrips-mediated self-pollination of two facultatively xenogamous wetland species. *American Journal of Botany* **78**, 959–963.

Baker, P. J., Bunyavejchewin, S., Oliver, C. D. and Ashton, P. S. (2005) Disturbance history and historical stand dynamics of a seasonal tropical forest in western Thailand. *Ecological Monographs* **75**, 317–343.

Baker, T. R., Swaine, M. D. and Burslem, D. (2003) Variation in tropical forest growth rates: Combined effects of functional group composition and resource availability. *Perspectives in Plant Ecology Evolution and Systematics*, **6**, 21–36.

Baker, W. J., Dransfield, J. and Hedderson, T. A. (2000a) Phylogeny, character evolution, and a new classification of the calamoid palms. *Systematic Botany* **25**, 297–322.

Baker, W. J., Hedderson, T. A. and Dransfield, J. (2000b) Molecular phylogenetics of subfamily Calamoideae (Palmae) based on nrDNA ITS and cpDNA rps16 intron sequence data. *Molecular Phylogenetics and Evolution* **14**, 195–217.

Balmford, A., Bruner, A., Cooper, P., Costanza, R., Farber, S., Green, R. E., Jenkins, M., Jefferiss, P., Jessamy, V., Madden, J., Munro, K., Myers, N., Naeem, S., Paavola, J., Rayment, M., Rosendo, S., Roughgarden, J., Trumper, K. and Turner, R. K. (2002) Economic reasons for conserving wild nature. *Science* **297**, 950–953.

Balmford, A., Gaston, K. J., Blyth, S., James, A. and Kapos, V. (2003) Global variation in terrestrial conservation costs, conservation benefits, and unmet conservation needs. *Proceedings of the National Academy of Sciences of the United States of America* **100**, 1046–1050.

Balmford, A., Green, R. E. and Scharlemann, J. P. W. (2005) Sparing land for nature: exploring the potential impact of changes in agricultural yield on the area needed for crop production. *Global Change Biology* **11**, 1594–1605.

Balmford, A. and Whitten, T. (2003) Who should pay for tropical conservation, and how could the costs be met? *Oryx* **37**, 238–250.

Baltzer, J. L., Davies, S. J., Bunyavejchewin, S. and Noor, N. S. M. (2008) The role of desiccation tolerance in determining tree species distributions along the Malay-Thai peninsula. *Functional Ecology*, **22**, 221–231.

Baltzer, J. L., Thomas, S. C., Nilus, R. and Burslem, D. (2005) Edaphic specialization in tropical trees: Physiological correlates and responses to reciprocal transplantation. *Ecology*, **86**, 3063–3077.

Baraloto, C. and Goldberg, D. E. (2004) Microhabitat associations and seedling bank dynamics in a neotropical forest. *Oecologia* **141**, 701–712.

Baraloto, C., Morneau, F., Bonal, D., Blanc, L. and Ferry, B. (2007) Seasonal water stress tolerance and habitat associations within four neotropical tree genera. *Ecology* **88**, 478–489.

Barber, R. T. and Chavez, F. P. (1983) Biological consequences of El-Niño. *Science* **222**, 1203–1210.

Barberis, I.M. and Tanner, E.V.J. (2005) Gaps and root trenching increase tree seedling growth in Panamanian semi-evergreen forest. *Ecology* **86**, 667–674.

Barbier, E. B. and Burgess, J. C. (1997) The economics of tropical forest land use options. *Land Economics* **73**, 174–195.

Barker, M. G. and Booth, W. E. (1996) Vertical profiles in a Brunei rain forest: II. Leaf characteristics of *Dryobalanops lanceolata*. *Journal of Tropical Forest Science* **9**, 52–66.

Barkman, T. J. and Simpson, B. B. (2001) Origin of high-elevation Dendrochilum species (Orchidaceae) endemic to Mount Kinabalu, Sabah, Malaysia. *Systematic Botany* **26**, 658–669.

Barlow, B. A. (1981) The loranthaceous mistletoes in Australia. *Ecological Biogeography of Australia*. Keast, A. Hague, Netherlands, W.Junk. Pp. 556–574.

Barlow, J., Gardner, T. A., Araujo, I. S., Avila-Pires, T. C., Bonaldo, A. B., Costa, J. E., Esposito, M. C., Ferreira, L. V., Hawes, J., Hernandez, M. M., Hoogmoed, M. S., Leite, R. N., Lo-Man-Hung, N. F., Malcolm, J. R., Martins, M. B., Mestre, L. A. M., Miranda-Santos, R., Nunes-Gutjahr, A. L., Overal, W. L., Parry, L., Peters, S. L., Ribeiro-Junior, M. A., da Silva, M. N. F., Motta, C. D. and Peres, C. A. (2007) Quantifying the biodiversity value of tropical primary, secondary, and plantation forests. *Proceedings of the National Academy of Sciences of the United States of America* **104**, 18555–18560.

Barlow, J., Lagan, B. O. and Peres, C. A. (2003) Morphological correlates of fire-induced tree mortality in a central Amazonian forest. *Journal of Tropical Ecology* **19**, 291–299.

Barlow, J. and Peres, C. A. (2008) Fire-mediated dieback and compositional cascade in an Amazonian forest. *Philosophical Transactions of the Royal Society B* **363**, 1787–1794.

Barone, J. A. (1998) Host-specificity of folivorous insects in a moist tropical forest. *Journal of Animal Ecology* **67**, 400–409.

Barone, J. A. (2000) Comparison of herbivores and herbivory in the canopy and understorey for two tropical tree species. *Biotropica* **32**, 307–317.

Barre, R. Y., Grant, M. and Draper, D. (2009) The role of taboos in conservation of sacred groves in Ghana's Tallensi-Nabdam district. *Social and Cultural Geography* **10**, 25–39.

Barrett, P. M. and Willis, K. J. (2001) Did dinosaurs invent flowers? Dinosaur-angiosperm coevolution revisited. *Biological Reviews* **76**, 411–447.

Basset, Y. (1991) The seasonality of arboreal arthropods foraging within an Australian rain forest tree. *Ecological Entomology* **16**, 265–278.

Basset, Y. (1999) Diversity and abundance of insect herbivores foraging on seedlings in a rainforest in Guyana. *Ecological Entomology* **24**, 245–259.

Basset, Y. (2001) Invertebrates in the canopy of tropical rain forests: How much do we really know? *Plant Ecology* **153**, 87–107.

Basset, Y., Aberlenc, H. P. and Delvare, G. (1992) Abundance and stratification of foliage arthropods in a lowland rain forest of Cameroon. *Ecological Entomology* **17**, 310–318.

Basset, Y., Novotny, V., Miller, S. E. and Pyle, R. (2000) Quantifying biodiversity: Experience with parataxonomists and digital photography in Papua New Guinea and Guyana. *Bioscience* **50**, 899–908.

Basuki, I. and Sheil, D. (2005) *Local Perspectives of Forest Landscapes: A Preliminary Evaluation of Land and Soils, and their Importance in Malinau, East Kalimantan, Indonesia.* CIFOR, Bogor, Indonesia.

Bates, H. W. (1863) *The Naturalist on the River Amazons: A record of adventures, habits of animals, sketches of Brazilian and Indian life and aspects of nature under the Equator during eleven years of travel.* J. Murray, London, UK.

Bawa, K. S. (1974) Breeding systems of tree species of a lowland tropical community. *Evolution* **28**, 85–92.

Bawa, K. S. (1979) Breeding systems of trees in a tropical wet forest. *New Zealand Journal of Botany* **17**, 521–524.

Bawa, K. S. (1980) Evolution of dioecy in flowering plants. *Annual Review of Ecology and Systematics* **11**, 15–39.

Bawa, K. S. (1992) Mating systems, genetic differentiation and speciation in tropical rain forest plants. *Biotropica* **24**, 250–255.

Bawa, K. S., Bullock, S. H., Perry, D. R., Coville, R. E. and Grayum, M. H. (1985) Reproductive biology of tropical lowland rain forest trees. 2. Pollination systems. *American Journal of Botany* **72**, 346–356.

Bawa, K. S. and Crisp, J. E. (1980) Wind-pollination in the understorey of a rainforest in Costa Rica. *Journal of Ecology* **68**, 871–876.

Bawa, K. S., Kang, H. S. and Grayum, M. H. (2003) Relationships among time, frequency, and duration of flowering in tropical rain forest trees. *American Journal of Botany* **90**, 877–887.

Bawa, K. S., Kress, W. J., Nadkarni, N. M. and Lele, S. (2004) Beyond paradise—Meeting the challenges in tropical biology in the 21st century. *Biotropica* **36**, 437–446.

Bawa, K. S. and Opler, P. A. (1975) Dioecism in tropical forest trees. *Evolution* **29**, 167–179.

Bayliss-Smith, T., Hviding, E. and Whitmore, T. (2003) Rainforest composition and histories of human disturbance in Solomon Islands. *Ambio* **32**, 346–352.

Bazzaz, F.A. and Pickett, S.T.A. (1980) Physiological ecology of tropical succession—a comparative review. *Annual Review of Ecology and Systematics* **11**, 287–310.

Beach, T. (1998) Soil catenas, tropical deforestation, and ancient and contemporary soil erosion in the Peten, Guatemala. *Physical Geography* **19**, 378–405.

Beaman, J. H. and Burley, J. S. (2003) Progress in the floristic inventory of Borneo. In: *Borneo in transition: people, forests, conservation and development*. Padoch, C. and Peluso, N. L. (eds.) Oxford University Press, New York, USA. Pp. 93–113.

Beaman, R. S., Decker, P. J. and Beaman, J. H. (1988) Pollination of *Rafflesia* (Rafflesiaceae). *American Journal of Botany* **75**, 1148–1162.

Beard, K. C. and Wang, B. Y. (1991) Phylogenetic and biogeographic significance of the tarsiiform primate *Asiomomys changbaicus* from the Eocene of Jilin Province, Peoples Republic of China. *American Journal of Physical Anthropology* **85**, 159–166.

Beath, D. D. N. (1996) Pollination of *Amorphophallus johnsonii* (Araceae) by carrion beetles (*Phaeochrous amplus*) in a Ghanaian rain forest. *Journal of Tropical Ecology* **12**, 409–418.

Beattie, A. (1985) *The Evolutionary Ecology of Ant-Plant Mutualisms*. Cambridge University Press, Cambridge, UK.

Beck, J., Muhlenberg, E. and Fiedler, K. (1999) Mud-puddling behavior in tropical butterflies: in search of proteins or minerals? *Oecologia* **119**, 140–148.

Becker, M. and Johnson, D. E. (2001) Cropping intensity effects on upland rice yield and sustainability in West Africa. *Nutrient Cycling in Agroecosystems* **59**, 107–117.

Becker, P., Moure, J. S. and Peralta, F. J. A. (1991) More about euglossine bees in Amazonian forest fragments. *Biotropica* **23**, 586–591.

Beehler, B. M., Sengo, J. B., Filardi, C. and Merg, K. (1995) Documenting the lowland rainforest avifauna in Papua New Guinea: effects of patchy distributions, survey effort and methodology. *Emu* **95**, 149–161.

Beekman, E. M., Ed. (1993) *The Poison Tree: Selected Writings of Rumphius on the Natural History of the Indies*. Oxford University Press, Oxford, UK.

Beerling, D. J., Lake, J. A., Berner, R. A., Hickey, L. J., Taylor, D. W. and Royer, D. L. (2002) Carbon isotope evidence implying high O_2/CO_2 ratios in the Permo-Carboniferous atmosphere. *Geochimica et Cosmochimica Acta* **66**, 3757–3767.

Belcher, B. M. (2005) Forest product markets, forests and poverty reduction. *International Forestry Review* **7**, 82–89.

Bell, C. D., Soltis, D. E. and Soltis, P. S. (2005) The age of the angiosperms: A molecular timescale without a clock. *Evolution* **59**, 1245–1258.

Bellingham, P. J., Kapos, V., Varty, N., Healey, J. R., Tanner, E. V. J., Kelly, D. L., Dalling, J. W., Burns, L. S., Lee, D. and Sidrak, G. (1992) Hurricanes need not cause high mortality: The effects of Hurricane Gilbert on forests in Jamaica. *Journal of Tropical Ecology* **8**, 217–223.

Bennett, E. L. and Robinson, J. G. (2000) *Hunting of Wildlife in Tropical Forests. Implications for Biodiversity and Forest Peoples*. Biodiversity Series, Impact Studies. Washington DC.

Bennett, E. L., Blencowe, E., Brandon, K., Brown, D., Burn, R. W., Cowlishaw, G., Davies, G., Dublin, H., Fa, J. E., Milner-Gulland, E. J., Robinson, J. G., Rowcliffe, J. M., Underwood, F. M. and Wilkie, D. S. (2007) Hunting for consensus: Reconciling bushmeat harvest, conservation, and development policy in west and central Africa. *Conservation Biology* **21**, 884–887.

Bennett, M. (2008) *Eco-Certification: Can It Deliver Conservation and Development in the Tropics?* Working Paper No. 65. World Agroforestry Centre, Bogor, Indonesia. Pp. 64.

Benzing, D. H. (1995) Vascular Epiphytes. *Forest Canopies*. Lowman, M. D. and Nadkri, N. M. Academic Press Inc., San Diego, California, USA. Pp. 225–254.

Benzing, D. H. (2000) *Bromeliaceae: profile of an adaptive radiation*. Cambridge University Press, Cambridge, UK.

Benzing, D. H. and Clements, M. A. (1991) Dispersal of the orchid *Dendrobium insigne* by the ant *Iridomyrmex cordatus* in Papua New Guinea. *Biotropica* **23**, 604–607.

Bereau, M., Bonal, D., Louisanna, E. and Garbaye, J. (2005) Do mycorrhizas improve tropical tree seedling performance under water stress and low light conditions? A case study with *Dicorynia guianensis* (Caesalpiniaceae). *Journal of Tropical Ecology* **21**, 375–381.

Berg, C. C. (1989) Classification and distribution of *Ficus*. *Experientia* **45**, 605–611.

Berger, L., Speare, R., Daszak, P., Green, D. E., Cunningham, A. A., Goggin, C. L., Slocombe, R., Ragan, M. A., Hyatt, A. D., McDonald, K. R., Hines, H. B., Lips, K. R., Marantelli, G. and Parkes, H. (1998) Chytridiomycosis causes amphibian mortality associated with population declines in the rain forests of Australia and Central America. *Proceedings of the National Academy of Sciences of the United States of America* **95**, 9031–9036.

Berkes, F. (2004) Rethinking community-based conservation. *Conservation Biology* **18**, 621–630.

Berkes, F. (2007) Community-based conservation in a globalized world. *Proceedings of the National Academy of Sciences of the United States of America* **104**, 15188–15193.

Berry, P. E., Huber, O. and Holst, B. K. (1995) Floristic analysis and phytogeography. *Flora of the Venezuelan Guayana*. Steyermark, J. A., Berry, P. E. and Holst, B. K. Timber Press, Portland, OR, USA. Pp. 161–191.

Bertram, J. E. A. (2004) New perspectives on brachiation mechanics. *American Journal of Physical Anthropology* **47**, 100–117.

Betts, R. A., Cox, P. M., Collins, M., Harris, P. P., Huntingford, C. and Jones, C. D. (2004) The role of ecosystem-atmosphere interactions in simulated Amazonian precipitation decrease and forest dieback under global climate warming. *Theoretical and Applied Climatology* **78**(1–3), 157–175.

Bhagwat, S. A. and Willis, K. J. (2008) Agroforestry as a solution to the oil-palm debate. *Conservation Biology* **22**, 1368–1369.

Bhagwat, S. A., Willis, K. J., Birks, H. J. B. and Whittaker, R. J. (2008) Agroforestry: a refuge for tropical biodiversity? *Trends in Ecology and Evolution* **23**, 261–267.

Bickford, D., Iskandar, D. and Barlian, A. (2008) A lungless frog discovered on Borneo. *Current Biology* **18**, R374–R375.

Biesmeijer, J. C., Giurfa, M., Koedam, D., Potts, S. G., Joel, D. M. and Dafni, A. (2005) Convergent evolution: floral guides, stingless bee nest entrances, and insectivorous pitchers. *Naturwissenschaften* **92**, 444–450.

Bignell, D. E. and Eggleton, P. (1995) On the elevated intestinal pH of higher termites (Isoptera, Termitidae). *Insectes Sociaux* **42**, 57–69.

Biju, S. D. and Bossuyt, F. (2003) New frog family from India reveals an ancient biogeographical link with the Seychelles. *Nature* **425**, 711–714.

Bininda-Emonds, O., Cardillo, M., Jones, K. E., MacPhee, R. D. E., Beck, R. M. D., Grenyer, R., Price, S. A., Vos, R. A., Gittleman, J. L. and Purvis, A. (2007) The delayed rise of present-day mammals. *Nature* **446**, 507–512.

Bird, M. I., Taylor, D. and Hunt, C. (2005) Environments of insular Southeast Asia during the Last Glacial Period: a savanna corridor in Sundaland? *Quaternary Science Reviews* **24**, 2228–2242.

Birkinshaw, C. R. and Colquhoun, I. C. (1998) Pollination of *Ravenala madagascariensis* and *Parkia madagascariensis* by *Eulemur macaco* in Madagascar. *Folia Primatologica* **69**, 252–259.

Bjorholm, S., Svenning, J. C., Baker, W. J., Skov, F. and Balslev, H. (2006) Historical legacies in the geographical diversity patterns of New World palm (Arecaceae) subfamilies. *Botanical Journal of the Linnean Society* **151**, 113–125.

Bjorholm, S., Svenning, J. C., Skov, F. and Balslev, H. (2005) Environmental and spatial controls of palm (Arecaceae) species richness across the Americas. *Global Ecology and Biogeography* **14**, 423–429.

Blackburn, D. C. and Measey, G. J. (2009) Dispersal to or from an African biodiversity hotspot? *Molecular Ecology* **18**, 1904–1915.

Blakemore, R. J., Csuzdi, C., Ito, M. T., Kaneko, N., Kawaguchi, T. and Schilthuizen, M. (2007b) Taxonomic status and ecology of Oriental *Pheretima darnleiensis* (Fletcher, 1886) and other earthworms (Oligochaeta: Megascolecidae) from Mt Kinabalu, Borneo. *Zootaxa* **487**, 23–44.

Blakemore, R. J., Csudzi, C., Ito, M. T., Kaneko, N., Paoletti, M. G., Spiridonov, S. E., Uchida, T. and Van Praagh, B. D. (2007a) *Megascolex (Promegascolex) mekongianus* Cognetti, 1922—its extent, ecology and allocation to *Amynthas* (Clitellata/Oligochaeta: Megascolecidae) *Opusc. Zool. Budapest* **36**, 19–30.

Blate, G. M. (2005) Modest trade-offs between timber management and fire susceptibility of a Bolivian semi-deciduous forest. *Ecological Applications* **15**, 1649–1663.

Bloor, J.M.G. and Grubb, P.J. (2004) Morphological plasticity of shade-tolerant tropical rainforest tree seedlings exposed to light changes. *Functional Ecology*, **18**, 337–348.

Boardman, N.K. (1989) Adaptation of plants to their light environment. *Current Contents/Agriculture Biology and Environmental Sciences*, 14–14.

Bobe, R. (2006) The evolution of arid ecosystems in eastern Africa. *Journal of Arid Environments* **66**, 564–584.

Boddy, L. (1993) Saprotrophic cord-forming fungi—warfare strategies and other ecological aspects. *Mycological Research* **97**, 641–655.

Bodley, J. H. and Benson, F. C. (1980) Stilt-root walking by an Iriarteoid palm in the Peruvian Amazon. *Biotropica* **12**, 67–71.

Boedhihartono, A. K., Gunarso, P., Levang, P. and Sayer, J. (2007) The principles of conservation and development: Do they apply in Malinau? *Ecology and Society* **12**(2).

Bohn, H. F. and Federle, W. (2004) Insect aquaplaning: *Nepenthes* pitcher plants capture prey with the peristome, a fully wettable water-lubricated anisotropic surface. *Proceedings of the National Academy of Sciences of the United States of America* **101**, 14138–14143.

Bolaños, O. and Schmink, M. (2005) Women's place is not in the forest: Gender issues in a timber management project in Bolivia. *The Equitable Forest: Diversity, Community and Resource Management*. Colfer, C. J. P. Resources for the Future/CIFOR, Washington, DC, USA.

Bonaccorso, E., Koch, I. and Peterson, A. T. (2006) Pleistocene fragmentation of Amazon species' ranges. *Diversity and Distributions* **12**, 157–164.

Bonan, G. B. (2008) Forests and climate change: Forcings, feedbacks, and the climate benefits of forests. *Science* **320**, 1444–1449.

Bongers, F., Poorter, L., Hawthorne, W. and Sheil, D. (2009) The intermediate disturbance hypothesis applies to tropical forests, but disturbance contributes little to tree diversity. *Ecology Letters* **12**, 798–805.

Bongers, F. and Sterck, F.J. (1998) Architecture and development of rainforest trees: Responses to light variation. *Dynamics of tropical communities* (ed. by D. M. Newbery, H. H. T. Prins and N. Brown) Blackwell Science, Oxford, UK. Pp. 125–162.

Boo, E. (1990) *Ecotourism: The Potentials and Pitfalls*. USA World Wildlife Fund, Washington DC.

Borchert, R., Renner, S. S., Calle, Z., Navarrete, D., Tye, A., Gautier, L., Spichiger, R. and von Hildebrand, P. (2005) Photoperiodic induction of synchronous flowering near the Equator. *Nature* **433**, 627–629.

Borchert, R., Rivera, G. and Hagnauer, W. (2002) Modification of vegetative phenology in a tropical semi-deciduous forest by abnormal drought and rain. *Biotropica* **34**, 27–39.

Borchert, R., Robertson, K., Schwartz, M.D. and Williams-Linera, G. (2005) Phenology of temperate trees in tropical climates. *International Journal of Biometeorology* **50**, 57–65.

Borgerhoff-Mulder, M. and Coppolillo, P. (2005) *Conservation: Ecology, Economics, and Culture*. Princeton University Press, Princeton, NJ, USA.

Borges, S. H. (2004) Species poor but distinct: bird assemblages in white sand vegetation in Jau National Park, Brazilian Amazon. *Ibis* **146**, 114–124.

Borggaard, O. K., Gafur, A. and Petersen, L. (2003) Sustainability appraisal of shifting cultivation in the Chittagong Hill Tracts of Bangladesh. *Ambio* **32**, 118–123.

Borland, A.M., Griffiths, H., Maxwell, C., Broadmeadow, M.S.J., Griffiths, N.M. and Barnes, J.D. (1992) On the ecophysiology of the Clusiaceae in Trinidad: expression of CAM in *Clusia minor* during the transition from wet to dry season and characterization of three endemic species. *New Phytologist*, **122**, 349–357.

Botta, A., Ramankutty, N. and Foley, J. A. (2002) Long-term variations of climate and carbon fluxes over the Amazon basin. *Geophysical Research Letters* **29**, 9.

Boucher, D. H., Vandermeer, J. H., de la Cerda, I. G., Mallona, M. A., Perfecto, I. and Zamora, N. (2001) Post-agriculture versus post-hurricane succession in southeastern Nicaraguan rain forest. *Plant Ecology* **156**, 131–137.

Boulter, S. L., Kitching, R. L. and Howlett, B. G. (2006) Family, visitors and the weather: patterns of flowering in tropical rain forests of northern Australia. *Journal of Ecology* **94**, 369–382.

Bowles, I. A., Rice, R. E., Mittermeier, R. A. and da Fonseca, G. A. B. (1998) Logging and tropical forest conservation. *Science* **280**, 1899–1900.

Bowman, D. (1998) Tansley Review No. 101: The impact of Aboriginal landscape burning on the Australian biota. *New Phytologist* **140**, 385–410.

Bowman, D. and Prior, L. D. (2005) Why do evergreen trees dominate the Australian seasonal tropics? *Australian Journal of Botany* **53**, 379–399.

Boyce, C. K., Hotton, C. L., Fogel, M. L., Cody, G. D., Hazen, R. M., Knoll, A. H. and Hueber, F. M. (2007) Devonian landscape heterogeneity recorded by a giant fungus. *Geology* **35**, 399–402.

Boyd, R. S. (2004) Ecology of metal hyperaccumulation. *New Phytologist* **162**, 563–567.

Boyd, R. S. and Jaffre, T. (2001) Phytoenrichment of soil Ni content by *Sebertia acuminata* in New Caledonia and the concept of elemental allelopathy. *South African Journal of Science* **97**, 535–538.

Brady, K.U., Kruckeberg, A.R. and Bradshaw, H.D. (2005) Evolutionary ecology of plant adaptation to serpentine soils. *Annual Review of Ecology Evolution and Systematics* **36**, 243–266.

Brandani, A., Hartshorn, G. S. and Orians, G. H. (1988) Internal heterogeneity of gaps and species richness in Costa Rican tropical wet forest. *Journal of Tropical Ecology* **4**, 99–119.

Brauman, A. (2000) Effect of gut transit and mound deposit on soil organic matter transformations in the soil feeding termite: A review. *European Journal of Soil Biology* **36**, 117–125.

Bray, D. B., Merino-Perez, L., Negreros-Castillo, P., Segura-Warnholtz, G., Torres-Rojo, J. M. and Vester, H. F. M. (2003) Mexico's community-managed forests as a global model for sustainable landscapes. *Conservation Biology* **17**, 672–677.

Brearley, F.Q., Press, M.C. and Scholes, J.D. (2003) Nutrients obtained from leaf litter can improve the growth of dipterocarp seedlings. *New Phytologist* **160**, 101–110.

Brehm, G., Sussenbach, D. and Fiedler, K. (2003) Unique elevational diversity patterns of geometrid moths in an Andean montane rainforest. *Ecography* **26**, 456–466.

Breitsprecher, A. and Bethel, J.S. (1990) Stem-growth periodicity of trees in a tropical wet forest of Costa Rica. *Ecology* **71**, 1156–1164.

Brewer, S. W. and Rejmanek, M. (1999) Small rodents as significant dispersers of tree seeds in a Neotropical forest. *Journal of Vegetation Science* **10**, 165–174.

Brienen, R. J. W. and Zuidema, P. A. (2005) Relating tree growth to rainfall in Bolivian rain forests: A test for six species using tree ring analysis. *Oecologia* **146**, 1–12.

Brienen, R. J. W. and Zuidema, P. A. (2006) Lifetime growth patterns and ages of Bolivian rain forest trees obtained by tree ring analysis. *Journal of Ecology* **94**, 481–493.

Briggs, J. C. (2003) The biogeographic and tectonic history of India. *Journal of Biogeography* **30**, 381–388.

Brncic, T. M., Willis, K. J., Harris, D. J. and Washington, R. (2007) Culture or climate? The relative influences of past processes on the composition of the lowland Congo rainforest. *Philosophical Transactions of the Royal Society B* **362**, 229–242.

Brokaw, N. and Busing, R. T. (2000) Niche versus chance and tree diversity in forest gaps. *Trends in Ecology and Evolution* **15**, 183–188.

Brokaw, N. V. L. and Walker, L. R. (1991) Summary of the effects of Caribbean hurricanes on vegetation. *Biotropica* **23**, 442–447.

Brook, B. W., Bradshaw, C. J. A., Koh, L. P. and Sodhi, N. S. (2006) Momentum drives the crash: Mass extinction in the tropics. *Biotropica* **38**, 302–305.

Brook, B. W., Sodhi, N. S. and Ng, P. K. L. (2003) Catastrophic extinctions follow deforestation in Singapore. *Nature* **424**, 420–423.

Brookfield, H., Potter, L. and Byron, Y. (1995) *In Place of the Forest: Environmental and Socio-economic Transformations in Borneo and the Eastern Malay Peninsula*. United Nations University Press, Tokyo, Japan.

Brooks, T. M. and Balmford, A. (1996) Atlantic forest extinctions. *Nature* **380**, 115–115.

Brooks, T. M., Bakarr, M. I., Boucher, T., Da Fonseca, G. A. B., Hilton-Taylor, C., Hoekstra, J. M., Moritz, T., Olivier, S., Parrish, J., Pressey, R. L., Rodrigues, A. S. L., Sechrest, W., Stattersfield, A., Strahm, W. and Stuart, S. N. (2004) Coverage provided by the global protected-area system: Is it enough? *Bioscience* **54**, 1081–1091.

Brooks, T. M., Mittermeier, R. A., da Fonseca, G. A. B., Gerlach, J., Hoffmann, M., Lamoreux, J. F., Mittermeier, C. G., Pilgrim, J. D. and Rodrigues, A. S. L. (2006) Global biodiversity conservation priorities. *Science* **313**, 58–61.

Brooks, T. M., Pimm, S. L. and Collar, N. J. (1997) Deforestation predicts the number of threatened birds in insular southeast Asia. *Conservation Biology* **11**, 382–394.

Brooks, T. M., Wright, S. J. and Sheil, D. (2009) Conserving tropical forest biodiversity: evidence for what works. *Conservation Biology*, in press.

Brosi, B. J., Daily, G. C., Shih, T. M., Oviedo, F. and Duran, G. (2008) The effects of forest fragmentation on bee communities in tropical countryside. *Journal of Applied Ecology* **45**, 773–783.

Brosius, J. P. (1991) Foraging in the tropical rainforests: the case of the case of the Penan of Sarawak, East Malaysia (Borneo) *Human Ecology* **19**, 123–150.

Brown, D. and Schreckenberg, K. (1998) *Shifting cultivators as agents of deforestation: assessing the evidence*. Overseas Development Institute (London), Natural Resources Perspective Number **29**, 10.

Brown, D. R. (2008) A spatiotemporal model of shifting cultivation and forest cover dynamics. *Environment and Development Economics* **13**, 643–671.

Brown, J. H. (2001) Mammals on mountainsides: elevational patterns of diversity. *Global Ecology and Biogeography* **10**, 101–109.

Brown, J. H. and Lomolino, M. V. (1998) *Biogeography*. Sinauer Associates, Sunderland, MA, USA. Pp. 560.

Brown, J. K. S. (2005) Geological, evolutionary, and ecological basis of the diversification of neotropical butterflies: Implications for conservation. In: *Tropical Rain Forests: Past, Present, and Future*. Bermingham, E., Dick, C. W. and Moritz, C. (eds.). University of Chicago Press, Chicago, IL, USA. Pp. 166–201.

Brown, K. S. and Hutchings, R. W. (1997) Disturbance, fragmentation and the dynamics of diversity in Amazonian forest butterflies. In: *Tropical Forest Remnants: Ecology, Management, and Conservation of Fragmented Communities*. Laurance, W. F. and Bierregaard, R. O. (eds.). University of Chicago Press, Chicago, IL, USA. Pp. 91–110.

Brown, N. (1993) The implications of climate and gap microclimate for seedling growth conditions in a Bornean lowland rain forest. *Journal of Tropical Ecology* **9**, 153–168.

Brown, N. D. and Jennings, S. (1998) Gap-size niche differentiation by tropical rainforest trees: a testable hypothesis or a broken-down bandwagon? In: *Dynamics of Tropical Communities*. Newbery, D. M., Prins H. H. T. and Brown N. D. (eds.). Blackwell Science, Oxford, UK. Pp. 79–94.

Brown, N. D. and Whitmore, T. C. (1992) Do dipterocarp seedlings really partition tropical rain forest gaps? *Philosophical Transactions of the Royal Society of London Series B* **335**, 369–378.

Brown, S., Hall, C. A. S., Knabe, W., Raich, J., Trexler, M. C. and Woomer, P. (1993) Tropical Forests: Their past, present, and potential future role in the terrestrial carbon budget. *Water, Air, and Soil Pollution* **70**, 71–94.

Brühl, C. A., Gunsalam, G. and Linsenmair, K. E. (1998) Stratification of ants (Hymenoptera, Formicidae) in a primary rain forest in Sabah, Borneo. *Journal of Tropical Ecology* **14**, 285–297.

Bruijnzeel, L. A. (2004) Hydrological functions of tropical forests: not seeing the soil for the trees? *Agriculture Ecosystems and Environment* **104**, 185–228.

Bruijnzeel, L. A. and Hamilton, L. S. (2000) *Decision time for cloud forests*. IHP Humid Tropics Programme Series No. 13. Unesco, Paris, France.

Bruna, E. (1999) Seed germination in rainforest fragments. *Nature* **402**, 139.

Brundrett, M. C. (2002) Coevolution of roots and mycorrhizas of land plants. *New Phytologist* **154**, 275–304.

Brune, A. (1998) Termite guts: the world's smallest bioreactors. *Trends in Biotechnology* **16**, 16–21.

Bruner, A. G., Gullison, R. E. and Balmford, A. (2004) Financial costs and shortfalls of managing and expanding protected-area systems in developing countries. *Bioscience* **54**, 1119–1126.

Bruner, A. G., Gullison, R. E., Rice, R. E. and da Fonseca, G. A. B. (2001) Effectiveness of parks in protecting tropical biodiversity. *Science* **291**, 125–128.

Bruun, T. B., Mertz, O. and Elberling, B. (2006) Linking yields of upland rice in shifting cultivation to fallow length and soil properties. *Agriculture Ecosystems and Environment* **113**, 139–149.

Bryant, R. L. (1992) *Forest problems in colonial Burma: historical variations on contemporary themes*. Workshop on the Political Ecology of Southeast Asian Forests: Transdisciplinary Discourses, London, England.

Bryant, R. L. (1996) Romancing colonial forestry: The discourse of 'forestry as progress' in British Burma. *Geographical Journal* **162**, 169–178.

Buchmann, S. L. (1987) The ecology of oil flowers and their bees. *Annual Review of Ecology and Systematics* **18**, 343–369.

Buckley, T. R., Attanayake, D. and Bradler, S. (2008) Extreme convergence in stick insect evolution: Phylogenetic placement of the Lord Howe Island tree lobster. *Proceedings of the Royal Society B* **276**, 1055–1062.

Bullock, S. H. (1994) Wind pollination of neotropical dioecious trees. *Biotropica* **26**, 172–179.

Bullock, S. H. (1997) Effects of seasonal rainfall on radial growth in two tropical tree species. *International Journal of Biometeorology* **41**, 13–16.

Bun, Y., King, T. and Shearman, P. (2004) *China's Impact on Papua New Guinea's Forestry Industry*. Forest Trends, Washington, DC, USA.

Bungard, R.A., Press, M.C. and Scholes, J.D. (2000) The influence of nitrogen on rain forest dipterocarp seedlings exposed to a large increase in irradiance. *Plant Cell and Environment* **23**, 1183–1194.

Bunker, D.E. and Carson, W.P. (2005) Drought stress and tropical forest woody seedlings: Effect on community structure and composition. *Journal of Ecology* **93**, 794–806.

Burd, M. (1994) Bateman's principle and plant reproduction: the role of pollen limitation in fruit and seed set. *Botanical Review* **60**, 83–139.

Burghouts, T. B. A., Van Straalen, N. M. and Bruijnzeel, L. A. (1998) Spatial heterogeneity of element and litter turnover in a Bornean rain forest. *Journal of Tropical Ecology* **14**, 477–505.

Burgoyne, P. M., van Wyk, A. E., Anderson, J. M. and Schrire, B. D. (2005) Phanerozoic evolution of plants on the African plate. *Journal of African Earth Sciences* **43**, 13–52.

Burke, A. (2003) Inselbergs in a changing world: Global trends. *Diversity and Distributions* **9**, 375–383.

Burkey, T. V. (1994) Tropical tree species-diversity: a test of the Janzen-Connell model. *Oecologia* **97**, 533–540.

Burney, D. A. and Flannery, T. F. (2005) Fifty millenia of catastrophic extinctions after human contact. *Trends in Ecology and Evolution* **20**, 395–401.

Burnham, R. J. and Graham, A. (1999) The history of neotropical vegetation: New developments and status. *Annals of the Missouri Botanical Garden* **86**, 546–589.

Burslem, D. F. R. P., Grubb, P.J. and Turner, I.M. (1995) Responses to nutrient addition among shade-tolerant tree seedlings of lowland tropical rain-forest in Singapore. *Journal of Ecology* **83**, 113–122.

Burslem, D. F. R. P., Grubb, P. J. and Turner, I. M. (1996) Responses to simulated drought and elevated nutrient supply among shade-tolerant tree seedlings of lowland tropical forest in Singapore. *Biotropica* **28**, 636–648.

Burslem, D. F. R. P., Pinard, M. and Hartley, S. E. (2005) *Biotic Interactions in the Tropics*. Cambridge University Press, Cambridge, UK. Pp. 564.

Burslem, D. F. R. P. and Swaine, M. D. (2002) Forest dynamics and regeneration. In: *Foundations of Tropical Forest Biology*. Chazdon, R. L. and Whitmore, T. (eds.) University of Chicago Press, Chicago, IL, USA. Pp. 577–583.

Burslem, D. F. R. P. and Whitmore, T. C. (1999) Species diversity, susceptibility to disturbance and tree population dynamics in tropical rain forest. *Journal of Vegetation Science* **10**, 767–776.

Burslem, D. F. R. P., Whitmore, T. C. and Brown, G. C. (2000) Short-term effects of cyclone impact and long-term recovery of tropical rain forest on Kolombangara, Solomon Islands. *Journal of Ecology* **88**, 1063–1078.

Bush, M. B. (1994) Amazonian speciation: A necessarily complex model. *Journal of Biogeography* **21**, 5–17.

Bush, M. B. and Silman, M. R. (2007) Amazonian exploitation revisited: ecological asymmetry and the policy pendulum. *Frontiers in Ecology and the Environment* **5**, 457–465.

Bush, M. B., Silman, M. R. and Listopad, C. (2007) *A regional study of Holocene climate change and human occupation in Peruvian Amazonia*. Annual Meeting of the Association for Tropical Biology and Conservation, Uberlandia, Brazil (24–28 July 2005).

Busing, R. T. and Brokaw, N. (2002) Tree species diversity in temperate and tropical forest gaps: the role of lottery recruitment. *Folia Geobotanica* **37**, 33–43.

Butler, B. J. and Chazdon, R. L. (1998) Species richness, spatial variation, and abundance of the soil seed bank of a secondary tropical rain forest. *Biotropica* **30**, 214–222.

Butler, C. D. (2008) Human health and forests: An overview. *Human Health and Forests: A Global Oveview of Issues, Practice and Policy* Colfer, C. J. P. Earthscan, London, UK. Pp. 13–33.

Butler, R. A., Koh, L. P. and Ghazoul, J. (2009) REDD in the red: palm oil could undermine carbon payment schemes. *Conservation Letters.* **2**, 67–73.

Butler, R. W. (1991) Tourism, environment, and sustainable development. *Environmental Conservation* **18**, 201–209.

Butynski, T. and Kalina, J. (1998) Gorilla tourism: A critical look. In: *Conservation of Biological Resources*. Milner-Gulland, E. J. and Mace, R. (eds.). Blackwell Science, Oxford, UK. Pp. 279–313.

Byrne, M. M. (1994) Ecology of twig-dwelling ants in a wet lowland tropical forest. *Biotropica* **26**, 61–72.

Cable, S. and Cheek, M. E. (1998) *The Plants of the Mount Cameroon: A Conservation Checklist*. Royal Botanic Gardens, Kew, London, UK. Pp. 198.

Cáceres, M. E. S., Maia, L. C. and Lücking, R. (2000) Foliicolous lichens and their lichenicolous fungi in the Atlantic rain forest of Brazil: diversity, ecogeography and conservation. *Bibliotheca Lichenologica* **75**, 47–70.

Cai, Z. Q., Slot, M. and Fan, Z. X. (2005) Leaf development and photosynthetic properties of three tropical tree species with delayed greening. *Photosynthetica* **43**, 91–98.

Calder, I. R. (2005) *The Blue Revolution: Land Use and Integrated Water Resources Management*. Earthscan, London, UK. Pp. 208.

Calder, I. R., Wright, I. R. and Murdiyarso, D. (1986) A study of evaporation from tropical rain forest—west Java. *Journal of Hydrology* **89**, 13–31.

Campbell, D.G., Richardson, P.M. and Rosas, A. (1989) Field screening for allelopathy in tropical forest trees, particularly *Duroia hirsuta*, in the Brazilian Amazon. *Biochemical Systematics and Ecology* **17**, 403–407.

Campos, M. T. and Ehringhaus, C. (2003) Plant virtues are in the eyes of the beholders: A comparison of known palm uses among indigenous and folk communities of Southwestern Amazonia. *Economic Botany* **57**, 324–344.

Canaday, C. (1996) Loss of insectivorous birds along a gradient of human impact in Amazonia. *Biological Conservation* **77**, 63–77.

Canadell, J., Jackson, R. B., Ehleringer, J. B., Mooney, H. A., Sala, O. E. and Schulze, E.-D. (1996) Maximum rooting depth of vegetation types at the global scale. *Oecologia* **108**, 583–595.

Canham, C. D., Denslow, J. S., Platt, W. J., Runkle, J. R., Spies, T. A. and White, P. S. (1990) Light Regimes beneath Closed Canopies and Tree-Fall Gaps in Temperate and Tropical Forests. *Canadian Journal of Forest Research* **20**, 620–631.

Canham, C. D., Finzi, A. C., Pacala, S. W. and Burbank, D. H. (1994) Causes and consequences of resource heterogeneity in forests: Interspecific variation in light transmission by canopy trees. *Canadian Journal of Forest Research* **24**, 337–349.

Cao, K.F. (2000) Water relations and gas exchange of tropical saplings during a prolonged drought in a Bornean heath forest, with reference to root architecture. *Journal of Tropical Ecology* **16**, 101–116.

Capers, R. S. and Chazdon, R. L. (2004) Rapid assessment of understory light availability in a wet tropical forest. *Agricultural and Forest Meteorology* **123**, 177–185.

Capers, R. S., Chazdon, R. L., Brenes, A. R. and Alvarado, B. V. (2005) Successional dynamics of woody seedling communities in wet tropical secondary forests. *Journal of Ecology* **93**, 1071–1084.

Capistrano, D. and Colfer, C. J. P. (2005) Decentralization: Issues, lessons, and reflections. In: *The Politics of Decentralization: Forests, Power, and People*. Capistrano, D. and Colfer, C. J. P. (eds.). Earthscan, London, UK. Pp. 296–313.

Caplan, M. L. and Bustin, R. M. (1999) Devonian-Carboniferous Hangenberg mass extinction event, widespread organic-rich mudrock and anoxia: causes and consequences. *Palaeogeography Palaeoclimatology Palaeoecology* **148**, 187–207.

Carnaval, A. C., Hickerson, M. J., Haddad, C. F. B., Rodrigues, M. T. and Moritz, C. (2009) Stability predicts genetic diversity in the Brazilian Atlantic Forest hotspot. *Science* **323**, 785–789.

Carpenter, R. J., Read, J. and Jaffre, T. (2003) Reproductive traits of tropical rain forest trees in New Caledonia. *Journal of Tropical Ecology* **19**, 351–365.

Carranza, S. and Arnold, E. N. (2003) Investigating the origin of transoceanic distributions: mtDNA shows *Mabuya* lizards (Reptilia, Scincidae) crossed the Atlantic twice. *Syst. Biodivers* **1**, 275–282.

Carthew, S. M. and Goldingay, R. L. (1997) Non-flying mammals as pollinators. *Trends in Ecology and Evolution* **12**, 104–108.

Carvajal, A. and Adler, G. H. (2005) Biogeography of mammals on tropical Pacific islands. *Journal of Biogeography* **32**, 1561–1569.

Carvajal G.d. (1970) *The Discovery of the Amazon*. AMS Press, New York, USA.

Case, R. J., Pauli, G. E. and Soejarto, D. D. (2005) Factors in maintaining indigenous knowledge among ethnic communities of Manus Islands. *Economic Botany* **59**, 356–365.

Cashdan, E. (2001) Ethnic diversity and its environmental determinants: Effects of climate, pathogens, and habitat diversity. *American Anthropologist* **103**, 968–991.

Caswell, H. (1976) Community structure: Neutral model analysis. *Ecological Monographs* **46**, 327–354.

Cavelier, J., Solis, D. and Jaramillo, M. A. (1996) Fog interception in montane forest across the Central Cordillera of Panama. *Journal of Tropical Ecology* **12**, 357–369.

Cavelier, J., Tanner, E. and Santamaria, J. (2000) Effect of water, temperature and fertilizers on soil nitrogen net transformations and tree growth in an elfin cloud forest of Colombia. *Journal of Tropical Ecology* **16**, 83–99.

CBD. (2001) *The Value of Forest Ecosystems*. Montreal, Canada, Secretariat of the Convention on Biological Diversity. CBD Technical Series No. 4: 67.

Ceballos, G. and Brown, J. H. (1995) Global patterns of mammalian diversity, endemism, and endangerment. *Conservation Biology* **9**, 559–568.

Chanderbali, A. S., van der Werff, H. and Renner, S. S. (2001) Phylogeny and historical biogeography of Lauraceae: Evidence from the chloroplast and nuclear genomes. *Annals of the Missouri Botanical Garden* **88**, 104–134.

Chape, S., Blyth, S., Fish, L., Fox, P. and Spalding, M. (2003) *2003 United Nations List of Protected Areas*. IUCN and UNEP-WCMC, Gland, Switzerland, and Cambridge, UK.

Chapin, F. S. (1980) The mineral nutrition of wild plants. *Annual Review of Ecology and Systematics* **11**, 233–260.

Chapin, F. S., Vitousek, P. M. and Vancleve, K. (1986) The nature of nutrient limitation in plant-communities. *American Naturalist* **127**, 48–58.

Chapman, C. A., Chapman, L. J., Struhsaker, T. T., Zanne, A. E., Clark, C. J. and Poulsen, J. R. (2005) A long-term evaluation

of fruiting phenology: importance of climate change. *Journal of Tropical Ecology* **21**, 31–45.

Chapman, C. A., Chapman, L. J., Vulinec, K., Zanne, A. and Lawes, M. J. (2003) Fragmentation and alteration of seed dispersal processes: An initial evaluation of dung beetles, seed fate, and seedling diversity. *Biotropica* **35**, 382–393.

Chapman, C. A., Kaufman, L. and Chapman, L. J. (1998) Buttress formation and directional stress experienced during critical phases of tree development. *Journal of Tropical Ecology* **14**, 341–349.

Charles-Dominique, P., Blanc, P., Larpin, D., Ledru, M. P., Riera, B., Sarthou, C., Servant, M. and Tardy, C. (1998) Forest perturbations and biodiversity during the last ten thousand years in French Guiana. *Acta Oecologica* **19**, 295–302.

Charles-Dominique, P., Chave, J., Dubois, M. A., De Granville, J.-J., Riera, B. and Vezzoli, C. (2003) Colonization front of the understorey palm *Astrocaryum sciophilum* in a pristine rain forest of French Guiana. *Global Ecology and Biogeography* **12**, 237–248.

Charrette, N. A., Cleary, D. F. R. and Mooers, A. O. (2006) Range-restricted, specialist Bornean butterflies are less likely to recover from ENSO-induced disturbance. *Ecology* **87**, 2330–2337.

Chase, M. W. and Hills, H. G. (1992) Orchid phylogeny, flower sexuality, and fragrance seeking. *Bioscience* **42**, 43–49.

Chazdon, R. L. (1992) Patterns of growth and reproduction of *Geonoma congesta*, a clustered understory palm. *Biotropica* **24**, 43–51.

Chazdon, R. L. (2008) Chance and determinism in tropical forest succession. *Tropical Forest Community Ecology*. Carson, W. P. and Schnitzer, S. A. Wiley-Blackwell, Oxford, UK. Pp. 384–408.

Chazdon, R. L., Brenes, A. R. and Alvarado, B. V. (2005) Effects of climate and stand age on annual tree dynamics in tropical second-growth rain forests. *Ecology* **86**, 1808–1815.

Chazdon, R. L. and Fetcher, N. (1984) Photosynthetic light environments in a lowland tropical rain-forest in Costa-Rica. *Journal of Ecology* **72**, 553–564.

Chazdon, R. L., Harvey, C. A., Komar, O., Griffith, D. M., Ferguson, B. G., Martinez-Ramos, M., Morales, H., Nigh, R., Soto-Pinto, L., van Breugel, M. and Philpott, S. M. (2009a) Beyond reserves: A research agenda for conserving biodiversity in human-modified tropical landscapes. *Biotropica* **41**, 142–153.

Chazdon, R. L., Letcher, S. G., van Breugel, M., Martinez-Ramos, M., Bongers, F. and Finegan, B. (2007) Rates of change in tree communities of secondary Neotropical forests following major disturbances. *Philosophical Transactions of the Royal Society B* **362**, 273–289.

Chazdon, R.L. and Pearcy, R.W. (1991) The importance of sunflecks for forest understory plants: Photosynthetic machinery appears adapted to brief, unpredictable periods of radiation. *Bioscience*, **41**, 760–766.

Chazdon, R. L., Peres, C. A., Dent, D., Sheil, D., Lugo, A. E., Lamb, D., Stork, N. E. and Miller, S. (2009b) Where are the wild things? Assessing the potential for species conservation in tropical secondary forests. *Conservation Biology* in press.

Cherrett, J. M. (1968) Foraging behaviour of *Atta cephalotes* L (Hymenoptera: Formicidae) I. Foraging pattern and plant species attacked in tropical rain forest. *Journal of Animal Ecology* **37**, 387–403.

Chesner, C. A., Rose, W. I., Deino, A., Drake, R. and Westgate, J. A. (1991) Eruptive History of Earths Largest Quaternary Caldera (Toba, Indonesia) Clarified. *Geology* **19**, 200–203.

Chiatante, D., Scippa, G. S., Iorio, A. d., Sarnataro, M., di Iorio, A. and Radoglou, K. (2001) The stability of trees growing on slope depends upon a particular conformational structure imposed by mechanical stress in their root system. *Proceedings International Conference Forest Research: a challenge for an integrated European approach, Thessaloniki, Greece, 27 August 1 September 2001, Volume II*. Thessaloniki, NAGEF—Forest Research Institute. Pp. 477–482.

Chomitz, K. (2007) *At Loggerheads? Agricultural Expansion, Poverty Reduction, and Environment in the Tropical Forests*. World Bank, Washington DC, USA.

Christ, C., Hillel, O., Matus, S. and Sweeting, J. (2003) *Tourism and Biodiversity: Mapping Tourism's Global Footprint*. Washington DC, UNEP and CI.

Christensen, J. H., Hewitson, B., Busuioc, A., Chen, A., Gao, X., Held, I., Jones, R., Kolli, R. K., Kwon, W.-T., Laprise, R., Magaña Rueda, V., Mearns, L., Menéndez, C. G., Räisänen, J., Rinke, A., A., S. and Whetton, P. (2007) Regional Climate Projections. *Climate Change 2007: The Physical Science Basis. Contribution of Working Group I to the Fourth Assessment Report of the Intergovernmental Panel on Climate Change*. Solomon, S., Qin, D., Manning, M., Chen, Z., Marquis, M., Averyt, K. B., Tignor, M. and Miller, H. L. Cambridge University Press, Cambridge, UK and New York, USA.

Chust, G., Chave, J., Condit, R., Aguilar, S., Lao, S. and Perez, R. (2006) Determinants and spatial modeling of tree beta-diversity in a tropical forest landscape in Panama. *Journal of Vegetation Science* **17**, 83–92.

Cibois, A., Slikas, B., Schulenberg, T. S. and Pasquet, E. (2001) An endemic radiation of Malagasy songbirds is revealed by mitochondrial DNA sequence data. *Evolution* **55**, 1198–1206.

Cione, A. L., Figini, A. J. and Tonni, E. P. (2001) Did the megafauna range to 4300 BP in South America? *Radiocarbon* **43**, 69–75.

Clair, B., Fournier, M., Prevost, F. M., Beauchene, J. and Bardet, S. (2003) Biomechanics of buttressed trees: bending strains and stresses. *American Journal of Botany* **90**, 1349–1356.

Clark, D. A. (2004) Tropical forests and global warming: slowing it down or speeding it up? *Frontiers in Ecology and the Environment* **2**, 73–80.

Clark, D. A., Brown, S., Kicklighter, D. W., Chambers, J. Q., Thomlinson, J. R. and Ni, J. (2001a) Measuring net primary production in forests: Concepts and field methods. *Ecological Applications* **11**, 356–370.

Clark, D. A., Brown, S., Kicklighter, D. W., Chambers, J. Q., Thomlinson, J. R., Ni, J. and Holland, E. A. (2001b) Net primary production in tropical forests: An evaluation and synthesis of existing field data. *Ecological Applications* **11**, 371–384.

Clark, D. A. and Clark, D. B. (1992) Life-history diversity of canopy and emergent trees in a neotropical rain-forest. *Ecological Monographs* **62**, 315–344.

Clark, D. A. and Clark, D. B. (1994) Climate-induced annual variation in canopy tree growth in a costa-rican tropical rain-forest. *Journal of Ecology* **82**, 865–872.

Clark, D. A. and Clark, D. B. (1999) Assessing the growth of tropical rain forest trees: Issues for forest modeling and management. *Ecological Applications*, **9**, 981–997.

Clark, D. A. and Clark, D. B. (2001) Getting to the canopy: Tree height growth in a neotropical rain forest. *Ecology* **82**, 1460–1472.

Clark, D. A., Piper, S. C., Keeling, C. D. and Clark, D. B. (2003) Tropical rain forest tree growth and atmospheric carbon dynamics linked to interannual temperature variation during 1984–2000. *Proceedings of the National Academy of Sciences of the United States of America* **100**, 5852–5857.

Clark, D. B. (1996) Abolishing virginity. *Journal of Tropical Ecology* **12**, 735–739.

Clark, D. B. and Clark, D. A. (1989) The role of physical damage in the seedling mortality regime of a neotropical rainforest. *Oikos* **55**, 225–230.

Clark, D. B. and Clark, D. A. (1991) The impact of physical damage on canopy tree regeneration in tropical rain forest. *Journal of Ecology* **79**, 447–457.

Clark, D. B., Clark, D. A. and Read, J. M. (1998) Edaphic variation and the mesoscale distribution of tree species in a neotropical rain forest. *Journal of Ecology* **86**, 101–112.

Clark, D. B., Palmer, M. W. and Clark, D. A. (1999) Edaphic factors and the landscape-scale distributions of tropical rain forest trees. *Ecology* **80**, 2662–2675.

Clark, D. L., Nadkarni, N. M. and Gholz, H. L. (1998) Growth, net production, litter decomposition, and net nitrogen accumulation by epiphytic bryophytes in a tropical montane forest. *Biotropica* **30**, 12–23.

Clarke, B. T. (1997) The natural history of amphibian skin secretions, their normal functioning and potential medical applications. *Biological Reviews of the Cambridge Philosophical Society* **72**, 365–379.

Clarke, H. L. (1894) The meaning of tree life. *American Naturalist*, **28**, 465–72.

Cleary, D. F. R. and Priadjati, A. (2005) Vegetation responses to burning in a rain forest in Borneo. *Plant Ecology* **177**, 145–163.

Cleaver, K. (1992) Deforestation in the western and central African forest: The agricultural and demographic causes, and some solutions. *Conservation of West and Central African Rainforests*. Cleaver, K., Munasinghe, M., Dyson, M., Egli, N., Peuker, A. and Wencelius, F. Washington, DC, World Bank. World Bank Environment Paper No. 1. Pp. 65–78.

Clement, C. R., McCann, J. M. and Smith, N. J. H. (2003) Agrobiodiversity in Amazonia and its relationship with Dark Earths. *Amazonian Dark Earths: Origin, Properties and Management*. Lehmann, J., Kern, D. C., Glaser, B. and Woods, W. I. Kluwer, Dordrecht. Pp. 159–178.

Clement, R. M. and Horn, S. P. (2001) Pre-Columbian land-use history in Costa Rica: a 3000-year record of forest clearance, agriculture and fires from Laguna Zoncho. *Holocene* **11**, 419–426.

Clements, D. B. (1916) *Plant Succession: An Analysis of the Development of Vegetation*. Carnegie Institute, Washington, DC, USA. Pp. 512.

Cleveland, C. C., Reed, S. C. and Townsend, A. R. (2006) Nutrient regulation of organic matter decomposition in a tropical rain forest. *Ecology* **87**, 492–503.

Cleveland, C. C., Townsend, A. R. and Schmidt, S. K. (2002) Phosphorus limitation of microbial processes in moist tropical forests: Evidence from short-term laboratory incubations and field studies. *Ecosystems* **5**, 680–691.

Clinebell, R. R., Phillips, O. L., Gentry, A. H., Stark, N. and Zuuring, H. (1995) Prediction of neotropical tree and liana species richness from soil and climatic data. *Biodiversity and Conservation* **4**, 56–90.

Cloutier, D., Hardy, O. J., Caron, H., Ciampi, A. Y., Degen, B., Kanashiro, M. and Schoen, D. J. (2007) Low inbreeding and high pollen dispersal distances in populations of two Amazonian forest tree species. *Biotropica* **39**, 406–415.

Coates, D. (1993) Fish ecology and management of the Sepik-Ramu, New Guinea, a large contemporary tropical river basin. *Environmental Biology of Fishes* **38**, 345–368.

Cochrane, M. A. and Schulze, M. D. (1999) Fire as a recurrent event in tropical forests of the eastern Amazon: Effects on forest structure, biomass, and species composition. *Biotropica* **31**, 2–16.

Cochrane, M. A., Alencar, A., Schulze, M. D., Souza, C. M., Nepstad, D. C., Lefebvre, P. and Davidson, E. A. (1999) Positive feedbacks in the fire dynamic of closed canopy tropical forests. *Science* **284**, 1832–1835.

Cockburn, P. F. (1978) The flora. In: *Kinabalu: Summit of Borneo*. Luping, D. M., Wen, C. and Dingley, E. R. (eds.). The Sabah Society, Kota Kinabalu, Sabah, Malaysia. Pp. 179–198.

Cohn-Haft, M., Whittaker, A. and Stouffer, P. C. (1997) A new look at the species poor central Amazon: The avifauna north of Manaus, Brazil. *Ornithological Monographs* **48**, 205–235.

Coleman, C.M., Boyd, R.S. and Eubanks, M.D. (2005) Extending the elemental defense hypothesis: Dietary metal concentrations below hyperaccumulator levels could harm herbivores. *Journal of Chemical Ecology*, **31**, 1669–1681.

Coley, P. D. and Barone, J. A. (1996) Herbivory and plant defenses in tropical forests. *Annual Review of Ecology and Systematics* **27**, 305–335.

Coley, P. D. and Kursar, T. A. (1996) Anti-herbivore defenses of young tropical leaves: physiological constraints and ecological tradeoffs. *Tropical Forest Plant Ecophysiology*. Smith, A. P., Mulkey, S. S. and Chazdon, R. L. Chapman and Hall, New York, USA. Pp. 305–336.

Coley, P. D., Kursar, T. A. and Machado J.-L. (1993) Colonization of tropical rain forest leaves by epiphylls: effects of site and host plant leaf lifetime. *Ecology* **74**, 619–623.

Coley, P. D., Bryant, J. P. and Chapin, F. S. (1985) Resource availability and plant antiherbivore defense. *Science* **230**, 895–899.

Coley, P. D., Lokvam, J., Rudolph, K., Bromberg, K., Sackett, T. E., Wright, L., Brenes-Arguedas, T., Dvorett, D., Ring, S., Clark, A., Baptiste, C., Pennington, R. T. and Kursar, T. A. (2005) Divergent defensive strategies of young leaves in two species of *Inga*. *Ecology* **86**, 2633–2643.

Coley, P.D. and Barone, J.A. (1996) Herbivory and plant defenses in tropical forests. *Annual Review of Ecology and Systematics* **27**, 305–335.

Colfer, C. J. P. (2008) *The Longhouse of the Tarsier: Changing Landscapes, Gender and Well Being in Borneo*. Borneo Research Council, UNESCO and CIFOR, Phillips, Maine.

Colfer, C. J. P., Dudley, R. G. and Gardner, R. (2008) Forest Women, Health and Childbearing. In: *Human Health and Forests: A Global, Interdisciplinary Overview*. Colfer, C. J. P. (ed.) Earthscan, London, UK. Pp. 113–133.

Colfer, C. J. P., Ed. (2005) *The Equitable Forest: Diversity, Community and Natural Resources*. RFF/CIFOR, Washington, DC, USA.

Colfer, C. J. P., Peluso, N. and Chung, C. S. (1997) *Beyond slash and burn: Lesson from the Kenyah in managing Borneo's tropical rain forests.* New York Botanical Gardens, New York, USA.

Colfer, C. J. P., Sheil, D. and Kishi, M. (2006) *Forests and human health: Assessing the evidence.* CIFOR Occasional Paper, No. 45. Center for International Forestry Research, Bogor, Indonesia. Pp. 119.

Colinvaux, P. A., De Oliveira, P. E. and Bush, M. B. (2000) Amazonian and neotropical plant communities on glacial time-scales: The failure of the aridity and refuge hypotheses. *Quaternary Science Reviews* **19**, 141–169.

Colinvaux, P. A., Irion, G., Rasanen, M. E., Bush, M. B. and de Mello, J. (2001) A paradigm to be discarded: Geological and paleoecological data falsify the Haffer and Prance refuge hypothesis of Amazonian speciation. *Amazoniana-Limnologia Et Oecologia Regionalis Systemae Fluminis Amazonas* **16**, 609–646.

Collins, M. (2005) El Niño- or La Niña-like climate change? *Climate Dynamics* **24**, 89–104.

Colwell, R. K. (2000) A barrier runs through it. Or maybe just a river. *Proceedings of the National Academy of Sciences of the United States of America* **97**, 13470–13472.

Colwell, R. K. and Coddington, J. A. (1994) Estimating terrestrial biodiversity through extrapolation. *Philosophical Transactions of the Royal Society of London, Series B* **345**, 101–118.

Colwell, R. K. and Hurtt, G. C. (1994) Nonbiological gradients in species richness and a spurious Rapoport effect. *American Naturalist* **144**, 570–595.

Colwell, R. K., Brehm, G., Cardelus, C. L., Gilman, A. C. and Longino, J. T. (2008) Global warming, elevational range shifts, and lowland biotic attrition in the wet tropics. *Science* **322**, 258–261.

Compton, S. G. (2002) *Sailing with the wind: dispersal by small flying insects. Dispersal ecology.* The 42nd Symposium of the British Ecological Society, University of Reading, UK.

Compton, S. G. and Ware, A. B. (1991) Ants disperse the eliaosome-bearing eggs of an African stick insect. *Psyche* **98**, 207–213.

Compton, S. G., Ellwood, M. D. F., Davis, A. J. and Welch, K. (2000) The flight heights of chalcid wasps (Hymenoptera, Chalcidoidea) in a lowland Bornean rain forest: Fig wasps are the high fliers. *Biotropica* **32**, 515–522.

Condit, R., Hubbell, S. P. and Foster, R. B. (1992) Recruitment near conspecific adults and the maintenance of tree and shrub diversity in a neotropical forest. *American Naturalist* **140**, 261–286.

Condit, R., Hubbell, S.P. and Foster, R.B. (1995) Mortality-rates of 205 neotropical tree and shrub species and the impact of a severe drought. *Ecological Monographs*, **65**, 419–439.

Condit, R., Pitman, N., Leigh, E. G., Chave, J., Terborgh, J., Foster, R. B., Nunez, P., Aguilar, S., Valencia, R., Villa, G., Muller-Landau, H. C., Losos, E. and Hubbell, S. P. (2002) Beta-diversity in tropical forest trees. *Science* **295**, 666–669.

Condit, R., Watts, K., Bohlman, S. A., Perez, R., Foster, R. B. and Hubbell, S. P. (2000) Quantifying the deciduousness of tropical forest canopies under varying climates. *Journal of Vegetation Science* **11**, 649–658.

Connell, J. H. (1970) *On the role of natural enemies in preventing competitive exclusion in some marine animals and in rain forest trees.* Proceedings of the Advanced Study Institute of Dynamics Numbers and Populations.

Connell, J. H. (1978) Diversity in tropical rain forests and coral reefs: High diversity of trees and corals is maintained only in a non-equilibrium state. *Science* **199**, 1302–1310.

Connell, J. H. and Lowman, M. D. (1989) Low-diversity tropical rain forests—some possible mechanisms for their existence. *American Naturalist* **134**, 88–119.

Connell, J. H. and Green, P. T. (2000) Seedling dynamics over thirty-two years in a tropical rain forest tree. *Ecology* **81**, 568–584.

Conservation International. (2007) Biodiversity Hotspots: Tropical Andes. Retrieved 15 March 2009, from http://www.biodiversityhotspots.org/xp/hotspots/andes/Pages/biodiversity.aspx.

Conti, E., Eriksson, T., Schonenberger, J., Sytsma, K. J. and Baum, D. A. (2002) Early tertiary out-of-India dispersal of Crypteroniaceae: Evidence from phylogeny and molecular dating. *Evolution* **56**, 1931–1942.

Coomes, D. A. and Grubb, P. J. (2000) Impacts of root competition in forests and woodlands: a theoretical framework and review of experiments. *Ecological Monographs* **70**, 171–207.

Coomes, D. A. and Grubb, P. J. (2003) Colonization, tolerance, competition and seed-size variation within functional groups. *Trends In Ecology and Evolution* **18**, 283–291.

Coomes, D. A., Allen, R. B., Bentley, W. A., Burrows, L. E., Canham, C. D., Fagan, L., Forsyth, D. M., Gaxiola-Alcantar, A., Parfitt, R. L., Ruscoe, W. A., Wardle, D. A., Wilson, D. J. and Wright, E. F. (2005) The hare, the tortoise and the crocodile: the ecology of angiosperm dominance, conifer persistence and fern filtering. *Journal of Ecology* **93**, 918–935.

Coomes, D. A. and Grubb, P. J. (1998) Responses of juvenile trees to above- and belowground competition in nutrient-starved Amazonian rain forest. *Ecology* **79**, 768–782.

Cordeiro, N. J. and Howe, H. F. (2001) Low recruitment of trees dispersed by animals in African forest fragments. *Conservation Biology* **15**, 1733–1741.

Cordeiro, N. J., Patrick, D. A. G., Munisi, B. and Gupta, V. (2004) Role of dispersal in the invasion of an exotic tree in an East

African submontane forest. *Journal of Tropical Ecology* **20**, 449–457.

Corlett, M. T. and Primack, R. B. (2006) Tropical rainforests and the need for cross-continental comparisons. *Trends in Ecology and Evolution* **21**, 104–110.

Corlett, R. T. (1998) Frugivory and seed dispersal by vertebrates in the Oriental (Indomalayan) Region. *Biological Reviews* **73**, 413–448.

Corlett, R. T. (2007) The impact of hunting on the mammalian fauna of tropical Asian forests. *Biotropica* **39**, 292–303.

Cornell, H. V. and Hawkins, B. A. (2003) Herbivore responses to plant secondary compounds: A test of phytochemical coevolution theory. *American Naturalist* **161**, 507–522.

Correa, S. B., Winemiller, K. O., Lopez-Fernandez, H. and Galetti, M. (2007) Evolutionary perspectives on seed consumption and dispersal by fishes. *Bioscience* **57**, 748–756.

Corry, S. (1993) *'Harvest moonshine' taking you for a ride: A critique of the 'rainforest harvest'*. Survival International, London. Pp. 17.

Costa, F. R. C. (2004) Structure and composition of the ground-herb community in a terra-firma Central Amazonian forest. *Acta Amazonica* **34**, 53–59.

Costa, F. R. C. (2006) Mesoscale gradients of herb richness and abundance in central Amazonia. *Biotropica* **38**, 711–717.

Costanza, R., d'Arge, R., de Groot, R., Farber, S., Grasso, M., Hannon, B., Limburg, K., Naeem, S., O'Meill, R. V., Paruelo, J., Raskin, R. G., Sutton, P. and van den Belt, M. (1997) The value of the world's ecosystem services and natural capital. *Nature* **387**, 253–260.

Costello, C. and Ward, M. (2006) Search, bioprospecting and biodiversity conservation. *Journal of Environmental Economics and Management* **52**, 615–626.

Cousins, S. H. (1989) Species richness and the energy theory. *Nature* **340**, 350–351.

Cowie, R. H. (1998) Patterns of introduction of non-indigenous non-marine snails and slugs in the Hawaiian Islands. *Biodiversity and Conservation* **7**, 349–368.

Cowie, R. H. and Holland, B. S. (2006) Dispersal is fundamental to biogeography and the evolution of biodiversity on oceanic islands. *Journal of Biogeography* **33**, 193–198.

Cowling, S. A., Betts, R. A., Cox, P. M., Ettwein, V. J., Jones, C. D., Maslin, M. A. and Spall, S. A. (2004) Contrasting simulated past and future responses of the Amazonian forest to atmospheric change. *Philosophical Transactions of the Royal Society B* **359**, 539–547.

Cox, P. A., Sperry, L. R., Tuominen, M. and Bohlin, L. (1989) Pharmacological activity of the Samoan ethnopharmacopoeia. *Economic Botany* **43**, 487–497.

Cox, P. M., Betts, R. A., Collins, M., Harris, P. P., Huntingford, C. and Jones, C. D. (2004) Amazonian forest dieback under climate-carbon cycle projections for the 21st century. *Theoretical and Applied Climatology* **78**, 137–156.

Cracraft, J. (1985) Historical biogeography and patterns of differentiation within the South American avifauna: areas of endemism. *Ornithological Monographs* **36**, 49–84.

Cramer, J. M., Mesquita, R. C. G. and Williamson, G. B. (2007) Forest fragmentation differentially affects seed dispersal of large and small-seeded tropical trees. *Biological Conservation* **137**, 415–423.

Cramer, W., Bondeau, A., Schaphoff, S., Lucht, W., Smith, B. and Sitch, S. (2004) Tropical forests and the global carbon cycle: impacts of atmospheric carbon dioxide, climate change and rate of deforestation. *Philosophical Transactions of the Royal Society of London Series B* **359**, 331–343.

Crampton, W. G. R., Castello, L. and Viana, J. P. (2004) Fisheries in the Amazon várzea: historical trends, current status, and factors affecting sustainability. In: *People and Nature: Wildlife Conservation in South and Central America*. Silvinus, K., Bodmer, R. and Fragoso, J. (eds.). Columbia University Press, New York, USA. Pp. 76–98.

Crayn, D.M., Winter, K. and Smith, J.A.C. (2004) Multiple origins of crassulacean acid metabolism and the epiphytic habit in the neotropical family bromeliaceae. *Proceedings of the National Academy of Sciences of the United States of America*, **101**, 3703–3708.

Cressler, W. L. (2001) Evidence of earliest known wildfires. *Palaios* **16**, 171–174.

Cristoffer, C. and Peres, C. A. (2003) Elephants versus butterflies: The ecological role of large herbivores in the evolutionary history of two tropical worlds. *Journal of Biogeography* **30**, 1357–1380.

Crook, M. J., Ennos, A. R. and Banks, J. R. (1997) The function of buttress roots: a comparative study of the anchorage systems of buttressed (*Aglaia* and *Nephelium ramboutan* species) and non-buttressed (*Mallotus wrayi*) tropical trees. *Journal of Experimental Botany* **48**, 1703–1716.

Crowley, T. J. and Berner, R. A. (2001) CO_2 and climate change. *Science* **292**, 870–872.

Cullen, L., Bodmer, R. E. and Padua, C. V. (2000) Effects of hunting in habitat fragments of the Atlantic forests, Brazil. *Biological Conservation* **95**, 49–56.

Culley, T. M. and Klooster, M. R. (2007) The cleistogamous breeding system: A review of its frequency, evolution, and ecology in angiosperms. *Botanical Review* **73**, 1–30.

Cunningham, A. B., Shanley, P. and Laird, S. (2008) Health, habitats and medicinal plant use. In: *Human Health and Forests: A Global Overview of Issues, Practice and Policy*. Colfer, C. J. P. (ed.). Earthscan, London, UK. Pp. 35–62.

Cunningham, S. A. (1996) Pollen supply limits fruit initiation by a rain forest understorey palm. *Journal of Ecology* **84**, 185–194.

Curran, L. M. and Leighton, M. (2000) Vertebrate responses to spatiotemporal variation in seed production of mast-fruiting dipterocarpaceae. *Ecological Monographs* **70**, 101–128.

Curran, L. M. and Webb, C. O. (2000) Experimental tests of the spatiotemporal scale of seed predation in mast-fruiting Dipterocarpaceae. *Ecological Monographs* **70**, 129–148.

Curran, L. M., Caniago, I., Paoli, G. D., Astianti, D., Kusneti, M., Leighton, M., Nirarita, C. E. and Haeruman, H. (1999) Impact of El Nino and logging on canopy tree recruitment in Borneo. *Science* **286**, 2184–2188.

Curran, L. M., Trigg, S. N., McDonald, A. K., Astiani, D., Hardiono, Y. M., Siregar, P., Caniago, I. and Kasischke, E. (2004) Lowland forest loss in protected areas of Indonesian Borneo. *Science* **303**, 1000–1003.

Curran, T. J., Brown, R. L., Edwards, E., Hopkins, K., Kelley, C., McCarthy, E., Pounds, E., Solan, R. and Wolf, J. (2008) Plant functional traits explain interspecific differences in immediate cyclone damage to trees of an endangered rainforest community in north Queensland. *Austral Ecology* **33**, 451–461.

Currie, C. R., Bot, A. N. M. and Boomsma, J. J. (2003) Experimental evidence of a tripartite mutualism: Bacteria protect ant fungus gardens from specialized parasites. *Oikos* **101**, 91–102.

Currie, C. R., Mueller, U. G. and Malloch, D. (1999) The agricultural pathology of ant fungus gardens. *Proceedings of the National Academy of Sciences of the United States of America* **96**, 7998–8002.

Currie, D. J. (1991) Energy and large-scale patterns of animal-species and plant-species richness. *American Naturalist* **137**, 27–49.

Curtis, T. P. and Sloan, W. T. (2005) Exploring microbial diversity: A vast below. *Science* **309**, 1331–1333.

Cushman, J.C. and Borland, A.M. (2002) Induction of crassulacean acid metabolism by water limitation. *Plant, Cell and Environment* **25**, 295–310.

Cushman, J.C. (2005) Crassulacean acid metabolism: Recent advances and future opportunities. *Functional Plant Biology* **32**, 375–380.

Cutrim, E., Martin, D. W. and Rabin, R. (1995) Enhancement of cumulus clouds over deforested lands in Amazonia. *Bulletin of the American Meteorological Society* **76**, 1801–1805.

da Costa, M. L. and Kern, D. C. (1999) Geochemical signatures of tropical soils with archaeological black earth in the Amazon, Brazil. *Journal of Geochemical Exploration* **66**, 369–385.

Da Silva, H. R., De Britto-Pereira, M. C. and Caramaschi, U. (1989) Frugivory and seed dispersal by *Hyla truncata*, a Neotropical tree-frog. *Copeia* 781–783.

Da Silva, N. J. and Sites, J. W. (1995) Patterns of diversity of neotropical squamate reptile species with emphasis on the Brazilian Amazon and the conservation potential of indigenous reserves. *Conservation Biology* **9**, 873–901.

Da Silva, R. P., Dos Santos, J., Tribuzy, E. S., Chambers, J. Q., Nakamura, S. and Higuchi, N. (2002) Diameter increment and growth patterns for individual tree growing in central Amazon, Brazil. *Forest Ecology and Management* **166**, 295–301.

Daily, G. C., Ceballos, G., Pacheco, J., Suzan, G. and Sanchez-Azofeifa, A. (2003) Countryside biogeography of neotropical mammals: Conservation opportunities in agricultural landscapes of Costa Rica. *Conservation Biology* **17**, 1814–1826.

Dalle, S. P., de Blois, S., Caballero, J. and Johns, T. (2006) Integrating analyses of local land-use regulations, cultural perceptions and land-use/land cover data for assessing the success of community-based conservation. *Forest Ecology and Management* **222**, 370–383.

Dalling, J. W. and Hubbell, S. P. (2002) Seed size, growth rate and gap microsite conditions as determinants of recruitment success for pioneer species. *Journal of Ecology* **90**, 557–568.

Dalling, J. W., Muller-Landau, H. C., Wright, S. J. and Hubbell, S. P. (2002) Role of dispersal in the recruitment limitation of neotropical pioneer species. *Journal of Ecology* **90**, 714–727.

Dalling, J. W., Swaine, M. D. and Garwood, N. C. (1998) Dispersal patterns and seed bank dynamics of pioneer trees in moist tropical forest. *Ecology* **79**, 564–578.

Dalling, J. W., Swaine, M. D. and Garwood, N. C. (1997) Soil seed bank community dynamics in seasonally moist lowland tropical forest, Panama. *Journal of Tropical Ecology* **13**, 659–680.

Dalling, J. W. and Hubbell, S. P. (2002) Seed size, growth rate and gap microsite conditions as determinants of recruitment success for pioneer species. *Journal of Ecology* **90**, 557–568.

Dalling, J. W., Winter, K., Nason, J. D., Hubbell, S. P., Murawski, D. A. and Hamrick, J. L. (2001) The unusual life history of *alseis blackiana*: A shade-persistent pioneer tree? *Ecology* **82**, 933–945.

D'Angelo, S. A., Andrade, A. C. S., Laurance, S. G., Laurance, W. F. and Mesquita, R. C. G. (2004) Inferred causes of tree mortality in fragmented and intact Amazonian forests. *Journal of Tropical Ecology* **20**, 243–246.

Danielsen, F. and Heegaard, M. (1994) *The impact of logging and foret conversion of lowland forest birds and other wildlife in Seberida, Riau Province, Sumatra.* Proceedings of

NORINDA Seminar, Indonesian Institute of Sciences, Jakarta, Indonesia.

Danielsen, F., Balete, D. S., Poulsen, M. K., Enghoff, M., Nozawa, C. M. and Jensen, A. E. (2000) A simple system for monitoring biodiversity in protected areas of a developing country. *Biodiversity and Conservation* **9**, 1671–1705.

Dargavel, J., Dixon, K. and Semple, N. (eds.) (1988) *Changing tropical forests: historical perspectives on today's challenges in Asia, Australia, and Oceania.* Centre for Resource and Environmental Studies, Australian National University, Canberra.

Darwin, C. (1845) *Journal of researches into the natural history and geology of the countries visited during the voyage of* HMS Beagle *round the world, under the command of Capt. Fitzroy, R.N.* 2nd edn. John Murray, London, UK.

Daszak, P., Cunningham, A. A. and Hyatt, A. D. (2003) Infectious disease and amphibian population declines. *Diversity and Distributions* **9**, 141–150.

Davidson, D. W., Cook, S. C., Snelling, R. R. and Chua, T. H. (2003) Explaining the abundance of ants in lowland tropical rainforest canopies. *Science* **300**, 969–972.

Davidson, D. W., Longino, J. T. and Snelling, R. R. (1988) Pruning of host plant neighbors by ants: an experimental approach. *Ecology* **69**, 801–808.

Davies, G. (2002) Bushmeat and international development. *Conservation Biology* **16**, 587–589.

Davies, R. G., Eggleton, P., Jones, D. T., Gathorne-Hardy, F. J. and Hernandez, L. M. (2003) Evolution of termite functional diversity: analysis and synthesis of local ecological and regional influences on local species richness. *Journal of Biogeography* **30**, 847–877.

Davis, A. J. (2000) Species richness of dung-feeding beetles (Coleoptera: Aphodiidae, Scarabaeidae, Hybosoridae) in tropical rainforest at Danum Valley, Sabah, Malaysia. *Coleopterists Bulletin* **54**, 221–231.

Davis, C. C., Bell, C. D., Fritsch, P. W. and Mathews, S. (2002a) Phylogeny of *Acridocarpus brachylophon* (Malpighiaceae): Implications for tertiary tropical floras and Afroasian biogeography. *Evolution* **56**, 2395–2405.

Davis, C. C., Bell, C. D., Mathews, S. and Donoghue, M. J. (2002b) Laurasian migration explains Gondwanan disjunctions: Evidence from Malpighiaceae. *Proceedings of the National Academy of Sciences of the United States of America* **99**, 6833–6837.

Davis, C. C., Fritsch, P. W., Bell, C. D. and Mathews, S. (2004) High-latitude tertiary migrations of an exclusively tropical clade: Evidence from Malpighiaceae. *International Journal of Plant Sciences* **165**, S107–S121.

Davis, C. C., Webb, C. O., Wurdack, K. J., Jaramillo, C. A. and Donoghue, M. J. (2005) Explosive radiation of malpighiales

supports a mid-Cretaceous origin of modern tropical rain forests. *American Naturalist* **165**, E36–E65.

Dawkins, H. C. (1958) The management of natural tropical high forest with special reference to Uganda. In: *Institute Paper 34,* Commonwealth Forestry Bureau, Oxford, UK.

Dawkins, H. C. and Philip, M. S. (1998) *Tropical Moist Forest Silviculture and Management: A History of Success and Failure.* CAB International, Oxford, UK. Pp. 386.

Daws, M. I., Burslem, D. F. R. P., Crabtree, L. M., Kirkman, P., Mullins, C. E. and Dalling, J. W. (2002) Differences in seed germination responses may promote coexistence of four sympatric *Piper* species. *Functional Ecology* **16**, 258–267.

de Almeida, W. R., Wirth, R. and Leal, I. R. (2008) Edge-mediated reduction of phorid parasitism on leaf-cutting ants in a Brazilian Atlantic forest. *Entomologia Experimentalis Et Applicata* **129**, 251–257.

de Beer, J. H. and McDermott, M. J. (1989) *The Economic Value of Non-Timber Forest Products in South East Asia.* Netherlands Committee for IUCN, Amsterdam, Netherlands. Pp. 197.

de Blas, D. E. and Perez, M. R. (2008) Prospects for reduced impact logging in Central African logging concessions. *Forest Ecology and Management* **256**, 1509–1516.

de Castro, M. C., Monte-Mor, R. L., Sawyer, D. O. and Singer, B. H. (2006) Malaria risk on the Amazon frontier. *Proceedings of the National Academy of Sciences of the United States of America* **103**, 2452–2457.

de Dijn, B. P. E. (2003) Vertical stratification of flying insects in a Surinam lowland rainforest. In: *Arthropods of tropical forests: spatio-temporal dynamics and resource use in the canopy.* Basset, Y., Novotny, V., Miller, S. E. and Kitching, R. L. (eds.). Cambridge University Press, Cambridge, UK. Pp. 110–122.

de Gouvenain, R. C. and Silander, J. A. (2003) Do tropical storm regimes influence the structure of tropical lowland rain forests? *Biotropica* **35**, 166–180.

de Jong, W., Chokkalingam, U. and Perera, G. A. D. (2001) The evolution of swidden fallow secondary forests in Asia. *Journal of Tropical Forest Science* **13**, 800–815.

de la Torre, S., Snowdon, C. T. and Bejarano, M. (2000) Effects of human activities on wild pygmy marmosets in Ecuadorian Amazonia. *Biological Conservation* **94**, 153–163.

de Merode, E., Homewood, K. and Cowlishaw, G. (2004) The value of bushmeat and other wild foods to rural households living in extreme poverty in Democratic Republic of Congo. *Biological Conservation* **118**, 573–581.

de Rouw, A. (1987) Tree management as part of two farming systems in the wet forest zone (Ivory Coast). *Acta Oecologica Oecologia Applicata* **8**, 39–51.

de Rouw, A. (1995) The fallow period as a weed break in shifting cultivation (tropical wet forests). *Agriculture Ecosystems and Environment* **54**, 31–43.

de Souza, R. P. and Valio, I. F. M. (2001) Seed size, seed germination, and seedling survival of Brazilian tropical tree species differing in successional status. *Biotropica* **33**, 447–457.

De Steven, D. (1989) Genet and ramet demography of *Oenocarpus mapora* ssp. *mapora*, a clonal palm of Panamanian tropical moist forest. *Journal of Ecology* **77**, 579–596.

De Steven, D. and Wright, S. J. (2002) Consequences of variable reproduction for seedling recruitment in three neotropical tree species. *Ecology* **83**, 2315–2327.

de Vivo, M. and Carmignotto, A. P. (2004) Holocene vegetation change and the mammal faunas of South America and Africa. *Journal of Biogeography* **31**, 943–957.

de Winter, A. J. and Gittenberger, E. (1998) The land snail fauna of a square kilometer patch of rainforest in southwestern Cameroon, high species richness, low abundance and seasonal fluctuations. *Malacologia* **40**, 231–250.

de Winter, W. P. and Amoroso, V. B. E. (2003) *Plant resources of South East Asia 15: Cryptograms: Ferns and fern allies.* Prosea Foundation and Backhuys Publishers, Bogor, Indonesia and Leiden, Netherlands.

DeFries, R., Hansen, A., Newton, A. C. and Hansen, M. C. (2005) Increasing isolation of protected areas in tropical forests over the past twenty years. *Ecological Applications* **15**, 19–26.

Delissio, L. J. and Primack, R. B. (2003) The impact of drought on the population dynamics of canopy-tree seedlings in an aseasonal Malaysian rain forest. *Journal of Tropical Ecology* **19**, 489–500.

Denham, T. P., Haberle, S. and Lentfer, C. (2004) New evidence and revised interpretations of early agriculture in Highland New Guinea. *Antiquity* **78**, 839–857.

Dennis, R. A., Mayer, J., Applegate, G., Chokkalingam, U., Colfer, C. J. P., Kurniawan, I., Lachowski, H., Maus, P., Permana, R. P., Ruchiat, Y., Stolle, F., Suyanto and Tomich, T. P. (2005) Fire, people and pixels: Linking social science and remote sensing to understand underlying causes and impacts of fires in Indonesia. *Human Ecology* **33**, 465–504.

Denslow, J. S. (1987) Tropical rainforest gaps and tree species diversity. *Annual Review of Ecology and Systematics* **18**, 431–451.

Denslow, J. S. and Guzman, S. (2000) Variation in stand structure, light and seedling abundance across a tropical moist forest chronosequence, Panama. *Journal of Vegetation Science* **11**, 201–212.

Denslow, J. S., Ellison, A. M. and Sanford, R. E. (1998) Treefall gap size effects on above- and below-ground processes in a tropical wet forest. *Journal of Ecology* **86**, 597–609.

Denslow, J.S., Newell, E. and Ellison, A.M. (1991) The effect of understory palms and cyclanths on the growth and survival of inga seedlings. *Biotropica* **23**, 225–234.

Develey, P. F. and Stouffer, P. C. (2001) Effects of roads on movements by understory birds in mixed- species flocks in central Amazonian Brazil. *Conservation Biology* **15**, 1416–1422.

DeVries, P. J. (2001) Butterflies. In: *Encyclopedia of Biodiversity.* Levin, S. A. (ed.). Academic Press, London, UK. **Vol 1**, 559–573.

Dew, J. L. and Boubli, J. P. (2002) *Tropical Fruits and Frugivores: The Search for Strong Interactors.* Springer, Dordrecht, The Netherlands. Pp. 243.

DeWalt, S. J., Denslow, J. S. and Ickes, K. (2004) Natural-enemy release facilitates habitat expansion of the invasive tropical shrub *Clidemia hirta. Ecology* **85**, 471–483.

DeWalt, S. J., Ickes, K., Nilus, R., Harms, K. E. and Burslem, D. F. R. P. (2006) Liana habitat associations and community structure in a Bornean lowland tropical forest. *Plant Ecology* **186**, 203–216.

Dezzeo, N., Hernandez, L. and Folster, H. (1997) Canopy dieback in lower montane forests of Alto Uriman, Venezuelan Guayana. *Plant Ecology* **132**, 197–209.

DFID. (2002) *Wildlife and poverty study.* Wildlife Advisory Group, Rural Livelihoods Department, London, UK.

Di Rosa, I., Simoncelli, F., Fagotti, A. and Pascolini, R. (2007) The proximate cause of frog declines? *Nature* **447**, E4–E5.

Dial, R. (2003) Energetic savings and the body size distributions of gliding mammals. *Evolutionary Ecology Research* **5**, 1151–1162.

Dial, R., Bloodworth, B., Lee, A., Boyne, P. and Heys, J. (2004) The distribution of free space and its relation to canopy composition at six forest sites. *Forest Science* **50**, 312–325.

Diamond, A. W. and Hamilton, A. C. (1980) The distribution of forest passerine birds and Quaternary climate change in Tropical Africa. *Journal of Zoology* **191**, 379–402.

Diamond, J. (2005) *Collapse: How Societies Choose to Fail or Succeed.* Viking Books, London, UK. Pp. 575.

Diamond, J. M. (1987) Did Komodo dragons evolve to eat pygmy elephants? *Nature* **326**, 832.

Diamond, J. M. (1999) Dirty eating for healthy living. *Nature* **400**, 120–121.

Dick, C. W. and Heuertz, M. (2008) The complex biogeographic history of a widespread tropical tree species. *Evolution* **62**, 2760–2774.

Dick, C. W., Abdul-Salim, K. and Bermingham, E. (2003a) Molecular systematic analysis reveals cryptic tertiary diversification of a widespread tropical rain forest tree. *American Naturalist* **162**, 691–703.

Dick, C. W., Bermingham, E., Lemes, M. R. and Gribel, R. (2007) Extreme long-distance dispersal of the lowland tropical rainforest tree *Ceiba pentandra* L. (Malvaceae) in Africa and the Neotropics. *Molecular Ecology* **16**, 3039–3049.

Dick, C. W., Etchelecu, G. and Austerlitz, F. (2003b) Pollen dispersal of tropical trees (*Dinizia excelsa*: Fabaceae) by native insects and African honeybees in pristine and fragmented Amazonian rainforest. *Molecular Ecology* **12**, 753–764.

Dick, C. W., Hardy, O. J., Jones, F. A. and Petit, R. J. (2008) Spatial scales of pollen and seed-mediated gene flow in tropical rain forest trees. *Tropical Plant Biology* **1**, 20–33.

Didham, R. K. (1997) The influence of edge effects and forest fragmentation on leaf litter invertebrates in central Amazonia. In: *Tropical Forest Remnants: Ecology, Management, and Conservation of Fragmented Communities.* Laurance, W. F. and Bierregaard, R. O. (eds.). University of Chicago Press, Chicago, IL, USA. Pp. 55–70.

Didham, R. K. and Lawton, J. H. (1999) Edge structure determines the magnitude of changes in microclimate and vegetation structure in tropical forest fragments. *Biotropica* **31**, 17–30.

Didham, R. K., Hammond, P. M., Lawton, J. H., Eggleton, P. and Stork, N. E. (1998) Beetle species responses to tropical forest fragmentation. *Ecological Monographs* **68**, 295–323.

Dietz, J., Leuschner, C., Holscher, D. and Kreilein, H. (2007) Vertical patterns and duration of surface wetness in an old-growth tropical montane forest, Indonesia. *Flora* **202**, 111–117.

Dirzo, R. and Miranda, A. (1990) Contemporary neotropical defaunation and forest structure, function, and diversity: a sequel. *Conservation Biology* **4**, 444–447.

Dodson, C. (1962) Pollination and variation in the subtribe Catasetinae (Orchidaceae). *Annals of the Missouri Botanical Garden* **49**, 35–57.

Doerr, S. H. (1999) Karst-like landforms and hydrology in quartzites of the Venezuelan Guyana shield: Pseudokarst or real karst? *Zeitschrift Fur Geomorphologie* **43**, 1–17.

Doligez, A. and Joly, H. I. (1997) Genetic diversity and spatial structure within a natural stand of a tropical fores tree species, *Carapa procera* (Meliaceae), in French Guiana. *Heredity* **79**, 72–82.

Dominy, N. J., Grubb, P. J., Jackson, R. V., Lucas, P. W., Metcalfe, D. J., Svenning, J. C. and Turner, I. M. (2008) In tropical lowland rain forests monocots have tougher leaves than dicots, and include a new kind of tough leaf. *Annals of Botany* **101**, 1363–1377.

Dominy, N. J., Lucas, P. W. and Wright, S. J. (2003) Mechanics and chemistry of rain forest leaves: Canopy and understorey compared. *Journal of Experimental Botany* **54**, 2007–2014.

Donoghue, M. J. (2008) A phylogenetic perspective on the distribution of plant diversity. *Proceedings of the National Academy of Sciences of the United States of America* **105**, 11549–1555.

Douglas, I., Bidin, K., Balamurugan, G., Chappell, N. A., Walsh, R. P. D., Greer, T. and Sinun, W. (1999) The role of extreme events in the impacts of selective tropical forestry on erosion during harvesting and recovery phases at Danum Valley, Sabah. *Philosophical Transactions of the Royal Society of London Series B* **354**, 1749–1761.

Douglas, I., Greer, T., Bidin, K. and Spilsbury, M. (1993) Impacts of rain-forest logging on river systems and communities in Malaysia and Kalimantan. *Global Ecology and Biogeography Letters* **3**, 245–252.

Dounias, E. and Froment, A. (2006) When forest-based hunter-gatherers become sedentary: consequences for diet and health. *Unasylva* **57**, 26–33.

Dounias, E., Selzner, A., Kishi, M., Kurniawan, I. and Siregar, R. (2007) Back to the trees? Diet and health as indicators of adaptive responses to environmental change. The case of the Punan Tubu in the Malinau Research Forest. In: *Managing forest resources in a decentralized environment: Lessons learnt from the Malinau Research Forest, East Kalimantan Indonesia.* Gunarso, P., Setyawati, T., Sunderland, T. and Shackleton, C. (eds.). CIFOR, Bogor, Indonesia. Pp. 157–180.

Dove, M. R. (1993) A revisionist view of tropical deforestation and development. *Environmental Conservation* **20**, 17–24.

Dove, M. R. (1994) Transition from native forest rubbers to *Hevea brasiliensis* (Euphorbiaceae) among tribal smallholders in Borneo. *Economic Botany* **48**, 382–396.

Downing, T. E., Hecht, S. B. and Pearson, H. A. (eds.) (1992) *Development or Destruction: The Conversion of Tropical Forest to Pasture in Latin America.* Westview Special Studies in Social, Political, and Economic Development. Westview Press, Boulder, CO, USA. Pp. 405.

Dransfield, S. and Widjaja, E. A. E. (1995) *Plant resources of South East Asia 7: Bamboos.* Prosea Foundation, Bogor, Indonesia. Pp. 189.

Dressler, R. L. (1968) Pollination by euglossine bees. *Evolution* **22**, 202–210.

Dressler, R. L. (1982) Biology of the orchid bees (Euglossini). *Annual Review of Ecology and Systematics* **13**, 373–394.

Ducousso, M., Bena, G., Bourgeois, C., Buyck, B., Eyssartier, G., Vincelette, M., Rabevohitra, R., Randrihasipara, L., Dreyfus, B. and Prin, Y. (2004) The last common ancestor of Sarcolaenaceae and Asian dipterocarp trees was ectomycorrhizal before the India-Madagascar separation, about 88 million years ago. *Molecular Ecology* **13**, 231–236.

Dudgeon, D. (2000) The ecology of tropical Asian rivers and streams in relation to biodiversity conservation. *Annual Review of Ecology and Systematics* **31**, 239–263.

Dudley, R. and Devries, P. (1990) Tropical rain forest structure and the geographical distribution of gliding vertebrates. *Biotropica* **22**, 429–431.

Duellman, W. E. (1990) Herpetofaunas in Neotropical rainforests: comparative composition, history and resource use. In: *Four Neotropical Rainforests*. Gentry, A. H. (ed.). Yale University Press, New Haven, CT, USA. Pp. 455–505.

Duellman, W. E. (1999) Global distribution of amphibian: patterns conservation and future challenges. *Patterns of distribution of amphibians: a global perspective*. Duellman, W. E. Johns Hopkins University Press Baltimore, Maryland, USA. Pp. 1–30.

Duffus, D. (1993) Tsitika to Baram: The myth of sustainability. *Conservation Biology* **7**, 440–442.

Duivenvoorden, J. F. (1996) Patterns of tree species richness in rain forests of the middle Caqueta area, Colombia, NW Amazonia. *Biotropica* **28**, 142–158.

Duivenvoorden, J. F., Svenning, J. C. and Wright, S. J. (2002) Beta diversity in tropical forests. *Science* **295**, 636–637.

Dumbacher, J. P., Beehler, B. M., Spande, T. F. and Garraffo, H. M. (1992) Homobatrachotoxin in the genus *Pitohui*: chemical defense in birds. *Science* **258**, 799–801.

Dumbacher, J. P., Wako, A., Derrickson, S. R., Samuelson, A., Spande, T. F. and Daly, J. W. (2004) Melyrid beetles (*Choresine*): A putative source for the batrachotoxin alkaloids found in poison-dart frogs and toxic passerine birds. *Proceedings of the National Academy of Sciences of the United States of America* **101**, 15857–15860.

Dunisch, O. and Morais, R.R. (2002) Regulation of xylem sap flow in an evergreen, a semi-deciduous, and a deciduous Meliaceae species from the Amazon. *Trees-Structure and Function* **16**, 404–416.

Dunn, R. R. (2005) Modern insect extinctions, the neglected majority. *Conservation Biology* **19**, 1030–1036.

Dupuy, J. M. and Chazdon, R. L. (1998) Long-term effects of forest regrowth and selective logging on the seed bank of tropical forests in NE Costa Rica. *Biotropica* **30**, 223–237.

Dussourd, D. E. and Eisner, T. (1987) Vein-cutting behaviour: Insect counterploy to the latex defense of plants. *Science* **237**, 898–901.

Duvall, C. S. (2006) On the origin of the tree *Spondias mombin* in Africa. *Journal of Historical Geography* **32**, 249–266.

Dyer, L. A., Dodson, C. D., Beihoffer, J. and Letourneau, D. K. (2001) Trade-offs in antiherbivore defenses in *Piper cenocladum*: Ant mutualists versus plant secondary metabolites. *Journal of Chemical Ecology* **27**, 581–592.

Dykes, A. P. (2002) Weathering-limited rainfall-triggered shallow mass movements in undisturbed steepland tropical rainforest. *Geomorphology* **46**, 73–93.

Dykstra, D. and Heinrich, R. (1996) *FAO model code of forest harvesting*. FAO, Rome, Italy.

Eamus, D. and Prior, L. (2001) Ecophysiology of trees of seasonally dry tropics: Comparisons among phenologies. *Advances in Ecological Research* **32**, 113–197.

Earl of Cranbrook (1991) *Mammals of South-East Asia*. Oxford University Press, Oxford, UK. Pp. 96.

Earl of Cranbrook and Edwards, D. S. (1994) *Belalong: A Tropical Rainforest*. The Royal Geographical Society, London, UK. Pp. 389.

Ebert, D. (1998) Behavioral asymmetry in relation to body weight and hunger in the tropical social spider *Anelosimus eximius* (Araneae, Theridiidae) *Journal of Arachnology* **26**, 70–80.

Edwards, W. and Gadek, P. (2002) Multiple resprouting from diaspores and single cotyledons in the Australian tropical tree species *Idiospermum australiense*. *Journal of Tropical Ecology* **18**, 943–948.

Eggeling, W. J. (1947) Observations on the ecology of the Budongo Rain Forest, Uganda. *Journal of Ecology* **34**, 20–87.

Eggleton, P., Homathevi, R., Jones, D. T., MacDonald, J. A., Jeeva, D., Bignell, D. E., Davies, R. G. and Maryati, M. (1999) Termite assemblages, forest disturbance and greenhouse gas fluxes in Sabah, East Malaysia. *Philosophical Transactions of the Royal Society of London Series B* **354**, 1791–1802.

Ehrlich, P. R. and Raven, P. H. (1964) Butterflies and Plants — a Study in Coevolution. *Evolution* **18**, 586–608.

Eisenberg, J. F. and Thorington, R. W. (1973) A preliminary analysis of a Neotropical mammal fauna. *Biotropica* **5**, 150–161.

Elbert, W., Taylor, P. E., Andreae, M. O. and Paschl, U. (2006) Contribution of fungi to primary biogenic aerosols in the atmosphere: active discharge of spores, carbohydrates, and inorganic ions by Asco- and Basidiomycota. *Atmospheric Chemistry and Physics Discussions* **6**, 11317–11355.

Ellwood, M. D. F. and Foster, W. A. (2004) Doubling the estimate of invertebrate biomass in a rainforest canopy. *Nature* **429**, 549–551.

Eltahir, E. A. B. and Bras, R. L. (1994) Precipitation Recycling in the Amazon Basin. *Quarterly Journal of the Royal Meteorological Society* **120**, 861–880.

Emberton, K. C. (1997) Diversities and distributions of 80 land snail species in southeastern-most Madagascan rainforests, with a report that lowlands are richer than highlands in endemic and rare species. *Biodiversity and Conservation* **6**, 1137–1154.

Emmons, L. H. and Gentry, A. H. (1983) Tropical forest structure and the distribution of gliding and prehensile-tailed vertebrates. *American Naturalist* **121**, 513–524.

Emsley, M. G. (1965) The geographical distribution of the color-pattern components of *Heliconius erato* and *Heliconius melpomene* with genetical evidence for the systematic relationship between the two species. *Zoologica* **49**, 245–286.

Endler, J. A. (1977) *Geographic variation, speciation, and clines.* Princeton University Press, Princeton, NJ, USA. Pp. 262.

Endress, P. K. (1994) *Diversity and Evolutionary Biology of Tropical Flowers.* Cambridge University Press, Cambridge, UK. Pp. 511.

Engelbrecht, B.M.J. and Kursar, T.A. (2003) Comparative drought-resistance of seedlings of 28 species of co-occurring tropical woody plants. *Oecologia*, **136**, 383–393.

Engelbrecht, B.M.J., Comita, L.S., Condit, R., Kursar, T.A., Tyree, M.T., Turner, B.L. and Hubbell, S.P. (2007) Drought sensitivity shapes species distribution patterns in tropical forests. *Nature*, **447**, 80–U2.

Enters, T. (2000) Rethinking stakeholder involvement in biodiversity conservation projects. In: *Forest Conservation Genetics: Principles and Practice.* Young, A., Boshier, D. and Boyle, T. (eds.). CSIRO Publishing, Collingwood, Australia. Pp. 263–273.

Erkens, R. H. J., Chatrou, L. W., Maas, J. W., Niet, T. v. d. and Savolainen, V. (2007) A rapid diversification of rainforest trees (Guatteria; Annonaceae) following dispersal from Central into South America. *Molecular Phylogenetics and Evolution* **44**, 399–411.

Ernest, K. A. (1989) Insect herbivory on a tropical understory tree: Effects of leaf age and habitat. *Biotropica* **21**, 194–199.

Erwin, D. H. (2008) Extinction as the loss of evolutionary history. *Proceedings of the National Academy of Sciences of the United States of America* **105** (Suppl. 1), 11520–11527.

Erwin, T. L. (1982) Tropical forests: their richness in Coleoptera and other arthropod species. *The Coleopterists Bulletin* **36**, 74–75.

Eshet, Y., Rampino, M. R. and Visscher, H. (1995) Fungal event and palynological record of ecological crisis and recovery across the Permian-Triassic boundary. *Geology* **23**, 967–70.

Esquivel, R. E. and Carranza, J. (1996) Pathogenicity of *Phylloporia chrysita* (Aphyllophorales: Hymenochaetaceae) on *Erythrochiton gymnanthus* (Rutaceae). *Revista de Biología Tropical* **44**(Suppl 4), 137–145.

Everham, E. M. and Brokaw, N. V. L. (1996) Forest damage and recovery from catastrophic wind. *Botanical Review* **62**, 113–185.

Ewers, R. M. and Rodrigues, A. S. L. (2008) Estimates of reserve effectiveness are confounded by leakage. *Trends in Ecology and Evolution* **23**, 113–116.

Eyre, L. (1998) The tropical rainforests of the eastern Caribbean: present status and conservation. *Caribbean Geography* **9**, 101–20.

Faegri, K. and van der Pijl, L. (1966) *The Principles of Pollination Ecology.* Pergamon Press, Oxford, UK.

Falconer, J. (1990) The major significance of minor forest products: examples from West Africa. *Appropriate Technology* **17**, 13–16.

Falcon-Lang, H. J. and Cantrill, D. J. (2001) Leaf phenology of some mid-Cretaceous polar forests, Alexander Island, Antarctica. *Geological Magazine* **138**, 39–52.

Falster, D. S. and Westoby, M. (2003) Leaf size and angle vary widely across species: What consequences for light interception? *New Phytologist* **158**, 509–525.

Falster, D. S. and Westoby, M. (2005) Alternative height strategies among 45 dicot rain forest species from tropical queensland, australia. *Journal of Ecology* **93**, 521–535.

FAO. (2000) *On Definitions of Forest and Forest Change.* Food and Agriculture Organization of the United Nations, Rome, Italy.

FAO. (2001) *Lecture notes on the major soils of the world.* Food and Agriculture Organization of the United Nations, Rome, Italy.

FAO. (2003) *Workshop on Tropical Secondary Forest Management in Africa: Reality and Perspectives,* Food and Agriculture Organization of the United Nations, Rome, Italy.

FAO. (2006) *Global Forest Resources Assessment: Progress towards sustainable forest management.* Food and Agriculture Organization of the United Nations, Rome, Italy.

FAO. (2007) *State of the World's Forests.* Food and Agriculture Organization of the United Nations, Rome, Italy. Pp. 157.

FAOSTAT. (2008) FAOSTAT Crops. Retrieved 15 December 2008, from http://faostat.fao.org/site/567/default.aspx#ancor.

Fargione, J., Hill, J., Tilman, D., Polasky, S. and Hawthorne, P. (2008) Land clearing and the biofuel carbon debt. *Science* **319**, 1235–1238.

Farji Brener, A. G., Valverde, O., Paolini, L., Torre, M. d. l. A. I., Quintero, E., Bonaccorso, E., Arnedo, L., Villalobos, R., de los, A. l. T. M. and la Torre, M. d. l. A. (2002) Function of the drip tips of leaves and their vertical distribution in a tropical rain forest in Costa Rica. *Revista de Biologia Tropical* **50**, 561–567.

Farjon, A. (1998) *World Checklist and Bibliography of Conifers.* Royal Botanical Gardens, Kew, London, UK.

Farnsworth, N. (1988) Screening plants for new medicines. *Biodiversity.* Wilson, E. O. National Academy Press, Washington, DC, USA. Pp. 83–97.

Farrell, B. D., Dussourd, D. E. and Mitter, C. (1991) Escalation of Plant Defense–Do Latex and Resin Canals Spur Plant Diversification? *American Naturalist* **138**, 881–900.

Farris-Lopez, K., Denslow, J. S., Moser, B. and Passmore, H. (2004) Influence of a common palm, Oenocarpus mapora, on seedling establishment in a tropical moist forest in Panama. *Journal of Tropical Ecology* **20**, 429–438.

Fearnside, P. M. (1997a) Environmental services as a strategy for sustainable development in rural Amazonia. *Ecological Economics* **20**, 53–70.

Fearnside, P. M. (1997b) Transmigration in Indonesia: Lessons from its environmental and social impacts. *Environmental Management* **21**, 553–570.

Fearnside, P. M. (1997c) Wood density for estimating forest biomass in Brazilian Amazonia. *Forest Ecology and Management* **90**, 59–87.

Fearnside, P. M. (2003) Conservation policy in Brazilian Amazonia: Understanding the dilemmas. *World Development* **31**, 757–779.

Fearnside, P. M. (2005) Deforestation in Brazilian Amazonia: History, rates, and consequences. *Conservation Biology* **19**, 680–688.

Fearnside, P. M. (2006a) Dams in the Amazon: Belo Monte and Brazil's hydroelectric development of the Xingu River basin. *Environmental Management* **38**, 16–27.

Fearnside, P. M. (2006b) Mitigation of climate change in the Amazon. In: *Emerging Threats to Tropical Forests*. Laurance, W. F. and Peres, C. A. (eds.). University of Chicago Press, Chicago, IL, USA. Pp. 353–375.

Fearnside, P. M. (2007) Brazil's Cuiaba-Santarem (BR-163) Highway: The environmental cost of paving a soybean corridor through the Amazon. *Environmental Management* **39**, 601–614.

Fearnside, P. M. (2008) The roles and movements of actors in the deforestation of Brazilian Amazonia. *Ecology and Society* **13**, 23.

Fearnside, P. M., Barbosa, R. I. and Graca, P. (2007) Burning of secondary forest in Amazonia: Biomass, burning efficiency and charcoal formation during land preparation for agriculture in Apiau, Roraima, Brazil. *Forest Ecology and Management* **242**, 678–687.

Feder, M. E. and Burggren, W. W. (1992) *Environmental Physiology of the Amphibians*. University of Chicago Press, Chicago, IL, USA. Pp. 472.

Fedorov, A. A. (1966) The structure of the tropical rain forest and speciation in the humid tropics. *Journal of Ecology* **54**, 1–11.

Feeley, K. J. and Terborgh, J. W. (2005) The effects of herbivore density on soil nutrients and tree growth in tropical forest fragments. *Ecology* **86**, 116–124.

Feild, T. S. and Arens, N. C. (2005) Form, function and environments of the early angiosperms: merging extant phylogeny and ecophysiology with fossils. *New Phytologist* **166**, 383–408.

Feild, T. S., Arens, N. C. and Dawson, T. E. (2003) The ancestral ecology of angiosperms: Emerging perspectives from extant basal lineages. *International Journal of Plant Sciences* **164**, S129–S142.

Feild, T. S., Arens, N. C., Doyle, J. A., Dawson, T. E. and Donoghue, M. J. (2004) Dark and disturbed: A new image of early angiosperm ecology. *Paleobiology* **30**, 82–107.

Fellows, L. and Scofield, A. (1995) Chemical diversity in plants. In: *Intellectual Property Rights and Biodiversity Conservation*. Swanson, T. M. (ed.). Cambridge University Press, Cambridge, UK. Pp. 19–44.

Fenner, M. and Thompson, J. (2005) *The Ecology of Seeds*. Cambridge University Press, Cambridge, UK. Pp. 250.

Ferguson, B. G., Vandermeer, J., Morales, H. and Griffith, D. M. (2003) Post-agricultural succession in El Peten, Guatemala. *Conservation Biology* **17**, 818–828.

Ferraro, P. J. and Kiss, A. (2002) Direct payments to conserve biodiversity. *Science* **298**, 1718–1719.

Ferraro, P. J., Uchida, T. and Conrad, J. M. (2005) Price premiums for eco-friendly commodities: Are 'Green' markets the best way to protect endangered ecosystems? *Environmental and Resource Economics* **32**, 419–438.

Ferraz, G., Russell, G. J., Stouffer, P. C., Bierregaard, R. O., Pimm, S. L. and Lovejoy, T. E. (2003) Rates of species loss from Amazonian forest fragments. *Proceedings of the National Academy of Sciences of the United States of America* **100**, 14069–14073.

Ferrer, A. and Gilbert, G. S. (2003) Effect of tree host species on fungal community composition in a tropical rain forest in Panama. *Diversity and Distributions* **9**, 455–468.

Fetcher, N., Haines, B. L., Cordero, R. A., Lodge, D. J., Walker, L. R., Fernandez, D. S. and Lawrence, W. T. (1996) Responses of tropical plants to nutrients and light on a landslide in Puerto Rico. *Journal of Ecology* **84**, 331–341.

Fichtler, E., Clark, D.A. and Worbes, M. (2003) Age and long-term growth of trees in an old-growth tropical rain forest, based on analyses of tree rings and C14. *Biotropica*, **35**, 306–317.

Fimbel, C. (1994) The relative use of abandoned farm clearings and old forest habitats by primates and a forest antelope at Tiwai, Sierra Leone, West Africa. *Biological Conservation* **70**, 277–286.

Fine, P. V. A. and Ree, R. H. (2006) Evidence for a time-integrated species-area effect on the latitudinal gradient in tree diversity. *American Naturalist* **168**, 796–804.

Fine, P. V. A., Daly, D. C., Munoz, G. V., Mesones, I. and Cameron, K. M. (2005) The contribution of edaphic heterogeneity to the evolution and diversity of Burseraceae trees in the western Amazon. *Evolution* **59**, 1464–1478.

Fine, P. V. A., Mesones, I. and Coley, P. D. (2004) Herbivores promote habitat specialization by trees in Amazonian forests. *Science* **305**, 663–665.

Fine, P. V. A., Miller, Z. J., Mesones, I., Irazuzta, S., Appel, H. M., Stevens, M. H. H., Saaksjarvi, I., Schultz, L. C. and Coley, P. D. (2006) The growth-defense trade-off and habitat specialization by plants in Amazonian forests. *Ecology,* **87**, S150–S162.

Finegan, B. (1996) Pattern and process in neotropical secondary rain forests: The first 100 years of succession. *Trends in Ecology and Evolution* **11**, 119–124.

Finegan, B. and Nasi, R. (2004) The biodiversity and conservation potential of shifting cultivation landscapes. In: *Agroforestry and Biodiversity Conservation in Tropical Landscapes.* Schroth, G., Da Fonseca, G. A. B., Harvey, C. A., Gascon, C., Vasconcelos, H. L. and Izac, A.-M. N. (eds.). Island Press, Washington, DC, USA. Pp. 153–197.

Firn, R. D. (2003) Bioprospecting—why is it so unrewarding? *Biodiversity and Conservation* **12**, 207–216.

Fischer, J., Brosi, B., Daily, G. C., Ehrlich, P. R., Goldman, R., Goldstein, J., Lindenmayer, D. B., Manning, A. D., Mooney, H. A., Pejchar, L., Ranganathan, J. and Tallis, H. (2008) Should agricultural policies encourage land sparing or wildlife-friendly farming? *Frontiers in Ecology and the Environment* **6**, 380–385.

Fischer, R. C., Wanek, W., Richter, A. and Mayer, V. (2003) Do ants feed plants? A N-15 labelling study of nitrogen fluxes from ants to plants in the mutualism of *Pheidole* and *Piper. Journal of Ecology* **91**, 126–134.

Fisher, B. L. (2003) Formicidae, ants. In: *The Natural History of Madagascar.* Goodman, S. M. and Benstead, J. P. (eds.). University of Chicago Press, Chicago, IL, USA. Pp. 811–819.

Fisher, J. B. (1982) A survey of buttresses and aerial roots of tropical trees for presence of reaction wood. *Biotropica* **14**, 56–61.

Fitter, A. H. and Moyersoen, B. (1996) Evolutionary trends in root-microbe symbioses. *Philosophical Transactions of the Royal Society of London Series B* **351**, 1367–1375.

Fitter, A. H., Gilligan, C. A., Hollingworth, K., Kleczkowski, A., Twyman, R. M. and Pitchford, J. W. (2005) Biodiversity and ecosystem function in soil. *Functional Ecology* **19**, 369–377.

Fittkau, E. J. and Klinge, H. (1973) On biomass and trophic structure of the central Amazonian rain forest ecosystem. *Biotropica* **5**, 2–14.

Fitton, J., Mahoney, J., Wallace, P. and Saunders, A., Eds. (2004) *Origin and evolution of the Ontong Java Plateau.* Geological Society of London Special Publication, London, UK.

Fjeldsa, J. (1994) Geographical patterns for relict and young species of birds in Africa and South America and implications for conservation priorities. *Biodiversity and Conservation* **3**, 207–226.

Fjeldsa, J. and Bowie, R. C. K. (2008) New perspectives on the origin and diversification of Africa's forest avifauna. *African Journal of Ecology* **46**, 235–247.

Fjeldsa, J. and Lovett, J. C. (1997a) Biodiversity and environmental stability. *Biodiversity and Conservation* **6**, 315–323.

Fjeldsa, J. and Lovett, J. C. (1997b) Geographical patterns of old and young species in African forest biota: The significance of specific montane areas as evolutionary centres. *Biodiversity and Conservation* **6**, 325–346.

Fleagle, J. G. (1999) *Primate Adaptation and Evolution.* Academic Press, San Diego, CA, USA. Pp. 596.

Fleagle, J. G. and Kay, R. F. (1997) Platyrrhines, catarrhines, and the fossil record. In: *New World Primates: Ecology, Evolution, and Behavior.* Kinzey, W. G. (ed.). Aldine de Gruyter, New York, USA. Pp. 3–23.

Fleagle, J. G., Janson, C. and Reed, C. K. (1999) *Primate Communities.* Cambridge University Press, Cambridge, UK.

Fleischbein, K., Wilcke, W., Goller, R., Boy, J., Valarezo, C., Zech, W. and Knoblich, K. (2005) Rainfall interception in a lower montane forest in Ecuador: effects of canopy properties. *Hydrological Processes* **19**, 1355–1371.

Fleming, T. H., Breitwisch, R. and Whitesides, G. H. (1987) Patterns of tropical vertebrate frugivore diversity. *Annual Review of Ecology and Systematics* **18**, 91–109.

Flenley, J. R. (1998) Tropical forests under the climates of the last 30,000 years. *Climatic Change* **39**, 177–197.

Foley, J. A., DeFries, R., Asner, G. P., Barford, C., Bonan, G., Carpenter, S. R., Chapin, F. S., Coe, M. T., Daily, G. C., Gibbs, H. K., Helkowski, J. H., Holloway, T., Howard, E. A., Kucharik, C. J., Monfreda, C., Patz, J. A., Prentice, I. C., Ramankutty, N. and Snyder, P. K. (2005) Global consequences of land use. *Science* **309**, 570–574.

Foody, G. M. (2003) Remote sensing of tropical forest environments: towards the monitoring of environmental resources for sustainable development. *International Journal of Remote Sensing* **24**, 4035–4046.

Foody, G. M. and Cutler, M. E. J. (2003) Tree biodiversity in protected and logged Bornean tropical rain forests and its measurement by satellite remote sensing. *Journal of Biogeography* **30**, 1053–1066.

Forest Trends (undated) *Reduced Impact Logging (RIL).* Washington, DC, USA.

Forget, P. M. (1990) Seed dispersal of *Vouacapoua americana* (Caesalpiniaceae) by caviomorph rodents in French Guiana. *Journal of Tropical Ecology* **6**, 459–468.

Forget, P. M. (1993) Postdispersal predation and scatterhoarding of *Dipteryx panamensis* (Papilionaceae) seeds by rodents in Panama. *Oecologia* **94**, 255–261.

Forget, P. M., Mercier, F. and Collinet, F. (1999) Spatial patterns of two rodent-dispersed rain forest trees *Carapa procera* (Meliaceae) and *Vouacapoua americana* (Caesalpiniaceae) at Paracou, French Guiana. *Journal of Tropical Ecology* **15**, 301–313.

Forget, P. M. and Wall, S. B. V. (2001) Scatter-hoarding rodents and marsupials: convergent evolution on diverging continents. *Trends in Ecology and Evolution* **16**, 65–67.

Forsberg, B. R., Araujolima, C., Martinelli, L. A., Victoria, R. L. and Bonassi, J. A. (1993) Autotrophic carbon sources for fish of the Central Amazon. *Ecology* **74**, 643–652.

Forte, J. (ed.) (1996) *Makusipe kamanto iseru: sustaining the Makushi way of life.* Makushi Research Unit, Business Print, Georgetown, Guyana.

Foster, D., Swanson, F., Aber, J., Burke, I., Brokaw, N., Tilman, D. and Knapp, A. (2003) The importance of land-use legacies to ecology and conservation. *Bioscience* **53**, 77–88.

Foster, R. B. (1977) *Tachigalia versicolor* is a suicidal neotropical tree. *Nature* **268**, 624–626.

Fostier, A. H., Forti, M. C., Guimaraes, J. R. D., Melfi, A. J., Boulet, R., Santo, C. M. E. and Krug, F. J. (2000) Mercury fluxes in a natural forested Amazonian catchment (Serra do Navio, Amapa State, Brazil) *Science of the Total Environment* **260**, 201–211.

Fox, J.E.D. (1973) Dipterocarp seedling behaviour in Sabah. *Malaysian Forester*, **36**, 205–214.

Fragoso, C. and Lavelle, P. (1992) Earthworm communities of tropical rain forests. *Soil Biology and Biochemistry* **24**, 1397–1408.

Fragoso, J. M. V., Silvius, K. M. and Correa, J. A. (2003) Long-distance seed dispersal by tapirs increases seed survival and aggregates tropical trees. *Ecology* **84**, 1998–2006.

Frahm, J. P. (1990) The ecology of epiphytic bryophytes on Mt. Kinabalu, Sabah (Malysia). *Nova Heldwigia* **51**, 121–132.

Frahm, J. P., Frey, W., Kuerschner, H. and Menzel, M. (1996) *Mosses and liverworts of Mount Kinabalu. Kota Kinabalu, Sabah, Malaysia.* Sabah Park Publications, Kota Kinabalu, Malaysia. Pp. 91.

Frahm, J. P. and Gradstein, S. R. (1991) An altitudinal zonation of tropical rain-forests using byrophytes. *Journal of Biogeography* **18**, 669–678.

Francis, A. P. and Currie, D. J. (2003) A globally consistent richness-climate relationship for angiosperms. *American Naturalist* **161**, 523–536.

Franco, W. and Dezzeo, N. (1994) Soils and soil-water regime in the terra-firme-caatinga forest complex near San Carlos De Rio Negro, State of Amazonas, Venezuela. *Interciencia* **19**, 305–316.

Frangi, J. L. and Lugo, A. E. (1991) Hurricane damage to a flood-plain forest in the Luquillo mountains of Puerto Rico. *Biotropica* **23**, 324–335.

Frangi, J. L. and Lugo, A. E. (1998) A flood plain palm forest in the Luquillo Mountains of Puerto Rico five years after Hurricane Hugo. *Biotropica* **30**, 339–348.

Fraver, S., Brokaw, N. V. L. and Smith, A. P. (1998) Delimiting the gap phase in the growth cycle of a Panamanian forest. *Journal of Tropical Ecology* **14**, 673–681.

Fredericksen, N. J. and Fredericksen, T. S. (2004) Impacts of selective logging on amphibians in a Bolivian tropical humid forest. *Forest Ecology and Management* **191**, 275–282.

Frederickson, M. E. and Gordon, D. M. (2007) The devil to pay: a cost of mutualism with *Myrmelachista schumanni* ants in 'devil's gardens' is increased herbivory on *Duroia hirsuta* trees. *Proceedings of the Royal Society B* **274**, 1117–1123.

Frederickson, M. E., Greene, M. J. and Gordon, D. M. (2005) 'Devil's gardens' bedevilled by ants. *Nature* **437**, 495–496.

Fredriksson, G. M. (2001) *Extinguishing the 1998 Forest Fires and Subsequent Coal Fires in the Sungai Wain Protection Forest, East Kalimantan, Indonesia.* TROPENBOS,Balikpapan, Kalimantan, Indonesia.

Fritsch, J. M. (1993) The hydrological effects of clearing tropical rain forest and of the implementation of alternative land uses. *International Association of Hydrological Sciences Publication* **216**, 53–66.

Fritts, T. H. and Rodda, G. H. (1998) The role of introduced species in the degradation of island ecosystems: A case history of Guam. *Annual Review of Ecology and Systematics* **29**, 113–140.

Frohlich, J. and Hyde, K. D. (1999) Biodiversity of palm fungi in the tropics: are global fungal diversity estimates realistic? *Biodiversity and Conservation* **8**, 977–1004.

Frohlich, J., Hyde, K. D. and Petrini, O. (2000) Endophytic fungi associated with palms. *Mycological Research* **104**, 1202–1212.

Frohlich, M. W. and Chase, M. W. (2007) Afer a dozen years of progress the origin of the angiosperms is still a great mystery. *Nature* **450**, 1184–1189.

FSC (2002) *The SLIMFs Inititive: A progress report.* Forest Stewardship Council, Oaxaca, Mexico.

FSC (2009) from http://www.fsc.org/77.html.

Gadgil, M. and Guha, R. (1994) Ecological conflicts and the environmental movement in India. *Development and Change* **25**, 101–136.

Gaither, J. C. (1994) Understory avifauna of a Bornean peat swamp forest: is it depauperate. *Wilson Bulletin* **106**, 381–390.

Gale, N. and Barfod, A. S. (1999) Canopy tree mode of death in a western Ecuadorian rain forest. *Journal of Tropical Ecology* **15**, 415–436.

Galeano, G., Suarez, S. and Balslev, H. (1998) Vascular plant species count in a wet forest in the Choco area on the Pacific coast of Colombia. *Biodiversity and Conservation* **7**, 1563–1575.

Galetti, M. (1993) Diet of the scaly-headed parrot (*Pionus maximiliani*) in a semideciduous forest in southeastern Brazil. *Biotropica* **25**, 419–425.

Galetti, M. (2001) Seed dispersal of mimetic fruits: Parasitism, mutualism, aposematism or exaptation? *Seed Dispersal and Frugivory: Ecology, Evolution and Conservation.* Levey, D. J., Silva, W. R. and Galetti, M. CABI Publishing, Wallingford, UK. Pp. 177–191.

Galetti, M., Donatti, C. I., Pizo, M. A. and Giacomini, H. C. (2008) Big fish are the best: Seed dispersal of Bactris glaucescens by the pacu fish (*Piaractus mesopotamicus*) in the Pantanal, Brazil. *Biotropica* **40**, 386–389.

Galetti, M., Laps, R. and Pizo, M. A. (2000) Frugivory by toucans (Ramphastidae) at two altitudes in the Atlantic forest of Brazil. *Biotropica* **32**(4B), 842–850.

Galvez, D. and Pearcy, R.W. (2003) Petiole twisting in the crowns of psychotria limonensis: Implications for light interception and daily carbon gain. *Oecologia*, **135**, 22–29.

Garcia-Guzman, G. and Dirzo, R. (2001) Patterns of leaf-pathogen infection in the understory of a Mexican rain forest: Incidence, spatiotemporal variation, and mechanisms of infection. *American Journal of Botany* **88**, 634–645.

Garcia-Moreno, J., Arctander, P. and Fjeldsa, J. (1999) Strong diversification at the treeline among Metallura hummingbirds. *Auk* **116**, 702–711.

Gargominy, O. and Ripken, T. (1998) Micro-pulmonates in tropical rainforest litter: a new bio-jewel? In *Abstracts of the World Congress of Malacology, Washington, D.C.* Bieler, R. and Nikkelsen, P. M. (eds.). Unitas Malacologica, Chicago, IL, USA. Pp. 116.

Gartner, B. L. (1989) Breakage and regrowth of *Piper* species in rainforest understorey. *Biotropica* **21**, 303–307.

Garwood, N. C. (1989) Tropical soil seed banks: a review. In: *Ecology of Soil Seed Banks.* Leck, M. A., Parker, V. T. and Simpson, R. L. (eds.). Academic Press, San Diego, CA, USA. Pp. 149–209.

Garwood, N. C., Janos, D. P. and Brokaw, N. (1979) Earthquake-caused landslides: A major disturbance to tropical forests. *Science* **205**, 997–999.

Gascon, C., Lovejoy, T. E., Bierregaard, R. O., Malcolm, J. R., Stouffer, P. C., Vasconcelos, H. L., Laurance, W. F., Zimmerman, B., Tocher, M. and Borges, S. (1999) Matrix habitat and species richness in tropical forest remnants. *Biological Conservation* **91**, 223–229.

Gash, J. H. C. and Nobre, C. A. (1997) Climatic effects of Amazonian deforestation: Some results from ABRACOS. *Bulletin of the American Meteorological Society* **78**, 823–830.

Gathorne-Hardy, F. J., Syaukani, Davies, R. G., Eggleton, P. and Jones, D. T. (2002) Quaternary rainforest refugia in south-east Asia: using termites (Isoptera) as indicators. *Biological Journal of the Linnean Society* **75**, 453–466.

Gaume, L., Perret, P., Gorb, E., Gorb, S., Labat, J. J. and Rowe, N. (2004) How do plant waxes cause flies to slide? Experimental tests of wax-based trapping mechanisms in three pitfall carnivorous plants. *Arthropod Structure and Development* **33**, 103–111.

Gautier-Hion, A. and Michaloud, G. (1989) Are figs always keystone resources for tropical frugivorous vertebrates: A test in Gabon. *Ecology* **70**, 1826–1833.

Gavin, D. G. and Peart, D. R. (1997) Spatial structure and regeneration of *Tetramerista glabra* in peat swamp rain forest in Indonesian Borneo. *Plant Ecology* **131**, 223–231.

Gavin, D. G. and Peart, D. R. (1999) Vegetative life history of a dominant rain forest canopy tree. *Biotropica* **31**, 288–294.

Gebo, D. L., Dagosto, M., Beard, K. C. and Qi, T. (2001) Middle Eocene primate tarsals from China: Implications for haplorhine evolution. *American Journal of Physical Anthropology* **116**, 83–107.

Genereux, D. P. and Jordan, M. (2006) Interbasin groundwater flow and groundwater interaction with surface water in a lowland rainforest, Costa Rica: A review. *Journal of Hydrology* **320**, 385–399.

Gentry, A. H. (1982) Patterns of neotropical plant species diversity. *Evolutionary Biology.* Hecht, M. K., Wallace, B. and Prance, G. T. Plenum Press, New York, USA. **15**. Pp. 1–84.

Gentry, A. H. (1988a) Changes in plant community diversity and floristic composition on environmental and geographic gradients. *Annals of the Missouri Botanical Gardens* **75**, 1–34.

Gentry, A. H. (1988b) Tree species richness of upper Amazonian forests. *Proceedings of the National Academy of Sciences of the United States of America* **85**, 156–159.

Gentry, A. H. (1991) The distribution and evolution of climbing plants. In: *The Biology of Vines.* Putz F. E. and Mooney H. A. (eds.). Cambridge University Press, Cambridge, UK. Pp. 3–49.

Gentry, A. H. and Dodson, C. (1987) Contribution of nontrees to species richness of a tropical rainforest. *Biotropica* **19**, 149–156.

Gerhardt, K. (1996) Effects of root competition and canopy openness on survival and growth of tree seedlings in a tropical seasonal dry forest. *Forest Ecology and Management* **82**, 33–48.

German, L. A. (2003) Historical contingencies in the coevolution of environment and livelihood: contributions to the debate on Amazonian Black Earth. *Geoderma* **111**, 307–331.

Geron, C., Owen, S., Guenther, A., Greenberg, J., Rasmussen, R., Bai, J. H., Li, Q. J. and Baker, B. (2006) Volatile organic compounds from vegetation in southern Yunnan Province, China: Emission rates and some potential regional implications. *Atmospheric Environment* **40**, 1759–1773.

Ghalambor, C. K., Huey, R. B., Martin, P. R., Tewksbury, J. J. and Wang, G. (2006) Are mountain passes higher in the tropics? Janzen's hypothesis revisited. *Integrative and Comparative Biology* **46**, 5–17.

Ghazoul, J. (2001a) Barriers to biodiversity conservation in forest certification. *Conservation Biology* **15**, 315–317.

Ghazoul, J. (2001b) Can floral repellents pre-empt potential ant-plant conflicts? *Ecology Letters* **4**, 295–299.

Ghazoul, J. (2002) Impact of logging on the richness and diversity of forest butterflies in a tropical dry forest in Thailand. *Biodiversity and Conservation* **11**, 521–541.

Ghazoul, J. (2005) Pollen and seed dispersal among dispersed plants. *Biological Reviews* **80**, 413–443.

Ghazoul, J. (2007) Recognising the complexities of ecosystem management and the ecosystem service concept. *Gaia* **16**, 215–221.

Ghazoul, J. (2009a) Bailing out creatures great and small. *Science* **323**, 460–460.

Ghazoul, J. (2009b) Bankers saved, environmentalists adrift, poor drowning. *Biotropica* **41**, 1–2.

Ghazoul, J. and McLeish, M. (2001) Reproductive ecology of tropical forest trees in logged and fragmented habitats in Thailand and Costa Rica. *Plant Ecology* **153**, 335–345.

Ghazoul, J. and Satake, A. (2009) Nonviable seed set enhances plant fitness: the sacrificial sibling hypothesis. *Ecology* **90**, 369–377.

Ghazoul, J., Garcia, C. A. and Kushalappa, C. G. (2009) Landscape labelling: a concept for next-generation payment for ecosystem service schemes. *Forest Ecology and Management* **258**, 1889–1895.

Ghazoul, J., Liston, K. A. and Boyle, T. J. B. (1998) Disturbance-induced density-dependent seed set in *Shorea siamensis* (Dipterocarpaceae), a tropical forest tree. *Journal of Ecology* **86**, 462–473.

Gilbert, G. S. (2002) Evolutionary ecology of plant diseases in natural ecosystems. *Annual Review of Phytopathology* **40**, 13–43.

Gilbert, G. S. (2005a) Nocturnal fungi: Airborne spores in the canopy and understory of a tropical rain forest. *Biotropica* **37**, 462–464.

Gilbert, G. S. (2005b) Dimensions of plant disease in tropical forests. In *Biotic Interactons in the Tropics: Their Role in the Maintenance of Species Diversity.* D. F. R. P. Burslem, M. A. Pinard and S. E. Hartley (eds.). Cambridge University Press, Cambridge, UK. Pp. 141–164.

Gilbert, G. S., Ferrer, A. and Carranza, A. (2002) Polypore fungal diversity and host density in a moist tropical forest. *Biodiversity and Conservation* **11**, 947–957.

Gilbert, G. S., Harms, K. E., Hamill, D. N. and Hubbell, S. P. (2001) Effects of seedling size, El Nino drought, seedling density, and distance to nearest conspecific adult on 6-year survival of *Ocotea whitei* seedlings in Panama. *Oecologia* **127**, 509–516.

Gilbert, G. S., Hubbell, S. P. and Foster, R. B. (1994) Density and distance-to-adult effects of a canker disease of trees in a moist tropical forest. *Oecologia* **98**, 100–108.

Gilbert, G. S. and Sousa, W. P. (2002) Host specialization among wood-decay polypore fungi in a caribbean mangrove forest. *Biotropica* **34**, 396–404.

Gilbert, L. E. (1972) Pollen feeding and reproductive biology of *Heliconius* butterflies. *Proceedings of the National Academy of Sciences of the United States of America* **69**, 1403–1407.

Gillespie, T. R., Chapman, C. A. and Greiner, E. C. (2005) Effects of logging on gastrointestinal parasite infections and infection risk in African primates. *Journal of Applied Ecology* **42**, 699–707.

Gillett, J. B. (1962) Pest pressure, an underestimated factor in evolution. *Taxonomy and Geography: A Symposium.* Nichols, D. (ed.). London Systematics Association, London, UK. Pp. 37–46.

Gillison, A. N. (2001) *Vegetation Survey and Habitat Assessment of the Tesso Nilo Forest Complex*, WWF-US. Pp. 76.

Gilmartin, A. J. and Brown, G. K. (1985) Cleistogamy in *Tillandsia capillaris* (Bromeliaceae). *Biotropica* **17**, 256–259.

Gilmore, D. P., Da Costa, C. P. and Duarte, D. P. F. (2001) Biology of the sloth. *Brazilian Journal of Medical and Biological Research* **34**, 9–24.

Givnish, T. J. (1988) Adaptation to sun and shade: a whole-plant perspective. *Australian Journal of Plant Physiology* **15**, 63–92.

Givnish, T. J. (1990) Leaf mottling: relation to growth form and leaf phenology and possible role as camouflage. *Functional Ecology* **4**, 463–474.

Givnish, T. J. (1999) On the causes of gradients in tropical tree diversity. *Journal of Ecology* **87**, 193–210.

Givnish, T. J. (2002) Adaptive significance of evergreen vs deciduous leaves: Solving the triple paradox. *Silva Fennica*, **36**, 703–743.

Givnish, T. J., Burkhardt, E. L., Happel, R. E. and Weintraub, J. D. (1984) Carnivory in the bromeliad *Brocchinia reducta*, with a cost-benefit model for the general restriction of carnivorous plants to sunny, moist, nutrient-poor habitats. *American Naturalist* **124**, 479–497.

Givnish, T. J., Pires, J. C., Graham, S. W., McPherson, M. A., Prince, L. M., Patterson, T. B., Rai, H. S., Roalson, E. H., Evans, T. M., Hahn, W. J., Millam, K. C., Meerow, A. W., Molvray, M., Kores, P. J., O'Brien, H. E., Hall, J. C., Kress, W. J. and Sytsma, K. J. (2005) Repeated evolution of net venation and fleshy fruits among monocots in shaded habitats confirms a priori predictions: evidence from an ndhF phylogeny. *Proceedings of the Royal Society B* **272**, 1481–1490.

Glaser, B., Haumaier, L., Guggenberger, G. and Zech, W. (2001) The 'Terra Preta' phenomenon: a model for sustainable agriculture in the humid tropics. *Naturwissenschaften* **88**, 37–41.

Gleason, H. A. (1926) The individualistic concept of the plant association. *Bulletin of the Torrey Botanical Club* **53**, 7–26.

Glew, L. and Hudson, M. D. (2007) Gorillas in the midst: the impact of armed conflict on the conservation of protected areas in sub-Saharan Africa. *Oryx* **41**, 140–150.

Global Witness (1996) *Corruption, War and Forest Policy: The Unsustainable Exploitation of Cambodia's Forests.* London, UK.

Godinez-Alvarez, H. (2004) Pollination and seed dispersal by lizards: a review. *Revista Chilena De Historia Natural* **77**, 569–577.

Godoy, J. R., Petts, G. and Salo, J. (1999) Riparian flooded forests of the Orinoco and Amazon basins: a comparative review. *Biodiversity and Conservation* **8**, 551–586.

Godoy, R., Wilkie, D., Overman, H., Cubas, A., Cubas, G., Demmer, J., McSweeney, K. and Brokaw, N. (2000) Valuation of consumption and sale of forest goods from a Central American rain forest. *Nature* **406**, 62–63.

Goldammer, J. G. (ed.) (1992) *Tropical Forests in Transition: Ecology of Natural and Anthropogenic Disturbance Processes.* Birkhäuser-Verlag, Basel, Switzerland. Pp. 270.

Goldingay, R. L., Carthew, S. M. and Whelan, R. J. (1991) The Importance of non-flying mammals in pollination. *Oikos* **61**, 79–87.

Goldstein, G., Andrade, J. L., Meinzer, F. C., Holbrook, N. M., Cavelier, J., Jackson, P. and Celis, A. (1998) Stem water storage and diurnal patterns of water use in tropical forest canopy trees. *Plant Cell Environ* **21**, 397–406.

Goller, R., Wilcke, W., Fleischbein, K., Valarezo, C. and Zech, W. (2006) Dissolved nitrogen, phosphorus, and sulfur forms in the ecosystem fluxes of a montane forest in Ecuador. *Biogeochemistry* **77**, 57–89.

Gomez-Pompa, A. (2004) The role of biodiversity scientists in a troubled world. *Bioscience* **54**, 217–225.

Gomez-Pompa, A., Vazquezy, C. and Guevara, S. (1972) Tropical rainforests: Non-renewable resource. *Science* **177**, 762–765.

Good, R. (1962) On the geographical relationships of the angiosperm flora of New Guinea. *Bulletin of the British Museum (Natural History), Botany* **12**, 205–226.

Goodman, S. M. and Benstead, J. P. (2003) *The Natural History of Madagascar.* University of Chicago Press, Chicago, IL, USA. Pp. 1728.

Goosem, M. (1997) Internal fragmentation: The effects of roads, highways, and powerline clearings on movements and mortality of rainforest vertebrates. In: *Tropical Forest Remnants: Ecology, Management, and Conservation of Fragmented Communities.* Laurance, W. F. and Bierregaard Jr, R. O. (eds.). University of Chicago Press, Chicago, IL, USA. Pp. 241–255.

Gorb, E., Kastner, V., Peressadko, A., Arzt, E., Gaume, L., Rowe, N. and Gorb, S. (2004) Structure and properties of the glandular surface in the digestive zone of the pitcher in the carnivorous plant *Nepenthes ventrata* and its role in insect trapping and retention. *Journal of Experimental Biology* **207**, 2947–2963.

Gordon, L. J., Steffen, W., Jonsson, B. F., Folke, C., Falkenmark, M. and Johannessen, A. (2005) Human modification of global water vapor flows from the land surface. *Proceedings of the National Academy of Sciences* **102**, 7612–7617.

Gordon, W. F., Vinsons, B., Newstrom, L. E., Barthell, J. F., Haber, W. A. and Frankie, J. K. (1990) Plant phenology, pollination ecology, pollinator behaviour and conservation of pollinators in Neotropical dry forest. In: *Reproductive ecology of tropical forest plants.* Bawa, K. S. and Hadley, M. (eds.). UNESCO, Paris, France. Pp. 37–48.

Gorelick, R. (2001) Did insect pollination cause increased seed plant diversity? *Biological Journal of the Linnean Society* **74**, 407–427.

Gottsberger, G. (1978) Seed dispersal by fish in inundated regions of Humaita, Amazonia. *Biotropica* **10**, 170–183.

Gotwald, W. (1995) *Army Ants: the Biology of Social Predation.* Cornell University Press, Ithaca, NY, USA. Pp. 302.

Gould, K.S. and Lee, D.W. (1996) Physical and ultrastructural basis of blue leaf iridescence in four malaysian understory plants. *American Journal of Botany,* **83**, 45–50.

Goulding, M. (1980) *The Fishes and the Forest: Explorations in Amazonian Natural History.* University of California Press, Berkeley, CA, USA. Pp. 250.

Gower, D. J. and Wilkinson, M. (2005) Conservation biology of caecilian amphibians. *Conservation Biology* **19**, 45–55.

Gradstein, S. R. and Pocs, T. (1989) Bryophytes. *Ecosystems of the World. 14B. Tropical forest Ecosystems. Biogeographical and Ecological Studies.* Lieth, H. and Werger, M. J. A. Elsevier, Amsterdam, Netherlands. Pp. 311–325.

Grafen, A. and Godfray, H. C. J. (1991) Vicarious selection explains some paradoxes in dioecious fig pollinator systems. *Proceedings of the Royal Society of London Series B* **245**, 73–76.

Graham, C. H., Moritz, C. and Williams, S. E. (2006) Habitat history improves prediction of biodiversity in rainforest fauna. *Proceedings of the National Academy of Sciences of the United States of America* **103**, 632–636.

Graham, E. A., Mulkey, S. S., Kitajima, K., Phillips, N. G. and Wright, S. J. (2003) Cloud cover limits net CO_2 uptake and growth of a rainforest tree during tropical rainy seasons. *Proceedings of the National Academy of Sciences of the United States of America*, **100**, 572–576.

Graham, R. M., Lee, D. W. and Norstog, K. (1993) Physical and ultrastructural basis of blue leaf iridescence in 2 neotropical ferns. *American Journal of Botany* **80**, 198–203.

Grainger, A. (2008) Difficulties in tracking the long-term global trend in tropical forest area. *Proceedings of the National Academy of Sciences of the United States of America* **105**, 818–823.

Grauel, W.T. and Putz, F.E. (2004) Effects of lianas on growth and regeneration of *Prioria copaifera* in Darien, Panama. *Forest Ecology and Management* **190**, 99–108.

Grayson, D. K. (2001) The Archaeological record of human impacts on animal populations. *Journal of World Prehistory* **15**, 1–68.

Green, R. E., Cornell, S. J., Scharlemann, J. P. W. and Balmford, A. (2005) Farming and the fate of wild nature. *Science* **307**, 550–555.

Greenberg, J. P., Guenther, A. B., Madronich, S., Baugh, W., Ginoux, P., Druilhet, A., Delmas, R. and Delon, C. (1999) Biogenic volatile organic compound emissions in central Africa during the Experiment for the Regional Sources and Sinks of Oxidants (EXPRESSO) biomass burning season. *Journal of Geophysical Research-Atmospheres* **104**, 30659–30671.

Greenberg, R. (1981) The abundance and seasonality of forest canopy birds on Barro Colorado Island, Panama. *Biotropica* **13**, 241–251.

Greenberg, R. (1995) Insectivorous migratory birds in tropical ecosystems: the Breeding Currency Hypothesis. *Journal of Avian Biology* **26**, 260–264.

Greene, S. (2004) Indigenous people incorporated? Culture as politics, culture as property in pharmaceutical bioprospecting. *Current Anthropology* **45**, 211–237.

Greig, N. (1993) Regeneration mode in neotropical *Piper*: habitat and species comparisons. *Ecology* **74**, 2125–2135.

Greig, N. (1993a) Pre-dispersal seed predation on five *Piper* species in tropical rain forest. *Oecologia* **93**, 412–420.

Greig, N. and Mauseth, J. D. (1991) Structure and function of dimorphic prop roots in *Piper auritum* L. *Bulletin of the Torrey Botanical Club* **118**, 176–183.

Grieg-Gran, M., Porras, I. and Wunder, S. (2005) How can market mechanisms for forest environmental services help the poor? Preliminary lessons from Latin America. *World Development* **33**, 1511–1527.

Griffiths, M. and van Schaik, C. P. (1993) The impact of human traffic on the abundance and activity periods of Sumatran rainforest wildlife. *Conservation Biology* **7**, 623–626.

Grifo, F., Newman, D., Fairfield, A. S., Bhattacharya, B. and Grupenhoff, J. T. (1997) The Origins of Presciption Drugs.

Biodiversity and Human Health. Grifo, F. and Rosenthal, J. Island Press, Washington, DC, USA. Pp. 131–163.

Grimaldi, C. and Pedro, G. (1996) Acidic hydrolysis as a soil formation mechanism in humid tropical regions. Role of the forest and consequences on the genesis of tropical white sands. *Comptes Rendus de l'Academie des Sciences Serie Ii Fascicule a Sciences de la Terre et des Planetes* **323**, 483–492.

Grimaldi, D. (1999) The co-radiations of pollinating insects and angiosperms in the Cretaceous. *Annals of the Missouri Botanical Garden* **86**, 373–406.

Griscom, B. W. and Ashton, P. M. S. (2003) Bamboo control of forest succession: *Guadua sarcocarpa* in Southeastern Peru. *Forest Ecology and Management* **175**, 445–454.

Griscom, B. W. and Ashton, P. M. S. (2006) A self-perpetuating bamboo disturbance cycle in a neotropical forest. *Journal of Tropical Ecology* **22**, 587–597.

Grison-Pige, L., Bessiere, J. M. and Hossaert-McKey, M. (2002) Specific attraction of fig-pollinating wasps: Role of volatile compounds released by tropical figs. *Journal of Chemical Ecology* **28**, 283–295.

Groombridge, B. and Jenkins, M. (1998) *Freshwater Biodiversity: A Preliminary Global Assessment*. World Conservation Press, Cambridge. WCMC Biodiversity Series No. 8. Available online from: http://www.unepwcmc.org/information_ services/publications/freshwater/toc.htm.

Grove, R. H. (1996) *Green Imperialism: Colonial Expansion, Tropical Island Edens and the Origins of Environmentalism, 1600–1860*. Cambridge University Press, Cambridge, UK. Pp. 560.

Grubb, P. J. (1977) Maintenance of species richness in plant communities: Importance of regeneration niche. *Biological Reviews of the Cambridge Philosophical Society* **52**, 107–145.

Grubb, P. J. (1998) A reassessment of the strategies of plants which cope with shortages of resources. *Perspectives in Plant Ecology, Evolution and Systematics*, **1**, 3–31.

Grubb, P. J., Jackson, R. V., Barberis, I. M., Bee, J. N., Coomes, D. A., Dominy, N. J., De la Fuente, M. A. S., Lucas, P. W., Metcalfe, D. J., Svenning, J. C., Turner, I. M. and Vargas, O. (2008) Monocot leaves are eaten less than dicot leaves in tropical lowland rain forests: Correlations with toughness and leaf presentation. *Annals of Botany* **101**, 1379–1389.

Grünmeier, R. (1990) Pollination by bats and non-flying mammals of the African tree *Parkia bicolor* (Mimosaceae). *Memoirs of the New York Botanical Garden* **55**, 83–104.

Guariguata, M. R. (1990) Landslide disturbance and forest regeneration in the upper Luquillo mountains of Puerto Rico. *Journal of Ecology* **78**, 814–832.

Guariguata, M. R. (1998) Response of forest tree saplings to experimental mechanical damage in lowland Panama. *Forest Ecology and Management* **102**, 103–111.

Guariguata, M. R. (2000) Seed and seedling ecology of tree species in neotropical secondary forests: Management implications. *Ecological Applications* **10**, 145–154.

Guariguata, M. R. and Ostertag, R. (2001) Neotropical secondary forest succession: changes in structural and functional characteristics. *Forest Ecology and Management* **148**, 185–206.

Guariguata, M. R., Chazdon, R. L., Denslow, J. S., Dupuy, J. M. and Anderson, L. (1997) Structure and floristics of secondary and old-growth forest stands in lowland Costa Rica. *Plant Ecology* **132**, 107–120.

Guenter, S., Stimm, B., Cabrera, M., Diaz, M. L., Lojan, M., Ordonez, E., Richter, M. and Weber, M. (2008) Tree phenology in montane forests of southern Ecuador can be explained by precipitation, radiation and photoperiodic control. *Journal of Tropical Ecology* **24**, 247–258.

Guenther, A. (2008) Are plant emissions green? *Nature* **452**, 701–702.

Guerrera, W., Sleeman, J. M., Jasper, S. B., Pace, L. B., Ichinose, T. Y. and Reif, J. S. (2003) Medical survey of the local human population to determine possible health risks to the mountain gorillas of Bwindi Impenetrable Forest National Park, Uganda. *International Journal of Primatology* **24**, 197–207.

Guevara, R. and Romero, I. (2004) Spatial and temporal abundance of mycelial mats in the soil of a tropical rain forest in Mexico and their effects on the concentration of mineral nutrients in soils and fine roots. *New Phytologist* **163**, 361–370.

Guha, R. (1989) *The Unquiet Woods: Ecological Change and Peasant Resistance in the Himalaya*. Oxford University Press, New Delhi, India. Pp. 244.

Gullison, R. E. (2003) Does forest certification conserve biodiversity? *Oryx* **37**, 153–165.

Gullison, R. E. and Losos, E. C. (1993) The role of foreign debt in deforestation in Latin America. *Conservation Biology* **7**, 140–147.

Gunnell, G. F. (1995) Omomyid primates (Tarsiiformes) from the Bridger Formation, Middle Eocene, Southern Green River Basin, Wyoming. *Journal of Human Evolution* **28**, 147–187.

Gupta, A. K. and Chivers, D. J. (1999) Biomass and use of resources in south and south-east Asian primate communities. In: *Primate Communities*. Fleagle, J. G., Janson, C. and Read, C. K. (eds.). Cambridge University Press, Cambridge, UK. Pp. 38–54.

Gustafsson, L., Nasi, R., Dennis, R., Nguyen Hoang Nghia, Sheil, D., Meijaard, E., Dykstra, D., Priyadi, H., Pham Quang Thu (2007) *Logging for the ark: Improving the conservation value of production forests in South East Asia*. CIFOR, Bogor, Indonesia. Pp. 74.

Guzman-Grajales, S. M. and Walker, L. R. (1991) Differential seedling responses to litter after Hurricane Hugo in the Luquillo experimental forest, Puerto Rico. *Biotropica* **23**, 407–413.

Gyasi, E., Agyepong, G. T., Ardayfio-Schandorf, E., Enu-Kwesi, L., Nabila, J. S. and Owusu-Bennoah, E. (1995) Production pressure and environmental change in the forest-savanna zone of Southern Ghana. *Global Environmental Change* **5**, 355–366.

Haber, W. A. (1984) Pollination by deceit in a mass-flowering tropical tree *Plumeria rubra* L. (Apocynaceae) *Biotropica* **16**, 269–275.

Hackel, J. D. (1999) Community conservation and the future of Africa's wildlife. *Conservation Biology* **13**, 726–734.

Haddad, C. F. B. and Prado, C. P. A. (2005) Reproductive modes in frogs and their unexpected diversity in the Atlantic forest of Brazil. *Bioscience* **55**, 207–217.

Hadfield, M. G. (1986) Extinction in Hawaiian achatinelline snails. *Malacologia* **27**, 67–81.

Haffer, J. (1969) Speciation in Amazonian forest birds. *Science* **165**, 131–137.

Haffer, J. (1974) *Avian speciation in tropical South America*. Nuttal Ornithological Club, Cambridge, UK. Pp. 390.

Haffer, J. (1997a) Alternative models of vertebrate speciation in Amazonia: An overview. *Biodiversity and Conservation* **6**, 451–476.

Haffer, J. (1997b) Contact zones between birds of southern Amazonia. *Ornithological Monographs* **48**, 281–305.

Haffer, J. and Prance, G. T. (2001) Climatic forcing of evolution in Amazonia during the Cenozoic: On the refuge theory of biotic differentiation. *Amazoniana-Limnologia Et Oecologia Regionalis Systemae Fluminis Amazonas* **16**, 579–605.

Haines, B. L. (1978) Element and energy flows through colonies of the leaf-cutting ant, *Atta colombica*, in Panama. *Biotropica* **10**, 270–277.

Hall, J. B. and Swaine, M. D. (1981) *Distribution and ecology of vascular plants in a tropical rain forest: forest vegetation in Ghana*. Junk, The Hague, Netherlands. Pp. 398.

Hall, J. P. W. (2005) Montane speciation patterns in *Ithomiola* butterflies (Lepidoptera: Riodinidae): are they consistently moving up in the world? *Proceedings of the Royal Society B* **272**, 2457–2466.

Hall, J. S., McKenna, J. J., Ashton, P. M. S. and Gregoire, T. G. (2004) Habitat characterizations underestimate the role of edaphic factors controlling the distribution of *Entandrophragma. Ecology* **85**, 2171–2183.

Hall, L. S. and Richards, G. C. (2000) *Flying foxes: fruit and blossom bats of Australia*. Krieger Publishing Company, Malabar, Australia. Pp. 160.

Hamer, K. C., Hill, J. K., Benedick, S., Mustaffa, N., Sherratt, T. N., Maryati, M. and Chey, V. K. (2003) Ecology of butterflies in natural and selectively logged forests of northern Borneo: The importance of habitat heterogeneity. *Journal of Applied Ecology* **40**, 150–162.

Hamer, K. C., Hill, J. K., Lace, L. A. and Langan, A. M. (1997) Ecological and biogeographical effects of forest disturbance on tropical butterflies of Sumba, Indonesia. *Journal of Biogeography* **24**, 67–75.

Hames, R. (2007) The ecologically noble savage debate. *Annual Review of Anthropology* **36**, 177–190.

Hamilton, A. C. (1974) Distribution patterns of forest trees in Uganda and their historical significance. *Vegetatio* **29**, 21–35.

Hamilton, A. C. (2004) Medicinal plants, conservation and livelihoods. *Biodiversity and Conservation* **13**, 1477–1517.

Hamilton, A. C. and Bensted-Smith, R. E. (1989) *Forest Conservation in the East Usambara Mountains, Tanzania.* IUCN, Gland. Switzerland. Pp. 392.

Hamilton, A. C. and Taylor, D. (1991) History of climate and forests in tropical Africa during the last 8 million years. *Climatic Change* **19**, 65–78.

Hammond, D. S. and Brown, V. K. (1995) Seed size of woody plants in relation to disturbance, dispersal, soil type in wet neotropical forests. *Ecology,* **76**, 2544–2561.

Hammond, D. S. and Brown, V. K. (1998) Disturbance, phenology and life-history characteristics: Factors influencing distance/density-dependent attack on tropical seeds and seedlings. In: *Dynamics of tropical communities.* Newbery, D. M., Prins, H. H. T. and Brown, N. (eds.). Blackwell Science, Oxford, UK. Pp. 51–78.

Hammond, D. S., Brown, V. K. and Zagt, R. (1999) Spatial and temporal patterns of seed attack and germination in a large-seeded neotropical tree species. *Oecologia* **119**, 208–218.

Hammond, D. S., Gond, V., de Thoisy, B., Forget, P. M. and DeDijn, B. P. E. (2007) Causes and consequences of a tropical forest gold rush in the Guiana Shield, South America. *Ambio* **36**, 661–670.

Hammond, P. M. (1990) Insect abundance and diversity in the Dumoga Bone National Park, N. Sulawesi, with special reference to the beetle fauna of lowland rain forest in the Toraut region. In: *Insects and the rain forests of South East Asia (Wallacea).* Knight, W. J. and Holloway, J. D. (eds.). The Royal Entomological Society, London, UK. Pp. 197–254.

Hamrick, J. L. and Murawski, D. A. (1990) The breeding structure of tropical tree populations. *Plant Species Biology* **5**, 157–165.

Hamrick, J. L., Murawski, D. A. and Nason, J. D. (1993) The influence of seed dispersal mechanisms on the genetic structure of tropical tree populations. *Vegetatio* **107/108**, 281–297.

Hansen, A. J. and Rotella, J. J. (2002) Biophysical factors, land use, and species viability in and around nature reserves. *Conservation Biology* **16**, 1112–1122.

Hansen, D. M., Beer, K. and Muller, C. B. (2006) Mauritian coloured nectar no longer a mystery: a visual signal for lizard pollinators. *Biology Letters* **2**, 165–168.

Hansen, D. M. and Galetti, M. (2009) The forgotten megafauna. *Science* **324**, 42–43.

Hansen, D. M. and Muller, C. B. (2009) Reproductive ecology of the endangered enigmatic Mauritian endemic *Roussea simplex* (Rousseaceae) *International Journal of Plant Sciences* **170**, 42–52.

Hansen, I., Brimer, L. and Molgaard, P. (2004) Herbivore-deterring secondary compounds in heterophyllous woody species of the Mascarene Islands. *Perspectives in Plant Ecology Evolution and Systematics* **6**, 187–203.

Hansen, J., Sato, M., Ruedy, R., Lo, K., Lea, D. W. and Medina-Elizade, M. (2006) Global temperature change. *Proceedings of the National Academy of Sciences of the United States of America* **103**, 14288–14293.

Hanson, T. R., Brunsfeld, S. J., Finegan, B. and Waits, L. P. (2008) Pollen dispersal and genetic structure of the tropical tree *Dipteryx panamensis* in a fragmented Costa Rican landscape. *Molecular Ecology* **17**, 2060–2073.

Harcourt, A. H. (1999) Biogeographic relationships of primates on South-East Asian islands. *Global Ecology and Biogeography* **8**, 55–61.

Harcourt, A. H. (2006) Rarity in the tropics: biogeography and macroecology of the primates. *Journal of Biogeography* **33**, 2077–2087.

Harcourt, A. H., Coppeto, S. A. and Parks, S. A. (2002) Rarity, specialization and extinction in primates. *Journal of Biogeography* **29**, 445–456.

Hardham, A. R. (2005) *Phytophthora cinnamomi. Molecular Plant Pathology* **6**, 589–604.

Hardin, G. (1968) The tragedy of the commons. *Science* **162**, 1243–1248.

Hardner, J., Frumhoff, P. and Goetze, D. (2000) Prospects for mitigating carbon, conserving biodiversity, and promoting socioeconomic development through the Clean Development Mechanism. *Mitigation and Adaptation Strategies for Global Change* **5**, 61–80.

Hardner, J. and Rice, R. (2002) Rethinking green consumerism. *Scientific American* **286**, 88–95.

Hardy, C. R. and Faden, R. B. (2004) *Plowmanianthus,* a new genus of Commelinaceae with five new species from tropical America. *Systematic Botany* **29**, 316–333.

Hardy, O. J., Maggia, L., Bandou, E., Breyne, P., Caron, H., Chevallier, M. H., Doligez, A., Dutech, C., Kremer, A., Latouche-Halle, C., Troispoux, V., Veron, V. and Degen, B.

(2006) Fine-scale genetic structure and gene dispersal inferences in 10 Neotropical tree species. *Molecular Ecology* **15**, 559–571.

Harms, K. E., Condit, R., Hubbell, S. P. and Foster, R. B. (2001) Habitat associations of trees and shrubs in a 50-ha neotropical forest plot. *Journal of Ecology* **89**, 947–959.

Harms, K. E. and Dalling, J. W. (2000) A bruchid beetle and a viable seedling from a single diaspore of *Attalea butyracea*. *Journal of Tropical Ecology* **16**, 319–325.

Harms, K. E., Powers, J. S. and Montgomery, R. A. (2004) Variation in small sapling density, understory cover, and resource availability in four neotropical forests. *Biotropica* **36**, 40–51.

Harms, K. E., Wright, S. J., Calderon, O., Hernandez, A. and Herre, E. A. (2000) Pervasive density-dependent recruitment enhances seedling diversity in a tropical forest. *Nature* **404**, 493–495.

Harper, D. B., Hamilton, J. T. G., Ducrocq, V., Kennedy, J. T., Downey, A. and Kalin, R. M. (2003) The distinctive isotopic signature of plant-derived chloromethane: Possible application in constraining the atmospheric chloromethane budget. *Chemosphere* **52**, 433–436.

Harper, K. N., Ocampo, P. S., Steiner, B. M., George, R. W., Silverman, M. S., Bolotin, S., Pillay, A., Saunders, N. J. and Armelagos, G. J. (2008) On the origin of the Treponematoses: A phylogenetic approach. *PLoS Neglected Tropical Diseases* **2**, 1.

Harrington, G. N., Freeman, A. N. D. and Crome, F. H. J. (2001) The effects of fragmentation of an Australian tropical rain forest on populations and assemblages of small mammals. *Journal of Tropical Ecology* **17**, 225–240.

Harrison, R. D. (2003) Fig wasp dispersal and the stability of a keystone plant resource in Borneo. *Proceedings of the Royal Society of London Series B* **270**, S76–S79.

Harrison, R. D. (2005) Figs and the diversity of tropical rainforests. *BioScience* **55**, 1053–1064.

Harrison, R. D. and Rasplus, J. Y. (2006) Dispersal of fig pollinators in Asian tropical rain forests. *Journal of Tropical Ecology* **22**, 631–639.

Harrison, R. D. and Yamamura, N. (2003) A few more hypotheses for the evolution of dioecy in figs (*Ficus*, Moraceae). *Oikos* **100**, 628–635.

Hart, J. A. (2000) Impact and sustainability of indigenous hunting in the Ituri Forest, Congo-Zaire: a comparison of unhunted and hunted duiker populations. In *Hunting for Sustainability in Tropical Forests*. Robinson, J. G. and Bennett, E. L. (eds.). Columbia University Press, New York, USA. Pp. 106–153.

Hart, T. B. and Hart, J. A. (1986) The ecological basis of hunter-gatherer subsistence in African rain forests: The Mbuti of eastern Zaire. *Human Ecology* **14**, 29–55.

Hart, T. B., Hart, J. A., Dechamps, R., Fournier, M. and Ataholo, M. (1994) *Changes in forest composition over the last 4000 years in the Ituri basin, Zaire.* XIVth Congress of the Association pour l'Etude Taxonomique de la Flore d'Afrique Tropicale (AETFAT), Wageningen, Netherlands.

Hart, T. B., Hart, J. A. and Murphy, P. G. (1989) Monodominant and species-rich forests of the humid tropics: Causes for their co-occurrence. *American Naturalist* **133**, 613–633.

Harvey, C. A., Komar, O., Chazdon, R., Ferguson, B. G., Finegan, B., Griffith, D. M., Martinez-Ramos, M., Morales, H., Nigh, R., Soto-Pinto, L., Van Breugel, M. and Wishnie, M. (2008) Integrating agricultural landscapes with biodiversity conservation in the Mesoamerican hotspot. *Conservation Biology* **22**, 8–15.

Hattenschwiler, S., Hagerman, A. E. and Vitousek, P. M. (2003) Polyphenols in litter from tropical montane forests across a wide range of soil fertility. *Biogeochemistry* **64**, 129–148.

Hattenschwiler, S. and Vitousek, P. M. (2000) The role of polyphenols in terrestrial ecosystem nutrient cycling. *Trends in Ecology and Evolution* **15**, 238–243.

Haugaasen, T. and Peres, C. A. (2005) Mammal assemblage structure in Amazonian flooded and unflooded forests. *Journal of Tropical Ecology* **21**, 133–145.

Hawkins, B. A. (2004) Are we making progress toward understanding the global diversity gradient? *Basic and Applied Ecology* **5**, 1–3.

Hawkins, B. A., Diniz, J. A. F., Jaramillo, C. A. and Soeller, S. A. (2006) Post-Eocene climate change, niche conservatism, and the latitudinal diversity gradient of New World birds. *Journal of Biogeography* **33**, 770–780.

Hawkins, B. A., Field, R., Cornell, H. V., Currie, D. J., Guegan, J. F., Kaufman, D. M., Kerr, J. T., Mittelbach, G. G., Oberdorff, T., O'Brien, E. M., Porter, E. E. and Turner, J. R. G. (2003) Energy, water, and broad-scale geographic patterns of species richness. *Ecology* **84**, 3105–3117.

Hawksworth, D. L. (1991) The fungal dimension of biodiversity: magnitude, significance, and conservation. *Mycological Research* **95**, 641–655.

Hawksworth, D. L. (2001) The magnitude of fungal diversity: The 1.5 million species estimate revisited. *Mycological Research* **105**, 1422–1432.

Hawksworth, D. L. (2004) Fungal diversity and its implications for genetic resource collections. *Studies in Mycology* **50**, 9–17.

Hawksworth, D. L. and Colwell, R. R. (1992) Microbial diversity 21: Biodiversity amongst microorganisms and its relevance. *Biodiversity and Conservation* **1**, 221–226.

Hawksworth, D. L. and Rossman, A. Y. (1997) Where are all the undescribed fungi? *Phytopathology* **87**, 888–891.

Hawthorne, W.D. (1995) *Ecological profiles of Ghanaian forest trees*. Oxford Forestry Institute, University of Oxford, Oxford, UK. Pp. 345.

Hawthorne, W.D. (1996) Holes and the sums of parts in Ghanaian forest: Regeneration, scale and sustainable use. *Proceedings of the Royal Society of Edinburgh* **104B**, 75-176.

Hawthorne, W. D. and Parren, M. P. E. (2000) How important are forest elephants to the survival of woody plant species in Upper Guinean forests? *Journal of Tropical Ecology* **16**, 133-150.

Hay, S. I., Guerra, C. A., Tatem, A. J., Noor, A. M. and Snow, R. W. (2004) The global distribution and population at risk of malaria: past, present and future. *The Lancet—Infectious Diseases* **4**, 327-336.

Hayes, F. E. and Sewlal, J. A. N. (2004) The Amazon River as a dispersal barrier to passerine birds: effects of river width, habitat and taxonomy. *Journal of Biogeography* **31**, 1809-1818.

Hazlett, D.L. (1987) Seasonal cambial activity for *Pentaclethra*, *Goelthalsia*, and *Carapa* trees in a Costa Rican lowland forest. *Biotropica* **19**, 357-360.

He, J., Chee, C. W. and Goh, C. J. (1996) 'Photoinhibition' of *Heliconia* under natural tropical conditions: The importance of leaf orientation for light interception and leaf temperature. *Plant Cell and Environment* **19**, 1238-1248.

He, X. H., C. Critchley and Bledsoe, C. (2003) Nitrogen transfer within and between plants through common mycorrhizal networks (CMNs). *Critical Reviews in Plant Sciences* **22**, 531-567.

Head, J. J., Bloch, J. I., Hastings, A. K., Bourque, J. R., Cadena, E. A., Herrera, F. A., Polly, P. D. and Jaramillo, C. A. (2009) Giant boid snake from the Palaeocene neotropics reveals hotter past equatorial temperatures. *Nature* **457**, 715-U4.

Headland, T. N. and Bailey, R. C. (1991) Introduction: Have hunter-gatherers ever lived in tropical rain forest independently of agriculture? *Human Ecology* **19**, 115-122.

Heads, M. (2006) Seed plants of Fiji: An ecological analysis. *Biological Journal of the Linnean Society* **89**, 407-431.

Heads, M. (2009) Globally basal centres of endemism: the Tasman-Coral Sea region (South-West Pacific), Latin America and Madagascar/South Africa. *Biological Journal of the Linnean Society* **96**, 222-245.

Heaney, L. R. (2001) Small mammal diversity along elevational gradients in the Philippines: An assessment of patterns and hypotheses. *Global Ecology and Biogeography* **10**, 15-39.

Hebant, C. and Lee, D.W. (1984) Ultrastructural basis and developmental control of blue iridescence in selaginella leaves. *American Journal of Botany*, **71**, 216-219.

Hecht, S. and Cockburn, A. (1989) *The Fate of the Forest*. Verso, New York, USA. Pp. 456.

Hecht, S. B. (1993) The logic of livestock and deforestation in Amazonia. *Bioscience* **43**, 687-695.

Hecht, S. B., Norgaard, R. B. and Possio, G. (1988) The economics of cattle ranching in eastern Amazonia. *Interciencia* **13**, 233-240.

Heckenberger, M. J., Kuikuro, A., Kuikuro, U. T., Russell, J. C., Schmidt, M., Fausto, C. and Franchetto, B. (2003) Amazonia 1492: Pristine forest or cultural parkland? *Science* **301**, 1710-1714.

Heckenberger, M. J., Peterson, J. B. and Neves, E. G. (1999) Village size and permanence in Amazonia: Two archaeological examples from Brazil. *Latin American Antiquity* **10**, 353-376.

Heckenberger, M. J., Russell, J. C., Toney, J. R. and Schmidt, M. J. (2007) The legacy of cultural landscapes in the Brazilian Amazon: implications for biodiversity. *Philosophical Transactions of the Royal Society B* **362**, 197-208.

Hegde, R. and Enters, T. (2000) Forest products and household economy: a case study from Mudumalai Wildlife Sanctuary, Southern India. *Environmental Conservation* **27**, 250-259.

Hejnowicz, Z. and Barthlott, W. (2005) Structural and mechanical peculiarities of the petioles of giant leaves of *Amorphophallus* (Araceae). *American Journal of Botany* **92**, 391-403.

Hellier, A., Newton, A. C. and Gaona, S. O. (1999) Use of indigenous knowledge for rapidly assessing trends in biodiversity: a case study from Chiapas, Mexico. *Biodiversity and Conservation* **8**, 869-889.

Helmer, E. H., Brandeis, T. J., Lugo, A. E. and Kennaway, T. (2008) Factors influencing spatial pattern in tropical forest clearance and stand age: Implications for carbon storage and species diversity. *Journal of Geophysical Research-Biogeosciences* **113**, G2.

Hemp, A. (2002) Ecology of the pteridophytes on the southern slopes of Mt. Kilimanjaro. I. Altitudinal distribution. *Plant Ecology* **159**, 211-239.

Henderson, A., Galeano, G. and R., B. (1995) *Field Guide to the Palms of the Americas*. Princeton University Press, Princeton, NJ, USA. Pp. 498.

Henkel, T. W., Mayor, J. R. and Woolley, L. P. (2005) Mast fruiting and seedling survival of the ectomycorrhizal, monodominant *Dicymbe corymbosa* (Caesalpiniaceae) in Guyana. *New Phytologist* **167**, 543-556.

Henry, O. (1999) Frugivory and the importance of seeds in the diet of the orange-rumped agouti (*Dasyprocta leporina*) in French Guiana. *Journal of Tropical Ecology* **15**, 291-300.

Henwood, A. A. (1993) Ecology and taphonomy of Dominican Republic amber and its inclusions. *Lethaia* **26**, 237-245.

Hernandez-Stefanoni, J. L., Pineda, J. B. and Valdes-Valadez, G. (2006) Comparing the use of indigenous knowledge with classification and ordination techniques for assessing the species composition and structure of vegetation in a tropical forest. *Environmental Management* **37**, 686–702.

Herrera, C. M. (2002) Seed dispersal by vertebrates. In: *Plant-Animal Interactions: An Evolutionary Approach.* Herrera, C. M. and Pellmyr, O. (eds.). Blackwell Science, Oxford, UK. Pp. 185–208.

Herrera, L. G. and Del Rio, C. M. (1998) Pollen digestion by New World bats: Effects of processing time and feeding habits. *Ecology* **79**, 2828–2838.

Hershkovitz, P. (1977) *Living New World Monkeys (Platyrrhini) with an Introduction to Primates.* University of Chicago Press, Chicago, IL, USA. Pp. 1132.

Herz, H., Beyschlag, W. and Holldobler, B. (2007) Herbivory rate of leaf-cutting ants in a tropical moist forest in Panama at the population and ecosystem scales. *Biotropica* **39**, 482–488.

Herzog, S. K., Kessler, M. and Bach, K. (2005) The elevational gradient in Andean bird species richness at the local scale: a foothill peak and a high-elevation plateau. *Ecography* **28**, 209–222.

Heydon, M. J. and Bulloh, P. (1996) The impact of selective logging on sympatric civet species in Borneo. *Oryx* **30**, 31–36.

Heydon, M. J. and Bulloh, P. (1997) Mousedeer densities in a tropical rainforest: The impact of selective logging. *Journal of Applied Ecology* **34**, 484–496.

Heying, H. E. (2001) Social and reproductive behaviour in the Madagascan poison frog, *Mantella laevigata*, with comparisons to the dendrobatids. *Animal Behaviour* **61**, 567–577.

Hikosaka, K. (2005) Leaf canopy as a dynamic system: Ecophysiology and optimality in leaf turnover. *Annals of Botany*, **95**, 521–533.

Hill, D. S. and Abang, F. (2005) *The Insects of Borneo (including South-East and East Asia).* University Malaysia, Kota Samarahan, Sarawak, Malaysia. Pp. 435.

Hill, J. K. (1999) Butterfly spatial distribution and habitat requirements in a tropical forest: impacts of selective logging. *Journal of Applied Ecology* **36**, 564–572.

Hill, J. K., Hamer, K. C., Lace, L. A. and Banham, W. M. T. (1995) Effects of selective logging on tropical forest butterflies on Buru, Indonesia. *Journal of Applied Ecology* **32**, 754–760.

Hill, K., McMillan, G. and Farina, R. (2003) Hunting-related changes in game encounter rates from 1994 to 2001 in the Mbaracayu Reserve, Paraguay. *Conservation Biology* **17**, 1312–1323.

Hill, K. and Padwe, J. (2000) Sustainability of Aché hunting in the Mbaracayu Reserve, Paraguay. In: *Hunting for Sustainability in Tropical Forests.* Robinson, J. G. and Bennett, E. L. (eds.). Columbia University Press, New York, USA. Pp. 79–105.

Hillebrand, H. (2004) On the generality of the latitudinal diversity gradient. *American Naturalist* **163**, 192–211.

Hilton-Taylor, C. (2002) *IUCN Red List of Threatened Species.* World Conservation Union, Gland, Switzerland.

Hobbie, S. E. and Vitousek, P. M. (2000) Nutrient limitation of decomposition in Hawaiian forests. *Ecology* **81**, 1867–1877.

Hobbs, J. J. (2004) Problems in the harvest of edible birds' nests in Sarawak and Sabah, Malaysian Borneo. *Biodiversity and Conservation* **13**, 2209–2226.

Hofer, U. and Bersier, L. F. (2001) Herpetofaunal diversity and abundance in tropical upland forests of Cameroon and Panama. *Biotropica* **33**, 142–152.

Hofstede, R. G. M., Wolf, J. H. D. and Benzing, D. H. (1993) Epiphytic biomass and nurient status of a Colombian upper montane rain forest. *Selbyana* **14**, 37–45.

Hoft, R. and Hoft, M. (1995) The differential effects of elephants on rain forest communities in the Shimba Hills, Kenya. *Biological Conservation* **73**, 67–79.

Holbrook, K. M. and Smith, T. B. (2000) Seed dispersal and movement patterns in two species of *Ceratogymna* hornbills in a West African tropical lowland forest. *Oecologia* **125**, 249–257.

Holbrook, K. M., Smith, T. B. and Hardesty, B. D. (2002) Implications of long-distance movements of frugivorous rain forest hornbills. *Ecography* **25**, 745–749.

Holbrook, N. M. and Putz, F. E. (1996a) From epiphyte to tree: Differences in leaf structure and leaf water relations associated with the transition in growth form in eight species of hemiepiphytes. *Plant Cell and Environment* **19**, 631–642.

Holbrook, N. M. and Putz, F. E. (1996b) Water relations of epiphytic and terrestrially rooted strangler figs in a Venezuelan palm savanna. *Oecologia* **106**, 424–431.

Holder, C. D. (2003) Fog precipitation in the Sierra de las Minas Biosphere Reserve, Guatemala. *Hydrological Processes* **17**, 2001–2010.

Holdridge, L. R. (1947) Determination of world plant formations from simple climatic data. *Science* **105**, 367–368.

Holldobler, B. and Wilson, E. O. (1990) *The Ants.* Springer Verlag, Berlin, Germany. Pp. 732.

Holmes, T. P., Blate, G. M., Zweede, J. C., Pereira Jr, R., Barreto, P., Boltz, F. and Bauch, R. (2000) *Financial Costs and Benefits of Reduced-Impact Logging Relative to Conventional Logging in the Eastern Amazon.* Tropical Forest Foundation, Washington, DC, USA. Pp. 48.

Holopainen, J. K. (2004) Multiple functions of inducible plant volatiles. *Trends in Plant Science* **9**, 529–533.

Holscher, D., Kohler, L., van Dijk, A. and Bruijnzeel, L. A. (2004) The importance of epiphytes to total rainfall interception by a tropical montane rain forest in Costa Rica. *Journal of Hydrology* **292**, 308–322.

Honigsbaum, M. (2001) *The Fever Trail: In Search of the Cure for Malaria*. Pan Macmillan, London, UK. Pp. 307.

Hood, L. A., Swaine, M. D. and Mason, P. A. (2004) The influence of spatial patterns of damping-off disease and arbuscular mycorrhizal colonization on tree seedling establishment in Ghanaian tropical forest soil. *Journal of Ecology* **92**, 816–823.

Hooper, E., Legendre, P. and Condit, R. (2005) Barriers to forest regeneration of deforested and abandoned land in Panama. *Journal of Applied Ecology* **42**, 1165–1174.

Hoorn, C. (2006) The birth of the mighty Amazon. *Scientific American* **295**, 52–59.

Hopkins, M. S. and Graham, A. W. (1984) Viable soil seed banks in disturbed lowland tropical rainforest sites in north Queensland. *Australian Journal of Ecology* **9**, 71–79.

Horner-Devine, M. C., Daily, G. C., Ehrlich, P. R. and Boggs, C. L. (2003) Countryside biogeography of tropical butterflies. *Conservation Biology* **17**, 168–177.

Houle, A. (1999) The origin of platyrrhines: An evaluation of the Antarctic scenario and the floating island model. *American Journal of Physical Anthropology* **109**, 541–559.

Houlton, B. Z., Sigman, D. M. and Hedin, L. O. (2006) Isotopic evidence for large gaseous nitrogen losses from tropical rainforests. *Proceedings of the National Academy of Sciences of the United States of America* **103**, 8745–8750.

House, S. M. (1993) Pollination success in a population of dioecious rain forest trees. *Oecologia* **96**, 555–561.

Houter, N. C. and Pons, T. L. (2005) Gap size effects on photoinhibition in understorey saplings in tropical rainforest. *Plant Ecology* **179**, 43–51.

Howard, P. C. (1991) *Nature Conservation in Uganda's Tropical Forest Reserves*. IUCN, Gland, Switzerland and Cambridge, UK.

Howe, H. F., Schupp, E. W. and Westley, L. C. (1985) Early consequences of seed dispersal for a Neotropical tree (*Virola surinamensis*). *Ecology* **66**, 781–791.

Hu, S., Dilcher, D. L., Jarzen, D. M. and Taylor, D. W. (2008) Early steps of angiosperm-pollinator coevolution. *Proceedings of the National Academy of Sciences of the United States of America* **105**, 240–245.

Huante, P., Rincon, E. and Chapin, F. S. (1995) Responses to phosphorus of contrasting successional tree-seedling species from the tropical deciduous forest of Mexico. *Functional Ecology* **9**, 760–766.

Hubbell, S. P. (1979) Tree dispersion, abundance, and diversity in a tropical dry forest. *Science* **203**, 1299–1309.

Hubbell, S. P. (2001) *The Unified Neutral Theory of Biodiversity and Biogeography*. Princeton University Press, Princeton, NJ, USA. Pp. 375.

Hubbell, S. P. and Foster, R. B. (1990) Structure, dynamics and equilibrium status of old-growth forest on Barro Colorado Island. In: *Four Neotropical Rain Forests*. Gentry, A. (ed.). Yale University Press, New Haven, CT, USA. Pp. 522–541.

Hubbell, S. P. and Foster, R. B. (1992) Short-term dynamics of a neotropical forest: why ecological research matters to tropical conservation and management. *Oikos* **63**, 48–61.

Hubbell, S. P., Foster, R. B., O'Brien, S. T., Harms, K. E., Condit, R., Wechsler, B., Wright, S. J. and de Lao, S. L. (1999) Light-gap disturbances, recruitment limitation, and tree diversity in a neotropical forest. *Science* **283**, 554–557.

Hubert, N. and Renno, J. F. (2006) Historical biogeography of South American freshwater fishes. *Journal of Biogeography* **33**, 1414–1436.

Hudson, W. H. (1904) *Green mansions: A romance of the tropical forest*. Duckworth, London, UK. Pp. 252.

Huete, A. R., Didan, K., Shimabukuro, Y. E., Ratana, P., Saleska, S. R., Hutyra, L. R., Yang, W. Z., Nemani, R. R. and Myneni, R. (2006) Amazon rainforests green-up with sunlight in dry season. *Geophysical Research Letters* **33**, 6405.

Hughes, J. M. and Baker, A. J. (1999) Phylogenetic relationships of the enigmatic hoatzin (*Opisthocomus hoazin*) resolved using mitochondrial and nuclear gene sequences. *Molecular Biology and Evolution* **16**, 1300–1307.

Hughes, W. O. H., Eilenberg, J. and Boomsma, J. J. (2002) Trade-offs in group living: transmission and disease resistance in leaf-cutting ants. *Proceedings of the Royal Society of London Series B* **269**, 1811–1819.

Hughes, W. O. H., Thomsen, L., Eilenberg, J. and Boomsma, J. J. (2004) Diversity of entomopathogenic fungi near leaf-cutting ant nests in a neotropical forest, with particular reference to *Metarhizium anisopliae* var *anisopliae*. *Journal of Invertebrate Pathology* **85**, 46–53.

Humphries, S. S. and Kainer, K. A. (2006) Local perceptions of forest certification for community-based enterprises. *Forest Ecology and Management* **235**, 30–43.

Hunt, T., Bergsten, J., Levkanicova, Z., Papadopoulou, A., John, O. S., Wild, R., Hammond, P. M., Ahrens, D., Balke, M., Caterino, M. S., Gomez-Zurita, J., Ribera, I., Barraclough, T. G., Bocakova, M., Bocak, L. and Vogler, A. P. (2007) A comprehensive phylogeny of beetles reveals the evolutionary origins of a superradiation. *Science* **318**, 1913–1916.

Hunt, T. L. (2007) Rethinking Easter Island's ecological catastrophe. *Journal of Archaeological Science* **34**, 485–502.

Hurles, M. E., Matisoo-Smith, E., Gray, R. D. and Penny, D. (2003) Untangling oceanic settlement: The edge of the knowable. *Trends in Ecology and Evolution* **18**, 531–540.

Hurtt, G. C. and Pacala, S. W. (1995) The consequences of recruitment limitation: Reconciling chance, history and competitive differences between plants. *Journal of Theoretical Biology* **176**, 1–12.

Huston, M. (1979) General hypothesis of species-diversity. *American Naturalist* **113**, 81-101.

Huston, M. (1980) Soil nutrients and tree species richness in Costa Rican forests. *Journal of Biogeography* **7**, 147-157.

Huston, M. (1994) *Biological Diversity: The Coexistence of Species on Changing Landscapes*. Cambridge University Press, Cambridge, UK. Pp. 681.

Hutto, R. L. (1980) Winter habitat distribution of migratory land birds in western Mexico, with special reference to small, foliage-gleaning insectivores. *Migrant Birds in the Neotropics: Ecology, Behavior, Distribution and Conservation*. Keast, A. and Morton, E. S. Washington, D.C., Smithsonian Institution Press.

Ichii, K., Hashimoto, H., Nemani, R. and White, M. (2005) Modeling the interannual variability and trends in gross and net primary productivity of tropical forests from 1982 to 1999. *Global and Planetary Change* **48**, 274-286.

ICRAF (2000) *Complex Agroforests*. International Centre for Research in Agroforestry, Bogor, Indonesia. Pp. 14.

ICRAF (2008) *Social Mobilization and Local Awareness of Rights and Opportunities for Environmental Services Market*. RUPES Synthesis Note. World Agroforestry Centre, Bogor, Indonesia. **2**.

Inderjit and Weiner, J. (2001) Plant allelochemical interference or soil chemical ecology? *Perspectives in Plant Ecology Evolution and Systematics* **4**, 3-12.

Inger, R. F. (1980a) Densities of floor-dwelling frogs and lizards in lowland forests of Southeast Asia and Central America. *American Naturalist* **115**, 761-770.

Inger, R. F. (1980b) Relative abundances of frogs and lizards in forests of Southeast Asia. *Biotropica* **12**, 14-22.

Intachat, J. and Holloway, J. D. (2000) Is there stratification in diversity or preferred flight height of geometroid moths in Malaysian lowland tropical forest? *Biodiversity and Conservation* **9**, 1417-1439.

International Ecotourism Society (2006) *Fact Sheet: Global Ecotourism*. The International Ecotourism Society, Washington, DC, USA. Pp. 6.

IPCC (2007) *Climate Change 2007: contribution of Working Group 1 to the fourth assessment report of the Intergovernmental Panel on Climate Change*. Cambridge University Press, Cambridge, UK.

Irion, G., Muller, J., deMello, J. N. and Junk, W. J. (1995) Quaternary geology of the Amazonian lowland. *Geo-Marine Letters* **15**, 172-178.

Irvine, A. and Armstrong, J. (1990) Beetle pollination in tropical forests of Australia. *Reproductive Ecology of Tropical Forest Plants*. Bawa, K. and Hadley, M. Carnforth, Parthenon and UNESCO. Pp. 135-148.

Ishida, A., Toma, T. and Marjenah (1999) Limitation of leaf carbon gain by stomatal and photochemical processes in the top canopy of macaranga conifera, a tropical pioneer tree. *Tree Physiology* **19**, 467-473.

Ishida, A., Yazaki, K. and Hoe, A.L. (2005) Ontogenetic transition of leaf physiology and anatomy from seedlings to mature trees of a rain forest pioneer tree, *Macaranga gigantea*. *Tree Physiology* **25**, 513-522.

Iskander, D. T. (1999) Amphibian declines monitoring in the Leuser Management Unit, Aceh, North Sumatra, Indonesia. *Froglog* **34**, 2.

Itioka, T., Inoue, T., Kaliang, H., Kato, M., Nagamitsu, T., Momose, K., Sakai, S., Yumoto, T., Mohamad, S. U., Hamid, A. A. and Yamane, S. (2001) Six-year population fluctuation of the giant honey bee *Apis dorsata* (Hymenoptera: Apidae) in a tropical lowland dipterocarp forest in Sarawak. *Annals of the Entomological Society of America* **94**, 545-549.

Itioka, T., Nomura, M., Inui, Y., Itino, T. and Inoue, T. (2000) Difference in intensity of ant defense among three species of *Macaranga myrmecophytes* in a southeast Asian dipterocarp forest. *Biotropica* **32**, 318-326.

Itioka, T. and Yamauti, M. (2004) Severe drought, leafing phenology, leaf damage and lepidopteran abundance in the canopy of a Bornean aseasonal tropical rain forest. *Journal of Tropical Ecology* **20**, 479-482.

ITTO (2005) *Status of Tropical Forest Management 2005*. International Tropical Timber Organization, Yokohama, Japan.

ITTO (2009) *Annual Review and Assessment of the World Timber Situation*. International Tropical Timber Organization, Yokohama, Japan.

Iturralde, R.B. (2001) The influence of ultramafic soils on plants in Cuba. *South African Journal of Science* **97**, 510-512.

IUCN/ITTO (2009) *Guidelines for the Conservation and Sustainable Use of Biodiversity in Tropical Timber Production Forests*. ITTO Policy Development Series No 17. The World Conservation Union (IUCN), Gland, Switzerland, and International Tropical Timber Organization, Yokohama, Japan. Pp. 120.

Iversen, S. T. (1991) The Usambara Mountains, Tanzania: Phytogeography of the vascular plant flora. *Acta Universitas Upsaliensis* **29**, 1-234.

Ivey, C. T. and DeSilva, N. (2001) A test of the function of drip tips. *Biotropica* **33**, 188-191.

Jablonski, D., Roy, K. and Valentine, J. W. (2006) Out of the tropics: Evolutionary dynamics of the latitudinal diversity gradient. *Science* **314**, 102-106.

Jackson, P.C., Cavelier, J., Goldstein, G., Meinzer, F.C. and Holbrook, N.M. (1995) Partitioning of water sources among plants of a lowland tropical forest. *Oecologia*, **101**, 197-203.

Jacobs, B. F. (2004) Palaeobotanical studies from tropical Africa: Relevance to the evolution of forest, woodland and savannah biomes. *Philosophical Transactions of the Royal Society of London Series B* **359**, 1573–1583.

Jago, L. C. F. and Boyd, W. E. (2005) How a wet tropical rainforest copes with repeated volcanic destruction. *Quaternary Research* **64**, 399–406.

Jans, L., Poorter, L., Vanrompaey, R. and Bongers, F. (1993) Gaps and forest zones in tropical moist forest in Ivory Coast. *Biotropica* **25**, 258–269.

Jansen, P. A., Van Der Meer, P. J. and Bongers, F. (2008) Spatial contagiousness of canopy disturbance in tropical rain forest: An individual-tree-based test. *Ecology* **89**, 3490–3502.

Jansen, S., Broadley, M.R., Robbrecht, E. and Smets, E. (2002) Aluminum hyperaccumulation in angiosperms: A review of its phylogenetic significance. *Botanical Review*, **68**, 235–269.

Janson, C. H. (1983) Adaptation of fruit morphology to dispersal agents in a neotropical forest. *Science* **219**, 187–189.

Janson, C. H., Terborgh, J. and Emmons, L. H. (1981) Non-flying mammals as pollinating agents in the Amazonian forest. *Biotropica* **13**, 1–6.

Jansson, R. (2003) Global patterns in endemism explained by past climatic change. *Proceedings of the Royal Society of London Series B* **270**, 583–590.

Janzen, D. H. (1967) Why mountain passes are higher in the tropics. *American Naturalist* **101**, 233–249.

Janzen, D. H. (1970) Herbivores and the number of tree species in tropical forests. *American Naturalist* **104**, 501–528.

Janzen, D. H. (1974a) Epiphytic myrmecophytes in Sarawak: Mutualism through the feeding of plants by ants. *Biotropica* **6**, 237–259.

Janzen, D. H. (1974b) Tropical blackwater rivers, animals, and mast fruiting by the Dipterocarpaceae. *Biotropica* **6**, 69–103.

Janzen, D. H. (1976) Depression of reptile biomass by large herbivores. *American Naturalist* **110**, 371–400.

Janzen, D. H. (1979) How to be a fig. *Annual Review of Ecology and Systematics* **10**, 13–51.

Janzen, D. H. (1983) No park is an island: Increase in interference from outside as park size decreases. *Oikos* **41**, 402–410.

Janzen, D. H. (1984) Dispersal of small seeds by big herbivores: Foliage is the fruit. *American Naturalist* **123**, 338–353.

Janzen, D. H. and Martin, P. S. (1982) Neotropical anachronisms: The fruits the gomphotheres ate. *Science* **215**, 19–27.

Janzen, D. H. (ed.) (1983) *Costa Rican Natural History*. University of Chicago Press, Chicago, IL, USA. Pp. 816.

Janzen, J. M. (1978) *The Quest for Therapy in Lower Zaire*. University of California Press, Berkeley, CA, USA. P. 266.

Jaramillo, C., Rueda, M. J. and Mora, G. (2006) Cenozoic plant diversity in the Neotropics. *Science* **311**, 1893–1896.

Jauhiainen, J., Takahashi, H., Heikkinen, J. E. P., Martikainen, P. J. and Vasander, H. (2005) Carbon fluxes from a tropical peat swamp forest floor. *Global Change Biology* **11**, 1788–1797.

Jenik, J. (1973) Root system of tropical trees. 8. Stilt-roots and allied adaptations. *Preslia* **45**, 250–264.

Jenik, J. (1978) Roots and root systems in tropical trees: morphologic and ecologic apsects. In *Tropical Trees as Living Systems*. Tomlinson, P. B. and Zimmermann, M. H. (eds.). Cambridge University Press, Cambridge, U.K. Pp. 323–349.

Jenkins, C. N. and Pimm, S. L. (2003) How big is the global weed patch? *Annals of the Missouri Botanical Garden* **90**, 172–178.

Jennings, S. B., Brown, N. D. and Sheil, D. (1999) Assessing forest canopies and understorey illumination: canopy closure, canopy cover and other measures. *Forestry* **72**, 59–73.

Jetz, W., Rahbek, C. and Colwell, R. K. (2004) The coincidence of rarity and richness and the potential signature of history in centres of endemism. *Ecology Letters* **7**, 1180–1191.

Ji, R. and Brune, A. (2006) Nitrogen mineralization, ammonia accumulation, and emission of gaseous NH_3 by soil-feeding termites. *Biogeochemistry* **78**, 267–283.

Jobbagy, E. G. and Jackson, R. B. (2001) The distribution of soil nutrients with depth: Global patterns and the imprint of plants. *Biogeochemistry* **53**, 51–77.

John, R., Dalling, J. W., Harms, K. E., Yavitt, J. B., Stallard, R. F., Mirabello, M., Hubbell, S. P., Valencia, R., Navarrete, H., Vallejo, M. and Foster, R. B. (2007) Soil nutrients influence spatial distributions of tropical tree species. *Proceedings of the National Academy of Sciences of the United States of America* **104**, 864–869.

Johns, A. D. (1986) Effects of selective logging on the behavioral ecology of West Malaysian primates. *Ecology* **67**, 684–694.

Johns, A. D. and Skorupa, J. P. (1987) Responses of rain-forest primates to habitat disturbance: a review. *International Journal of Primatology* **8**, 157–191.

Johns, A. G. (1997) *Timber production and biodiversty conservation in tropical rainforests*. Cambridge University Press, Cambridge, UK.

Johns, N. D. (1999) Conservation in Brazil's chocolate forest: The unlikely persistence of the traditional cocoa agroecosystem. *Environmental Management* **23**, 31–47.

Johns, R. J. (1986) The instability of the tropical ecosystem in New Guinea. *Blumea* **31**, 341–371.

Johns, R. J. (1989) The influence of drought on tropical rainforest vegetation in Papua New Guinea. *Mountain Research and Development* **9**, 248–251.

Johnson, D. E. and Group, I. P. S. (1996) *Palms: their conservation and sustained utilization*. IUCN, Gland, Switzerland and Cambridge, UK. Pp. 116.

Johnson, K. R. and Ellis, B. (2002) A tropical rainforest in Colorado 1.4 million years after the Cretaceous-Tertiary boundary. *Science* **296**, 2379–2383.

Johnson, M. D., Sherry, T. W., Strong, A. M. and Medori, A. (2005) Migrants in Neotropical bird communities: An assessment of the breeding currency hypothesis. *Journal of Animal Ecology* **74**, 333–341.

Jones, E. W. (1956) Ecological studies on the rainforest of southern Nigeria. IV. The plateau forest of the Okomu forest reserve. *Journal of Ecology* **44**, 83–117.

Jones, M. J., Sullivan, M. S., Marsden, S. J. and Linsley, M. D. (2001) Correlates of extinction risk of birds from two Indonesian islands. *Biological Journal of the Linnean Society* **73**, 65–79.

Jones, M. M., Tuomisto, H., Clark, D. B. and Olivas, P. (2006) Effects of mesoscale environmental heterogeneity and dispersal limitation on floristic variation in rain forest ferns. *Journal of Ecology* **94**, 181–195.

Jones, T., Ehardt, C. L., Butynski, T. M., Davenport, T. R. B., Mpunga, N. E., Machaga, S. J. and De Luca, D. W. (2005) The highland mangabey *Lophocebus kipunji*: a new species of African monkey. *Science* **308**, 1161–1164.

Jordano, P. (1992) Fruits and fugivory. *Seeds: The Ecology of Regeneration in Plant Communities*. Fenner, M. Wallingford (eds.). CAB International, Wallingford, UK. Pp. 105–156.

Jordano, P. (1995) Angiosperm fleshy fruits and seed dispersers: a comparative analysis of adaptation and constraints in plant-animal interactions. *American Naturalist* **145**, 163–191.

Jørgensen, P. M. and León-Yánez, S., Eds. (1999) *Catalogue of the Vascular Plants of Ecuador*. Monographs in Systematic Botany of the Missouri Botanical Gardens. Pp. 1181.

Joshi, L., Wibawa, G., Vincent, G., Boutin, D., Akiefnawati, R., Manurung, G., van Noordwijk, M. and Williams, S. (2002) *Jungle Rubber: a traditional agroforestry system under pressure*. ICRAF: World Agroforestry Centre, Bogor, Indonesia. Pp. 44.

Judziewicz, E. J., Clark, L. G., Londono, X. and Stern, M. J. (1999) *American Bamboos*. Smithsonian Institution Press, Washington DC, USA. Pp. 392.

Jullien, M. and Thiollay, J. M. (1998) Multi-species territoriality and dynamic of neotropical forest understorey bird flocks. *Journal of Animal Ecology* **67**, 227–252.

Junk, W. J. (1989) The use of Amazonian floodplains under an ecological perspective. *Interciencia* **14**, 317–322.

Kabakoff, R. P. and Chazdon, R. L. (1996) Effects of canopy species dominance on understorey light availability in low-elevation secondary forest stands in Costa Rica. *Journal of Tropical Ecology* **12**, 779–788.

Kaimowitz, D. (2004) Forests and water: a policy perspective. *Journal of Forest Research* **9**, 289–291.

Kaimowitz, D. and Angelsen, A. (1998) *Economic models of tropical deforestation: a review*. CIFOR, Bogor, Indonesia. Pp.153.

Kaimowitz, D. and Sheil, D. (2007) Conserving what and for whom? Why conservation should help meet basic human needs in the tropics. *Biotropica* **39**, 567–574.

Kaimowitz, D. and Smith, J. (2001) Soybean technology and the loss of natural vegetation in Brazil and Bolivia. *Agricultural technologies and tropical deforestation*. Angelsen, A. and Kaimowitz, D. (eds.). CAB International, Wallingford, Oxford, UK. Pp. 195–211.

Kainer, K. A., Schmink, M., Leite, A. C. P. and Fadell, M. J. D. (2003) Experiments in forest-based development in Western Amazonia. *Society and Natural Resources* **16**, 869–886.

Kainer, K. A., Wadt, L. H. O., Gomes-Silva, D. A. P. and Capanu, M. (2006) Liana loads and their association with *Bertholletial excelsa* fruit and nut production, diameter growth and crown attributes. *Journal of Tropical Ecology* **22**, 147–154.

Kalema-Zikusoka, G., Rothman, J. M. and Fox, M. T. (2005) Intestinal parasites and bacteria of mountain gorillas (*Gorilla beringei beringei*) in Bwindi Impenetrable National Park, Uganda. *Primates* **46**, 59–63.

Kalko, E. K. V. (1998) Organisation and diversity of tropical bat communities through space and time. *Zoology* **101**, 281–297.

Kalko, E. K. V., Herre, E. A. and Handley, C. O. (1996) Relation of fig fruit characteristics to fruit-eating bats in the New and Old World tropics. *Journal of Biogeography* **23**, 565–576.

Kammesheidt, L. (1998) The role of tree sprouts in the restorations of stand structure and species diversity in tropical moist forest after slash-and-burn agriculture in Eastern Paraguay. *Plant Ecology* **139**, 155–165.

Kapan, D. D. (2001) Three-butterfly system provides a field test of mullerian mimicry. *Nature* **409**, 338–340.

Kaplin, B. A. and Lambert, J. E. (2002) Effectiveness of seed dispersal by *Cercopithecus* monkeys: Implications for seed input into degraded areas. In: *Seed Dispersal and Frugivory: Ecology, Evolution and Conservation*. Levey, D. J., Silva, W. R. and Galetti, M. (eds.). CAB International, Wallingford, UK. Pp. 351–364.

Kapos, V. (1989) Effects of isolation on the water status of forest patches in the Brazilian Amazon. *Journal of Tropical Ecology* **5**, 173–185.

Kappelman, J., Rasmussen, D. T., Sanders, W. J., Feseha, M., Bown, T., Copeland, P., Crabaugh, J., Fleagle, J., Glantz, M., Gordon, A., Jacobs, B., Maga, M., Muldoon, K., Pan, A., Pyne, L., Richmond, B., Ryan, T., Seiffert, E. R., Sen, S., Todd, L., Wiemann, M. C. and Winkler, A. (2003) Oligocene mammals from Ethiopia and faunal exchange between Afro-Arabia and Eurasia. *Nature* **426**, 549–552.

Karl, T., Potosnak, M., Guenther, A., Clark, D., Walker, J., Herrick, J. D. and Geron, C. (2004) Exchange processes of volatile organic compounds above a tropical rain forest: Implications for modeling tropospheric chemistry above dense vegetation. *Journal of Geophysical Research-Atmospheres* **109**, D18306.

Karr, J. R. (1971) Structure of avian communies in selected Panama and Illinois habitats. *Ecological Monographs* **41**, 207–233.

Karr, J. R. (1980) Geographical variation in the avifaunas of tropical forest undergrowth. *Auk* **97**, 283–298.

Karr, J. R. and Roth, R. R. (1971) Vegetation structure and avian diversity in several New World areas. *American Naturalist* **105**, 423–435.

Karsenty, A. (2007a) *The architecture of proposed REDD schemes after Bali: facing critical choices.* International Workshop on the International Regime: Avoided Deforestation and the Evolution of Public and Private Policies Towards Forests in Developing Countries, Paris, France.

Karsenty, A. (2007b) Questioning rent for development swaps: New market-based instruments for biodiversity acquisition and the land-use issue in tropical countries. *International Forestry Review* **9**, 503–513.

Karsenty, A. (2008) REDD and the evolution of an international forest regime. *International Forestry Review* **10**, 423–423.

Kartawinata, K., Adisoemarto, S., Riswan, S. and Vayda, A. P. (1981) The impact of man on a tropical forest in Indonesia. *Ambio* **10**, 115–119.

Kay, K. M. and Schemske, D. W. (2003) Pollinator assemblages and visitation rates for 11 species of neotropical *Costus* (Costaceae) *Biotropica* **35**, 198–207.

Kay, R. F., Madden, R. H., VanSchaik, C. and Higdon, D. (1997) Primate species richness is determined by plant productivity: Implications for conservation. *Proceedings of the National Academy of Sciences of the United States of America* **94**, 13023–13027.

Kays, R. and Allison, A. (2001) Arboreal tropical forest vertebrates: current knowledge and research trends. *Plant Ecology* **153**, 109–120.

Keller, R. (1996) *Identification of tropical woody plants in the absence of flowers and fruits: a field guide.* Birkhäuser, Basel, Switzerland. Pp. 229.

Kelly, D. L., Tanner, E. V. J., Lughadha, E. M. N. and Kapos, V. (1994) Floristics and Biogeography of a Rain-Forest in the Venezuelan Andes. *Journal of Biogeography* **21**, 421–440.

Kelt, D. A. and Van Vuren, D. H. (2001) The ecology and macroecology of mammalian home range area. *American Naturalist* **157**, 637–645.

Kennedy, D. N. and Swaine, M. D. (1992) Germination and growth of colonizing species in artificial gaps of different sizes in dipterocarp rain forest. *Philosophical Transactions of the Royal Society of London, Series B* **335**, 357–366.

Kershaw, P. and Wagstaff, B. (2001) The southern conifer family Araucariaceae: History, status, and value for paleoenvironmental reconstruction. *Annual Review of Ecology and Systematics* **32**, 397–414.

Kesselmeier, J., Ciccioli, P., Kuhn, U., Stefani, P., Biesenthal, T., Rottenberger, S., Wolf, A., Vitullo, M., Valentini, R., Nobre, A., Kabat, P. and Andreae, M. O. (2002) Volatile organic compound emissions in relation to plant carbon fixation and the terrestrial carbon budget. *Global Biogeochemical Cycles* **16**, 1126.

Kessler, M. (2000) Elevational gradients in species richness and endemism of selected plant groups in the central Bolivian Andes. *Plant Ecology* **149**, 181–193.

Kessler, M. (2001) Patterns of diversity and range size of selected plant groups along an elevational transect in the Bolivian Andes. *Biodiversity and Conservation* **10**, 1897–1921.

Kessler, M. (2002) The elevational gradient of Andean plant endemism: varying influences of taxon-specific traits and topography at different taxonomic levels. *Journal of Biogeography* **29**, 1159–1165.

Khumbongmayum, A. D., Khan, M. L. and Tripathi, R. S. (2005) Sacred groves of Manipur, northeast India: biodiversity value, status and strategies for their conservation. *Biodiversity and Conservation* **14**, 1541–1582.

Kiers, E. T., Lovelock, C. E., Krueger, E. L. and Herre, E. A. (2000) Differential effects of tropical arbuscular mycorrhizal fungal inocula on root colonization and tree seedling growth: Implications for tropical forest diversity. *Ecology Letters* **3**, 106–113.

Kiesecker, J. M., Blaustein, A. R. and Belden, L. K. (2001) Complex causes of amphibian population declines. *Nature* **410**, 681–684.

Killmann, W., Bull, G. Q., Schwab, O. and Pulkki, R. E. (2001) *Reduced impact logging: does it cost or does it pay?* Applying Reduced Impact Logging to Advance Sustainable Forest Management, Kuching, Malaysia, Food and Agriculture Organization of the United Nations, Regional Office for Asia and the Pacific.

King, D. A. and Clark, D. B. (2004) Inferring growth rates from leaf display in tropical forest saplings. *Journal of Tropical Ecology*, **20**, 351–354.

King, D. A., Davies, S. J. and Noor, N. S. M. (2006) Growth and mortality are related to adult tree size in a Malaysian mixed dipterocarp forest. *Forest Ecology and Management*, **223**, 152–158.

King, D. A., Davies, S. J., Supardi, M. N. N. and Tan, S. (2005) Tree growth is related to light interception and wood density in two mixed dipterocarp forests of Malaysia. *Functional Ecology*, **19**, 445–453.

King, D. A. and Maindonald, J. H. (1999) Tree architecture in relation to leaf dimensions and tree stature in temperate and tropical rain forests. *Journal of Ecology*, **87**, 1012–1024.

Kingston, T., Francis, C. M., Akbar, Z. and Kunz, T. H. (2003) Species richness in an insectivorous bat assemblage from Malaysia. *Journal of Tropical Ecology* **19**, 67–79.

Kinnaird, M. F. and O'Brien, T. G. (1998) Ecological effects of wildfire on lowland rainforest in Sumatra. *Conservation Biology* **12**, 954–956.

Kinsman, S. (1990) Regeneration by fragmentation in tropical montane forest shrubs. *American Journal of Botany* **77**, 1626–1633.

Kiss, A. (2004) Is community-based ecotourism a good use of biodiversity conservation funds? *Trends in Ecology and Evolution* **19**, 232–237.

Kitajima, K. (1994) Relative importance of photosynthetic traits and allocation patterns as correlates of seedling shade tolerance of 13 tropical trees. *Oecologia*, **98**, 419–428.

Kitajima, K. and Hogan, K.P. (2003) Increases of chlorophyll a/b ratios during acclimation of tropical woody seedlings to nitrogen limitation and high light. *Plant Cell and Environment*, **26**, 857–865.

Kitajima, K., Mulkey, S. S. and Wright, S. J. (2005) Variation in crown light utilization characteristics among tropical canopy trees. *Annals of Botany* **95**, 535–547.

Klein, B. C. (1989) Effects of forest fragmentation on dung and carrion beetle communities in central Amazonia. *Ecology* **70**, 1715–1725.

Klinger, L. F., Greenberg, J., Guenther, A., Tyndall, G., Zimmerman, P., M'Bangui, M. and Moutsambote, J. M. (1998) Patterns in volatile organic compound emissions along a savanna-rainforest gradient in central Africa. *Journal of Geophysical Research-Atmospheres* **103**, 1443–1454.

Klooster, D. (1999) Community-based forestry in Mexico: Can it reverse processes of degradation? *Land Degradation and Development* **10**, 365–381.

Klooster, D. (2000) Community forestry and tree theft in Mexico: Resistance or complicity in conservation? *Development and Change* **31**, 281–305.

Knapp, S. and Mallet, J. (2003) Refuting refugia? *Science* **300**, 71–72.

Koch, P. L. and Barnosky, A. D. (2006) Late quaternary extinctions: State of the debate. *Annual Review of Ecology Evolution and Systematics* **37**, 215–250.

Koh, L. P. (2007) Potential habitat and biodiversity losses from intensified biodiesel feedstock production. *Conservation Biology* **21**, 1373–1375.

Koh, L. P., Dunn, R. R., Sodhi, N. S., Colwell, R. K., Proctor, H. C. and Smith, V. S. (2004a) Species coextinctions and the biodiversity crisis. *Science* **305**, 1632–1634.

Koh, L. P. and Ghazoul, J. (2010) Accounting for the matrix in predicting biodiversity losses due to land use change. *Conservation Biology*, **in press**.

Koh, L. P. and Ghazoul, J. (2008) Biofuels, biodiversity and people: understanding the conflicts and finding opportunities. *Biological Conservation* **141**, 2450–2460.

Koh, L. P., Levang, P. and Ghazoul, J. (2009) Designer landscapes for sustainable biofuels. *Trends in Ecology and Evolution* **24**, 431–438.

Koh, L. P., Sodhi, N. S. and Brook, B. W. (2004b) Co-extinctions of tropical butterflies and their hostplants. *Biotropica* **36**, 272–274.

Koh, L. P., Sodhi, N. S. and Brook, B. W. (2004c) Ecological correlates of extinction proneness in tropical butterflies. *Conservation Biology* **18**, 1571–1578.

Kohyama, T., Suzuki, E., Partomihardjo, T., Yamada, T. and Kubo, T. (2003) Tree species differentiation in growth, recruitment and allometry in relation to maximum height in a Bornean mixed dipterocarp forest. *Journal of Ecology*, **91**, 797–806.

Kokou, K., Couteron, P., Martin, A. and Caballe, G. (2002) Taxonomic diversity of lianas and vines in forest fragments of southern Togo. *Revue d'Ecologie—La Terre et la Vie* **57**, 3–18.

Komarudin, H., Siagian, Y. and Colfer, C. P. (2008) *Collective action to secure property rights for the poor: A case study in Jambi province, Indonesia.* CAPRi Working Paper, Centre for International Forestry Research, Bogor, Indonesia. **90**. Pp. 49.

Kondgen, S., Kuhl, H., N'Goran, P. K., Walsh, P. D., Schenk, S., Ernst, N., Biek, R., Formenty, P., Maetz-Rensing, K., Schweiger, B., Junglen, S., Ellerbrok, H., Nitsche, A., Briese, T., Lipkin, W. I., Pauli, G., Boesch, C. and Leendertz, F. H. (2008) Pandemic human viruses cause decline of endangered great apes. *Current Biology* **18**, 260–264.

Koniger, M., Harris, G.C., Virgo, A. and Winter, K. (1995) Xanthophyll-cycle pigments and photosynthetic capacity in tropical forest species: A comparative field-study on canopy, gap and understory plants. *Oecologia*, **104**, 280–290.

Koptur, S. and Lee, M. A. B. (1993) Plantlet formation in tropical montane ferns: a preliminary investigation. *American Fern Journal* **83**, 60–66.

Kottelat, M. and Lim, K. K. P. (1994) Diagnoses of two new genera and three new species of earthworm eels from the Malay Peninsula and Borneo (Teleostei: Chaudhuriidae). *Ichthyological Exploration of Freshwaters* **5**, 181–190.

Kottelat, M. and Whitten, A. J. (1996) *Freshwater biodiversity in Asia with special reference to fish.* World Bank Technical Paper 343. World Bank, Washington, DC, USA. Pp. 59.

Kottelat, M., Whitten, A., Kartikasari, S. and Wirjoatmodjo, S. (1993) *Freshwater fishes of western Indonesia and Sulawesi*. Periplus Editions Ltd, Hong Kong. P. 344.

Kraus, T. E. C., Dahlgren, R. A. and Zasoski, R. J. (2003) Tannins in nutrient dynamics of forest ecosystems. *Plant and Soil* **256**, 41–66.

Krause, G.H., Grube, E., Koroleva, O.Y., Barth, C. and Winter, K. (2004) Do mature shade leaves of tropical tree seedlings acclimate to high sunlight and UV radiation? *Functional Plant Biology*, **31**, 743–756.

Krause, G.H. and Winter, K. (1996) Photoinhibition of photosynthesis in plants growing in natural tropical forest gaps a chlorophyll fluorescence study. *Botanica Acta*, **109**, 456–462.

Krauss, M. (1992) The world's languages in crisis. *Language* **68**, 4–10.

Kress, W. and Beach, J. (1994) Flowering plant reproductive systems. In: *La Selva: Ecology and Natural History of a Tropical Rainforest*. Mcdade, L., Bawa, K., Hespenheide, H. and Hartshorn, G. S. (eds.). University of Chicago Press, Chicago, IL, USA. Pp. 161–176.

Kress, W. J. (1986) A symposium: The biology of tropical epiphytes. *Selbyana* **9**, 1–22.

Kress, W. J. (1993) Coevolution of plants and animals: Pollination of flowers by primates in Madagascar. *Current Science* **65**, 253–257.

Kress, W. J., Schatz, G. E., Andrianifahanana, M. and Morland, H. S. (1994) Pollination of *Ravenala madagascariensis* (Strelitziaceae) by lemurs in Madagascar: Evidence for an archaic coevolutionary system. *American Journal of Botany* **81**, 542–551.

Kriger, K. M. and Hero, J. M. (2007) The chytrid fungus *Batrachochytrium dendrobatidis* is non-randomly distributed across amphibian breeding habitats. *Diversity and Distributions* **13**, 781–788.

Kriger, K. M., Pereoglou, F. and Hero, J. M. (2007) Latitudinal variation in the prevalence and intensity of chytrid (*Batrachochytrium dendrobatidis*) infection in Eastern Australia. *Conservation Biology* **21**, 1280–1290.

Krijger, C. L., Opdam, M., Thery, M. and Bongers, F. (1997) Courtship behaviour of manakins and seed bank composition in a French Guianan rain forest. *Journal of Tropical Ecology* **13**, 631–636.

Krings, M., Kerp, H., Taylor, T. N. and Taylor, E. L. (2003) How Paleozoic vines and lianas got off the ground: On scrambling and climbing Carboniferous-early Permian pteridosperms. *Botanical Review* **69**, 204–224.

Krishnamani, R. and Mahaney, W. C. (2000) Geophagy among primates: adaptive significance and ecological consequences. *Animal Behaviour* **59**, 899–915.

Kromer, T., Kessler, M., Gradstein, S. R. and Acebey, A. (2005) Diversity patterns of vascular epiphytes along an elevational gradient in the Andes. *Journal of Biogeography* **32**, 1799–1809.

Kruijt, B., Malhi, Y., Lloyd, J., Norbre, A. D., Miranda, A. C., Pereira, M. G. P., Culf, A. and Grace, J. (2000) Turbulence statistics above and within two Amazon rain forest canopies. *Boundary-Layer Meteorology* **94**, 297–331.

Kubitzki, K. and Ziburski, A. (1994) Seed dispersal in floodplain forests of Amazonia. *Biotropica* **26**, 30–43.

Kuhn, U., Rottenberger, S., Biesenthal, T., Wolf, A., Schebeske, G., Ciccioli, P., Brancaleoni, E., Frattoni, M., Tavares, T. M. and Kesselmeier, J. (2004) Seasonal differences in isoprene and light-dependent monoterpene emission by Amazonian tree species. *Global Change Biology* **10**, 663–682.

Kummer, D. M. and Turner, B. L. (1994) The human causes of deforestation in Southeast Asia. *Bioscience* **44**, 323–328.

Kupfer, A., Nabhitabhata, J. and Himstedt, W. (2004) Reproductive ecology of female caecilian amphibians (genus *Ichthyophis*): a baseline study. *Biological Journal of the Linnean Society* **83**, 207–217.

Kuppers, M., Timm, H., Orth, F., Stegemann, J., Stober, R., Schneider, H., Paliwal, K., Karunaichamy, K. and Ortiz, R. (1996) Effects of light environment and successional status on lightfleck use by understory trees of temperate and tropical forests. *Tree Physiology*, **16**, 69–80.

Kurokawa, H., Yoshida, T., Nakamura, T., Lai, J.H. and Nakashizuka, T. (2003) The age of tropical rain-forest canopy species, Borneo ironwood (*Eusideroxylon zwageri*), determined by C14 dating. *Journal of Tropical Ecology*, **19**, 1–7.

Kursar, T. A., Caballero-George, C. C., Capson, T. L., Cubilla-Rios, L., Gerwick, W. H., Gupta, M. P., Ibanez, A., Linington, R. G., McPhail, K. L., Ortega-Barria, E., Romero, L. I., Solis, P. N. and Coley, P. D. (2006) Securing economic benefits and promoting conservation through bioprospecting. *Bioscience* **56**, 1005–1012.

Kursar, T. A. and Coley, P. D. (1992) Delayed greening in tropical leaves: an antiherbivore defense. *Biotropica* **24**, 256–262.

Kursar, T.A. and Coley, P. D. (1993) Photosynthetic induction times in shade-tolerant species with long and short-lived leaves. *Oecologia*, **93**, 165–170.

Kursar, T.A. and Coley, P. D. (1999) Contrasting modes of light acclimation in two species of the rainforest understory. *Oecologia* **121**, 489–498.

Kursar, T. A. and Coley, P. D. (2003) Convergence in defense syndromes of young leaves in tropical rainforests. *Biochemical Systematics and Ecology* **31**, 929–949.

Kusters, K., de Foresta, H., Ekadinata, A. and van Noordwijk, M. (2007) Towards solutions for state vs. local community conflicts over forestland: The impact of formal recognition of user rights in Krui, Sumatra, Indonesia. *Human Ecology* **35**, 427–438.

Kyereh, B., Swaine, M.D. and Thompson, J. (1999) Effect of light on the germination of forest trees in ghana. *Journal of Ecology,* **87**, 772–783.

Ladiges, P. Y. and Cantrill, D. (2007) New Caledonia–Australian connections: biogeographic patterns and geology. *Australian Systematic Botany* **20**, 383–389.

LaFrankie, J. V., Ashton, P. S., Chuyong, G. B., Co, L., Condit, R., Davies, S. J., Foster, R., Hubbell, S. P., Kenfack, D., Lagunzad, D., Losos, E. C., Nor, N. S. M., Tan, S., Thomas, D. W., Valencia, R. and Villa, G. (2006) Contrasting structure and composition of the understory in species-rich tropical rain forests. *Ecology* **87**, 2298–2305.

LaFrankie, J. V. and Saw, L. G. (2005) The understorey palm *Licuala* (Arecaceae) suppresses tree regeneration in a lowland forest in Asia. *Journal of Tropical Ecology* **21**, 703–706.

Lahm, S. A. (2001) Hunting and wildlife in Northeastern Gabon. Why conservation should extend beyond protected areas. *African Rain Forest Ecology and Conservation: An Interdisciplinary Perspective.* Weber, W., White, L. J. T., Vedder, A. and Naughton-Treves, L. (eds.). Yale University Press, New Haven, CT, USA. Pp. 344–354.

Laidlaw, R. K. (2000) Effects of habitat disturbance and protected areas on mammals of Peninsular Malaysia. *Conservation Biology* **14**, 1639–1648.

Lakatos, M., Lange-Bertalot, H. and Budel, B. (2004) Diatoms living inside the thallus of the green algal lichen *Coenogonium linkii* in neotropical lowland rain forests. *Journal of Phycology* **40**, 70–73.

Lal, R. (1996) Deforestation and land-use effects on soil degradation and rehabilitation in western Nigeria .1. Soil physical and hydrological properties. *Land Degradation and Development* **7**, 19–45.

Lambert, F. R. and Collar, N. J. (2002) The future for Sundaic lowland forest birds: long-term effects of commercial logging and fragmentation. *Forktail* **18**, 127–146.

Lambert, J. E. (2001) Red-tailed guenons (*Cercopithecus ascanius*) and *Strychnos mitis*: Evidence for plant benefits beyond seed dispersal. *International Journal of Primatology* **22**, 189–201.

Langenheim, J. H. (1995) Biology of amber-producing trees: Focus on case studies of Hymenaea and Agathis. *Amber, Resinite, and Fossil Resins* **617**, 1–31.

Langhe, E. D. and Maret, P. d. (1999) Tracking the banana: its significance in early agriculture. In: *The Prehistory of Food: Appetites for Change.* Gosden, C. and Hather, J. (eds.). Routledge, London, UK. Pp. 377–396.

Lanly, J. (1982) *Tropical forest resources.* FAO Forestry Paper, Rome, Italy. **30**.

Lapied, E. (2002) Rhinodrilus saulensis *Nov. Sp. (Oligochaeta, Glossoscolecidae), a new giant earthworm species from French Guiana.* the 7th International Symposium on Earthworm Ecology. Cardiff, UK.

Laporte, N. T., Stabach, J. A., Grosch, R., Lin, T. S. and Goetz, S. J. (2007) Expansion of industrial logging in Central Africa. *Science* **316**, 1451–1451.

Large Mammal Conservation in India? *Hunting for Sustainability in Tropical Forests.* Robinson, J. G. and Bennett, E. L. Columbia University Press New York, USA. Pp. 339–355.

Larsen, H. O. and Olsen, C. S. (2007) Unsustainable collection and unfair trade? Uncovering and assessing assumptions regarding Central Himalayan medicinal plant conservation. *Biodiversity and Conservation* **16**, 1679–1697.

Larsen, M. C., Torres-Sanchez, A. J. and Concepcion, I. M. (1999) Slopewash, surface runoff and fine-litter transport in forest and landslide scars in humid-tropical steeplands, Luquillo Experimental Forest, Puerto Rico. *Earth Surface Processes and Landforms* **24**, 481–502.

Lasso, E. and Ackerman, J. D. (2004) The flexible breeding system of *Werauhia sintenisii*, a cloud forest bromeliad from Puerto Rico. *Biotropica* **36**, 414–417.

Lathiere, J., Hauglustaine, D. A., Friend, A. D., De Noblet-Ducoudre, N., Viovy, N. and Folberth, G. A. (2006) Impact of climate variability and land use changes on global biogenic volatile organic compound emissions. *Atmospheric Chemistry and Physics* **6**, 2129–2146.

Laube, S. and Zotz, G. (2003) Which abiotic factors limit vegetative growth in a vascular epiphyte? *Functional Ecology* **17**, 598–604.

Laurance, W. F. (2003) Slow burn: the insidious effects of surface fires on tropical forests. *Trends in Ecology and Evolution* **18**, 209–212.

Laurance, W. F. (2004) The perils of payoff: corruption as a threat to global biodiversity. *Trends in Ecology and Evolution* **19**, 399–401.

Laurance, W. F. (2005) Forest–climate interactions in fragmented tropical landscapes. In *Tropical Forests and Global Atmospheric Change.* Malhi, Y. and Phillips, O. (eds.). Oxford University Press, Oxford, UK. Pp. 31–38.

Laurance, W. F., Croes, B. M., Guissouegou, N., Buij, R., Dethier, M. and Alonso, A. (2008) Impacts of roads, hunting, and habitat alteration on nocturnal mammals in African rainforests. *Conservation Biology* **22**, 721–732.

Laurance, W. F., Delamonica, P., Laurance, S. G., Vasconcelos, H. L. and Lovejoy, T. E. (2000) Rainforest fragmentation kills big trees. *Nature* **404**, 836–836.

Laurance, W. F., Fearnside, P. M., Laurance, S. G., Delamonica, P., Lovejoy, T. E., Rankin de Merona, J., Chambers, J. Q. and Gascon, C. (1999) Relationship between soils and Amazon forest biomass: A landscape-scale study. *Forest Ecology and Management* **118**, 127-138.

Laurance, W. F., Nascimento, H. E. M., Laurance, S. G., Andrade, A., Ribeiro, J., Giraldo, J. P., Lovejoy, T. E., Condit, R., Chave, J., Harms, K. E. and D'Angelo, S. (2006a) Rapid decay of tree-community composition in Amazonian forest fragments. *Proceedings of the National Academy of Sciences of the United States of America* **103**, 19010-19014.

Laurance, W. F., Nascimento, H. E. M., Laurance, S. G., Andrade, A. C., Fearnside, P. M., Ribeiro, J. E. L. and Capretz, R. L. (2006b) Rain forest fragmentation and the proliferation of successional trees. *Ecology* **87**, 469-482.

Laurance, W. F., Nascimento, H. E. M., Laurance, S. G., Condit, R., D'Angelo, S. and Andrade, A. (2004) Inferred longevity of Amazonian rainforest trees based on a long-term demographic study.

Laurance, W. F., Oliveira, A. A., Laurance, S. G., Condit, R., Nascimento, H. E. M., Sanchez-Thorin, A. C., Lovejoy, T. E., Andrade, A., D'Angelo, S., Ribeiro, J. E. and Dick, C. W. (2004) Pervasive alteration of tree communities in undisturbed Amazonian forests. *Nature* **428**, 171-175.

Laurance, W. F., Perez-Salicrup, D., Delamonica, P., Fearnside, P. M., D'Angelo, S., Jerozolinski, A., Pohl, L. and Lovejoy, T. E. (2001) Rain forest fragmentation and the structure of Amazonian liana communities. *Ecology* **82**, 105-116.

Laurance, W. F. and Williamson, G. B. (2001) Positive feedbacks among forest fragmentation, drought, and climate change in the Amazon. *Conservation Biology* **15**, 1529-1535.

Lawrence, D. (2003) The response of tropical tree seedlings to nutrient supply: Meta-analysis for understanding a changing tropical landscape. *Journal of Tropical Ecology*, **19**, 239-250.

Lawton, J. H., Bignell, D. E., Bolton, B., Bloemers, G. F., Eggleton, P., Hammond, P. M., Hodda, M., Holt, R. D., Larsen, T. B., Mawdsley, N. A., Stork, N. E., Srivastava, D. S. and Watt, A. D. (1998) Biodiversity inventories, indicator taxa and effects of habitat modification in tropical forest. *Nature* **391**, 72-76.

Lawton, R. O., Nair, U. S., Pielke, R. A. and Welch, R. M. (2001) Climatic impact of tropical lowland deforestation on nearby montane cloud forests. *Science* **294**, 584-587.

Lawton, R. O. and Putz, F. E. (1988) Natural disturbance and gap-phase regeneration in a wind-exposed tropical cloud forest. *Ecology* **69**, 764-777.

Leach, M. and Mearns, R. (eds.) (1996) *The Lie of the Land: Challenging Received Wisdom on the African Environment.* James Curry, Portsmouth, NH, USA. Pp. 256.

Leakey, A.D.B., Scholes, J.D. and Press, M.C. (2005) Physiological and ecological significance of sunflecks for dipterocarp seedlings. *Journal of Experimental Botany*, **56**, 469-482.

LeBauer, D. S. and Treseder, K. K. (2008) Nitrogen limitation of net primary productivity in terrestrial ecosystems is globally distributed. *Ecology* **89**, 371-379.

Lebot, V. (1999) Biomolecular evidence for plant domestication in Sahul. *Genetic Resources and Crop Evolution* **46**, 619-628.

Lecorff, J. (1993) Effects of light and nutrient availability on chasmogamy and cleistogamy in an understory tropical herb, *Calathea micans* (Marantaceae) *American Journal of Botany* **80**, 1392-1399.

Lecorff, J. and Horvitz, C. C. (1995) Dispersal of seeds from chasmogamous and cleistogamous flowers in an ant-dispersed neotropical herb. *Oikos* **73**, 59-64.

Lecorff, J. and Horvitz, C. C. (2005) Population growth versus population spread of an ant-dispersed neotropical herb with a mixed reproductive strategy. *Ecological Modelling* **188**, 41-51.

Lee, D.W., Lowry, J.B. and Stone, B.C. (1979) Abaxial anthocyanin layer in leaves of tropical rain-forest plants: Enhancer of light capture in deep shade. *Biotropica*, **11**, 70-77.

Lee, J. E., Oliveira, R. S., Dawson, T. E. and Fung, I. (2005) Root functioning modifies seasonal climate. *Proceedings of the National Academy of Sciences of the United States of America* **102**, 17576-17581.

Lee, S. L., Wickneswari, R., Mahani, M. C. and Zakri, A. H. (2000) Genetic diversity of a tropical tree species, *Shorea leprosula* Miq. (Dipterocarpaceae), in Malaysia: Implications for conservation of genetic resources and tree improvement. *Biotropica* **32**, 213-224.

Lees, A., Garnett, M. and Wright, S. (1991) *A Representative Forest System for the Solomon Islands.* Maruia Society, Nelson, New Zealand. Pp. 185.

Lees, D. C., Kremen, C. and Andriamampianina, L. (1999) A null model for species richness gradients: bounded range overlap of butterflies and other rainforest endemics in Madagascar. *Biological Journal of the Linnean Society* **67**, 529-584.

Legouallec, J.L., Cornic, G. and Blanc, P. (1990) Relations between sunfleck sequences and photoinhibition of photosynthesis in a tropical rain-forest understory herb. *American Journal of Botany*, **77**, 999-1006.

Lehmann, J., Kern, D. C., Glaser, B. and Woods, W. I. (2003) *Amazonian Dark Earths: Origin, Properties and Management.* Kluwer, Dordrecht, The Netherlands. Pp. 523.

Leigh, E. G. (1999) *Tropical forest ecology: A View from Barro Colorado Island.* Oxford University Press, New York, USA. Pp. 264.

Leigh, E. G., Davidar, P., Dick, C. W., Puyravaud, J. P., Terborgh, J., ter Steege, H. and Wright, S. J. (2004) Why do some tropical forests have so many species of trees? *Biotropica* **36**, 447-473.

Leigh, E. G., Hladik, A., Hladik, C. M. and Jolly, A. (2007) The biogeography of large islands, or how does the size of the ecological theater affect the evolutionary play? *Revue Écologique (Terre Vie)* **62**, 105–168.

Lejju, B. J., Robertshaw, P. and Taylor, D. (2006) Africa's earliest bananas? *Journal of Archaeological Science* **33**, 102–113.

Lelieveld, J., Butler, T. M., Crowley, J. N., Dillon, T. J., Fischer, H., Ganzeveld, L., Harder, H., Lawrence, M. G., Martinez, M., Taraborrelli, D. and Williams, J. (2008) Atmospheric oxidation capacity sustained by a tropical forest. *Nature*, 737–740.

Lentz, D. L. (2000) *Imperfect Balance: Landscape Transformations in the Precolumbian Americas.* Columbia University Press, New York, USA.

Leonard, K. L. (2003) African traditional healers and outcome-contingent contracts in health care. *Journal of Development Economics* **71**, 1–22.

Leon-Vargas, Y., Engwald, S. and Proctor, M. C. F. (2006) Microclimate, light adaptation and desiccation tolerance of epiphytic bryophytes in two Venezuelan cloud forests. *Journal of Biogeography* **33**, 901–913.

Leps, J., Novovotny, V., Cizek, L., Molem, K., Isua, B., Boen, W., Kutil, R., Auga, J., Kasbal, M., Manumbor, M. and Hiuk, S. (2002) Successful invasion of the neotropical species *Piper aduncum* in rain forests in Papua New Guinea. *Applied Vegetation Science* **5**, 255–262.

Lerdau, M. and Slobodkin, K. (2002) Trace gas emissions and species-dependent ecosystem services. *Trends in Ecology and Evolution* **17**, 309–312.

Leroy, E. M., Kumulungui, B., Pourrut, X., Rouquet, P., Hassanin, A., Yaba, P., Delicat, A., Paweska, J. T., Gonzalez, J. P. and Swanepoel, R. (2005) Fruit bats as reservoirs of Ebola virus. *Nature* **438**, 575–576.

Lesack, L. F. W. (1993) Export of nutrients and major ionic solutes from a rainforest catchment in the Central Amazon Basin. *Water Resources Research* **29**, 743–758.

Letourneau, D. K. (1998) Ants, stem borers and fungal pathogens: experimental tests of a fitness advantage in *Piper* ant-plants. *Ecology* **79**, 593–603.

Letourneau, D. K. and Dyer, L. A. (1998) Experimental test in lowland tropical forest shows top-down effects through four trophic levels. *Ecology* **79**, 1678–1687.

Letts, M.G. and Mulligan, M. (2005) The impact of light quality and leaf wetness on photosynthesis in north-west Andean tropical montane cloud forest. *Journal of Tropical Ecology*, **21**, 549–557.

Levang, P., Sitorus, S. and Dounias, E. (2007) City life in the midst of the forest: A Punan Hunter-Gatherer's vision of conservation and development. *Ecology and Society* **12**, 1.

Levey, D. J. and Byrne, M. M. (1993) Complex ant plant interactions: Rainforest ants as secondary dispersers and postdispersal seed predators. *Ecology* **74**, 1802–1812.

Levins, R. and Culver, D. (1971) Regional coexistence of species and competition between rare species (mathematical model/ habitable patches) *Proceedings of the National Academy of Sciences of the United States of America* **68**, 1246–&.

Lev-Yadun, S. (2003) Weapon (thorn) automimicry and mimicry of aposematic colorful thorns in plants. *Journal of Theoretical Biology* **224**, 183–188.

Lev-Yadun, S., Dafni, A., Flaishman, M. A., Inbar, M., Izhaki, I., Katzir, G. and Ne'eman, G. (2004) Plant coloration undermines herbivorous insect camouflage. *BioEssays* **26**, 1126–1130.

Lewis, A. R. (1988) Buttress arrangement in *Pterocarpus officinalis* (Fabaceae): Effects of crown asymmetry and wind. *Biotropica* **20**, 280–285.

Lewis, O. T., Memmott, J., Lasalle, J., Lyal, C. H. C., Whitefoord, C. and Godfray, H. C. J. (2002) Structure of a diverse tropical forest insect-parasitoid community. *Journal of Animal Ecology* **71**, 855–873.

Lewis, S. L. (2006) Tropical forests and the changing earth system. *Philosophical Transactions of the Royal Society B* **361**, 195–210.

Lewis, S. L., Lopez-Gonzalez, G., Sonke, B., Affum-Baffoe, K., Baker, T. R., Ojo, L. O., Phillips, O. L., Reitsma, J. M., White, L., Comiskey, J. A., Djuikouo, M. N., Ewango, C. E. N., Feldpausch, T. R., Hamilton, A. C., Gloor, M., Hart, T., Hladik, A., Lloyd, J., Lovett, J. C., Makana, J. R., Malhi, Y., Mbago, F. M., Ndangalasi, H. J., Peacock, J., Peh, K. S. H., Sheil, D., Sunderland, T., Swaine, M. D., Taplin, J., Taylor, D., Thomas, S. C., Votere, R. and Woll, H. (2009) Increasing carbon storage in intact African tropical forests. *Nature* **457**, 1003–U3.

Lewis, S. L., Malhi, Y. and Phillips, O. L. (2004) Fingerprinting the impacts of global change on tropical forests. *Philosophical Transactions of the Royal Society of London Series B* **359**, 437–462.

Lewis, S.L. and Tanner, E.V.J. (2000) Effects of above- and belowground competition on growth and survival of rain forest tree seedlings. *Ecology*, **81**, 2525–2538.

Lewis, T. (1973) *Thrips: Their Biology, Ecology and Economic Importance.* Academic Press, London, UK. Pp. 349.

Lieberman, D., Lieberman, M., Peralta, R. and Hartshorn, G. S. (1996) Tropical forest structure and composition on a large-scale altitudinal gradient in Costa Rica. *Journal of Ecology* **84**, 137–152.

Lieberman, M., Lieberman, D. (1994) Patterns of density and dispersion of forest trees. In: *La Selva: ecology and natural history of a neotropical rainforest.* Mcdade, L. A., Bawa, K. S., Hespenheide, H. and Hartshorn, G. S. (eds.). University of Chicago Press, Chicago, IL, USA. Pp. 106–119.

Lieberman, M., Lieberman, D. and Peralta, R. (1989) Forests are not just Swiss cheese: Canopy stereogeometry of non-gaps in tropical forests. *Ecology* **70**, 550–552.

Lieberman, M., Lieberman, D., Peralta, R. and Hartshorn, G. S. (1995) Canopy closure and the distribution of tropical forest tree species at La Selva, Costa Rica. *Journal of Tropical Ecology* **11**, 161–178.

Lima, H. N., Schaefer, C. E. R., Mello, J. W. V., Gilkes, R. J. and Ker, J. C. (2002) Pedogenesis and pre-Colombian land use of Terra Preta Anthrosols (Indian black earth) of Western Amazonia. *Geoderma* **110**(1–2), 1–17.

Linder, H. P. and Rudall, P. J. (2005) Evolutionary history of Poales. *Annual Review of Ecology Evolution and Systematics* **36**, 107–124.

Linkie, M., Smith, R. J., Zhu, Y., Martyr, D. J., Suedmeyer, B., Pramono, J. and Leader-Williams, N. (2008) Evaluating biodiversity conservation around a large Sumatran protected area. *Conservation Biology* **22**, 683–690.

Lips, K. R., Brem, F., Brenes, R., Reeve, J. D., Alford, R. A., Voyles, J., Carey, C., Livo, L., Pessier, A. P. and Collins, J. P. (2006) Emerging infectious disease and the loss of biodiversity in a Neotropical amphibian community. *Proceedings of the National Academy of Sciences of the United States of America* **103**, 3165–3170.

Lips, K. R., Burrowes, P. A., Mendelson, J. R. and Parra-Olea, G. (2005a) Amphibian declines in Latin America: Widespread population declines, extinctions, and impacts. *Biotropica* **37**, 163–165.

Lips, K. R., Burrowes, P. A., Mendelson, J. R. and Parra-Olea, G. (2005b) Amphibian population declines in Latin America: A synthesis. *Biotropica* **37**, 222–226.

Lips, K. R., Reeve, J. D. and Witters, L. R. (2003) Ecological traits predicting amphibian population declines in Central America. *Conservation Biology* **17**, 1078–1088.

Listabarth, C. (1993) Insect-induced wind pollination of the palm *Chamaedorea pinnatifrons* and pollination in the related *Wendlandiella* sp. *Biodiversity and Conservation* **2**, 39–50.

Liu, W. J., Meng, F. R., Zhang, Y. P., Liu, Y. H. and Li, H. M. (2004) Water input from fog drip in the tropical seasonal rain forest of Xishuangbanna, South-West China. *Journal of Tropical Ecology* **20**, 517–524.

Lloyd, J. and Taylor, J. A. (1994) On the temperature-dependence of soil respiration. *Functional Ecology* **8**, 315–323.

Lodge, D. J., Hawksworth, D. L. and Ritchie, B. J. (1996) Microbial diversity and tropical forest functioning. In: *Biodiversity and Ecosystem Processes in Tropical Forests*. Orians, G. H., Dirzo, R. and Cushman, J. H. (eds.). Springer-Verlag, Berlin, Germany. **122**. Pp. 69–100.

Lodge, D. J., McDowell, W. H. and McSwiney, C. P. (1994) The importance of nutrient pulses in tropical forests. *Trends in Ecology and Evolution* **9**, 384–387.

Londono, A. C., Alvarez, E., Forero, E. and Morton, C. M. (1995) A new genus and species of Dipterocarpaceae from the neotropics. 1. Introduction, taxonomy, ecology, and distribution. *Brittonia* **47**, 225–236.

Lopez, L. and Terborgh, J. (2007) Seed predation and seedling herbivory as factors in tree recruitment failure on predator-free forested islands. *Journal of Tropical Ecology* **23**, 129–137.

Lopez, O. R., Kursar, T. A., Cochard, H. and Tyree, M. T. (2005) Interspecific variation in xylem vulnerability to cavitation among tropical tree and shrub species. *Tree Physiology*, **25**, 1553–1562.

Lopez-Portillo, J., Ewers, F. W., Angeles, G. and Fisher, J. B. (2000) Hydraulic architecture of *Monstera acuminata*: evolutionary consequences of the hemiepiphytic growth form. *New Phytologist* **145**, 289–299.

Lord, J., Egan, J., Clifford, T., Jurado, E., Leishman, M., Williams, D. and Westoby, M. (1997) Larger seeds in tropical floras: Consistent patterns independent of growth form and dispersal mode. *Journal of Biogeography* **24**, 205–211.

Losos, E. C. and Leigh, J., E. G., Eds. (2004) *Tropical Forest Diversity and Dynamism: Findings from a Large-Scale Plot Network*. University of Chicago Press, Chicago, IL, USA. Pp. 688.

Lovejoy, N. R., Albert, J. S. and Crampton, W. G. R. (2006) Miocene marine incursions and marine/freshwater transitions: Evidence from Neotropical fishes. *Journal of South American Earth Sciences* **21**, 5–13.

Lovell, W. G. (1992) Heavy shadows and black night: Disease and depopulation in colonial Spanish-America. *Annals of the Association of American Geographers* **82**, 426–443.

Lovelock, C. E., Jebb, M. and Osmond, C.B. (1994) Photoinhibition and recovery in tropical plant-species: Response to disturbance. *Oecologia*, **97**, 297–307.

Lovelock, C. E., Kyllo, D., Popp, M., Isopp, H., Virgo, A. and Winter, K. (1997) Symbiotic vesicular-arbuscular mycorrhizae influence maximum rates of photosynthesis in tropical tree seedlings grown under elevated CO_2. *Australian Journal of Plant Physiology*, **24**, 185–194.

Lovelock, C. E., Wright, S. F., Clark, D. A. and Ruess, R. W. (2004) Soil stocks of glomalin produced by arbuscular mycorrhizal fungi across a tropical rain forest landscape. *Journal of Ecology* **92**, 278–287.

Lovett, J. C. and Wasser, S. K. (1993) *Biogeography and Ecology of the Rain Forests of Eastern Africa*. Cambridge University Press, Cambridge, UK. Pp. 351.

Lucas, P. W., Darvell, B. W., Lee, P. K. D., Yuen, T. D. B. and Choong, M. F. (1998) Colour cues for leaf food selection by long-tailed macaques (*Macaca fascicularis*) with a new suggestion for the evolution of trichromatic colour vision. *Folia Primatologica* **69**, 139–152.

Lucking, R. and Bernecker-Lucking, A. (2005) Drip-tips do not impair the development of epiphyllous rain-forest lichen communities. *Journal of Tropical Ecology* **21**, 171–177.

Lucking, R. and Matzer, M. (2001) High foliicolous lichen alpha-diversity on individual leaves in Costa Rica and Amazonian Ecuador. *Biodiversity and Conservation* **10**, 2139–2152.

Lucky, A., Erwin, T. L. and Witman, J. D. (2002) Temporal and spatial diversity and distribution of arboreal Carabidae (Coleoptera) in a western Amazonian rain forest. *Biotropica* **34**, 376–386.

Lugo, A. E., Applefield, M., Pool, D. J. and McDonald, R. B. (1983) The impact of Hurricane David on the forests of Dominica. *Canadian Journal of Forest Research* **13**, 201–211.

Lugo, A. E., Gonzalezliboy, J.A., Cintron, B. and Dugger, K. (1978) Structure, productivity, and transpiration of a sub-tropical dry forest in Puerto Rico. *Biotropica*, **10**, 278–291.

Lugo, A. E. and Helmer, E. (2004) Emerging forests on abandoned land: Puerto Rico's new forests. *Forest Ecology and Management* **190**, 145–161.

Lugo, A. E., Parrotta, J. A. and Brown, S. (1993) Loss of species caused by tropical deforestation and their recovery through management. *Ambio* **22**, 106–109.

Lugo, A. E. and Scatena, F. N. (1996) Background and catastrophic tree mortality in tropical moist, wet, and rain forests. *Biotropica* **28**, 585–599.

Lumer, C. and Schoer, R. D. (1986) Pollination of *Blakea austin-smithii* and *Blakea penduliflora* (Melastomataceae) by small rodents in Costa Rica. *Biotropica* **18**, 363–364.

Lusk, C. H. and Piper, F. I. (2007) Seedling size influences relationships of shade tolerance with carbohydrate-storage patterns in a temperate rainforest. *Functional Ecology* **21**, 78–86.

Luteyn, J. L. (2002) Diversity, adaptation, and endemism in Neotropical Ericaceae: Biogeographical patterns in the Vaccinieae. *The Botanical Review* **68**, 55–87.

Luttge, U. (2004) Ecophysiology of crassulacean acid metabolism (CAM). *Annals of Botany* **93**, 629–652.

Lynam, T., de Jong, W., Sheil, D., Kusumanto, T. and Evans, K. (2007) A review of tools for incorporating community knowledge, preferences, and values into decision making in natural resources management. *Ecology and Society* **12**, 1.

Lynch, O. J. and Harwell, E. (2002) *Whose Resources? Whose Common Good? Towards a New Paradigm of Environmental Justice and the National Interest in Indonesia.* Lembaga Studi dan Advokasi Masyarakat (ELSAM), Jakarta.

Maass, J. M., Martinez-Yrizar, A., Patino, C. and Sarukhan, J. (2002) Distribution and annual net accumulation of above-ground dead phytomass and its influence on throughfall quality in a Mexican tropical deciduous forest ecosystem. *Journal of Tropical Ecology* **18**, 821–834.

Macdonald, D. W. (1996) Dangerous liaisons and disease. *Nature* **379**, 400–401.

MacDonald, G. E. (2004) Cogongrass (*Imperata cylindrica*): Biology, ecology, and management. *Critical Reviews in Plant Sciences* **23**, 367–380.

MacDonald, J. A., Jeeva, D., Eggleton, P., Davies, R., Bignell, D. E., Fowler, D., Lawton, J. and Maryati, M. (1999) The effect of termite biomass and anthropogenic disturbance on the CH_4 budgets of tropical forests in Cameroon and Borneo. *Global Change Biology* **5**, 869–879.

Macedo, M. and Prance, G. T. (1978) Notes on the vegetation of Amazonia II. The dispersal of plants in Amazonian white sands Campinas: the Campinas as functional islands. *Brittonia* **30**, 203–215.

Mack, A. L. (1993) The sizes of vertebrate-dispersed fruits: a Neotropical-Paleotropical comparison. *American Naturalist* **142**, 840–856.

Mack, A. L. (1998) The potential impact of small-scale physical disturbance on seedlings in a papuan rainforest. *Biotropica* **30**, 547–552.

Mackensen, J., Bauhus, J. and Webber, E. (2003) Decomposition rates of coarse woody debris—A review with particular emphasis on Australian tree species. *Australian Journal of Botany* **51**, 27–37.

Madhusudan, M. D. and Karanth, K. U. (2000) Hunting for an answer: Is local hunting compatible with large mammal conservation in India? In: *Hunting for Sustainability in Tropical Forests* Robinson. Robinson J. G. and Bennett, E. L. (eds.). Columbia University Press, New York, USA. Pp. 339–355.

Maffi, L. (2005) Linguistic, cultural, and biological diversity. *Annual Review of Anthropology* **34**, 599–617.

Mahar, D. J. (1989) *Government Policies and Deforestation in Brazil's Amazon Region.* World Bank, Washington, DC, USA. Pp. 56.

Makarieva, A. M. and Gorshkov, V. G. (2007) Biotic pump of atmospheric moisture as driver of the hydrological cycle on land. *Hydrology and Earth System Sciences* **11**, 1013–1033.

Maley, J. (2002) A catastrophic destruction of African forests about 2,500 years ago still exerts a major influence on present vegetation formations. *IDS Bulletin-Institute of Development Studies* **33**, 13–30.

Malhi, Y., Baker, T. R., Phillips, O. L., Almeida, S., Alvarez, E., Arroyo, L., Chave, J., Czimczik, C. I., Di Fiore, A., Higuchi, N., Killeen, T. J., Laurance, S. G., Laurance, W. F., Lewis, S. L., Montoya, L. M. M., Monteagudo, A., Neill, D. A., Vargas, P. N., Patino, S., Pitman, N. C. A., Quesada, C. A., Salomao, R., Silva, J. N. M., Lezama, A. T., Martinez, R. V., Terborgh, J., Vinceti, B. and Lloyd, J. (2004) The above-ground coarse wood productivity of 104 Neotropical forest plots. *Global Change Biology* **10**, 563–591.

Malhi, Y. and Wright, J. (2004) Spatial patterns and recent trends in the climate of tropical rainforest regions. *Philosophical Transactions of the Royal Society of London Series B* **359**, 311–329.

Malhi, Y. and Wright, J. (2005) Late twentieth-century patterns and trends in the climate of tropical forest regions. *Tropical Forests and Global Atmospheric Change*. Malhi, Y. and Phillips, O. Oxford University Press, Oxford, UK. Pp. 3–16.

Mallet, J. and Turner, J. (1998) Biotic drift or the shifting balance: did forest islands drive the diversity of warningly coloured butterflies? In: *Evolution on Islands*. Grant, P. (ed.). Oxford University Press, Oxford, UK. Pp. 262–280.

Mancina, C. A., Balseiro, F. and Herrera, L. G. (2005) Pollen digestion by nectarivorous and frugivorous Antillean bats. *Mammalian Biology* **70**, 282–290.

Mandon-Dalger, I., Clergeau, P., Tassin, J., Riviere, J. N. and Gatti, S. (2004) Relationships between alien plants and an alien bird species on Reunion Island. *Journal of Tropical Ecology* **20**, 635–642.

Manfroi, O. J., Koichiro, K., Nobuaki, T., Masakazu, S., Nakagawa, M., Nakashizuka, T. and Chong, L. (2004) The stemflow of trees in a Bornean lowland tropical forest. *Hydrological Processes* **18**, 2455–2474.

Mann, C. C. (2008) Ancient earthmovers of the Amazon. *Science* **321**, 1148–1152.

Manne, L. L. (2003) Nothing has yet lasted forever: current and threatened levels of biological and cultural diversity. *Evolutionary Ecology Research* **5**, 517–527.

Mannheimer, S., Bevilacqua, G., Caramaschi, E. P. and Scarano, F. R. (2003) Evidence for seed dispersal by the catfish *Auchenipterichthys longimanus* in an Amazonian lake. *Journal of Tropical Ecology* **19**, 215–218.

Marcoy, Paul (1872) *A Journey Across South America from the Pacific Ocean to the Atlantic Ocean*. Blackie and Son, Glasgow, UK.

Marengo, J. A., Nobre, C. A., Tomasella, J., Cardoso, M. F. and Oyama, M. D. (2008) Hydro-climatic and ecological behaviour of the drought of Amazonia in 2005. *Philosophical Transactions of the Royal Society B* **363**, 1773–1778.

Margules, C. R. and Pressey, R. L. (2000) Systematic conservation planning. *Nature* **405**, 243–253.

Markesteijn, L. and Poorter, L. (2009 in press) Seedling root morphology and biomass allocation of 62 tropical tree species in relation to drought- and shade-tolerance. *Journal of Ecology*.

Marquis, R. J. (1984) Leaf herbivores decrease fitness of a tropical plant. *Science* **226**, 537–539.

Marquis, R. J. and Braker, H. E. (1994) Plant-herbivore interactions: Diversity, specificity, and impact. In *La Selva: Ecology and Natural History of a Tropical Rainforest* Mcdade,

L. A., Bawa, K. S., Hespenheide, H. A. and Hartshorn, G. S. (eds.). University of Chicago Press, Chicago, IL, USA. Pp. 261–281.

Marquis, R. J., Newell, E. A. and Villegas, A. C. (1997) Non-structural carbohydrate accumulation and use in an understorey rain forest shrub and relevance for the impact of leaf herbivory. *Functional Ecology*, **11**, 636–643.

Marquis, R. J., Young, H. J. and Braker, H. E. (1986) The influence of understory vegetation cover on germination and seedling establishment in a tropical lowland wet forest. *Biotropica* **18**, 273–278.

Marsden, S. J. (1998) Changes in bird abundance following selective logging on Seram, Indonesia. *Conservation Biology* **12**, 605–611.

Marsh, C., Johns, A. and Ayres, J. (1987) Effects of habitat disturbance on rain forest primates. In: *Primate Conservation in the Tropical Rain Forest*. Marsh, C. and Mittermeier, R. (eds.). Alan R. Liss, Inc., New York, USA. Pp. 83–107.

Marshall, L. G., Webb, S. D., Sepkoski, J. J. and Raup, D. M. (1982) Mammalian evolution and the Great American Interchange. *Science* **215**, 1351–1357.

Martin, A. P. and Palumbi, S. R. (1993) Body size, metabolic-rate, generation time, and the molecular clock. *Proceedings of the National Academy of Sciences of the United States of America* **90**, 4087–4091.

Martin, P. A. W. and Travers, R. S. (1989) Worldwide abundance and distribution of *Bacillus thuringiensis* isolates. *Applied and Environmental Microbiology* **55**, 2437–2442.

Martinez-Ramos, M. and Alvarez-Buylla, E.R. (1998) How old are tropical rain forest trees? *Trends in Plant Science*, **3**, 400–405.

Martinez-Ramos, M., Alvarez-Buylla, E., Sarukhan, J. and Pinero, D. (1988) Treefall age-determination and gap dynamics in a tropical forest. *Journal of Ecology* **76**, 700–716.

Martin-Smith, K. M. (1998) Effects of disturbance caused by selective timber extraction on fish communities in Sabah, Malaysia. *Environmental Biology of Fishes* **53**, 155–167.

Martius, C. (1994) Diversity and ecology of termites in Amazonian forests. *Pedobiologia* **38**, 407–428.

Mas, A. H. and Dietsch, T. V. (2004) Linking shade coffee certification to biodiversity conservation: Butterflies and birds in Chiapas, Mexico. *Ecological Applications* **14**, 642–654.

Matelson, T. J., Nadkarni, N. M. and Solano, R. (1995) Tree damage and annual mortality in a montane forest in Monteverde, Costa Rica. *Biotropica* **27**, 441–447.

Mateo, N. (1997) *Wild biodiversity: the last frontier?* In: *Proceedings of the international workshop on genetic resources, Oberursel, Germany, 25–26 August*, BMU, Bonn. Pp. 14–15.

Matson, P. A. and Vitousek, P. M. (2006) Agricultural intensification: will land be spared from farming be land spared for nature? *Conservation Biology* **20**, 709–710.

Mawdsley, N. A., Compton, S. G. and Whittaker, R. J. (1998) Population persistence, pollination mutualisms, and figs in fragmented tropical landscapes. *Conservation Biology* **12**, 1416–1420.

May, R. M. (1990) How many species. *Philosophical Transactions of the Royal Society of London Series B* **330**, 293–304.

May, R. M., Lawton, J. H. and Stork, N. E. (1995) Assessing extinction rates. In: *Extinction Rates*. Lawton, J. H. and May, R. M. (eds.). Oxford University Press, Oxford, UK. Pp. 1–24.

Mayor, J. R. and Henkel, T. W. (2006) Do ectomycorrhizas alter leaf-litter decomposition in monodominant tropical forests of Guyana? *New Phytologist* **169**, 579–588.

Mayr, E. and Diamond, J. M. (2001) *The Birds of Northern Melanesia: Speciation, Ecology, and Biogeography*. Oxford University Press, Oxford, UK. Pp. 548.

Mbida, C. M., Doutrelepont, H., Vrydaghs, L., Swennen, R. L., Swennen, R. J., Beeckman, H., Langhe, E. D. and Maret, P. D. (2001) First evidence of banana cultivation in central Africa during the third millennium before present. *Vegetation History and Archaeobotany* **10**, 1–6.

McCain, C. M. (2004) The mid-domain effect applied to elevational gradients: species richness of small mammals in Costa Rica. *Journal of Biogeography* **31**, 19–31.

McCarthy, J. (2001) Gap dynamics of forest trees: A review with particular attention to boreal forests. *Environmental Reviews* **9**, 1–59.

McCarthy-Neumann, S. and Kobe, R.K. (2008) Tolerance of soil pathogens co-varies with shade tolerance across species of tropical tree seedlings. *Ecology* **89**, 1883–1892.

McCay, M. G. (2003) Winds under the rain forest canopy: The aerodynamic environment of gliding tree frogs. *Biotropica* **35**, 94–102.

McClure, F. A. (1966) *The Bamboos: A Fresh Perspective*. Harvard University Press, Cambridge, MA, USA. Pp. 347.

McCoy, E. D. (1990) The distribution of insects along elevational gradients. *Oikos* **58**, 313–322.

McDade, L. A., Bawa, K. S., Hespenheide H. A. and Hartshorn G. S. (eds.) (1994) *La Selva: Ecology and Natural History of a Neotropical Rain Forest*. University of Chicago Press, Chicago, IL, USA. Pp. 493.

McDaniel, J., Kennard, D. and Fuentes, A. (2005) Smokey the Tapir: Traditional fire knowledge and fire prevention campaigns in lowland Bolivia. *Society and Natural Resources* **18**, 921–931.

McDowell, T., Volovsek, M. and Manos, P. (2003) Biogeography of *Exostema* (Rubiaceae) in the Caribbean region in light of molecular phylogenetic analyses. *Systematic Botany* **28**, 431–441.

McDowell, W. H. (1998) Internal nutrient fluxes in a Puerto Rican rain forest. *Journal of Tropical Ecology* **14**, 521–536.

McElwain, J. C. and Punyasena, S. W. (2007) Mass extinction events and the plant fossil record. *Trends in Ecology and Evolution* **22**, 548–557.

Mcguire, K.L. (2007) Common ectomycorrhizal networks may maintain monodominance in a tropical rain forest. *Ecology*, **88**, 567–574.

McMahon, T. A. and Kronauer, R. E. (1976) Tree structures: deducing the principles of mechanical design. *Journal of Theoretical Biology* **59**, 443–466.

McShane, T. O. and Newby, S. A. (2004) Expecting the unattainable: The assumptions behind ICDPs. In: *Getting Biodiversity Projects to Work: Towards more effective conservation and development*. Mcshane, T. O. and Wells, M. P. (eds.). Columbia University Press, New York, USA. Pp. 49–74.

McSweeney, K. (2005) Natural insurance, forest access, and compounded misfortune: Forest resources in smallholder coping strategies before and after Hurricane Mitch, northeastern Honduras. *World Development* **33**, 1453–1471.

Md. Nor, S. (2001) Elevational diversity patterns of small mammals on Mount Kinabalu, Sabah, Malaysia. *Global Ecology and Biogeography* **10**, 41–62.

Mediavilla, S. and Escudero, A. (2003) Photosynthetic capacity, integrated over the lifetime of a leaf, is predicted to be independent of leaf longevity in some tree species. *New Phytologist*, **159**, 203–211.

Meegaskumbura, M., Bossuyt, F., Pethiyagoda, R., Manamendra-Arachchi, K., Bahir, M., Milinkovitch, M. C. and Schneider, C. J. (2002) Sri Lanka: An amphibian hot spot. *Science* **298**, 379–379.

Meijaard, E. (2004) Biogeographic history of the Javan leopard *Panthera pardus* based on a craniometric analysis. *Journal of Mammalogy* **85**, 302–310.

Meijaard, E. and Sheil, D. (2007) A logged forest in Borneo is better than none at all. *Nature* **446**, 974.

Meijaard, E. and Sheil, D. (2008) The persistence and conservation of Borneo's mammals in lowland rain forests managed for timber: observations, overviews and opportunities. *Ecological Research* **23**, 21–34.

Meijaard, E., Sheil, D., Marshall, A. J. and Nasi, R. (2008) Phylogenetic age is positively correlated with sensitivity to timber harvest in Bornean mammals. *Biotropica* **40**, 76–85.

Meijaard, E., Sheil, D., Nasi, R., Augeri, D., Rosenbaum, B., Iskandar, D., Setyawati, T., Lammertink, A., Rachmatika, I., Wong, A., Soehartono, T., Stanley, S. and O'Brien, T. (2005) *Life after logging: reconciling wildlife conservation and production forestry in Indonesian Borneo*. CIFOR, UNESCO and ITTO, Bogor, Indonesia.

Meijaard, E. and van der Zon, A. P. M. (2003) Mammals of south-east Asian islands and their Late Pleistocene environments. *Journal of Biogeography* **30**, 1245–1257.

Meimberg, H., Wistuba, A., Dittrich, P. and Heubl, G. (2001) Molecular phylogeny of Nepenthaceae based on cladistic analysis of plastid trnK intron sequence data. *Plant Biology* **3**, 164–175.

Mena, P. V., Stallings, J. R., Regalado, J. B. and Cueva, R. L. (2000) The sustainability of current hunting practices by the Huaorani. In: *Hunting for Sustainability in Tropical Forests*. Robinson, J. G. and Bennett, E. L. (eds.). Columbia University Press, New York, USA. Pp. 57–78.

Mertz, O., Wadley, R. L., Nielsen, U., Bruun, T. B., Colfer, C. J. P., Neergaard, A. d., Jepsen, M. R., Martinussen, T., Zhao, Q., Noweg, G. T. and Magid, J. (2008) A fresh look at shifting cultivation: Fallow length an uncertain indicator of productivity. *Agricultural Systems* **96**, 75–84.

Mesquita, R. C. G., Delamonica, P. and Laurance, W. F. (1999) Effect of surrounding vegetation on edge-related tree mortality in Amazonian forest fragments. *Biological Conservation* **91**, 129–134.

Meyer, C. F. J. and Zotz, G. (2004) Do growth and survival of aerial roots limit the vertical distribution of hemiepiphytic aroids? *Biotropica* **36**, 483–491.

Meyer-Berthaud, B. and Decombeix, A. (2007) Palaeobotany: A tree without leaves. *Nature* **446**, 861–862.

Michon, G. and de Foresta, H. (1999) Agro-forests: Incorporating a forest vision in agroforestry. In: *Agroforestry in Sustainable Agricultural Systems*. Buck, L. E., Lassoie, J. P. and Fernandez, E. C. M. (eds.). CRC Press and Lewis Publishers, New York, USA. Pp. 381–406.

Midgley, J. J. (2003) Is bigger better in plants? The hydraulic costs of increasing size in trees. *Trends in Ecology and Evolution* **18**, 5–6.

Miller, E. R., Gunnell, G. F. and Martin, R. D. (2005) Deep time and the search for anthropoid origins. *Yearbook of Physical Anthropology* **48**. Pp. 60–95.

Miller, R. P. and Nair, P. K. R. (2006) Indigenous agroforestry systems in Amazonia: from prehistory to today. *Agroforestry Systems* **66**, 151–164.

Mindzie, C. M., Doutrelepont, H., Vrydaghs, L., Swennen, R. L., Swennen, R. J., Beeckman, H., de Langhe, E. and de Maret, P. (2001) First archaeological evidence of banana cultivation in central Africa during the third millennium before present. *Vegetation History and Archaeobotany* **10**, 1–6.

Mirmanto, E., Proctor, J., Green, J., Nagy, L. and Suriantata (1999) Effects of nitrogen and phosphorus fertilization in a lowland evergreen rainforest. *Philosophical Transactions of the Royal Society of London Series B* **354**, 1825–1829.

Mittelbach, G. G., Schemske, D. W., Cornell, H. V., Allen, A. P., Brown, J. M., Bush, M. B., Harrison, S. P., Hurlbert, A. H., Knowlton, N., Lessios, H. A., McCain, C. M., McCune, A. R., McDade, L. A., McPeek, M. A., Near, T. J., Price, T. D., Ricklefs, R. E., Roy, K., Sax, D. F., Schluter, D., Sobel, J. M. and Turelli, M. (2007) Evolution and the latitudinal diversity gradient: speciation, extinction and biogeography. *Ecology Letters* **10**, 315–331.

Mittermeier, R. A. and Konstant, W. R. (2001) Conservation of Primate Populations. In: *Encyclopedia of Biodiversity*. Levin, S. A. (ed.). Academic Press, San Diego, CA, USA. Vol. 4. Pp. 879–890.

Mittermeier, R. A., Myers, N., Thomsen, J. B., da Fonseca, G. A. B. and Olivieri, S. (1998) Biodiversity hotspots and major tropical wilderness areas: Approaches to setting conservation priorities. *Conservation Biology* **12**, 516–520.

Mittermeier, R. A., Robles Gil, P., Hoffmann, M., Pilgrim, J., Brooks, T., Mittermeier, C. G., Lamoreux, J. and da Fonseca, G. A. B. (2004) *Hotspots: Revisited*. CEMEX, Mexico.

Miyamoto, K., Suzuki, E., Kohyama, T., Seino, T., Mirmanto, E. and Simbolon, H. (2003) Habitat differentiation among tree species with small-scale variation of humus depth and topography in a tropical heath forest of Central Kalimantan, Indonesia. *Journal of Tropical Ecology* **19**, 43–54.

Moegenburg, S. M. and Levey, D. J. (2002) Prospects for conserving biodiversity in Amazonian extractive reserves. *Ecology Letters* **5**, 320–324.

Möhler, O., DeMott, P. J., Vali, G. and Levin, Z. (2007) Microbiology and atmospheric processes: the role of biological particles in cloud physics. *Biogeosciences Discussions* **4**, 2559–2591.

Moles, A. T., Ackerly, D. D., Tweddle, J. C., Dickie, J. B., Smith, R., Leishman, M. R., Mayfield, M. M., Pitman, A., Wood, J. T. and Westoby, M. (2007) Global patterns in seed size. *Global Ecology and Biogeography* **16**, 109–116.

Moles, A. T., Ackerly, D. D., Webb, C. O., Tweddle, J. C., Dickie, J. B. and Westoby, M. (2005) A brief history of seed size. *Science* **307**, 576–580.

Moles, A. T. and Westoby, M. (2003) Latitude, seed predation and seed mass. *Journal of Biogeography* **30**, 105–128.

Molino, J. F. and Sabatier, D. (2001) Tree diversity in tropical rain forests: A validation of the intermediate disturbance hypothesis. *Science* **294**, 1702–1704.

Moll, D. and Jansen, K. P. (1995) Evidence for a role in seed dispersal by two tropical herbivorous turtles. *Biotropica* **27**, 121–127.

Molnar, A. (2003) *Forest Certification and Communities: Looking Forward to the Next Decade*. Forest Trends, Washington DC.

Molnar, A., Scherr, S. J. and Khare, A. (2004) *Who Conserves the World's Forests? A New Assessment of Conservation and Investment Trends*. Forest Trends, Washington, DC, USA.

Momose, K., Nagamitsu, T. and Inoue, T. (1998a) Thrips cross-pollination of *Popowia pisocarpa* (Annonaceae) in a lowland dipterocarp forest in Sarawak. *Biotropica* **30**, 444–448.

Momose, K., Yumoto, T., Nagamitsu, T., Kato, M., Nagamasu, H., Sakai, S., Harrison, R. D., Itioka, T., Hamid, A. A. and Inoue, T. (1998b) Pollination biology in a lowland dipterocarp forest in Sarawak, Malaysia. I. Characteristics of the plant-pollinator community in a lowland dipterocarp forest. *American Journal of Botany* **85**, 1477–1501.

Monagle, C. (2001) *Biodiversity and Intellectual Property Rights: Reviewing Intellectual Property Rights in Light of the Objectives of the Convention on Biological Diversity.* Geneva, The Center for International Environmental Law (CIEL) and World Wide Fund For Nature (WWF). Pp. 34.

Monteith, J.L. and Unsworth, M.H. (1990) *Principles of Environmental Physics.* Chapman and Hall, New York, USA. Pp. 291.

Montgomery, R. (2004) Relative importance of photosynthetic physiology and biomass allocation for tree seedling growth across a broad light gradient. *Tree Physiology* **24**, 155–167.

Montgomery, R. A. (2004) Effects of understory foliage on patterns of light attenuation near the forest floor. *Biotropica* **36**, 33–39.

Montgomery, R. A. and Chazdon, R. L. (2001) Forest structure, canopy architecture, and light transmittance in tropical wet forests. *Ecology* **82**, 2707–2718.

Montgomery, R. A. and Chazdon, R. L. (2002) Light gradient partitioning by tropical tree seedlings in the absence of canopy gaps. *Oecologia* **131**, 165–174.

Montgomery, R. A. and Givnish, T. J. (2008) Adaptive radiation of photosynthetic physiology in the hawaiian lobeliads: Dynamic photosynthetic responses. *Oecologia*, **155**, 455–467.

Moog, J., Fiala, B., Werner, M., Weissflog, A., Saw, L. G. and Maschwitz, U. (2003) Ant-plant diversity in Peninsular Malaysia, with special reference to the Pasoh Forest Reserve. In: *Pasoh: Ecology of a Lowland Rain Forest in Southeast Asia.* T. Okuda, N. Manokaran, Y. Matsumoto, K. Niiyama, S.C. Thomas and Ashton, P. S. (eds.). Springer Press, Tokyo, Japan. Pp. 459–494.

Moore, D. L. and Stephenson S. L. (2003) Microhabitat distribution of protostelids in a tropical wet forest in Costa Rica. *Mycologia* **95**, 11–18.

Moran, J. A., Booth, W. E. and Charles, J. K. (1999) Aspects of pitcher morphology and spectral characteristics of six Bornean *Nepenthes* pitcher plant species: Implications for prey capture. *Annals of Botany* **83**, 521–528.

Moreau, C. S., Bell, C. D., Vila, R., Archibald, S. B. and Pierce, N. E. (2006) Phylogeny of the ants: Diversification in the age of angiosperms. *Science* **312**, 101–104.

Moritz, C., Hoskin, C. J., MacKenzie, J. B., Phillips, B. L., Tonione, M., Silva, N., VanDerWal, J., Williams, S. E. and Graham, C. H. (2009) Identification and dynamics of a cryptic suture zone in tropical rainforest. *Proceedings of the Royal Society B: Biological Sciences* **276**, 1235–1244.

Moritz, C., Patton, J. L., Schneider, C. J. and Smith, T. B. (2000) Diversification of rainforest faunas: An integrated molecular approach. *Annual Review of Ecology and Systematics* **31**, 533–563.

Morley, R. J. (1998) Palynological evidence for Tertiary plant dispersals in the SE Asian region in relation to plate tectonics and climate. *Biogeography and Geological Evolution of SE Asia.* Hall, R. and Holloway, J. Leiden, Netherlands, Bakhuys Publishers. Pp. 177–200.

Morley, R. J. (2000) *Origin and Evolution of Tropical Rain Forests.* John Wiley and Sons, Ltd., Chichester, UK. Pp. 362.

Morley, R. J. (2003) Interplate dispersal paths for megathermal angiosperms. *Perspectives in Plant Ecology Evolution and Systematics* **6**, 5–20.

Morris, M. H., Negreros-Castillo, P. and Mize, C. (2000) Sowing date, shade, and irrigation affect big-leaf mahogany (*Swietenia macrophylla* King). *Forest Ecology and Management,* **132**, 173–181.

Morris, R. J., Lewis, O. T. and Godfray, H. C. J. (2004) Experimental evidence for apparent competition in a tropical forest food web. *Nature* **428**, 310–313.

Mortensen, H. S., Dupont, Y. L. and Olesen, J. M. (2008) A snake in paradise: Disturbance of plant reproduction following extirpation of bird flower-visitors on Guam. *Biological Conservation* **141**, 2146–2154.

Morton, C. M., Dayanandan, S. and Dissanayake, D. (1999) Phylogeny and biosystematics of *Pseudomonotes* (Dipterocarpaceae) based on molecular and morphological data. *Plant Systematics and Evolution* **216**, 197–205.

Morwood, M. J., Sutikna, T., Saptorno, E. W., Westaway, K. E., Jatmiko, Due, R. A., Moore, M. W., Yuniawati, D. Y., Hadi, P., Zhao, J. X., Turney, C. S. M., Fifield, K., Allen, H. and Soejono, R. P. (2008) Climate, people and faunal succession on Java, Indonesia: Evidence from Song Gupuh. *Journal of Archaeological Science* **35**, 1776–1789.

Moyersoen, B., Becker, P. and Alexander, I. J. (2001) Are ectomycorrhizas more abundant than arbuscular mycorrhizas in tropical heath forests? *New Phytologist* **150**, 591–599.

Muchhala, N. (2006) Nectar bat stows huge tongue in its rib cage. *Nature* **444**, 701–702.

Mueller, U. G., Gerardo, N. M., Aanen, D. K., Six, D. L., Schultz, T. R. (2005) The evolution of agriculture in insects. *Annual Review of Ecology Evolution and Systematics* **36**, 563–595.

Mueller-Dombois, D. (2002) Forest vegetation across the tropical Pacific: A biogeographically complex region with many analogous environments. *Plant Ecology* **163**, 155–176.

Muir, J. (1911) *My First Summer in the Sierra.* Houghton Mifflin Co., Boston, New York, USA.

Muller-Landau, H. C., Condit, R. S., Chave, J., Thomas, S. C., Bohlman, S. A., Bunyavejchewin, S., Davies, S., Foster, R., Gunatilleke, S., Gunatilleke, N., Harms, K. E., Hart, T., Hubbell, S. P., Itoh, A., Kassim, A. R., LaFrankie, J. V., Lee, H. S., Losos, E., Makana, J.-R., Ohkubo, T., Sukumar, R., Sun, I.-F., N., N. S. M., Tan, S., Thompson, J., Valencia, R., Muñoz, G. V., Wills, C., Yamakura, T., Chuyong, G., Dattaraja, H. S., Esufali, S., Hall, P., Hernandez, C., Kenfack, D., Kiratiprayoon, S., Suresh, H. S., Thomas, D., Vallejo, M. I. and Ashton, P. (2006) Testing metabolic ecology theory for allometric scaling of tree size, growth and mortality in tropical forests. *Ecology Letters* **9**, 575–588.

Mullner, A., Linsenmair, K. E. and Wikelski, M. (2004) Exposure to ecotourism reduces survival and affects stress response in hoatzin chicks (*Opisthocomus hoazin*) *Biological Conservation* **118**, 549–558.

Munoz-Pina, C., Guevara, A., Torres, J. M. and Brana, J. (2008) Paying for the hydrological services of Mexico's forests: Analysis, negotiations and results. *Ecological Economics* **65**, 725–736.

Murakami, N. and Yatabe, Y. (2001) Recognition of biological species in *Asplenium nidus* complex using molecular data and crossing experiments. 5th International Flora Malesiana Symposium 9–14 September 2001, Sydney, Australia. From http://www.anbg.gov.au/fm/fms5/fms5_fern_abstracts.html.

Muraoka, H., Takenaka, A., Tang, Y., Koizumi, H. and Washitani, I. (1998) Flexible leaf orientations of arisaema heterophyllum maximize light capture in a forest understorey and avoid excess irradiance at a deforested site. *Annals of Botany*, **82**, 297–307.

Murawski, D. A., Dayanandan, B. and Bawa, K. S. (1994) Outcrossing rates of two endemic *Shorea* species from Sri Lankan tropical rain forests. *Biotropica* **26**, 23–29.

Murray, K. G. (1988) Avian seed dispersal of three neotropical gap-dependent plants. *Ecological Monographs* **58**, 271–298.

Muslin, E. H. and Homann, P. H. (1992) Light as a hazard for the desiccation-resistant resurrection fern *Polypodium polypodioides* l. *Plant Cell and Environment*, **15**, 81–89.

Musselman, L. J. (1980) The biology of *Striga, Orobanche*, and other root-parasitic weeds. *Annual Review of Phytopathology* **18**, 463–489.

Muth, C. C. and Bazzaz, F. A. (2003) Tree canopy displacement and neighborhood interactions. *Canadian Journal of Forest Research* **33**, 1323–1330.

Mutke, J. and Barthlott, W. (2005) Patterns of vascular plant diversity at continental to global scales. *Biologiske Skrifter Kongelige Danske Videnskabernes Selskab* **55**, 521–531.

Myers, J. A. and Kitajima, K. (2007) Carbohydrate storage enhances seedling shade and stress tolerance in a neotropical forest. *Journal of Ecology*, **95**, 383–395.

Myers, N. (1992) Tropical forests: The policy challenge. *The Environmentalist* **12**, 15–27.

Myers, N. (1993) The main deforestation fronts. *Environmental Conservation* **20**, 9–16.

Myers, N. (2000) Shifting versus shifted cultivators. *Bioscience* **50**, 845–846.

Myers, N., Mittermeier, R. A., Mittermeier, C. G., da Fonseca, G. A. B. and Kent, J. (2000) Biodiversity hotspots for conservation priorities. *Nature* **403**, 853–858.

Myneni, R. B., Yang, W. Z., Nemani, R. R., Huete, A. R., Dickinson, R. E., Knyazikhin, Y., Didan, K., Fu, R., Juarez, R. I. N., Saatchi, S. S., Hashimoto, H., Ichii, K., Shabanov, N. V., Tan, B., Ratana, P., Privette, J. L., Morisette, J. T., Vermote, E. F., Roy, D. P., Wolfe, R. E., Friedl, M. A., Running, S. W., Votava, P., El-Saleous, N., Devadiga, S., Su, Y. and Salomonson, V. V. (2007) Large seasonal swings in leaf area of Amazon rainforests. *Proceedings of the National Academy of Sciences of the United States of America* **104**, 4820–4823.

Nabe-Nielsen, J. (2001) Diversity and distribution of lianas in a neotropical rain forest, Yasuni National Park, Ecuador. *Journal of Tropical Ecology* **17**, 1–19.

Nabe-Nielsen, J. and Hall, P. (2002) Environmentally induced clonal reproduction and life history traits of the liana *Machaerium cuspidatum* in an Amazonian rain forest, Ecuador. *Plant Ecology* **162**, 215–226.

Nagamitsu, T. and Inoue, T. (1997) Cockroach pollination and breeding system of *Uvaria elmeri* (Annonaceae) in a lowland mixed-dipterocarp forest in Sarawak. *American Journal of Botany* **84**, 208–213.

Nagamitsu, T. and Inoue, T. (2005) Floral resource utilization by stingless bees (Apidae, Meliponini). In: *Pollination Ecology and the Rain Forest: Sarawak Studies*. Roubik, D. W., Sakai, S. and Hamid Karim, A. A. (eds.). Springer, New York, USA. Pp. 73–88.

Nagendra, H. (2008) Do parks work? Impact of protected areas on land cover clearing. *Ambio* **37:**, 330–337.

Naidoo, R., Balmford, A., Ferraro, P. J., Polasky, S., Ricketts, T. H. and Rouget, M. (2006) Integrating economic costs into conservation planning. *Trends in Ecology and Evolution* **21**, 681–687.

Naito, Y., Kanzaki, M., Iwata, H., Obayashi, K., Lee, S. L., Muhammad, N., Okuda, T. and Tsumura, Y. (2008) Density-dependent selfing and its effects on seed performance in a tropical canopy tree species, *Shorea acuminata* (Dipterocarpaceae). *Forest Ecology and Management* **256**, 375–383.

Naito, Y., Konuma, A., Iwata, H., Suyama, Y., Seiwa, K., Okuda, T., Lee, S. L., Muhammad, N. and Tsumura, Y. (2005) Selfing and inbreeding depression in seeds and seedlings of

Neobalanocarpus heimii (Dipterocarpaceae). *Journal of Plant Research* **118**, 423–430.

Naka, L. N. (2004) Structure and organization of canopy bird assemblages in central Amazonia. *Auk* **121**, 88–102.

Nakagawa, M., Takeuchi, Y., Kenta, T. and Nakashizuka, T. (2005) Predispersal seed predation by insects vs. vertebrates in six dipterocarp species in Sarawak, Malaysia. *Biotropica* **37**, 389–396.

Namara, A., Gray, M. and McNeilage, A. (2000) *People and Bwindi Forest: A historical account as given by local community members.* Institute of Tropical Forest Conservation, University of Mbarara, Uganda. Unpublished Report. Pp. 34.

Nascimento, H. E. M. and Laurance, W. F. (2004) Biomass dynamics in Amazonian forest fragments. *Ecological Applications* **14**, S127–S138.

Nascimento, H. E. M., Laurance, W. F., Condit, R., Laurance, S. G., D'Angelo, S. and Andrade, A. C. (2005) Demographic and life-history correlates for Amazonian trees. *Journal of Vegetation Science*, **16**, 625–634.

Nasi, R., Brown, D., Wilkie, D., Bennett, E., Tutin, C., van Tol, G. and Christophersen, T. (2008) *Conservation and Use of Wildlife-Based Resources: The Bushmeat Crisis*. Technical Series. **33**. Pp. 50.

Nason, J. D., Herre, E. A. and Hamrick, J. L. (1996) Paternity analysis of the breeding structure of strangler fig populations: Evidence for substantial long-distance wasp dispersal. *Journal of Biogeography* **23**, 501–512.

Nathan, R. and Katul, G. G. (2005) Foliage shedding in deciduous forests lifts up long-distance seed dispersal by wind. *Proceedings of the National Academy of Sciences of the United States of America* **102**, 8251–8256.

Naughton-Treves, L., Holland, M. B. and Brandon, K. (2005) The role of protected areas in conserving biodiversity and sustaining local livelihoods. *Annual Review of Environment and Resources* **30**, 219–252.

Nebel, G., Quevedo, L., Jacobsen, J. B. and Helles, F. (2005) Development and economic significance of forest certification: the case of FSC in Bolivia. *Forest Policy and Economics* **7**, 175–186.

Nelson, B. W., Ferreira, C. A. C., Dasilva, M. F. and Kawasaki, M. L. (1990) Endemism centers, refugia and botanical collection density in Brazilian Amazonia. *Nature* **345**, 714–716.

Nelson, B. W., Kapos, V., Adams, J. B., Oliveira, W. J., Braun, O. P. G. and Doamaral, I. L. (1994) Forest disturbance by large blowdowns in the Brazilian Amazon. *Ecology* **75**, 853–858.

Nepstad, D., Carvalho, G., Barros, A. C., Alencar, A., Capobianco, J. P., Bishop, J., Moutinho, P., Lefebvre, P., Silva, U. L. and Prins, E. (2001) Road paving, fire regime feedbacks,

and the future of Amazon forests. *Forest Ecology and Management* **154**, 395–407.

Nepstad, D., McGrath, D., Alencar, A., Barros, A. C., Carvalho, G., Santilli, M. and Diaz, M. D. V. (2002) Frontier governance in Amazonia. *Science* **295**, 629.

Nepstad, D., Schwartzman, S., Bamberger, B., Santilli, M., Ray, D., Schlesinger, P., Lefebvre, P., Alencar, A., Prinz, E., Fiske, G. and Rolla, A. (2006) Inhibition of Amazon deforestation and fire by parks and indigenous lands. *Conservation Biology* **20**, 65–73.

Nepstad, D. C., de Carvalho, C. R., Davidson, E. A., Jipp, P. H., Lefebvre, P. A., Negreiros, G. H., da Silva, E. D., Stone, T. A., Trumbore, S. E. and Vieira, S. (1994) The role of deep roots in the hydrological and carbon cycles of Amazonian forests and pastures. *Nature* **372**, 666–669.

Nettle, D. (1998) Explaining global patterns of language diversity. *Journal of Anthropological Archaeology* **17**, 354–374.

Neumann, R. P. and Hirsch, E. (2000) *Commercialisation of non-timber forest products: Review and analysis of research.* Bogor, Indonesia, Centre for International Forestry Research.

Newbery, D. M. (1991) Floristic variation within kerangas (heath) forest: reevaluation of data from Sarawak and Brunei. *Vegetatio* **96**, 43–86.

Newbery, D. M., Alexander, I. J. and Rother, J. A. (1997) Phosphorus dynamics in a lowland African rain forest: The influence of ectomycorrhizal trees. *Ecological Monographs* **67**, 367–409.

Newbery, D. M., Chuyong, G. B., Green, J. J., Songwe, N. C., Tchuenteu, F. and Zimmermann, L. (2002) Does low phosphorus supply limit seedling establishment and tree growth in groves of ectomycorrhizal trees in a central African rainforest? *New Phytologist*, **156**, 297–311.

Newbery, D. M., Kennedy, D. N., Petol, G. H., Madani, L. and Ridsdale, C. E. (1999) Primary forest dynamics in lowland dipterocarp forest at Danum Valley, Sabah, Malaysia, and the role of the understorey. *Philosophical Transactions of the Royal Society of London Series B* **354**, 1763–1782.

Newbery, D. M. and Lingenfelder, M. (2004) Resistance of a lowland rain forest to increasing drought intensity in Sabah, Borneo. *Journal of Tropical Ecology*, **20**, 613–624.

Newbery, D. M., van der Burgt, X. M. and Moravie, M. A. (2004) Structure and inferred dynamics of a large grove of *Microberlinia bisulcata* trees in central African rain forest: the possible role of periods of multiple disturbance events. *Journal of Tropical Ecology* **20**, 131–143.

Newmark, W. D. (1991) Tropical forest fragmentation and the local extinction of understorey birds in the Eastern Usambara Mountains, Tanzania. *Conservation Biology* **5**, 67–78.

Newson, L. A. (1996) The population of the Amazon basin in 1492: A view from the Ecuadorian headwaters. *Transactions of the Institute of British Geographers* **21**, 5–26.

Newton-Fisher, N. E. (1999) The diet of chimpanzees in the Budongo Forest Reserve, Uganda. *African Journal of Ecology* **37**, 344–354.

Ng, F. S. P. (1980) Germination ecology of Malaysian woody plants. *Malaysian Forester* **43**, 406–437.

Ng, P. (1994a) Diversity and conservation of blackwater fishes in Peninsular Malaysia, particularly in the North Selangor peat swamp forest. *Hydrobiologia* **285**, 203–218.

Ng, P. K. L. (1994b) Peat swamp fishes of Southeast Asia—Diversity under threat. *Wallaceana* **73**, 1–5.

Ng, P. K. L. and Lim, K. K. P. (1993) The Southeast Asian catfish genus *Encheloclarias* (Teleostei: Clariidae), with descriptions of four new species. *Ichthyol. Explor. Freshwaters* **4**, 21–37.

Ngabo, C. K. M. and Dranzoa, C. (2001) Bird communities in gaps of Budongo Forest Reserve, Uganda. *Ostrich* **Supplement 15**, 38–43.

NHM. (2008) World's longest insect revealed. Retrieved 24 November 2008, from http://www.nhm.ac.uk/about-us/news/2008/october/worlds-longest-insect-revealed.html.

Nicholls, H. (2004) The conservation business. *PloS Biology* **2**, 1256–1259.

Nickrent, D. L. (2002) Plantas parásitas en el mundo. *Plantas Parásitas de la Península Ibérica e Islas Baleares.* López-Sáez, J. A., Catalán, P. and Sáez, L. Madrid, Spain, Mundi-Prensa Libros, S. A.: 7–27.

Nicotra, A. B., Chazdon, R. L. and Iriarte, S. V. B. (1999) Spatial heterogeneity of light and woody seedling regeneration in tropical wet forests. *Ecology* **80**, 1908–1926.

Nieder, J., Prospera, J. and Michaloud, G. (2001) Epiphytes and their contribution to canopy diversity. *Plant Ecology* **153**, 51–63.

Pp.Niesten, E., Frumhoff, P. C., Manion, M. and Hardner, J. J. (2002) Designing a carbon market that protects forests in developing countries. *Philosophical Transactions of the Royal Society of London Series A* **360**, 1875–1888.

Niesten, E. T. and Rice, R. E. (2006) Conservation incentic agreements as an alternative to tropical forest exploitation. In: *Emerging Threats to Tropical Forests.* Laurance, W. F. and Peres, C. A. (eds.). University of Chicago Press, Chicago, IL, USA. Pp. 337–352.

Niinemets, U. and Valladares, F. (2004) Photosynthetic acclimation to simultaneous and interacting environmental stresses along natural light gradients: Optimality and constraints. *Plant Biology* **6**, 254–268.

Niklas, K. J. (1995) Size-dependent allometry of tree height, diameter and trunk-taper. *Annals of Botany* **75**, 217–227.

Nilsson, L. A., Rabakonandrianina, E., Pettersson, B. and Grunmeier, R. (1993) Lemur pollination in the Malagasy rainforest liana *Strongylodon craveniae* (Leguminosae) *Evolutionary Trends in Plants* **7**, 49–56.

Normand, S., Vormisto, J., Svenning, J. C., Grandez, C. and Balslev, H. (2006) Geographical and environmental controls of palm beta diversity in paleo-riverine terrace forests in Amazonian Peru. *Plant Ecology* **186**, 161–176.

Norton, A., English-Loeb, G. and Belden, E. (2001) Host plant manipulation of natural enemies: leaf domatia protect beneficial mites from insect predators. *Oecologia* **126**, 535–542.

Noss, A. J. (2000) Cable snares and nets in the Central African Republic. In: *Hunting for Sustainability in Tropical Forests.* Robinson, J. G. and Bennett, E. L. (eds.). Columbia University Press, New York, USA. Pp. 282–304.

Novacek, M. J. (1999) 100 million years of land vertebrate evolution: The Cretaceous-Early Tertiary transition. *Annals of the Missouri Botanical Garden* **86**, 230–258.

Novick, R. R., Dick, C. W., Lemes, M. R., Navarro, C., Caccone, A. and Bermingham, E. (2003) Genetic structure of mesoamerican populations of big-leaf mahogany (*Swietenia macrophylla*) inferred from microsatellite analysis. *Molecular Ecology* **12**, 2885–2893.

Novotny, V. and Basset, Y. (2005) Host specificity of insect herbivores in tropical forests. *Proceedings of the Royal Society B* **272**, 1083–1090.

Novotny, V., Basset, Y., Miller, S. E., Drozd, P. and Cizek, L. (2002) Host specialization of leaf-chewing insects in a New Guinea rainforest. *Journal of Animal Ecology* **71**, 400–412.

Novotny, V., Basset, Y., Miller, S. E., Weiblen, G. D., Bremer, B., Cizek, L. and Drozd, P. (2002) Low host specificity of herbivorous insects in a tropical forest. *Nature* **416**, 841–844.

Novotny, V., Miller, S. E., Basset, Y., Cizek, L., Darrow, K., Kaupa, B., Kua, J. and Weiblen, G. D. (2005) An altitudinal comparison of caterpillar (Lepidoptera) assemblages on *Ficus* trees in Papua New Guinea. *Journal of Biogeography* **32**, 1303–1314.

Novotny, V., Miller, S. E., Cizek, L., Leps, J., Janda, M., Basset, Y., Weiblen, G. D. and Darrow, K. (2003) Colonising aliens: caterpillars (Lepidoptera) feeding on *Piper aduncum* and *P. umbellatum* in rainforests of Papua New Guinea. *Ecological Entomology* **28**, 704–716.

Novozhilov, Y. K., Schnittler, M., Rollins, A. W. and Stephenson, S. L. (2001) Myxomycetes from different forest types in Puerto Rico. *Mycotaxon* **77**, 285–299.

Numata, S., Kachi, N., Okuda, T. and Manokaran, N. (2004) Delayed greening, leaf expansion, and damage to sympatric Shorea species in a lowland rain forest. *Journal of Plant Research* **117**, 19–25.

Nunez-Iturri, G. and Howe, H. F. (2007) Bushmeat and the fate of trees with seeds dispersed by large primates in a lowland rain forest in western Amazonia. *Biotropica* **39**, 348–354.

NyHagen, D. F., Kragelund, C., Olesen, J. M. and Jones, C. G. (2001) Insular interactions between lizards and flowers: flower visitation by an endemic Mauritian gecko. *Journal of Tropical Ecology* **17**, 755–761.

Oates, J. F. (1999) *Myth and reality in the rain forest: How conservation strategies are failing in West Africa.* University of California Press, Berkeley, CA, USA.

Oates, J. F., Whitesides, G. H., Davies, A. G., Waterman, P. G., Green, S. M., Dasilva, G. L. and Mole, S. (1990) Determinants of variation in tropical forest primate biomass: New evidence from West Africa. *Ecology* **71**, 328–343.

Oberbauer, S.F., Strain, B.R. and Riechers, G.H. (1987) Field water relations of a wet-tropical forest tree species, pentaclethra-macroloba (mimosaceae) *Oecologia*, **71**, 369–374.

Oberhuber, W. and Edwards, G.E. (1993) Temperature-dependence of the linkage of quantum yield of photosystem-II to CO_2 fixation in C_4 and C_3 plants. *Plant Physiology*, **101**, 507–512.

O'Brien, T. G., Kinnaird, M. F., Dierenfeld, E. S., Conklin-Brittain, N. L., Wrangham, R. W. and Silver, S. C. (1998) What's so special about figs? *Nature* **392**, 668–668.

Odegaard, F. (2000) How many species of arthropods? Erwin's estimate revised. *Biological Journal of The Linnean Society* **71**, 583–597.

O'Dowd, D. J., Green, P. T. and Lake, P. S. (2003) Invasional 'meltdown' on an oceanic island. *Ecology Letters* **6**, 812–817.

O'Dowd, D. J. and Willson, M. F. (1989) Leaf domatia and mites on Australasian plants: Ecological and evolutionary implications. *Biological Journal of the Linnean Society* **37**, 191–236.

O'sDowd, D. J. and Willson, M. F. (1991) Associations between mites and leaf domatia. *Trends in Ecology and Evolution* **6**, 179–182.

Ogden, C. (1990) *Who is Dependent on Forest and Tree Foods? A Review of the Literature.* Rome, The Community Forestry Unit, Food and Agriculture Organization of the United Nations. Pp. 36.

Oguchi, R., Hikosaka, K. and Hirose, T. (2005) Leaf anatomy as a constraint for photosynthetic acclimation: Differential responses in leaf anatomy to increasing growth irradiance among three deciduous trees. *Plant Cell and Environment*, **28**, 916–927.

Ohlemüller, R., Anderson, B. J., Araujo, M. B., Butchart, S. H. M., Kudrna, O., Ridgely, R. S. and Thomas, C. D. (2008) The coincidence of climatic and species rarity: high risk to small-range species from climate change. *Biology Letters* **4**, 568–572.

Ohsawa, M. (1991) Structural Comparison of Tropical Montane Rain-Forests Along Latitudinal and Altitudinal Gradients in South and East-Asia. *Vegetatio* **97**, 1–10.

Ohsawa, M. (1993) Latitudinal pattern of mountain vegetation zonation in southern and eastern Asia. *Journal of Vegetation Science* **4**, 13–18.

Ohsawa, M. (1995) Latitudinal comparison of altitudinal changes in forest structure, leaf-type, and species richness in humid monsoon Asia. *Vegetatio* **121**, 3–10.

Olander, L. P., Gibbs, H. K., Steininger, M., Swenson, J. J. and Murray, B. C. (2008) Reference scenarios for deforestation and forest degradation in support of REDD: a review of data and methods. *Environmental Research Letters* **3**(2).

Oldroyd, B. P. and Wongsiri, S. (2006) *Asian Honey Bees: Biology, Conservation, and Human Interactions.* Harvard University Press, Cambridge, MA, USA.

Olesen, J. M. and Jordano, P. (2002) Geographic patterns in plant-pollinator mutualistic networks. *Ecology* **83**, 2416–2424.

Olesen, J. M. and Valido, A. (2003) Lizards as pollinators and seed dispersers: an island phenomenon. *Trends in Ecology and Evolution* **18**, 177–181.

Oliveira, P. J. C., Asner, G. P., Knapp, D. E., Almeyda, A., Galvan-Gildemeister, R., Keene, S., Raybin, R. F. and Smith, R. C. (2007) Land-use allocation protects the Peruvian Amazon. *Science* **317**, 1233–1236.

Ollerton, J. and Cranmer, L. (2002) Latitudinal trends in plant-pollinator interactions: are tropical plants more specialised? *Oikos* **98**, 340–350.

Ollerton, J., Johnson, S. D. and Hingston, A. B. (2006) Geographic variation in diversity and specificity of pollination systems. In: *Plant-pollinator interactions: From specialization to generalization.* Waser, N. M. and Ollerton, J. (eds.). University of Chicago Press, Chicago, IL, USA. Pp. 283–308.

Olson, D. M. and Dinerstein, E. (2000) *The Global 200: Priority ecoregions for global conservation.* 47th Annual Systematics Symposium of the Missouri Botanical Garden, St Louis, Missouri, USA.

Onguene, N. A. and T. W. Kuyper (2002) Importance of the ectomycorrhizal network for seedling survival and ectomycorrhiza formation in rain forests of south Cameroon. *Mycorrhiza* **12**, 13–17.

Ortiz, E. G. (2002) Brazil nut (*Bertholletia excelsa*). In: *Tapping the Green Market: Certification and Management of Non-Timber Forest Products.* Shanley, P., Pierce, A. R., Laird, S. A. and Guillen, A. (eds.). Earthscan, London, UK. Pp. 61–74.

Osada, N., Takeda, H., Furukawa, A. and Awang, M. (2001) Fruit dispersal of two dipterocarp species in a Malaysian rain forest. *Journal of Tropical Ecology* **17**, 911–917.

Osada, N., Takeda, H., Furukawa, A. and Awang, M. (2002) Changes in shoot allometry with increasing tree height in a tropical canopy species, *Elateriospermum tapos*. *Tree Physiology*, **22**, 625-632.

Osmond, C.B. (1994) What is photoinhibition? Some insights from comparisons of shade and sun plants. In: *Photoinhibition of photosynthesis: from molecular mechanisms to the field*. Baker, N. R. and Bowyer, J. R. (eds.). BIOS Scientific, Oxford, UK. Pp. 1-24.

Osorio, D., Smith, A. C., Vorobyev, M. and Buchanan-Smith, H. M. (2004) Detection of fruit and the selection of primate visual pigments for color vision. *American Naturalist* **164**, 696-708.

Ostertag, R. (1998) Belowground effects of canopy gaps in a tropical wet forest. *Ecology*, **79**, 1294-1304.

Ostertag, R. (2001) Effects of nitrogen and phosphorus availability on fine-root dynamics in Hawaiian montane forests. *Ecology*, **82**, 485-499.

Ostertag, R., Scatena, F. N. and Silver, W. L. (2003) Forest floor decomposition following hurricane litter inputs in several Puerto Rican forests. *Ecosystems* **6**, 261-273.

Ostertag, R., Silver, W. L. and Lugo, A. E. (2005) Factors affecting mortality and resistance to damage following hurricanes in a rehabilitated subtropical moist forest. *Biotropica* **37**, 16-24.

Osunkoya, O. O. (1996) Light requirements for regeneration in tropical forest plants: Taxon-level and ecological attribute effects. *Australian Journal of Ecology*, **21**, 429-441.

Otero-Arnaiz, A. and Oyama, K. (2001) Reproductive phenology, seed-set and pollination in *Chamaedorea alternans*, an understorey dioecious palm in a rain forest in Mexico. *Journal of Tropical Ecology* **17**, 745-754.

Owiunji, I. (2000) Changes in avian communities of Budongo Forest Reserve after 70 years of selective logging. *Ostrich* **71**, 216-219.

Paciorek, C. J., Condit, R., Hubbell, S. P. and Foster, R. B. (2000) The demographics of resprouting in tree and shrub species of a moist tropical forest. *Journal of Ecology* **88**, 765-777.

Padmanaba, M. and Sheil, D. (2007) Finding and promoting a local conservation consensus in a globally important tropical forest landscape. *Biodiversity and Conservation* **16**, 137-151.

Page, C. N. (2002) Ecological strategies in fern evolution: a neopteridological overview. *Review of Palaeobotany and Palynology* **119**(1-2), 1-33.

Page, S. E., Rieley, J. O., Shotyk, O. W. and Weiss, D. (1999) Interdependence of peat and vegetation in a tropical peat swamp forest. *Philosophical Transactions of the Royal Society of London Series B* **354**, 1885-1897.

Page, S. E., Siegert, F., Rieley, J. O., Boehm, H. D. V., Jaya, A. and Limin, S. (2002) The amount of carbon released from peat and forest fires in Indonesia during 1997. *Nature* **420**, 61-65.

Pagiola, S. (2008) Payments for environmental services in Costa Rica. *Ecological Economics* **65**, 712-724.

Paini, D. R. (2004) Impact of the introduced honey bee (*Apis mellifera*) (Hymenoptera: Apidae) on native bees: A review. *Austral Ecology* **29**, 399-407.

Pan, A. D., Jacobs, B. F., Dransfield, J. and Baker, W. J. (2006) The fossil history of palms (Arecaceae) in Africa and new records from the Late Oligocene (28-27 Mya) of north-western Ethiopia. *Botanical Journal of the Linnean Society* **151**, 69-81.

Paoli, G. D. and Curran, L. M. (2007) Soil nutrients limit fine litter production and tree growth in mature lowland forest of southwestern Borneo. *Ecosystems*, **10**, 503-518.

Paoli, G. D., Curran, L. M. and Slik, J. W. F. (2008) Soil nutrients affect spatial patterns of aboveground biomass and emergent tree density in southwestern Borneo. *Oecologia*, **155**, 287-299.

Paoli, G. D., Curran, L. M. and Zak, D. R. (2005) Phosphorus efficiency of Bornean rain forest productivity: Evidence against the unimodal efficiency hypothesis. *Ecology*, **86**, 1548-1561.

Paoli, G. D., Curran, L. M. and Zak, D. R. (2006) Soil nutrients and beta diversity in the Bornean Dipterocarpaceae: evidence for niche partitioning by tropical rain forest trees. *Journal of Ecology* **94**, 157-170.

Paoli, G. D., Peart, D. R., Leighton, M. and Samsoedin, I. (2001) An ecological and economic assessment of the nontimber forest product gaharu wood in Gunung Palung National Park, West Kalimantan, Indonesia. *Conservation Biology* **15**, 1721-1732.

Parmentier, I., Malhi, Y., Senterre, B., Whittaker, R. J., Alonso, A., Balinga, M. P. B., Bakayoko, A., Bongers, F., Chatelain, C., Comiskey, J. A., Cortay, R., Kamdem, M. N. D., Doucet, J. L., Gautier, L., Hawthorne, W. D., Issembe, Y. A., Kouame, F. N., Kouka, L. A., Leal, M. E., Lejoly, J., Lewis, S. L., Nusbaumer, L., Parren, M. P. E., Peh, K. S. H., Phillips, O. L., Sheil, D., Sonke, B., Sosef, M. S. M., Sunderland, T. C. H., Stropp, J., Ter Steege, H., Swaine, M. D., Tchouto, M. G. P., van Gemerden, B. S., van Valkenburg, J. and Woll, H. (2007) The odd man out? Might climate explain the lower tree alpha-diversity of African rain forests relative to Amazonian rain forests? *Journal of Ecology* **95**, 1058-1071.

Parmentier, I., Stevart, T. and Hardy, O. J. (2005) The inselberg flora of Atlantic Central Africa. I. Determinants of species assemblages. *Journal of Biogeography* **32**, 685-696.

Parolin, P., De Simone, O., Haase, K., Waldhoff, D., Rottenberger, S., Kuhn, U., Kesselmeier, J., Kleiss, B., Schmidt, W., Piedade, M. T. F. and Junk, W. J. (2004a) Central Amazonian floodplain forests: Tree adaptations in a pulsing system. *Botanical Review* **70**, 357-380.

Parolin, P., Ferreira, L.V., Albernaz, A. and Almeida, S.S. (2004b) Tree species distribution in varzea forests of Brazilian Amazonia. *Folia Geobotanica*, **39**, 371-383.

Passos, L. and Oliveira, P. S. (2002) Ants affect the distribution and performance of seedlings of *Clusia criuva*, a primarily bird-dispersed rain forest tree. *Journal of Ecology* **90**, 517–528.

Passos, L. and Oliveira, P. S. (2004) Interaction between ants and fruits of *Guapira opposita* (Nyctaginaceae) in a Brazilian sandy plain rainforest: Ant effects on seeds and seedlings. *Oecologia* **139**, 376–382.

Patel, A., Anstett, M. C., Hossaertmckey, M. and Kjellberg, F. (1995) Pollinators entering female dioecious figs: Why commit suicide. *Journal of Evolutionary Biology* **8**, 301–313.

Pattanayak, S. K. and Yasuoka, J. (2008) Deforestation and malaria: revisiting the human ecology perspective. In: *Human Health and Forests: A Global Overview of Issues, Practice and Policy*. Colfer, C. J. P. (ed.). Earthscan, London, UK. Pp. 197–217.

Patterson, B. D. (2000) Patterns and trends in the discovery of new Neotropical mammals. *Diversity and Distributions* **6**, 145–151.

Patterson, B. D., Stotz, D. F., Solari, S., Fitzpatrick, J. W. and Pacheco, V. (1998) Contrasting patterns of elevational zonation for birds and mammals in the Andes of southeastern Peru. *Journal of Biogeography* **25**, 593–607.

Patz, J. A., Daszak, P., Tabor, G. M., Aguirre, A. A., Pearl, M., Epstein, J., Wolfe, N. D., Kilpatrick, A. M., Foufopoulos, J., Molyneux, D., Bradley, D. J. (2004) Unhealthy landscapes: Policy recommendations on land use change and infectious disease emergence. *Environmental Health Perspectives* **112**, 1092–1098.

Paul, M.J. and Pellny, T.K. (2003) Carbon metabolite feedback regulation of leaf photosynthesis and development. *Journal of Experimental Botany* **54**, 539–547.

Pauw, A., Van Bael, S. A., Peters, H. A., Allison, S. D., Camargo, J. L. C., Cifuentes-Jara, M., Conserva, A., Restom, T. G., Heartsill-Scalley, T., Mangan, S. A., Nunez-Iturri, G., Rivera-Ocasio, E., Rountree, M., Vetter, S. and de Castilho, C. V. (2004) Physical damage in relation to carbon allocation strategies of tropical forest tree saplings. *Biotropica* **36**, 410–413.

Payne, J. (1995) Links between vertebrates and the conservation of Southeast Asian rainforests. In: *Ecology, Conservation and Managemnet of Southeast Asian Rainforests*. Primack, R. B. and Lovejoy, T. E. (eds.). Yale University Press, New Haven, CT, USA. Pp. 54–65.

Paz, H. (2003) Root/shoot allocation and root architecture in seedlings: Variation among forest sites, microhabitats, and ecological groups. *Biotropica*, **35**, 318–332.

Pearce, D. and Puroshothaman, S. (1995) The economic value of plant-based pharmaceuticals. In: *Intellectual Property Rights and Biodiversity Conservation*. Swanson, T. M. (ed.). Cambridge University Press, Cambridge, UK. Pp. 127–138.

Pearce, D., Putz, F. E. and Vanclay, J. K. (2003) Sustainable forestry in the tropics: panacea or folly? *Forest Ecology and Management* **172**, 229–247.

Pearcy, R. W. (1983) The light environment and growth of C_3 and C_4 tree species in the understorey of a Hawaiian forest. *Oecologia*, **58**, 19–25.

Pearcy, R. W., Chazdon, R. L., Gross, L. J. and Mott, K. A. (1994) Photosynthetic utilization of sunflecks: A temporally patchy resource on a time scale of seconds to minutes. In: *Exploitation of environmental heterogeneity by plants: Ecophysiological processes above and below ground*. Caldwell, M. M. and Pearcy, R. W. (eds.). Academic Press, San Diego, CA, USA. Pp. 175–208.

Pearcy, R. W., Muraoka, H. and Valladares, F. (2005) Crown architecture in sun and shade environments: Assessing function and trade-offs with a three-dimensional simulation model. *New Phytologist*, **166**, 791–800.

Pearson, D. L. (1977) Pan-tropical comparison of bird community structure on six lowland forest sites. *Condor* **79**, 232–244.

Pearson, T. R. H., Burslem, D. F. R. P., Mullins, C. E. and Dalling, J. W. (2002) Germination ecology of neotropical pioneers: Interacting effects of environmental conditions and seed size. *Ecology* **83**, 2798–2807.

Pearson, T. R. H., Burslem, D. F. R. P., Mullins, C. E. and Dalling, J. W. (2003) Functional significance of photoblastic germination in neotropical pioneer trees: a seed's eye view. *Functional Ecology* **17**, 394–402.

Pelissier, R., Dray, S. and Sabatier, D. (2002) Within-plot relationships between tree species occurrences and hydrological soil constraints: an example in French Guiana investigated through canonical correlation analysis. *Plant Ecology* **162**, 143–156.

Peluso, N. L. (1993) Traditions of forest control in Java: implications for social forestry and sustainability. *Global Ecology and Biogeography Letters* **3**, 138–157.

Peña-Claros, M., Fredericksen, T. S., Alarcon, A., Blate, G. M., Choque, U., Leano, C., Licona, J. C., Mostacedo, B., Pariona, W., Villegas, Z. and Putz, F. E. (2008) Beyond reduced-impact logging: Silvicultural treatments to increase growth rates of tropical trees. *Forest Ecology and Management* **256**, 1458–1467.

Penalosa, J. (1984) Basal branching and vegetative spread in two tropical rain forest lianas. *Biotropica* **16**, 1–9.

Pendry, C. A. and Proctor, J. (1996) The causes of altitudinal zonation of rain forests on Bukit Belalong, Brunei. *Journal of Ecology* **84**, 407–418.

Pendry, C. A. and Proctor, J. (1997) Altitudinal zonation of rain forest on Bukit Belalong, Brunei: Soils, forest structure and floristics. *Journal of Tropical Ecology* **13**, 221–241.

Pennington, R. T. and Dick, C. W. (2004) *The role of immigrants in the assembly of the South American rainforest tree flora.* Discussion Meeting on Plant Phylogeny and the Origin of Major Biomes, London, UK.

Penuelas, J. and Munne-Bosch, S. (2005) Isoprenoids: an evolutionary pool for photoprotection. *Trends in Plant Science* **10**, 166–169.

Pereira, I. M., Andrade, L. A., Sampaio, E. and Barbosa, M. R. V. (2003) Use-history effects on structure and flora of caatinga. *Biotropica* **35**, 154–165.

Pereira, O. L., Kasuya, M. C. M., Borges, A. C. and Araújo, E. F. (2005) Morphological and molecular characterization of mycorrhizal fungi isolated from neotropical orchids in Brazil. *Canadian Journal of Botany* **83**, 54–65.

Peres, C. A. (1994) Indigenous reserves and nature conservation in Amazonian forests. *Conservation Biology* **8**, 586–588.

Peres, C. A. (1997) Primate community structure at twenty western Amazonian flooded and unflooded forests. *Journal of Tropical Ecology* **13**, 381–405.

Peres, C. A. (1999) Effects of subsistence hunting and forest types on the structure of Amazonian primate communities. In: *Primate Communities.* Fleagle, J. G., Janson, C. and Read, C. K. (eds.). Cambridge University Press, Cambridge, UK. Pp. 268–283.

Peres, C. A. (2000) Effects of subsistence hunting on vertebrate community structure in Amazonian forests. *Conservation Biology* **14**, 240–253.

Peres, C. A. and Baider, C. (1997) Seed dispersal, spatial distribution and population structure of Brazilnut trees (*Bertholletia excelsa*) in southeastern Amazonia. *Journal of Tropical Ecology* **13**, 595–616.

Peres, C. A. and Palacios, E. (2007) Basin-wide effects of game harvest on vertebrate population densities in Amazonian forests: Implications for animal-mediated seed dispersal. *Biotropica* **39**, 304–315.

Peres, C. A. and van Roosmalen, M. (2001) Primate frugivory in two species-rich neotropical forests: Implications for the demography of large seeded plants in overhunted areas. In *Seed Dispersal and Frugivory: Ecology, Evolution and Conservation.* Levey, D. J., Silva, W. R. and Galetti, M. (eds.). CABI Publishing, Wallingford, UK. Pp. 407–421.

Peres, C. A. and van Roosmalen, M. G. M. (1996) Avian dispersal of 'mimetic seeds' of *Ormosia lignivalvis* by terrestrial granivores: Deception or mutualism? *Oikos* **75**, 249–258.

Perez, V. R., Godfrey, L. R., Nowak-Kemp, M., Burney, D. A., Ratsimbazafy, J. and Vasey, N. (2005) Evidence of early butchery of giant lemurs in Madagascar. *Journal of Human Evolution* **49**, 722–742.

Perfecto, I., Vandermeer, J., Mas, A. and Pinto, L. S. (2005) Biodiversity, yield, and shade coffee certification. *Ecological Economics* **54**, 435–446.

Perreijn, K. (2002) *Symbiotic Nitrogen Fixation by Leguminous Trees in Tropical Rain Forests in Guyana.* Tropenbos-Guyana Series 11.PhD Thesis, University of Utrecht, The Netherlands.

Peters, C. M., Gentry, A. H. and Mendelsohn, R. O. (1989) Valuation of an Amazonian rainforest. *Nature* **339**, 655–656.

Peters, H. A. (2001) *Clidemia hirta* invasion at the Pasoh Forest Reserve: An unexpected plant invasion in an undisturbed tropical forest. *Biotropica* **33**, 60–68.

Peters, H. A. (2003) Neighbour-regulated mortality: the influence of positive and negative density dependence on tree populations in species-rich tropical forests. *Ecology Letters* **6**, 757–765.

Peters, H. A., Pauw, A., Silman, M. R. and Terborgh, J. W. (2004) Failing palm fronds structure Amazonian rainforest sapling communities. *Proceedings of the Royal Society of London Series B* **271**: S367–S369.

Peters, W. J. and Neuenschwander, L. F. (1988) *Slash and Burn: Farming in the Third World Forest.* University of Idaho Press, Moscow, ID, USA. Pp. 113.

Petit, R. J. and Hampe, A. (2006) Some evolutionary consequences of being a tree. *Annual Review of Ecology Evolution and Systematics* **37**, 187–214.

Pfannes, K.R. and Baier, A.C. (2002) Devil's gardens in the Ecuadorian Amazon: Association of the allelopathic tree *Duroia hirsuta* (Rubiaceae) and its gentle ants. *Revista De Biologia Tropical* **50**, 293–301.

Pfeiffer, M. and Linsenmair, K. E. (2001) Territoriality in the Malaysian giant ant *Camponotus gigas* (Hymenoptera/Formicidae). *Journal of Ethology* **19**, 75–85.

Pfund, J.-L. (2000) *Culture sur brulis et gestion des ressources naturelles, evolution et perspectives de trois terroirs ruraux du versant est de Madagascar.* Thesis. ETH Zurich, Zurich, Switzerland.

Phillips, J. and Firth, A. (1990) *Introduction to Intellectual Property Law.* Butterworths, London, UK. Pp. 488.

Phillips, O. L., Aragao, L., Lewis, S. L., Fisher, J. B., Lloyd, J., Lopez-Gonzalez, G., Malhi, Y., Monteagudo, A., Peacock, J., Quesada, C. A., van der Heijden, G., Almeida, S., Amaral, I., Arroyo, L., Aymard, G., Baker, T. R., Banki, O., Blanc, L., Bonal, D., Brando, P., Chave, J., de Oliveira, A. C. A., Cardozo, N. D., Czimczik, C. I., Feldpausch, T. R., Freitas, M. A., Gloor, E., Higuchi, N., Jimenez, E., Lloyd, G., Meir, P., Mendoza, C., Morel, A., Neill, D. A., Nepstad, D., Patino, S., Penuela, M. C., Prieto, A., Ramirez, F., Schwarz, M., Silva, J., Silveira, M., Thomas, A. S., ter Steege, H., Stropp, J., Vasquez, R., Zelazowski, P., Davila, E. A., Andelman, S., Andrade, A., Chao, K. J., Erwin, T., Di Fiore, A., Honorio, E., Keeling, H., Killeen, T. J., Laurance, W. F., Cruz, A. P., Pitman, N. C. A., Vargas, P. N., Ramirez-Angulo, H., Rudas, A., Salamao, R., Silva, N.,

Terborgh, J. and Torres-Lezama, A. (2009) Drought sensitivity of the Amazon rainforest. *Science* **323**, 1344–1347.

Phillips, O. L., Baker, T. R., Arroyo, L., Higuchi, N., Killeen, T., Laurance, W. F., Lewis, S. L., Lloyd, J. J., Malhi, Y., Monteagudo, A., Neill, D., Nunez Vargas, P., Silva, N., J., T., Vasquez, M., Alexiades, M., Almeida, S., Brown, S., J., C., Comiskey, J., Czimczik, C., Di Fiore, A., Erwin, T., C., K., Laurance, S. G., Nascimento, H. E. M., Olivier, J., Palacios, W., Patino, S., Pitman, N., C.A., Q., Saldias, M., Torres Lezama, A. and Vincenti, B. (2005) Late twentieth-century patterns and trends in Amazon tree turnover. In *Tropical Forests and Global Atmospheric Change*. Malhi, Y. and Phillips, O. L. (eds.). Oxford University Press, Oxford, UK. Pp. 107–128.

Phillips, O. L., Baker, T. R., Arroyo, L., Higuchi, N., Killeen, T. J., Laurance, W. F., Lewis, S. L., Lloyd, J., Malhi, Y., Monteagudo, A., Neill, D. A., Vargas, P. N., Silva, J. N. M., Terborgh, J., Martinez, R. V., Alexiades, M., Almeida, S., Brown, S., Chave, J., Comiskey, J. A., Czimczik, C. I., Di Fiore, A., Erwin, T., Kuebler, C., Laurance, S. G., Nascimento, H. E. M., Olivier, J., Palacios, W., Patino, S., Pitman, N. C. A., Quesada, C. A., Salidas, M., Lezama, A. T. and Vinceti, B. (2004) Pattern and process in Amazon tree turnover, 1976–2001. *Philosophical Transactions of the Royal Society of London Series B* **359**, 381–407.

Phillips, O. L. and Gentry, A. H. (1994) Increasing turnover through time in tropical forests. *Science* **263**, 954–958.

Phillips, O., Gentry, A., Hall, P., Sawyer, S. and Vasquez, R. (1994) Dynamics and species richness of tropical rain forests. *Proceedings of the National Academy of Science, USA.* **91**, 2805–2809.

Phillips, O., Hall, P., Sawyer, S. and Vasquez, R. (1997) Species richness, tropical forest dynamics, and sampling: Response. *Oikos* **79**, 183–187.

Phillips, O. L., Lewis, S. L., Baker, T. R., Chao, K. J. and Higuchi, N. (2008) The changing Amazon forest. *Philosophical Transactions of the Royal Society B* **363**, 1819–1827.

Phillips, O. L., Martinez, R. V., Arroyo, L., Baker, T. R., Killeen, T., Lewis, S. L., Malhi, Y., Mendoza, A. M., Neill, D., Vargas, P. N., Alexiades, M., Ceron, C., Di Fiore, A., Erwin, T., Jardim, A., Palacios, W., Saldias, M. and Vinceti, B. (2002) Increasing dominance of large lianas in Amazonian forests. *Nature* **418**, 770–774.

Phillips, O. L., Vargas, P. N., Monteagudo, A. L., Cruz, A. P., Zans, M. E. C., Sanchez, W. G., Yli-Halla, M. and Rose, S. (2003) Habitat association among Amazonian tree species: a landscape-scale approach. *Journal of Ecology* **91**, 757–775.

Phillips, T. (2007) *Invisible but all too real: the illegal roads speeding destruction of the rainforest*. The Guardian, 21 April 2007. Manchester. Available online at: http://www.guardian.co.uk/environment/2007/apr/21/Brazil.conservationandendangeredspecies

Pickett, S. T. A. and Kempf, J. S. (1980) Branching Patterns in Forest Shrubs and Understory Trees in Relation to Habitat. *New Phytologist* **86**, 219–228.

Pierce, S., Maxwell, K., Griffiths, H. and Winter, K. (2001) Hydrophobic trichome layers and epicuticular wax powders in bromeliaceae. *American Journal of Botany* **88**, 1371–1389.

Piippo, S., Tan, B. C., Murphy, D. H., Juslen, A. and Meng-Shyan, C. (2002) *A guide to the common liverworts and hornworts of Singapore*. Singapore Science Centre.

Pimentel, D., McNair, M., Duck, L., Pimentel, M. and Kamil, J. (1997) The value of forests to world food security. *Human Ecology* **25**, 91–120.

Pimm, S., Raven, P., Peterson, A., Sekercioglu, C. H. and Ehrlich, P. R. (2006) Human impacts on the rates of recent, present, and future bird extinctions. *Proceedings of the National Academy of Sciences of the United States of America* **103**, 10941–10946.

Pimm, S. L. (1986) Community stability and structure. *Conservation Biology: The Science of Scarcity and Diversity*. Soulé, M. E. Sunderland, Sinauer Associates, MA, USA. Pp. 309–329.

Pimm, S. L. (1991) *The Balance of Nature: Ecological Issues in the Conservation of Species and Communities*. University of Chicago Press, Chicago, IL, USA.

Pimm, S. L. and Lawton, J. H. (1977) Number of trophic levels in ecological communities. *Nature* **268**, 329–331.

Pimm, S. L. and Lawton, J. H. (1978) Feeding on more than one trophic level. *Nature* **275**, 542–544.

Pimm, S. L. and Raven, P. (2000) Extinction by numbers. *Nature* **403**, 843–845.

Pimm, S. L., Russell, G. J., Gittleman, J. L. and Brooks, T. M. (1995) The future of biodiversity. *Science* **269**, 347–350.

Piperno, D. R. and Pearsall, D. M. (1998) *The Origins of Agriculture in the Lowland Neotropics*. Academic Press, San Diego.

Pitman, N. C. A., Ceron, C. E., Reyes, C. I., Thurber, M. and Arellano, J. (2005) Catastrophic natural origin of a species-poor tree community in the world's richest forest. *Journal of Tropical Ecology* **21**, 559–568.

Pitman, N. C. A. and Jorgensen, P. M. (2002) Estimating the size of the world's threatened flora. *Science* **298**, 989.

Pitman, N. C. A., Terborgh, J., Silman, M. R. and Nuez, P. (1999) Tree species distributions in an upper Amazonian forest. *Ecology* **80**, 2651–2661.

Pizo, M. A. (1997) Seed dispersal and predation in two populations of *Cabralea canjerana* (Meliaceae) in the Atlantic forest of southeastern Brazil. *Journal of Tropical Ecology* **13**, 559–578.

Pizo, M. A. (2008) The use of seeds by a twig-dwelling ant on the floor of a tropical rain forest. *Biotropica* **40**, 119–121.

Pizo, M. A. and Oliveira, P. S. (2000) The use of fruits and seeds by ants in the Atlantic forest of southeast Brazil. *Biotropica* **32**, 851–861.

Pizo, M. A. and Oliveira, P. S. (2001) Size and lipid content of nonmyrmecochorous diaspores: effects on the interaction with litter-foraging ants in the Atlantic Rain Forest of Brazil. *Plant Ecology* **157**, 37–52.

Poffenberger, M. and McGean, B. (1996) *Village Voices, Forest Choices: Joint Forest Management in India.* Oxford University Press, Delhi.

Poinar, G. (2005) *Culex malariager*, N. sp (Diptera: Culicidae) from Dominican amber: The first fossil mosquito vector of plasmodium. *Proceedings of the Entomological Society of Washington* **107**, 548–553.

Poinar, G. O. and Poinar, R. (1999) *The amber forest: reconstruction of a vanished world.* Princeton University Press, Princeton, NJ, USA.

Poinar, G. O., Jr and Danforth, B. N. (2006) A fossil bee from early Cretaceous Burmese amber. *Science* **314**, 614.

Pons, T., Alexander, E., Houter, N., Rose, S. and Rijkers, T. (2005) Ecophysiological patterns in guianan forest plants. *Tropical Forests of the Guiana Shield.* Hammond, D. S. (ed.). CAB International, Wallingford, UK. Pp. 195–231.

Pons, T.L. and Welschen, R.A.M. (2004) Midday depression of net photosynthesis in the tropical rainforest tree eperua grandiflora: Contributions of stomatal and internal conductances, respiration and rubisco functioning. *Tree Physiology*, **24**, 599–599.

Pontzer, H. and Wrangham, R. W. (2004) Climbing and the daily energy cost of locomotion in wild chimpanzees: implications for hominoid locomotor evolution. *Journal of Human Evolution* **46**, 317–335.

Poorter, H., Pepin, S., Rijkers, T., De Jong, Y., Evans, J.R. and Korner, C. (2006a) Construction costs, chemical composition and payback time of high- and low-irradiance leaves. *Journal of Experimental Botany* **57**, 355–371.

Poorter, L. (1999) Growth responses of 15 rain forest tree species to a light gradient: The relative importance of morphological and physiological traits. *Functional Ecology* **13**, 396–410.

Poorter, L. (2005) Resource capture and use by tropical forest tree seedlings and their consequences for competition. *Biotic Interaction in the Tropics: Their Role in the Maintenance of Species Diversity.* Burslem, D. F. R. P., Pinard M. A. and Hartley S. E. (eds.). Cambridge University Press, Cambridge, UK.

Poorter, L. (2007) Are species adapted to their regeneration niche, adult niche, or both? *American Naturalist* **169**, 433–442.

Poorter, L. (2009) Leaf traits show different relationships with species shade tolerance in moist versus dry tropical forests. *New Phytologist* **181**, 890–900.

Poorter, L. and Arets, E. (2003) Light environment and tree strategies in a Bolivian tropical moist forest: an evaluation of the light partitioning hypothesis. *Plant Ecology* **166**, 295–306.

Poorter, L. and Bongers, F. (2006) Leaf traits are good predictors of plant performance across 53 rain forest species. *Ecology* **87**, 1733–1743.

Poorter, L., Bongers, L. and Bongers, F. (2006b) Architecture of 54 moist-forest tree species: Traits, trade-offs, and functional groups. *Ecology* **87**, 1289–1301.

Poorter, L. and Kitajima, K. (2007) Carbohydrate storage and light requirements of tropical moist and dry forest tree species. *Ecology* **88**, 1000–1011.

Poorter, L. and Markesteijn, L. (2008) Seedling traits determine drought tolerance of tropical tree species. *Biotropica* **40**, 321–331.

Poorter, L. and Rozendaal, D. M. A. (2008) Leaf size and leaf display of thirty-eight tropical tree species. *Oecologia* **158**, 35–46.

Poorter, L., Wright, S.J., Paz, H., Ackerly, D.D., Condit, R., Ibarra-Manriques, G., Harms, K.E., Licona, J.C., Martinez-Ramos, M., Mazer, S.J., Muller-Landau, H.C., Peña-Claros, M., Webb, C.O. and Wright, I.J. (2008) Are functional traits good predictors of demographic rates? Evidence from five neotropical forests. *Ecology* **89**, 1908–1920.

Poorter, L., Zuidema, P. A., Peña-Claros, M. and Boot, R. G. A. (2005) A monocarpic tree species in a polycarpic world: How can *Tachigali vasquezii* maintain itself so successfully in a tropical rain forest community? *Journal of Ecology* **93**, 268–278.

Porembski, S. and Barthlott, W. (2000) Granitic and gneissic outcrops (inselbergs) as centers of diversity for desiccation-tolerant vascular plants. *Plant Ecology* **151**, 19–28.

Porembski, S., Szarzynski, J., Mund, J. P. and Barthlott, W. (1996) Biodiversity and vegetation of small-sized inselbergs in a West African rain forest (Tai, Ivory Coast) *Journal of Biogeography* **23**, 47–55.

Portnoy, S. and Willson, M. F. (1993) Seed dispersal curves: Behavior of the tail of the distribution. *Evolutionary Ecology* **7**, 25–44.

Posey, D. A. (2002) Commodification of the sacred through intellectual property rights. *Journal of Ethnopharmacology* **83**, 3–12.

Posey, D. A., Dutfield, G. and Plenderleith, K. (1995) Collaborative research and intellectual property rights. *Biodiversity and Conservation* **4**, 892–902.

Poszwa, A., Dambrine, E., Ferry, B., Pollier, B. and Loubet, M. (2002) Do deep tree roots provide nutrients to the tropical rainforest? *Biogeochemistry* **60**, 97–118.

Potter, C. S., Alexander, S. E., Coughlan, J. C. and Klooster, S. A. (2001) Modeling biogenic emissions of isoprene: exploration of model drivers, climate control algorithms, and use of global satellite observations. *Atmospheric Environment* **35**, 6151–6165.

Potter, L. M. (1997) A forest product out of control: *Gutta percha* in Indonesia and the wider Malay world, 1845–1915. In: *Paper Landscapes: Exploration in the Environmental History of Indonesia*. Boomgard, P., Colombijn, F. and Henley, D. (eds.) KITLV Press, Leiden, The Netherlands. Pp. 281–308.

Potts, M. D., Ashton, P. S., Kaufman, L. S. and Plotkin, J. B. (2002) Habitat patterns in tropical rain forests: A comparison of 105 plots in northwest Borneo. *Ecology* **83**, 2782–2797.

Poulsen, A. D. and Balslev, H. (1991) Abundance and cover of ground herbs in an Amazonian rainforest. *Journal of Vegetation Science* **2**, 315–322.

Poulsen, A. D., Tuomisto, H. and Balslev, H. (2006) Edaphic and floristic variation within a 1-ha plot of lowland Amazonian rain forest. *Biotropica* **38**, 468–478.

Poulsen, J. R., Clark, C. J., Connor, E. F. and Smith, T. B. (2002) Differential resource use by primates and hornbills: Implications for seed dispersal. *Ecology* **83**, 228–240.

Pounds, J. A., Bustamante, M. R., Coloma, L. A., Consuegra, J. A., Fogden, M. P. L., Foster, P. N., La Marca, E., Masters, K. L., Merino-Viteri, A., Puschendorf, R., Ron, S. R., Sanchez-Azofeifa, G. A., Still, C. J. and Young, B. E. (2006) Widespread amphibian extinctions from epidemic disease driven by global warming. *Nature* **439**, 161–167.

Poux, C., Chevret, P., Huchon, D., de Jong, W. W. and Douzery, E. J. P. (2006) Arrival and diversification of caviomorph rodents and platyrrhine primates in South America. *Systematic Biology* **55**, 228–244.

Powell, A. H. and Powell, G. V. N. (1987) Population dynamics of male euglossine bees in Amazonian forest fragments. *Biotropica* **19**, 176–179.

Prance, G. T. (1980) A note on the probable pollination of *Combretum* by *Cebus* Monkeys. *Biotropica* **12**, 239–239.

Prance, G. T. (1985) The pollination of Amazonian plants. *Key Environments: Amazonia*. Prance, G. T. and Lovejoy, T. E. (eds.) Pergamon Press, New York. Pp. 166–191.

Prance, G. T., Beentje, H., J., D. and Johns, R. (2000) The tropical flora remains under collected. *Annals of the Missouri Botanical Gardens* **87**, 67–71.

Prather, M., Ehalt, D., Dentener, F., Derwent, R., Dlugokencky, E., Holland, E., Isaksen, I., Katima, J., Kirchhoff, V., Matson, P., Midgley, P. and Wang, M. (2001) Atmospheric chemistry and greenhouse gases. *Climate change 2001: the scientific basis. Third Assessment Report of the Intergovernmental Panel on Climate Change*. Houghton J.T., D. Y., Griggs D.J., Noguer M., Van Der Linden P.J., Dai X., Maskell K. And Johnson C.A. Cambridge University Press, Cambridge, NY, USA.

Prentice, I. C. (1998) *Interactions of climate change and the terrestrial biosphere*. Vatican Conference on Geosphere-Biosphere Interactions and Climate, Vatican City, Rome, Italy.

Press, M. C. and Phoenix, G. K. (2005) Impacts of parasitic plants on natural communities. *New Phytologist* **166**, 737–751.

Pretty, J. (2002) *Agri-culture; Reconnecting people, land and nature*. Earthscan, London, UK.

Price, J. P. (2004) Floristic biogeography of the Hawaiian Islands: influences of area, environment and paleogeography. *Journal of Biogeography* **31**, 487–500.

Price, J. P. and Wagner, W. L. (2004) Speciation in Hawaiian angiosperm lineages: Cause, consequence, and mode. *Evolution* **58**, 2185–2200.

Priess, J., Then, C. and Folster, H. (1999) Litter and fine-root production in three types of tropical premontane rain forest in SE Venezuela. *Plant Ecology* **143**, 171–187.

Primack, R. B. and Corlett, R. T. (2005) *Tropical Rainforests: An Ecological and Biogeographical Comparison*. Blackwell, Oxford, UK.

Pringle, E. G., Alvarez-Loayza, P. and Terborgh, J. (2007) Seed characteristics and susceptibility to pathogen attack in tree seeds of the Peruvian Amazon. *Plant Ecology* **193**, 211–222.

Proctor, J. (2003) Vegetation and soil and plant chemistry on ultramafic rocks in the tropical far east. *Perspectives in Plant Ecology Evolution and Systematics*, **6**, 105–124.

Proctor, J., Brearley, F. Q., Dunlop, H., Proctor, K., Supramono and Taylor, D. (2001) Local wind damage in Barito Ulu, Central Kalimantan: a rare but essential event in a lowland dipterocarp forest? *Journal of Tropical Ecology* **17**, 473–475.

Prugnolle, F., Rousteau, A. and Belin-Depoux, M. (2001) Spatial occupation of *Cyathea muricata* Willd. (Cyatheaceae) in Guadeloupean tropical rain forest. II. At the population level. *Acta Botanica Gallica*, **148**, 81–91.

Pueyo, S., He, F. and Zillio, T. (2007) The maximum entropy formalism and the idiosyncratic theory of biodiversity. *Ecology Letters* **10**, 1017–1028.

Puhakka, M., Kalliola, R., Rajasilta, M. and Salo, J. (1992) River Types, Site Evolution and Successional Vegetation Patterns in Peruvian Amazonia. *Journal of Biogeography* **19**, 651–665.

Purseglove, J. W. (1972) *Tropical crops: Monocotyledons 2*. John Wiley and Sons, New York, USA.

Purves, D. and Pacala, S. (2008) Predictive models of forest dynamics. *Science* **320**, 1452–1453.

Puttker, T., Meyer-Lucht, Y. and Sommer, S. (2008) Fragmentation effects on population density of three rodent species in secondary Atlantic Rainforest, Brazil. *Studies on Neotropical Fauna and Environment* **43**, 11–18.

Putz, F. E. (1984) The natural history of lianas on Barro Colorado Island, Panama. *Ecology* **65**, 1713–1724.

Putz, F. E. and Brokaw, N. V. L. (1989) Sprouting of broken trees on Barro Colorado Island, Panama. *Ecology* **70**, 508–512.

Putz, F. E., Coley, P. D., Lu, K., Montalvo, A. and Aiello, A. (1983) Uprooting and snapping of trees: structural determinants and ecological consequences. *Canadian Journal of Forest Research* **13**, 1011–1020.

Putz, F. E. and Mooney, H. A. (1991) *The Biology of Vines.* Cambridge University Press, Cambridge, UK.

Putz, F. E., Parker, G. G. and Archibald, R. M. (1984) Mechanical abrasion and intercrown spacing. *American Midland Naturalist* **112**, 24–28.

Putz, F. E., Sist, P., Fredericksen, T. and Dykstra, D. (2008) Reduced-impact logging: Challenges and opportunities. *Forest Ecology and Management* **256**, 1427–1433.

Pyke, C. R., Condit, R., Aguilar, S. and Lao, S. (2001) Floristic composition across a climatic gradient in a neotropical lowland forest. *Journal of Vegetation Science* **12**, 553–566.

Quintana-Ascencio, P. F., Gonzalez-Espinosa, M., Ramirez-Marcial, N., Dominguez-Vazquez, G. and Martinez-Ico, M. (1996) Soil seed banks and regeneration of tropical rain forest from milpa fields at the Silva Lacandona, Chiapas, Mexico. *Biotropica* **28**, 192–209.

Quintero, I. and Roslin, T. (2005) Rapid recovery of dung beetle communities following habitat fragmentation in central Amazonia. *Ecology* **86**, 3303–3311.

Raaimakers, D. and Lambers, H. (1996) Response to phosphorus supply of tropical tree seedlings: A comparison between a pioneer species *Tapirira obtusa* and a climax species *Lecythis corrugata*. *New Phytologist*, **132**, 97–102.

Raaimakers, D., Boot, R.G.A., Dijkstra, P., Pot, S. and Pons, T. (1995) Photosynthetic rates in relation to leaf phosphorus content in pioneer versus climax tropical rainforest trees. *Oecologia*, **102**, 120–125.

Raffauf, R. F. (1996) *Plant Alkaloids: A Guide To Their Discovery and Distribution.* Food Products Press and The Haworth Press, London, UK.

Rahbek, C. (1995) The elevational gradient of species richness: A uniform pattern. *Ecography* **18**, 200–205.

Rahbek, C. (1997) The relationship among area, elevation, and regional species richness in neotropical birds. *American Naturalist* **149**, 875–902.

Raich, J. W., Russell, A. E., Kitayama, K., Parton, W. J. and Vitousek, P. M. (2006) Temperature influences carbon accumulation in moist tropical forests. *Ecology* **87**, 76–87.

Rametsteiner, E. and Simula, M. (2003) Forest certification: An instrument to promote sustainable forest management? *Journal of Environmental Management* **67**, 87–98.

Ramirez, S. R., Gravendeel, B., Singer, R. B., Marshall, C. R. and Pierce, N. E. (2007) Dating the origin of the Orchidaceae from a fossil orchid with its pollinator. *Nature* **448**, 1042–1045.

Rampino, M. R. and Self, S. (1992) Volcanic winter and accelerated glaciation following the Toba super-eruption. *Nature* **359**, 50–52.

Rao, M. (2000) Variation in leaf-cutter ant (*Atta* sp.) densities in forest isolates: the potential role of predation. *Journal of Tropical Ecology* **16**, 209–225.

Rao, M., Terborgh, J. and Nunez, P. (2001) Increased herbivory in forest isolates: Implications for plant community structure and composition. *Conservation Biology* **15**, 624–633.

Ratiarison, S. and Forget, P. M. (2005) Frugivores and seed removal at *Tetragastris altissima* (Burseraceae) in a fragmented forested landscape of French Guiana. *Journal of Tropical Ecology* **21**, 501–508.

Raunkiaer, C. (1934) *The Life Forms of Plants and Statistical Plant Geography.* Clarendon Press, Oxford, UK.

Raven, P. H. and Axelrod, D. I. (1974) Angiosperm biogeography and past continental movements. *Annals of the Missouri Botanical Garden* **61**, 539–673.

Raymond, A. and Metz, C. (2004) Ice and its consequences: Glaciation in the Late Ordovician, Late Devonian, Pennsylvanian-Permian, and Cenozoic compared. *Journal of Geology* **112**, 655–670.

Read, J. and Sanson, G.D. (2003) Characterizing sclerophylly: The mechanical properties of a diverse range of leaf types. *New Phytologist*, **160**, 81–99.

Reagan, D. P., Camilo, G. R. and Waide, R. B. (1996) The community food web: Major properties and patterns of organisation. *The Food Web of a Tropical Rain Forest.* Reagan, D. P. and Waide, R. B. (eds.) University of Chicago Press, Chicago, IL, USA. Pp. 462–510.

Reagan, D. P. and Waide, R. B. (1996) *The Food Web of a Tropical Rain Forest.* University of Chicago Press, Chicago, IL, USA.

Reay, D. S., Dentener, F., Smith, P., Grace, J. and Feely, R. A. (2008) Global nitrogen deposition and carbon sinks. *Nature Geoscience* **1**, 430–437.

Rebelo, C. F. and Williamson, G. B. (1996) Driptips vis-à-vis soil types in central Amazonia. *Biotropica* **28**, 159–163.

Rebertus, A. J. (1988) Crown shyness in a tropical cloud forest. *Biotropica* **20**, 338–339.

Reed, K. E. and Bidner, L. R. (2004) Primate communities: Past, present, and possible future. *American Journal of Physical Anthropology* **Suppl. 39**, 2–39.

Rees, P. M., Noto, C. R., Parrish, J. M. and Parrish, J. T. (2004) Late Jurassic climates, vegetation, and dinosaur distributions. *Journal of Geology* **112**, 643–653.

Regal, P. J. (1982) Pollination by wind and animals: ecology of geographic patterns. *Annual Review of Ecology and Systematics* **13**, 497–524.

Reich, P. B., Ellsworth, D. S. and Uhl, C. (1995) Leaf carbon and nutrient assimilation and conservation in species of differing successional status in an oligotrophic Amazonian forest. *Functional Ecology* **9**, 65–76.

Reich, P. B., Hobbie, S. E., Lee, T., Ellsworth, D. S., West, J. B., Tilman, D., Knops, J. M. H., Naeem, S. and Trost, J. (2006)

Nitrogen limitation constrains sustainability of ecosystem response to CO_2. *Nature* **440**, 922–925.

Reich, P. B., Uhl, C., Walters, M. B. and Ellsworth, D. S. (1991) Leaf life-span as a determinant of leaf structure and function among 23 Amazonian tree species. *Oecologia* **86**, 16–24.

Reich, P. B., Walters, M. B. and Ellsworth, D. S. (1997) From tropics to tundra: Global convergence in plant functioning. *Proceedings of the National Academy of Sciences of the United States of America* **94**, 13730–13734.

Reich, P. B., Walters, M. B., Ellsworth, D. S. and Uhl, C. (1994) Photosynthesis-nitrogen relations in Amazonian tree species. 1. Patterns among species and communities. *Oecologia* **97**, 62–72.

Reich, P. B., Walters, M. B., Ellsworth, D. S., Vose, J. M., Volin, J. C., Gresham, C. and Bowman, W. D. (1998) Relationships of leaf dark respiration to leaf nitrogen, specific leaf area and leaf life-span: A test across biomes and functional groups. *Oecologia* **114**, 471–482.

Reich, P. B., Wright, I. J., Cavender-Bares, J., Craine, J. M., Oleksyn, J., Westoby, M. and Walters, M. B. (2003) The evolution of plant functional variation: Traits, spectra, and strategies. *International Journal of Plant Sciences* **164**, S143–S164.

Reid, N., Smith, M. S. and Yan, Z. (1995) Ecology and Population Ecology of Mistletoes. *Forest Canopies*. Lowman, M. D. and Nadkarni, N. M. (eds.). Academic Press, San Diego, California, USA. Pp. 285–310.

Reilly, W. K. (1990) Debt-for-nature swaps: The time has come. *International Environmental Affairs* **2**, 134–140.

Reinhard, J. and Rowell, D. M. (2005) Social behaviour in an Australian velvet worm, *Euperipatoides rowelli* (Onychophora: Peripatopsidae) *Journal of Zoology* **267**, 1–7.

Remington, C. L. (1968) Suture-zones of hybrid interaction between recently joined biotas. *Evolutionary Biology*. Dobzhansky, T., Hecht, M. K. and Steere, C. W. (eds.) Plenum Press, New York. Pp. 321–428.

Remsen, J. V., Hyde, M. A. and Chapman, A. (1993) The diets of Neotropical trogons, motmots, barbets and toucans. *Condor* **95**, 178–192.

Renner, S. (2004) Plant dispersal across the tropical Atlantic by wind and sea currents. *International Journal of Plant Sciences* **165**, S23–S33.

Renner, S. S. and Feil, J. P. (1993) Pollinators of tropical dioecious angiosperms. *American Journal of Botany* **80**, 1100–1107.

Report of the Intergovernmental Panel on Climate Change. Cambridge University Press, Cambridge, UK, and New York, USA.

Restrepo, C. and Alvarez, N. (2006) Landslides and their contribution to land-cover change in the mountains of Mexico and Central America. *Biotropica* **38**, 446–457.

Retallack, G. J. (2001) Cenozoic expansion of grasslands and climatic cooling. *Journal of Geology* **109**, 407–426.

Retallack, G. J. and Germanheins, J. (1994) Evidence from Paleosols for the geological antiquity of rainforest. *Science* **265**, 499–502.

Reys, P., Sabino, J. and Galetti, M. (2009) Frugivory by the fish *Brycon hilarii* (Characidae) in western Brazil. *Acta Oecologica* **35**, 136–141.

Rice, R. E., Sugal, C. A., Ratay, S. M. and de Fonseca, G. A. B. (2001) *Sustainable forest management: A review of conventional wisdom*. Advances in Applied Biodiversity Science, CABS/Conservation International, Washington, DC, USA. **3**. Pp. 1–29.

Richards, P. W. (1973) The tropical rain forest. *Scientific American* **229**, 58–67.

Richards, P. W. (1996) *The Tropical Rain Forest: An Ecological Study*. Cambridge University Press, Cambridge, UK.

Richardson, J. E., Chatrou, L. W., Mols, J. B., Erkens, R. H. J. and Pirie, M. D. (2004) Historical biogeography of two cosmopolitan families of flowering plants: Annonaceae and Rhamnaceae. *Philosophical Transactions of the Royal Society of London Series B* **359**, 1495–1508.

Richardson, J. E., Pennington, R. T., Pennington, T. D. and Hollingsworth, P. M. (2001) Rapid diversification of a species-rich genus of neotropical rain forest trees. *Science* **293**, 2242–2245.

Ricketts, T. H. (2001) The matrix matters: Effective isolation in fragmented landscapes. *American Naturalist* **158**, 87–99.

Ricketts, T. H., et al. (2005) Pinpointing and preventing imminent extinctions. *Proceedings of the National Academy of Scences of the United States of America* **102**, 18497–18501.

Ricklefs, R. E. (2005) Phylogenetic perspectives on patterns of regional and local species richness. *Tropical Rainforests: Past, Present and Future*. Bermingham, E., Dick, C. W. and Moritz, C. (eds.) University of Chicago Press, Chicago, IL, USA. Pp. 16–40.

Ricklefs, R. E. and Bermingham, E. (2007) The causes of evolutionary radiations in archipelagoes: Passerine birds in the Lesser Antilles. *American Naturalist* **169**, 285–297.

Ricklefs, R. E. and Schluter, D. (1993) *Species Diversity in Ecological Communities: Historical and Geographic Perspectives*. University of Chicago Press, Chicago, IL, USA.

Rico-Gray, V. and Oliveira, P. S. (2007) *The Ecology and Evolution of Ant-Plant Interactions*. University of Chicago Press, Chicago, IL, USA.

Ridley, P., Howse, P. E. and Jackson, C. W. (1996) Control of the behaviour of leaf-cutting ants by their 'symbiotic' fungus. *Experientia* **52**, 631–635.

Riede, K. (1997) Bioacoustic monitoring of insect communities in a Bornean rainforest canopy. *Canopy Arthropods*. N.E., S.,

Adis, J. A. and Didham, R. K. Chapman and Hall, London, UK. Pp. 442–452.

Rieley, J. O., Page, S. E. and Shepherd, P. A. (1997) Tropical Bog Forests of South East Asia. *Conserving Peatlands*. Arkyn, L., Stoneman, R. E. and Ingram, H. A. P. CAB International Wallingford, UK. Pp. 35–41.

Rijkers, T., Jan De Vries, P.J., Pons, T.L. and Bongers, F. (2000) Photosynthetic induction in saplings of three shade-tolerant tree species: Comparing understorey and gap habitats in a french guiana rain forest. *Oecologia*, **125**, 331–340.

Roberts, A. F. (1997) Anarchy, abjection, and absurdity: A case of metaphoric medicine among the Tabwa of Zaire. *The Anthropology of Medicine: From Culture to Method*. Romanucci-Ross, L., Moerman, D. E. and Tancredi, L. R. Bergin and Garvey, Westport, Connecticut, USA. Pp. 224–239.

Robinson, D. and A. Fitter (1999) The magnitude and control of carbon transfer between plants linked by a common mycorrhizal network. *Journal of Experimental Botany* **50**, 9–13.

Robinson, J. G. and Bennett, E. L. (2002) Will alleviating poverty solve the bushmeat crisis? *Oryx* **36**, 332–332.

Robinson, J. G. and Bennett, E. L. (2004) Having your wildlife and eating it too: an analysis of hunting sustainability across tropical ecosystems. *Animal Conservation* **7**, 397–408.

Rockwell, C., Kainer, K. A., Marcondes, N. and Baraloto, C. (2007) Ecological limitations of reduced-impact logging at the smallholder scale. *Forest Ecology and Management* **238**, 365–374.

Rodrigues, A. S. L., Akcakaya, H. R., Andelman, S. J., Bakarr, M. I., Boitani, L., Brooks, T. M., Chanson, J. S., Fishpool, L. D. C., Da Fonseca, G. A. B., Gaston, K. J., Hoffmann, M., Marquet, P. A., Pilgrim, J. D., Pressey, R. L., Schipper, J., Sechrest, W., Stuart, S. N., Underhill, L. G., Waller, R. W., Watts, M. E. J. and Yan, X. (2004) Global gap analysis: Priority regions for expanding the global protected-area network. *Bioscience* **54**, 1092–1100.

Rodrigues, E. (2006) Plants and animals utilized as medicines in the Jau National Park (JNP), Brazilian Amazon. *Phytotherapy Research* **20**, 378–391.

Rogers, R. W., Barnes, A. and Conran, J. G. (1994) Lichen succession on *Wilkiea macrophylla* leaves. *Lichenologist* **26**, 135–147.

Rohde, K. (1992) Latitudinal gradients in species-diversity: The search for the primary cause. *Oikos* **65**, 514–527.

Romero, G. Q. and Benson, W. W. (2004) Leaf domatia mediate mutualism between mites and a tropical tree. *Oecologia* **140**, 609–616.

Roosevelt, A. C., da Costa, M. L., Machado, C. L., Michab, M., Mercier, N., Valladas, H., Feathers, J., Barnett, W., da Silveira,

M. I., Henderson, A., Sliva, J., Chernoff, B., Reese, D. S., Holman, J. A., Toth, N. and Schick, K. (1996) Paleoindian cave dwellers in the Amazon: The peopling of the Americas. *Science* **272**, 373–384.

Rosenzweig, M. (1995) *Species Diversity in Space and Time*. Cambridge University Press, Cambridge, UK.

Ros-Tonen, M. A. F., van Andel, T., Morsello, C., Otsuki, K., Rosendo, S. and Scholz, I. (2008) Forest-related partnerships in Brazilian Amazonia: There is more to sustainable forest management than reduced impact logging. *Forest Ecology and Management* **256**, 1482–1497.

Roth-Nebelsick, A., Uhl, D., Mosbrugger, V. and Kerp, H. (2001) Evolution and function of leaf venation architecture: A review. *Annals of Botany*, **87**, 553–566.

Rottenberger, S., Kuhn, U., Wolf, A., Schebeske, G., Oliva, S. T., Tavares, T. M. and Kesselmeier, J. (2004) Exchange of short-chain aldehydes between Amazonian vegetation and the atmosphere. *Ecological Applications* **14**, S247–S262.

Roubik, D. W. (1989) *Ecology and Natural History of Tropical Bees*. Cambridge University Press, Cambridge, UK.

Roubik, D. W. (2001) Ups and downs in pollinator populations: when is there a decline? *Conservation Ecology* **5**, 2.

Roy, H. E., Steinkraus, D. C. Eilenberg, J., Hajek, A. E. and Pell, J. K. (2006) Bizarre interactions and endgames: Entomopathogenic fungi and their arthropod hosts. *Annual Review of Entomology* **51**, 331–357.

Roy, J. and Salager, J.L. (1992) Midday depression of net CO_2 exchange of leaves of an emergent rain-forest tree in french-guiana. *Journal of Tropical Ecology*, **8**, 499–504.

Roy, M. S. (1997) Recent diversification in African greenbuls (Pycnonotidae: *Andropadus*) supports a montane speciation model. *Proceedings of the Royal Society of London Series B* **264**, 1337–1344.

Royer, D. L., Wilf, P., Janesko, D. A., Kowalski, E. A. and Dilcher, D. L. (2005) Correlations of climate and plant ecology to leaf size and shape: Potential proxies for the fossil record. *American Journal of Botany* **92**, 1141–1151.

Rudel, T. K., Coomes, O. T., Moran, E., Achard, F., Angelsen, A., Xu, J. C. and Lambin, E. (2005) Forest transitions: towards a global understanding of land use change. *Global Environmental Change* **15**, 23–31.

Ruiz-Perez, M., Almeida, M., Dewi, S., Costa, E. M. L., Pantoja, M. C., Puntodewo, A., Postigo, A. D. and de Andrade, A. G. (2005) Conservation and development in Amazonian extractive reserves: The case of Alto Jurua. *Ambio* **34**, 218–223.

Ruiz-Perez, M., Belcher, B., Achdiawan, R., Alexiades, M., Aubertin, C., Caballero, J., Campbell, B., Clement, C., Cunningham, T., Fantini, R., de Foresta, H., Fernandez, C. G., Gautam, K. H., Martinez, P. H., de Jong, W., Kusters, K., Kutty,

M. G., Lopez, C., Fu, M. Y., Alfaro, M. A. M., Nair, T. K. R., Ndoye, O., Ocampo, R., Rai, N., Ricker, M., Schreckenberg, K., Shackleton, S., Shanley, P., Sunderland, T. and Youn, Y. C. (2004) Markets drive the specialization strategies of forest peoples. *Ecology and Society* **9**:2.

Russell, A. E., Raich, J. W. and Vitousek, P. M. (1998) The ecology of the climbing fern *Dicranopteris linearis* on windward Mauna Loa, Hawaii. *Journal of Ecology* **86**, 765–779.

Russell-Smith, A. and Stork, N. E. (1995) Composition of spider communities in the canopies of rain forest trees in Borneo. *Journal of Tropical Ecology* **11**, 223–235.

Russo, S.E., Davies, S.J., King, D.A. and Tan, S. (2005) Soil-related performance variation and distributions of tree species in a Bornean rain forest. *Journal of Ecology*, **93**, 879–889.

Rykiel, E. J. (1985) Towards a definition of ecological disturbance. *Australian Journal of Ecology* **10**, 361–365.

Saakov, S. G. (1983) Parallel variation in palms. *Botanicheskii Zhurnal* **68**, 1235–1241.

Sack, L. and Frole, K. (2006) Leaf structural diversity is related to hydraulic capacity in tropical rain forest trees. *Ecology*, **87**, 483–491.

Sack, L. and Grubb, P.J. (2003) Crossovers in seedling relative growth rates between low and high irradiance: Analyses and ecological potential. *Functional Ecology*, **17**, 281–287.

Sack, L., Tyree, M.T. and Holbrook, N.M. (2005) Leaf hydraulic architecture correlates with regeneration irradiance in tropical rainforest trees. *New Phytologist*, **167**, 403–413.

Sage, R. F. and McKown, A. D. (2006) Is C_4 photosynthesis less phenotypically plastic than C_3 photosynthesis? *Journal of Experimental Botany* **57**, 303–317.

Sagers, C. L. (1993) Reproduction in neotropical shrubs: the occurrence and some mechanisms of asexuality. *Ecology* **74**, 615–618.

Sagers, C. L. and Coley, P. D. (1995) Benefits and costs of defense in a neotropical shrub. *Ecology* **76**, 1835–1843.

Sagers, C. L., Ginger, S. M. and Evans, R. D. (2000) Carbon and nitrogen isotopes trace nutrient exchange in an ant-plant mutualism. *Oecologia* **123**, 582–586.

Saint-Paul, U., Zuanon, J., Correa, M. A. V., Garcia, M., Fabre, N. N., Berger, U. and Junk, W. J. (2000) Fish communities in central Amazonian white- and blackwater floodplains. *Environmental Biology of Fishes* **57**, 235–250.

Sakai, S. (2001) Thrips pollination of androdioecious *Castilla elastica* (Moraceae) in a seasonal tropical forest. *American Journal of Botany* **88**, 1527–1534.

Sakai, S. (2002) General flowering in lowland mixed dipterocarp forests of Southeast Asia. *Biological Journal of the Linnean Society* **75**, 233–247.

Sakai, S., Harrison, R. D., Momose, K., Kuraji, K., Nagamasu, H., Yasunari, T., Chong, L. and Nakashizuka, T. (2006) Irregular droughts trigger mass flowering in aseasonal tropical forests in Asia. *American Journal of Botany* **93**, 1134–1139.

Sakai, S. and Inoue, T. (1999) A new pollination system: Dung-beetle pollination discovered in *Orchidantha inouei* (Lowiaceae, Zingiberales) in Sarawak, Malaysia. *American Journal of Botany* **86**, 56–61.

Salafsky, N., Salzer, D., Stattersfield, A. J., Hilton-Taylor, C., Neugarten, R., Butchart, S. H. M., Collen, B., Cox, N., Master, L. L., O'Connor, S. and Wilkie, D. (2008) A standard lexicon for biodiversity conservation: unified classifications of threats and actions. *Conservation Biology* **22**, 897–911.

Salafsky, N. and Wollenberg, E. (2000) Linking livelihoods and conservation: A conceptual framework and scale for assessing the integration of human needs and biodiversity. *World Development* **28**, 1421–1438.

Salisbury, D. S. and Schmink, M. (2007) Cows versus rubber: Changing livelihoods among Amazonian extractivists. *Geoforum* **38**, 1233–1249.

Salleo, S. and Nardini, A. (2000) Sclerophylly: Evolutionary advantage or mere epiphenomenon? *Plant Biosystems*, **134**, 247–259.

Salm, R. (2006) Invertebrate and vertebrate seed predation in the Amazonian palm *Attalea maripa*. *Biotropica* **38**, 558–560.

Sanchez-Cordero, V. (2001) Elevation gradients of diversity for rodents and bats in Oaxaca, Mexico. *Global Ecology and Biogeography* **10**, 63–76.

Sandbrook, C. and Semple, S. (2006) The rules and the reality of mountain gorilla Gorilla beringei beringei tracking: how close do tourists get? *Oryx* **40**, 428–433.

Sanderson, M. G. (1996) Biomass of termites and their emissions of methane and carbon dioxide: A global database. *Global Biogeochemical Cycles* **10**, 543–557.

Sanford, R. L. J. (1987) Apogeotropic Roots in an Amazon Rain Forest. *Science* **235**, 1062–1064.

Santiago, L. S. (2000) Use of coarse woody debris by the plant community of a Hawaiian montane cloud forest. *Biotropica* **32**(4a), 633–641.

Santiago, L.S., Kitajima, K., Wright, S.J. and Mulkey, S.S. (2004) Coordinated changes in photosynthesis, water relations and leaf nutritional traits of canopy trees along a precipitation gradient in lowland tropical forest. *Oecologia*, **139**, 495–502.

Sastry, C. B. (2002) Rattan in the twenty-first century: an outlook. *Rattan: Current Research Issues and Prospects for Conservation and Sustainable Development*. Food and Agriculture Organization of the United Nations, Rome, Italy.

Saulei, S. M. and Swaine, M. D. (1988) Rainforest seed dynamics during succession at Gogol, Papua New Guinea. *Journal of Ecology* **76**, 1133–1152.

Savenije, H. H. G. (2004) The importance of interception and why we should delete the term evapotranspiration from our vocabulary. *Hydrological Processes* **18**, 1507–1511.

Sayer, J., Elliott, C., Barrow, E., Gretzinger, S., Maginnis, S., McShane, T. and Shepherd, G. (2005) Implications for biodiversity conservation of decentralized forest resources management. *The politics of decentralization: forests, people and power.* Colfer, C. J. P. and Capistrano, D. London, Earthscan. Pp. 121–137.

Sayer, J., McNeely, J., Maginnis, S., Boedhihartono, I., Shepherd, G. and Fisher, B. (2007) *Local rights and tenure for forests: opportunity or threat for conservation.* Rights and Resources Initiative, Washington, DC, USA. Pp. 16.

Scariot, A. (2000) Seedling mortality by litterfall in Amazonian forest fragments. *Biotropica* **32**, 662–669.

Scatena, F. N. and Larsen, M. C. (1991) Physical Aspects of Hurricane Hugo in Puerto-Rico. *Biotropica* **23**, 317–323.

Schatz, G. E., Williamson, G. B., Cogswell, C. M. and Stam, A. C. (1985) Stilt roots and growth of arboreal palms. *Biotropica* **17**, 206–209.

Schelkle, M. and R. L. Peterson (1996) Suppression of common root pathogens by helper bacteria and ectomycorrhizal fungi in vitro. *Mycorrhiza* **6**, 481–485.

Schemske, D. W. and Pautler, L. P. (1984) The effects of pollen composition on fitness components in a neotropical herb. *Oecologia* **62**, 31–36.

Schenka, H. J. and Jackson, R. B. (2005) Mapping the global distribution of deep roots in relation to climate and soil characteristics. *Geoderma* **126**, 129–140.

Schilthuizen, M., Chai, H. N., Kimsin, T. E. and Vermeulen, J. J. (2003) Abundance and diversity of land snails (Mollusca: Gastropoda) on limestone hills in Borneo. *Raffles Bulletin of Zoology* **51**, 35–42.

Schilthuizen, M. and Rutjes, H. A. (2001) Land snail diversity in a square kilometre of tropical rainforest in Sabah, Malaysian Borneo. *Journal of Molluscan Studies* **67**, 417–423.

Schlesinger, W. H. and Andrews, J. A. (2000) Soil respiration and the global carbon cycle. *Biogeochemistry* **48**, 7–20.

Schmidt, G., Stuntz, S. and Zotz, G. (2001) Plant size: an ignored parameter in epiphyte ecophysiology? *Plant Ecology* **153**, 65–72.

Schneider, C. and Moritz, C. (1999) Rainforest refugia and evolution in Australia's Wet Tropics. *Proceedings of the Royal Society of London Series B* **266**, 191–196.

Schneider, C. J., Cunningham, M. and Moritz, C. (1998) Comparative phylogeography and the history of endemic vertebrates in the Wet Tropics rainforests of Australia. *Molecular Ecology* **7**, 487–498.

Schneider, H., Schuettpelz, E., Pryer, K. M., Cranfill, R., Magallon, S. and Lupia, R. (2004) Ferns diversified in the shadow of angiosperms. *Nature* **428**, 553–557.

Schnittler, M. and S. L. Stephenson (2000) Myxomycete biodiversity in four different forest types in Costa Rica. *Mycologia* **92**, 626–637.

Schnitzer, S. A. (2005) A mechanistic explanation for global patterns of liana abundance and distribution. *American Naturalist* **166**, 262–276.

Schnitzer, S. A. and Bongers, F. (2002) The ecology of lianas and their role in forests. *Trends in Ecology and Evolution* **17**, 223–230.

Schnitzer, S. A., Dalling, J. W. and Carson, W. P. (2000) The impact of lianas on tree regeneration in tropical forest canopy gaps: evidence for an alternative pathway of gap-phase regeneration. *Journal of Ecology* **88**, 655–666.

Schnitzer, S. A., Kuzee, M. E. and Bongers, F. (2005) Disentangling above- and below-ground competition between lianas and trees in a tropical forest. *Journal of Ecology*, **93**, 1115–1125.

Schnitzler, H. U., Kalko, E. K. V., Kaipf, I. and Grinnell, A. D. (1994) Fishing and echolocation behavior of the greater bulldog bat, *Noctilio leporinus*, in the field. *Behavioral Ecology and Sociobiology* **35**, 327–345.

Schreuder, M. D. J., Brewer, C. A. and Heine, C. (2001) Modelled influences of non-exchanging trichomes on leaf boundary layers and gas exchange. *Journal of Theoretical Biology*, **210**, 23–32.

Schroth, G. and Burkhardt, J. (2003) Nutrient exchange with the atmosphere. *Trees, Crops and Soil Fertility.* Schroth, G. and Sinclair, F. L. CAB International, Wallingford, Oxon, UK.

Schuepp, P.H. (1993) Tansley review no 59 leaf boundary-layers. *New Phytologist*, **125**, 477–507.

Schulze, C. H., Linsenmair, K. E. and Fiedler, K. (2001) Understorey versus canopy: patterns of vertical stratification and diversity among Lepidoptera in a Bornean rain forest. *Plant Ecology* **153**(1–2), 133–152.

Schulze, M., Grogan, J., Uhl, C., Lentini, M. and Vidal, E. (2008a) Evaluating ipe (*Tabebuia*, Bignoniaceae) logging in Amazonia: Sustainable management or catalyst for forest degradation? *Biological Conservation* **141**, 2071–2085.

Schulze, M., Grogan, J. and Vidal, E. (2008b) Forest certification in Amazonia: standards matter. *Oryx* **42**, 229–239.

Schupp, E. W. and Feener, D. H. J. (1991) Phylogeny, lifeform, and habitat dependence of ant-defended plants in a Panamanian forest. *Ant-Plant Interactions.* Huxley, C. R. and Cutler, D. F. (eds.) Oxford University Press, Oxford, UK. Pp. 175–197.

Schupp, E. W. and Frost, E. J. (1989) Differential predation of *Welfia georgii* seeds in treefall gaps and the forest understory. *Biotropica* **21**, 200–203.

Secrett, C. (1986) The environmental impact of transmigration. *The Ecologist* **16**, 77–88.

Seidel, D. J., Fu, Q., Randel, W. J. and Reichler, T. J. (2008) Widening of the tropical belt in a changing climate. *Nature Geoscience* **1**, 21–24.

Seidler, T. G. and Plotkin, J. B. (2006) Seed dispersal and spatial pattern in tropical trees. *PloS Biology* **4**, 2132–2137.

Sekercioglu, C. H., Daily, G. C. and Ehrlich, P. R. (2004) Ecosystem consequences of bird declines. *Proceedings of the National Academy of Sciences of the United States of America* **101**, 18042–18047.

Sekercioglu, C. H., Ehrlich, P. R., Daily, G. C., Aygen, D., Goehring, D. and Sandi, R. F. (2002) Disappearance of insectivorous birds from tropical forest fragments. *Proceedings of the National Academy of Sciences of the United States of America* **99**, 263–267.

Sekercioglu, C. H., Loarie, S. R., Brenes, F. O., Ehrlich, P. R. and Daily, G. C. (2007) Persistence of forest birds in the Costa Rican agricultural countryside. *Conservation Biology* **21**, 482–494.

Sekhsaria, P. (2007) Conservation in India and the need to think beyond 'Tiger vs. Tribal'. *Biotropica* **39**, 575–577.

Setoguchi, H., Asakawa Osawa, T., Pintaud, J.-C., Jaffre, T. and Veillon, J.-M. (1998) Phylogenetic relationships within Araucariaceae based on rbcL gene sequences. *American Journal of Botany* **85**, 1507–1516.

Shanahan, M., Harrison, R. D., Yamuna, R., Boen, W. and Thornton, I. W. B. (2001a) Colonization of an island volcano, Long Island, Papua New Guinea, and an emergent island, Motmot, in its caldera lake. V. Colonization by figs (*Ficus* spp.), their dispersers and pollinators. *Journal of Biogeography* **28**(11–12), 1365–1377.

Shanahan, M., So, S., Compton, S. G. and Corlett, R. (2001b) Fig-eating by vertebrate frugivores: A global review. *Biological Reviews* **76**, 529–572.

Shanley, P. and Luz, L. (2003) The impacts of forest degradation on medicinal plant use and implications for health care in eastern Amazonia. *Bioscience* **53**, 573–584.

Sharkey, T. D. (2005) Effects of moderate heat stress on photosynthesis: importance of thylakoid reactions, rubisco deactivation, reactive oxygen species, and thermotolerance provided by isoprene. *Plant Cell and Environment* **28**, 269–277.

Sharkey, T. D. and Yeh, S. S. (2001) Isoprene emission from plants. *Annual Review of Plant Physiology and Plant Molecular Biology*, **52**, 407–436.

Sharpe, B. (1998) 'First the forest': Conservation, 'community' and 'participation' in southwest Cameroon. *Africa* **68**, 25–45.

Sheil, D. (1995) A critique of permanent plot methods and analysis with examples from Budongo Forest, Uganda. *Forest Ecology and Management*, **77**, 11–34.

Sheil, D. (1998) A half century of permanent plot observation in Budongo Forest, Uganda: Histories, highlights and hypotheses. *Forest Biodiversity Research, Monitoring and Modeling: Conceptual background and Old World case studies.* Dallmeier, F. and Comiskey, J. A. (eds.) M.A.B. UNESCO, Paris. Pp. 399–428.

Sheil, D. (1999a) Developing tests of successional hypotheses with size-structured populations, and an assessment using long-term data from a Ugandan rain forest. *Plant Ecology* **140**, 117–127.

Sheil, D. (1999b) Tropical forest diversity, environmental change and species augmentation: After the intermediate disturbance hypothesis. *Journal of Vegetation Science* **10**, 851–860.

Sheil, D. (2003) Growth assessment in tropical trees: large daily diameter fluctuations and their concealment by dendrometer bands. *Canadian Journal of Forest Research* **33**, 2027–2035.

Sheil, D. and Burslem, D. F. R. P. (2003) Disturbing hypotheses in tropical forests. *Trends in Ecology and Evolution* **18**, 18–26.

Sheil, D. and Ducey, M. (2002) An extreme-value approach to detect clumping and an application to tropical forest gap-mosaic dynamics. *Journal of Tropical Ecology* **18**, 671–686.

Sheil, D., Jennings, S. and Savill, P. (2000) Long-term permanent plot observations of vegetation dynamics in Budongo, a Ugandan rain forest. *Journal of Tropical Ecology*, **16**, 765–800.

Sheil, D. and Lawrence, A. (2004) Tropical biologists, local people and conservation: new opportunities for collaboration. *Trends in Ecology and Evolution* **19**, 634–638.

Sheil, D. and Murdiyarso, D. (2009) How forests attract rain: An examination of a new hypothesis. *Bioscience* **59**, 341–347.

Sheil, D., Puri, R., Wan, M., Basuki, I., van Heist, M., Liswanti, N., Rukmiyati, Rachmatika, I. and Samsoedin, I. (2006) Recognizing local people's priorities for tropical forest biodiversity. *Ambio* **35**, 17–24.

Sheil, D. and Salim, A. (2004) Forest tree persistence, elephants, and stem scars. *Biotropica* **36**, 505–521.

Sheil, D., Salim, A., Chave, J. R., Vanclay, J. and Hawthorne, W. D. (2006) Illumination-size relationships of 109 coexisting tropical forest tree species. *Journal of Ecology*, **94**, 494–507.

Sheil, D., van Heist, M., Liswanti, N., Basuki, I. and Wan, M. (2009) Biodiversity landscapes and livelihoods: a local perspective. *The Decentralization of Forest Governance: Politics, Economics and the Fight for Control of Forests in Indonesian Borneo.* Moeliono, M., Wollenberg, E. and G. L. Earthscan, London, UK. Pp. 61–87.

Sheil, D. and Wunder, S. (2002) The value of tropical forest to local communities: Complications, caveats, and cautions. *Conservation Ecology* **6**(2).

Pp.Shelford, Robert W. (1916) *A Naturalist in Borneo.* First published by T. Fisher Unwin Ltd, London, UK; reprinted (1985) Oxford University Press, Oxford, UK.

Shepard, G. (2005) Psychoactive botanicals in ritual, religion and shaminism. *Ethnopharmacology*. Elisabetsky, E. and Etkin, N. Oxford, UK, UNESCO/EOLSS Publishers.

Shephard, G., Yu, D., Lizarralde, M. and Italiano, M. (2001) Rain forest habitat classification among the Matsigenka of the Peruvian Amazon. *Journal of Ethnobiology* **21**, 1–38.

Sherman, P. M. (2003) Effects of land crabs on leaf litter distributions and accumulations in a mainland tropical rain forest. *Biotropica* **35**, 365–374.

Shimamura, T. and Momose, K. (2005) Organic matter dynamics control plant species coexistence in a tropical peat swamp forest. *Proceedings of the Royal Society B* **272**, 1503–1510.

Sidle, R. C., Tani, M. and Ziegler, A. D. (2006a) Catchment processes in Southeast Asia: Atmospheric, hydrologic, erosion, nutrient cycling, and management effects. *Forest Ecology and Management* **224**(1–2), 1–4.

Sidle, R. C., Ziegler, A. D., Negishi, J. N., Nik, A. R., Siew, R. and Turkelboom, F. (2006b) Erosion processes in steep terrain: Truths, myths, and uncertainties related to forest management in Southeast Asia. *Forest Ecology and Management* **224**(1–2), 199–225.

Siegert, F., Ruecker, G., Hinrichs, A. and Hoffmann, A. (2001) Increased damage from fires in logged forests during droughts caused by El Nino. *Nature* **414**, 437–440.

Silman, M. R., Terborgh, J. W. and Kiltie, R. A. (2003) Population regulation of a dominant-rain forest tree by a major seed-predator. *Ecology* **84**, 431–438.

Silvera, K., Santiago, L.S. and Winter, K. (2005) Distribution of crassulacean acid metabolism in orchids of panama: Evidence of selection for weak and strong modes. *Functional Plant Biology*, **32**, 397–407.

Simpson, B. and Neff, J. (1983) Evolution and diversity of floral rewards. *Handbook of Experimental Pollination Ecology*. Jones, C. and Little, R. (eds.). Van Nostrand Reinhold, New York, USA. Pp. 277–293.

Simpson, B. B. and Haffer, J. (1978) Speciation patterns in Amazonian forest biota. *Annual Review of Ecology and Systematics* **9**, 497–518.

Sist, P. and Ferreira, F. N. (2007) Sustainability of reduced-impact logging in the Eastern Amazon. *Forest Ecology and Management* **243**(2–3), 199–209.

Sist, P., Sheil, D., Kartawinata, K. and Priyadi, H. (2003) Reduced-impact logging in Indonesian Borneo: Some results confirming the need for new silvicultural prescriptions. *Forest Ecology and Management* **179**, 415–427.

Sizer, N. and Tanner, E. V. J. (1999) Responses of woody plant seedlings to edge formation in a lowland tropical rainforest, Amazonia. *Biological Conservation* **91**(2–3), 135–142.

Skillman, J.B., Garcia, M. and Winter, K. (1999) Whole-plant consequences of crassulacean acid metabolism for a tropical forest understory plant. *Ecology*, **80**, 1584–1593.

Slater, C. (2003) *Entangled Edens: Visions of the Amazon*. University of California Press, Berkeley and Los Angeles, USA.

Slik, J. W. F. and Eichhorn, K. A. O. (2003) Fire survival of lowland tropical rain forest trees in relation to stem diameter and topographic position. *Oecologia* **137**, 446–455.

Slik, J. W. F., Poulsen, A. D., Ashton, P. S., Cannon, C. H., Eichhorn, K. A. O., Kartawinata, K., Lanniari, I., Nagamasu, H., Nakagawa, M., van Nieuwstadt, M. G. L., Payne, J., Purwaningsih, Saridan, A., Sidiyasa, K., Verburg, R. W., Webb, C. O. and Wilkie, P. (2003) A floristic analysis of the lowland dipterocarp forests of Borneo. *Journal of Biogeography* **30**, 1517–1531.

Smith, A. P. (1972) Buttressing of tropical trees: a descriptive model and new hypotheses. *American Naturalist* **106**, 32–46.

Smith, J., Obidzinski, K., Subarudi and Suramenggala, I. (2003) Illegal logging, collusive corruption and fragmented governments in Kalimantan, Indonesia. *International Forestry Review* **5**, 293–302.

Smith, K. R. (2008) Wood: The Fuel that Warms You Thrice. *Human Health and Forests: A Global Overview of Issues, Practice and Policy*. Colfer, C. J. P. Earthscan/CIFOR, London, UK. Pp. 97–112.

Smith, T. and Huston, M. (1989) A theory of the spatial and temporal dynamics of plant-communities. *Vegetatio*, **83**, 49–69.

Smith, T. B., Calsbeek, R., Wayne, R. K., Holder, K. H., Pires, D. and Bardeleben, C. (2005) Testing alternative mechanisms of evolutionary divergence in an African rain forest passerine bird. *Journal of Evolutionary Biology* **18**, 257–268.

Smith, T. B., Wayne, R. K., Girman, D. J. and Bruford, M. W. (1997) A role for ecotones in generating rainforest biodiversity. *Science* **276**, 1855–1857.

Snow, D. W. (1966) A possible selective factor in evolution of fruiting seasons in tropical forest. *Oikos* **15**, 274–&.

Snow, D. W. (1981) Tropical frugivorous birds and their food plants: A world survey. *Biotropica* **13**, 1–14.

Soares-Filho, B. S., Nepstad, D. C., Curran, L. M., Cerqueira, G. C., Garcia, R. A., Ramos, C. A., Voll, E., McDonald, A., Lefebvre, P. and Schlesinger, P. (2006) Modelling conservation in the Amazon basin. *Nature* **440**, 520–523.

Soderstrom, T. and Calderon, C. (1971) Insect pollination in tropical rain forest grasses. *Biotropica* **3**, 1–16.

Sodhi, N. S., Brooks, T. M., Koh, L. P., Acciaioli, G., Erb, M., Tan, A. K. J., Curran, L. M., Brosius, P., Lee, T. M., Patlis, J. M., Gumal, M. and Lee, R. J. (2006) Biodiversity and human livelihood crises in the Malay archipelago. *Conservation Biology* **20**, 1811–1813.

Sodhi, N. S., Liow, L. H. and Bazzaz, F. A. (2004) Avian extinctions from tropical and subtropical forests. *Annual Review of Ecology Evolution and Systematics* **35**, 323–345.

Soejarto, D. D. and Farnsworth, N. R. (1989) Tropical rain forests: potential source of new drugs. *Perspectives in Biology and Medicine* **32**, 244–256.

Sohn, J. (2007) *Protecting the Peruvian Amazon and its People From the Risks of Oil and Gas Development*. World Resources Institute, Washington, DC, USA.

Solds, D. E., Bell, C. D., Kim, S. and Soltis, P. S. (2008) Origin and early evolution of angiosperms. *Year in Evolutionary Biology 2008*. **1133**, 3–25.

Solem, A., Climo, F. M. and Roscoe, D. J. (1981) Sympatric species diversity of New Zealand land snails. *New Zealand Journal of Zoology* **8**, 453–485.

Sollins, P. (1998) Factors influencing species composition in tropical lowland rain forest: Does soil matter? *Ecology*, **79**, 23–30.

Solomona, S., Plattner, G.-K., Knutti, R. and Friedlingstein, P. (2009) Irreversible climate change due to carbon dioxide emissions. *Proceedings of the National Academy of Sciences* **106**, 1704–1709.

Sommer, A. (1976) Attempt at an assessment of the world's tropical forests. *Unasylva* **28**, 5–25.

Sorenson, M. D., Oneal, E., Garcia-Moreno, J. and Mindell, D. P. (2003) More taxa, more characters: The hoatzin problem is still unresolved. *Molecular Biology and Evolution* **20**, 1484–1498.

Sousa, W. P. (1984) The role of disturbance in natural communities. *Annual Review of Ecology and Systematics* **15**, 353–391.

Southgate, D. (1998) *Tropical Forest Conservation*. Oxford University Press, New York, USA.

Spiteri, A. and Nepal, S. K. (2006) Incentive-based conservation programs in developing countries: A review of some key issues and suggestions for improvements. *Environmental Management* **37**, 1–14.

Stadtmüller, T. and Agudelo, N. (1990) Amounts and variability of cloud moisture input in a tropical cloud forest. *International Association of Hydrology Scientific Publication* **193**, 25–32.

Stark, N. M. and Jordan, C. F. (1978) Nutrient Retention by Root Mat of an Amazonian Rain-Forest. *Ecology* **59**, 434–437.

Stattersfield, A. J., Crosby, M. J., Long, A. J. and Wege, D. C. (1998) *Endemic Bird Areas of the World: Priorities for Biodiversity Conservation*. BirdLife International.

Stebbins, G. L. (1974) *Flowering Plants: Evolution Above the Species Level*. Harvard University Press, Cambridge, MA, USA.

Stefenon, V. M., Gailing, O. and Finkeldey, R. (2006) Phylogenetic relationship within genus *Araucaria* (Araucariaceae) assessed by means of AFLP fingerprints. *Silvae Genetica* **55**, 45–52.

Steffen, W., Crutzen, P. J. and McNeill, J. R. (2007) The Anthropocene: Are humans now overwhelming the great forces of nature. *Ambio* **36**, 614–621.

Stein, W. E., Mannolini, F., Hernick, L. V., Landing, E. and Berry, C. M. (2007) Giant cladoxylopsid trees resolve the enigma of the Earth's earliest forest stumps at Gilboa. *Nature* **446**, 904–907.

Steinberg, P. F. (2005) From public concern to policy effectiveness: civic conservation in developing countries. *Journal of International Wildlife Law and Policy* **8**, 341–365.

Steiner, K. E. (1985) The Role of nectar and oil in the pollination of *Drymonia serrulata* (Gesneriaceae) by *Epicharis* bees (Anthophoridae) in Panama. *Biotropica* **17**, 217–229.

Stephenson, N. L. and van Mantgem, P. J. (2005) Forest turnover rates follow global and regional patterns of productivity. *Ecology Letters* **8**, 524–531.

Stephenson, S. L. and J. C. Landolt (1998) Dictyostelid cellular slime molds in canopy soils of tropical forests. *Biotropica* **30**, 657–661.

Stephenson, S. L., Schnittler, M. and Lado, C. (2004) Ecological characterization of a tropical myxomycete assemblage: Maquipucuna Cloud Forest Reserve, Ecuador. *Mycologia* **96**, 488–497.

Stepp, J. R. (2004) The role of weeds as sources of pharmaceuticals. *Journal of Ethnopharmacology* **92**(2–3), 163–166.

Sterck, F.J., Poorter, L. and Schieving, F. (2006) Leaf traits determine the growth-survival trade-off across rain forest tree species. *American Naturalist*, **167**, 758–765.

Stern, N. (2007) *The Economics of Climate Change*. Cambridge University Press, Cambridge, UK.

Stevens, G.C. (1987) Lianas as structural parasites: The *Bursera simaruba* example. *Ecology*, **68**, 77–81.

Steyermark, J. A. (1979) Flora of the Guayana Highland: Endemicity of the generic flora of the summits of the Venezuela tepuis. *Taxon* **28**, 45–54.

Stiles, F. G. (1977) Coadapted competitors: Flowering seasons of hummingbird-pollinated plants in a tropical forest. *Science* **198**, 1177–1178.

Stocker, G. C. and Irvine, A. K. (1983) Seed dispersal by cassowaries (*Casuarius casuarius*) in North Queenslands rainforests. *Biotropica* **15**, 170–176.

Stoner, K. E., Riba-Hernandez, P., Vulinec, K. and Lambert, J. E. (2007) The role of mammals in creating and modifying seedshadows in tropical forests and some possible consequences of their elimination. *Biotropica* **39**, 316–327.

Stork, N. E. (1988) Insect diversity: facts, fiction and speculation. *Biological Journal of the Linnean Society* **35**, 321–337.

Stork, N. E. (1993) How many species are there. *Biodiversity and Conservation* **2**, 215–232.

Stork, N. E. and Blackburn, T. M. (1993) Abundance, body size and biomass of arthropods in tropical forest. *Oikos* **67**, 483–489.

Stouffer, P. C. and Bierregaard, R. O. (1995) Use of Amazonian forest fragments by understory insectivorous birds. *Ecology* **76**, 2429-2445.

Strasberg, D., Faloya, V. and Lepart, J. (1995) Patterns of tree mortality in an island tropical rainforest subjected to recurrent windstorms. *Acta Oecologica* **16**, 237-248.

Strassburg, B., Turner, R. K., Fisher, B., Schaeffer, R. and Lovett., A. (2009) Reducing emissions from deforestation: The 'combined incentives' mechanism and empirical simulations. *Global Environmental Change* **19**, 265-278.

Struhsaker, T. T. (1975) *The Red Colobus Monkey*. University of Chicago Press, Chicago, IL, USA.

Struhsaker, T. T. (1997) *Ecology of an African Rain Forest: Logging in Kibale and the Conflict between Conservation and Exploitation*. University Press of Florida, Gainesville, FL, USA.

Stuart, S. N., Chanson, J. S., Cox, N. A., Young, B. E., Rodrigues, A. S. L., Fischman, D. L. and Waller, R. W. (2004) Status and trends of amphibian declines and extinctions worldwide. *Science* **306**, 1783-1786.

Stuart, T. S. (1968) Revival of respiration and photosynthesis in dried leaves of *Polypodium polypodioides*. *Planta* **83**, 185-206.

Styger, E., Rakotondramasy, H. M., Pfeffer, M. J., Fernandes, E. C. M. and Bates, D. M. (2007) Influence of slash-and-burn farming practices on fallow succession and land degradation in the rainforest region of Madagascar. *Agriculture, Ecosystems and Environment* **119**, 257-269.

Sugimoto, A., Inoue, T., Tayasu, I., Miller, L., Takeichi, S. and Abe, T. (1998) Methane and hydrogen production in a termite-symbiont system. *Ecological Research* **13**, 241-257.

Suh, S. O., McHugh, J. V., Pollock, D. D. and Blackwell, M. (2005) The beetle gut: a hyperdiverse source of novel yeasts. *Mycological Research* **109**, 261-265.

Sun, C. and Moermond, T. C. (1997) Foraging ecology of three sympatric *Turacos* in a montane forest in Rwanda. *Auk* **114**, 396-404.

Sunderland, T. C. H. and Dransfield, J. (2002) Rattan (various spp.) *Tapping the Green Market: Certification and Management of Non-Timber Forest Products*. Shanley, P., Pierce, A. R., Laird, S. A. and Guillen, A. Earthscan, London, UK. Pp. 225-239.

Sunderlin, W. D., Angelsen, A., Belcher, B., Burgers, P., Nasi, R., Santoso, L. and Wunder, S. (2005) Livelihoods, forests, and conservation in developing countries: An overview. *World Development* **33**, 1383-1402.

Sunderlin, W. D., Hatcher, J. and Liddle, M. (2008) *From Exclusion to Ownership? Challenges and Opportunities in Advancing Forest Tenure Reform*. Rights and Resources Initiative, Washington, DC, USA.

Sussman, R. W. and Raven, P. H. (1978) Pollination by lemurs and marsupials: Archaic coevolutionary system. *Science* **200**, 731-736.

Svenning, J. C. (2000) Growth strategies of clonal palms (Arecaceae) in a neotropical rainforest, Yasuni, Ecuador. *Australian Journal of Botany* **48**, 167-178.

Svenning, J. C., Engelbrecht, B. M. J., Kinner, D. A., Kursar, T. A., Stallard, R. F. and Wright, S. J. (2006) The relative roles of environment, history and local dispersal in controlling the distributions of common tree and shrub species in a tropical forest landscape, Panama. *Journal of Tropical Ecology* **22**, 575-586.

Svenning, J. C., Kinner, D. A., Stallard, R. F., Engelbrecht, B. M. J. and Wright, S. J. (2004) Ecological determinism in plant community structure across a tropical forest landscape. *Ecology* **85**, 2526-2538.

Swaine, M. D. (1996) Rainfall and soil fertility as factors limiting forest species distributions in Ghana. *Journal of Ecology* **84**, 419-428.

Swaine, M. D. and Hall, J. B. (1983) Early succession on cleared forest land in Ghana. *Journal of Ecology* **71**, 601-627.

Swaine, M. D. and Whitmore, T. C. (1988) On the definition of ecological species groups in tropical rain forests. *Vegetatio* **75**, 81-86.

Swanborough, P., Doley, D., Keenan, R. and Yates, D. (1998) Photosynthetic characteristics of *flindersia brayleyana* and *castanospermum australe* from tropical lowland and upland sites. *Tree Physiology*, **18**, 341-347.

Swanson, A. R., Vadell, E. M. and Cavender, J. C. (1999) Global distribution of forest soil dictyostelids. *Journal of Biogeography* **26**, 133-148.

Swenson, U., Backlund, A., McLoughlin, S. and Hill, R. S. (2001a) Nothofagus Biogeography Revisited with Special Emphasis on the Enigmatic Distribution of Subgenus Brassospora in New Caledonia. *Cladistics* **17**(1 %R doi:10.1111/j.1096-0031.2001.tb00109.x), 28-47.

Swenson, U., Hill, R. S. and McLoughlin, S. (2001b) Biogeography of Nothofagus Supports the Sequence of Gondwana Break-up *Taxon* **50**, 1025-1041.

Synnott, T. J. (1985) *A checklist of the flora of the Budongo Forest Reserve, Uganda, with notes on ecology and phenology*. Commonwealth Forestry Institute, University of Oxford, Oxford, UK.

Szarzynski, J. and Anhuf, D. (2001) Micrometeorological conditions and canopy energy exchanges of a neotropical rain forest (Surumoni-Crane Project, Venezuela) *Plant Ecology* **153**, 231-239.

Tait, N. N. (1998) The Onychophora and Tardigrada. *Invertebrate Zoology*. Anderson, D. T. Oxford University Press, Oxford, UK. Pp. 204-220.

Takacs, D. (1996) *The Idea of Biodiversity: Philosophies of Paradise*. Johns Hopkins University Press, Baltimore, USA.

Takamura, K. (2001) Effects of termite exclusion on decay of heavy and light hardwood in a tropical rain forest of Peninsular Malaysia. *Journal of Tropical Ecology* **17**, 541–548.

Takasaki, Y., Barham, B. L. and Coomes, O. T. (2004) Risk coping strategies in tropical forests: floods, illnesses, and resource extraction. *Environment and Development Economics* **9**, 203–224.

Takenaka, A., Takahashi, K. and Kohyama, T. (2001) Optimal leaf display and biomass partitioning for efficient light capture in an understorey palm, licuala arbuscula. *Functional Ecology*, **15**, 660–668.

Takyu, M., Kubota, Y., Aiba, S., Seino, T. and Nishimura, T. (2005) Pattern of changes in species diversity, structure and dynamics of forest ecosystems along latitudinal gradients in East Asia. *Ecological Research* **20**, 287–296.

Talley, S. M., Setzer, W. N. and Jackes, B. R. (1996) Host associations of two adventitious-root-climbing vines in a north Queensland tropical rain forest. *Biotropica*, **28**, 356–366.

Tang, H. T. and Chang, P. F. (1979) Sudden mortality in a regenerated stand of *Shorea curtissii* Senaling Inas forest reserve, Negri Sembilan. *Malay Forester*, **42**, 240–254.

Tang, Y. H., Kachi, N., Furukawa, A. and Awang, M. B. (1999) Heterogeneity of light availability and its effects on simulated carbon gain of tree leaves in a small gap and the understory in a tropical rain forest. *Biotropica*, **31**, 268–278.

Tanner, E. V. J., Kapos, V. and Franco, W. (1992) Nitrogen and phosphorus fertilization effects on venezuelan montane forest trunk growth and litterfall. *Ecology*, **73**, 78–86.

Tanner, E. V. J., Teo, V. K., Coomes, D. A. and Midgley, J. J. (2005) Pair-wise competition-trials amongst seedlings of ten dipterocarp species; the role of initial height, growth rate and leaf attributes. *Journal of Tropical Ecology*, **21**, 317–328.

Tansley, A. G. (1935) The use and abuse of vegetational concepts and terms. *Ecology* **16**, 284–307.

Tawaraya, K., Takaya, Y., Turjaman, M., Tuah, S. J., Limin, S. H., Tamai, Y., Cha, J. Y., Wagatsuma, T. and Osaki, M. (2003) Arbuscular mycorrhizal colonization of tree species grown in peat swamp forests of Central Kalimantan, Indonesia. *Forest Ecology and Management* **182**, 381–386.

Taylor, D., Saksena, P., Sanderson, P. G. and Kucera, K. (1999) Environmental change and rain forests on the Sunda shelf of Southeast Asia: drought, fire and the biological cooling of biodiversity hotspots. *Biodiversity and Conservation* **8**, 1159–1177.

Taylor, P. L. (2005) A Fair Trade approach to community forest certification? A framework for discussion. *Journal of Rural Studies* **21**, 433–447.

Taylor, T. N. and Osborn, J. M. (1996) The importance of fungi in shaping the paleoecosystem. *Review of Palaeobotany and Palynology* **90**, 249–262.

Tchouto, M. G. P., de Wilde, J., de Boer, W. F., van der Maesen, L. J. G. and Cleef, A. M. (2009) Bio-indicator species and Central African rain forest refuges in the Campo-Ma'an area, Cameroon. *Systematics and Biodiversity* **7**, 21–31.

Tejerina-Garro, F. L., Fortin, R. and Rodriguez, M. A. (1998) Fish community structure in relation to environmental variation in floodplain lakes of the Araguaia River, Amazon Basin. *Environmental Biology of Fishes* **51**, 399–410.

Temple, S. A. (1977) Plant-animal mutualism: coevolution with dodo leads to near extinction of plant. *Science* **197**, 885–886.

Teo, D. H. L., Tan, H. T. W., Corlett, R. T., Wong, C. M. and Lum, S. K. Y. (2000) *Continental rain forest fragments in Singapore resist invasion by exotic plants*. Conference on the Biogeography of Southeast Asia 2000, Netherlands.

ter Steege, H. and Hammond, D. S. (2000) *An analysis at the ecosystem level: community characteristics, diversity and disturbance*. Plant diversity in Guyana. With recommendation for a protected areas strategy. Ter Steege, H. Wageningen, The Netherlands, Tropenbos Foundation. Pp. 101–117.

ter Steege, H. and Hammond, D. S. (2001) Character convergence, diversity, and disturbance in tropical rain forest in Guyana. *Ecology* **82**, 3197–3212.

ter Steege, H., Jetten, V. G., Polak, A. M. and Werger, M. J. A. (1993) Tropical rain forest types and soil factors in a watershed area in Guyana. *Journal of Vegetation Science* **4**, 705–716.

ter Steege, H., Pitman, N., Sabatier, D., Castellanos, H., Van der Hout, P., Daly, D. C., Silveira, M., Phillips, O., Vasquez, R., Van Andel, T., Duivenvoorden, J., De Oliveira, A. A., Ek, R., Lilwah, R., Thomas, R., Van Essen, J., Baider, C., Maas, P., Mori, S., Terborgh, J., Vargas, P. N., Mogollon, H. and Morawetz, W. (2003) A spatial model of tree alpha-diversity and tree density for the Amazon. *Biodiversity and Conservation* **12**, 2255–2277.

ter Steege, H., Sabatier, D., Castellanos, H., Van Andel, T., Duivenvoorden, J., De Oliveira, A. A., Ek, R., Lilwah, R., Maas, P. and Mori, S. (2000) An analysis of the floristic composition and diversity of Amazonian forests including those of the Guiana Shield. *Journal of Tropical Ecology* **16**, 801–828.

Terborgh, J. (1973) On the notion of favorableness in plant ecology. *American Naturalist* **107**, 481–501.

Terborgh, J. (1986) Keystone plant resources in the tropical forest. *Conservation Biology: The Science of Scarcity and Diversity*. Soule, M. E. Sinauer Associates, Sunderland, MA, USA. Pp. 330–344.

Terborgh, J. (1992a) *Diversity and the tropical rain forest*. Scientific American Library, Salt Lake City, Utah, USA.

Terborgh, J. (1992b) Maintenance of diversity in tropical forests. *Biotropica* **24**(2b), 283–292.

Terborgh, J. (1999) *Requiem for Nature*. Island Press, Washington DC.

Terborgh, J., Feeley, K., Silman, M., Nunez, P. and Balykjian, B. (2006) Vegetation dynamics of predator-free land-bridge islands. *Journal of Ecology* **94**, 253–263.

Terborgh, J., Foster, R. B. and Nunez, P. (1996) Tropical tree communities: A test of the nonequilibrium hypothesis. *Ecology* **77**, 561–567.

Terborgh, J., Lopez, L., Nunez, P., Rao, M., Shahabuddin, G., Orihuela, G., Riveros, M., Ascanio, R., Adler, G. H., Lambert, T. D. and Balbas, L. (2001) Ecological meltdown in predator-free forest fragments. *Science* **294**, 1923–1926.

Terborgh, J., Robinson, S. K., Parker, T. A., Munn, C. A. and Pierpont, N. (1990) Structure and organization of an Amazonian forest bird community. *Ecological Monographs* **60**, 213–238.

Tewksbury, J. J., Huey, R. B. and Deutsch, C. A. (2008) Putting the heat on tropical animals. *Science* **320**, 1296–1297.

Thapa, B. (1998) Debt-for-nature swaps: an overview. *International Journal of Sustainable Development and World Ecology* **5**, 249–262.

Thapa, B. (1999) *The Relationship Between Debt-For-Nature Swaps and Protected Area Tourism: A Plausible Strategy for Developing Countries*. Wilderness Science in a Time of Change Conference, Missoula, MT. USDA Forest Service Proceedings RMRS-P-15-VOL-2. 2000.

Thiollay, J. M. (1992) Influence of selective logging on bird species diversity in a Guianan rain forest. *Conservation Biology* **6**, 47–63.

Thiollay, J. M. (1995) The role of traditional agroforests in the conservation of rain forest bird diversity in Sumatra. *Conservation Biology* **9**, 335–353.

Thiollay, J. M. (1997) Disturbance, selective logging and bird diversity: A Neotropical forest study. *Biodiversity and Conservation* **6**, 1155–1173.

Thiollay, J. M. (1999) Frequency of mixed species flocking in tropical forest birds and correlates of predation risk: an intertropical comparison. *Journal of Avian Biology* **30**, 282–294.

Thiollay, J. M. (2002) Avian diversity and distribution in French Guiana: patterns across a large forest landscape. *Journal of Tropical Ecology* **18**, 471–498.

Thomas, C. D., Cameron, A., Green, R. E., Bakkenes, M., Beaumont, L. J., Collingham, Y. C., Erasmus, B. F. N., de Siqueira, M. F., Grainger, A., Hannah, L., Hughes, L., Huntley, B., van Jaarsveld, A. S., Midgley, G. F., Miles, L., Ortega-Huerta, M. A., Peterson, A. T., Phillips, O. L. and Williams, S. E. (2004) Extinction risk from climate change. *Nature* **427**, 145–148.

Thomas, S. C. (1991) Population densities and patterns of habitat use among anthropoid primates of the Ituri Forest, Zaire. *Biotropica* **23**, 68–83.

Thomas, S. C. (1997) Geographic parthenogenesis in a tropical forest tree. *American Journal of Botany* **84**, 1012–1015.

Thomas, S. C. (1996a) Asymptotic height as a predictor of growth and allometric characteristics Malaysian rain forest trees. *American Journal of Botany* **83**, 556–566.

Thomas, S. C. (1996b) Reproductive allometry in malaysian rain forest trees: Biomechanics versus optimal allocation. *Evolutionary Ecology* **10**, 517–530.

Thomas, S. C. and Bazzaz, F. A. (1999) Asymptotic height as a predictor of photosynthetic characteristics in Malaysian rain forest trees. *Ecology* **80**, 1607–1622.

Thompson, H. (1910) *Gold Coast: Report on Forests*. Colonial Reports-Miscellaneous No. 66. HMSO, London, UK.

Thompson, J. N. (2005) *The Geographic Mosaic of Coevolution*. University of Chicago Press, Chicago, IL, USA.

Thomson, J. D., Herre, E.A., Hamrick, J.L., Stone, J.L. (1991) Genetic mosaics in strangler fig trees: implications for tropical conservation. *Science* **254**, 1214–1216.

Tian, H. Q., Melillo, J. M., Kicklighter, D. W., McGuire, A. D., Helfrich, J. V. K., Moore, B. and Vorosmarty, C. J. (1998) Effect of interannual climate variability on carbon storage in Amazonian ecosystems. *Nature* **396**, 664–667.

Tiessen, H., Cuevas, E. and Chacon, P. (1994) The role of soil organic-matter in sustaining soil fertility. *Nature* **371**, 783–785.

Tilman, D., Cassman, K. G., Matson, P. A., Naylor, R. and Polasky, S. (2002) Agricultural sustainability and intensive production practices. *Nature* **418**, 671–677.

Timm, H.C., Stegemann, J. and Kuppers, M. (2002) Photosynthetic induction strongly affects the light compensation point of net photosynthesis and coincidentally the apparent quantum yield. *Trees-Structure and Function*, **16**, 47–62.

Titus, J.H., Holbrook, N.M. and Putz, F.E. (1990) Seed germination and seedling distribution of *Ficus pertusa* and *F. tuerkheimii*: Are strangler figs autotoxic? *Biotropica*, **22**, 425–428.

Torquebiau, E. F. (1988) Photosynthetically Active Radiation Environment, Patch Dynamics and Architecture in a Tropical Rainforest in Sumatra. *Australian Journal of Plant Physiology* **15**, 327–342.

Torti, S. D. and Coley, P. D. (1999) Tropical monodominance: A preliminary test of the ectomycorrhizal hypothesis. *Biotropica* **31**, 220–228.

Torti, S. D., Coley, P. D. and Kursar, T. A. (2001) Causes and consequences of monodominance in tropical lowland forests. *American Naturalist* **157**, 141–153.

Toy, R. J., Marshall, A. G. and Pong, T. Y. (1992) Fruiting phenology and the survival of insect fruit predators: a case-study from the South-East Asian Dipterocarpaceae. *Philosophical Transactions of the Royal Society of London Series B* **335**, 417–423.

Travers, S. E., Gilbert, G. S. and Perry, E. F. (1998) The effect of rust infection on reproduction in a tropical tree (*Faramea occidentalis*) *Biotropica* **30**, 438–443.

Tschapka, M. and von Helversen, O. (1999) Pollinators of syntopic *Marcgravia* species in Costa Rican lowland rain forest: Bats and opossums. *Plant Biology* **1**, 382–388.

Tsukaya, H., Okada, H. and Mohamed, M. (2004) A novel feature of structural variegation in leaves of the tropical plant *Schismatoglottis calyptrata*. *Journal of Plant Research* **117**, 477.

Tucker, R. P. and Richards, J. F., Eds. (1983) *Global Deforestation and the Nineteenth Century World Economy*. Duke University Press, Durham, NC, USA.

Tuomisto, H. and Ruokolainen, K. (1997) The role of ecological knowledge in explaining biogeography and biodiversity in Amazonia. *Biodiversity and Conservation* **6**, 347–357.

Tuomisto, H. and Ruokolainen, K. (2006) Analyzing or explaining beta diversity? Understanding the targets of different methods of analysis. *Ecology* **87**, 2697–2708.

Tuomisto, H., Ruokolainen, K., Kalliola, R., Linna, A., Danjoy, W. and Rodriguez, Z. (1995) Dissecting Amazonian biodiversity. *Science* **269**, 63–66.

Tuomisto, H., Ruokolainen, K. and Yli-Halla, M. (2003) Dispersal, environment, and floristic variation of western Amazonian forests. *Science* **299**, 241–244.

Turner, B. L. (1974) Prehistoric intensive agriculture in Mayan lowlands. *Science* **185**, 118–124.

Turner, B. L. (1976) Population density in the classic Maya lowlands: New evidence for old approaches. *Geographical Review* **66**, 73–82.

Turner, I.M. (2001) *The Ecology of Trees in the Tropical Rain Forest*. Cambridge University Press, Cambridge, UK.

Turner, J. (2004) *Spice: the history of temptation*. Vintage Books, Random House, New York, USA.

Tyler, B. M., S. Tripathy, et al. (2006) *Phytophthora* genome sequences uncover evolutionary origins and mechanisms of pathogenesis. *Science* **313**, 1261–1266.

Tyree, M. T. (2003) Hydraulic limits on tree performance: Transpiration, carbon gain and growth of trees. *Trees-Structure and Function*, **17**, 95–100.

Tyree, M. T., Patino, S. and Becker, P. (1998) Vulnerability to drought-induced embolism of Bornean heath and dipterocarp forest trees. *Tree Physiology*, **18**, 583–588.

Uhl, C., Clark, K., Clark, H. and Murphy, P. (1981) Early plant succession after cutting and burning in the upper Rio Negro region of the Amazon basin. *Journal of Ecology* **69**, 631–649.

Uhl, C., Clark, K., Dezzeo, N. and Maquirino, P. (1988) Vegetation dynamics in Amazonian treefall gaps. *Ecology* **69**, 751–763.

Uhl, C. and Vieira, I. C. G. (1989) Ecological impacts of selective logging in the Brazilian Amazon:a case study from the Paragominas region of the state of Para. *Biotropica* **21**, 98–106.

Uma Shaanker, R., Ganeshaiah, K. N., Krishnan, S., Ramya, R., Meera, C., Aravind, N. A., Kumar, A., Rao, D., Vanaraj, G., Ramachandra, J., Gauthier, R., Ghazoul, J., Poole, N. and Reddy, B. V. C. (2004) Livelihood gains and ecological costs of non-timber forest product dependence: Assessing the roles of dependence, ecological knowledge and market structure in three contrasting human and ecological settings in south India. *Environmental Conservation* **31**, 242–253.

UNDP-GEF.(2008) *Biodiversity: Delivering results*. United Nations Development Programme and Global Environemt Facility Unit, New York, USA. Pp. 60.

United Nations (2008) *The Millennium Development Goals Report 2008*. United Nations, New York, USA. Pp. 56.

Urbas, P., Araujo, M. V., Leal, I. R. and Wirth, R. (2007) Cutting more from cut forests: Edge effects on foraging and herbivory of leaf-cutting ants in Brazil. *Biotropica* **39**, 489–495.

Utzurrum, R. C. B. and Heideman, P. D. (1991) Differential ingestion of viable vs nonviable *Ficus* seeds by fruit bats. *Biotropica* **23**, 311–312.

Vaculik, A., Kounda-Kiki, C., Sarthou, C. and Ponge, J. F. (2004) Soil invertebrate activity in biological crusts on tropical inselbergs. *European Journal of Soil Science* **55**, 539–549.

Vadell, E. M., Holmes, M. T. and Cavender, J. C. (1995) *Dictyostelium citrinum*, *D. medusoides* and *D. granulophorum*: Three new members of the Dictyosteliaceae from forest soils of Tikal, Guatemala. *Mycologia* **87**, 551–559.

Val, A. L. and de Almeida-Val, V. M. F. (1995) *Fishes of the Amazon and their Environment: Physiological and Biochemical Aspect*. Springer-Verlag, Berlin.

Valencia, R., Balslev, H. and Mino, G. P. Y. (1994) High tree alpha-diversity In Amazonian Ecuador. *Biodiversity and Conservation* **3**, 21–28.

Valencia, R., Foster, R. B., Villa, G., Condit, R., Svenning, J. C., Hernandez, C., Romoleroux, K., Losos, E., Magard, E. and Balslev, H. (2004) Tree species distributions and local habitat variation in the Amazon: large forest plot in eastern Ecuador. *Journal of Ecology* **92**, 214–229.

Valladares, F., Skillman, J.B. and Pearcy, R.W. (2002) Convergence in light capture efficiencies among tropical forest understory plants with contrasting crown architectures: A case of morphological compensation. *Am. J. Bot.*, **89**, 1275–1284.

Valladares, F., Wright, S.J., Lasso, E., Kitajima, K. and Pearcy, R.W. (2000) Plastic phenotypic response to light of 16 congeneric shrubs from a panamanian rainforest. *Ecology* **81**, 1925–1936.

Vallan, D., Andreone, F., Raherisoa, V. H. and Dolch, R. (2004) Does selective wood exploitation affect amphibian diversity? The case of An'Ala, a tropical rainforest in eastern Madagascar. *Oryx* **38**, 410–417.

Van Bael, S. A., Aiello, A., Valderrama, A., Medianero, E., Samaniego, M. and Wright, S. J. (2004) General herbivore outbreak following an El Niño-related drought in a lowland Panamanian forest. *Journal of Tropical Ecology* **20**, 625–633.

Van Bael, S. A., Brawn, J. D. and Robinson, S. K. (2003) Birds defend trees from herbivores in a Neotropical forest canopy. *Proceedings of the National Academy of Sciences of the United States of America* **100**, 8304–8307.

van der Hammen, T. and Hooghiemstra, H. (2000) Neogene and Quaternary history of vegetation, climate, and plant diversity in Amazonia. *Quaternary Science Reviews* **19**, 725–742.

van der Pijl, L. (1982) *Principles of Dispersal in Higher Plants.* Springer-Verlag, New York, USA.

Van Gelder, H.A., Poorter, L. and Sterck, F.J. (2006) Wood mechanics, allometry, and life-history variation in a tropical rain forest tree community. *New Phytologist*, **171**, 367–378.

Van Houtan, K. S., Pimm, S. L., Halley, J. M., Bierregaard, R. O. and Lovejoy, T. E. (2007) Dispersal of Amazonian birds in continuous and fragmented forest. *Ecology Letters* **10**, 219–229.

Van Nieuwstadt, M. G. L. and Sheil, D. (2005) Drought, fire and tree survival in a Borneo rain forest, East Kalimantan, Indonesia. *Journal of Ecology* **93**, 191–201.

van Roosmalen, M. G. M., van Roosmalen, T. and Mittermeier, R. A. (2002) A taxonomic review of the Titi monkeys, genus *Callicebus* Thomas, 1903, with the description of two new species, *Callicebus bernhardi* and *Callicebus stephennashi*, from Brazilian Amazonia. *Neotropical Primates* **10** (Suppl.), 1–52.

Van Schaik, C. P. and Mirmanto, E. (1985) Spatial variation in the structure and litterfall of a Sumatran rain forest. *Biotropica* **17**, 196–205.

Van Schaik, C. P., Terborgh, J. W. and Wright, S. J. (1993) The phenology of tropical forests: Adaptive significance and consequences for primary consumers. *Annual Review of Ecology and Systematics*, **24**, 353–377.

Vanclay, J.K. (1991) Aggregating tree species to develop diameter increment equations for tropical rain-forests. *Forest Ecology and Management*, **42**, 143–168.

Vande weghe, J. P. (2004) *Forests of Central Africa, Nature and Man.* ECOFAC and Lannoo Publishers, Tielt, Belgium.

Vander Wall, S. B., Kuhn, K. M. and Beck, M. J. (2005) Seed removal, seed predation, and secondary dispersal. *Ecology* **86**, 801–806.

Vandermaarel, E. (1993) Some remarks on disturbance and its relations to diversity and stability. *Journal of Vegetation Science* **4**, 733–736.

Vandermeer, J. (1996) Disturbance and neutral competition theory in rain forest dynamics. *Ecological Modelling* **85**, 99–111.

Vandermeer, J., Boucher, D., Perfecto, I. and de la Cerda, I. G. (1996) A theory of disturbance and species diversity: Evidence from Nicaragua after Hurricane Joan. *Biotropica* **28**, 600–613.

Vandermeer, J., de la Cerda, I. G., Perfecto, I., Boucher, D., Ruiz, J. and Kaufmann, A. (2004) Multiple basins of attraction in a tropical forest: Evidence for nonequilibrium community structure. *Ecology* **85**, 575–579.

Vandermeer, J., Mallona, M. A., Boucher, D., Yih, K. and Perfecto, I. (1995) Three years of ingrowth following catastrophic hurricane damage on the caribbean coast of Nicaragua: evidence in support of the direct regeneration hypothesis. *Journal of Tropical Ecology* **11**, 465–471.

Vandermeer, J., Zamora, N., Yih, K. and Boucher, D. (1990) Initial regeneration of a tropical forest in the Caribbean coast of Nicaragua after Hurricane Joan. *Revista De Biologia Tropical* **38**(2B), 347–359.

Vasconcelos, H. L. and Cherrett, J. M. (1997) Leaf-cutting ants and early forest regeneration in central Amazonia: Effects of herbivory on tree seedling establishment. *Journal of Tropical Ecology* **13**, 357–370.

Vasconcelos, H. L. and Luizao, F. J. (2004) Litter production and litter nutrient concentrations in a fragmented Amazonian landscape. *Ecological Applications* **14**, 884–892.

Vasquez, R. (1991) *Caraipa* (Guttiferae) In Peru. *Annals of the Missouri Botanical Garden* **78**, 1002–1008.

Vazquez-Yanes, C. and Orozco-Segovia, A. (1993) Patterns of seed longevity and germination in the tropical rainforest. *Annual Review of Ecology and Systematics* **24**, 69–87.

Vazquez-Yanes, C., Orozco-Segovia, A., Rincon, E., Sanchez-Coronado, M. E., Huante, P., Toledo, J. R. and Barradas, V. L. (1990) Light beneath the litter in a tropical forest: effect on seed germination. *Ecology* **71**, 1952–1958.

Vedeld, P., Angelsen, A., Sjaastad, E. and Berg, G. K. (2004) *Counting on the Environment. Forest Incomes and the Rural Poor.* Environment Department Papers, World Bank, Washington, DC, USA. **98**. Pp. 94.

Veenendaal, E. M., Abebrese, I. K., Walsh, M. F. and Swaine, M. D. (1996) Root hemiparasitism in a West African rainforest tree *Okoubaka aubrevillei* (Santalaceae) *New Phytologist* **134**, 487–493.

Veenendaal, E. M., Swaine, M. D., Blay, D., Yelifari, N. B. and Mullins, C. E. (1996) Seasonal and long-term soil water regime in West African tropical forest. *Journal of Vegetation Science* **7**, 473–482.

Vega, F. E. and P. F. Dowd (2005) The role of yeasts as insect endosymbionts. In: *Insect-Fungal Associations: ecology and evolution*. F. E. Vega and M. Blackwell (eds.). Oxford University Press, New York, USA.

Veldkamp, E., Purbopuspito, J., Corre, M. D., Brumme, R. and Murdiyarso, D. (2008) Land use change effects on trace gas fluxes in the forest margins of Central Sulawesi, Indonesia. *Journal of Geophysical Research* **113**, 1–11.

Vences, M., Vieites, D. R., Glaw, F., Brinkmann, H., Kosuch, J., Veith, M. and Meyer, A. (2003) Multiple overseas dispersal in amphibians. *Proceedings of the Royal Society of London Series B* **270**, 2435–2442.

Veneklaas, E. J. and Poorter, L. (1998) Growth and carbon partitioning of tropical tree seedlings in contrasting light environments. In: *Inherent variation in plant growth: Physiological mechanisms and ecological consequences*. M.V. Vuuren (ed.). Backhuys Publishers, Leiden, The Netherlands. Pp. 337–361.

Veríssimo, A. and Lentini, M. (2007) The Brazilian Amazon. Retrieved 15 December 2008, from http://forestryencyclopedia.jot.com/WikiHome/The%20Brazilian%20Amazon?themePage=printable.

Vernooy, R., Ed. (2006) *Social and Gender Analysis in Natural Resource Management: Learning Studies and Lessons from Asia*. Sage Publications, London, UK.

Viana, V. M., Tabanez, A. A. and Batista, J. (1997) Dynamics and restoration of forest fragments in the Brazilian Atlantic moist forest. In: *Tropical Forest Remnants: Ecology, Management, and Conservation of Fragmented Communities*. Laurance, W. F. and Bierregaard, R. O. (eds.). University of Chicago Press, Chicago, IL, USA. Pp. 351–365.

Vieira, I. C. G., de Almeida, A. S., Davidson, E. A., Stone, T. A., de Carvalho, C. J. R. and Guerrero, J. B. (2003) Classifying successional forests using Landsat spectral properties and ecological characteristics in eastern Amazonia. *Remote Sensing of Environment* **87**, 470–481.

Vieites, D. R., Chiari, Y., Vences, M., Andreone, F., Rabemananjara, F., Bora, P., Nieto-Roman, S. and Meyer, A. (2006) Mitochondrial evidence for distinct phylogeographic units in the endangered Malagasy poison frog *Mantella bernhardi*. *Molecular Ecology* **15**, 1617–1625.

Vierling, L. A. and Wessman, C. A. (2000) Photosynthetically active radiation heterogeneity within a monodominant Congolese rain forest canopy. *Agricultural and Forest Meteorology* **103**, 265–278.

Visser, D. R. and Mendoza, G. A. (1994) Debt-for-nature swaps in Latin America. *Journal of Forestry* **92**, 13–16.

Vitousek, P. M. (2004) *Nutrient cycling and limitation: Hawai'i as a model system*. Princeton University Press, New Jersey, USA.

Vitousek, P. M., Cassman, K., Cleveland, C., Crews, T., Field, C. B., Grimm, N. B., Howarth, R. W., Marino, R., Martinelli, L., Rastetter, E. B. and Sprent, J. I. (2002) Towards an ecological understanding of biological nitrogen fixation. *Biogeochemistry* **57**, 1–45.

Vitt, L. J., Avila-Pires, T. C. S., Caldwell, J. P. and Oliveira, V. R. L. (1998) The impact of individual tree harvesting on thermal environments of lizards in Amazonian rain forest. *Conservation Biology* **12**, 654–664.

Vitt, L. J. and Zani, P. A. (1998) Prey use among sympatric lizard species in lowland rain forest of Nicaragua. *Journal of Tropical Ecology* **14**, 537–559.

Vittor, A. Y., Gilman, R. H., Tielsch, J., Glass, G., Shields, T., Lozano, W. S., Pinedo-Cancino, V. and Patz, J. A. (2006) The effect of deforestation on the human-biting rate of *Anopheles darlingi*, the primary vector of falciparum malaria in the Peruvian Amazon. *American Journal of Tropical Medicine and Hygiene* **74**, 3–11.

Voeks, R. A. (2004) Disturbance pharmacopoeias: Medicine and myth from the humid tropics. *Annals of the Association of American Geographers* **94**, 868–888.

Vogelmann, T.C., Nishio, J.N. and Smith, W.K. (1996) Leaves and light capture: Light propagation and gradients of carbon fixation within leaves. *Trends in Plant Science*, **1**, 65–70.

Voigt, C. C. (2004) The power requirements (Glossophaginae: Phyllostomidae) in nectar-feeding bats for clinging to flowers. *Journal of Comparative Physiology B* **174**, 541–548.

Volkov, I., Banavar, J. R., Hubbell, S. P. and Maritan, A. (2007) Patterns of relative species abundance in rainforests and coral reefs. *Nature* **450**, 45–49.

Vonesh, J. R. (2001) Patterns of richness and abundance in a tropical African leaf-litter herpetofauna. *Biotropica* **33**, 502–510.

von Humboldt, Alexander (1849) *Aspects of Nature*. Longman, Brown, Green, London, UK; originally published as *Ansichten der Natur* (Stuttgart and Tübingen, 1808).

Vonuexkull, H.R. and Mutert, E. (1995) Global extent, development and economic impact of acid soils. *Plant and Soil*, **171**, 1–15.

Voris, H. K. (2000) Maps of Pleistocene sea levels in Southeast Asia: shorelines, river systems and time durations. *Journal of Biogeography* **27**, 1153–1167.

Vormisto, J. (2002) Palms as rainforest resources: how evenly are they distributed in Peruvian Amazonia? *Biodiversity and Conservation* **11**, 1025–1045.

Vormisto, J., Svenning, J. C., Hall, P. and Balslev, H. (2004) Diversity and dominance in palm (Arecaceae) communities in terra firme forests in the western Amazon basin. *Journal of Ecology* **92**, 577–588.

Vulinec, K. (2002) Dung beetle communities and seed dispersal in primary forest and disturbed land in Amazonia. *Biotropica* **34**, 297–309.

Vulinec, K., Mellow, D. J. and da Fonseca, C. R. V. (2007) Arboreal foraging height in a common neotropical dung beetle, *Canthon subhyalinus* Harold (Coleoptera: Scarabaeidae) *Coleopterists Bulletin* **61**, 75–81.

Wadley, R. L. and Colfer, C. J. P. (2004) Sacred forest, hunting, and conservation in West Kalimantan, Indonesia. *Human Ecology* **32**, 313–338.

Wagner, T. (2001) Seasonal changes in the canopy arthropod fauna in *Rinorea beniensis* in Budongo Forest, Uganda. *Plant Ecology* **153**, 169–178.

Wahlberg, A. (2006) Bio-politics and the promotion of traditional herbal medicine in Vietnam. *Health* **10**, 123–147.

Wahungu, G. M., Muoria, P. K., Moinde, N. N., Oguge, N. O. and Kirathe, J. N. (2005) Changes in forest fragment sizes and primate population trends along the River Tana floodplain, Kenya. *African Journal of Ecology* **43**, 81–90.

Waide, R. B. and Reagan, D. P. (1996) The Rain Forest Setting. In: *The Food Web of a Tropical Rain Forest*. Reagan, D. M. and Waide, R. B. (eds.) University of Chicago Press, Chicago, IL, USA. Pp. 1–16.

Walker, L. R. (1991) Tree damage and recovery from Hurricane Hugo in Luquillo experimental forest, Puerto Rico. *Biotropica* **23**, 379–385.

Walker, L. R. (1994) Effects of fern thickets on woodland development on landslides in Puerto Rico. *Journal of Vegetation Science* **5**, 525–532.

Walker, T. W. and Syers, J. K. (1976) Fate of phosphorus during pedogenesis. *Geoderma* **15**, 1–19.

Wallace, A. R. (1869 reprint 1996) *The Malay Archipelago*. [Macmillan London] Oxford University Press, Kuala Lumpur.

Wallace, A. R. (1852) On the monkeys of the Amazon. *Proceedings of the Zoological Society of London* **20**, 107–110.

Wallace, A. R. (1853) *A Narrative of Travels on the Amazon and Rio Negro with an Account of the Native Tribes, and Observations on the Climate, Geology, and Natural History of the Amazon Valley*. Ward, Lock and Co., London, UK.

Wallace, A. R. (1863) On the physical geography of the Malay Archipelago. *Journal of the Royal Geographical Society of London* **33**, 217–234.

Wallace, A. R. (1876) *The Geographical Distribution of Animals: With a study of the relations of living and extinct faunas as elucidating the past changes of the Earth's surface*. Harper and Brothers, New York, USA.

Walters, R.G. (2005) Towards an understanding of photosynthetic acclimation. *Journal of Experimental Botany*, **56**, 435–447.

Wanek, W., Huber, W., Arndt, S.K. and Popp, M. (2002) Mode of photosynthesis during different life stages of hemiepiphytic *Clusia* species. *Functional Plant Biology*, **29**, 725–732.

Wang, B. C., Sork, V. L., Leong, M. T. and Smith, T. B. (2007) Hunting of mammals reduces seed removal and dispersal of the afrotropical tree *Antrocaryon klaineanum* (Anacardiaceae). *Biotropica* **39**, 340–347.

Wang, Y. H. and Augspurger, C. (2004) Dwarf palms and cyclanths strongly reduce Neotropical seedling recruitment. *Oikos* **107**, 619–633.

Wanga, H., Moore, M. J., Soltis, P. S., Bell, C. D., Brockington, S. F., Alexandre, R., Davis, C. C., Latvis, M., Manchester, S. R. and Soltis, D. E. (2009) Rosid radiation and the rapid rise of angiosperm-dominated forests. *Proceedings of the National Academy of Sciences of the United States of America* **106**, 3853–3858.

Wanntorp, L. and Wanntorp, H. E. (2003) The biogeography of *Gunnera* L.: vicariance and dispersal. *Journal of Biogeography* **30**, 979–987.

Ward, M., Dick, C. W., Gribel, R. and Lowe, A. J. (2005) To self, or not to self. A review of outcrossing and pollen-mediated gene flow in neotropical trees. *Heredity* **95**, 246–254.

Waser, N. M., Chittka, L., Price, M. V., Williams, N. M. and Ollerton, J. (1996) Generalization in pollination systems, and why it matters. *Ecology* **77**, 1043–1060.

Watling, J. R., Robinson, S. A., Woodrow, I. E. and Osmond, C. B. (1997) Responses of rainforest understorey plants to excess light during sunflecks. *Australian Journal of Plant Physiology*, **24**, 17–25.

Watling, R. and Harper, D. B. (1998) Chloromethane production by wood-rotting fungi and an estimate of the global flux to the atmosphere. *Mycological Research* **102**, 769–787.

Watts, D. P. and Mitani, J. C. (2002) Hunting behavior of chimpanzees at Ngogo, Kibale National Park, Uganda. *International Journal of Primatology* **23**, 1–28.

Webb, C. O. and Peart, D. R. (2000) Habitat associations of trees and seedlings in a Bornean rain forest. *Journal of Ecology* **88**, 464–478.

Webb, L. J. (1959) A physiognomic classification of Australian rain forests. *Journal of Ecology* **47**, 551–570.

Webb, L. J., Tracey, J. G. and Haybrook, K. P. (1967) A factor toxic to seedlings of the same species associated with roots of the non-gregarious subtropical rain forest tree *Grevillea robusta*. *Journal of Applied Ecology*, **14**, 13–25.

Weiblen, G., Flick, B. and Spencer, H. (1995) Seed set and wasp predation in dioecious *Ficus variegata* from an Australian wet tropical forest. *Biotropica* **27**, 391–394.

Weiblen, G. D. (2002) How to be a fig wasp. *Annual Review of Entomology* **47**, 299–330.

Weiblen, G. D., Yu, D. W. and West, S. A. (2001) Pollination and parasitism in functionally dioecious figs. *Proceedings of the Royal Society of London Series B* **268**, 651–659.

Weir, A. (1998) Notes on the Laboulbeniales of Sulawesi. The genus *Rickia*. *Mycological Research* **102**, 327–343.

Weir, A. and Hammond, P. M. (1997) Laboulbeniales on beetles: Host utilization patterns and species richness of the parasites. *Biodiversity and Conservation* **6**, 701–719.

Welden, C. W., Hewett, S. W., Hubbell, S. P. and Foster, R. B. (1991) Sapling survival, growth, and recruitment: relationship to canopy height in a neotropical forest. *Ecology* **72**, 35–50.

Wells, M. and McShane, T. O. (2004) Integrating protected area management with local needs and aspirations. *Ambio* **33**, 513–519.

Wenny, D. G. and Levey, D. J. (1998) Directed seed dispersal by bellbirds in a tropical cloud forest. *Proceedings of the National Academy of Sciences of the United States of America* **95**, 6204–6207.

Westcott, D. A., Bradford, M. G., Dennis, A. J. and Lipsett-Moore, G. (2002) *Keystone fruit resources and Australia's tropical rain forests*. Symposium on Tropical Fruits and Frugivores: The Search for Strong Interactors, Panama City, Panama.

Westoby, M., Leishman, M. and Lord, J. (1996) Comparative ecology of seed size and dispersal. *Philosophical Transactions of the Royal Society of London Series B* **351**, 1309–1317.

Wheelwright, N. T. (1985) Fruit size, gape width, and the diets of fruit-eating birds. *Ecology* **66**, 808–818.

Whigham, D. F., Olmsted, I., Cano, E. C. and Harmon, M. E. (1991) The impact of Hurricane Gilbert on trees, litterfall, and woody debris in a dry tropical forest in the northeastern Yucatan peninsula. *Biotropica* **23**, 434–441.

Whinnett, A., Zimmermann, M., Willmott, K. R., Herrera, N., Mallarino, R., Simpson, F., Joron, M., Lamas, G. and Mallet, J. (2005) Strikingly variable divergence times inferred across an Amazonian butterfly 'suture zone'. *Proceedings of the Royal Society B* **272**, 2525–2533.

White, A. and Martin, A. (2005) Who owns the world's forests? Forest tenure and public forests in transition. *Forestry and Development*. Sayer, J. (ed.) London, Earthscan. Pp. 72–103.

White, L. J. T. and Oates, J. F. (1999) New data on the history of the plateau forest of Okomu, southern Nigeria: an insight into how human disturbance has shaped the African rain forest. *Global Ecology and Biogeography* **8**, 355–361.

White, P. S. and Jentsch, A. (2001) The search for generality in studies of disturbance and ecosystem dynamics. *Progress in Botany* **62**, 399–450.

White, T. C. R. (2007) Flooded forests: Death by drowning, not herbivory. *Journal of Vegetation Science* **18**, 147–148.

Whitehead, D. R. (1983) Wind pollination: some ecological and evolutionary perspectives. *Pollination Biology*. Real, L. Academic Press, Orlando, Florida, USA. Pp. 97–108.

Whitehouse, A. E. and Mulyana, A. A. S. (2004) Coal fires in Indonesia. *International Journal of Coal Geology* **59**, 91–97.

Whitman, A. A., Hagan, J. M. and Brokaw, N. V. L. (1998) Effects of selection logging on birds in northern Belize. *Biotropica* **30**, 449–457.

Whitmore, T. C. (1966a) *Guide to the Forests of the British Solomon Islands*. Oxford University Press, Oxford, UK.

Whitmore, T. C. (1966b) Social status of *Agathis* in a rain forest in Melanesia. *Journal of Ecology*, **54**, 285–301.

Whitmore, T. C. (1978) Gaps in the forest canopy. *Tropical Trees as Living Systems*. Tomlinson, P. B. and Zimmerman, M. M. Cambridge University Press, New York, USA. Pp. 639–655.

Whitmore, T. C. (1998) *An Introduction to Tropical Rain Forests (Second Edition)* Oxford University Press, Oxford, UK.

Whitmore, T. C. and Brown, N. D. (1996) Dipterocarp seedling growth in rain forest canopy gaps during six and a half years. *Philosophical Transactions of the Royal Society of London Series B*, **351**, 1195–1203.

Whitmore, T. C. and Burslem, D. F. R. P. (1998) Major disturbances in tropical rainforests. In: *Dynamics of Tropical Communities*. Newbery, D. M., Prins, H. H. T. and Brown, N. (eds) Blackwell Science. Pp. 549–565.

Whitmore, T. C., Peralta, R. and Brown, K. (1985) Total species count in a Costa Rican rain forest. *Journal of Tropical Ecology* **1**, 375–378.

Whitney, K. D., Fogiel, M. K., Lamperti, A. M., Holbrook, K. M., Stauffer, D. J., Hardesty, B. D., Parker, V. T. and Smith, T. B. (1998) Seed dispersal by *Ceratogymna* hornbills in the Dja Reserve, Cameroon. *Journal of Tropical Ecology* **14**, 351–371.

Whittaker, R. H. (1975) *Communities and Ecosystems*. Macmillan, London, UK.

Whittaker, R. J., Bush, M. B. and Richards, K. (1989) Plant recolonization and vegetation succession on the Krakatau Islands, Indonesia. *Ecological Monographs* **59**, 59–123.

Whittaker, R. J., Field, R. and Partomihardjo, T. (2000) How to go extinct: lessons from the lost plants of Krakatau. *Journal of Biogeography* **27**, 1049–1064.

Whittaker, R. J., Schmitt, S. F., Jones, S. H., Partomihardjo, T. and Bush, M. B. (1998) Stand biomass and tree mortality from permanent forest plots on Krakatau, Indonesia, 1989–1995. *Biotropica* **30**, 519–529.

WHO. (1978) *Report of the International Conference on Primary Health Care*. International Conference on Primary Health Care, Alma-Ata, USSR, World Health Organization and United Nations Children's Fund.

WHO. (2005) World Malaria Report. Retrieved on 12 April 2009 from http://www.globalpolicy.org/socecon/develop/africa/2005/05malariareport.pdf.

Wiens, J. J. and Donoghue, M. J. (2004) Historical biogeography, ecology and species richness. *Trends in Ecology and Evolution* **19**, 639–644.

Wier, A., Dolan, M., Grimaldi, D., Guerrero, R., Wagensberg, J. and Margulis, L. (2002) Spirochete and protist symbionts of a termite (*Mastotermes electrodominicus*) in Miocene amber. *Proceedings of the National Academy of Sciences of the United States of America* **99**, 1410–1413.

Wiersum, K. F. (1997) Indigenous exploitation and management of tropical forest resources: An evolutionary continuum in forest-people interactions. *Agriculture Ecosystems and Environment* **63**, 1–16.

Wikstrom, N., Savolainen, V. and Chase, M. W. (2001) Evolution of the angiosperms: calibrating the family tree. *Proceedings of the Royal Society of London Series B* **268**, 2211–2220.

Wilcke, W., Amelung, W., Martius, C., Garcia, M. V. B. and Zech, W. (2000) Biological sources of polycyclic aromatic hydrocarbons (PAHs) in the Amazonian Rain Forest. *Journal of Plant Nutrition and Soil Science* **163**, 27–30.

Wilcke, W., Krauss, M. and Amelung, W. (2002) Carbon isotope signature of polycyclic aromatic hydrocarbons (PAHs): Evidence for different sources in tropical and temperate environments? *Environmental Science and Technology* **36**, 3530–3535.

Wilcke, W., Valladarez, H., Stoyan, R., Yasin, S., Valarezo, C. and Zech, W. (2003) Soil properties on a chronosequence of landslides in montane rain forest, Ecuador. *Catena* **53**, 79–95.

Wilcox, B. A. and Ellis, B. (2006) Forests and emerging infectious diseases of humans. *Unasylva* **224**, 11–18.

Wilde, V. L. and Vainio-Mattila, A. (1995) *Gender Analysis and Forestry*. FAO, Rome, Italy.

Wiles, G. J., Bart, J., Beck, R. E. and Aguon, C. F. (2003) Impacts of the brown tree snake: Patterns of decline and species persistence in Guam's avifauna. *Conservation Biology* **17**, 1350–1360.

Wilf, P., Cuneo, N. R., Johnson, K. R., Hicks, J. F., Wing, S. L. and Obradovich, J. D. (2003) High plant diversity in Eocene South America: Evidence from Patagonia. *Science* **300**, 122–125.

Wilf, P., Labandeira, C. C., Johnson, K. R. and Ellis, B. (2006) Decoupled plant and insect diversity after the end-Cretaceous extinction. *Science* **313**, 1112–1115.

Wilhere, G. F. (2008) The how-much-is-enough myth. *Conservation Biology* **22**, 514–517.

Wilkie, D. S. and Carpenter, J. F. (1999) Bushmeat hunting in the Congo Basin: an assessment of impacts and options for mitigation. *Biodiversity and Conservation* **8**, 927–955.

Williams, J. S. and Cooper, R. M. (2003) Elemental sulphur is produced by diverse plant families as a component of defence against fungal and bacterial pathogens. *Physiological and Molecular Plant Pathology* **63**, 3–16.

Williams, M. (2003) *Deforesting the Earth: from prehistory to global crisis*. University of Chicago Press, Chicago, IL, USA.

Williams, P. H. and Gaston, K. J. (1994) Measuring more of biodiversity: can higher-taxon richness predict wholesale species richness? *Biological Conservation* **67**, 211–217.

Williams, P. H., Humphries, C. J. and Gaston, K. J. (1994) Centers of seed-plant diversity: The family way. *Proceedings of the Royal Society of London Series B* **256**, 67–70.

Williams, P. H., Prance, G. T., Humphries, C. J. and Edwards, K. S. (1996) Promise and problems in applying quantitative complementary areas for representing the diversity of some neotropical plants (families Dichapetalaceae, Lecythidaceae, Caryocaraceae, Chrysobalanaceae and Proteaceae) *Biological Journal of the Linnean Society* **58**, 125–157.

Williams, S. E. (1997) Patterns of mammalian species richness in the Australian tropical rainforests: Are extinctions during historical contractions of the rainforest the primary determinants of current regional patterns in biodiversity? *Wildlife Research* **24**, 513–530.

Williams, S. E. and Hero, J. M. (2001) Multiple determinants of Australian tropical frog biodiversity. *Biological Conservation* **98**, 1–10.

Williams, S. E. and Pearson, R. G. (1997) Historical rainforest contractions, localized extinctions and patterns of vertebrate endemism in the rainforests of Australia's wet tropics. *Proceedings of the Royal Society of London Series B* **264**, 709–716.

Williams-Linera, G. and Lawton, R. O. (1995) The ecology of hemiepiphytes in forest canopies. *Forest Canopies*. Lowman, M. D. and Nadkarni, N. M. (eds.). Academic Press, San Diego, California, USA. Pp. 255–283.

Williamson, G. B. and Costa, F. (2000) Dispersal of Amazonian trees: Hydrochory in *Pentaclethra macroloba*. *Biotropica* **32**, 548–552.

Williamson, G. B., Laurance, W. F., Oliveira, A. A., Delamonica, P., Gascon, C., Lovejoy, T. E. and Pohl, L. (2000) Amazonian tree mortality during the 1997 El Niño drought. *Conservation Biology*, **14**, 1538–1542.

Willis, E. O. (1980) Ecological roles of migratory and resident birds on Barro Colorado Island, Panama. *Migrant Birds in the Neotropics*. Keast, A. and Morton, E. S. Washington, Smithsonian Institute Press. Pp. 205–225.

Willott, S. J. (1999) The effects of selective logging on the distribution of moths in a Bornean rainforest. *Philosophical Transactions of the Royal Society of London Series B* **354**, 1783–1790.

Wills, C., Harms, K. E., Condit, R., King, D., Thompson, J., He, F. L., Muller-Landau, H. C., Ashton, P., Losos, E., Comita, L., Hubbell, S., LaFrankie, J., Bunyavejchewin, S., Dattaraja, H. S., Davies, S., Esufali, S., Foster, R., Gunatilleke, N., Gunatilleke, S., Hall, P., Itoh, A., John, R., Kiratiprayoon, S., de Lao, S. L., Massa, M., Nath, C., Noor, M. N. S., Kassim, A. R., Sukumar, R., Suresh, H. S., Sun, I. F., Tan, S., Yamakura, T. and Zimmerman, E. (2006) Nonrandom processes maintain diversity in tropical forests. *Science* **311**, 527–531.

Willson, M. F., Irvine, A. K. and Walsh, N. G. (1989) Vertebrate dispersal syndromes in some Australian and New Zealand plant communities, with geographic comparisons. *Biotropica* **21**, 133–147.

Wilshusen, P. R., Brechin, S. R., Fortwangler, C. L. and West, P. C. (2002) Reinventing a square wheel: Critique of a resurgent protection paradigm in international biodiversity conservation. *Society and Natural Resources* **15**, 17–40.

Wilson, E. O. (1987a) The arboreal ant fauna of Peruvian Amazon forests: a first assessment. *Biotropica* **19**, 245–251.

Wilson, E. O. (1987b) The little things that run the world (the importance and conservation of invertebrates) *Conservation Biology* **1**, 344–346.

Wilson, E. O. (1988) *Biodiversity*. National Academy Press, Washington, DC, USA.

Wilson, E. O. (1992) *The Diversity of Life*. Belknap Press, Harvard University Press, Cambridge, MA, USA.

Wilson, E. O. (2002) What is nature worth? There's a powerful economic argument for preserving our living natural environment. 2nd National Conference on Science, Policy and the Environment, Sustainable Communities: Science and Solutions. Retrieved 17 July, 2005, from http://www.ncseonline.org/NCSEconference/2001conference/page.cfm?FID=1965.

Winemiller, K. O. and Jepsen, D. B. (2004) Migratory neotropical fishes subsidize food webs of oligotrophic blackwater rivers. In: *Food Webs at the Landscape Level*. Polis, G. A., Power, M. E. and Huxel, G. R. (eds.). University of Chicago Press, Chicago, IL, USA. Pp. 115–132.

Winn, A. A. (1999) The functional significance and fitness consequences of heterophylly. *International Journal of Plant Sciences* **160**, S113–S121.

Wirth, R., Beyschlag, W., Ryel, R., Herz, H. and Hölldobler, B. (2003) *The herbivory of leaf-cutting ants. A case study on* Atta colombica *in the tropical rainforest of Panama*. Springer Verlag, Berlin, Germany.

Wirth, R., Beyschlag, W., Ryel, R. J. and Hölldobler, B. (1997) Annual foraging of the leaf-cutting ant *Atta colombica* in a semideciduous rain forest in Panama. *Journal of Tropical Ecology* **13**, 741–757.

Wirth, R., Meyer, S. T., Almeida, W. R., Araujo, M. V., Barbosa, V. S. and Leal, I. R. (2007) Increasing densities of leaf-cutting ants (*Atta* spp.) with proximity to the edge in a Brazilian Atlantic forest. *Journal of Tropical Ecology* **23**, 501–505.

Wirth, R., Weber, B. and Ryel, R. J. (2001) Spatial and temporal variability of canopy structure in a tropical moist forest. *Acta Oecologica* **22**, 235–244.

Witmer, M. C. and Cheke, A. S. (1991) The dodo and the tambalacoque tree: An obligate mutualism reconsidered. *Oikos* **61**, 133–137.

Wittmann, F. and Junk, W. J. (2003) Sapling communities in Amazonian white-water forests. *Journal of Biogeography* **30**, 1533–1544.

Wittmann, F., Junk, W. J. and Piedade, M. T. F. (2004) The varzea forests in Amazonia: flooding and the highly dynamic geomorphology interact with natural forest succession. *Forest Ecology and Management* **196**, 199–212.

Wittmann, F. and Parolin, P. (2005) Aboveground roots in Amazonian floodplain trees. *Biotropica* **37**, 609–619.

Wittmann, F., Schongart, J., Parolin, P., Worbes, M., Piedade, M.T.F. and Junk, W.J. (2006) Wood specific gravity of trees in Amazonian white-water forests in relation to flooding. *Iawa Journal*, **27**, 255–268.

Wolda, H. (1992) Trends in abundance of tropical forest insects. *Oecologia* **89**, 47–52.

Wolfe, J. A. (1972) An interpretation of Alaskan Tertiary floras. *Floristics and Paleofloristics of Asia and Eastern North America*. Graham, A. (ed.). Elsevier, Amsterdam, The Netherlands. Pp. 201–233.

Wolfe, J. A. (1978) Paleobotanical interpretation of Tertiary climates in northern hemisphere. *American Scientist* **66**, 694–703.

Wolfe, N. D., Daszak, P., Kilpatrick, A. M. and Burke, D. S. (2005) Bushmeat hunting deforestation, and prediction of zoonoses emergence. *Emerging Infectious Diseases* **11**, 1822–1827.

Wollenberg, E., Campbell, B., Dounias, E., Gunarso, P., Moeliono, M. and Sheil, D. (2009) Interactive land use planning in Indonesian rainforest landscapes: Reconnecting plans to practice. *Ecology and Society* **14**, 35.

Won, H. and Renner, S. S. (2006) Dating dispersal and radiation in the gymnosperm *Gnetum* (Gnetales): Clock calibration when outgroup relationships are uncertain. *Systematic Biology* **55**, 610–622.

Wong, M. (1984) Understorey foliage arthropods in the virgin and regenerating habitats of Pasoh forest reserve, West Malaysia. *The Malaysian Forester* **47**, 43–69.

Wood, J. J., Beaman, R. S. and Beaman, J. H. (1993) *The Plants of Mount Kinabalu, 2. Orchids.* Royal Botanical Garden, Kew, Kew, UK.

Wood, T. E., Lawrence, D., Clark, D. A. and Chazdon, R. L. (2009) Rain forest nutrient cycling and productivity in response to large-scale litter manipulation. *Ecology* **90**, 109–121.

Woods, P. (1989) Effects of logging, drought, and fire on structure and composition of tropical forests in Sabah, Malaysia. *Biotropica* **21**, 290–298.

Worbes, A., Staschel, R., Roloff, A. and Junk, W. J. (2003) Tree ring analysis reveals age structure, dynamics and wood production of a natural forest stand in Cameroon. *Forest Ecology and Management,* **173**, 105–123.

Worbes, M. (1999) Annual growth rings, rainfall-dependent growth and long-term growth patterns of tropical trees from the caparo forest reserve in venezuela. *Journal of Ecology*, **87**, 391–403.

Worbes, M. and Junk, W. J. (1989) Dating tropical trees by means of C14 from bomb tests. *Ecology,* **70**, 503–507.

World Bank (1990) *Indonesia: sustainable development of forest, land and water.* World Bank, Washington, DC, USA. Pp. 252.

World Bank (2002) *A Revised Forest Strategy for the World Bank Group.* World Bank, Washington, DC, USA.

World Bank (2004) *Sustaining Forests: A Development Strategy.* World Bank, Washington, DC, USA.

World Bank (2008) Commodity price data (pink sheet) Retrieved 19 December, 2008, from http://siteresources. worldbank.org/INTDAILYPROSPECTS/Resources/Pnk1208R. pdf.

WRI. (2003) Watersheds of the World: Global Maps 02. Freshwater Fish Species Richness by Basin. from http://www. earthtrends.wri.org/pdf_library/maps/watersheds/gm2.pdf.

Wright, D., Currie, D. and Maurer, B. (1993) Energy supply and patterns of species richness on local and regional scales. In. *Species diversity in ecological communities: historical and geographical perspectives.* Ricklefs, R. and Schluter, D. (eds.). University of Chicago Press, Chicago, IL, USA. Pp. 66–74.

Wright, D. H. (1983) Species-energy theory: an extension of species-area theory. *Oikos* **41**, 496–506.

Wright, I. R., Gash, J. H. C., da Roche, H. R. and Roberts, J. M. (1996) Modelling surface conductance for Amazonian pasture and forest. *Amazonian Deforestation and Climate.* Gash, J. H. C., Nobre, C. A., Roberts, J. M. and Victoria, R. L. (eds.). John Wiley and Sons, San Francisco, CA, USA. Pp. 437–458.

Wright, I. J., Reich, P. B., Westoby, M., Ackerly, D. D., Baruch, Z., Bongers, F., Cavender-Bares, J., Chapin, T., Cornelissen, J. H. C., Diemer, M., Flexas, J., Garnier, E., Groom, P. K., Gulias, J., Hikosaka, K., Lamont, B. B., Lee, T., Lee, W., Lusk, C., Midgley, J. J., Navas, M. L., Niinemets, U., Oleksyn, J., Osada, N.,

Poorter, H., Poot, P., Prior, L., Pyankov, V. I., Roumet, C., Thomas, S. C., Tjoelker, M. G., Veneklaas, E. J. and Villar, R. (2004) The worldwide leaf economics spectrum. *Nature,* **428**, 821–827.

Wright, I. J. and Westoby, M. (1999) Differences in seedling growth behaviour among species: Trait correlations across species, and trait shifts along nutrient compared to rainfall gradients. *Journal of Ecology,* **87**, 85–97.

Wright, S. J. (2002) Plant diversity in tropical forests: a review of mechanisms of species coexistence. *Oecologia* **130**, 1–14.

Wright, S. J. (2005) Tropical forests in a changing environment. *Trends in Ecology and Evolution* **20**, 553–560.

Wright, S. J. and Calderon, O. (2006) Seasonal, El Niño and longer term changes in flower and seed production in a moist tropical forest. *Ecology Letters* **9**, 35–44.

Wright, S. J. and Muller-Landau, H. C. (2006) The future of tropical forest species. *Biotropica* **38**, 287–301.

Wright, S. J., Muller-Landau, H. C. and Schipper, J. (2009) The future of tropical species on a warmer planet. *Conservation Biology* **in press**.

Wronski, T. and Hausdorf, B. (2008) Distribution patterns of land snails in Ugandan rain forests support the existence of Pleistocene forest refugia. *Journal of Biogeography* **35**, 1759–1768.

Wuebbles, D. J. and Hayhoe, K. (2002) Atmospheric methane and global change. *Earth-Science Reviews* **57**, 177–210.

Wunder, S. (2001) Poverty alleviation and tropical forests: What scope for synergies? *World Development* **29**, 1817–1833.

Wunder, S. (2008) Payments for environmental services and the poor: concepts and preliminary evidence. *Environment and Development Economics* **13**, 279–297.

Wunder, S. and Alban, M. (2008) Decentralized payments for environmental services: The cases of Pimampiro and PROFAFOR in Ecuador. *Ecological Economics* **65**, 685–698.

Wunder, S. and Sunderlin, W. D. (2004) Oil, macroeconomics, and forests: Assessing the linkages. *World Bank Research Observer* **19**, 231–257.

Wurth, M.K.R., Pelaez-Riedl, S., Wright, S.J. and Korner, C. (2005) Non-structural carbohydrate pools in a tropical forest. *Oecologia,* **143**, 11–24.

Yamada, T., Suzuki, E., Yamakura, T. and Tan, S. (2005) Tap-root depth of tropical seedlings in relation to species-specific edaphic preferences. *Journal of Tropical Ecology* **21**, 155–160.

Yanoviak, S. P., Kaspari, M., Dudley, R. and Poinar, G. (2008) Parasite-induced fruit mimicry in a tropical canopy ant. *American Naturalist* **171**, 536–544.

Yap, S. W., Chack, C. V., Majuakim, L., Anuar, M. and Putz, F. E. (1995) Climbing bamboo (*Dinochloa* spp.) in Sabah: biomechanical characteristics, mode of ascent, and abundance in logged-over forest. *Journal of Tropical Forest Science* **8**, 96–202.

Yasuda, M., Miura, S. and Nor Azman, H. (2000) Evidence for food hoarding behaviour in terrestrial rodents in Pasoh Forest Reserve, a Malaysian lowland rain forest. *Journal of Tropical Forest Science* **12**, 164–173.

Yavitt, J. B., Wright, S. J. and Wieder, R. K. (2004) Seasonal drought and dry-season irrigation influence leaf-litter nutrients and soil enzymes in a moist, lowland forest in Panama. *Austral Ecology* **29**, 177–188.

Yeates, D. K., Bouchard, P. and Monteith, G. B. (2002) Patterns and levels of endemism in the Australian Wet Tropics rainforest: evidence from flightless insects. *Invertebrate Systematics* **16**, 605–619.

Yesson, C., Russell, S. J., Parrish, T., Dalling, J. W. and Garwood, N. C. (2004) Phylogenetic framework for *Trema* (Celtidaceae) *Plant Systematics and Evolution* **248**, 85–109.

Yih, K., Boucher, D. H., Vandemeer, J. H. and Zamora, N. (1991) Recovery of the rain forest of southeastern Nicaragua after destruction by Hurricane Joan. *Biotropica* **23**, 106–113.

Yoda, K. (1978) Light climate within the forest. *Biological Production in a Warm-Temperate Evergreen Oak Forest of Japan*. Kira, T., Ono, Y. and Hosokawa, T. Tokyo, University of Tokyo Press. Pp. 46–54.

Yokouchi, Y., Ikeda, M., Inuzuka, Y. and Yukawa, T. (2002) Strong emission of methyl chloride from tropical plants. *Nature* **416**, 163–165.

Young, A. M. (1982) Effects of shade cover and availability of midge breeding sites on pollinating midge populations and fruit set in two cocoa farms. *Journal of Applied Ecology* **19**, 47–63.

Young, B. E., Lips, K. R., Reaser, J. K., Ibanez, R., Salas, A. W., Cedeno, J. R., Coloma, L. A., Ron, S., La Marca, E., Meyer, J. R., Munoz, A., Bolanos, F., Chaves, G. and Romo, D. (2001) Population declines and priorities for amphibian conservation in Latin America. *Conservation Biology* **15**, 1213–1223.

Young, K. R., Ewel, J. J. and Brown, B. J. (1987) Seed dynamics during forest succession in Costa Rica. *Vegetatio* **71**, 157–173.

Young, O. R. (2002) Institutional interplay: the environmental consequences of cross-scale interactions. *The Drama of the Commons*. Ostrom, E., Dietz, T., Dolsak, N., Stern, P. C., Stonich, S. and Weber, E. U. National Academy Press, Washington, DC, USA. Pp. 263–292.

Young, T. P. and Hubbell, S. P. (1991) Crown asymmetry, treefalls, and repeat disturbance of broad-leaved forest gaps. *Ecology* **72**, 1464–1471.

Yu, D. W. (1994) The structural role of epiphytes in ant gardens. *Biotropica* **26**, 222–226.

Yu, D. W., Terborgh, J. W. and Potts, M. D. (1998) Can high tree species richness be explained by Hubbell's null model? *Ecology Letters* **1**, 193–199.

Yusop, Z., Douglas, I. and Nik, A. R. (2006) Export of dissolved and undissolved nutrients from forested catchments in Peninsular Malaysia. *Forest Ecology and Management* **224**, 26–44.

Zagt, R. J. (1997) *Tree demography in the tropical rain forest in Guyana. Tropenbos-Guyana Series 3*. Tropenbos-Guyana programme, Georgetown, Guyana.

Zarin, D. J., Schulze, M. D., Vidal, E. and Lentini, M. (2007) Beyond reaping the first harvest: Management objectives for timber production in the Brazilian Amazon. *Conservation Biology* **21**, 916–925.

Zech, W., Senesi, N., Guggenberger, G., Kaiser, K., Lehmann, J., Miano, T. M., Miltner, A. and Schroth, G. (1997) Factors controlling humification and mineralization of soil organic matter in the tropics. *Geoderma* **79**, 117–161.

Zerega, N. J. C., Clement, W. L., Datwyler, S. L. and Weiblen, G. D. (2005) Biogeography and divergence times in the mulberry family (Moraceae) *Molecular Phylogenetics and Evolution* **37**, 402–416.

Zhang, D. Y. and Lin, K. (1997) The effects of competitive asymmetry on the rate of competitive displacement: How robust is Hubbell's community drift model? *Journal of Theoretical Biology* **188**, 361–367.

Zhang, L. M., Yu, G. R., Sun, X. M., Wen, X. F., Ren, C. Y., Song, X., Liu, Y. F., Guan, D. X., Yan, J. H. and Zhang, Y. P. (2006) Seasonal variation of carbon exchange of typical forest ecosystems along the eastern forest transect in China. *Science in China Series D-Earth Sciences* **49**, 47–62.

Zhang, Q. F., Justice, C. O. and Desanker, P. V. (2002) Impacts of simulated shifting cultivation on deforestation and the carbon stocks of the forests of central Africa. *Agriculture Ecosystems and Environment* **90**, 203–209.

Zhou, G., Minakawa, N., Githeko, A. K. and Yan, G. Y. (2004) Association between climate variability and malaria epidemics in the East African highlands. *Proceedings of the National Academy of Sciences of the United States of America* **101**, 2375–2380.

Ziegler, A. M., Eshel, G., McAllister Rees, P., Rothfus, T. A., Rowley, D. B. and Sunderlin, D. (2003) Tracing the tropics across land and sea: Permian to present. *Lethaia* **36**, 227–254.

Zimmerman, J. K., Everham, E. M., Waide, R. B., Lodge, D. J., Taylor, C. M. and Brokaw, N. V. L. (1994) Responses of tree species to hurricane winds in subtropical wet forest in

Puerto Rico: implications for tropical tree life-histories. *Journal of Ecology* **82**, 911–922.

Zinn, T. L. and Humphrey, S. R. (1981) Seasonal food resources and prey selection of the southeastern brown bat (*Myotis austroriparius*) in Florida. *Florida Scientist* **44**, 81–90.

Zipperlen, S. W. and Press, M. C. (1996) Photosynthesis in relation to growth and seedling ecology of two dipterocarp rain forest tree species. *Journal of Ecology*, **84**, 863–876.

Zotz, G. (1995) How fast does an epiphyte grow? *Selbyana* **16**, 150–154.

Zotz, G. (2005) Vascular epiphytes in the temperate zones: a review. *Plant Ecology* **176**, 173–183.

Zotz, G. and Winter, K. (1993) Short-term photosynthesis measurements predict leaf carbon balance in tropical rain forest canopy plants. *Planta*, **191**, 409–412.

Zou, X., Zucca, C. P., Waide, R. B. and McDowell, W. H. (1995) Long-term influence of deforestation on tree species composition and litter dynamics of a tropical rain forest in Puerto Rico. *Forest Ecology and Management* **78**, 147–157.

Index